Modern Welding

by

Andrew D. Althouse
Technical-Vocational Education Consultant
Member of the American Welding Society

Carl H. Turnquist
Career Education Consultant
Member of the American Welding Society

William A. Bowditch
Career Education Consultant
Gaylord, Michigan
Life Member of the American Welding Society
Member of the American Vocational Association

Kevin E. Bowditch
Welding Engineer
Production Manager
Vista Industrial Products, Inc.
Vista, California
Member of the American Welding Society
Member of the American Vocational Association

Publisher
THE GOODHEART-WILLCOX COMPANY, INC.
Tinley Park, Illinois

Library of Congress Card Catalog Number 96-22368
International Standard Book Number 1-56637-330-1

3 4 5 6 7 8 9 0 97 00 99 98

William A. Bowditch has an extensive teaching and welding background. He has been a teacher, department head, and supervisor of special needs and vocational programs. In addition to his formal college training in preparation for teaching, Bill has taken several specialized courses in industry, such as the Hobart Welding School and American Welding Society courses. He is a member of the American Vocational Association and a life member of the American Welding Society. As a coauthor of *Modern Welding*, he has guided this Goodheart-Willcox book through many revisions to keep it up to date and technically correct in order to maintain its value as an authoritative welding text. He is also a coauthor of *Welding Technology Fundamentals.*

Kevin E. Bowditch is the production manager of a precision sheet metal firm that produces fabricated metal parts and weldments. His welding experience includes working for an automotive firm, a construction company (building two nuclear plants), and two aerospace firms. While working for one aerospace firm, Kevin designed resistance welding and soldering equipment, special equipment for custom applications, and worked to develop correct welding and soldering schedules for customers. He has a Bachelor's degree in welding engineering from The Ohio State University and has attended specialized conferences and courses sponsored by the American Welding Society, American Society of Mechanical Engineers, and American National Standards Institute. Kevin joined his father as a coauthor of *Modern Welding*, beginning with the 1984 edition, and has been a coauthor of *Welding Technology Fundamentals* since its first edition was published in 1991.

Library of Congress Cataloging in Publication Data
Modern welding: complete coverage of the welding field
in one
easy-to-use volume! / by Andrew D. Althouse ... [et al.].

p. cm.
Includes index.
ISBN 1-56637-330-1
1. Welding. I. Althouse, Andrew Daniel.
TS227.M557 1997
671.5'2--dc20   96-22368
                                                                      CIP

INTRODUCTION

Modern Welding is an authoritative and technically correct text written for use by students and teachers in secondary schools, colleges and universities, technical, and trade schools. It is also suitable for use by individuals who wish to learn independently about contemporary processes and techniques used in welding.

Modern Welding will provide you with an understanding of virtually all the welding and cutting processes used in production and repair today. This book covers the theory, fundamentals of operation, equipment used, and techniques recommended for all of the welding and cutting processes that are commercially used. The many tables and charts that are included in **Modern Welding** provide information regarding recommended settings for the variables involved in the various welding processes. These tables include the recommended gas pressures, tip sizes, welding speeds, amperage and voltage settings, wire feed rates, electrode sizes, inert gas selections, and much, much more.

General shop safety, safety attitudes, and specific welding shop safety practices are covered in Chapter 1. Safety information and cautions are also printed in red throughout the text wherever they apply. Chapters 2 and 3 cover print reading and the interpretation of the American Welding Society welding symbols found on welding prints. The text is divided into nine sections. Each section covers a different welding process or group of related welding processes. Chapter 33 covers the subject of getting and holding a job in the welding industry. Technical information that may be required while working in the welding industry is included in Chapter 34. An extensive *Glossary of Welding Terms* is provided at the end of the book.

Modern Welding may be studied in order from Chapter 1 to Chapter 34, or it may be used to study welding processes in any desired order. Each section stands alone, and does not rely on previously acquired knowledge. **Modern Welding** is extensively illustrated with full-color and black-and-white photographs, as well as drawings that have been color-coded to help you better understand each welding process and its equipment. All figure captions should be read, since they may contain details that are not included in the text. Measurements are generally shown in dual form: US Conventional followed by SI Metric. Welding terms used throughout conform to the usage in the American National Standard *ANSI/AWS A3.0 Standard Welding Terms and Definitions.* Nonstandard or "trade" terms, when given, are clearly identified. The topics covered in **Modern Welding** are presented in a logical sequence designed to make learning and teaching the technology of welding easier and more effective.

William A. Bowditch
Kevin E. Bowditch

HOW TO USE THE COLOR KEY
Colors are used throughout **Modern Welding** to help show the flow of different gases and to indicate various materials or equipment features. The following key shows what each color represents.

Gases

Oxygen	high-pressure		low-pressure
Fuel gas	high-pressure		low-pressure
Shielding gas (1)	high-pressure		low-pressure
Shielding gas (2)	high-pressure		low-pressure
Air	high-pressure		vacuum

Weld

Base metal or plastic	edge	surface
Molten metal or plastic		
Welding rod		
Weld bead or surfacing material		
Flame, arc, or plasma		
Electrode		
Flux		
Slag		
Fumes		
Direction or motion		
Welding machines/equipment		
Wires, leads, graph curves		
Water		

Other Materials
Special features, materials or components not otherwise color-coded

TABLE OF CONTENTS

PART 1 Welding Fundamentals

PART 7 Special Processes

PART 8 Metal Technology

Part 1

WELDING FUNDAMENTALS

Safety is a vital part of any welding activity. This welder is wearing proper protective clothing and a safety harness while working high above the ground on skyscraper construction. (Lincoln Electric Co.)

Welders on construction sites often must be prepared for severe weather conditions. This welder has dressed for warmth while working in winter weather.

Chapter 1

SAFETY IN THE WELDING SHOP

LEARNING OBJECTIVES

After studying this chapter, you will be able to:
* Identify several common causes of accidents.
* Recognize possible safety hazards in the welding shop or other work environments.
* Select and properly use safety equipment appropriate to working conditions.
* Recognize and evaluate potential safety hazards and react appropriately to prevent accidents.

1.1 ACCIDENTS

The *American Heritage Dictionary* defines **accident** as "An unexpected undesirable event."

Many of the experts who study accidents and encourage safe practices believe that there is some personal or physical factor responsible in every accident. Some of the *personal factors* that may be responsible for causing accidents are:

* *Stress:* People who are under stress may be distracted from their work by thoughts of worry, anger, sorrow, love, or hate.
* *Illness:* When a person is ill, he or she may not be able to give all the attention necessary to a task.
* *Fatigue:* If a person does not get sufficient sleep, for whatever reason, he or she may be less alert to the requirements of the job.
* *Lack of Job Knowledge:* When a person is not sufficiently trained for the job or task and its safety hazards, accidents can more readily occur.
* *Age:* There are more accidental deaths at age 18 than at any other age. As people age, they normally have a reduction in vision and strength and an increase in reaction time. Wisdom and a

concern for the results of personal actions, normally increase with age. However, the age at which these changes take place varies considerably.
* *Lack of Wisdom:* This is the lack of intelligence and experience to know when an action may cause an accident. "Horseplay" and practical joking have no place in a shop. There are too many hard and sharp items to bump into or fall on.
* *Poor Attitude:* How you feel about yourself and your job has much to do with a good attitude toward safety rules, housekeeping, and the wearing of safety equipment and guards.
* *Drugs:* The use of alcohol, street drugs, and some prescription drugs will affect reflexes, perception, coordination, and judgment.

Some of the *physical factors* responsible or involved in shop accidents are:

* *Equipment Failure:* Hazardous equipment that is poorly maintained is an accident waiting to happen.
* *Time of Day:* Starting and quitting times may be the most congested and dangerous periods in the shop. After lunch, some people become less attentive.
* *Housekeeping:* Metal pieces, hoses, cable, dirt and oil on the shop floor, and poorly stored cylinders and combustibles often are factors in shop accidents.

Most of the conditions listed above are in the control of the worker. Even if you cannot control them, you should be aware of their effects on safety. Three of the most important factors in safety on the job are:

* Staying healthy in mind and body.
* Becoming well-trained in the required job or task and its possible hazards involved.
* Having a good attitude toward safety rules, equipment, and training on the job.

1.2 GENERAL SHOP SAFETY

Working and moving about in a welding shop or welding environment can present many dangers such as heat, sparks, fumes, radiation, high voltage, hot metal, moving vehicles, hazardous machinery, and moving overhead cranes and their loads. However, through training and the use of adequate safety equipment, all these safety hazards can be controlled.

1.2.1 Clothing and Personal Safety

Workers in the shop should wear clothing appropriate to the job. **Everyone, including office personnel and visitors, should wear goggles or approved safety glasses while in the shop.** Hard hats are necessary if there are overhead hazards.

Welders should wear work clothes or coveralls. The shirt or coveralls should have covered pockets and have buttons at the neck. Trousers and coveralls should not have cuffs that could catch hot metal spatter. A cap of some kind should be worn to protect the hair from hot metal spatter. Gloves should be worn to protect against hot and sharp metal. Leather gloves with gauntlet-type cuffs are recommended for welders. Figure 1-1 shows a welder adequately dressed for welding.

Steel-toed safety shoes are recommended for welders and other workers who handle heavy articles. Oxyfuel gas welders and cutters should wear approved goggles with the correct shade lens.

Anyone performing arc welding should wear an arc welding helmet with a lens of the correct shade for the process and amperage being used. See Figure 5-48 for a chart of recommended welding lens shades. Some arc welders also wear a pair of goggles under their arc welding helmet to protect them from arc rays and metal spatter that might reflect off the inside of the helmet.

For his or her own welfare, every worker in the shop should be constantly alert for safety and health hazards.

1.2.2 Housekeeping

One of the most important factors in shop safety is *housekeeping*. The floors and workbenches should be kept clean of dirt, scrap metal, grease, oil, and anything that is not essential to the job at hand. *Combustibles* such as wood, paper, rags, and flammable liquids must be kept clear of all areas where sparks or hot metal may fly. Aisles must be kept clear of hoses and electric cables, which can cause tripping accidents. Hoses and cables also could be run over and possibly damaged. Every worker has a responsibility to help keep the shop clean and clear of hazards and to report safety problems and hazards to the shop supervisor.

1.2.3 Fire Hazards

Paint, oil, cleaning chemicals, and other possible combustibles must be kept in steel cabinets designed for such storage, in accordance with the local fire codes.

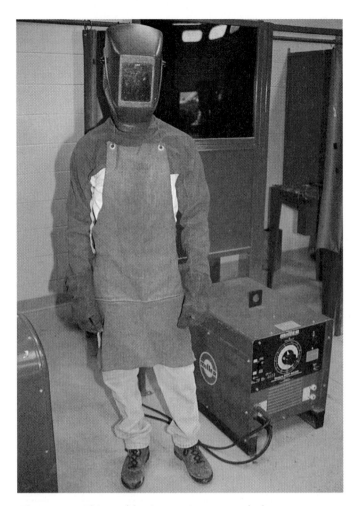

Figure 1-1. *This welder is wearing an arc helmet, cap, gloves, leather apron, leather cape, cuffless trousers, and high-top shoes as protection from hot metal and harmful arc rays.*

Fire exits, fire blankets, and extinguishers should be clearly marked. Fire extinguishers and blankets are usually mounted on a bright red surface for visibility. All workers should know how to use the extinguishers and blankets. They should also know where the exits are and which one to use in an emergency. Periodic fire drills are a good idea. It is also important to know what equipment should be shut down before leaving the work area to reduce fire hazards.

1.2.4 Electrical Hazards

Electrical devices of various kinds are common in welding shops. All electrical devices are hazardous, but some use extremely high and dangerous voltages. All equipment and areas where 220 volts or more are used must be well-marked. Usually, this is done with a sign stating "**Danger: High Voltage,**" or stating the specific voltage that is involved. **Electrical equipment must be installed and repaired only by well-trained and competent technicians.**

1.2.5 Machinery Hazards

Machinery must operated only after thorough training on how the machine operates, its safety hazards, its safety features, the correct placement of hands and feet, and the proper sequence of operation.

The pedestal grinder, a simple piece of equipment, can be one of the must dangerous pieces of machinery in the shop! For example, failing to adjust the tool rest to its closest safe point and securely tighten it can result in an accident. So can pushing hard on the work and attempting to grind it in a downward direction on the wheel. This is extremely dangerous: it may cause the workpiece to flip up and be pulled between the tool rest and the grinding wheel. It can result in a thumb that is ground on the wheel, a painful injury. See Figure 1-2.

1.2.6 Fumes and Ventilation

Dust, fumes, and metal particles can be a hazard to health. For this reason, shops should normally have high

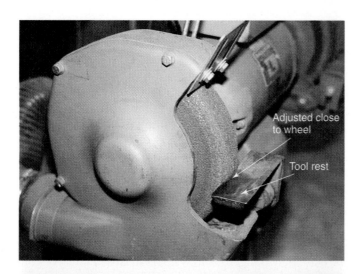

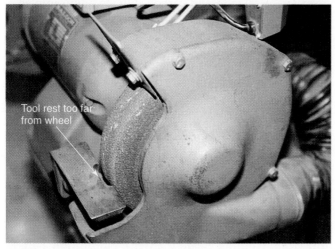

Figure 1-2. Pedestal grinder safety. A—The tool rest must be adjusted to be as close as possible to the grinding wheel without touching, and should be slightly above the centerline of the wheel. B—This tool rest is dangerously far away from the wheel.

ceilings for greater air volume and well-designed ventilation systems to provide adequate volume of air movement in accordance with OSHA requirements.

1.2.7 Lifting

Lifting is always a hazard to the body, but training in how to lift objects safely can reduce the chance of injury. Always keep your back straight and lift with your legs. Limits should be set on how much weight a worker may lift. The maximum weight to be lifted may vary from shop to shop and by worker classification. In many shops, any worker who may have to lift anything is required to wear a lifting belt or back brace.

1.2.8 Hazardous Obstacles

Safety hazards or obstacles in the shop should be well-marked. Signs, fences, or barriers should be erected while *temporary hazards* are present, so that all workers are fully aware of them. *Permanent hazards* are often painted with wide yellow and black stripes to create high visibility.

1.2.9 Hand and Power Tools

All hand and power tools should be examined prior to use for loose parts that might fly off and injure the operator or a worker nearby. Power cords should be checked for cut or frayed insulation and exposed wires. Return and report any defective equipment to the shop supervisor or to maintenance.

1.3 SAFETY IN THE WELDING ENVIRONMENT

Many of the safety practices to be observed in the welding environment apply equally to all welding processes.

1.3.1 Designated Welding and Cutting Areas

All welding and cutting should be done in designated areas of the shop, if possible. These areas should be made safe for welding and cutting operations with concrete floors, arc filter screens, protective drapes or curtains, and fire extinguishers. No combustibles should be stored nearby.

1.3.2 Ventilation

Adequate and approved *ventilation* is required in welding and cutting areas to ensure the proper removal of toxic fumes, dust, and dirt. The air also may need to go through filters and cleaners before it is recirculated. Ventilation pickups should be located so that fumes are picked up below the level of the welder's nose, before he/she has to inhale them.

Toxic fumes may be created when welding or brazing operations heat fluxes and metals containing such chemicals as cadmium, chromium, lead, zinc, or beryllium. When these chemicals are present, or are suspected to be present, welding or brazing must be done with excellent ventilation or while wearing a *purified air breathing apparatus*. Figures 1-3 and 1-4 show two types of helmets or hoods equipped with air purifying systems. A *powered air purifier* provides cleaned air to the helmet from a small electrically powered purifier that is worn around the

Figure 1-3. *A small, electrically powered air purifier worn around the waist supplies clean air to the welding helmet and hood. (Hornell Speedglas, Inc.)*

Figure 1-4. An air-supplied purifier which supplies clean air to the welding helmet and hood through a flexible hose is being worn by this welder during cutting operations. (3M Occupational Health and Environmental Safety Division)

waist. An *air-supplied purifier* provides cleaned air to the helmet through a long hose from a central source.

Two types of *air-purifying respirators* are shown in Figures 1-5 and 1-6. Shop air flows through one or two air filters on the respirator before being breathed by the welder. The respirator shown in Figure 1-5 may be made of vinyl or silicone rubber. A full face respirator is shown in Figure 1-6.

Figure 1-5. *This welder is cutting a pipe using an oxyfuel gas cutting torch. The welder is wearing a air-purifying respirator which filters dust and other particles from the air that he or she breathes. It will not protect against toxic fumes, however. (3M Occupational Health and Environmental Safety Division)*

1.3.3 Suffocation Hazards

Gases that are heavier than air or lighter than air can be extremely dangerous to welders in closed tanks or confined spaces. Argon and carbon dioxide are examples of heavier-than-air gases. Helium is an example of a lighter-than-air gas. These gases are colorless and odorless and will displace the oxygen in a closed space. Argon, for example, can asphyxiate a person in about seven seconds.

Confined spaces must be well ventilated when heavier-than-air or lighter-than-air gases are to be used. If proper ventilation cannot be obtained, the welder must go into the space using an air-supplied purifier.

1.3.4 Welding on Hazardous Containers

Unless you are trained in the proper procedures for doing so, never try to weld or cut a container that has held flammable or hazardous materials. Such a container may explode. Proper procedures are described in the American Welding Society (AWS) Publication F4.1-94, *Recommended Safe Practices for the Preparation for Welding and Cutting of Containers That Have Held Hazardous Substances*. You should also consult with the local fire marshal before welding or cutting such containers.

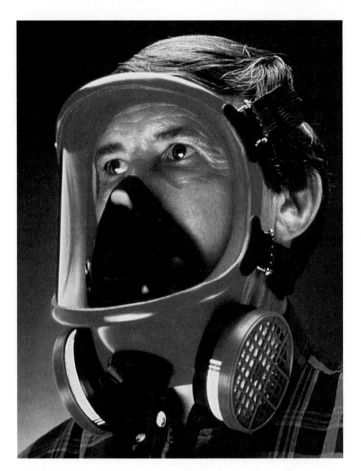

Figure 1-6. This air-purifying respirator is equipped with a full face mask.

1.4 OXYFUEL GAS WELDING AND CUTTING SAFETY

Welders should be well-trained in the function and operation of each part of an oxyfuel gas welding or cutting station. Welders also should be trained and tested in the correct methods of starting, testing for leaks, and shutting down an oxyfuel gas welding or cutting station. They must be aware of the hazards involved in the use of fuel gas, oxygen, and shielding gas cylinders and how these cylinders are safely stored.

1.4.1 Arc Ray and Spark Hazards

Other hazards to be aware of when welding are damage to the eyes and exposed skin from ultraviolet and infrared rays and from hot metal sparks. Approved welding goggles of the recommended shade number will protect the eyes from these hazards. Leather, wool, or flame-retardant-treated cotton clothing can be worn to protect the body. All clothing (such as coveralls, shirts, or capes) must be buttoned at the neck to prevent sparks from entering. When welding out of position, the welder should wear leather clothing, commonly called *leathers.* "Leathers" may be jackets, capes, leggings, or aprons. See

Figure 1-1. A cap of some kind should be worn when welding to protect the hair from burning. Ear plugs are worn by some welders to prevent hot metal sparks from entering their ears..

Trousers must not have cuffs that could catch sparks or hot molten metal; all shirt and jacket pockets must be covered for the same reason. No matches, lighters, or other flammables should be carried in jacket, shirt, or trouser pockets. Trousers should be long enough to overlap the tops of the shoes to keep hot metal out of the shoes. High-top, hard-toed shoes are recommended in a welding shop. Full gauntlet-type gloves of supple leather, or gauntlet gloves with at least leather palms, are recommended.

1.4.2 Hot and Sharp Metal Hazards

All metal parts in the shop should be considered hot. Metal must be marked "HOT" if it is left to cool. Pliers or tongs should be used to pick up all metal parts. When handling sheet metal or metal that may have sharp edges, it is wise to wear cut-proof gloves.

1.4.3 Hearing Hazards

Ear muffs or plugs should also be worn for hearing protection when plasma arc welding, cutting, or arc gouging. Ear protection is also necessary to protect against the high noise levels of other equipment in the shop or on the job site.

1.4.4 High-Pressure Cylinder and Flammable Gas Hazards

Cylinders of acetylene, oxygen, and other high-pressure gases should be stored in an upright position, in an approved area or room, and fastened to a rigid post or wall with steel bands or chains. Cylinders must always be stored with their safety caps in place. When in use at a station, cylinders must be fastened by the same methods to a wall or post, or mounted on an approved cylinder truck.

When cylinders are moved, they must always have their safety caps in place. Cylinders should be moved using a cylinder truck, or tipped and rolled on edge by a trained operator using both hands.

Acetylene gas must never be used at a working pressure on the gauge above 15 psig or 103 kPa. Oxygen and other gases should not be used at pressures greater than half of the highest reading on the gauge used. For example, if the highest reading on the oxygen working pressure gauge is 40 psig, oxygen should not be used at a pressure greater than 20 psig.

When welding is completed and the flame is extinguished, the system must be *bled* (emptied) of all gases from the cylinder outlet to torch tip. This is done by closing both cylinder tank valves, opening the torch valves, and turning in the regulator adjusting screws until all the gauges read zero. Then, close the torch valves and turn out the regulator adjusting screws until they feel loose.

When a regulator is not in use, the adjusting screw must always be turned out until it feels loose. This is the off position for a regulator.

1.4.5 Fire Hazards While Welding or Cutting

Always light the oxyfuel gas torch flame using an approved *torch lighter* to avoid burning your fingers. See Figure 1-7. Never point the torch tip at anyone when lighting it or using it. Never point the torch at the cylinders, regulators, hoses, or anything else that may be damaged and cause a fire or explosion. Never lay a lighted torch down on the bench or workpiece, and do not hang it up while lighted. **If the torch is not in the welder's hand, it must be off.**

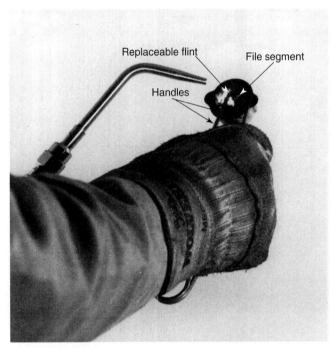

Figure 1-7. *Using a spark lighter to ignite the oxyfuel gas torch flame. Note how the gauntlet-type gloves protect the welder's wrist from sparks.*

To prevent possible fires and explosions, check valves and flashback arrestors must be installed in all oxyfuel gas welding and cutting outfits.

When cutting with oxyfuel gas equipment, clear the area of all combustible materials. It is recommended that at least one fire guard be posted with an extinguisher to watch for possible fires. More welding environment fires are caused while performing oxyfuel gas or arc cutting than by any other means.

1.4.6 Toxic Fume Hazards

The fumes that result from welding, brazing, braze welding, or cutting metals that are coated with or contain zinc, chromium, lead, cadmium, and beryllium are toxic. Cadmium fumes can kill!! Fumes from some fluxes used in soldering and brazing (those containing fluorides and sodium cyanide compounds) also can be toxic. Excellent ventilation or air-purifying breathing apparatus must be worn when working with these metals or fluxes. Refer to Heading 1.3.

Note: Air-purifying respirators will not filter out toxic fumes. They are designed to filter out dust, dirt, and metal particles.

The American Welding Society Standard Z49.1-94, *Safety in Welding and Cutting*, suggests placing a notice similar to the one shown in Figure 1-8 on major equipment used in the oxyfuel gas process. Labels like those shown in Figures 1-9 and 1-10 are required on all welding products containing hazardous materials.

1.5 ARC WELDING AND CUTTING SAFETY

Wherever appropriate, the safety precautions mentioned in the general shop and oxyfuel gas safety sections of this chapter should be applied to arc welding and cutting operations.

Many of the safety hazards involved in arc welding and cutting deal with the dangers of work with and around electricity. The AWS publication, *Safety in Welding and Cutting*, suggests that a notice similar to the one shown in Figure 1-11 be placed on major equipment used in an arc welding process.

1.5.1 Electrical Hazards

A *safety inspection* of the arc welding station should be made regularly by the welder. With the welding machine power off, the cables should be checked over their entire length for worn or cut insulation or exposed wires. Cable connectors should be checked for tightness. The electrode holder or welding gun should be checked for cracked insulators, loose contacts, and worn or cut hoses.

To avoid electrical shocks, a welder should never work with wet or damp gloves or while standing on a wet or damp floor. If work must be done under such conditions, a floor of nonconducting material should be built up above the wet areas. Welders who must weld in a seated or lying position should sit or lie on an insulated material or mat.

Electrical repairs should be attempted only by qualified electricians. Repairs to feed wire drives should be made only with the power off. Nothing other than the base metal, electrodes, and approved welding cables should ever be used to carry welding current.

1.5.2 Arc Ray Hazards

An arc welding face shield with the correct shade number lens must be worn while welding. This is done to protect the eyes, face, and neck from burns caused by the ultraviolet or infrared rays emitted by the electric arc.

1.5.3 Breathing Hazards

Arc welding must be done with good ventilation, with proper protective clothing, and with the same

WARNING: PROTECT yourself and others. Read and understand this label.

FUMES AND GASES can be dangerous to your health.
HEAT RAYS (INFRARED RADIATION from flame or hot metal) can injure eyes.

- Before use, read and understand the manufacturer's instructions, Material Safety Data Sheets (MSDSs), and your employer's safety practices.

- Keep your head out of the fumes.

- Use enough ventilation, exhaust at the flame, or both, to keep fumes and gases from your breathing zone and the general area.

- Wear correct eye, ear, and body protection.

- See American National Standard Z49.1, *Safety in Welding and Cutting*, published by the American Welding Society, 550 N.W. LeJeune Rd., P.O. Box 351040, Miami, Florida 33135; OSHA Safety and Health Standards, 29 CFR 1910, available from U.S. Government Printing Office, Washington, DC 20402.

DO NOT REMOVE THIS LABEL.

Figure 1-8. *An example of the AWS-recommended label to be used on oxyfuel gas welding and cutting equipment.*

DANGER: CONTAINS CADMIUM. Protect yourself and others. Read and understand this label.

FUMES ARE POISONOUS AND CAN KILL.

- Before use, read, understand and follow the manufacturer's instructions, Material Safety Data Sheets (MSDSs), and your employer's safety practices.

- Do not breathe fumes. Even brief exposure to high concentrations should be avoided.

- Use enough ventilation, exhaust at the work, or both, to keep fumes from your breathing zone and the general area. If this cannot be done, use air supplied respirators.

- Keep children away when using.

- See American National Standard Z49.1, *Safety in Welding and Cutting*, published by the American Welding Society, 550 N.W. LeJeune Rd., P.O. Box 351040, Miami, Florida 33135; OSHA Safety and Health Standards, 29 CFR 1910, available from U.S. Government Printing Office, Washington, DC 20402.

If chest pain, shortness of breath, cough, or fever develop after use, obtain medical help immediately.

DO NOT REMOVE THIS LABEL.

Figure 1-9. *An example of the AWS-recommended label to be placed on materials that contain cadmium. Similar labels are recommended for other hazardous materials.*

WARNING: CONTAINS FLUORIDES. Protect yourself and others. Read and understand this label.

FUMES AND GASES CAN BE DANGEROUS TO YOUR HEALTH. BURNS EYES AND SKIN ON CONTACT. CAN BE FATAL IF SWALLOWED.

- Before use, read, understand and follow the manufacturer's instructions, Material Safety Data Sheets (MSDSs), and your employer's safety practices.

- Keep your head out of the fumes.

- Use enough ventilation, exhaust at the work area, or both, to keep fumes and gases from your breathing zone and the general area.

- Avoid contact of flux with eyes and skin.

- Do not take internally.

- Keep out of reach of children.

- See American National Standard Z49.1, *Safety in Welding and Cutting*, published by the American Welding Society, 550 N.W. LeJeune Rd., P.O. Box 351040, Miami, Florida 33135; OSHA Safety and Health Standards, 29 CFR 1910, available from U.S. Government Printing Office, Washington, DC 20402.

First Aid: If contact in eyes, flush immediately with clean water for at least 15 minutes. If swallowed, induce vomiting. Never give anything by mouth to an unconscious person. Call a physician.

DO NOT REMOVE THIS LABEL.

Figure 1-10. *This AWS warning label is recommended for use on containers for brazing and welding fluxes that contain fluorides.*

WARNING: PROTECT yourself and others. Read and understand this label.

FUMES AND GASES can be dangerous to your health.
ARC RAYS can injure eyes and burn skin.
ELECTRIC SHOCK can KILL.

- Before use, read and understand the manufacturer's instructions, Material Safety Data Sheets (MSDSs), and your employer's safety practices.

- Keep your head out of the fumes.

- Use enough ventilation, exhaust at the arc, or both, to keep fumes and gases from your breathing zone and the general area.

- Wear correct eye, ear, and body protection

- Do not touch live electrical parts.

- See American National Standard Z49.1, *Safety in Welding and Cutting*, published by the American Welding Society, 550 N.W. LeJeune Rd., P.O. Box 351040, Miami, Florida 33135; OSHA Safety and Health Standards, 29 CFR 1910, available from U.S. Government Printing Office, Washington, DC 20402.

DO NOT REMOVE THIS LABEL.

Figure 1-11. *An example of the AWS warning label to be placed on arc welding equipment.*

precautions described in the general and oxyfuel gas welding safety sections. Ventilation should always be arranged so that the fumes are picked up before they reach the welder's eyes, nose, or mouth. The hazards from toxic fumes are the same for the arc welding processes as they are for all other forms of welding. Refer to Heading 1-3.

1.5.4 High-Pressure Cylinder Hazards

When high-pressure *shielding gases* or other gases are used, cylinders must be stored, moved, and fastened down as described in the oxyfuel gas safety section. Regulators must be operated and cared for in the same way as oxygen or acetylene gas regulators.

1.6 RESISTANCE WELDING SAFETY

Resistance welding processes generally present fewer safety hazards than do the arc or oxyfuel gas processes. However, hazards from sparks, moving machinery, rotating machinery, and electrical shocks are present when resistance welding. All general shop safety rules should be followed.

1.6.1 Molten and Sharp Metal Hazards

Resistance welders must wear clear goggles, lightly shaded welding goggles, or a full plastic face shield to protect their eyes from flying sparks. The welder also should wear leather or cut-proof gloves when handling sheet metal, because of its sharp edges. Comfortable clothing may be worn, but all collars and pockets should be buttoned, and shirt cuffs and pant cuffs rolled down to prevent sparks from entering these areas. A cap is recommended to protect the hair from sparks or hot metal. To prevent accidental burns, matches, lighters, and combustibles should not be carried in the pockets of work clothes.

1.6.2 Electrical and Mechanical Hazards

A resistance welding operator should never work on a damp or wet floor. If the floor is damp, the welding operator should work on a platform of wood or other insulated flooring material.

The welding operator should not open control panels of resistance welding machines because of the hazards of high voltages within the control cabinets. The operator also must be careful to avoid being pinched when the electrodes of the resistance welder close. These electrodes are brought together with considerable pressure by hydraulic or pneumatic (air) force.

Operating and safety instructions should be given to all operators regarding the hazards and operating sequences of resistance welding machines. They should be taught where to place their hands during machine operation.

1.7 SAFETY AROUND WELDING ROBOTS

All the general instructions on welding and cutting safety apply to working with welding robots, but additional hazards are present in a robotic work station.

1.7.1 Robot Motion Hazards

Robots are generally rather large and heavy, but often move very rapidly from one weld location to the next. Because of the force and speed involved, an operator must never enter the *work envelope* (working volume) of a robot when it is in operation. The robot could hit and severely injure a worker who enters that space.

Normally, the robot's work cell is fenced or equipped with an electric eye "kill switch" to prevent workers from entering the space while the machine is operating. When the work envelope must be entered to repair or adjust the robot, positioner, or job-holding fixtures, **the power to the robot must be shut off and the switch locked or tagged to prevent it being mistakenly turned on.**

1.7.2 Spark Hazards

Since the welding operator normally is stationed away from the actual weld site, spark hazards are less severe than in other forms of welding. Comfortable work clothes may be worn. However, goggles are suggested to protect against possible flying sparks.

1.8 SPECIAL WELDING PROCESS SAFETY

1.8.1 Eye Hazards

Laser welding and electron beam welding present eye hazards to the welding operator and anyone else in the beam radiation area. **Everyone working in the area affected by laser or electron beam radiation must wear special filter lenses, since the rays are so much more intense than those involved with other forms of welding.**

1.8.2 Hearing Hazards

The plasma arc welding and cutting, arc gouging, and explosion welding processes produce very high sound levels that can cause hearing damage. Approved ear plugs or ear muffs should be worn for protection.

1.8.3 Explosion Hazards

The welding process known as *explosion welding* presents special hazards due to the explosives that are used. Workers who are involved in this process must be trained in the use of explosive materials, correct safety procedures, and the proper operation of safety devices.

SAFETY TEST (GENERAL SHOP SAFETY)

Write your answers on a separate sheet of paper. In multiple-choice questions, you may select more than one answer. Do not write in this book.

1. A person who is under stress may be distracted by thoughts of _____.
    a. love
    b. worry
    c. hate
    d. illness
    e. fatigue
2. How you feel about yourself and your job has much to do with your _____.
    a. age
    b. health
    c. attitude
    d. job skill
    e. all the above
3. Which of the following is not considered a physical factor in accidents?
    a. Health.
    b. Equipment failure.
    c. Time of day.
    d. Housekeeping.
    e. All the above.
4. Which of the following items should be stored in an approved steel cabinet?
    a. Paint.
    b. Oil.
    c. Cleaning chemicals.
    d. Welding rods.
    e. Leathers.
5. Fire extinguishers and fire blankets should be mounted on a surface that is painted _____.
    a. orange
    b. yellow
    c. yellow and black
    d. red
    e. light blue
6. All hand tools should be inspected for _____.
    a. size
    b. loose parts
    c. cleanliness
    d. worn power cords
    e. all the above
7. Which of the items listed should be covered during job or task training?
    a. Safety hazards.
    b. Location of fire extinguishers.
    c. Correct placement of hands and feet.
    d. Machine safety features.
    e. All the above.
8. When lifting, which is *not* a good practice?
    a. Bending your back slightly.
    b. Lifting with your leg muscles.
    c. Wearing a back brace.
    d. Keeping your back straight.
    e. Limiting lifting to 35 pounds.
9. Which of these factors promote safety in the shop?
    a. Staying healthy in mind and body.
    b. Becoming well-trained in the required job or task.
    c. Having a good attitude toward safety and safety rules.
    d. Engaging in friendly games in the shop.
    f. Thinking about pleasant things to pass the time.
10. Which of the following are good safety practices when working around electricity?
    a. Allowing only trained technicians to do repairs on electrical equipment.
    b. Always checking for loose connections and worn insulation with the power off.
    c. Assuming that all electricity is dangerous.
    d. Clearly marking voltages of 220V or higher with "danger" signs.
    e. All the above.
11. *True or False?* Housekeeping is considered an important factor in shop safety.
12. *True or False?* Voltages of 220V or higher are usually marked "Danger," and the voltage stated, such as "Danger: High Voltages."
13. *True or False?* People are never under stress when thinking about love.
14. *True or False?* Grinders and wire wheels are not considered hazardous machines like punch presses and metal shears.
15. *True or False?* Well-designed shops have low ceilings to improve ventilation and air flow.
16. *True or False?* To prevent back injuries, the back should always be kept straight and the object lifted with the leg muscles.
17. *True or False?* Some companies require workers to wear a back brace or support when lifting anything.
18. *True or False?* Corners or stairways that stick out into the shop are always painted bright orange or red for visibility.
19. *True or False?* Welding shops normally have a high ceiling to provide better air flow.
20. *True or False?* Engaging in horseplay in the shop may be considered a lack of wisdom or intelligence.

SAFETY TEST (OXYFUEL GAS WELDING AND CUTTING)

Write your answers on a separate sheet of paper. In multiple choice questions, you may select more than one answer. Do not write in this book.

1. Which of the following are dangerous light rays developed when oxyfuel gas welding or cutting?
    a. Incandescent.
    b. Strobe.
    c. Ultraviolet.
    d. Infrared.
    e. Gaseous.

2. Recommended fabrics for clothing worn while oxyfuel gas welding or cutting are _____.
   a. cotton
   b. wool
   c. flame-retardant treated cotton
   d. leather
   e. All the above.
3. *True or False?* Trousers with narrow cuffs are preferred for welding or cutting.
4. *True or False?* Gloves made from flame-retardant-treated cotton are recommended for use in welding or cutting.
5. Acetylene gas must never be used at a gauge pressure above _____.
6. *True or False?* Metal that has been welded and is left somewhere to cool should be marked "HOT" so nobody is burned by touching it.
7. Which of the following statements about cylinder handling are correct?
   a. Store cylinders in an upright position.
   b. Fasten cylinders to a wall or column.
   c. Secure cylinders with chains or steel band.
   d. Always move cylinders with the safety cap in place.
   e. Store cylinders in a special room or cage.
8. *True or False?* If the highest pressure marked on the face of a pressure gauge is 400 psig, it is recommended that this gauge should not be used at pressures above 200 psig.
9. *True or False?* To avoid possible burns, pliers should be used to pick up metal in the welding shop.
10. *True or False?* A pressure regulator is "off" when the regulator adjusting screw is turned out.
11. The oxyfuel gas flame should be lighted using _____.
    a. a wooden match
    b. a book match
    c. a spark lighter
    d. another torch
    e. an electric arc
12. *True or False?* Whenever the torch is not in your hand, it should be turned off.
13. *True or False?* When oxyfuel gas cutting, the area should be cleared of combustible materials and at least one fire guard be posted with a fire extinguisher.
14. Which of the following metals may create toxic or deadly fumes when welded?
    a. Beryllium.
    b. Lead.
    c. Chromium.
    d. Cadmium.
    e. Sodium cyanide compounds.
15. *True or False?* The American Welding Society suggests that a label be attached to the major equipment in an oxyfuel gas welding or cutting station to warn of hazards.

16. *True or False?* When an oxyfuel gas station is shut down, the gases should be bled from the cylinder outlet to the torch tip.

SAFETY TEST (ARC WELDING AND CUTTING)

Write your answers on a separate sheet of paper. In multiple-choice questions, you may select more than one answer. Do not write in this book.

1. *True or False?* It is possible to receive a sunburn (ultraviolet ray burn) while arc welding.
2. *True or False?* A high-pressure shielding gas cylinder may be moved by slightly tipping the cylinder and rolling it on its edge.
3. *True or False?* The AWS suggests that a label which points out the hazards of arc welding be placed on all major arc welding equipment.
4. Which of the following statements about arc welding is *not* true?
   a. Never weld on a damp floor.
   b. Never weld lying down unless you are lying on an insulated material.
   c. Never attempt to make repairs to welding equipment.
   d. Never make adjustments or repairs to the wire feed mechanism.
   e. Always make safety inspections with the power on.
5. Ultraviolet and _____ rays are the only harmful rays emitted while arc welding.
6. Toxic fumes should always be picked up before they reach the welder's _____. .
7. Which of the metals listed may give off toxic or deadly fumes when welded?
   a. Cadmium.
   b. Chromium.
   c. Beryllium.
   d. Zinc.
   e. Lead.
8. *True or False?* High-pressure shielding gas cylinders should be fastened securely to a wall or column when empty or stored.
9. *True or False?* When welding on metals that may emit toxic fumes, it may be necessary to wear an air-supplied purifier breathing apparatus.
10. When arc cutting, it is always necessary to clear the area of combustible materials and post a _____ .

SAFETY TEST (RESISTANCE AND SPECIAL WELDING PROCESSES)

Write your answers on a separate sheet of paper. In multiple-choice questions, you may select more than one answer. Do not write in this book.

1. Which type of face and eye protection is suggested while resistance welding?
   a. Clear grinding goggles.
   b. Clear unbreakable plastic goggles.
   c. Unbreakable plastic face shields.
   d. A #14 shade lens.
   e. Any of the above.
2. *True or False?* Hot metal may fly about while spot welding.
3. It is recommended that the following be worn while resistance welding:
   a. Comfortable clothing buttoned at the neck.
   b. Eye protection.
   c. Cut-proof gloves.
   d. Hat or cap.
   e. Clothing with covered pockets.
4. Which of the following items may be safely carried in your pockets while resistance welding?
   a. Butane lighter.
   b. Book matches.
   c. Plastic comb.
   d. Wood matches.
   e. Metal rule.
5. *True or False?* While resistance welding, standing on an uninsulated platform is a good safety practice.
6. Why is it dangerous to work within the work envelope of the robot?
7. Name two welding processes that produce sounds so loud that they could damage the welder's hearing.
8. Special training in the use of explosives is required to do which type of welding?
9. The greatest hazard while laser welding is possible damage to the welder's _____.
10. *True or False?* It is recommended that the operator of a welding robot wear leathers, a hat or cap, and safety glasses.

Chapter 2

PRINT READING

Virtually everything that is or manufactured for sale or made by an artisan at home is first drawn. The drawing may be a hand sketch or what is called a *mechanical drawing*. A mechanical drawing was traditionally made using drafting instruments; now, it is often done on a computer. By making a drawing, the designer can see what the object will look like. All parts may be checked to see if they fit properly. Rotating or sliding parts can be checked to make certain that they clear all other parts. Drawings also serve to accurately communicate the designer's ideas to the people who will manufacture the part or assembly. All information required to make the parts must be on the drawing, so that the parts will be made exactly as the designer wants them. Persons using the drawings must be able to read them properly, so that they can accurately make the parts shown. The sizes of the various features of the part are given on the drawing as dimensions. The materials to be used and other needed information and special notes are also given on a mechanical drawing.

2.1 TYPES OF DRAWINGS

Drawings that are used in a shop to produce objects are known as *working drawings*. Working drawings exist in two forms. They are the assembly drawing and the detail drawing. These drawings can be made on a *translucent* material (a material that light can travel through) so that copies may be made of them using a light source. For this reason, the drawings are usually made on a special drafting paper or a treated cloth or a plastic film. Copies made from the original drawing are called *prints*. Up to a few years ago, most prints made were produced using a process that resulted in white lines on a blue background. Today, most prints are on white paper, but are still sometimes referred to as "blueprints." The proper term, however, is merely "prints." Drawings can also be made on standard white paper, rather than a translucent material. Copies of these drawings often can be made on standard office copying machines.

2.1.1 Assembly Drawings

The *assembly drawing* shows the object to be made as it would appear in a fully assembled, ready to use form. Important location dimensions and overall size dimensions are given on the assembly drawing. However, not all the dimensions that are required to make each piece in the assembly are shown on the assembly drawing.

2.1.2 Detail Drawings

Individual *detail drawings* are made of each different part in the assembly. These drawings may be made on separate sheets, but sometimes, a number of detail drawings may be made on one larger sheet of drafting paper. Every view required to make a part is shown on the detail drawing, along with every dimension required to produce it.

To be able to read a print, the theory of *orthographic projection* must be understood. Almost all assembly and detail drawing are made using the theory of orthographic projection. Understanding how a drawing is made helps a person to find various areas, lines, holes, welds, and other details on a drawing. This knowledge also helps in finding the sizes of parts on a detail drawing or the type of weld to make and where to place it.

2.2 THE THEORY OF ORTHOGRAPHIC PROJECTION

To completely describe an object and show all its holes, depressions, spaces, curves, and angles, it may be necessary to view it from various sides. All these sides and views must be shown, however, on a flat sheet of drawing material. To make all these views possible in a flat plane, the orthographic projection method of drawing was developed.

To make an orthographic drawing, the object to be drawn is imagined as being located and viewed inside a clear glass box. See Figure 2-1. All points and lines are imagined as projected onto the six sides of the box, just as they would be seen by a person viewing the object at each of the sides. The object is then drawn on each of the six sides of the box, Figure 2-2.

When the sides of the clear box are folded out until they are flat, all sides of the object are seen in one flat plane. See Figure 2-3.

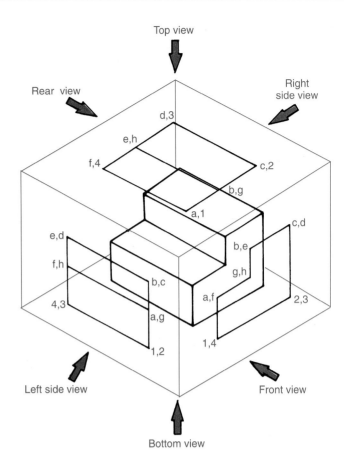

Figure 2-2. *Six views of an object are projected onto the sides of the orthographic box. For clarity, only three of the projected views are shown here.*

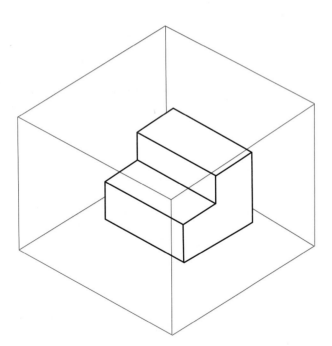

Figure 2-1. *The orthographic projection theory is illustrated by imagining the part enclosed in a clear glass box.*

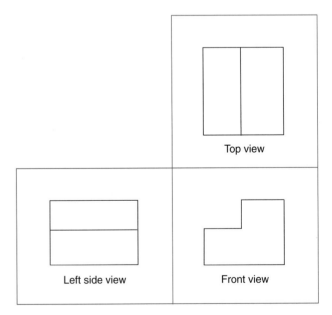

Figure 2-3. *The developed views can be folded out from the orthographic box to form a flat surface.*

2.3   VIEWS

When discussing orthographic drawings, the *direction* from which the object is viewed is given a name. Generally, the view that shows the most about the overall shape of the object is called the ***front view***. The view to the right of the front view (as the orthographic box is unfolded in your imagination), is known as the ***right side view***.

The view to the left of the front view is the *left side view*. The view above the front view is called the *top view* and the view below the front view is the *bottom view*. See Figure 2-2. A *back view* is possible on an orthographic drawing, but it is very seldom used.

To produce an object, it is seldom necessary to show the views of all six sides. A working drawing typically shows three views of an object, which is sufficient to describe its shape and give all the required dimensions. It is even possible with some objects (such as spheres, cubes, or cylinders) to describe and dimension the object with only *one* view.

2.4 USING A WORKING DRAWING

To find the size of a line, hole, surface, or other feature, it is necessary for the person reading a print to be able to follow a given point or line from one view to another. In Figure 2-4, each point on the object is designated by a number or letter. Each point on the upper surface has been given a letter; each point on the lower surface has been given a number. Lines are defined by the letter or number at each end, such as (*h-i*). The number or letter to the left side or closest to the viewer is stated first. Points are defined as (*a,b*) and surfaces by all points that define the surface as (*b,a,l,c*). It should be noted that a line in one view (*k-d*) will be a line in one other view (*k-d*) and a point or corner in the third view (*k,d*). Also, that a point or corner in one view (*1,2*) will be a line (*1-2*) in the other two views. And finally, that a surface in one view (*b,a,l,c*) will become a line in the other two views (*1-a*) and (*a-b*).

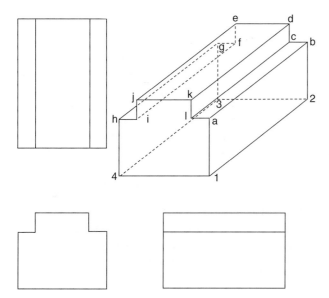

Figure 2-4. *An object with all corners identified by letters or numbers.*

2.4.1 An Exercise in Defining Points on a Drawing

In the three-dimensional view of the object in Figure 2-4, each end of each line has been given a letter or number. Working with a friend or in class, mentally and verbally define each line, point, and surface in each of the three orthographic views. *Do not mark in the textbook.* See your instructor for the correct answers.

A surface is identified by all the letters on the corners of that surface. The surface letters or numbers are given consecutively in either a clockwise or counterclockwise order, starting at any point. However, *lines* first must be given the letter or number to the left or closest to the viewer, then the point to the right or farthest away is given. Example: (*k-d*) in the right side view. Note that the numbers or letters are separated by a dash. *Points* are defined in the same way (the closest letter or number first), but separated by a comma. Example: (*a,b*) in the front view.

2.5 TYPES OF LINES USED ON A DRAWING

Refer to Figure 2-5 to see the various types of lines that may be used on a drawing. Lines on an original drawing are always made as black as possible. The thickness and shape of the various lines, however, will vary as their purpose varies on the drawing.

2.5.1 Object Lines and Hidden Lines

The outline of the object is a solid black line called an *object line*. It has a thickness of about 0.030" (0.70mm). Lines hidden from view by material in front of them are called *hidden edge lines*. They are lines that are made up of a series of 1/8" (3.2mm) dashes. These hidden lines are usually drawn about 0.020" (0.50mm) thick.

2.5.2 Centerlines

The center of a radius, circle, or cylinder is marked by a *center line*. This is a line about 0.010" (0.30mm) thick. It is made up of a series of 3/4" (19mm) long and 1/8" (3.2mm) short dashes and it runs through the exact center of a circular part and hole.

2.5.3 Dimension Lines

Dimension lines and dimensions are placed about 1/2" (12.7mm) away from the outer edge of the object. When a large number of dimensions is required, each dimension line is placed farther from the object at 1/2" (12.7mm) intervals. Dimension lines are 0.010" (0.30mm) thick. A long, thin arrowhead appears at each end of a *dimension line*.

2.5.4 Extension Lines

The ends or terminal points of a dimensioned line are marked by lines called *extension lines*. They are the same

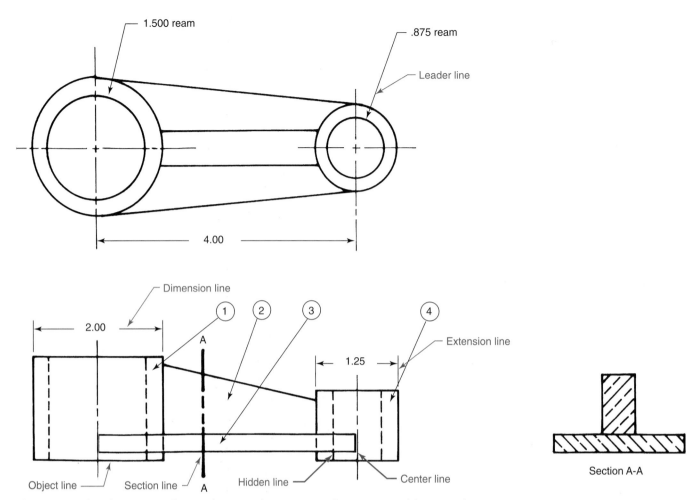

Figure 2-5. *This dimensioned parts drawing shows correct line types and line weights.*

thickness of a centerline, but are solid lines. The size of a given line or feature on the object is placed on a dimension line between a set of extension lines.

2.5.5 Section Lines

Occasionally, it is necessary to show the inside features of an object. To do so, it may be necessary to *theoretically* cut through the object at some point. The direction of the cutting line is shown with a line called a **section line.** This is the thickest line that may appear on a drawing. It is made up of a series of 3/4" long (19mm) and double 1/8" short (3.2mm) lines that are about 0.040" (0.90mm) thick.

2.5.6 Leader Lines

The size of a corner radius or size for a hole is given by using a **leader line** that points to the edge of the circle and "leads" the reader out to a clear area on the drawing where a diameter or dimension is given.

The difference in line types is used so that a person looking at a drawing may first see the object that stands out because of the heavy lines used. Then, by following lines and points from view to view, the person reading the print can find the extension lines and dimension lines, and finally the dimension that he or she is looking for.

2.6 DRAWINGS MADE TO A SCALE

If an object is drawn full size, it is said to be drawn *full scale.*

Large objects like airplanes, automobiles, houses, ships, or skyscrapers cannot be easily drawn full size (although this *is* occasionally done). To get some objects to fit on the size paper being used, however, the object must be *drawn to scale.* The object may be drawn one-half its actual size, or *half-scale.* This means that every actual 1" (25.4mm) on the object is drawn only 1/2" (12.7mm). Half-scale is shown as follows: 1/2 = 1. Objects may be drawn to any scale. Some very small parts, such as those used in a watch or an electronic device, may have to be increased in size on a drawing to be seen and dimensioned more easily. The scale for such a drawing might be 10 = 1. A 1/8" (3.2mm) part at a 10 = 1 scale would be drawn 1 1/4" (31.8mm) in size.

The scale of a drawing may be shown somewhere near the actual part drawing or may be indicated in the drawing's title block.

2.7   THE TITLE BLOCK

Large companies often have drafting paper or material printed in large quantities with their own specially designed title block imprinted on each sheet. Figure 2-6 shows a typical drawing title block, which contains a great deal of information that is required to make a part or an assembly.

The information given in a title block may includes the following:

- *Drawing Number.* Each drawing has a unique number to identify it.
- *Dash Number.* A dash followed by a number may be added to the drawing number to distinguish between right- and left-hand parts or assembly and detail drawings.
- *Permitted Part Size Tolerances.* Tolerance defining how accurately a part must be made.
- *Name Given the Part or Assembly.* A descriptive name that describes the part or its function.
- *Sheet Number.* Many sheets may be required to make all the parts required for a single assembly. If so, each page is numbered as follows: "Sheet 1 of 20," "Sheet 2 of 20," and so on.
- *Drawing Sheet Size.* Drawings are made on a variety of sheet sizes. The assembly drawing is normally on a large sheet with detail drawing on smaller sheets. The sheet size is shown as a letter code running from A to E. An "A" sheet is about 9" by 11" and "E" is about 36" x 48".
- *Scale of the Drawing.* Drawings may be made actual size, or they may be made smaller or larger than the actual size of the object. A drawing that is twice the actual size of the part would show a scale of 2 = 1 or 2/1. A drawing made half the actual size of a part would be in a scale of 1/2 = 1 or simply 1/2.
- *Responsible Signatures Area.* Spaces are often provided for signatures of the part designer, draftsperson, drawing checker, supervisor, inspector, or others involved in the production of a part.
- *Material.* The material to be used in making the part is shown.
- *Change Orders.* When requested changes have been made, the change is briefly described and a date of the change is given. Parts must be made according to the latest drawing change.

Other information may be given in a title block, depending on the company that designs and draws the part or assembly.

Drawings that are made on a computer may be printed out on paper that has the title block at one side, as shown in Figure 2-7. Often the title block is stored on the computer, and the drawing and title block are printed onto blank paper.

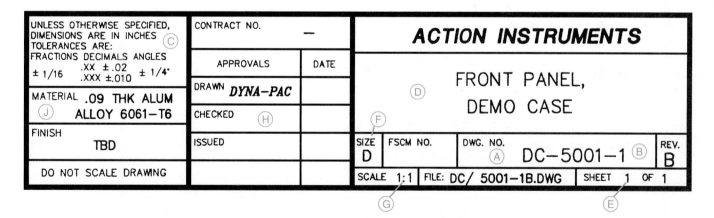

Figure 2-6. *A typical title block with areas or sections identified. A—Drawing Number. B—Dash Number. C—Permitted Part Size Tolerances. D—Name Given the Part or Assembly. E—Sheet Number. F—Drawing Sheet Size. G—Scale of the Drawing. H—Responsible Signatures Area. J—Material.*

Figure 2-7. *A computer-type title block.*

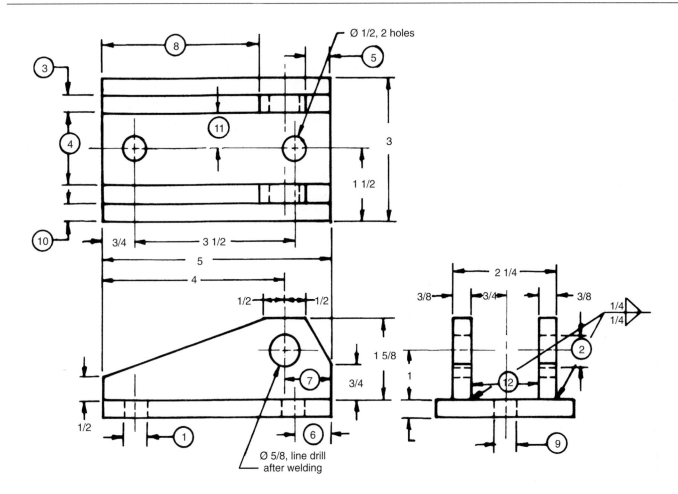

Ø 1/2, 2 holes

Ø 5/8, line drill after welding

TEST YOUR KNOWLEDGE

Answer these questions on a separate sheet of paper. Do not write in this book.

Refer to the figure above and read or calculate the following dimensions:

1. What is the distance **1**?
2. What is the distance **2**?
3. What is the distance **3**?
4. What is the distance **4**?
5. What is the distance **5**?
6. What is the distance **6**?
7. What is the distance **7**?
8. What is the distance **8**?
9. What is the distance **9**?
10. What is the distance 10?
11. What is the distance **11?**
12. What is the distance **12?**

Refer to Figure 2-6 for the answers to the following questions:

13. What is the name of the draftsperson or company who drew this part?
14. What is the scale of this drawing ?
15. What size paper is this drawing made on ?
16. What is the required accuracy of a size given as a three-place decimal?
17. What is this drawing number ?
18. How many other sheets are there for this part
19. What metal and metal thickness is used for this part?
20. What is the name of this part ?

Whether working in the shop or on a job site, a welder must know how to correctly read prints and be able to make accurate measurements.

Chapter 3

READING WELDING SYMBOLS

LEARNING OBJECTIVES

After studying this chapter, you will be able to:
* Identify the five basic welding joints.
* Identify and describe the various welds that may be used in each welding joint.
* Label the parts or areas of a grooved butt weld and a fillet weld.
* Locate and apply required weld and joint information from the AWS welding symbol.

3.1 BASIC WELDING JOINTS

There are five basic types of welded joints. They are the *butt joint, corner joint , tee joint, lap joint*, and *edge joint*. See Figure 3-1. The five basic types of welded joints can be made in the four different welding positions shown in Figure 3-2. For illustration purposes, most of the welds shown in this chapter will be in the flat welding position. Remember that you may be called upon to weld in any position.

The names or labels given to the various parts of a groove-type weld are shown in Figure 3-3. Study these names and make them a part of your vocabulary.

The edges of the *base metal* (metal to be welded) are often prepared for welding by cutting, machining, or grinding. This preparation is done to ensure that the base metal is welded through its entire thickness. Various edge shapes are created on thick metal sections to open up the joint area. This provides a space large enough to permit welding at the bottom of the joint.

The butt joint may be welded from the top, the bottom, or from both sides. A butt joint may be welded using one of the *edge shapes* shown in Figure 3-4: square-groove, bevel-groove, V-groove, U-groove, J-groove, flare-bevel-groove, flare-V-groove, or flanged-butt.

When the edge of one piece of base metal is placed on the surface of another piece at the very edge, this joint is

called a corner joint. Corner joints may be welded from the inside or outside of the corner. Occasionally, the corner may be welded from both sides. Figure 3-5 shows the methods used to prepare an inside or outside corner joint for welding.

The T-joint obtains its name from the placement of the base metals to form a "T" shape. If one piece of metal is placed on the surface of the other in any location, other than at the very edge, the joint is called a T-joint. See Figure 3-6 for methods of preparing the metal edges for welding several types of T-joints.

The metals that form the lap joint are seldom altered in preparation for welding. Edge joints may be prepared for welding in a number of ways, as shown in Figure 3-7.

3.2 THE WELDING SYMBOL

The welding symbol described on the following pages was developed by The American National Standards Institute (ANSI) and the American Welding Society (AWS). It is described in detail in the publication ANSI/AWS A2.4-93, *Standard Symbols for Welding, Brazing, and Nondestructive Examination.*

The entire symbol, as shown in Figure 3-8 with all of its numbers and other symbols, is called the *welding symbol.* Welding symbols are used on drawings of parts and assemblies which are joined together by welding. A welding symbol may appear in any view on the drawing. Whenever two or more pieces are joined by welding, the assembled item is called a *weldment*. When the pieces of a weldment are assembled, the lines along which their edges and surfaces come in contact are called *joints*.

The drawing of a weldment seldom shows how the edges are to be prepared or how the completed weld appears. The drawing shows only how the parts come together and what type of joint they will form. Occasionally, when an unusual or very complex weld joint is to be made, a "detail" drawing of the joint may be drawn with the joint preparation and weld shape shown and dimensioned. Refer to Figure 3-1 for the types of welded joints and the types of welds used on the various joints.

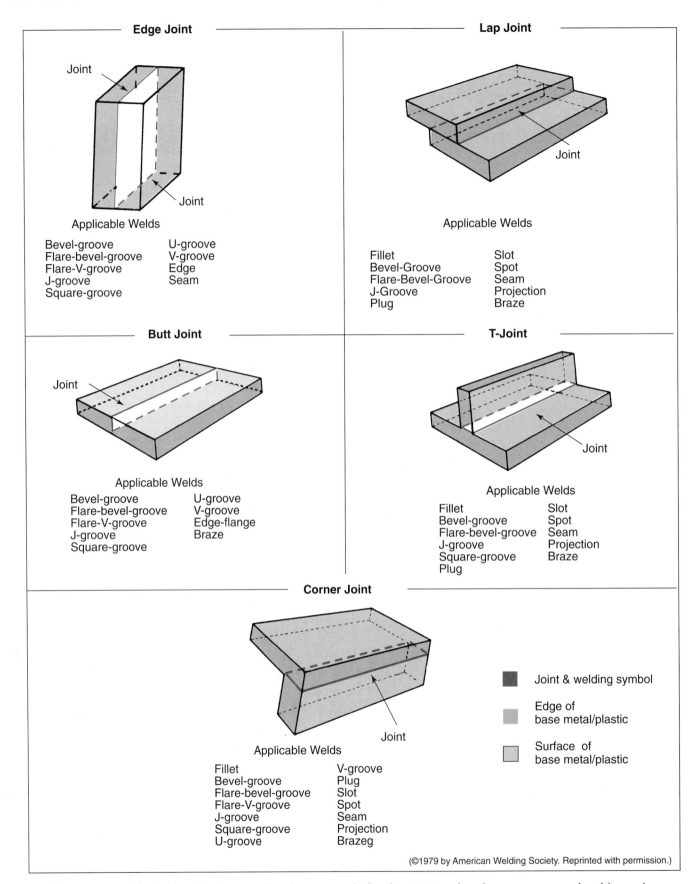

Figure 3-1. *Basic welding joint designs. Note that there are only five basic joints, but that many types of welds can be placed on each type of joint. (ANSI/AWS A3.0-94)*

Within the figure:

Edge Joint

Joint

Joint

Applicable Welds

Bevel-groove	U-groove
Flare-bevel-groove	V-groove
Flare-V-groove	Edge
J-groove	Seam
Square-groove	

Lap Joint

Joint

Applicable Welds

Fillet	Slot
Bevel-Groove	Spot
Flare-Bevel-Groove	Seam
J-Groove	Projection
Plug	Braze

Butt Joint

Joint

Applicable Welds

Bevel-groove	U-groove
Flare-bevel-groove	V-groove
Flare-V-groove	Edge-flange
J-groove	Braze
Square-groove	

T-Joint

Joint

Applicable Welds

Fillet	Slot
Bevel-groove	Spot
Flare-bevel-groove	Seam
J-groove	Projection
Square-groove	Braze
Plug	

Corner Joint

Joint

Applicable Welds

Fillet	V-groove
Bevel-groove	Plug
Flare-bevel-groove	Slot
Flare-V-groove	Spot
J-groove	Seam
Square-groove	Projection
U-groove	Brazeg

Joint & welding symbol

Edge of base metal/plastic

Surface of base metal/plastic

(©1979 by American Welding Society. Reprinted with permission.)

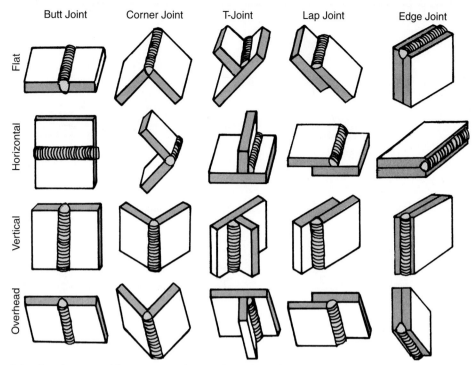

Butt Joint   Corner Joint   T-Joint   Lap Joint   Edge Joint

Flat

Horizontal

Vertical

Overhead

Figure 3-2. *Each of the five basic weld joints may be made in four different welding positions.*

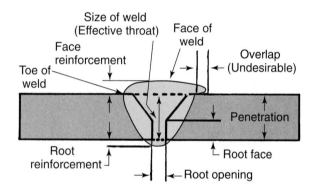

Size of weld
(Effective throat)

Face of
weld

Face
reinforcement

Overlap
(Undesirable)

Toe of
weld

Penetration

Root face

Root
reinforcement

Root opening

Figure 3-3. *Approved terms used to describe various parts of a completed groove-type weld.*

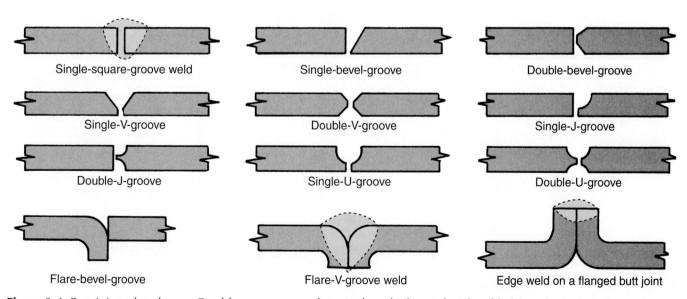

Single-square-groove weld   Single-bevel-groove   Double-bevel-groove

Single-V-groove   Double-V-groove   Single-J-groove

Double-J-groove   Single-U-groove   Double-U-groove

Flare-bevel-groove   Flare-V-groove weld   Edge weld on a flanged butt joint

Figure 3-4. *Butt joint edge shapes. Double grooves are often used on thick metal and welded from both sides. Note that the metal is bent to form the flare-groove, flare-V-groove, and edge-flange butt joints.*

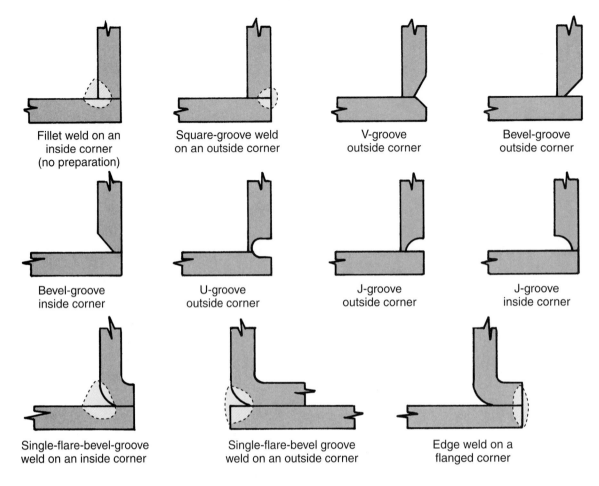

Figure 3-5. *Corner joint edge shapes. The weld is shown on five joints where the metal was not cut or ground.*

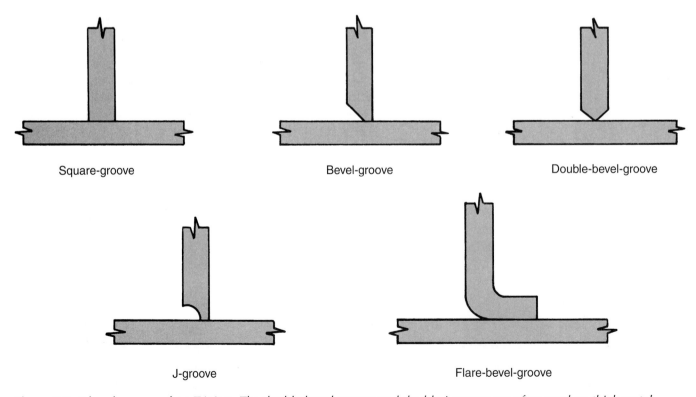

Figure 3-6. *Edge shapes used on T-joints. The double-bevel-groove and double-J-groove are often used on thick metal.*

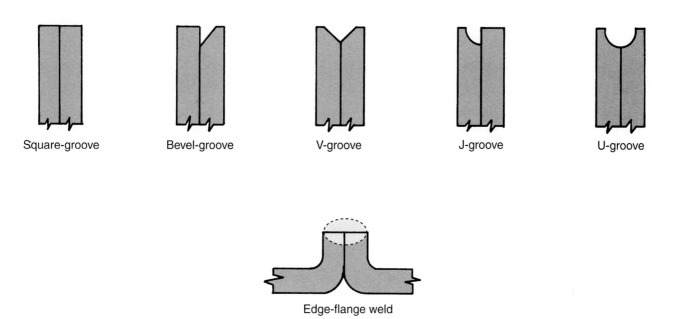

Square-groove   Bevel-groove   V-groove   J-groove   U-groove

Edge-flange weld

Figure 3-7. *Edge shapes used on edge joints.*

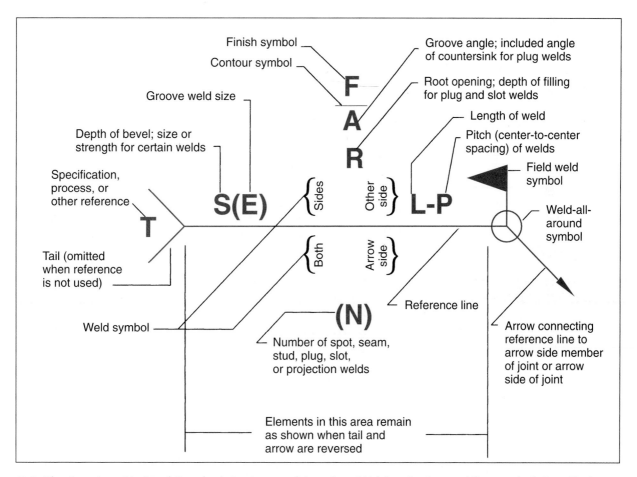

Figure 3-8. *The American National Standards Institute and American Welding Society welding symbol. Specific locations have been assigned on the welding symbol for various information and sizes. (ANSI/AWS A2.4-93)*

A complete welding symbol contains all the information about a welded joint. The welding symbol may appear in any view of the drawing. *A welding symbol applies to only one joint and applies to that joint only until it changes direction.* There are a few exceptions to this rule—they will be discussed later in this chapter. Some of the following information is given to the welder on the welding symbol:

- How to prepare the edges of the base metal prior to welding.
- What welding process to use.
- What type of weld to make.
- Where to place the weld.
- The size of the weld.
- The shape of the weld face.
- How to finish the weld surface after welding is completed.

Much more information regarding the weld is also given on the welding symbol. Dimensions on a welding symbol may be in SI Metric units or US Conventional units of measurement.

Part of the welding symbol is the *weld symbol*, which shows what type of weld is to be placed in a joint. Figure 3-9 shows the weld symbols used on the ANSI/AWS welding symbol.

Information given in each part or area of the welding symbol will be explained in later paragraphs. A number of weld drawings, with their corresponding welding symbols, will be shown to illustrate the information given in the various areas of the complete welding symbol. The edges of the weld joint, as they would be prepared and fitted up prior to welding, will be shown in blue. A completed weld for the welding symbol will also be shown.

3.3 THE REFERENCE LINE, ARROWHEAD, AND TAIL

The *reference line*, shown in Figure 3-10, is always drawn as a *horizontal* line. It is placed on the drawing near the joint to be welded. All information to be given on the

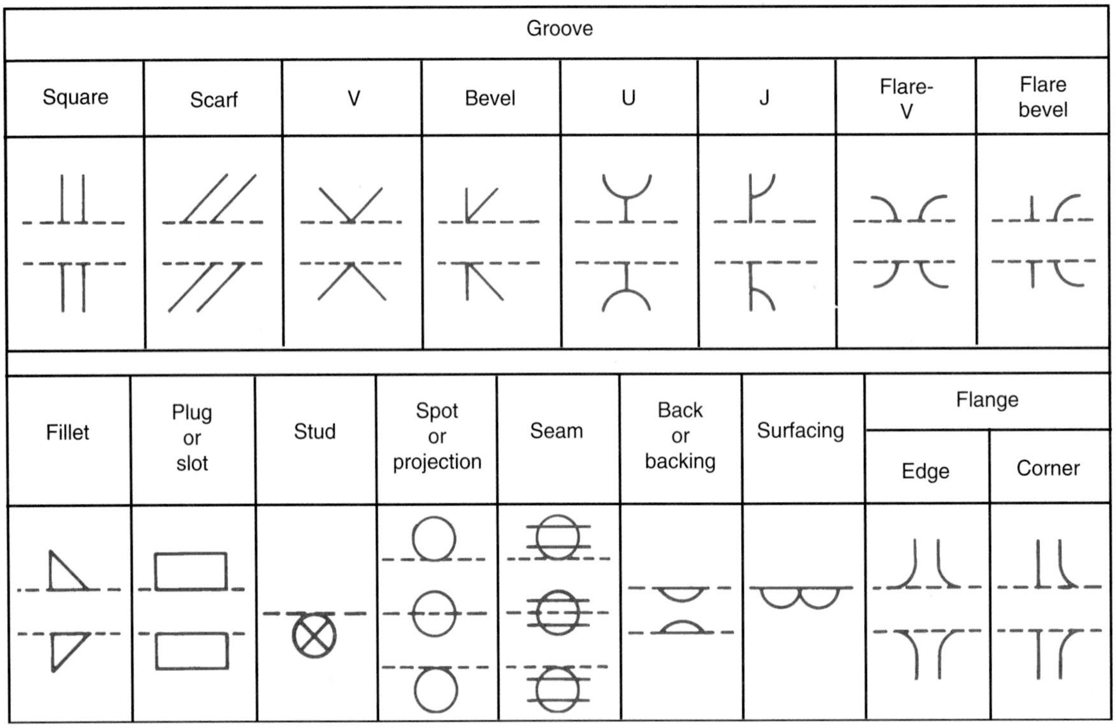

Figure 3-9. *Weld symbols. These are a part of the complete welding symbol. (ANSI/AWS A2.4-93)*

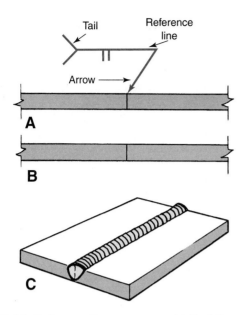

Figure 3-10. *Reference line, arrow, and tail of the welding symbol. A—Drawing and welding symbol. B—The edge shape of the metal. C—Completed weld.*

welding symbol is shown either above or below this horizontal reference line. Information given on a welding symbol is always shown in the same location, as indicated in Figure 3-8. The information is read from left to right whether it is above or below the reference line.

The *arrow* may be drawn from either end of the reference line. The arrow always touches the line that represents the welded joint.

The *tail* is used only when necessary. It may be used to give information on specifications, the welding process used, or other details required but not shown on the welding symbol. A number (such as 1, 2, or 3) may be used in the tail to refer the user to a note elsewhere on the drawing. A company may use its own number or letter codes in

the tail to indicate the welding process, procedure, finishing method, or company specification.

If no tail is used, a note such as "Unless otherwise specified, all welds will be made in accordance with Specification No. XXXX," will be displayed somewhere on the drawing.

3.4 WELD SYMBOLS

The *weld symbol* shown on the complete welding symbol indicates the type of weld to be made on a weld joint. It is a miniature drawing that shows how the edge or edges of the joint are prepared before welding.

Figure 3-11 illustrates how some of the weld symbols shown in Figure 3-9 are used on a welding symbol. The vertical line used with a fillet, bevel-groove, or J-groove weld is *always* drawn to the left.

3.5 THE ARROW SIDE AND OTHER SIDE

On the drawing of a welded part, the arrow of the welding symbol touches the line to be welded. The metal has two sides. The side of the metal which the arrow touches is always the *arrow side*. The opposite surface from the arrow is called the *other side*.

Because of the joint position on many weldments, there is no inside or outside, top or bottom, left or right. To simplify identifying the location of the weld, the terms arrow side and other side are used. On the welding symbol, the arrow side weld information is always shown *below* the reference line. The other side weld information is always shown *above* the reference line.

It is not always possible to place the welding symbol on the side to be welded. The drawing is sometimes too

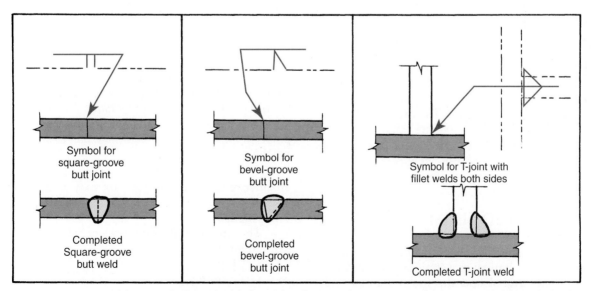

Figure 3-11. *Comparing welding symbols and actual welds. Phantom lines are not shown on a weld symbol. They are used here, however, to illustrate that the weld symbol is a miniature drawing of the edge shape and the type of weld used. The vertical line in the bevel-groove and fillet weld symbols is always drawn to the left.*

crowded and complicated. See Figure 3-12 for examples of the use of the arrow side and other side on the welding symbol.

If the weld is not to remain in an "as welded" condition, a *finish symbol* is used on the welding symbol. See Figure 3-14. The finish symbol indicates the method of finishing. A surface texture or degree of finishing may also be added if required. If all welds are to be finished in the same

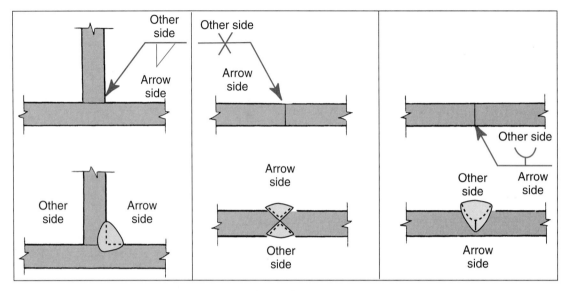

Figure 3-12. The side of the metal that the arrow touches is the arrow side. Placement of the weld symbol on the welding symbol determines whether the weld will be made on the arrow side or on the other side from the arrow.

3.6 ROOT OPENING AND GROOVE ANGLE

The *root opening* is the space between the pieces at the bottom or root of the joint. Prior to welding, the two pieces are spaced apart the distance indicated by the root opening. This root opening may be specified on the drawing in metric units, in fractions of an inch, or as a single-place decimal of an inch. The root opening size appears inside the weld symbol on the complete welding symbol.

The included angle or total angle of a groove weld is shown beyond the weld symbol. See Figure 3-13. When preparing the edges for a V-groove welding, half the groove angle is cut on each piece. When placed together, the combined angles will total the angle shown.

When a bevel-groove or J-groove weld is used, only one piece of metal is cut or ground. The arrow at the end of the welding symbol is bent to along its length the left or right to point to the piece that is to be cut or ground. See Figure 3-13, view D, and Figure 3-14, view A.

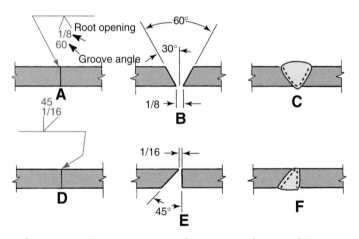

Figure 3-13. Root opening and groove angle. A and D show the weld symbol for a groove weld. B and E show the pieces cut and set up for welding. C and F show the completed weld. Note that the bend in the arrow at D points to the left piece, which is the part to be cut or machined.

3.7 CONTOUR AND FINISH SYMBOLS

The shape or contour of the completed weld bead is shown on the welding symbol as a straight or curved line between the weld symbol and the finish symbol. The straight contour line indicates that the weld bead is to be made as flat as possible. The curved contour line indicates a normal convex or concave weld bead. See Figure 3-14.

manner, a note on the drawing may indicate the finish used. Users of the finish symbol may create their own finish symbols.

The American Welding Society lists the following finish symbols:
C: Chipping
G: Grinding
M: Machining
R: Rolling
H: Hammering

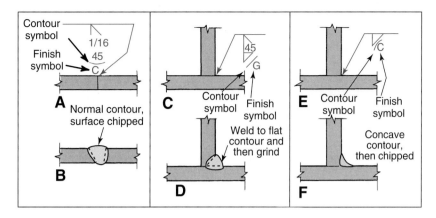

Figure 3-14. *Weld contour and finish symbols. In A, C, and E, contour and finish symbols are shown on the welding symbol. B, D, and F illustrate the shape and finish of the completed weld face.*

3.8 DEPTH OF BEVEL, SIZE OR STRENGTH, AND GROOVE WELD SIZE

The "S" position on the welding symbol indicates the ***depth of bevel*** or the depth of preparation for a groove-type joint. Information in the "S" position may also indicate the *size* or *strength* of certain welds.

The size of a fillet weld is the length of the legs (sides) of the triangle that can be drawn inside the cross section of the finished weld, as shown in Figure 3-15. The length of the legs on a typical fillet weld is equal, so only one dimension is given in the "S" area of the welding symbol. If the fillet weld has unequal legs, the size of each leg is given in parentheses ( ). The weld shape of this fillet weld may be shown on the part drawing to indicate which is the short leg. See Figure 3-16.

Weld size or strength information is also required for resistance spot or seam welds, electron beam welds, and plug and slot welds. This information also appears in the "S" position.

Groove weld size is given in parentheses in the "E" position on the welding symbol. Groove weld size is the depth to which the weld penetrates into the base metal. See Figure 3-17 for examples of depth of bevel and groove weld size. In Figure 3-17A, the depth of bevel is 1/4" and the groove weld size is 3/8". In Figure 3-17B, the "S" dimension states that the metal is cut to a depth of 3/8" on both sides. The "E" dimension shows that the depth of the weld is 5/8" on both sides.

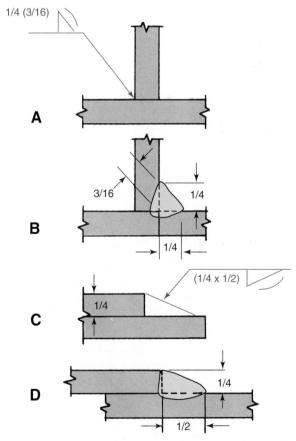

Figure 3-16. *Fillet weld size and shape. A—The single dimension indicates both legs are equal and 1/4" in size. The depth of the weld is 3/16". B—How the finished weld will look. C—Two dimensions in parentheses indicate unequal legs. Relative size of the legs is shown on the working drawing. D—How the finished weld will look.*

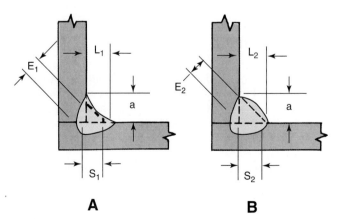

Figure 3-15. *Comparing the weld size and strength of a concave and a convex fillet weld. The fillet welds at A and B appear to be the same size (dimension "a"). The leg size (L_1 and L_2) appears the same in both welds. However, the actual weld size in A (S_1) is smaller than the actual weld size (S_2) in B. Also, effective throat E_1 is smaller than effective throat E_2.*

The groove weld size (depth of weld) and depth of bevel (depth of preparation) are generally determined by welding codes or specifications, or by a welding engineer.

When no groove weld size is shown for a single-groove or double-groove weld, complete penetration is required. See Figures 3-17 and 3-18 for examples of depth of bevel and groove weld size.

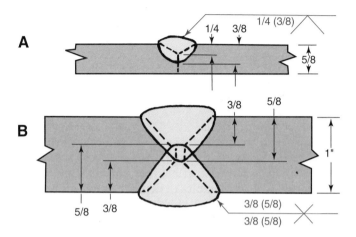

Figure 3-17. *The depth of the edge shape and groove weld size. A—the edge shape is cut 1/4" deep. The weld size or depth of penetration is 3/8". B—The edge shape is a double-V-groove. The edge shape is cut 3/8" deep and the depth of penetration is 5/8" on both sides of the joint.*

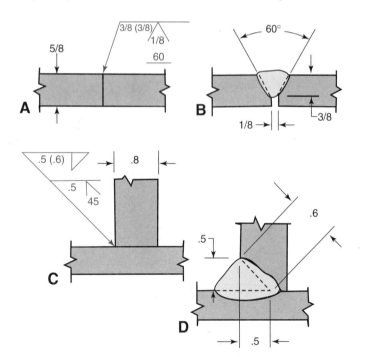

Figure 3-18. *Groove weld size. Groove weld size or depth of penetration is shown in parentheses at A and C. Note, at B, that the groove weld size is less than the metal thickness. At D, it is greater than the depth of bevel.*

3.9 LENGTH AND PITCH OF THE WELD

In many welded parts, it is not necessary to weld continuously from one end of the joint to the other. To save time and expense, where strength is not affected, short sections of weld may be spaced across the joint. This is called *intermittent welding.*

On intermittent welds, the *length dimension* is used to indicate the length of each weld. The *pitch dimension* indicates the distance from the center of one weld segment to the center of the next. See Figure 3-19 for examples of such welds. The length and pitch dimensions are always shown to the right of the basic weld symbol on the welding symbol.

When intermittent fillet welds are required on *both* sides of a welding joint, they may be one of two types. One type is chain intermittent welding; the other is staggered intermittent welding. The welds on either side of a *chain intermittent weld* begin and end at the same spot. The welds line up with each other on each side of the joint. The weld symbols also line up on each side of the reference line. *Staggered intermittent welds* are offset so the welded segments do not line up on each side of the joint. This is shown on the welding symbol by offsetting the fillet weld symbols. See Figure 3-19, views C and D.

Continuous and intermittent welds may be made on the same joint. In such a case, the drawing will use dimensions to show where each weld symbol's effectiveness begins and ends. See Figure 3-20, view A.

A spacing different from the regular pitch is used between the end of the continuous weld and the beginning of the intermittent weld. See the 4" dimension in Figure 3-20, view B. This spacing is equal to the intermittent pitch minus the length of one intermittent weld. The spacing between the continuous and intermittent welds in Figure 3-20, view B, equals the pitch minus the length, or 6" – 2" = 4", as shown.

3.10 BACKING WELDS AND MELT-THROUGH SYMBOLS

Weld joints that require complete penetration may be welded from both sides. A *stringer bead* (single pass weld without a weaving motion) may be all that is required on the side opposite a groove weld to ensure complete penetration. In such cases, a *backing weld symbol* may be used, Figure 3-21.

The *melt-through symbol* is used when 100% penetration is required on one-side welds, Figure 3-21.

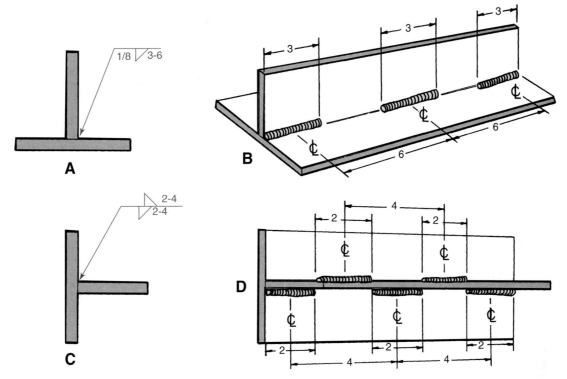

Figure 3-19. *Length and pitch dimensions of weld. A—Note placement of length (3) and pitch (6) on the welding symbol. B—Weld shows a series of 3" long welds which are 6" apart from center-to-center of the welds. C and D—Staggered weld. Notice staggered fillet symbols in C.*

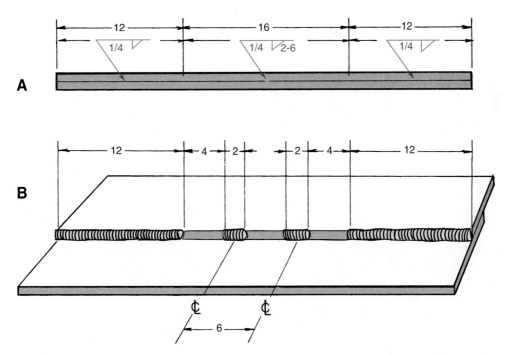

Figure 3-20. *Continuous and intermittent welds. Note that the dimensions on the top welding drawing limits use of the welding symbol to the distance shown. Note also that the spacing between continuous and intermittent weld is equal to the pitch minus the length of one intermittent weld (4" in this application).*

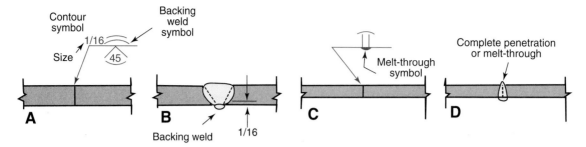

Figure 3-21. *Backing welds and melt-through symbols. The melt-through symbol is used on welds that are welded from one side only and require 100% penetration, as shown in C and D. A backing weld may be used to obtain 100% penetration when welding is possible on both sides, as shown in A and B. Note that contour symbols and dimensions may be used with these symbols, as in A.*

3.11 WELD-ALL-AROUND AND FIELD WELD SYMBOLS

Directions given on a welding symbol are no longer of any value when the weld joint makes a sharp change in direction, such as going around a corner. When the joint changes direction sharply, either a new welding symbol or a weld-all-around symbol may be used. The *weld-all-around symbol* indicates that the same type weld joint is to be used on all edges of a box or cylindrical part. See Figure 3-22.

Some parts are assembled and welded in the shop. It is often necessary to take parts to the job site or into the field to make final assembly and welds. When welds are to be made away from the shop, a *field weld symbol* is used, as in Figure 3-22, view D. On the print, the field weld symbol is a small black flag located at the arrow end of the reference line. If a weld is to be made in the shop, the field weld symbol is not used. Refer to Figure 3-8.

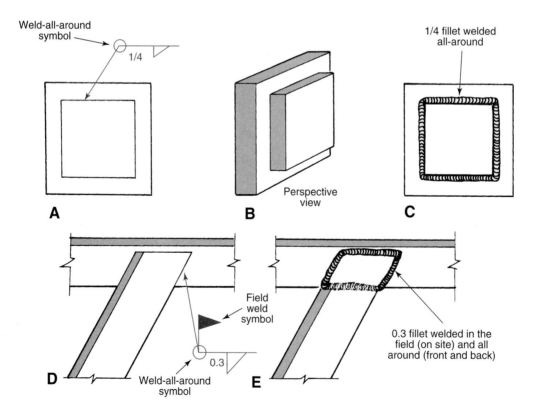

Figure 3-22. *Weld-all-around and field weld symbols. Note the 0.3" fillet in D and E is welded in the field. It is welded all around the angle iron both front and back.*

3.12 MULTIPLE REFERENCE LINES

When a sequence of operations is to be performed, two or more reference lines may be used. The reference line nearest the arrow indicates the first operation. The last operation to be performed is on the reference line furthest from the arrow. See Figure 3-23.

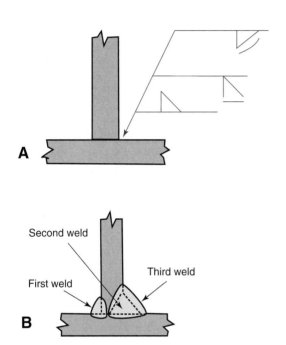

Figure 3-23. Use of multiple reference lines. In A, three reference lines are used. The first weld is a backing weld. The second operation is to perform the bevel-groove weld. The final operation is a contour fillet weld. Note that no sizes were used in this example. The completed joint is shown at B.

3.13 PLUG AND SLOT WELDS

Occasionally, it is necessary to weld two pieces together at points that are not located on the edges. This is done by creating a hole in one piece and welding the two pieces together through this hole, as shown in Figure 3-24. The holes are usually round, but may be of any shape. Holes may be drilled, flame-cut, or machined.

If the hole used is round, the weld is called *plug weld*. A weld made in a hole that is *not* round is known as a *slot weld*. The weld symbol used for the plug weld is shown in Figure 3-25. The size of the plug is shown to the left of the weld symbol. In either a plug or slot weld, the sides of the hole may be *countersunk* (slanted). If a countersink is used, its angle is indicated beyond the weld symbol, as shown in Figure 3-25. (See also location "A" in Figure 3-8.) The depth of the weld is shown *inside* the weld symbol. See location "R" in Figures 3-8 and 3-25. A plug weld's exact location on the weldment is shown on the assembly drawing.

The length, width, angle of the countersink, and the location and spacing of the slots for a slot weld, are not

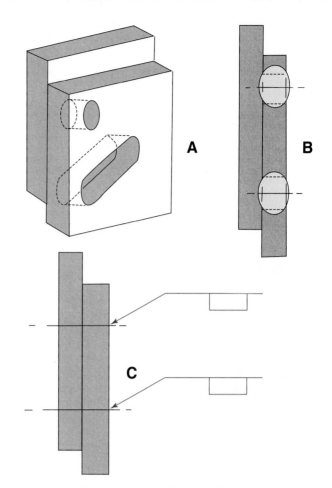

Figure 3-24. Plug and slot welds. A—The hole and slot are shown as cut in one piece. B—Cross-sections of the completed welds. C—The welding symbol for the plug and slot weld.

shown on the basic weld symbol. These dimensions are shown on the assembly drawing. The depth of filling for a slot or plug weld is shown inside the weld symbol. Figure 3-25 shows that the depth of filling for this weld is 1/4". If there is a series of plug or slot welds, the center-to-center distance is shown to the right of the weld symbol. See Figure 3-8 for the location of the *pitch* ("P") or center-to-center distance of welds; also see the weld symbol in Figure 3-26.

3.14 SPOT WELDS

A spot weld may be accomplished using resistance welding, gas tungsten arc welding, gas metal arc welding, electron beam welding, and ultrasonic welding. The *spot weld symbol* is a small circle. The circle may be on either side of the reference line, or it may straddle the reference line. If the weld is accomplished from the arrow side, the weld symbol should be below the reference line, as in all other welding symbols. If the welding is done on both sides, as in resistance spot welding, the circle straddles the reference line. See Figure 3-27.

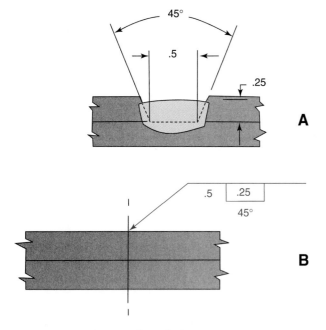

weld size is given to the left of the weld symbol. Weld strength is also shown to the left in pounds or newtons per spot. To the right of the weld symbol is found the weld spacing. The number of welds desired is shown in parentheses, centered above or below the spot welding symbol.

The welding process to be used is shown in the tail of the welding symbol. See Figure 3-28 for examples of the welding symbols used for spot welding. The welding symbol may be placed in any view of a drawing, as shown in Figure 3-28.

3.15 SEAM WELDS

Seam welds may be made with a number of processes, such as electron beam welding, resistance welding, or gas tungsten arc welding. The process to be used is shown in the tail. The size (width) of the weld and strength of the weld are shown to the left of the weld symbol. The strength is given in pounds per linear inch or in newtons per millimeter. The length of the seam may be shown to the right of the weld symbol.

Figure 3-25. *Plug weld symbol. A—Cross-section of a completed plug weld. The desired dimensions are also shown. B—The weld symbol used to complete the desired weld.*

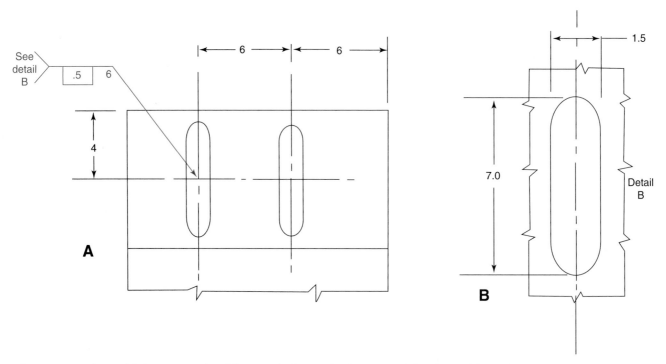

Figure 3-26. *Slot weld drawing and welding symbol. A—The location of the slot welds and the welding symbol are shown on the assembly drawing. B—Size dimensions for the slot weld may be shown in a detail drawing on the assembly drawing.*

Projection welding is another process used to produce spot welds. To indicate which piece has the projections on it, the circle is placed above or below the reference line.

The following information is given for a spot weld: size, strength, spacing, and the number of spot welds. The

The weld symbol may straddle the reference line if welded from both sides, as in resistance seam welding. For electron beam welding and gas tungsten arc seam welding, the symbol is placed above or below the reference line. This indicates from *which side* of the part the weld is to be made. See Figure 3-29.

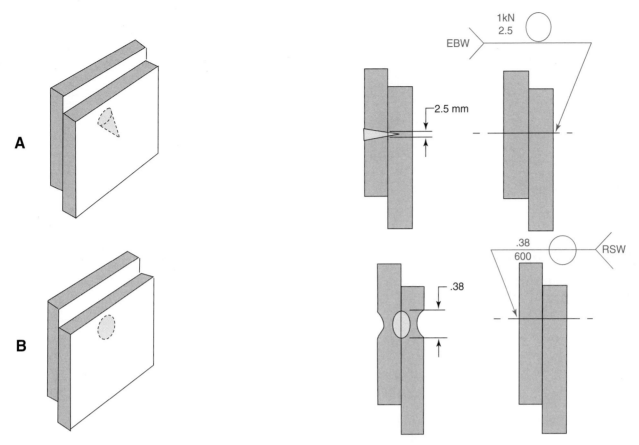

Figure 3-27. *The weld at A is an electron beam spot weld. Its size at the point of fusion is 2.5mm. Its required strength is 1 kilonewton. The weld is made from the other side. The weld at B is a resistance spot weld. Its size is 3/8" and its strength is 600 pounds. The weld is made from both sides and the symbol straddles the reference line.*

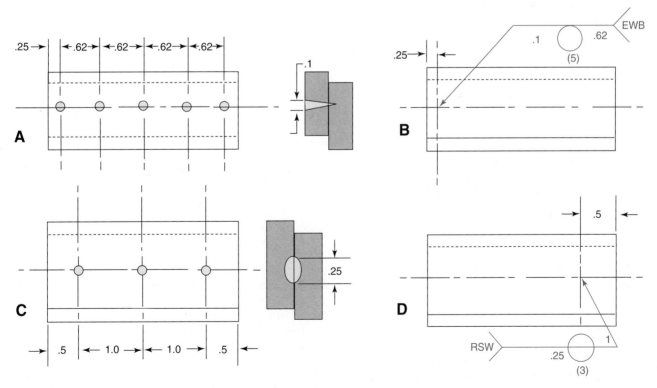

Figure 3-28. *A series of electron beam spot welds are shown in A and B. A—The desired weld and spacing is shown. B—The working drawing and welding symbol. C—Appearance of finished welds. D—Three .25 diameter resistance spot welds at 1" spacing are shown on the welding symbol on the working drawing.*

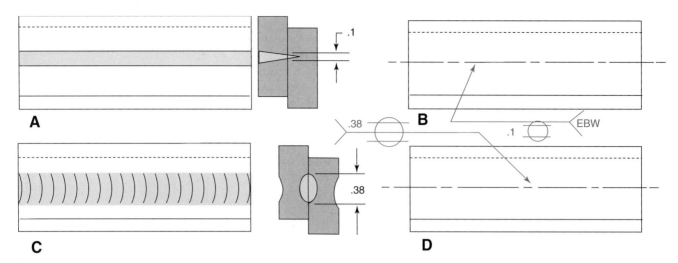

Figure 3-29. *Electron beam seam weld. A—An illustration of a seam weld made with the electron beam. Its size at the fusion point is 1". B—The welding symbol and weld symbol for this electron beam seam weld. The welding drawing and weld symbol shown in D will produce the weld shown in C.*

3.16   REVIEW OF WELDING SYMBOLS

Figure 3-30 shows examples of various welds and positions of welding. The ability to visualize the resulting weldment after reading the welding symbol is a very important skill for the welder to master. If you are not sure of the information given on the welding symbol or the location of needed information on the symbol, go back over this chapter. Reviewing this chapter periodically as you move through the material covered in this book will help build your print-reading confidence. You may also wish to refer to ANSI/AWS A2.4-93, *Standard Symbols for Welding, Brazing, and Nondestructive Testing.*

TEST YOUR KNOWLEDGE

Answer these questions on a separate sheet of paper. Do not write in this book.
1. List the five basic welding joints.
2. List five things the welding symbol will tell the welder about the weld that is to be made.
3. Why is the tail used on the welding symbol?
4. When used on the welding symbol, what does the weld symbol tell the welder?
5. The information below the reference line refers to the _____ side of the weld.
6. Where does the root opening appear on a groove weld symbol?
7. What information is shown in the "E" position or within the parentheses?

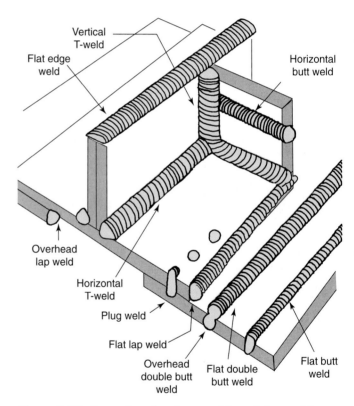

Figure 3-30. *Examples of various welds and the positions in which they may be made.*

8. On a chain intermittent welding symbol, are the fillet weld symbols aligned or offset?
9. An intermittent fillet weld with the length and pitch dimensions of 4-10 has how much space between the weld segments?

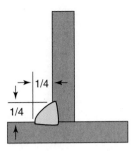

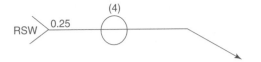

Refer to the welding symbol above when answering Questions 17-19.
17. What type of weld is to be made?
18. How many welds will be made?
19. What size are the welds?
20. What kind of weld symbol(s) may straddle the reference line?

10. On a separate sheet of paper, sketch the complete welding symbol for the weld sketched in the space above. The metal part shown is not to be ground or cut prior to welding, and the weld is to be continuous rather than intermittent.

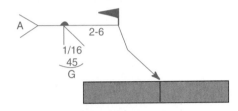

Refer to the welding symbol above when answering Questions 11 through 16:
11. Which piece of metal is to be ground, machined, or cut prior to welding?
12. What does the small, half-round symbol mean?
13. What shape is the weld face to be and what method will be used to finish it?
14. Is the weld made in the shop or on site (in the field)?
15. At what angle is the one piece ground, and how far apart are the pieces at the root of the weld?
16. Is the weld made continuously, or is it intermittent? If intermittent, how long is each weld? How far apart are the welds, center-to-center?

Oxyfuel gas cutting is often used on construction sites. This welder is piercing a bolt hole in an angle iron bracket that will be used to attach precast concrete panels to a building foundation.

Chapter 4

WELDING AND CUTTING PROCESSES

LEARNING OBJECTIVES

After studying this chapter, you will be able to:
* Cite the advantages of welding over other joining methods.
* Name and identify the various welding and cutting processes currently used in industry.
* Recognize the American Welding Society abbreviation for each of the welding and cutting processes currently in use in industry.
* Describe the operation of each welding and cutting process.

Welding is a general term for a number of *processes* (methods) used to join metallic or nonmetallic materials in a small area. The most common welding processes join metals or plastics by heating relatively small areas of the materials to their *welding temperature* (point at which the material melts). *Filler material* is normally added to the weld area while the materials are molten. However, welds can be made without adding filler material. Other welding processes join the materials using heat and pressure; some processes use only pressure. Not *all* materials can be welded, but most metals and a number of plastics are *weldable.*

This introductory section illustrates the many welding and cutting processes used in modern industry. American Welding Society (AWS) standard terminology is used in these descriptions. Drawings showing the parts and materials used in each process are color-keyed for clarity. In each illustration, the welding station is diagrammed. In some cases, details have been enlarged to help explain the process. To more easily understand the material in this chapter, you should refer to the drawings frequently while reading the text.

Each welding process description includes:
* Its application and purpose.
* The energy source used to produce heat.
* Controls used in the process.
* Operation of the process.
* Safety considerations involved in using the process.
* References to text chapters and paragraphs where more detailed explanations are given.

4.1 OXYACETYLENE WELDING (OAW)

The oxyacetylene welding process combines oxygen and acetylene gases to provide a high-temperature flame for welding. This flame provides enough heat to melt most metals. The oxyacetylene flame may also be used for all types of brazing.

An oxyacetylene outfit, as shown in Figure 4-1, is used in this process. Oxyacetylene welding is a manual process. The welder must personally control the torch movement and welding rod. Acetylene is supplied from one cylinder; compressed oxygen is supplied from another cylinder. Both cylinders must be equipped with a pressure-reducing regulator. Each regulator is fitted with two gauges.

One pressure gauge *(high)* indicates the pressure in the cylinder. The other pressure gauge *(low)* indicates the pressure of the gas being fed to the torch.

Separate flexible hoses carry the gases to the torch. The torch has two needle valves. One torch valve controls the rate of flow of the oxygen, the other controls the rate of flow of the acetylene to the torch tip. The mixed gases burn at the torch tip *orifice* (opening).

As shown in Figure 4-2, acetylene burns in the atmosphere with a yellow-red flame. A *carburizing flame* (excess acetylene with oxygen) is blue with an orange-and-red end. It may release black smoke. A *neutral flame* (perfect mixture of oxygen and acetylene) has a quiet, blue-white inner cone. This is the flame used in most welding processes. An *oxidizing flame* (excess of oxygen) results in a short, noisy, hissing inner cone. It tends to burn the metal being welded.

Other fuel gases can be used in place of acetylene. These include LP (liquefied petroleum), MAPP (methylacetylene propadiene), natural gas, and hydrogen.

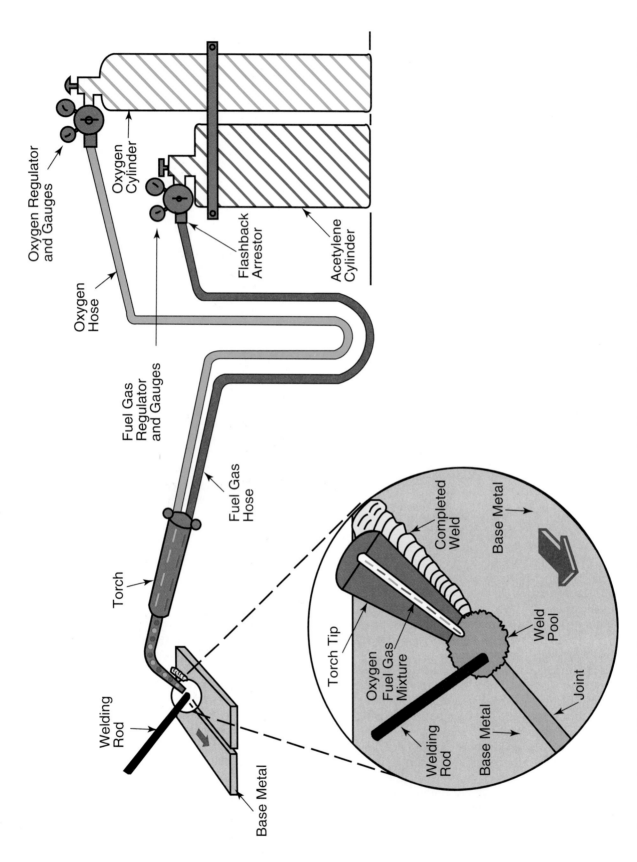

Figure 4-1. *Oxyacetylene welding (OAW). The oxygen and acetylene gas are mixed in a torch. The mixture burns at the torch tip. The heat from this flame is used to melt the base metal and welding rod. This melted material forms a welded joint.*

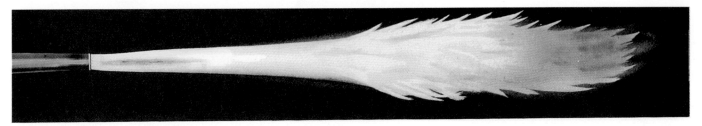

Acetylene Burning in Atmosphere
Open fuel gas valve until smoke clears from flame.

Carburizing Flame
(Excess acetylene with oxygen) Used for hard-facing and welding white metal.

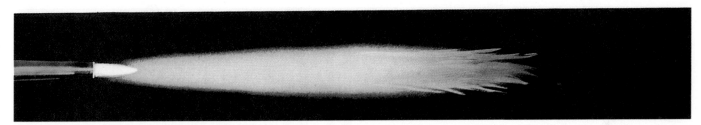

Neutral Flame
(Acetylene and oxygen.) Temperature 5589°F (3087°C). For fusion welding of steel and cast iron.

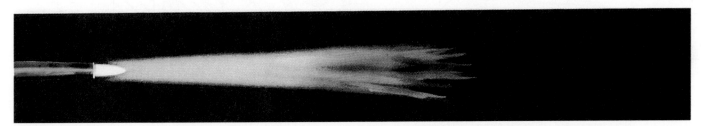

Oxidizing Flame
(Acetylene and excess oxygen) For braze welding with bronze rod.

Figure 4-2. Color appearance of oxyacetylene flames. (Smith Equipment, Division of Tescom Corp.)

Welding goggles should be worn for eye protection. Gloves, nonflammable clothing, and all other required safety clothing should be worn to protect against burns. Good fire safety and prevention techniques should be employed. Proper ventilation must be provided.

See Chapters 5 and 6 for additional oxyacetylene welding information.

4.2 OXYFUEL GAS CUTTING (OFC)

It is possible to *burn* (rapidly oxidize) iron or steel. The oxyfuel gas flame raises the temperature of the metal to a cherry-red color at 1472°F to 1832°F (800°C to 1000°C). Then, a high-pressure jet of oxygen from the cutting torch is directed at the metal. This causes the metal to burn and blow away very rapidly. For this reason, the term "burning" is sometimes used in connection with the oxyfuel gas cutting process.

The process requires cylinders of oxygen and a fuel gas, as shown in Figure 4-3. Each cylinder has a regulator and two pressure gauges. One pressure gauge indicates the pressure in the cylinder. The other gauge indicates the pressure of the gas being fed to the torch. Flexible hoses carry the gases to the torch. Construction details of a typical oxyfuel gas cutting torch are shown in Figure 4-4.

Several different fuel gases may be used in this process. The flame adjustments used when acetylene is the fuel gas are shown in Figure 4-5. The flame adjustments for liquefied petroleum (LP) gas are shown in Figure 4-6.

Welding goggles should be worn for eye protection. Approved gloves and proper clothing must be worn to prevent burns. It is important that the area of work be cleared of combustible material. Good fire prevention practices should be followed. Proper ventilation should be provided.

See Chapters 7 and 8 for more detailed instructions concerning oxyfuel gas cutting.

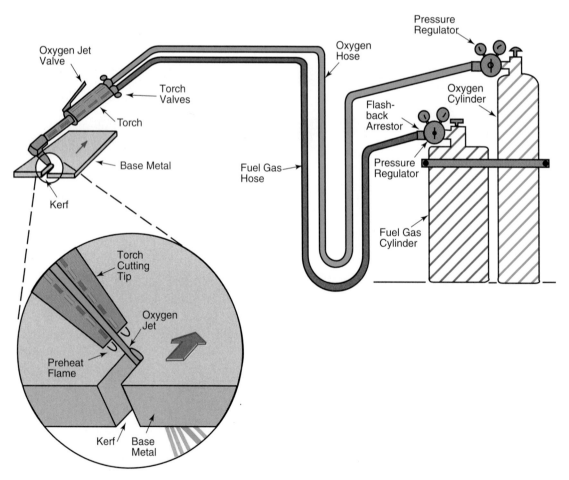

Figure 4-3. *Oxyfuel gas cutting (OFC). Oxygen and a fuel gas are mixed in the torch. The mixture burns at several orifices (openings) in the torch tip. When the flame has heated the base metal to a dull cherry red, a lever on the torch is pressed. This allows a jet of oxygen to rush out of an orifice. The oxygen jet quickly oxidizes the heated base metal and blows it away. This removal of material leaves a kerf (cut) in the base metal.*

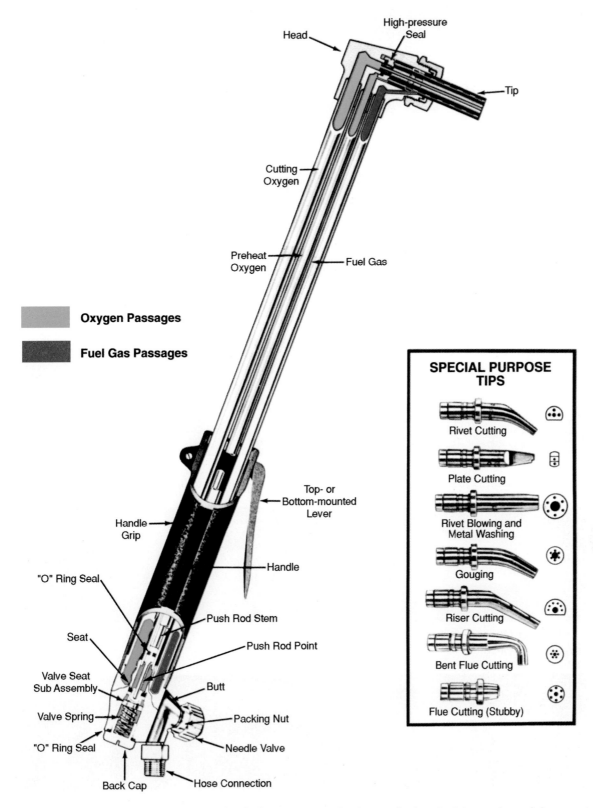

Oxygen Passages

Fuel Gas Passages

Figure 4-4. Cutting torch. Oxygen and a fuel gas are mixed, often in the head of the torch, and then carried to the tip orifices to form preheat flames. The cutting oxygen, turned off and on by the cutting oxygen lever, oxidizes the metal and blows it away to form the kerf. (Smith Equipment, Division of Tescom Corp.)

Acetylene Burning in Atmosphere
Open fuel gas valve until smoke clears from flame.

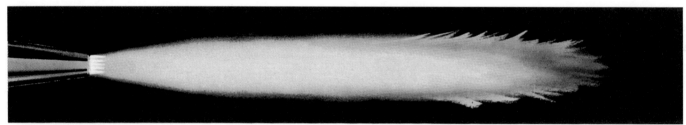

Carburizing Flame
(Excess acetylene with oxygen) Preheat flames require more oxygen.

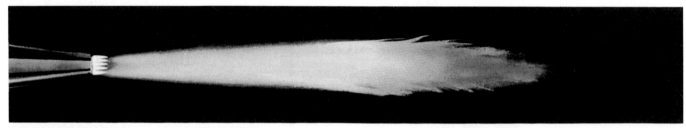

Neutral Flame
(Acetylene with oxygen) Temperature 5589°F (3087°C). Proper preheat adjustment for cutting.

Neutral Flame with Cutting Jet Open
Cutting jet must be straight and clean.

Oxidizing Flame
(Acetylene with excess oxygen) Not recommended for average cutting.

Figure 4-5. *Conditions of the oxyacetylene cutting flame when adjusting the torch. (Smith Equipment, Division of Tescom Corp.)*

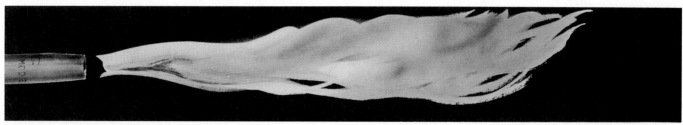

LP Gas Burning in Atmosphere
Open fuel gas valve until flame begins to leave tip end.

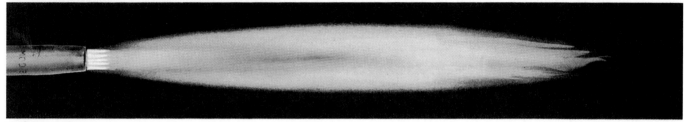

Reducing Flame
(Excess LP-gas with oxygen) Not hot enough for cutting.

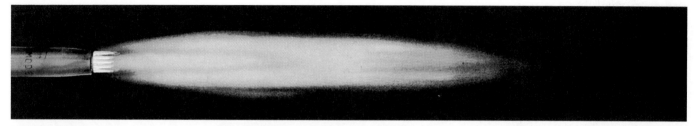

Neutral Flame
(LP-gas with oxygen) For preheating 1/8" and under prior to cutting.

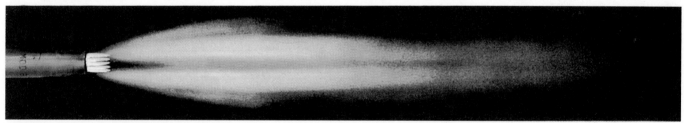

Oxidizing Flame with Cutting Jet Open
Cutting jet stream must be straight and clean.

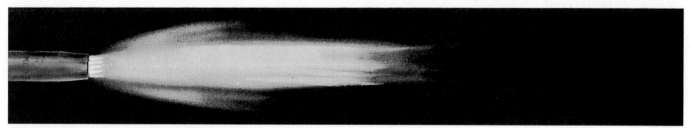

Oxidizing Flame without Cutting Jet Open
(LP-gas with excess oxygen) The highest temperature flame for fast starts and high cutting speeds.

Figure 4-6. *Conditions of the oxygen-LP gas cutting flame when adjusting the torch. (Smith Equipment, Division of Tescom Corp.)*

4.3 TORCH SOLDERING (TS)

Torch soldering as shown in Figure 4-7 uses an air-fuel gas flame. The oxyfuel gas flame referred to in Figure 4-1 may also be used for soldering. The base metal is heated enough to allow the solder to melt and then bond to the base metal. Soldering occurs at temperatures below 840°F (449°C). This is a popular method of joining metals in manufacturing and service operations. Torch soldering is used to fill a seam or to make an airtight joint.

The air-fuel or oxyfuel gas torch can be used for soft soldering. It can also be used to braze small metal parts.

In the soldering process, Figure 4-7, the amount of heat is controlled by the amounts of the gases flowing through the torch. When more heat is needed, larger torch tip orifices are used. The rate of gas flow is usually controlled by a needle valve on the torch. In air-fuel soldering, atmospheric air is drawn into the torch through holes at the base of the torch tip, as shown in Figure 4-7. A regulator is mounted on the cylinder to control the fuel gas pressure to the torch. The final flame adjustment is made with the torch valve.

MAPP gas (methylacetylene propadiene) or LP (liquefied petroleum) may be used as the fuel gas. Some torches may burn any of these gases. Other torches, however, can be used with only one type of fuel. Be sure to check the recommended gas for the type of torch being used.

Safety goggles or flash goggles are recommended to protect the eyes. Use pliers to handle the hot metal. Keep moisture away from molten solder. Moisture in contact with molten solder instantly changes to steam. This may cause molten solder to fly in all directions.

See Chapter 9 for more detailed instructions about soldering.

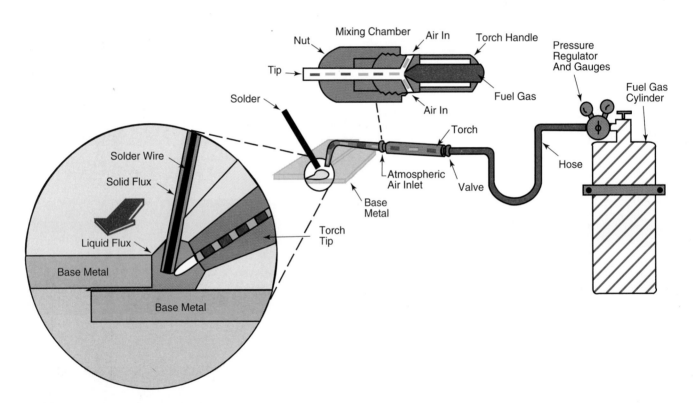

Figure 4-7. *Torch soldering (TS) with the air-fuel gas flame. A mixture of air and fuel gas is burned at the end of tip. The flame provides the heat to the base metal, which in turn melts the solder. A flux is needed to keep the base metal and solder clean enough to allow the solder to adhere (stick). The oxyfuel gas station may also be used for torch soldering. Refer to Figure 4-1.*

4.4 TORCH BRAZING (TB)

In torch brazing, an oxyfuel gas flame heats the base metal, and the heated base metal melts the brazing rod. Brazing is done at temperatures above 840°F (449°C). Brass and bronze, which have melting temperatures much lower than that of steel, are used as brazing rods.

When torch brazing, a thin layer of brass or bronze is used. The process of brazing is similar to soldering in this respect. There is usually less warping of the base metal when brazing than when welding. This is due to the lower temperature involved in brazing or braze welding. Figure 4-8 illustrates a typical torch brazing station.

The oxyfuel gas flame is adjusted to provide a neutral, oxidizing, or reducing (carburizing) flame, depending on the behavior of the metal in the joint. This means that flame adjustment is very important. See Figure 4-9.

Metal parts to be brazed are first carefully cleaned. A flux material is used to keep the metal clean. Flux is added in powder or paste form or as a coating on the brazing filler metal.

The oxyfuel gas flame is used to heat the base metal. The brazing filler metal is touched to the base metal, where it melts and flows into the joint. If the mating surfaces were properly cleaned, a good brazed joint should result.

Torch brazing can also be used to produce thick, beveled joints similar to those made with oxyfuel gas welding. The process used on these thicker joints is called braze welding. The base metal is not melted.

The welder should wear goggles, gloves, and fire-resistant clothing. There may be a severe health hazard if brazing filler metal contains zinc, cadmium, phosphorous, or beryllium alloys. Overheating of the brazing alloy should, therefore, be avoided. The brazing operation must be well-ventilated to remove the fumes from the brazing alloys and flux.

See Chapter 10 for more detailed instructions about brazing.

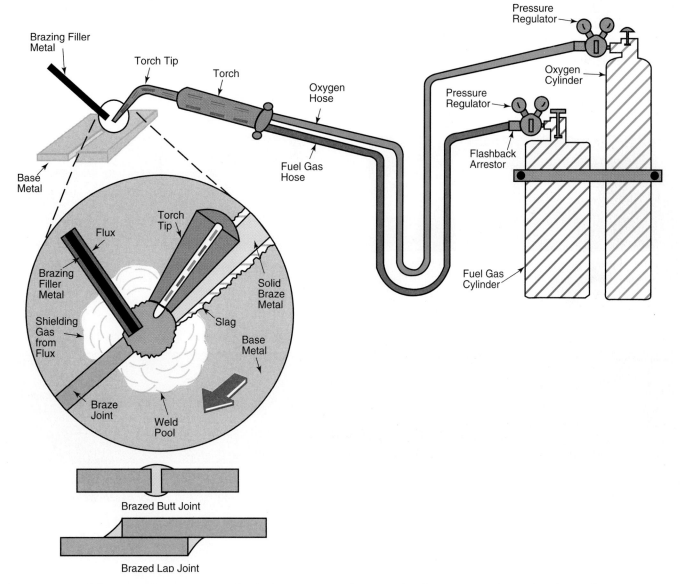

Figure 4-8. *Torch brazing (TB). Oxygen and a fuel gas are mixed in a torch. The mixture is burned at the torch tip. Heat raises the temperature of a spot on the base metal until the brass or bronze filler metal melts and adheres to it. A flux is used to keep the base metal and filler metal clean during the operation.*

LP Gas Burning in Atmosphere
Open fuel gas valve until flame begins to leave tip end.

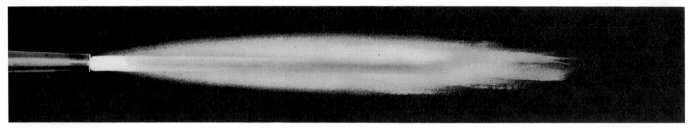

Reducing Flame
(Excess LP-gas with oxygen) For heating and soft soldering or silver brazing.

Neutral Flame
(LP-gas with oxygen) Temperature 4579°F (2526°C). For brazing light material.

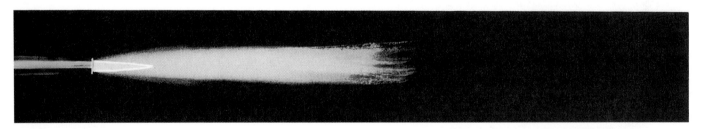

Oxidizing Flame
(LP-gas with excess oxygen) Hottest flame about 5300°F (2927°C). For fusion welding and heavy braze welding.

Figure 4-9. *Color appearance of the oxygen-LP gas flame. (Smith Equipment, Division of Tescom Corp.)*

4.5 SHIELDED METAL ARC WELDING (SMAW)

The shielded metal arc welding process uses an electric arc between a flux-covered metal electrode and the *base metal* (metal being welded). Heat from the electric arc melts both the end of the electrode and the base metal to be joined. This process is often used for maintenance work and small production welding. Heavy pipe welding is done almost exclusively with shielded metal arc welding.

Equipment used in this welding process provides an electric current for welding. The electric current may be either *alternating current (ac)* or *direct current (dc)*. Current adjustment controls on the welding machine allow the welder to set the desired current. Movement of the hand-held electrode holder is controlled by the welder. The electrode is a flux-covered metal wire. An electrical lead (cable) connects the electrode holder to the power source. Another lead connects the work to the power source to complete the circuit. Figure 4-10 illustrates a typical station for shielded metal arc (stick) welding.

The heat of the electric arc may be controlled by the current setting and by the arc length. Electrode diameter and flux material will determine the type (ac or dc) and amount of welding current required. The arc between the welding electrode and the base metal is *struck* (initiated) by the welder. Correct arc length must also be controlled by the welder.

Some of the covering on the electrode turns into a protective gas shield which surrounds the arc as the electrode melts. Some of the covering melts to form a slag which covers the completed weld. The slag layer protects the hot metal from oxidizing while it cools.

Welders must wear an approved helmet, gloves, and protective clothing. The welding workstation must be well ventilated.

Chapters 11 and 12 give more detailed information on shielded metal arc welding.

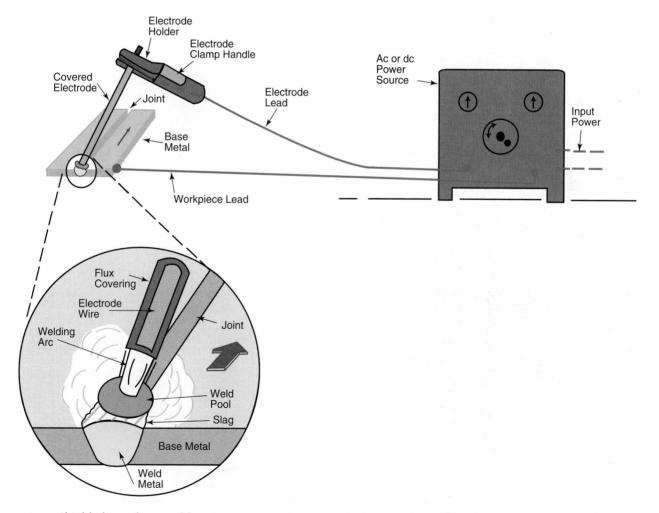

Figure 4-10. *Shielded metal arc welding (SMAW). An electric arc is drawn between the covered electrode and the base metal. The heat of the arc melts the end of the electrode and the base metal where the arc contacts it. The metal from the electrode provides the filler metal for the weld.*

4.6 GAS TUNGSTEN ARC WELDING (GTAW)

Gas tungsten arc welding uses the heat of an electric arc between a tungsten electrode and the base metal. A separate welding filler rod is fed into the molten base metal, if needed. A shielding gas flows around the arc to keep away air and other harmful materials. Gas tungsten arc welding is particularly desirable when welding stainless steel, aluminum, titanium, and many other nonferrous metals. GTAW may be done in any position with excellent results. The process is also called *TIG* (Tungsten Inert Gas) welding.

Figure 4-11 illustrates a typical station for gas tungsten arc welding. An ac-dc welding machine may be used with a regulated flow of a shielding gas, such as argon or helium. The shielding gas flows from a cylinder through a regulator, flowmeter, and a hose to the GTAW torch.

The welder manually operates the torch (tungsten electrode holder) and the filler metal rod. A heat-resistant gas flow cup or nozzle surrounds the electrode. Some small-capacity torches are air-cooled. Large torches are water-cooled.

Heating properties of the arc may be controlled by changing current and arc length. The diameter of the tungsten electrode, and the thickness and kind of base metal, will determine welding amperage.

Gas tungsten arc welding generates intense heat and light, with no metal spatter. The welder must wear an approved welding helmet, gloves, and welder's clothing.

See Chapters 13 and 14 for additional information.

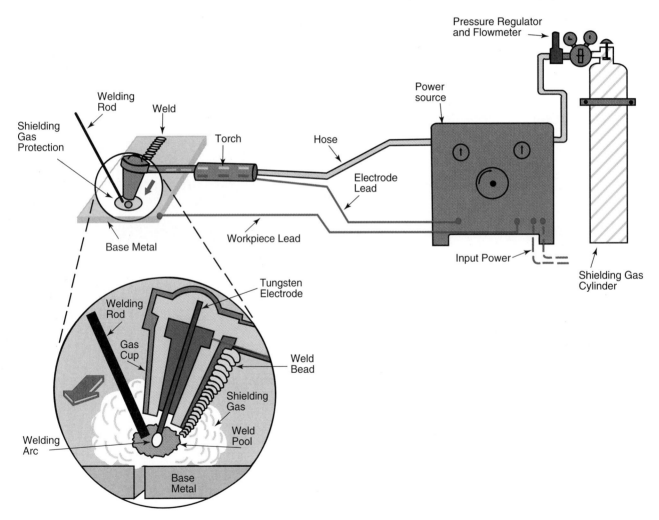

Figure 4-11. *Gas tungsten arc welding (GTAW). An electric arc is drawn between the end of the tungsten electrode and a spot on the base metal. Only the base metal melts. If filler metal is needed in the joint, a separate welding rod is used. The welding rod is added to the molten weld pool as needed. Shielding gas flows out a nozzle around the tungsten electrode.*

4.7 GAS METAL ARC WELDING (GMAW)

In *gas metal arc welding,* an electric arc between a continuously fed metal electrode and the base metal produces heat. The arc is shielded by a gas. This process is popular in production and repair shops. It is often called *MIG* (Metal Inert Gas) welding.

As shown in Figure 4-12, a shielding gas cylinder, regulator, flowmeter, and hose provide a flow of shielding gas to the arc. Shielding gases such as carbon dioxide, argon, or helium may be used. An electrode-feeding device continuously supplies metal electrode. A cable carries the electrode wire, current, and shielding gas to the torch and arc. The torch usually has a trigger-type switch for starting and stopping the electrode feed and gas flow.

A constant voltage dc welder is used with this process. The desired voltage is set on the welding machine. Current is changed by adjusting the feed speed of the wire.

Speed controls for the wire are usually mounted in the wire-feed mechanism. Shielding gas volume adjustments are made at a gas flowmeter on the regulator. The kind of shielding gas used usually depends on the metals being welded.

When performing GMAW, the welder:
1. Selects the electrode size.
2. Sets the desired voltage.
3. Adjusts the shielding gas flow.
4. Adjusts the rate of electrode feed.
5. Controls the torch movement and electrode extension. (The electrode extension is the distance from the torch tip to the arc.)

The welder must wear an approved helmet, gloves, and welder's clothing. The welding area must have good ventilation.

See Chapters 13 and 15 for additional information.

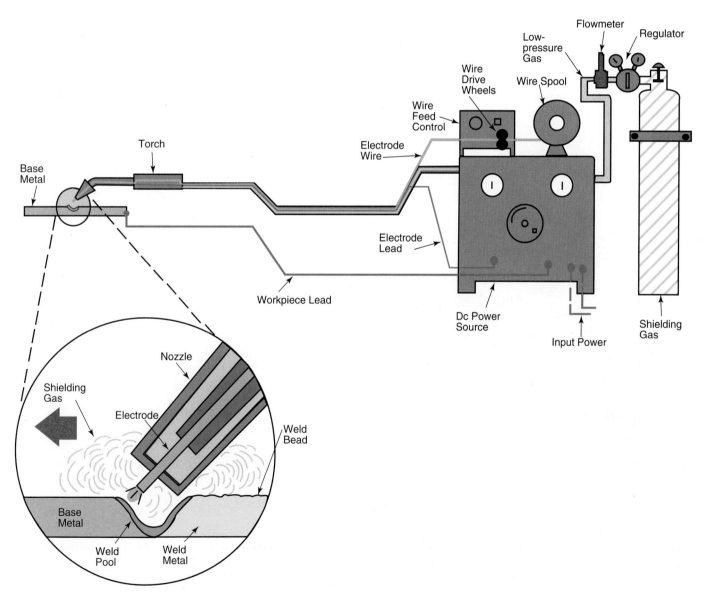

Figure 4-12. *Gas metal arc welding (GMAW). An electric arc is drawn between a metal electrode and the base metal. Heat from the arc melts the end of the electrode wire and a spot on the base metal. A shielding gas flows out of the torch nozzle. This gas keeps the oxygen and impurities in the air from contacting the weld.*

4.8 FLUX CORED ARC WELDING (FCAW)

In flux cored arc welding, Figure 4-13, heat is produced by an arc between a flux cored electrode and the base metal. This process is particularly desirable for welding structural steel and in other low-carbon applications. A *constant voltage* (constant potential) dc arc welding machine is used.

The electrode is a hollow metal tube with its center (core) filled with a flux material. The electrode is fed to the torch from a large spool on the electrode drive mechanism. Some flux cored wires are used without CO_2 shielding gas. Others use CO_2 gas.

The heat of the arc depends upon the arc length, voltage setting, and the wire feed speed. The speed of the electrode feed mechanism may be adjusted. Adjusting the wire speed varies the current at the arc. The higher the wire feed speed, the higher the current and, thus, the greater the heat of the arc.

The welder working with FCAW equipment is exposed to heat and light from the arc. An arc welder's helmet, leather gloves, and protective clothing should be worn. Excellent ventilation should also be provided.

Chapters 13 and 15 give more detailed information concerning the operation of this process.

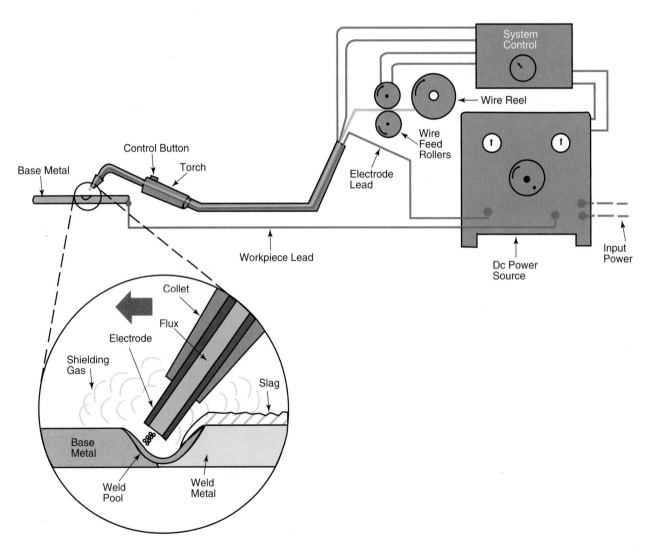

Figure 4-13. *Flux cored arc welding (FCAW). The heat energy comes from an electric arc. The end of the electrode and a spot on the base metal are melted to form the weld. The flux core, as it is heated, forms a gaseous shield around the arc and also provides a slag covering to keep air away from the weld until it cools.*

4.9 AIR CARBON ARC CUTTING (CAC-A)

The air carbon arc cutting process uses an electric arc to melt the base metal. A jet of air then blows the melted metal away. Air carbon arc cutting may be used on many metals.

Figure 4-14 shows a typical station for air carbon arc cutting and gouging. The electrical supply may be either direct current (dc) or alternating current (ac). An *electrode lead* (flexible cable) connects the electrode holder to the welding machine. A *workpiece lead* (ground cable) connects the base metal to the welding machine.

The air jet may be supplied from either a compressed air cylinder or an air compressor. The air line is attached to the electrode holder. A lever-operated valve in the electrode holder controls the air flow. The welder operates the electrode holder manually. This process can be used for either cutting or gouging metal.

Current is regulated by adjustments on the welding machine. The arc length is controlled by the welder. The length of the carbon electrode between the air jet nozzle and the arc must be maintained at such a distance that the air jet will be effective in blowing away the molten metal.

This cutting process produces considerable sparking. The welder must be protected by gloves, helmet, and clothing. Excellent ventilation is needed. Good fire prevention practices must be followed.

See Chapters 16 and 17 for more details.

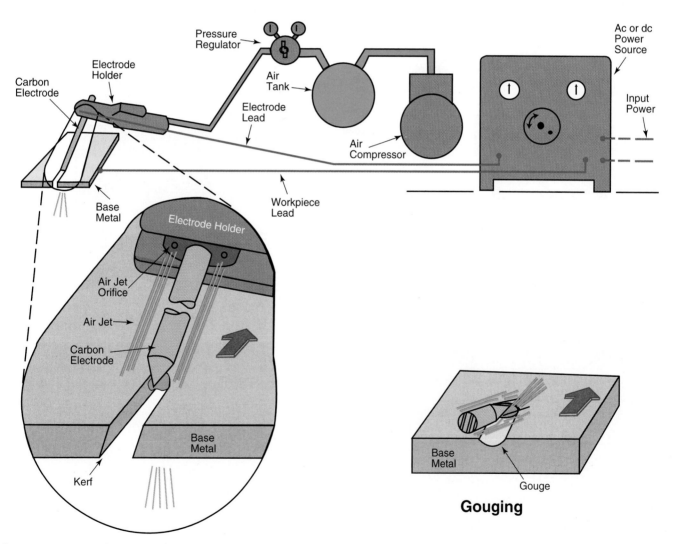

Figure 4-14. *Air carbon arc cutting (CAC-A). The electric arc between the carbon electrode and the base metal melts the base metal. Manually operated air jets attached to the electrode holder blow the molten metal away. This process is used for gouging base metal as well as for cutting.*

4.10 OXYGEN ARC CUTTING (AOC)

The oxygen arc cutting process uses an electric arc to heat the base metal. Then, a jet of oxygen is used to cut the heated metal. AOC is a rapid metal-cutting process. It is used for cutting cast iron, steel, and many other metals. Figure 4-15 illustrates a typical station for oxygen arc cutting.

The equipment includes a special electrode holder and a hollow metal electrode. An oxygen cylinder and regulator provide a controlled flow of pressurized oxygen through the hollow electrode. A flexible hose carries the oxygen to the electrode holder. The welder controls the oxygen through a hand valve on the electrode holder. An ac or dc welding machine supplies an arc current to the electrode. Current is conducted to the electrode holder through the electrode lead. A workpiece lead connects the metal *workpiece* (base metal) to the welding machine.

To make the cut, the welder strikes an arc and, as soon as the metal surface to be cut has a molten spot, opens the oxygen valve. The flow of oxygen into the arc and against the metal very rapidly oxidizes and blows away the metal. The heat of the arc is adjusted by a current adjustment control on the welding machine. This process may be used either in air or under water.

This cutting process may produce a great shower of sparks. There is also a danger from sparks falling around the legs and feet. The welder must wear a helmet, gloves, and protective clothing. Good ventilation and fire protection are important.

See Chapters 16 and 17 for more details on this process.

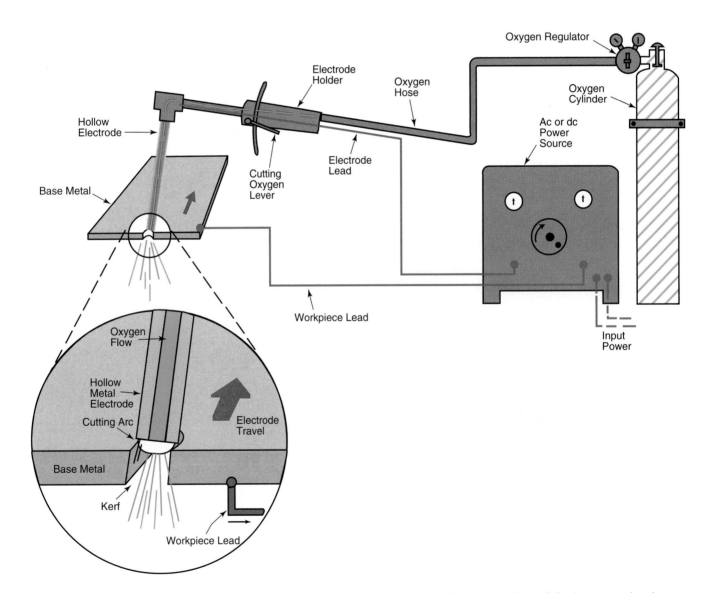

Figure 4-15. Oxygen arc cutting (AOC). An electric arc is drawn between a hollow electrode and the base metal. A lever on the holder allows a jet of oxygen to flow through the electrode. This jet cuts the base metal by rapid oxidization (combining the metal with oxygen).

4.11 GAS TUNGSTEN ARC CUTTING (GTAC)

In gas tungsten arc cutting, an arc between a tungsten electrode and the base metal heats the metal. The shielding gas then blows the melted metal away. This process is used on aluminum, stainless steel, nickel, and many other metals. Since the electrode being used to melt the metal is tungsten, the electrode does not burn away rapidly.

Figure 4-16 illustrates a typical station for gas tungsten arc cutting. Equipment includes a water-cooled electrode holder and a current supply. A regulated supply of shielding gas is required. Gases used are argon, helium, or hydrogen. The current supply is usually *direct current straight polarity (DCSP)*. The current is furnished by an arc welding generator or rectifier, and may be adjusted to meet job requirements.

The welder adjusts the flow of current, water, and shielding gas. The cutting torch can be either manually or automatically operated. Shielding gas serves two purposes. It blows the molten metal away from the cutting area and keeps the surfaces of the cut from oxidizing.

Gloves, helmet, and protective clothing must be worn. There is considerable sparking with this type of cutting.

See Chapters 16 and 17 for more details.

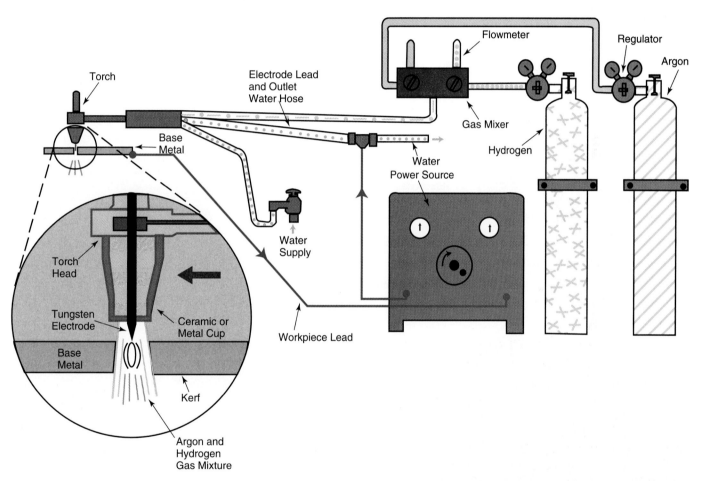

Figure 4-16. *Gas tungsten arc cutting (GTAC). An arc between a tungsten electrode and the base metal heats and melts the base metal. A mixture of argon and hydrogen gases is used to blow the melted metal away and form a kerf (cut) in the base metal.*

4.12 SHIELDED METAL ARC CUTTING (SMAC)

The shielded metal arc cutting process uses an arc between a metal electrode and the base metal. This arc melts the base metal. The electrode is heavily covered with flux. The molten metal drops away from the base metal to form a kerf (cut). This process is used mainly for small maintenance jobs.

The arc cutting process requires a current source, either ac or dc. A manually operated electrode holder provides a grip for controlling the electrode. Electric current from the arc welding machine flows through the electrode holder lead and forms an arc between the electrode and the workpiece. A workpiece lead (ground cable) between the workpiece and the power source completes the circuit in this process.

Figure 4-17 illustrates an arc cutting station. Heat from the arc is controlled by the arc length, current, and electrode material. In operation, both electrical and mechanical equipment is adjusted to provide the desired arc. The welder strikes the arc between the electrode and the base metal and the cutting is started. The welder moves the electrode as the cut progresses through the base metal.

Protection is needed from the intense heat, the light of the arc, and the sparks. This requires wearing an approved helmet, gloves, and welder's clothing. The equipment used must include the necessary shielding and safety devices. Excellent ventilation is needed. Good fire prevention practices must be followed.

Refer to Chapters 16 and 17 for more detailed information on SMAC.

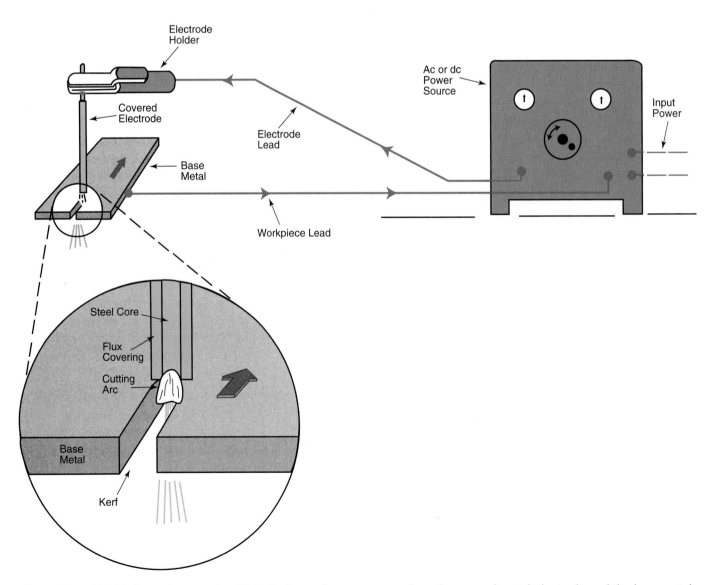

Figure 4-17. *Shielded metal arc cutting (SMAC). An arc between a very heavily covered metal electrode and the base metal melts the end of the electrode and the base metal. The metal electrode melts far back into the covering, producing a jet action of gases that blows the molten base metal away.*

4.13 PLASMA ARC CUTTING (PAC)

Plasma arc cutting uses an electric arc and fast-flowing ionized gases to melt and cut metals. This process cuts aluminum, stainless steel, and most other metals rapidly. Plasma arc cutting also can be used to cut nonmetals, such as concrete. Figure 4-18 illustrates a station for plasma arc cutting. The metal is melted by the heat of the plasma arc. Then, the molten metal is blown away by the high velocity of the shielding gas.

Plasma arc cutting requires a special water-cooled cutting nozzle. It makes use of a tungsten electrode connected to a source of dc power, compressed gas, and suitable controls.

Plasma arc cutting is usually used along with automatic cutting devices. The current is controlled by devices on the power source. Waterflow to cool the torch is usually manually adjusted by the welder.

Plasma arc cutting is a very noisy process. The operator or welder must be protected from the noise by the use of earplugs or industrial "ear muffs." It is sometimes necessary to use a "walkie-talkie" type of communication system where plasma arc cutting is being done. The welder must also be protected with an approved helmet, gloves, protective clothing, and other required safety equipment.

See Chapters 16 and 17 for more details.

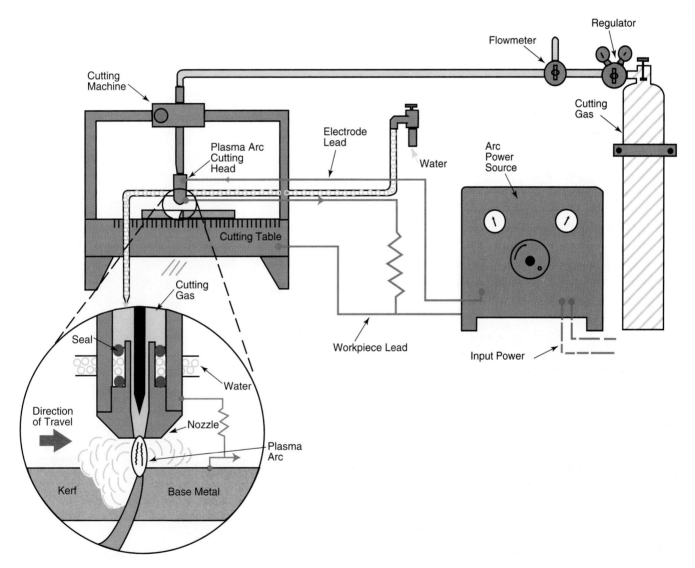

Figure 4-18. *Plasma arc cutting (PAC). An electric arc between a tungsten electrode and the base metal ionizes some of the cutting gas. This ionized gas (plasma) leaves the torch and hits the base metal. The very high plasma temperature superheats the base metal and very rapidly forms a kerf (cut) in the base metal.*

4.14 RESISTANCE SPOT WELDING (RSW)

Resistance spot welding passes an electric current through the metal being welded. Resistance to the electrical flow heats the metal to welding temperature. The process is used to weld together two or more overlapping pieces. It is well-suited to automatic welding. Spot welding is commonly used to join auto body sections, cabinets, and other sheet metal assemblies.

As shown in Figure 4-19, a step-down transformer converts fairly high-voltage/low-amperage current to a low-voltage/high-amperage current. The weld is made between two electrodes that press the metals together. A large electrical current flows through the metals being welded from one electrode to the second electrode.

These electrodes are special copper alloys that can carry the high current and still have physical strength to operate under high pressures. The electrodes on small spot welders, used to weld thin materials, may be air-cooled. Electrodes for welding thicker metals are water-cooled.

Resistance spot welding is controlled by the amperage, the electrode pressure, and the length of time the current flows. In an automatic spot welder, the welder sets the electrode current, electrode pressure, and the timing. The electronic controller repeats the desired weld cycle each time the start switch is pushed.

The welder must wear flash goggles. If the metals must be manually handled, gloves are needed to prevent cuts and burns to the hands.

See Chapters 18 and 19 for detailed information concerning spot welding.

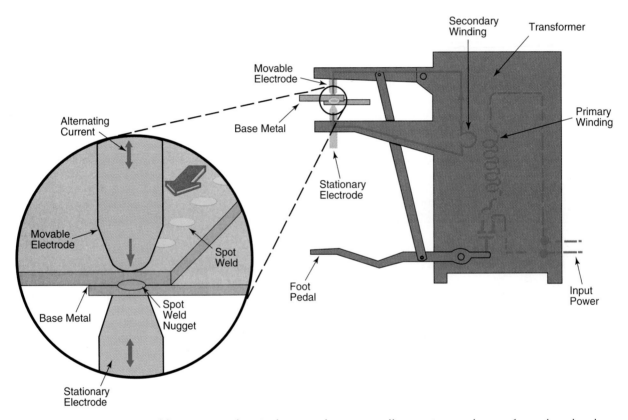

Figure 4-19. *Resistance spot welding (RSW). Electrical current heats a small area on two sheets of metal as the sheets are pressed together between the spot welding electrodes. The metals become hot enough to fuse together.*

4.15 PROJECTION WELDING (PW)

The projection welding process, Figure 4-20, uses resistance to the flow of electricity to create heat for welding. It is similar to spot welding, and is commonly used in production welding.

Prior to welding, one of the two pieces of metal is run through a machine that makes bumps or ***projections*** of a designed shape and size in the metal.

The welding machine electrodes are flat plates called ***platens***. The two pieces of base metal are placed together between the platens. They touch only at the projections or bumps on the one piece. Welding current is supplied by a resistance welder transformer. The welding current flows through the pieces to be welded while they are clamped between the platen plates. Due to the projections, the current is concentrated at the points of contact. These points heat up and fuse.

The welding current flows for a short time as pressure is applied between the two platens. This completes the weld. Timing of the current flow and application of welding pressure are important parts of this welding process.

Some flash and sparking may take place during the welding. The welder should wear flash goggles, approved clothing, safety shoes, and leather gloves.

A more complete explanation of the projection welding process is given in Chapters 18 and 19.

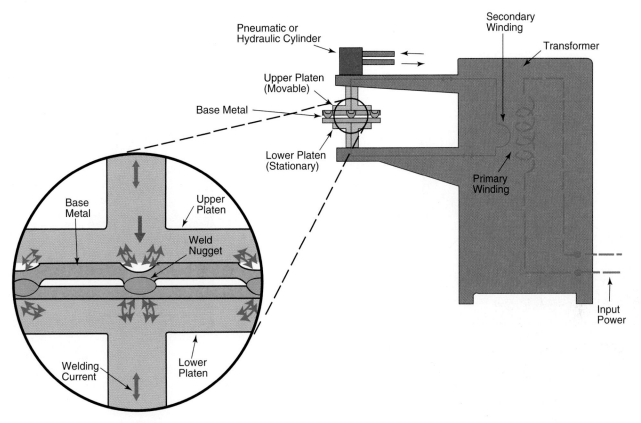

Figure 4-20. *Projection welding (PW). Electrical energy heats the projections in one sheet of base metal as they touch another sheet of base metal. Both pieces become hot enough at the contact spots to fuse together as pressure is applied.*

4.16 FLASH WELDING (FW)

The flash welding process uses an electric arc to heat the base metals. It combines resistance welding, arc welding, and pressure welding. Flash welding provides a strong, clean weld joint, and is chiefly used in production welding. The process, as shown in Figure 4-21, uses a step-down transformer to provide the welding current. The two pieces to be welded are held in current-conducting movable clamps.

To make a flash weld, the workpieces are brought together under light pressure. A low-voltage/high-amperage current travels between the base metals.

As soon as the current is established, the two pieces of metal are drawn apart very slightly. This causes an electric arc to form between them. The arc heats the surfaces of the two metals. When sufficiently heated, they are forced together under very high pressure. This pressure causes a slight outward flow of heated and somewhat dirty metal from the joining surfaces. The clean, heated subsurface metal brought into contact produces a good weld. The finished weld will have a flash or enlargement at the joint.

Welding heat is controlled by the current adjustment setting. The quality of the weld is controlled by the current, the length of time of the arc, and finally, the pressure at the time the two surfaces are brought together. When flash welding manually, the welder must be able to judge the metal temperatures, the time that the welding surfaces must be brought together, and the proper welding pressure.

It is necessary to wear protective clothing, a face shield, and gloves. Flying sparks (metal expelled at the joint) are produced during this process.

See Chapters 18 and 19 for more detailed information.

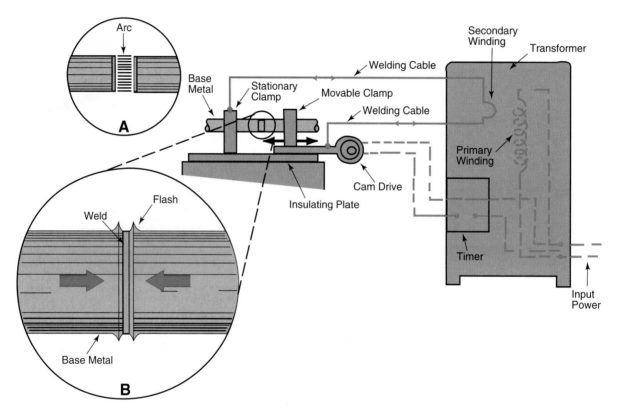

Figure 4-21. Flash welding (FW). A—Electrical energy creates an arc that melts the ends of two pieces of base metal. B—When the ends are molten, they are pushed together to fuse into one piece.

4.17 RESISTANCE SEAM WELDING (RSEW)

Resistance seam welding, Figure 4-22, is a special application of spot welding. It is often used to weld joints in containers and other products that require an airtight or vapor-tight seam.

The electrodes are in the form of wheels, with the work to be welded passed between them as they revolve. A timing device turns on the welding current at controlled, rapidly repeated intervals. The rapidly repeating current flow makes a series of overlapping spot welds that appear to be a continuous line of welding. These machines are usually automatic. The current, pressure on the electrodes, and sequence times are set by a welding operator. The timing of the process is regulated by an electronic controller.

The operator must wear flash goggles and all other required safety clothing. If the metal must be handled, the operator should wear approved gloves.

Refer to Chapters 18 and 19 for additional information on resistance seam welding.

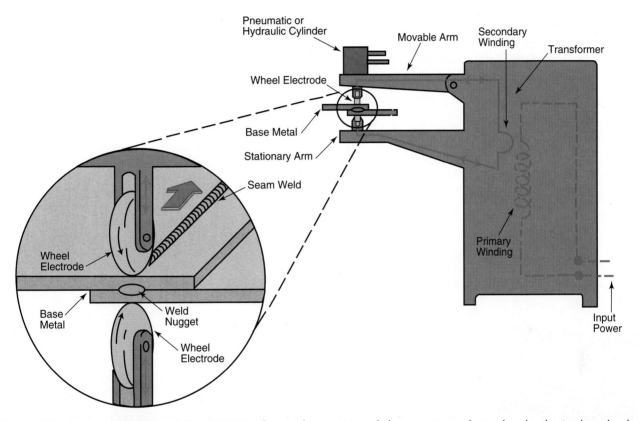

Figure 4-22. *Resistance seam welding (RSEW). Electrical energy travels between two electrode wheels. As the wheels travel, they clamp two sheets of base metal together. Electricity travels through the base metal, making it hot enough to fuse the two sheets together and form a seam weld.*

4.18 SUBMERGED ARC WELDING (SAW)

In submerged arc welding, an electric arc between an electrode and the base metal produces welding heat. The arc is submerged in a granular flux. Some of the flux melts and forms a shielding slag over the weld. This process is often used when welding thick plate joints. The equipment is usually automatically or semiautomatically operated. The electrode also feeds automatically into the arc. See Figure 4-23.

A hopper and feeding mechanism is used to provide a flow of flux over the joint being welded. The arc, generated by ac or dc current, is submerged in the flux. The chemical composition of the flux will affect the composition of the completed weld. Alloy elements can be added to the weld by controlling the chemical composition of the flux. As the weld progresses, some of the flux melts and forms a slag over the weld. A vacuum machine may be used to pick up the flux for reuse.

Welding heat is regulated by changing the voltage and wire feed speed on the welding machine. The electrode, which is usually power-fed and made of mild steel, extends through the flux to a point just above the base metal. The flux material shields the base metal from oxidation while it cools. The correct arc length is automatically maintained underneath the shielding material.

Since the arc is submerged in the shielding material, it cannot be seen during welding. This reduces, to some extent, the hazard of burns and flying sparks. Still, the welder should wear approved goggles, gloves, clothing, and provide good ventilation.

See Chapter 20 for more details on submerged arc welding.

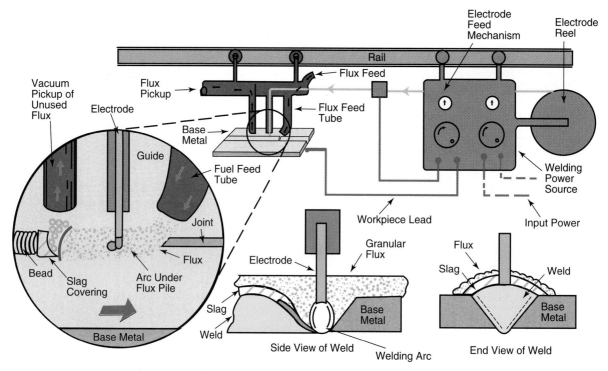

Figure 4-23. *Submerged arc welding (SAW). Electric current creates an arc to heat and melt the end of the electrode wire and the base metal. The arc is covered with a layer of flux. Some flux melts and forms a slag over the weld.*

4.19 ELECTROSLAG WELDING (ESW)

Electroslag welding is used to weld various joints on thick metal. The weld is done vertically, moving upward.

Prior to starting the weld, a flux material several inches deep is placed between the two base metals. The flux is able to conduct electricity.

Figure 4-24 illustrates a typical setup for electroslag welding of a butt joint. To start the weld, an electric arc is struck between one or more electrodes and the base metal. The electrodes are continuously fed by a drive mechanism. The electrodes are often automatically *oscillated* (moved back and forth) when extremely thick plates are welded. The flux is melted by the heat of the arc. After the flux becomes molten, the arc is stopped, but electricity contin-ues to flow. The flux (slag) is kept molten by its resistance to the flow of the electricity through the electrode and flux to the base metal.

The molten flux melts the base metal and the contin-uously fed filler metal to form a weld. Movable, water-cooled molds (shoes) are used on each side of the joint. The shoes are water-cooled to prevent them from melting into the weld. These shoes keep the molten flux, filler metal, and base metal in place as the weld moves upward. The molten flux shields the weld from the atmosphere.

The heat source is controlled by the amount of current and the physical characteristics of the flux. Operators must wear protective clothing. The heat developed in this process is very intense.

See Chapter 20 for more details.

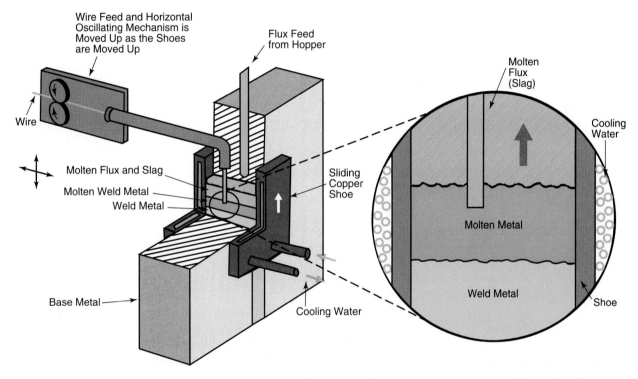

Figure 4-24. *Electroslag welding (ESW). The energy for welding is an electric current. This current keeps the flux molten. The current and the molten flux melts the base metal; the melting electrode wire provides the filler metal. The weld is formed vertically. Water-cooled shoes contain the molten metal until it solidifies.*

4.20 ARC STUD WELDING (SW)

Arc stud welding is a semiautomatic welding process. It is used to attach metal fastening devices to metal plates or beams without drilling and tapping. Bolts, screws, rivets, and spikes may be attached in this way.

Figure 4-25 shows an arc stud welding station. The heat source is an electric arc. The enlarged view shows the stud, stud chuck, and ceramic ferrule. Another sequence of drawings shows the four steps in making the welded joint.

The energy source is an electric welding transformer. The control on the welding machine determines the current in the electric arc. Current settings vary with the size of the stud and the kind of metal. The control unit has a timer that controls the duration of the arc.

The welder installs the stud in the gun. The gun is then positioned on the base metal. A switch on the gun starts the stud welding operation cycle.

The welder must wear gloves, a face shield with flash goggles, and fire-resistant clothing.

A more complete explanation of arc stud welding is given in Chapter 20.

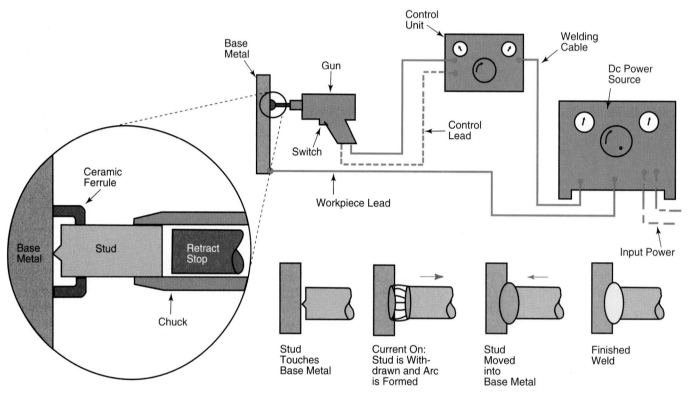

Figure 4-25. *Stud arc welding (SW). Electrical energy creates an arc that melts the end of a special electrode (a bolt, a pin, a clip, or a stud) and the base metal. The special electrode is then moved into the base metal weld pool. The assembly is held until the weld metal cools.*

4.21 COLD WELDING (CW)

No outside heat source is used in this process. Cold welding uses very high pressure to force metals together. Only the surface molecules are heated and fused to form a weld. The method is used mainly to join softer metals, such as aluminum-to-aluminum, copper-to-copper, and aluminum-to-copper. Good fusion occurs, resulting in a strong weld. Butt welds and lap welds may be made with this process. Metal surfaces being joined must be very, very clean.

The source of energy to produce the weld is tremendous pressure, usually produced by using hydraulic cylinders. See Figure 4-26. The weld is controlled by the size of the die surfaces in contact with the metal and the amount of hydraulic pressure.

The operator should wear gloves, a face shield or safety goggles, and approved clothing.

See Chapter 20 for more details on the cold welding process.

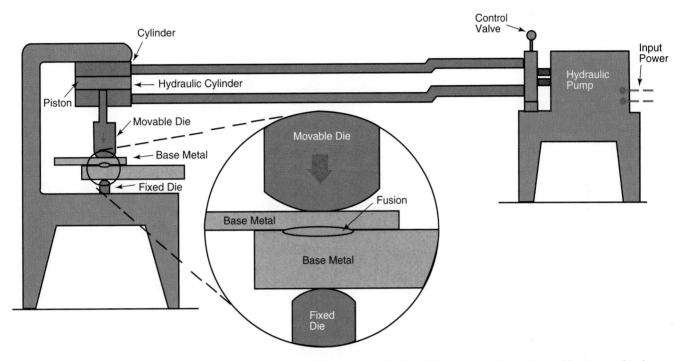

Figure 4-26. *Cold welding (CW). Pressure creates the welding energy. In this illustration, a thin sheet of base metal is lap-welded to a thicker sheet of base metal. Note that the upper die is larger than the bottom die when the top base metal is softer than the bottom base metal. A hydraulic piston is used to force the upper die against the metal. The hydraulic pressure used is only enough to fuse the surfaces of the base metals being welded.*

4.22 EXPLOSION WELDING (EXW)

Explosion welding, Figure 4-27, joins metals together by using a powerful shock wave. This creates enough pressure between two metals to cause surface flow and cohesion. It is often used to weld large sheets together. In one common application, EXW is used to weld thin stainless steel sheet to mild steel sheet. It is also used for joining aluminum and molybdenum.

The energy source is the tremendous shock wave caused by detonating an explosive material. The operation requires very careful setup. Bonding takes place in an instant. Such welding is done either in a safety chamber, or less frequently, under water.

Safety is very important! Both the explosives and fixtures must conform to approved written specifications. The welder must be protected from the sound of the explosion by wearing industrial "ear muffs" and/or earplugs. Face shields, safety helmets, and approved clothing should be used. Special permits are required from government authorities because of the explosives used.

Chapter 20 explains this special type of welding in more detail.

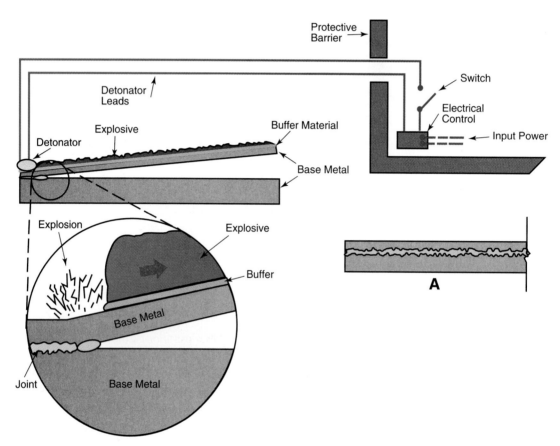

Figure 4-27. Explosion welding (EXW). A buffer material is applied to the top surface of one of the metal sheets. Welding energy comes from explosive material placed on top of buffer material. An ignitor (detonator) is operated from behind a barrier. The explosion proceeds from left to right and welds the top plate to the bottom plate almost at once, without deforming either piece of metal. The completed weld is shown at "A."

4.23 FORGE WELDING (FOW)

In forge welding, a furnace is used to heat two pieces of metal to a plastic temperature. *Forging* (hammer blows) is used to fuse the two pieces together. This process is used to join wrought iron, low-carbon steel, and medium-carbon steel workpieces. It can be used in places where there is no electricity or fuel gas available. Figure 4-28 illustrates a typical blacksmith's forge welding station.

The forge (furnace) has a cast iron pan or tub. A blower supplies air at the bottom of the forge through an opening called a *tuyere*. The fuel is usually a good grade of soft coal. Heat changes the burning coal to coke near the center of the fire. The fire is carefully tended by feeding coal from the outside toward the center. *Coking* means driving out the combustible gases from the coal. As this coal is being coked, the gas that is released will burn in an open flame over the fire. A thick bed of coked coal should be made before any welding is attempted by this process. Heat is controlled by the airflow through the tuyere. Increasing airflow increases the rate of combustion. Heating may also be done in a furnace burning a fuel gas.

Metal parts to be welded are placed down in the coke at the point of combustion. When the parts reach forging temperature (indicated by the bright red color of the metal), they are withdrawn. The heated parts are placed together on an anvil and hammered to make a weld.

The blacksmith/welder should wear eye protection, safety shoes, and flame-resistant clothing and gloves. This process produces a great deal of heat. Sparks are often created when the hot metal is pounded.

See Chapter 20 for more details.

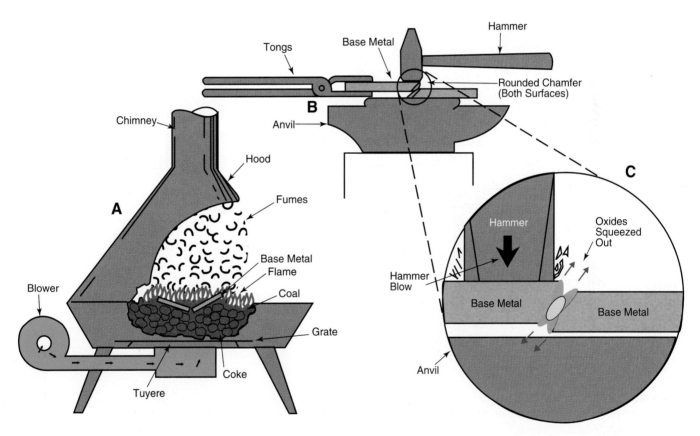

Figure 4-28. *A forge welding (FOW) station. A—Cross section through blacksmith's coal-fired forge. B—Hammer and anvil being used to make a forged weld. C—Enlarged view of the welding action.*

4.24 FRICTION WELDING (FRW)

Friction welding uses heat generated by friction (rubbing) to fuse two pieces of metal together. This process is used chiefly in butt welding fairly large, round rods or cylinders. Figure 4-29 illustrates a friction welding station.

No outside heat is supplied. One of the pieces is made to revolve. The ends of the pieces to be joined are then brought together under a light pressure. The resulting friction between the stationary and revolving part develops the heat needed to form the weld. As the metal surfaces reach the plastic state, they are forced together under a much greater pressure. The process creates a clean metal-to-metal welded surface.

Equipment includes the necessary clamping devices, a mechanism for revolving one part and a method for applying high pressure to the friction surfaces. The control of this type of welding is based on the speed at which the surfaces rotate against one another and the pressure that is applied.

Friction welding produces considerable sparking. The operator needs to wear approved clothing and goggles or a face shield to avoid injury.

See Chapter 20 for additional information.

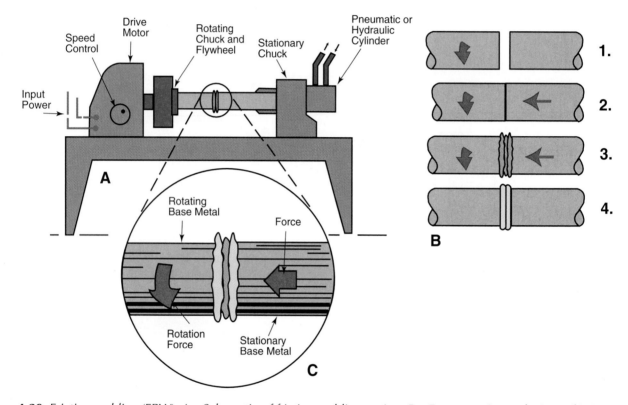

Figure 4-29. Friction welding (FRW). A—Schematic of friction welding station. B—Four steps in producing a friction weld. C—Close-up of a friction weld being made.

4.25 ULTRASONIC WELDING (USW)

In ultrasonic welding, very-high-frequency vibrations excite metal surface molecules. This movement among the molecules produces fusion. This process is most often used to join very light materials. For example, ultrasonic welding is used for attaching fine wires to foil or wires to wires.

Figure 4-30 illustrates the ultrasonic welding station. Since no outside heat is applied, this process is particularly desirable where the control of the heat-affected zone is important.

The horn and anvil contact the two materials to be welded. One of these rods vibrates at a very high (*ultra-sonic* or "above the normal sound range") frequency. This causes the materials being joined to vibrate at a corresponding rate. During the vibrations, some molecules of the two surfaces become intermixed and form a strong joint.

This type of welding is controlled by the rate of vibration and the pressure exerted by the vibrating elements on the parts being welded. Surfaces to be joined by ultrasonic welding must be very clean and free of oxidation.

Operators should wear gloves and goggles, since fine particles could be thrown off.

See Chapter 20 for more details.

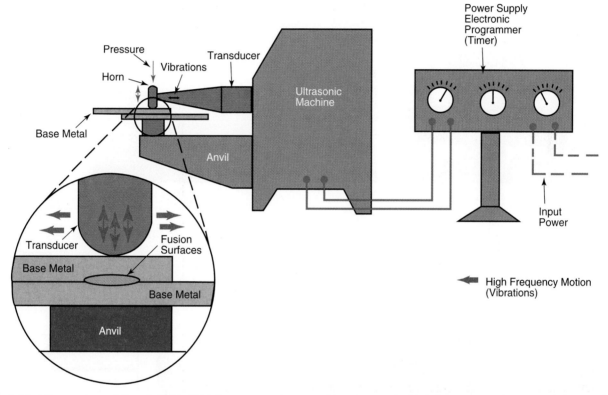

Figure 4-30. *Ultrasonic welding (USW). Welding energy is created by extremely high-frequency vibration of the machine horn. The molecules at the surfaces of the two pieces of base metal are made to move so rapidly that the surfaces fuse.*

4.26 LASER BEAM WELDING (LBW)

Laser beam welding uses a single-frequency light beam. This beam puts energy into a metal, causing it to heat up to its melting temperature. The laser beam is useful in welding small, light materials, particularly in locations where it is very difficult to weld with any other process. Figure 4-31 shows a typical laser beam welding set-up.

The beam is created by putting light or heat energy into a single molecule of a substance (typically ruby or carbon dioxide). The single-frequency energy of the single molecule increases in intensity by traveling between two mirrors until it passes through the less reflective of the two. The release of the laser beam is controlled by the operator.

The heat of the laser beam is very intense and is easily directed to the spot where the heat is desired. Since the laser beam can be reflected, it can be directed to the weld joint by a combination of mirrors. It can be either a continuous heat source or a pulsed beam. There is an instantaneous heat release when the beam contacts the base metals. Controlling the input to the laser beam source controls the resulting heat.

Laser beams must be very carefully guarded against being directed toward any part of the human body or anything of a heat-sensitive nature. Operators must wear special goggles designed for laser operations.

See Chapter 20 for more details.

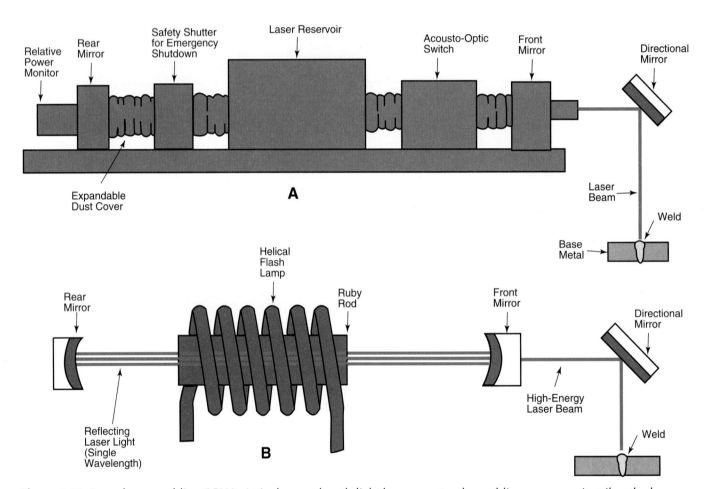

Figure 4-31. *Laser beam welding (LBW). A single-wavelength light beam creates the welding energy as it strikes the base metal. The laser beam can also be used for cutting and piercing base metals. A—Laser beam station. B—Laser beam light source.*

4.27 ELECTRON BEAM WELDING (EBW)

Electron beam welding uses energy from a focused stream of electrons to heat and fuse metals. It is a good process for welding thick parts when the distance between them is small. Welds can be made in deep, narrow spaces with an extremely narrow weld zone. Figure 4-32 shows an electron beam welding set-up. This process is used to join metals that are difficult to weld by other processes. It may also be used to weld metals at very high speeds. Because the equipment is large and expensive, it is used only where other processes cannot do the job.

The machine uses a filament that *emits* (gives off) electrons. These streams of electrons are controlled (focused and concentrated) by electromagnets called a *magnetic lens*. The electron beam is generated in much the same way as the light beam in a television receiver. It can be bent or directed by a magnetic field as in the television set. Weld energy is regulated by the current in the electron gun filament.

The electron beam weld is usually made in a vacuum, since air molecules tend to interfere with the beam. The vacuum chambers are also a shield against radiation.

The welder watches the weld through a safe optical system and directs the beam with remote controls. Some

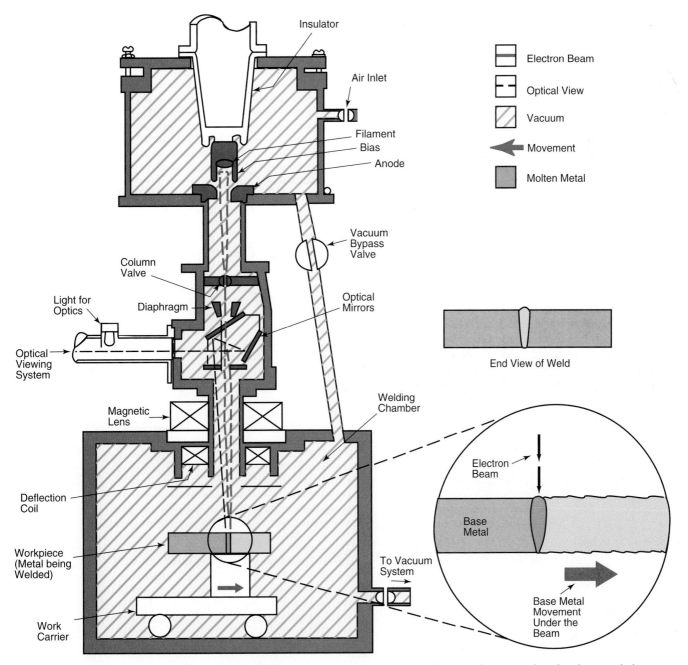

Figure 4-32. *Electron beam welding (EBW). Welding energy is created by striking the base metal with a beam of electrons. The heat energy is very intense. Very narrow welds are formed.*

modern electron beam welding equipment may be used to produce welds under nonvacuum conditions. The surfaces to be joined should be cleaned prior to welding. The parts being joined must fit together very tightly.

The welder and other persons near the machine must be protected from the radiation given off by the beam. Most machines use lead as a shielding material

See Chapter 20 for more on electron beam welding processes.

4.28 TORCH PLASTIC WELDING

In torch plastic welding, heated air or heated shielding gas melts and fuses together plastic base materials and plastic filler materials The use of plastics in modern industry makes considerable amounts of plastic welding necessary. Figure 4-33 shows a typical torch plastic welding set-up.

The gas or air is heated by an electric heating coil. The heated gas flows through the nozzles and heats the parts to be welded. The welding tip is designed to press the heated plastic filler material into the weld area. Plastics are joined at between 400°F and 800°F (204°C and 427°C). The heat source is controlled by adjusting the resistance unit and/or gas flow rate. The welder must manipulate both the torch and the filler material.

The temperature of the filler material during welding is hot enough to cause severe burns. It is advisable to wear gloves and safety goggles. Good ventilation is recommended.

See Chapter 22 for additional details.

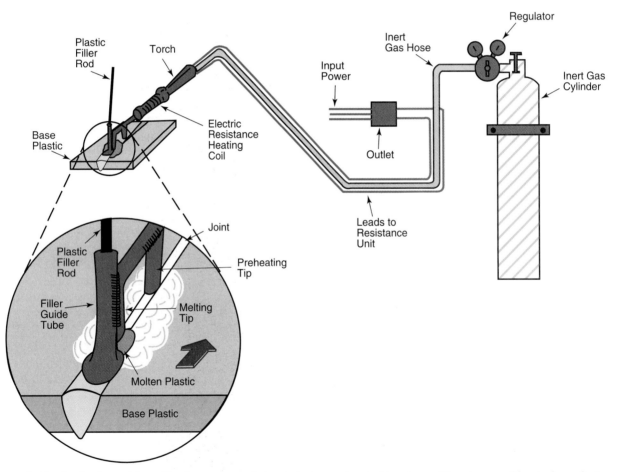

Figure 4-33. *Torch plastic welding. The heat source is heated air or a heated inert gas. The stream of heated gas first preheats the joint, then heated gas from a second orifice further heats the plastic as a plastic filler rod is fed through a tube to the joint.*

4.29 OXYGEN LANCE CUTTING (LOC)

In the oxygen lance cutting process, an oxyfuel gas flame heats the base metal while a jet of oxygen is directed at the heated metal to cut (burn) it. This method has been used for many years to cut heavy steel sections. Figure 4-34 illustrates a typical oxygen lance cutting station.

The oxygen lance is a straight piece of iron pipe with a hand valve. It is attached by a hose to one or more oxygen cylinders equipped with regulators and gauges.

The oxygen lance is used along with an oxyfuel gas cutting torch (see Heading 4.2). After the cut is started with the torch, the oxygen lance valve is opened. The stream of oxygen from the iron pipe flows into the kerf started by the cutting torch. It can cut sections of great thickness. The lance is gradually melted away during the cutting and must be replaced frequently.

As with all cutting operations, the welder must wear proper eye protection, gloves, and clothing to prevent personal injury. It is also important to cover or clear the work area of combustible materials. Follow all fire prevention practices. Good ventilation is required.

See Chapter 24 for more information.

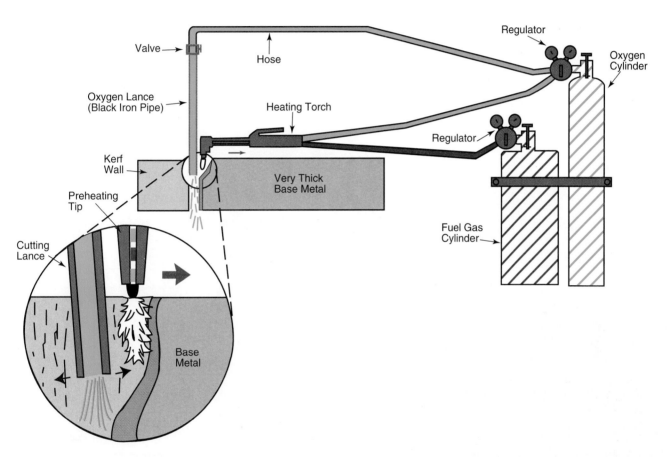

Figure 4-34. *Oxygen lance cutting (LOC). An oxyfuel gas torch is used to heat the base metal. A valve on the metal pipe lance is then opened and a jet of oxygen rapidly oxidizes the heated base metal and blows it away. The lance is gradually consumed as cutting progresses.*

4.30 OXYFUEL GAS UNDERWATER CUTTING (OFC)

To heat the base metal, this underwater cutting process uses an oxyfuel gas flame surrounded by compressed air. An oxygen jet oxidizes the heated metal and blows it away. The oxyhydrogen flame plus an oxygen jet has long been used to cut steel under water. The process is used for salvage operations and underwater construction.

Figure 4-35 shows an underwater cutting outfit. Acetylene gas cannot be used as a fuel gas, because it is unstable at pressures above 15 *psig* (pounds per square inch gauge) or 103.4 *kilopascals* (kPa). This means that acetylene cannot be used safely in water deeper than 10' to 12' (3.05m to 3.66m). Oxyhydrogen can be used at any depth up to 200' (61m).

The proprietary gas called MAPP (methylacetylene propadiene) may also be used for this purpose. It is stable under high pressures. The equipment for using either oxy-hydrogen or oxygen and MAPP is much the same. However, different cutting tips are used.

In addition to oxygen and a fuel gas, compressed air is required for underwater cutting. This air forms an air pocket in which the flame burns under water. The oxygen and the fuel gas cylinders are connected to the torch with regulators and suitable lengths of hose. The valves on the underwater cutting torch control the torch flame.

The fuel gas is usually lighted under water by an electric ignitor built into the air jacket. Gas cutting under water requires suitable diving gear and extensive training as a diver.

See Chapter 24 for more details.

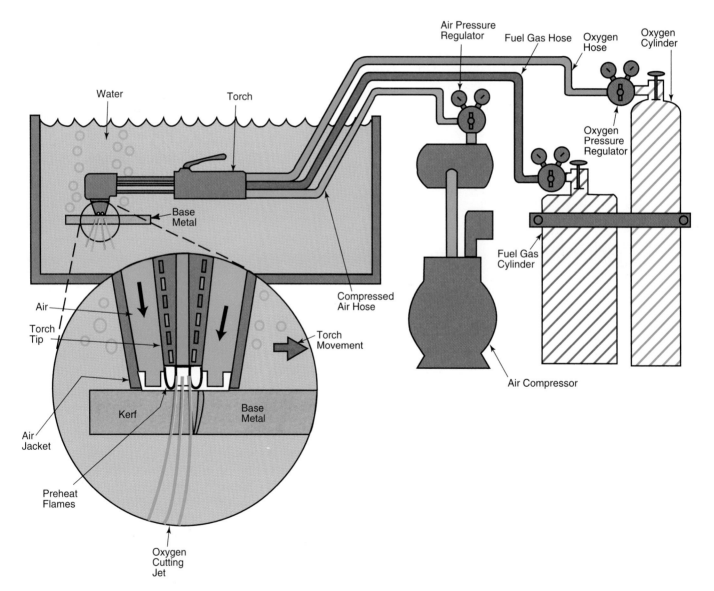

Figure 4-35. Oxyfuel gas underwater cutting. The oxyfuel gas flames burn in a "bubble" of compressed air which keeps the water away . The heated metal is rapidly oxidized and blown away by the oxygen jet.

4.31 FLUX CUTTING (FOC)

Flux cutting uses an oxyfuel gas torch to heat the base metal. This process is used for cutting alloy steels, cast iron, and nonferrous metals which form oxides with a very high melting temperature. Figure 4-36 shows a basic flux cutting station.

Flux cutting requires an oxygen cylinder and a fuel gas cylinder, both fitted with regulators and gauges. Compressed air carries the flux powder to the torch tip. An oxygen jet flows against the heated metal to cut it.

In flux cutting, flux powder is introduced into an oxyfuel gas cutting torch flame. The flux powder decreases the formation of *refractory oxides* (solids). The molten metal is then easily removed to form a kerf.

A very similar process uses an iron powder. The iron powder increases the total heat of the flame. The iron in the flux powder also absorbs the oxygen in the area of the cut, reducing the alloy oxides.

Because of the hazard of sparks, the welder must wear goggles, gloves, high shoes, and protective clothing. Be sure to follow all fire safety practices. The area must be cleared of combustible material.

See Chapter 24 for more details.

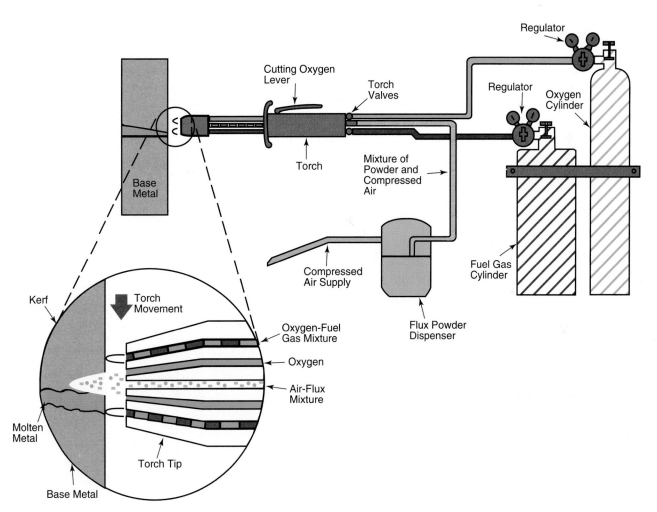

Figure 4-36. *Flux cutting (FOC). A flame, fed by the fuel gas orifices in the tip, heats the metal. The torch lever is then pressed, releasing both an oxygen jet and a jet of compressed air mixed with a powdered flux. This rapidly melts the base metal. The flux makes this melted metal very fluid. The force of the air jet and the oxygen jet blows the molten metal and flux away to form a kerf in the base metal. This process is used mainly for alloy steels.*

4.32 EXOTHERMIC (ULTRATHERMIC) CUTTING

Exothermic cutting is also known as ultrathermic cutting or arc-started LOC. It was developed in 1978 for underwater cutting on offshore oil platforms. Exothermic cutting is also used to make cuts on land.

This process uses a special cutting torch, a special rod, an oxygen cylinder, regulator and hose, a 12-volt battery, a copper striker plate, and two electrical leads. See Fig. 4-37.

The oxygen hose is connected to the regulator at one end and to the cutting torch at the other. The special rod is a copper-coated steel tube which is filled with fuel wires that will burn and create a great deal of heat. These fuel wires, once ignited, will continue to burn with oxygen to support the burning. This self-consuming process is called an *exothermic process*. The temperature of the burning exothermic tube and wires in oxygen reaches over 10,000°F (5537°C).

The torch holds the rod in a collet that is tightened when the collet nut is turned. One lead of the battery is connected to the torch collet. The other lead of the battery is connected to the copper striker plate.

To make a cut, the rod is scratched across the copper plate. When it begins to spark, the oxygen valve on the torch is opened. The burning rod is brought close to the base metal and the cut or gouge begins. The torch is then moved along the metal surface to complete the cut.

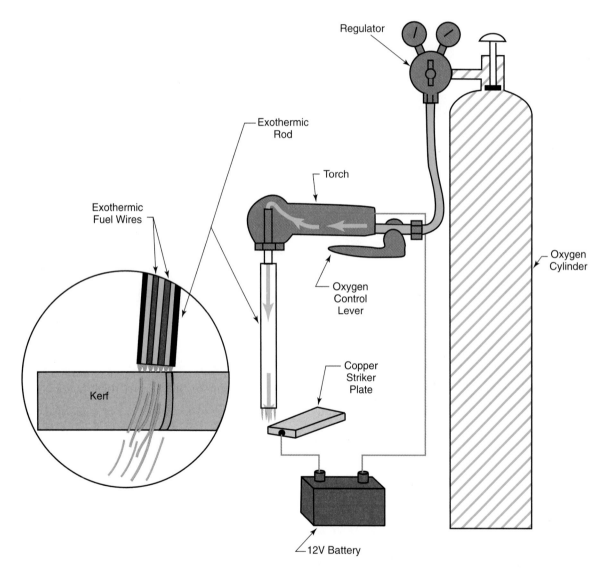

Figure 4-37. *Exothermic or ultrathermic cutting. Oxygen is fed to the cutting area through the torch and hollow rod. Fuel wires inside the larger exothermic rod are self-consuming. They are ignited by striking them on the copper striker plate. The fuel wires provide the heat to melt the base metal, while the oxygen blows the metal out of the kerf.*

4.33 STEEL TEMPERATURE AND COLOR RELATIONSHIPS

It is important to understand what happens when metals are heated, melted, and then cooled. The physical properties of steel are determined by these temperatures and actions.

There are thermometers that can accurately measure surface temperatures of steels (melting crayons, pyrometers, and thermocouples). However, the color of the steel is a very popular and quite accurate way to judge the temperature. Figure 4-38 illustrates the color range for steel.

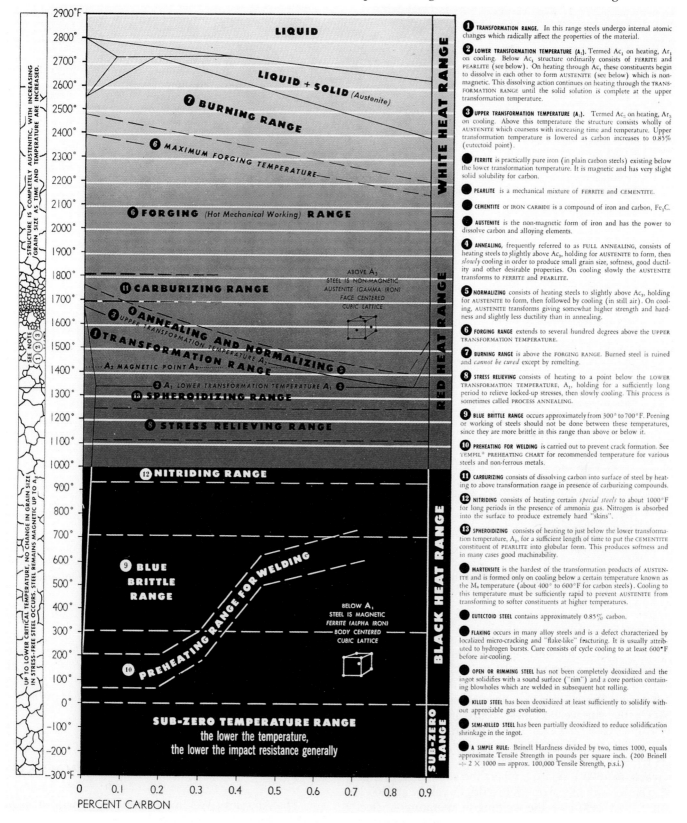

Figure 4-38. *Color-temperature chart of 0 to 90 point carbon steel. This chart shows the change in color of the steel surface as the temperature changes from 1000°F to 2900°F (538°C to 1539°C). (Tempil° Division, Air Liquide America Corp.)*

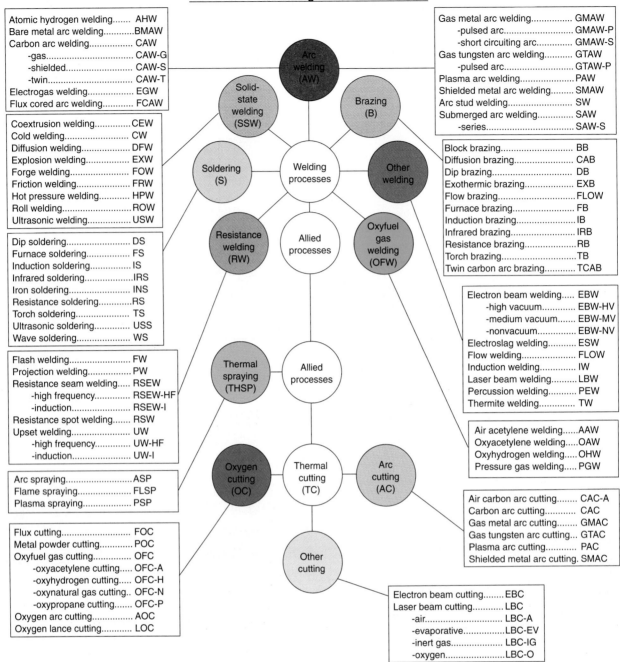

Figure 4-39. *Welding processes defined by the American Welding Society (AWS) and the American National Standards Institute. (ANSI/AWS A3.0-94)*

At low temperatures, steel looks gray to the naked eye. When heated, the steel surface starts to oxidize. First, the oxide coating is gray from 0°F to 100°F (-18°C to 38°C). From 400°F to 475°F (205°C to 245°C), the color changes from a faint straw color to a deep straw color. At 520°F (270°C), the surface becomes bronze in color. Between 540°F and 700°F (280°C to 370°C), the surface oxide turns purple, blue, then black. The surface is dark red at 1000°F (538°C). The red becomes brighter as the temperature increases to 1500°F (815°C), where the color is an orange red. The color finally changes to a bright yellow at 2400°F (1316°C). At 2400°F — the welding temperature — steel begins to melt. Note that temperature affects steel properties differently as the steel's carbon content varies.

Because metals are good conductors of heat, always use extreme caution when handling heated pieces of steel. Temperatures above 120°F (49°C) will cause severe burns. Protect the body by wearing approved goggles, leather gloves, and protective clothing. Handle hot metals with pliers or tongs.

See Chapter 27 for more details.

4.34 WELDING AND ALLIED PROCESSES

The American Welding Society recognizes 11 major headings for welding and allied processes. See Figure 4-39. Each process is explained in detail elsewhere in this text. Refer to Chapter 3 for information on welding symbols

TEST YOUR KNOWLEDGE

Write your answer to these questions on a separate sheet of paper. Do not write in this book.

1. What color is the neutral oxyacetylene flame?
2. In the OFC (oxyfuel gas cutting) process, what causes the rapid oxidation or burning of the steel after the metal is heated to a melting temperature?
3. What is the temperature of the oxidizing, oxygen-LP gas flame?
4. Why is submerged arc welding referred to as "submerged?"
5. In ESW (electroslag welding), what prevents the shoes from becoming welded to the base metal?
6. In ultrasonic welding, what creates the heat required for welding?
7. Why is most EBW (electron beam welding) done in a vacuum?
8. While cutting under water with oxyfuel gas, what keeps the water away from the preheating flames?
9. In the oxygen cutting process designated as OFC-H, what fuel gas is used?
10. What process uses the initials AOC?

In industrial applications, welding and cutting processes are often computer-controlled for improved accuracy and efficiency. This battery of oxyfuel gas cutting torches makes four identical cuts under computer numerical control (CNC). (ESAB Group)

Part 2

OXYFUEL GAS PROCESSES

Oxyfuel gas cutting is used extensively in fabrication shops and on construction sites. (American Welding Society)

Excellent ventilation is needed for all types of welding and cutting process to protect the welder and other workers from smoke and fumes. In many applications, such as this truck repair facility, a portable fume extractor can be moved to the welding location and its intake tube placed immediately over the work area. (Nederman, Inc.)

OXYFUEL GAS WELDING EQUIPMENT AND SUPPLIES

A welder should have a thorough knowledge of the purpose, design, construction, and operation of welding equipment and supplies that he or she is using. This knowledge is necessary to ensure the safe, conscientious use of the equipment and supplies that are used regularly.

5.1 COMPLETE OXYFUEL GAS WELDING OUTFIT

The term *oxyfuel gas welding outfit* refers to the basic equipment needed to weld. An *oxyfuel gas station* includes the oxyfuel gas welding outfit, welding table, ventilation, lighting, and other equipment necessary to welding and the welder's comfort.

The equipment necessary for an oxyfuel gas welding station may vary, depending upon the welding operations performed.

In general, a welding outfit will consist of the following:

* *Gas supplies,* consisting of an oxygen cylinder (compressed gas) or tank (bulk liquid), and a fuel gas cylinder or generator.
* *Regulators* with high- and low-pressure indicators or gauges. Cylinder and hose fittings are also required. One regulator with gauges and fittings is required for the oxygen cylinder and one regulator with gauges for the fuel gas cylinder.
* *Hoses* with appropriate fittings.
* Complete *welding torch* with welding tip, mixing chamber, needle valves, and hose fittings.
* *Welding goggles* with the recommended tinted filter shade lens. For oxyfuel gas welding, a number 4, 5, or 6 lens is normally recommended.
* A flame or *spark lighter* to ignite the welding flame.

The equipment for a complete oxyacetylene welding outfit is shown in Figure 5-1.

5.2 OXYGEN SUPPLY

Oxygen used for welding is stored in cylinders of various sizes, which are usually painted. Since there is no standard national code, the cylinders are painted a color selected by the manufacturer.

Oxygen is stored in cylinders at a pressure of from 2000 psig to 2640 psig (pounds per square inch gauge), or 13 790kPa to 18 202kPa (kilopascals). The pressure varies, depending on the cylinder material and size. The pressure in a cylinder also will vary according to room temperature.

Oxygen is obtained by three different processes. One process consists of the liquefying of atmospheric air by compression and cooling. The atmospheric air consists of approximately 21% oxygen, 78% nitrogen, 1% other gases (by volume).

Oxygen and nitrogen have different boiling temperatures, so it is easy to separate the oxygen from the nitrogen. Nitrogen boils at -320°F (-196°C) and oxygen

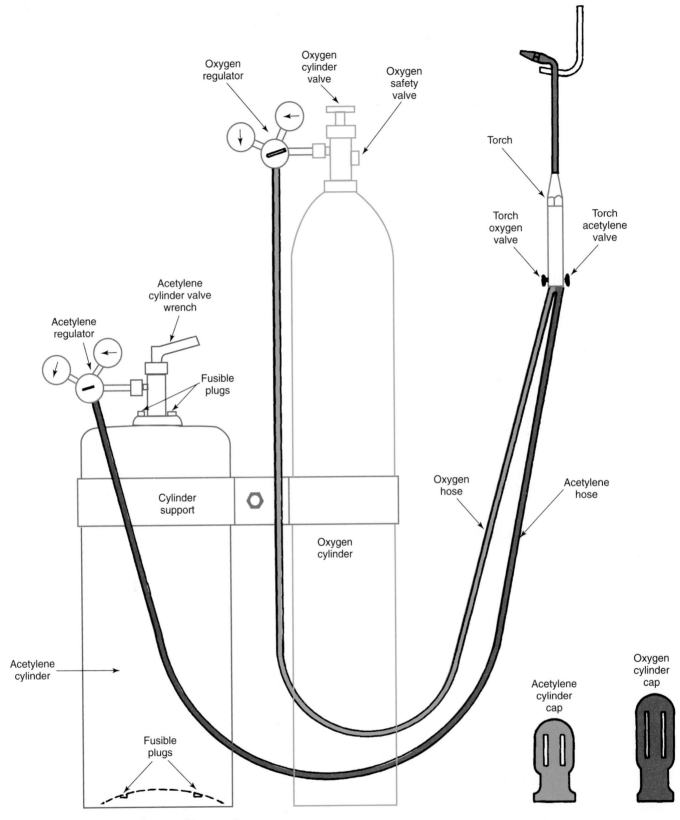

Figure 5-1. *An oxyacetylene welding outfit.*

boils at -297°F (-183°C). The oxygen will boil off first as air is cooled, since it boils at a higher temperature than nitrogen. Nitrogen and oxygen at these low temperatures are in a liquid form. Liquid oxygen and nitrogen may be placed in Dewar flasks for storage and shipment. As they are warmed, they become gaseous and they may then be stored under pressure in thick-walled cylinders.

Air contains 1% by volume of other gases. These gases are mainly water vapor, carbon dioxide, argon, helium, hydrogen, and neon. The water vapor and carbon dioxide are removed during the compression and liquefying process. Argon, neon, helium, and hydrogen each has its own boiling point. As oxygen and nitrogen are being separated, these other *elements* (gases) are also boiled off at different temperatures and stored as a liquid or as a gas.

A second method of producing oxygen is by electrolysis of water. In this process, an electric current is passed through water causing the water (H_2O) to separate into its elements, which are oxygen (O) and hydrogen (H). In the electrolytic process, oxygen will collect at the positive electrode, and hydrogen will collect at the negative electrode. Figure 5-2 illustrates an experimental schematic of the electrolysis of water.

A third method used to produce oxygen is to heat an *oxygen-bearing pellet*. This method is generally used only by individual crafters who do not need enough oxygen to warrant leasing or buying an oxygen cylinder.

Commercial oxygen is close to 100% pure. A popular method of distribution consists of shipping oxygen as a liquid. Liquid oxygen installations are chiefly used in steel mills and by steel fabricators who use large quantities of oxygen.

Oxygen sold in liquid form is in large thermos bottle-like tanks or vessels. These vessels are also known as *Dewar flasks*. Liquid oxygen is not held under very high pressure. The pressure in a liquid oxygen vessel is seldom greater than 240 psig (1655kPa). The evaporation of some of the liquid keeps the temperature of the liquid very low, approximately -297°F (-183°C). At this low temperature, the oxygen remains a liquid under normal atmospheric pressure. Liquid oxygen in a container will rapidly evaporate if the pressure is reduced or if the temperature is increased.

Some portable liquid oxygen vessels have been developed and experience indicates that their use will increase. The advantage of liquid oxygen storage is chiefly in the relatively small size and large gaseous volume of a Dewar flask. Figure 5-3 illustrates such a liquid oxygen storage vessel.

5.2.1 Oxygen Cylinders

Most oxyacetylene welders will be using oxygen from pressurized cylinders. Therefore, they must know how to use these cylinders and the various *safety precautions* which must be taken. The oxygen stored in cylinders is under very high pressure. The gauge pressure is about 2200 pounds per square inch of tank area (15 170kPa). For safety, cylinders must be of very sturdy construction. The Interstate Commerce Commission (ICC) has prepared specifications for the construction of these cylinders. They are forged in one piece. No part of the cylinder is less than 1/4" thick. The steel used is armor plate type, high-carbon steel. Figure 5-4 illustrates an oxygen cylinder with a section cut away to show construction of the cylinder and the valve. Oxygen cylinders are tested regularly. They must withstand hydrostatic (water) pressure of over 3,300 psig (22 753kPa). These cylinders are periodically heated and annealed. The annealing relieves stresses created during on-the-job handling. They are also periodically cleaned using a caustic solution that chemically cleans the cylinder inside and out.

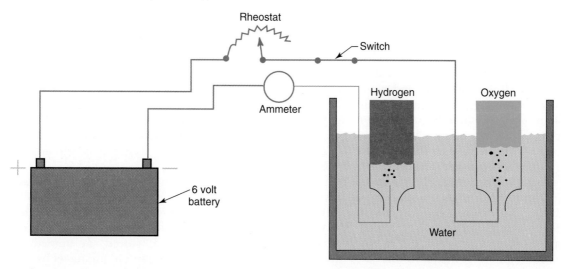

Figure 5-2. *A schematic illustrating the electrolysis of water. With dc current flowing, bubbles of oxygen will appear in one bottle and hydrogen in the other bottle. Oxygen is collected at the positive (+) terminal and hydrogen at the negative (-) terminal.*

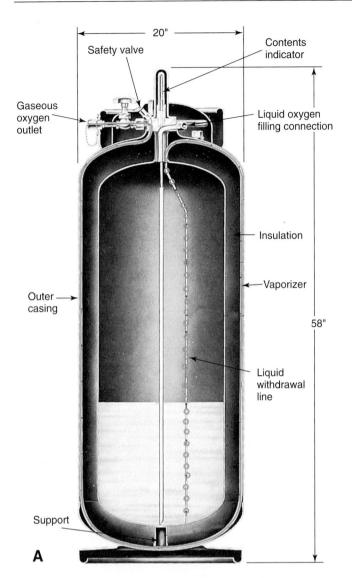

Figure 5-3. A liquid oxygen vessel or Dewar flask. A—In this sectional view, note that liquid oxygen drawn from the inner cylinder vaporizes as it passes through the withdrawal line in the insulated space surrounding the inner cylinder. B—This Dewar flask will furnish 250 cfh (7079L/hr) of oxygen at a constant flow. (Taylor-Wharton Cylinders)

The cylinder valve is of special design to withstand high pressure and is made of forged brass. The oxygen cylinder valve is a "back-seating valve." When the valve is turned all the way open, the stem is sealed to prevent the leakage of oxygen around it. When in use, this valve should be turned all the way open **and never used in a partly open position.** Figure 5-5 shows the internal construction of an oxygen cylinder valve. The cylinder valve is threaded into the top of the cylinder using a 1/2" (12.7mm) or 3/4" (19.1mm) pipe thread. The 3/4" pipe thread size is the most popular. The cylinder valve incorporates a pressure safety device. This device is a thin disc, as shown in Figure 5-5. The thin metal disc will burst, as shown in Figure 5-6, before the pressure in the cylinder becomes great enough to rupture the cylinder. The valve outlet fitting is a standard male thread to which all standard pressure regulators made in the U.S. may be attached. A handwheel for operating the valve is securely fastened to the valve stem.

Threads on the upper body of the cylinder are provided for the **cylinder (safety) cap**. The heavy steel cylinder cap is screwed onto the top of the cylinder to protect the valve from damage during movement or shipment. The thread size is 3 1/8" diameter (79.4mm) with 7 or 11 threads per inch. (In SI Metric units, the thread pitch is 3.62mm or 2.31mm.) If the cylinder valve should ever be broken off, the very high pressure oxygen will escape rapidly, possibly giving the cylinder "rocket propulsion." Also, the escaping oxygen will help any material it contacts to burn.

Because of these dangers, a cylinder truck should be used when moving cylinders. To reduce the danger of being tipped over or dropped, cylinders should be secured by chain or steel straps in a vertical position while being used, stored, or moved. When full or partially full, cylinders should never be allowed to stand by themselves without being secured to an adequate support.

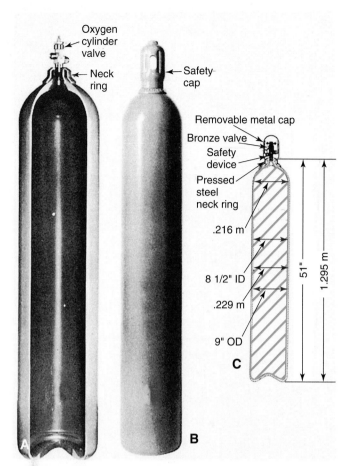

Figure 5-4. *A typical oxygen cylinder with a 244 ft³ (6909L) capacity. A—Internal construction of cylinder. Note the one-piece forged construction. B—Exterior of the cylinder. Note the safety cap over the valve. C—Dimensions of a 244 ft³ (6909L) cylinder. (Pressed Steel Tank Co.)*

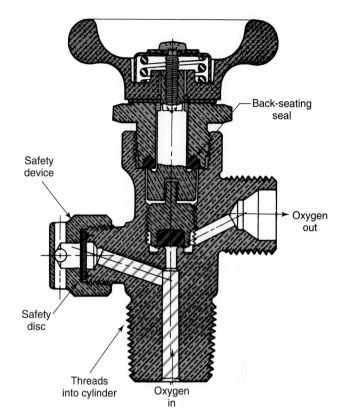

Figure 5-5. *Internal construction of an oxygen cylinder valve. Note that the valve must be opened completely to engage the back-seating seal.*

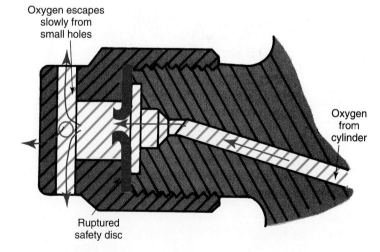

Figure 5-6. *A schematic of the oxygen cylinder safety plug. When the disc ruptures, oxygen escapes from several small, drilled holes.*

If it should become necessary to move an oxygen, acetylene, or any other gas cylinder by hand, proceed as follows:

1. **Make sure that a safety cap is securely screwed in place over the cylinder valve.**

2. Tip the cylinder to a small angle from vertical. Keep one hand at the top of the safety cap to keep the cylinder from falling.

3. With the other hand slowly rotate the cylinder. As it is rotated it will move across the floor. **Proceed slowly**

keeping the cylinder nearly vertical with one hand and rotating it with the other to move it to where it is to be used or stored. See Figure 5-7.

Cylinders should always be kept valve end up. The valve should be closed when the cylinder is not in use, whether the cylinder is full or empty.

Oxygen is purchased by the cubic foot (ft^3), measured at atmospheric pressure. The cylinders usually remain the property of the manufacturer and are loaned to the consumer. The price of an empty oxygen cylinder is quite high. It is generally to the advantage of the consumer to use the rental system. Most companies will not charge any rental for a cylinder if it is returned within thirty days from the date of delivery. After thirty days, a fee is charged on a per day basis, often referred to as a demurrage charge.

Oxygen cylinders come in many sizes and shapes. They are made from steel or aluminum under very strict specifications. Some common sizes of steel cylinders are: 55 ft^3 (1557L), 80 ft^3 (2265L), 125 ft^3 (3540L), 150 ft^3

(4248L), 250 ft^3 (7079L) as shown in Figure 5-8. Figure 5-9 shows other steel and aluminum cylinder sizes and shapes. Under full pressure, approximately 1 ft^3 (28L) of gas is stored in each 10 in^3 (0.164L) of space in the cylinder. As the pressure in an oxygen cylinder decreases, the amount of oxygen remaining in the cylinder also decreases.

Figure 5-8. *Oxygen cylinders of five different capacities. They range in capacity from 55 ft^3 to 250 ft^3 (1557L to 7078L). Under certain conditions they may be charged with 10% more oxygen. (Pressed Steel Tank Co., Inc.)*

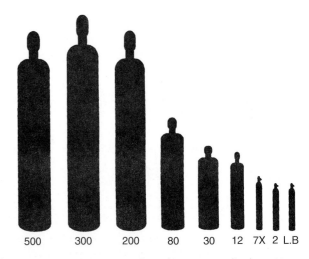

Figure 5-9. *Various sizes of steel oxygen cylinders. Note that the smallest three have post-type valves.*

Figure 5-7. *A cylinder being moved by tilting it with one hand and rolling it into position with the other hand. Notice that the safety cap is in place.*

5.2.2 The Oxygen Manifold

In many school welding shops and many industries, an individual oxygen cylinder is not a part of each welding station. Instead, one or more oxygen cylinders is attached to a manifold, and the oxygen is piped from the manifold to the welding stations. The oxygen leaving the manifold is reduced in pressure. Line pressures between 30 psig and 100 psig (206.8kPa and 689.5kPa) are common. The length and size of the pipe and the amount of oxygen being used will determine the pressure needed.

The oxygen manifold is usually located in the oxygen cylinder storage room. The manifold cylinders are generally outside the work area. The shop is safer and space is saved by not having the cylinders in the shop.

With a manifold system, an *oxygen line regulator* is used at each welding station to control the oxygen pressure to each torch. A line regulator has only one pressure gauge; this gauge indicates the pressure on the delivery (torch) side of the system. Figure 5-10 shows a typical oxygen manifold installation. Note that the manifold is pressurized up to the manual shut-off valve. This valve is kept closed when the oxygen system is not in use. If the manifold is to be shut down completely, it is recommended that the cylinder valves be closed as well. A *master regulator* controls the pressure of the oxygen in the piping system after it leaves the manifold. This regulator has two gauges. One gauge shows the pressure in the manifold, and the other shows the pressure in the line (piping). See Heading 5.4.4 for information concerning the construction and operation of these regulators.

In manifold installations, the copper tubing (pigtail) connecting the cylinders to the manifold should be frequently *annealed* (heated and cooled to soften it and make it less brittle). This is done because the tubing is subjected to high cylinder pressure, and it therefore may become brittle and more subject to breakage. **Local and national safety and fire codes must be checked to ensure proper installation of an oxygen manifold.**

5.2.3 Oxygen Safety Precautions

- If a label on a cylinder is missing or is not legible, do not assume that the cylinder contains a particular gas. Return it to the supplier.
- Do not permit smoking or open flames in any area where oxygen is stored, handled, or used.
- Liquid oxygen at -297°F (-183°C) may cause freeze burns to the eyes or skin if it comes in contact with them.
- Keep all organic materials such as oil, grease, kerosene, cloth, wood, tar, or coal dust away from contact with oxygen.
- Do not place liquid oxygen equipment on an asphalt surface or any surface with oil or grease deposits.
- Remove all clothing which has been splashed or otherwise saturated with oxygen gas. Such clothing is highly flammable. It is not safe to wear for at least 30 minutes or until no oxygen remains.

The following are some safety codes and standards for your reference:

- AWS Z49.1-88. *Safety in Welding and Cutting.*
- NFPA (National Fire Protection Association) Standard No. 50. *Bulk Oxygen Systems at Consumer Sites.*
- NFPA Standard No. 51. *Oxygen-Fuel Gas Systems for Welding, Cutting, and Allied Systems.*
- NFPA Standard No. 51B. *Cutting and Welding Processes.*
- Local and National Safety Codes.

5.3 ACETYLENE SUPPLY

The majority of oxyfuel gas welding and cutting is done using acetylene as the fuel gas. Acetylene will be the fuel gas which is referred to in most of the following sections of this chapter when the term "fuel gas" is used.

Acetylene is produced by the chemical combination of calcium carbide with water. The chemical formula for acetylene is C_2H_2. (See Heading 34.4 for further technical information concerning the chemical structure and nature of acetylene.)

Acetylene is made available for oxyacetylene welding using either an acetylene storage cylinder or an acetylene generator.

5.3.1 Acetylene Cylinders

Acetylene gas may be stored in cylinders specially designed for this purpose. The gas is first passed through filters and purifiers. **Storing acetylene in its gaseous form under pressure is not safe at pressures above 15 psig (103.4kPa).**

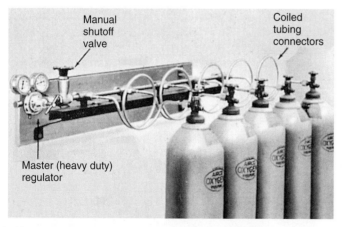

Figure 5-10. *An oxygen manifold for five cylinders. A master shutoff valve is located between the regulator and the cylinders. Be sure to check local and national fire safety codes prior to installation. (Airco Welding Products)*

The method used to safely store acetylene in cylinders is as follows:

1. New empty cylinders are filled with a *monolithic* (massively uniform) filler material that cures to a porosity of 85%, as required by federal safety regulations.
2. Acetylene gas is then pumped into the cylinder, where it is stored in the small spaces or pores in the monolithic filler. This prevents large acetylene gas accumulations.
3. The cylinder is then charged with acetone, which absorbs acetylene. The theory is that the acetylene molecules fit in between the acetone molecules. Using both of these techniques (monolithic filler and acetone) prevents the accumulation of a pocket of high-pressure acetylene.

Acetylene cylinders, like oxygen cylinders, are fabricated according to Interstate Commerce Commission (ICC) specifications.

The base of the acetylene cylinder is concave, and it usually has two plugs threaded into it. These pipe-threaded *fuse plugs* have a center made of a special metal alloy that will melt at a temperature of approximately 212°F (100°C). Fuse plugs also may be threaded into the top of the cylinder. Figures 5-11 and 5-12 illustrate the fuse plug. If the cylinder is subjected to a high temperature, the plugs will melt and allow the gas to escape before the pressure builds up enough to burst the cylinder. In a fire, the

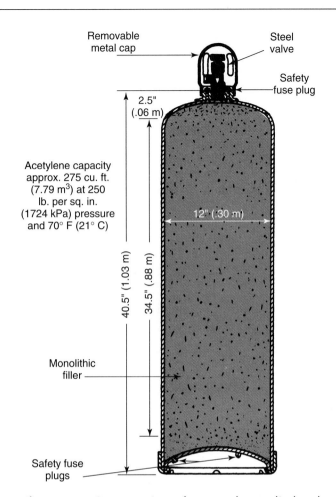

Figure 5-12. Cutaway view of an acetylene cylinder, showing the porous filler. Note the safety fuse plugs at the top and bottom of cylinder.

Figure 5-11. Acetylene cylinder fuse plugs. The body of the plug is usually made of brass. The center is a fusible material that melts at about 212°F (100°C). Note that the regulator attaches to a male thread.

acetylene escaping from the fuse plug will burn, but at a relatively slow rate.

These precautions are needed because the pressure in an acetylene cylinder builds up rapidly with an increase of temperature. Figure 5-12 shows the construction of an acetylene cylinder.

Acetylene cylinder valves come in two types. One has a handwheel which should remain attached at all times. Figure 5-13 shows the construction of the handwheel type of acetylene cylinder valve. The other type is provided

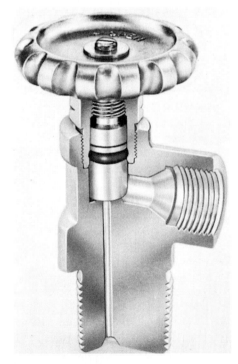

Figure 5-13. A cutaway view of an acetylene cylinder valve. Note that the regulator attaches to a female thread.

with a 3/8" (9.5mm) square shank. The square shank is turned by means of a 3/8" (9.5mm) square box-end wrench. The regulator fitting on the cylinder may have a male or female thread.

It is recommended that the cylinder valve be opened only 1/4 to 1/2 turn. **The wrench should be left in place on the valve stem whenever the acetylene cylinder valve is open. This is done so that the valve may be closed quickly in case a hose or some other part catches fire.**

The amount of acetylene in a cylinder cannot be estimated by the pressure in the cylinder, because the pressure of the acetylene gas coming out of the acetone solution will remain fairly constant (depending on the temperature) until most of the gas is consumed. The amount of acetylene in a cylinder can be determined accurately by weighing the cylinder. The *tare weight* (weight of the cylinder without the acetylene gas) is always stamped on the cylinder. Subtracting the tare weight from the actual cylinder weight will give the weight of the gas remaining. The gas weighs one pound per 14 1/2 ft^3 (1.10kg per m^3).

Acetylene cylinders should always be stored and used in an upright position. If they are not, some of the acetone is likely to escape with the acetylene and contaminate the equipment and the flame. Each time that an acetylene cylinder is refilled, the tare weight is checked and acetone is added if necessary.

Acetylene cylinders are available in a variety of sizes from 10 ft^3 to 380 ft^3 (283L to 10 762L). Some of the most common sizes used in welding are: 130 ft^3 (3682L), 290 ft^3 (8213L), and 330 ft^3 (9346L).

Two small-size acetylene cylinders are available for portable welding and cutting equipment. These are *B size,* 40 ft^3 (1133L), and *MC size,* 10 ft^3 (283L). The fittings on these two small size cylinders (sometimes called tanks) are called *Presto-O-Lite (POL)* fittings.

All acetylene cylinders are designed and constructed according to Interstate Commerce Commission Specification No. 8. Each design must be tested by the Bureau of Explosives and must pass these tests before it can be used commercially.

As noted, acetylene is dissolved in acetone. Therefore, acetylene cannot be drawn from the cylinder any faster than it can be released from the acetone. This release is a kind of boiling action that occurs efficiently at room temperature. The maximum safe rate for drawing acetylene from a cylinder is one-seventh of the cylinder's capacity per hour. Thus, a single 290 ft^3 (8213L) cylinder can supply acetylene at a rate of about 41 ft^3 (1161L) per hour. If acetylene is drawn out of a cylinder too rapidly, a considerable amount of acetone may be drawn from the cylinder along with the acetylene.

An oxyacetylene flame consuming some acetone will burn with a purple color. Acetone in the flame is not desirable, since it lowers the flame temperature and increases gas consumption. Moreover, the quality of the weld is affected.

Low or freezing temperatures can cause the flow of acetylene to decrease, since heat is needed to boil off the acetylene. Another factor that will cause a slowdown in acetylene flow is a nearly exhausted cylinder. As the cylinder approaches the empty or discharged condition,

acetylene is released more slowly. If a greater flow rate is needed, a number of cylinders may be connected to a manifold. This kind of installation makes it possible to use up more of the acetylene from each cylinder. As a result, the cost of the acetylene used may be decreased.

The rate at which acetylene may be drawn off depends on:

- The temperature of the cylinder.
- The amount of charge remaining in the cylinder.
- The number of cylinders providing the flow.

Too-rapid removal of acetylene can be dangerous. A pressure drop could occur which might cause a *flashback*. A flashback is a condition in which the torch flame burns inside the torch or hoses. If this occurs, shut off the gas supply immediately.

Figure 5-14 illustrates several sizes of acetylene cylinders. The cylinder on the extreme right is the MC size, while the one next to it is the B size. The valves in the larger cylinders shown in Figure 5-14 are known as POL Commercial. Some acetylene cylinders have a recessed

Figure 5-14. *Acetylene cylinders come in a variety of sizes to meet the needs of most users. Note the post-type valves on the smaller cylinders.*

top. This recess protects the cylinder valve, which has a female regulator connection or fitting.

5.3.2 Acetylene Manifold

As explained in Heading 5.2.2, there are several advantages in a manifold system in which the fuel gas and oxygen cylinders are not located at the welding stations in the shop. In the case of acetylene or other fuel gases, fire safety is one additional reason for having a manifold installation.

Figure 5-15 illustrates an acetylene cylinder manifold installation. It should be noted that the acetylene manifold system incorporates some features not included in an oxygen manifold system. In addition to regulator and control valves, the acetylene manifold also has a waterseal-type *flashback arrestor* to prevent flashback. Local and national fire safety codes must be checked to properly install an acetylene or fuel gas manifold.

Brass piping may be used with acetylene, but copper piping *must not* be used in the presence of acetylene. Copper and acetylene will form *copper acetylide*, an unstable compound that disassociates violently (explodes) at the slightest shock.

5.3.3 Acetylene Generator

Acetylene is produced by the chemical action of water with calcium carbide. An *acetylene generator* brings the proper amounts of water and calcium carbide together at the rate required to safely generate the acetylene needed.

Acetylene generators are available in two types: a low-pressure type, and a medium-pressure type. The low-pressure type generates acetylene at approximately 1/4 psig (4" to 6" of water column) or 1.72kPa. With this type of generator, it is necessary to use an injector-type torch. See Heading 5.6.2.

The medium-pressure type acetylene generator will produce acetylene at up to 15 psig (103kPa), the maximum safe pressure at which free acetylene gas may be stored. With this type of generator, positive-pressure type torches may be used. See Heading 5.6.1 for a description of the positive-pressure type torch. Figure 5-16 illustrates an acetylene generator. The use and installation of an acetylene generator is controlled by safety codes which should be carefully studied and rigidly followed. See Heading 5.3.4. A generator installation will require regular care, which includes checking and replenishing the water and calcium carbide supplies, and removal of the calcium hydroxide sludge. The removal and disposal of the sludge is perhaps the greatest problem in the use of an acetylene generator.

Referring to Figure 5-16, note that the generator is partially filled with water. The calcium carbide is stored in the hopper at the top of the generator. A feed mechanism meters the calcium carbide into the water. Acetylene is generated by the chemical action of the water on the calcium carbide. As soon as the predetermined acetylene pressure is reached, the feed of calcium carbide is stopped. No more can be fed into the water until the acetylene has been drawn off and the acetylene pressure in the generator decreases. A water-level sensing device stops the feeding of calcium carbide if there is insufficient water in the base of the generator.

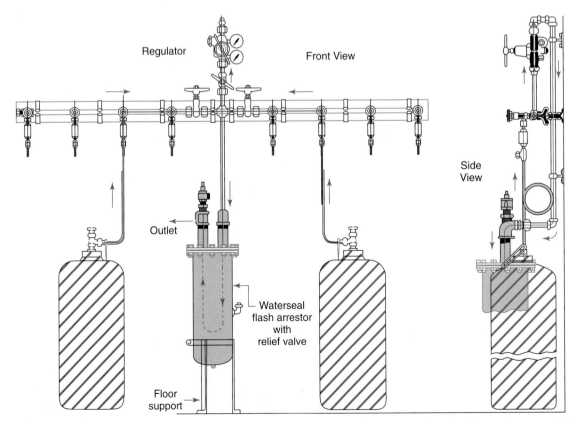

Figure 5-15. *A typical acetylene manifold installation. Be sure to check local or national codes for proper installation.*

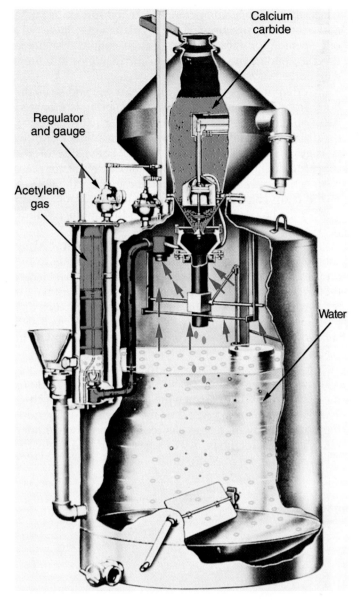

Figure 5-16. *A cutaway view of an acetylene generator. (Rexarc, Inc.)*

5.3.4 Acetylene Safety Precautions

- Concentrations of acetylene between 2.5% and 81% by volume in air are easy to ignite and may cause an explosion. Smoking, open flames, unapproved electrical equipment, or other sources of ignition must not be permitted in acetylene storage areas.
- Under certain conditions, acetylene forms readily explosive compounds with copper, silver, and mercury. Acetylene must be kept away from these metals, their salts, compounds, and high-concentration alloys.
- Keep cylinders away from overhead welding or cutting. Hot slag may fall on a cylinder and melt a fusible plug. The fusible plugs used on acetylene cylinders will melt at about 212°F

(100°C). Keep the torch flame away from the cylinder and fusible plugs.
- Adequate ventilation in welding areas is essential, since acetylene may displace air in a poorly ventilated space. Acetylene has a garlic-like odor. Any atmosphere that does not contain at least 18% oxygen may cause dizziness, unconsciousness, or even death.
- Leave the handwheel, wrench, or key on the cylinder valve for emergency shutoff.

The following are some safety codes and standards for your reference:
- AWS Z49.1-88. *Safety in Welding and Cutting.*
- NFPA Standard No. 51. *Oxygen-Fuel Gas Systems for Welding, Cutting, and Allied Systems.*
- NFPA Standard No. 51B. *Cutting and Welding Processes.*

Also, it is important to follow the equipment manufacturer's instructions.

5.4 PRESSURE REGULATOR PRINCIPLES

Gases are commonly stored in cylinders at pressures considerably above the working or flame pressures. Most welding torches operate at pressures of between 0 psig and 30 psig (0kPa and 207kPa). Therefore, it is necessary to provide a pressure-regulating mechanism to reduce and otherwise regulate the pressure from the cylinder. This mechanism is called a *pressure regulator*. Every system that requires the control of gas pressure uses such a regulator.

The regulator performs two functions:
- It reduces the high storage cylinder pressure to an adjusted, lower working pressure.
- It maintains a constant working pressure at the torch even though the cylinder pressure may vary.

Figure 5-17 shows a pressure regulator complete with gauges and fittings.

Figure 5-17. *A single-stage acetylene regulator. Note that both gauges are calibrated in US Conventional (psig) and SI Metric (kPa) units. (Smith Equipment, Division of Tescom Corp.)*

There are two basic types of regulator mechanisms, the nozzle-type and the stem-type. Figure 5-18 shows the construction of the nozzle-type regulator. Figure 5-20 illustrates a cross section of a stem-type regulator.

Most regulators have two gauges. The *high-pressure gauge* shows the cylinder pressure. The *low-pressure gauge* shows the working pressure, or the pressure of the gas being delivered to the torch.

Regulators are produced as:
*   Single-stage regulators, which reduce the cylinder pressure to a working pressure in one step. Examples—2000 psig to 5 psig (13 790kPa to 34.5kPa).
*   Two-stage regulators, which reduce the cylinder pressure to working pressure in two steps or stages. Example—2000 psig to 200 psig (13 790kPa to 1379kPa), then 200 psig 5 psig (1379kPa to 34.5kPa).

Regulators are generally forged or cast and made of brass, aluminum, or stainless steel.

A pressure relief valve is also threaded into the regulator body. If the pressure within the regulator becomes too high, the relief valve will vent the pressure so that the regulator does not rupture. See Figure 5-19.

5.4.1 Nozzle-Type Pressure Regulator

See Figure 5-18 for an illustration of a nozzle-type pressure regulator. A fitting to attach the regulator to the cylinder is threaded into the body. The regulator body generally has threaded openings for a high-pressure gauge and a low-pressure gauge. Line regulators may have only one opening for a low-pressure gauge. A third opening is used to connect the torch hose to the regulator low-pressure outlet.

A diaphragm or flexible disc separates and seals the body from the bonnet. The diaphragm spring is mounted between the diaphragm and the adjusting screw. Force on the diaphragm is adjusted by means of an adjusting screw. A cage or carrier is attached to the body side of the diaphragm. This cage curves down into the regulator body chamber. The seat is attached to the cage near the bottom of the body chamber. This seat is constantly pushed up against the nozzle by the body spring to stop the gas flow into the regulator.

The opening of the nozzle and seat is controlled by the position of the diaphragm. The line that leads to the nozzle comes from the cylinder and to this line is attached the high-pressure gauge. A fine mesh screen or ceramic filter is commonly located in this line. This filter keeps dirt from entering and damaging the regulator. The screen also serves as a flame arrestor and should be always left in place.

If the adjusting screw in the body is turned "in" (clockwise), the diaphragm spring on the bonnet side of the diaphragm will overcome the force of the body spring. The diaphragm spring will then move the seat away from the nozzle, allowing some gas to pass from the cylinder into the regulator body. As this gas enters the regulator body, it builds up pressure in the body. The force created

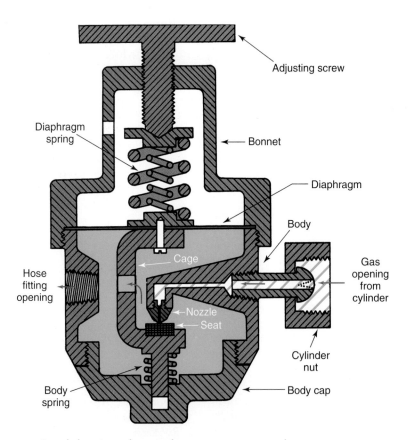

Figure 5-18. Schematic cross-sectional drawing of a nozzle-type pressure regulator.

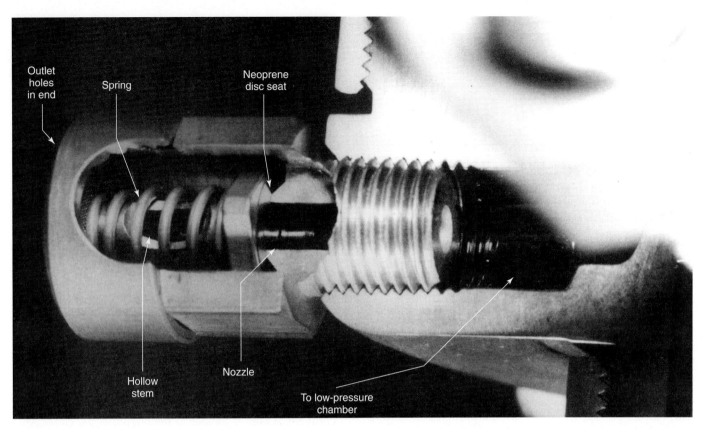

Figure 5-19. *A cutaway view of a regulator pressure relief valve. Overpressure may open and reset this valve. A sudden high-pressure or fire will rupture the neoprene seat and vent the regulator through the hollow stem. (Smith Equipment, Division of Tescom Corp.)*

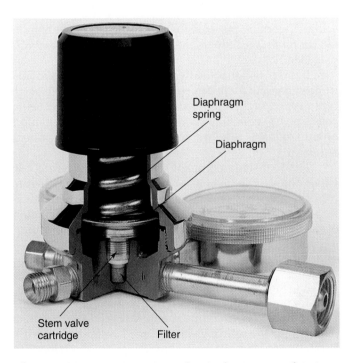

Figure 5-20. *A cutaway view of a single-stage regulator. Note the sintered metal filter below the stem-type valve cartridge. (CONCOA)*

by this pressure pushes the diaphragm up against the diaphragm spring. The upward movement of the diaphragm moves the cage and the seat up. This closes off the nozzle opening. The pressure will fall as the gas is released from the regulator and flows through the hose to the torch. This action allows the diaphragm spring to move the diaphragm down slightly. The nozzle and seat valve open and allow more gas to come into the regulator body. The balance of the diaphragm spring pushing the diaphragm and seat open, and the pressure under the diaphragm pushing the seat closed, keeps the working pressure flowing through the regulator relatively constant.

Note: When the adjusting screw on the regulator is turned all the way out, the flow of gas from the cylinder is stopped completely.

These regulators come in various gas-flow capacities and nozzle orifice sizes. The diaphragm, seat orifice, and spring size are designed to provide the volume of gas desired. Master regulators are designed to allow a large amount of gas to flow. Line regulators allow the flow of relatively small amounts of gas.

The springs are made of a good grade of spring steel, while the diaphragm may be made of brass, phosphor bronze, sheet spring steel, or stainless steel.

The diaphragm is sealed at the joint between the diaphragm and the regulator body by means of suitable gaskets and the clamping action between the body and the bonnet.

The nozzle is usually made of bronze, while the seat may be made of various materials such as nylon, Teflon®, neoprene, or some proprietary materials.

5.4.2 Stem-Type Pressure Regulator

The stem-type regulator works on the same principle as the nozzle-type, but instead of using a nozzle and seat, it employs a poppet valve and seat. Figure 5-20 illustrates a typical stem-type regulator. The operation of this type of regulator is described as follows.

High-pressure gas enters the chamber below the seat when the cylinder valve is opened. The construction is such that the high pressure and the body spring force the valve against its seat. When the pressure adjusting screw is turned in, it forces the diaphragm spring and diaphragm down. The diaphragm pushes the stem down. This downward movement forces the valve away from the seat.

As pressure builds below the diaphragm, the diaphragm moves up, closing the valve. Gas from the low-pressure chamber is fed to the torch.

As the pressure in the low-pressure chamber drops, the diaphragm spring forces the diaphragm, stem, and valve down again. This allows more high-pressure gas to enter the low-pressure chamber and shut off the valve again. This constant opening and closing of the valve as the diaphragm moves up and down controls the working (outlet) pressure within close preset limits.

The stem-type regulator lends itself to installations requiring a rather high rate of flow, and is commonly used on manifolds and flame cutting machines.

The materials in a stem-type pressure regulator are similar to those used in the nozzle-type. The seat may be constructed of neoprene, Teflon®, or special proprietary materials. The stem (pin) is usually made of stainless steel. The stem and seat may be designed to enable the complete assembly to be removed as a cartridge, permitting easy servicing. Figure 5-21 shows a stem-type valve built as a cartridge. Figure 5-22 shows an exploded view of a stem-type regulator.

5.4.3 Two-Stage Pressure Regulator

The two-stage pressure regulator may be considered to be two regulators in one. In the first stage, the high pressure is reduced and regulated to an intermediate pressure. This is done by a valve and diaphragm mechanism, which has a fixed pressure adjustment. Figure 5-23 illustrates an external view of a two-stage regulator. Figure 5-24 shows a two-stage regulator that has been cut away to show the two diaphragms and two needle valve cartridges. The second stage, which is adjustable, regulates the working pressure and gas flow to the torch. A two-stage regulator operates in the same manner as a single-stage regulator.

In a two-stage *oxygen* regulator, the pressure is reduced through the first stage to about 200 psig

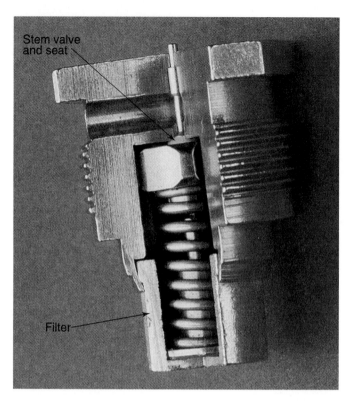

Figure 5-21. *A stem-type regulator valve in a capsule or cartridge. A filter is located at the bottom of the cartridge. (CONCOA)*

(1379kPa). This means that in the second stage, the pressure need only be regulated from 200 psig (1379kPa) to the selected working pressure. In a two-stage *acetylene* regulator, the first stage reduces the pressure to approximately 50 psig (344.7kPa).

It is claimed that the two-stage regulator will provide a more constant working pressure than the single-stage type, especially when large volumes of gas are being consumed. Figure 5-25 illustrates, in color, the mechanical operation and the pressure conditions in this type regulator.

The two-stage regulator may be constructed to use either the nozzle-type or stem-type mechanism. Figure 5-26 illustrates a two-stage stem-type regulator. Some two-stage regulators have been made which are a combination of the two styles, as shown in Figure 5-27.

5.4.4 Master Service Regulator

As stated in Headings 5.2.2 and 5.3.2, manifold systems for both oxygen and acetylene require the use of large master regulators to control the flow of the gases from the manifold to the welding station line. These regulators are basically of the same construction as described in Headings 5.4.1, 5.4.2, and 5.4.3; however, some modifications are made to adapt the regulator to its particular job. The regulator must be capable of controlling large volumes of gas, even when the difference between the cylinder pressure and the line pressure is small. These regulators always have two gauges; a high-pressure gauge

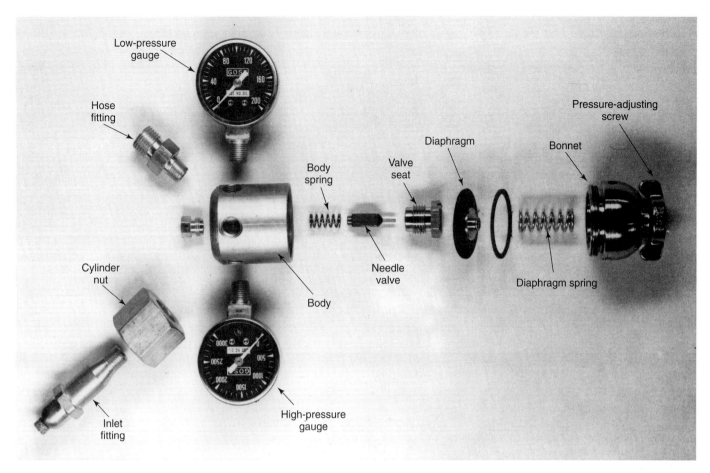

Figure 5-22. An exploded view of a single-stage regulator showing all the parts in the correct relative positions. (Goss, Inc.)

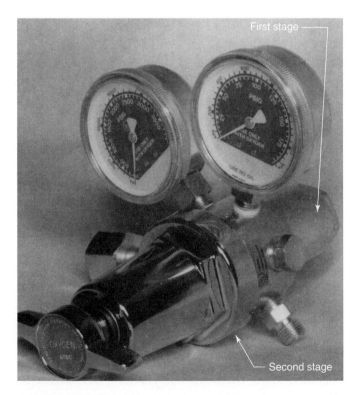

Figure 5-23. A two-stage regulator. Adjustment is on the second stage.

that indicates the manifold (cylinder) pressure, and a low-pressure gauge that indicates the line or working pressure. The typical cylinder regulator usually has insufficient gas-flow capacity to be used as a master regulator on a manifold installation.

5.4.5 Line or Station Regulator

Line or station regulators are used in connection with manifold systems. These regulators have the usual adjustment to control the working pressure to the torch. They should be handled in the same manner as other regulators. They are usually equipped with only one gauge. This gauge is connected to the discharge side of the regulator and indicates the pressure of the gas being delivered to the torch.

Note: Fuel gas and oxygen regulators and gauges must not be interchanged—a fire or explosion may result.

Since the inlet pressure is lower than with regulators attached to cylinders, the line regulator will have larger orifice nozzles and a more flexible, sensitive diaphragm. These regulators usually connect to the gas distribution pipe, so the inlet connection may be a standard pipe fitting, instead of the usual tank fitting. The discharge fitting will be the same as the usual regulator. *A right-hand-thread fitting is used for oxygen and a left-hand-thread fitting for acetylene or fuel gases.*

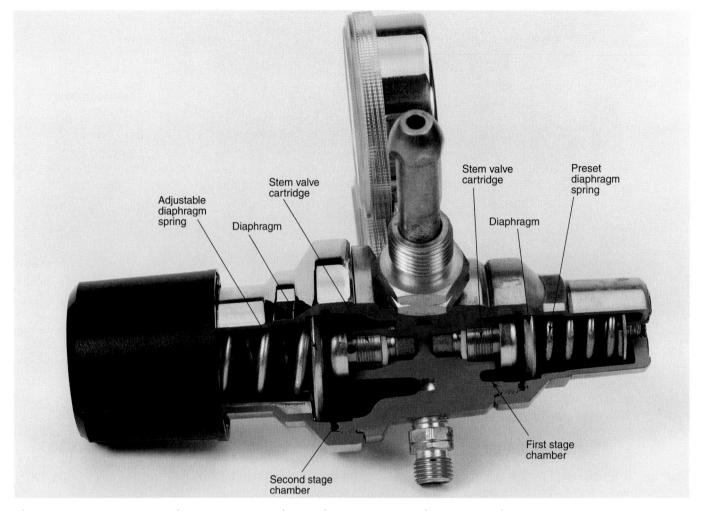

Figure 5-24. *A two-stage regulator (cutaway). Each stage has a stem-type valve in a cartridge. (CONCOA)*

5.4.6 Pressure Gauges

As mentioned in a previous paragraph, the gauges are mounted on the regulator. The high-pressure gauge is connected into the regulator between the regulator nozzle and the cylinder valve. It registers the cylinder pressure when the cylinder valve is opened. The low-pressure gauge is connected into the diaphragm chamber of the regulator, and registers the pressure of the gas flowing to the torch. Gauges are usually built with very small, delicate gears and springs similar to mechanical (nonelectronic) watches and clocks. Since they are of delicate construction, they must be handled accordingly. Refer to Figure 5-28 for details of pressure gauge construction.

The basic principle of operation of the pressure gauge depends on the *Bourdon tube*. This tube, usually made of phosphor bronze or stainless steel, is usually flat in cross section and is bent to fit inside the circular case. It is closed at one end, and connected with the pressure to be measured at the other. As the pressure increases, the tube tends to straighten. As it straightens, it operates the gear and pointer mechanisms. The dial is calibrated to indicate corresponding pressures. The gauge is usually fastened to the regulator body using a 1/4" NPT (National Pipe Thread) fitting.

Heavy glass or plastic covers the dial face and needle, and is held in the body by means of a large threaded clamp ring, called a **bezel**. The gauges come in various sizes. The 2 1/2" (63.5mm), 3" (76.2mm), and 3 1/2" (88.9mm) diameter dials are the most popular. The calibration of the gauges depends entirely upon the pressure to be used. It is generally recommended that the gauge scale read 50% higher than the highest pressure to be used. If 100 psig (689.5kPa) is the highest pressure to be used, a gauge with a 150 psig (1034.2kPa) or greater reading should be purchased.

The oxygen high-pressure gauge is usually calibrated from 0 psig to 3000 psig (0kPa to 20 684kPa) or 0 psig to 4000 psig (0kPa to 27 579kPa). There are sometimes two scales on a pressure gauge dial. One reads the pressure in pounds per square inch (psig) and the other in kilopascals (kPa). Oxygen is stored under direct pressure, and the amount of gas remaining in the cylinder is proportional to the pressure (Boyle's Law). Figure 5-29 illustrates the calibrations on the high- and low-pressure gauges on an acetylene cylinder.

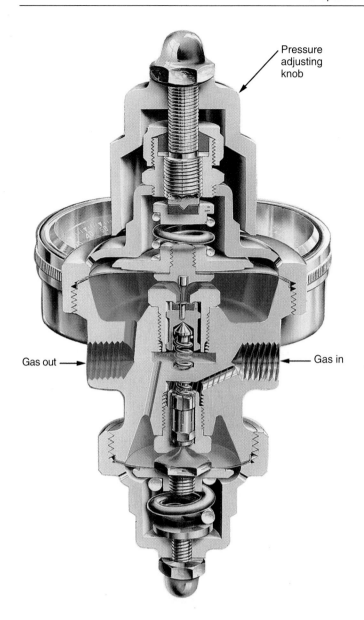

Figure 5-25. The two-stage regulator provides two diaphragms, two needles, and two seats. The first stage reduces the high gas pressure (hatched red) as it comes from the cylinder to some intermediate pressure (pink). The second stage is the low-pressure stage which reduces the intermediate pressure to some constant pressure needed by the torch (red).

The oxygen low-pressure gauge has a variety of dial calibrations. For light welding, the dial is calibrated up to 50 psig (345kPa). For heavy welding and for cutting, the gauge may read as high as 200 psig (1379kPa), 400 psig (2758kPa), or even 1000 psig (689 5kPa).

The acetylene high-pressure gauge is usually calibrated up to 400 psig or 500 psig (2758 or 3447kPa). The acetylene low-pressure gauge is usually only calibrated from 0 psig to 30 psig (0kPa to 207kPa) or from 0 psig to 50 psig (0kPa to 345kPa). Most acetylene low-pressure gauges are calibrated only up to 15 psig (103kPa), leaving the pressures above 15 psig (103kPa), red in color and unmarked.

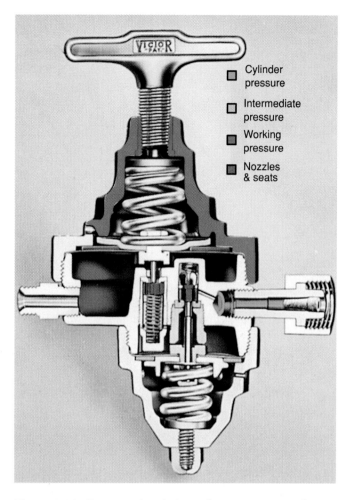

Figure 5-26. Cross-sectional view of a two-stage regulator that uses stem-type valves in both stages. (Victor Equipment Co.)

Remember, it is dangerous to use acetylene at pressures above 15 psig (103kPa).

To protect gauges from damage, a guard may be placed around the regulator and gauges, as shown in Figure 5-30.

5.5 WELDING HOSE

Hoses are built in several layers for strength. The inner layer is generally a very good grade of gum rubber. This layer is often surrounded by several layers of reinforced materials. The outer layer or covering is made of a colored rubber or plastic, which may be ribbed to provide a long-wearing surface. See Figure 5-31. Hoses are manufactured in three common colors: black, green, and red. The use of these colors is not standardized. However, the red is usually used for carrying acetylene or other fuel gases. Either the green or the black hose is used to carry the oxygen. The hose is specified according to its *inside diameter (ID)*, and comes in several sizes. The most common sizes are 3/16", 1/4", and 5/16" (4.8mm, 6.4mm, and 7.9mm). The size to be used depends on the size of the

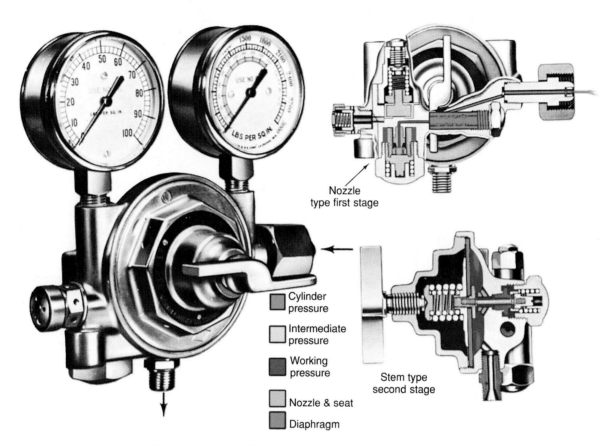

▦	Cylinder pressure
▢	Intermediate pressure
▦	Working pressure
▢	Nozzle & seat
▦	Diaphragm

Nozzle
type first stage

Stem type
second stage

Figure 5-27. *Cross-sectional view of a two-stage regulator which has a nozzle-type first stage and a stem-type second stage.*

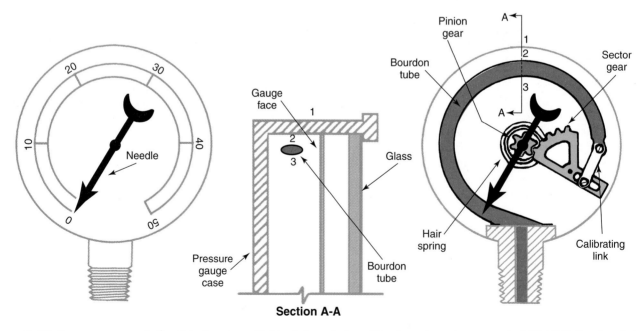

Figure 5-28. *Pressure gauge. Left—Exterior view. Right—Internal view. The thread on the gauge is 1/4" NPT. Section A-A shows the shape of the Bourdon tube.*

Figure 5-29. *A welder setting a 5 psig (34.5kPa) working pressure on the low-pressure gauge. Both gauges have US Conventional and SI Metric calibrations.*

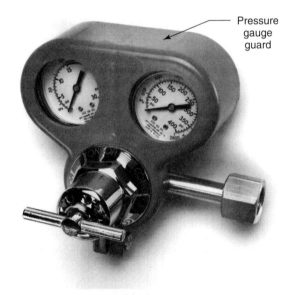

Pressure gauge guard

Figure 5-30. *A gauge guard installed to protect the pressure gauges from damage. (CONCOA)*

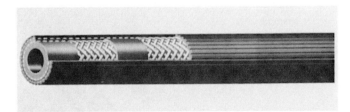

Figure 5-31. *A cutaway of a single welding hose. Note the rubber lining, the two layers of fabric, and the ribbed outer rubber cover. (Anchor Swan Corp.)*

torch, and the length of hose needed. The 3/16" (4.8mm) ID hose is very flexible and light, and is used extensively for light duty welding. The size (inner diameter) of the hoses should be increased if the hose is excessively long. To minimize entanglement, hoses can also be obtained as a double hose, as shown in Figure 5-32. Suppliers usually furnish hoses in 25' (7.62m) lengths.

Figure 5-32. *A double welding hose. One hose (green) carries the oxygen and the other hose (red) carries the fuel gas. (Anchor Swan Corp.)*

Hoses should never be interchanged, carrying first one gas and then another gas. If oxygen were to pass down a used acetylene hose, a combustible mixture might form. To prevent this, special precautions are used when attaching the hose to the regulators and torch. The hose is clamped to a nipple by means of a hose clamp. This nipple is fastened to a regulator or torch by means of a nut.

The nut and fitting have *right-hand threads* when they are to be used with oxygen. The oxygen nut may also be marked "OXY." The six sides of the oxygen nut are flat and smooth.

The fitting and nut used for acetylene have *left-hand threads*, and the nut may be marked "ACE." The acetylene nut also has a groove machined around its six sides. Both the oxygen nipple and the acetylene nipple use a rounded face as a sealing surface. Figure 5-33 illustrates some typical hose fittings.

The hose must be carefully handled to prevent accidents. It should not be allowed to come in contact with any flame or hot metal. Care should be taken that the hose is not kinked sharply. Kinking the hose might crack the fabric and permit the pressure to burst the hose. A kink will also hinder the gas flow. When the equipment is not being used, the hose should be hung away from the floor and away from things that might injure it. When welding, the hose should be protected from falling articles, from vehicles running over it, and from being stepped on, as these actions might injure the hose.

Hose reels are available. They are usually spring-loaded and roll up the hose into the container when the station is not being used.

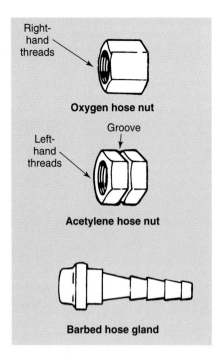

Figure 5-33. *Hose fittings. The fuel gas (acetylene) hose nut has an identifying groove cut into it. It also has a left-hand thread. (Airco Welding Products, Div. of Airco, Inc.)*

5.6 OXYACETYLENE TORCH TYPES

The welding torch (sometimes called a "blowpipe") and the *cutting torch* (sometimes incorrectly called a "burning torch") are each designed for a specific purpose. These torches are somewhat similar in construction, but the cutting torch is provided with a separate control valve to regulate an oxygen jet which does the cutting. These torches come in a variety of sizes and designs, depending on their intended use. See Chapters 7 and 8 for additional information.

There are two types of welding and cutting torches in common use. They are the positive-pressure type torch (also known as an equal-pressure type or medium-pressure type torch), and the injector-type torch.

5.6.1 Positive-Pressure Type Welding Torch

The gases are mixed in the positive-pressure type welding torch and are burned at the end of the torch tip. The torch consists of four main parts, as shown in Figure 5-34.

- Body.
- Hand torch valves.
- Mixing chamber.
- Tip.

The positive-pressure torch is used with cylinder gases. Its construction necessitates that each gas be supplied under enough pressure to force it into the mixing chamber, as illustrated schematically in Figure 5-35. The torch shown in Figure 5-36 has been cut away to show the mixing chamber and gas passages. Torches are made of various materials, including brass, aluminum, and/or stainless steel. The various parts are threaded and silver-brazed together.

The hand valves are located either at the end of the torch where the hoses attach to the handle, as shown in Figure 5-36, or at the tip end of the handle as shown in Figure 5-37. The torch valves are generally either a needle-and-seat or ball-and-seat design, as shown in Figure 5-38. These hand valves are used chiefly for shutting off the gas and turning it on; however, many welders use them to *throttle* (make the final flow adjustment to) the gases being fed to the torch.

Figures 5-34 and 5-39 illustrate the construction of a medium-duty positive-pressure type welding torch, while Figure 5-37 shows a light-duty torch with two different sets of tips.

The *mixing chamber* is usually located inside the torch body, although some torches incorporate the mixing chamber in the torch tube. Gases are fed to this chamber through two brass or stainless steel tubes leading from the torch valves. The size and design of the mixing chamber depends on the size of the torch. Some torch designs change the size and shape of the mixing chamber at the same time the tip is changed. The size and shape of the torch tip holes and the chambers should never be altered, nor should the parts be abused. Figure 5-36 illustrates a well-designed mixing chamber. The gases, after being

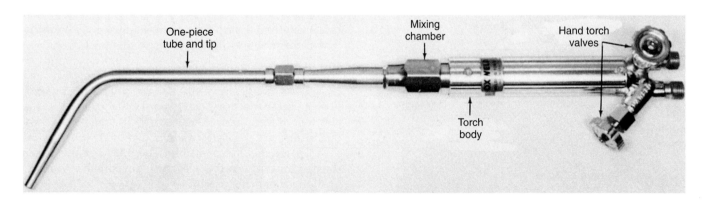

Figure 5-34. *An oxyacetylene positive-pressure welding torch. (ESAB Welding and Cutting Products)*

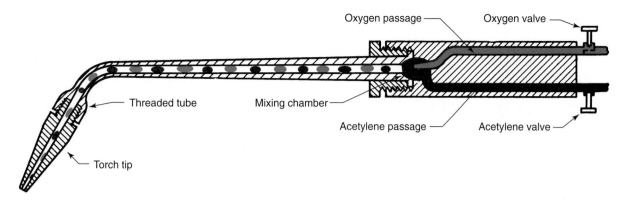

Figure 5-35. A schematic drawing of an oxyacetylene welding torch.

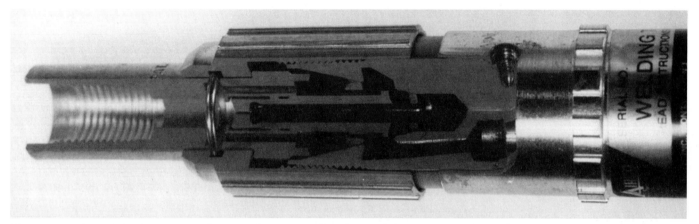

Figure 5-36. A torch body cutaway showing the mixing chamber and the oxygen and acetylene passages. (CONCOA)

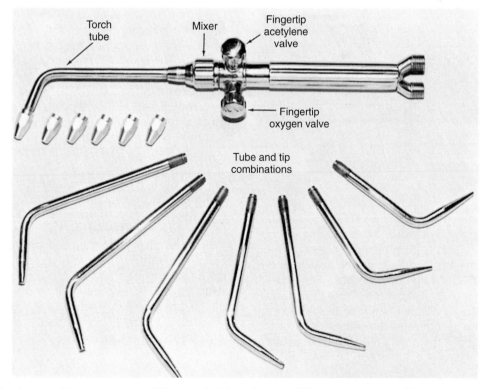

Figure 5-37. A light-duty positive-pressure welding torch. Note the two different types of welding tips and the location of the torch valves. The same mixer is used for all tips. (Smith Equipment, Division of Tescom Corp.)

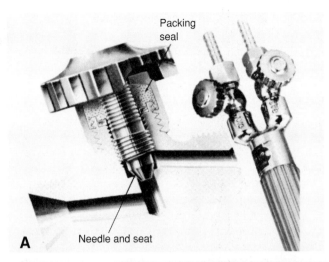

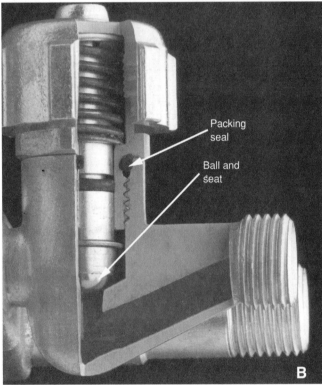

Figure 5-38. *Torch hand valves. A—A needle-and-seat type torch valve in cross-section. (Veriflo Corp.) B—A cutaway of a ball-and-seat type torch valve. (CONCOA)*

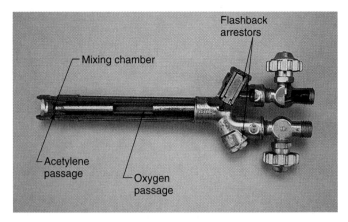

Figure 5-39. *This cutaway shows two flashback arrestors built into the torch body. Note the mixing chamber area. (Victor Equipment Co.)*

mixed, are fed through the tube or barrel of the body to the tip where combustion takes place. Most welding torches use a one-piece tube and tip combination. Separate tube and tip combinations are also used. In this case, both the tube and tip are threaded to permit assembly.

The *orifice*, or hole drilled in the tip, must be of accurate size. The tip size or number is normally stamped on the tip. The tip number indicates the orifice size and the relative volume of gases that it will allow to flow through the tip. The tips are usually made of copper, but some are nickel-plated to reflect heat and stay cooler. Heading 6.5.3 explains various systems used for indicating torch tip size. In this text, torch tip size is indicated in number drill sizes.

Refer to Figure 6-7 for the drill size and tip number used by various tip manufacturers.

5.6.2 Injector-Type Welding Torch

The injector-type (low-pressure) welding torch looks much like the positive-pressure type torch. However, the internal construction of the injector-type torch is somewhat different. The chief characteristic of the injector-type torch is its ability to operate using very low acetylene pressure. In general, the acetylene pressure remains practically constant regardless of the size tip or thickness of the metal being welded.

The ability of this torch to operate on low acetylene pressure has certain advantages. It is particularly desirable for use in connection with the low-pressure acetylene generator that supplies acetylene at a pressure of 1/4 psig (1.7kPa). The injector-type torch also has the advantage of being able to more completely draw the charge from acetylene cylinders.

Figure 5-40 shows the internal construction of the mixing chamber portion of an injector-type welding torch. It should be noted that the oxygen line enters the mixing chamber through a jet which is surrounded by the acetylene passage. As the oxygen flows from the jet, it draws (injects) the acetylene along with it. Handling the valves and the other operations of the torch is much the same as with the positive-pressure type torch. It should be noted that the oxygen pressure used in these torches is considerably higher than with the positive-pressure type torches. Follow the torch adjustments recommended by the torch manufacturer. The materials of injector-type torch construction are usually the same as those materials used in the positive-pressure torch.

5.6.3 Welding Tips

The oxyfuel gas welding flame is maintained at the end of a solid copper *welding tip*. Since copper conducts heat rapidly, there is little danger of overheating and backfiring. Tip size and condition are most important for proper welding. There are two types of tips, as shown in Figure 5-41:

- One-piece (tip and tip tube are one piece).
- Two-piece (tip and tip tube are separate).

The tip is subjected to both mechanical wear and flame erosion. As tips are removed and installed in a tip tube, the attaching threads may be subjected to considerable wear and abuse. Wrenches used on these tips should be of the box-end type to minimize damage (pliers should never be used). Do not try to remove a hot tip from a tip tube. Allow the tip and tip tube to cool first. Also, do not install a cold tip in a hot tip tube.

During welding, the molten metal may "pop" and throw molten droplets into the tip orifice, where they may remain until removed with a tip cleaner (described under Heading 5.6.4).

mechanically damaged by coming into contact with the welding work, the bench, or firebricks. This damage can roughen the end of the tip and cause the flame to burn with a "fishtail" appearance.

5.6.4 Welding Tip Cleaners

As mentioned before, welding tips are sometimes subject to considerable abuse. The orifice must be kept smooth and clean if the tip is to perform satisfactorily. If good performance is expected, carbon deposits and slag must be removed regularly. When cleaning a welding tip, exercise care to avoid enlarging or scarring the orifice.

Special welding tip cleaners have been developed to satisfactorily perform this service operation. The cleaner consists of a series of broach-like wires (a **broach** is a tool used to shape a hole) that corresponds to the diameters of

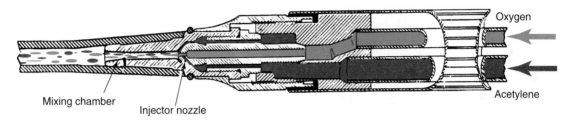

Figure 5-40. *Cross-sectional view of the mixing chamber area of an injector-type welding torch. The acetylene is injected (drawn) into the mixing chambers by the venturi (pulling) action of the oxygen jet. Injector torches are particularly adaptable for use with acetylene generators, which operate under low-pressure.*

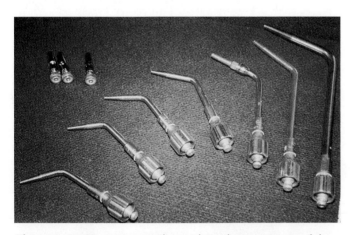

Figure 5-41. *Two commonly used tip designs. Most of the tips shown are of the one-piece design with their mixers. The third from the right, however, is a torch tube and mixer with a separate tip that screws into the threaded tube.*

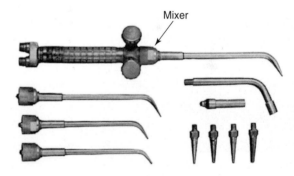

Figure 5-42. *A light-duty positive-pressure torch. A separate mixer is provided with each one-piece tip. (CONCOA)*

Always use tips which are made for a particular torch, as shown in Figure 5-42. Some makes of torches use a pliable, heat-resistant synthetic gasket to seal the joint between the tip tube and the torch. With this type of torch, no wrench is needed to install or service a tip. Hand-tightening is sufficient.

Avoid dropping a tip, since this could damage the seat which seals the joint. The flame end of the tip may be

tip orifices. See Figure 5-43. These wires are packaged in a holder, which makes their use safe and convenient. Figure 5-44 illustrates a tip cleaner in use.

Some welders prefer to use a number drill of the correct size to clean welding tip orifices. If a number drill is used, cleaning must be done very carefully, so that the orifice is not enlarged, bellmouthed, reamed out-of-round, or otherwise damaged.

The flame end of the tip must be clean and smooth. If a correctly shaped flame is desired, the tip's surface must be at a right angle to the center line of the tip orifice. A 4" mill file is commonly used to recondition the orifice end of the tip, as shown in Figure 5-45.

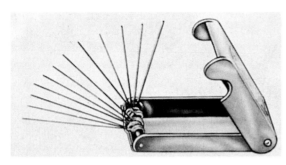

Figure 5-43. Welding tip orifice cleaner. Notice the various sizes of broaching wires. (Thermacote-Welco Co.)

Figure 5-45. Reconditioning the orifice end of a cutting torch tip. (Maitlen & Benson)

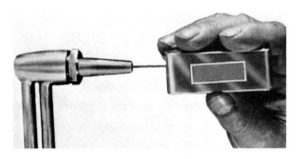

Figure 5-44. A tip cleaner used to clean cutting tip orifices. (Maitlen & Benson)

5.7 AIR-ACETYLENE TORCH

The air-acetylene torch, Figure 5-46, is often used where a light portable flame with a medium temperature of 2500°F (1370°C) is required. This torch is used extensively on copper plumbing (soft soldering) and refrigera-

tion lines (silver brazing). It is also used to solder or braze small parts. If large parts are to be silver-brazed, the oxyacetylene torch is recommended.

The air-acetylene torch receives its acetylene from a cylinder through a regulator and hose. As the acetylene flows through the torch, air is drawn in from the atmosphere to supply the oxygen needed for combustion. The torch operates on the same principle as the Bunsen burner used in chemistry laboratories.

The same safety precautions observed when handling the oxyacetylene torch should be used when working with the air-acetylene torch.

Cylinders for small portable air-acetylene torches come in various capacities. Refer to Figure 5-14. Regular sizes are 10 ft³ (0.28m³) and 40 ft³ (1.14m³). The 10 ft³ (0.28m³) cylinder is called the MC size. The valve fittings on both of these cylinders are Prest-O-Lite (POL) fittings.

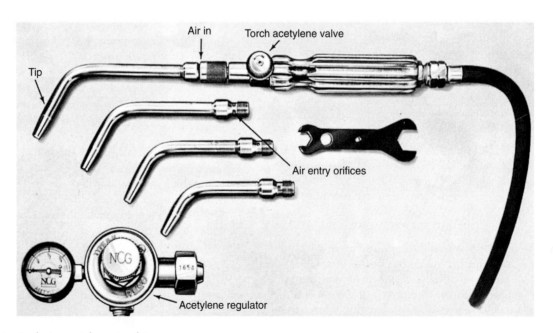

Figure 5-46. A typical air-acetylene torch.

5.8 WELDING GOGGLES AND PROTECTIVE CLOTHING

The welder must wear suitable goggles when doing oxyacetylene welding. The flame and pool of molten metal emit both ultraviolet and infrared rays. Both types of rays may cause eye injury if the welding work is viewed at a close distance. Goggles also protect eyes from flying sparks. Glare is reduced as well, so the welder can see the weld pool more clearly.

The common welding goggle has a sparkproof frame for the lenses. An elastic band holds the goggles securely on the welder's head. The welding lenses may be round or rectangular. If the lenses are round, they are 50mm (2") in diameter. If the lens is rectangular, it measures 2" x 4 1/4" (50mm x 108mm). Welding goggles, Figure 5-47, are often designed to fit over glasses.

The *filter lenses* are tinted either green or brown and are made in a variety of shade intensities. A more recent filter lens, which has a gold reflective layer on the outside, has been found to filter harmful rays exceptionally well. The degree of darkness of a filter lens is indicated by the shade number. The shade numbers range from 1 to 14; the higher the number, the darker the shade. The shade number is generally etched in the corner of the lens.

Filter lenses must conform to the American National Standards Institute requirements for eye protection for welding (ANSI Z87.1). Figure 5-48 lists recommended shades for various welding applications. If operations are of short duration, the lighter shade indicated may be used. For longer or continuous operation, the darker shade should be used. In general, shade number should increase with tip size.

A clear *cover lens* is used to protect the filter lens from metal spatter. The cover lens is optical quality clear glass or plastic. It is generally 3/64" to 1/16" (1.19mm to 1.59mm) thick. These cover lenses need to be replaced frequently, so that the welder's view of the weld will not be dimmed.

Cover lenses are rather inexpensive. Filter lenses are quite costly. For this reason, it is necessary that the filter lenses *always* be protected by cover lenses. Many cover lenses are protected by a thin layer of transparent plastic which keeps metal spatter from pitting and adhering to them. The life of these plastic-coated lenses is greater than

Figure 5-47. Two types of 50mm (2") round welding goggles. Both will fit over prescription glasses. The goggles at right have a soft flexible one-piece frame. (Jackson Products, Div. Thermadyne Industries, Inc.)

Process	Plate thickness		Suggested* shade no. (Comfort)
	in.	mm	
Gas welding			
Light	Under 1/8	Under 3.2	4 or 5
Medium	1/8 to 1/2	3.2 to 12.7	5 or 6
Heavy	Over 1/2	Over 12.7	6 or 8
Oxygen cutting			
Light	Under 1	Under 25	3 or 4
Medium	1 to 6	25 to 150	4 or 5
Heavy	Over 6	Over 150	5 or 6

*As a rule of thumb, start with a shade that is too dark to see the weld zone. Then go to a lighter shade which gives sufficient view of the weld zone without going below the minimum. In oxyfuel gas welding or cutting where the torch produces a high yellow light, it is desirable to use a filter lens that absorbs the yellow or sodium line in the visible light of the (spectrum) operation.

Figure 5-48. This chart is from A Guide to Recommended Welding Lens Shades for Oxyfuel Gas Welding and Cutting. (ANSI Z49.1-88)

them. The life of these plastic-coated lenses is greater than that of clear glass.

Some oxyfuel gas welders prefer to use the rectangular eyeshield type of eye protection. The lenses are the same size as those used in arc welding helmets. These rectangular eye shields fit over prescription glasses well, and give a good range of vision. Figure 5-49 illustrates the eyeshield type of eye protection. Figure 5-50 shows an eyeshield that permits the filter lens and cover lens to be flipped up for a view of the weld through a clear lens. Face shields with large, flexible, filter quality lenses also are available, Figure 5-51. **Always wear eye protection. Eyes can never be replaced.**

Figure 5-49. *This woman is wearing goggles with a rectangular lens as she makes an overhead weld.*

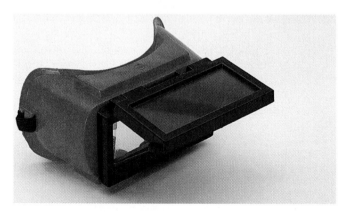

Figure 5-50. *Welding goggles with a 2" x 4 1/4" (50mm x 108mm) lens. The cover lens and filter lens may be flipped up to see through another clear lens. CAUTION: None of these lenses is suitable for grinding or chipping. (Jackson Products, Div. Thermadyne Industries, Inc.)*

Figure 5-51. *A filter quality face shield like this one worn by this welder will provide excellent visibility when welding.*

The welder must wear protective clothing. The hands should be protected with leather or fabric gloves. The cuffs on the gloves should be either the gauntlet type, or have an elastic band that makes a tight seal between the glove and the coat sleeve. Jackets should be made from either leather or from a fabric treated to be slow-burning or nonflammable. Trousers should be without cuffs, and made from a fabric treated to resist burning.

5.9 TORCH LIGHTERS AND ECONOMIZERS

Matches or burning paper should never be used for lighting a welding torch. Do not carry matches or other combustible items, such as combs or pens, in your pockets while welding. If a spark should enter the pocket, a serious burn might result before the fire could be extinguished.

A flint and steel spark lighter is perhaps the most popular type of torch lighter. See Figure 5-52. Pistol-grip spark lighters are also available.

Establishments that use a number of gas welding stations often provide pilot lights fueled by either natural gas or acetylene. The gas is piped to an outlet near the welding station, and a small flame is kept burning continuously. The flame outlet should be located overhead, where it will not have any chance of igniting anything on the working level of the room.

Figure 5-53 shows a combination *economizer* and lighter. It consists of a mechanism through which the oxygen and acetylene are fed before going to the torch. This mechanism is also used to hold the torch when it is not being used.

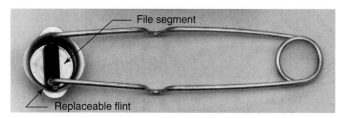

Figure 5-52. *Flint-and-steel spark lighter. The steel cup tends to trap the gas. When the flint is rubbed on the file segment, the spark quickly and safely ignites the fuel gas.*

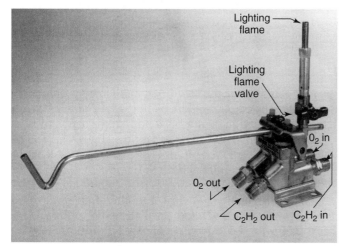

Figure 5-53. *A combination gas economizer and torch lighter. A small flame to light the torch is maintained at the tip of the vertical tube when the small valve is turned on. (CONCOA)*

Before the torch is placed in the holder, it is lighted and adjusted. When put in the holder, it presses a lever that turns off both the oxygen and acetylene, leaving a very small acetylene flame burning at a special outlet, or pilot light. When the torch is lifted from this holder, gas flow starts, the gas is ignited by the pilot light, and the torch is ready to be used for welding. The device saves considerable gas and time. Safety is also improved, since the chance of having the torch laid aside while still lighted is minimized.

5.10 OXYFUEL GAS WELDING SUPPLIES

Many supply items are needed in order to perform the usual oxyfuel gas welding operations. The more common supplies needed are:
- Welding gases, consisting of oxygen and fuel gas.
- Welding rod (filler metal) for steel, stainless steel, cast iron, aluminum, and hardfacing. (See Chapter 28)
- Fluxes for cast iron welding, aluminum welding, stainless steel welding, or soldering and brazing. (See Chapters 9 and 10).

- Firebrick.
- Carbon paste and forms.
- Noncombustible materials (sheet and powder).
- Glycerine.
- Litharge.

5.10.1 Other Fuel Gases

Handling oxygen and acetylene is explained in Headings 5.2 and 5.3. Other fuel gases in common use are:
- Hydrogen.
- LP (liquefied petroleum), propane, and butane.
- Natural gas.
- Methylacetylene propadiene (MAPP).
- Polypropylene-based fuel gas (FG-2).

No fuel gas mixed with oxygen has a flame temperature as high as the oxyacetylene flame. The *oxyhydrogen* flame, however, is very clean and is recommended for welding aluminum and magnesium. Because it can be used at a higher pressure than acetylene, it is also recommended for underwater welding and cutting. Since hydrogen is a reducing agent, this flame minimizes oxidation, if properly adjusted. A regular oxyacetylene torch may be used with hydrogen as the fuel gas. Hydrogen is supplied in cylinders, as is oxygen. The pressures in oxygen and hydrogen cylinders are about the same. The standard sizes of hydrogen cylinders are 200 ft^3 (5663L) and 100 ft^3 (2832L).

Hydrogen cylinders are fitted with special fittings, so the regulators used on these cylinders must be provided with proper mating attachments. **Hydrogen has no odor, and when combined with either air or oxygen in the proper proportions, it forms a powerful explosive mixture. Hydrogen connections should be regularly checked for leaks using a soap-and-water solution.**

Liquefied petroleum (LP) is sold under a variety of names. It may have some variations in chemical analysis. For welding use, the general title of liquefied petroleum (LP) gas is used. This fuel is supplied in liquid form, and is under a positive pressure which varies with the temperature. LP gas is used mostly for cutting, soldering, and brazing. Most oxygen welding or cutting torches can use LP fuel, but special LP cutting tips are required. An air-fuel gas type torch using LP gas is commonly used for general heating purposes. Refer to Figure 5-46.

Liquefied petroleum gas is sold by the pound (or kilogram). Common sizes of tanks are: 4 1/4 lb. (1.9kg), 11 lb. (5kg), 20 lb. (9.1kg), 30 lb. (13.6kg), 40 lb. (18.1kg), 60 lb. (27.2kg), and 100 lb. (45.4kg).

LP gas is also provided by the gallon. Sizes are: 120 gal. (454L), 250 gal. (946L), 500 gal. (1893L), and 1000 gal. (3785L).

The customer usually purchases the required tank size and returns it to the dealer for refilling. The larger sizes are usually leased. Industries that use large quantities of LP gas provide their own bulk storage. The fuel is delivered to them from bulk tank cars or trucks.

When using LP gas as a fuel, the pressure must be regulated using an LP gas regulator. These regulators are most often supplied with two gauges. One gauge shows the pressure in the storage tank, and the other indicates the pressure in the torch or burner line. These regulators are usually attached to the tank using a standard POL Commercial fitting.

Natural gas, piped to most communities, is an excellent fuel for certain uses. It is particularly adaptable for cutting, soldering, brazing, and preheating.

Because natural gas is delivered at a rather low pressure, injector-type torches are used both for cutting and for general heating. Some small torches have been developed which use compressed air and natural gas, particularly for soldering and brazing. Natural gas serves very nicely for many preheating operations. Natural gas systems should be protected by a waterseal or a blowback valve to keep air and oxygen from backfiring into the gas supply line. Refer to Figure 5-15 for the waterseal flash arrestor. Always consult all required safety authorities and regulations on this matter before proceeding.

Stabilized *methylacetylene propadiene* is a fuel gas sold under the trade name of **MAPP**. The fuel has the safety and ease of handling of liquefied petroleum gas (LP) with a heating value approaching that of acetylene.

MAPP may be stored and shipped in the liquefied state. The cylinders are available in various sizes. MAPP gas cylinders have the same thread as acetylene cylinders. The usual acetylene regulator may be used with this gas.

This fuel has some advantages over acetylene when used for cutting. It has a narrower explosive range. The explosive range is 3.4% to 10.8% in air, compared to a range of 2.5% to 80% for acetylene.

MAPP gas is used for underwater cutting, since—unlike acetylene—it may be used at pressures higher than 15 psig (103.4kPa). Oxyacetylene cutting torches may be used with this fuel gas. However, special tips made for MAPP gas must be used. These tips are available for all common cutting torches. OxyMAPP cutting is rapid and forms a very liquefied slag that flows away, leaving a clean cut.

The customer usually owns the storage cylinders, so there are no *demurrage* (charge per day) costs. Since the fuel is in the liquid form, a cylinder of MAPP fuel contains many more cubic feet or cubic meters of gas than an acetylene cylinder of equal size.

5.10.2 Check Valves and Flashback Arrestors

Check valves are used to prevent the reverse flow of gases through the torch, hoses, or regulators. See Figure 5-54. Figure 5-55 shows a check valve in the open and closed positions. It also explains how the check valve works.

A *check valve* allows a gas to flow in only one direction. Gas pressure opens the valve to allow gas flow. The valve closes when the gas stops flowing, or attempts to flow in the reverse direction. Check valves are generally installed at the torch inlet or placed at the regulator outlet. The valves may also be placed at *both* the torch and regulator.

Flashback arrestors are designed to eliminate the possibility of an explosion in the regulator or cylinder.

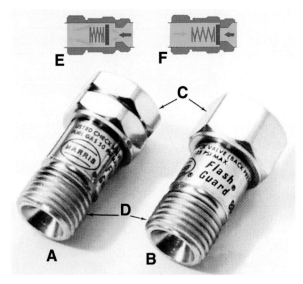

Figure 5-54. Check valves prevent mixtures of fuel gas and oxygen from backing up and possibly burning in the welding hose and regulator. A—Fuel gas check valve (note groove on hex nut). B—Oxygen check valve. C—Torch or regulator connection. D—Welding hose connection. E—Normal flow; valve open. F—Reverse flow; valve closed. (Harris Calorific Div. of the Lincoln Electric Co.)

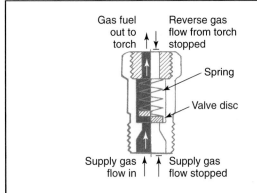

Gas fuel out to torch / Reverse gas flow from torch stopped / Spring / Valve disc / Supply gas flow in / Supply gas flow stopped

Check Valve
In normal operation, left, gas from the supply hose is at pressure higher than gases in the torch, so pressure in the hose lifts the valve disc against spring pressure and gas flows into the torch. If pressure inside the torch plus force of the spring on the disc exceed pressure in the supply hose, the disc seats and shuts off flow of gas from the torch into the hose. Not reusable after a flashback.

Figure 5-55. A check valve. (Welding Design & Fabrication)

Figure 5-56 shows a flashback arrestor that has a check valve and stainless steel flame arrestor to prevent flashbacks. Built into the flashback arrestor are the following safety valves:

- A reverse-flow check valve. This valve stops any flow of gas in the wrong direction.
- A pressure-sensitive cut-off valve that cuts off gas flow if an explosion occurs.

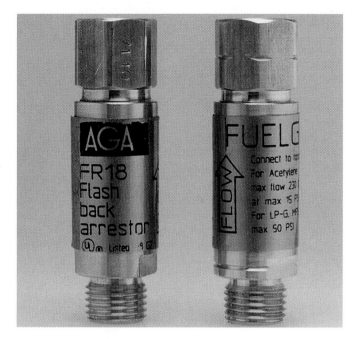

Figure 5-56. A torch-mounted, two-component flashback arrestor. This unit contains a check valve and a stainless steel flame arrestor. (AGA Gas, Inc.)

- A stainless steel filter that stops the flame.
- A heat-sensitive check valve that stops the gas flow if the arrestor reaches 220°F (104°C).

Figure 5-57 shows a flashback arrestor in the open and closed positions. An explanation of how the flashback arrestor works is also included. Another type of flashback arrestor is shown in Figure 5-58. This arrestor is also mounted on the regulator. It has the four components of a flashback arrestor that are listed above.

5.10.3 Gas Welding Rod (Filler Metal)

The American Welding Society defines welding rod as *"A form of filler metal used for welding or brazing which does not conduct the electric current."*

Some common welding rods are:
- Mild steel.
- Cast iron.
- Stainless steel.
- Braze welding alloys.
- Aluminum.
  - ◆ Drawn.
  - ◆ Extruded.
  - ◆ Cast.

Mild steel, braze welding alloys, stainless steel, and some aluminum rods are made in 36" (0.91m) lengths and are available in the following diameters: 1/16" (1.6mm), 3/32" (2.38mm), 1/8" (3.18mm), 5/32" (7.94mm), and 3/8"

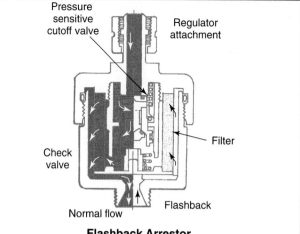

Flashback Arrestor

In normal operation, left, gas flows through the open cut-off valves and check valve through the flame arrestor filter into the hose.

In the event of a flashback, right, the stainless steel filter stops the flame and the pressure wave activates the cut-off valve, stopping the flow of gas to extinguish the flame. The check valve operates when gas flows towards the cylinder. If the arrestor is exposed to fire, the thermal cut-off valve shuts the gas supply. Reusable after a flashback.

Figure 5-57. Flashback arrestor. (Welding Design & Fabrication)

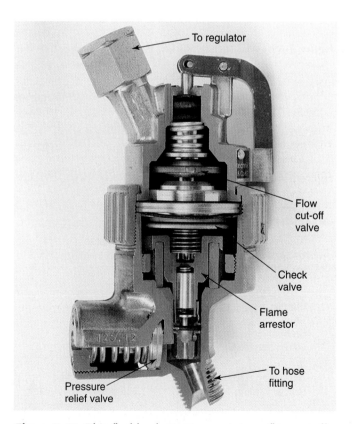

Figure 5-58. This flashback arrestor contains a flow cutoff valve, a check valve, a pressure relief valve, and a flame arrestor. (CONCOA)

(9.53mm). Metric-sized welding rods may not be available in the sizes shown above.

They are packaged in 50 lb. (22.7kg) bundles. Mild steel welding rods are copper-coated to keep them from rusting. Aluminum welding rods or wires are packed in 36" (0.91m) lengths or in coils. Some aluminum rods are flux-coated. The coated rods are sold in 28" (0.71m) lengths.

Iron and steel welding rod specifications are sometimes confusing. The problem has been somewhat simplified by the fact that the old Army Air Corps, Navy, and Federal specifications are now combined into one name and one series of numbers under military (MIL) specifications. However, both the MIL and the American Welding Society (AWS) numbers are still used. In this text, AWS specification numbers are used.

The AWS numbers for oxyacetylene steel welding rods are shown in Figure 5-59. In these numbers, the letter **R** stands for a welding rod. The letter **G** stands for gas welding. The numbers **45**, **60**, and **65** indicate the approximate tensile strength of the weld in thousands of pounds per square inch (45 = 45,000 psi). Figure 5-60 lists the characteristics of the RG 45 welding rod.

5.10.4 Welding Fluxes

The American Welding Society defines *flux* as follows: "*Material used to prevent, dissolve, or facilitate removal of oxides and other undesirable surface substances.*"

Fluxes are required when heating certain metals with the intent to join them with solder, silver alloys, or copper alloys. Fluxes are also used in many welding operations. The composition of a flux is determined by the specific application for which it is made. The general classification for fluxes is shown in Figure 5-61.

Note from the flux classifications in Figure 5-61 that *no* flux is required for mild steel. Protection of the weld from oxidation is not so critical with mild steel. The iron oxides melt at a much lower temperature than the mild steel and float to the surface during welding.

5.10.5 Firebricks

Firebricks are used to form welding table tops, to build forms around articles, to aid with preheating and stress relieving. They are useful for building up supports for articles to be welded or brazed. They can be used to

Number	Use	Tensile strength
RG 45	Mild steel rod	45,000 psi (310.3 MPa)
RG 60	Low-alloy steel rod	65,000 psi (448.2 MPa)
RG 65	Low-alloy high-strength steel rod	105,000 psi (723.9 MPa)

Figure 5-59. AWS numbers used for oxyacetylene steel welding rods.

Tensile strength lbs./sq. in.	Carbon max. percent	Manganese max. percent	Sulphur max. percent	Phosphorus max. percent	Silicon max. percent
45,000	0.06	0.25	0.035	0.025	0.03

Figure 5-60. Characteristics of the AWS RG 45 gas welding rod.

Soldering	
Aluminum	Sheet steel
Copper alloys	Stainless steel
Galvanized sheet	
Brazing	
Aluminum	Cast iron
Steel	Copper alloy
Steel alloy (such as stainless steel)	
Welding	
Aluminum welding	Cast iron welding
Braze welding	Stainless steel welding

Figure 5-61. Fluxes are available to solder, braze, or weld the metals shown.

protect combustible materials which may become heated when exposed to the welding or cutting flame.

Firebricks are made of *refractory* (difficult to burn) materials. The size of firebricks in common use is 8 3/4"x 4 1/2"x 2 1/2" (222mm x 111mm x 63.5mm).

5.10.6 Carbon Paste and Forms

There are many places where the welder will find carbon paste materials useful. A few typical uses for carbon paste or carbon forms are:

- For building dams to contain the molten metal when building up a broken section.
- For protecting drilled or threaded holes which are in or adjacent to a weld or braze.
- For building up a support for an uneven surface that is to be welded.

- As a protective cover over metal that is adjacent to a weld, and which might be injured by spatter or heat.

The paste form is available in various size cans. Formed carbon is available as round or square rods and in plate form.

5.10.7 Pipe Sealing Compounds

To make an airtight seal on threaded pipe joints, certain *proprietary* (private brand) compounds are recommended. **Apply these sealing compounds on the male threads only.** In this way, there is less danger of compound entering the pipe.

Sealing compounds are available in either paste or tape form. **You should avoid using sealing compounds that have a lead content. Sealing compounds with an oil content should never be used on oxygen lines.**

5.10.8 Clamps And Clamping Fixtures

The success or failure of many welding operations depends on how well the metals are held in place during the welding operation. Many varieties of clamps and clamping devices have been developed for this purpose.

Pliers with clamping jaws of special design for holding and aligning parts are shown in Figure 5-62. The pliers have deep jaws to allow them to clamp around obstructions. Figure 5-63 shows various applications for this type

of clamping pliers. Very complex shapes may be clamped for welding with the chain-clamp pliers, shown in Figure 5-64. C-clamp pliers have also been developed to hold parts firmly and in alignment, as shown in Figure 5-65.

Special fixtures are very convenient for aligning, holding, and positioning metals of various shapes, as shown in Figure 5-66. A special double fixture with a protractor scale which makes quick and accurate aligning of the parts possible is shown at No. 1 in the illustration. Figure 5-67 shows three other types of fast-acting clamps that may be used to hold weldments in alignment while they are welded.

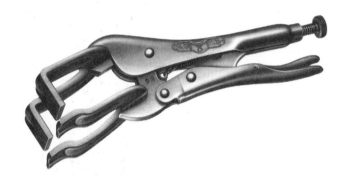

Figure 5-62. *A unique type of clamping pliers for holding materials being welded or brazed. (American Tool Companies, Inc.)*

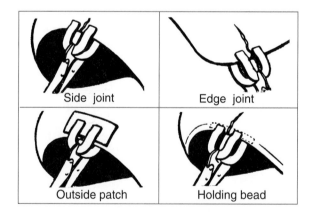
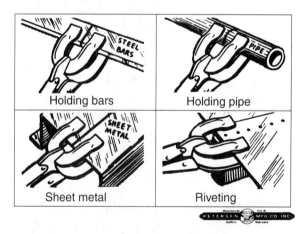

Figure 5-63. *Application of the clamping pliers shown in Figure 5-62. (American Tool Companies, Inc.)*

Figure 5-64. *A chain-type pliers. This pliers can clamp almost any shape material. (American Tool Companies, Inc.)*

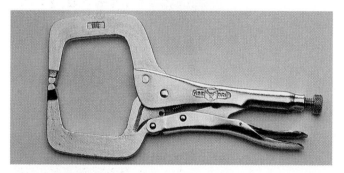

Figure 5-65. *A C-clamp pliers. This type pliers is available with longer jaws to clamp hard-to-reach areas. (American Tool Companies, Inc.)*

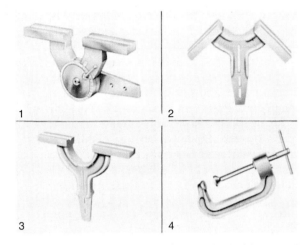

Figure 5-66. Special alignment fixtures for holding stock that is being welded. (Strippit, Inc./IDEX Corp.)

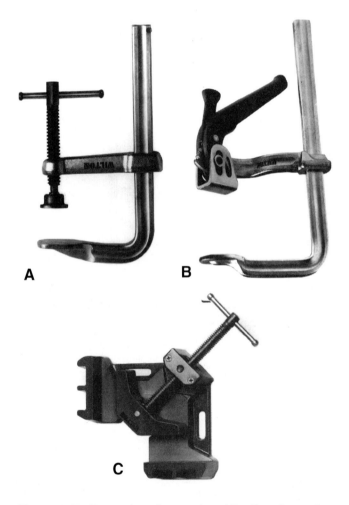

Figure 5-67. Fast-acting clamps. A and B—Two fast acting C-clamps. C—This clamp is used when parts are forming a 90° corner. (Wilton Corp.)

TEST YOUR KNOWLEDGE

Write your answers on a separate sheet of paper. Do not write in this book.

1. What parts and equipment make up a complete oxyacetylene welding station?
2. What type safety device is used in an oxygen cylinder?
3. What type safety device is used in an acetylene cylinder?
4. Completely describe the acetylene hose and regulator nut and its threads.
5. What materials are used in regulator diaphragms?
6. What factors determine the rate at which acetylene gas can be drawn from the cylinders?
7. Why must oil not be used on welding station fittings?
8. What colors are normally used on oxygen and acetylene hoses?
9. What is the maximum pressure that should be used with a gauge whose highest number is 400 psig (2758kPa)?
10. What is a two-stage regulator?
11. Name two types of welding torches and describe the difference between them.
12. What is the purpose of a flux?
13. What is the maximum working gauge pressure for acetylene gas?
14. Some flashback arrestors have four components. Name them.
15. What is MAPP?
16. List five (5) standard sizes for oxygen cylinders.
17. Why must clothing saturated with oxygen be removed?
18. What is the tensile strength of the RG 60 welding rod?
19. At what temperature, in degrees Fahrenheit and degrees Celsius, does the acetylene cylinder fuse plug melt?
20. Is the regulator open or closed when the regulator adjusting screw is turned out?

Chapter 6

OXYFUEL GAS WELDING

6.1 DEFINITION OF WELDING

Oxyfuel gas welding may be described as a metalworking process in which metals are joined by heating them to the melting point, and allowing the molten portions to fuse or flow together. Review Heading 4.1. Before attempting to make any welds, study Chapter 1 and Heading 6.14.

6.2 SOLDERING AND BRAZING

Two other metal joining processes that are often confused with oxyfuel gas welding are soldering and brazing.

Soldering occurs when two metals, which are not melted, are joined by a third metal that has a melting point *below* 840°F (450°C). Review Heading 4.3. An example of soldering is the joining of copper to steel using a tin-lead alloy. For more information on soldering, see Chapter 9.

Brazing is done when two metals, which are not melted, are joined with a third metal that melts at temperatures *above* 840°F (450°C). Review Heading 4.4. An example of brazing is the joining of two pieces of steel with a silver alloy. For more information on brazing see Chapter 10.

Manual welding or soldering may be defined as an art. The skill to weld and solder metals together can only be obtained after a diligent study of the methods and after careful and correct practice. In classifying welding as an art, it is meant that some persons can do welding better than others. This is due to a seemingly natural gift. However, it has been found that any person can become a successful welder with good instruction and by following correct procedures. Continuous practice is necessary to maintain a high standard of skill as a welder. For this reason, only proper equipment and metals should be used while learning welding.

A thorough, fundamental procedure should be followed. To correct any possible mistakes early, trainees should work under close supervision.

6.3 DIFFERENT TYPES OF WELDING AND CUTTING

The most common types of welding are: oxyfuel gas welding, shielded metal arc welding, gas tungsten arc welding, gas metal arc welding, and resistance welding. Other types include cold welding, ultrasonic welding, electron beam welding, friction welding, laser beam welding, and electroslag welding. Refer to the various welding procedures described in Chapter 4. Two popular types of thermal cutting are oxygen cutting and arc cutting. All of these welding and cutting processes will be explained in detail in later chapters.

The oxyfuel gas process will be studied first because:
- The fundamentals of gas welding include basic skills important to most other forms of welding.
- The oxyfuel gas process is a popular manual welding process. It is slower and easier to control than some other processes.
- It is easier for a beginner to observe the fusion process. The shape and flow of the oxyfuel gas weld pool and bead is similar to the various arc welding pools and beads.

6.4 OXYFUEL GAS WELDING

One of the most popular welding methods is to use an oxyfuel gas flame as the source of heat. This flame is produced by burning a fuel gas in the presence of the oxygen from the air, from a pure oxygen source, or from both. Oxyfuel gas flames may receive their oxygen in three ways:

From the surrounding atmosphere, which:
- gives lowest flame temperature.
- is the least clean.
- produces the least heat.

From air that is drawn in from the atmosphere through holes in the torch. This process:
- gives a higher flame temperature.
- is cleaner.
- gives more heat.

From a supply of pure oxygen under pressure that is mixed with the fuel gases before they burn. This:
- gives the highest flame temperature.
- is cleanest.
- gives the most heat.

Figure 6-1 illustrates the three methods of adding oxygen to a fuel gas to support combustion.

6.4.1 Welding Flames

In oxyfuel gas welding, metals are joined by melting and fusing. A very intense, concentrated flame is applied to the metal until a spot under the flame becomes molten and forms a pool of liquid metal. When two metals melt, and the molten pools run together and *fuse* (solidify), the edges of the two pieces become one. This process must be performed carefully to minimize damage to the metals.

Certain conditions are necessary for a good weld. They are:
- The temperature of the flame must be high enough to melt the metals.
- Enough heat must be supplied to overcome the heat losses.
- The flame must not *oxidize* (burn) the metal.
- The flame must not add dirt or foreign material to the metal.
- The flame must not add carbon or other harmful materials to the metal.
- The products of combustion should not be *toxic* (poisonous).

The quantity of heat is determined by the type and number of cubic feet per hour (liters per minute) of gases burned. To obtain more heat, a torch tip with a larger *orifice* (hole) is used. To feed sufficient gas to the larger tip, more pressure must also be supplied. Whether a large torch tip or a small torch tip is used, the temperature of the flame is the *same* with a given fuel gas.

It should be remembered that the amount of heat generated, and therefore the thickness of the metal that may be welded, will depend on the amount of fuel gas burned per unit of time. Therefore, the amount of heat depends on the size of the torch orifice.

There are several commercially used oxyfuel gas welding and cutting flames. They are:
- *Oxyacetylene,* (oxygen and acetylene) used for welding and cutting.
- *Oxyhydrogen* (oxygen and hydrogen) used for welding and cutting.

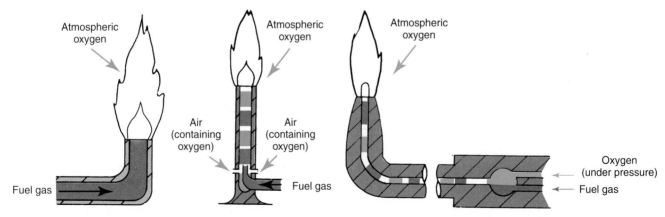

Figure 6-1. Three methods of combining oxygen and fuel gases to produce fuel gas flames.

- *Oxynatural gas* (oxygen and natural gas), used for cutting.
- *Oxypropane* (oxygen and propane), a liquefied petroleum gas, used for cutting.
- *MAPP* (oxygen and methylacetylene propadiene) used for welding and cutting.

For flame temperatures of various oxygen and fuel gas mixtures, see Heading 34.3.

6.4.2 The Oxyfuel Gas Flame

Harmful oxidizing or carburizing flames result when the wrong proportions of the two gases (oxygen and a fuel gas) are used. If there is too much oxygen, the result is an oxidizing flame that will burn the metal. If there is too much fuel gas, a carburizing flame results. This flame can add harmful carbon to the metal being welded. The oxy-hydrogen flame will not become carburizing. Figure 6-2 shows the various flame adjustments. See also Figures 6-3 and 6-4. The two harmful flames are easily recognized. See Headings 6.5.4 and 6.5.6 for flame adjustments. The correct flame heats the metal and does not carburize (add carbon) or oxidize (burn) it. The correct flame is called a neutral flame. A neutral flame is the result of a perfect proportion and mixture of the fuel gas and oxygen. In a neutral flame, these two gases unite so that the oxygen burns up the carbon and the hydrogen in the fuel gas, releasing only heat and harmless gases. The colors of the flames are shown in Figure 6-3. (Refer to Heading 33.4 for more on the chemistry of the welding flame.)

Chemically, when oxygen and acetylene burn they create carbon dioxide, water vapor, and heat. Oxygen and all fuel gases (with the exception of hydrogen) will form the following gases when burned: carbon dioxide (CO_2)

and water vapor (H_2O). These are considered harmless gases.

Oxygen in the air surrounding the flame is used to complete the burning process. When welding in corners, where the air has difficulty getting to the flame, additional cylinder oxygen must be fed to the flame.

The effect of an improper mixture of gases in the welding flame is easily recognized, and the final test for a neutral flame is determined by the manner in which the melting metal reacts to the flame. Dirt in a welding flame may come from two sources:

- Dirty gases.
- Dirty equipment.

Good quality gases should always be used. The purity of the gases made by a manufacturer should be noted and taken into consideration. A neutral oxyacetylene welding flame will produce a temperature of approximately 5589°F (3087°C). A neutral oxypropane flame reaches 4579°F (2526°C). An oxidizing flame will produce a slightly higher temperature. Temperatures required to melt various metals are listed in Figure 6-5.

An oxyacetylene welding flame is the hottest oxyfuel gas flame. It is hot enough to melt the common metals. Acetylene is generally the only fuel gas used for welding. Propylene gas and MAPP gas (methylacetylene propadiene) are used for preheating, cutting, torch brazing, flame spraying, and flame hardening. Propane is used for preheating, soldering, and brazing. Natural gas (methane) is generally used for heating, cutting, soldering, and brazing. Hydrogen gas is used in welding nonferrous metals and in some brazing operations.

6.5 THE OXYFUEL GAS WELDING OUTFIT

Before the procedures used to make a good weld are discussed, it is advisable to know the welding equipment required, its limitations, and its proper use.

Basically, the *oxyfuel gas welding outfit* consists of the oxygen and fuel gas cylinders; regulators, hoses, and hose fittings for each gas, and a welding torch. A *welding station* includes the welding outfit plus a welding table, ventilation, and possibly a booth. Figure 6-6 shows a complete oxyacetylene gas welding outfit. In the *order of flow* of the gases through the outfit, the following items will be found in common use:

- Gas cylinders (oxygen cylinder, fuel gas cylinder).
- Pressure regulators and gauges (oxygen regulator, fuel gas regulator).
- Hoses and fittings (oxygen hose, fuel gas hose).
- Welding torch.

There are two types of oxyfuel gas welding torches in common use. They are the *positive-pressure torch* and the injector-type torch. The positive-pressure torch normally operates under a pressure of from 1 psig to 10 psig (6.9kPa to 69kPa), depending on the torch tip size. The positive-pressure type torch is the most commonly used. When

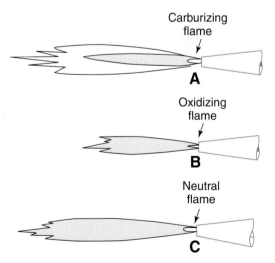

Figure 6-2. *Three oxyfuel gas flame adjustments. Top—A carburizing flame has three distinct flame sections. Center—The inner cone on an oxidizing flame is pointed, noisy, and has a bluish-white color. Bottom—The inner cone on a neutral flame is smooth and rounded and a brilliant white in color. The neutral flame is the correct flame for most welding operations.*

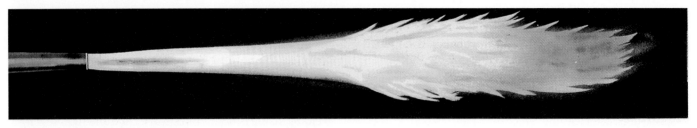

Acetylene Burning in Atmosphere
Open fuel gas valve until smoke clears from flame.

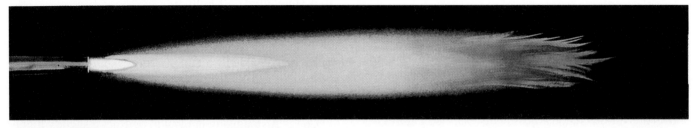

Carburizing Flame
(Excess acetylene with oxygen) Used for hard-facing and welding white metal.

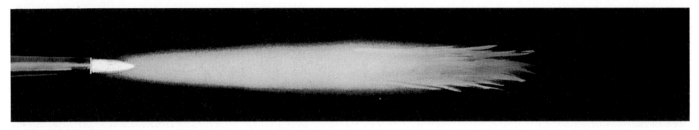

Neutral Flame
(Acetylene and oxygen) Temperature 5589°F (3087°C). For fusion welding of steel and cast iron.

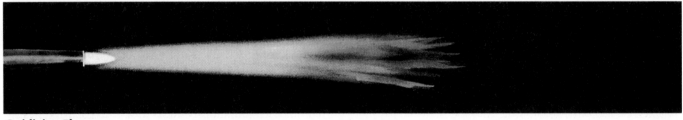

Oxidizing Flame
(Acetylene and excess oxygen) For braze welding with bronze rod.

Figure 6-3. *Color appearance of oxyacetylene flames. (Smith Equipment, Division of Tescom Corp.)*

LP Gas Burning in Atmosphere
Open fuel gas valve until flame begins to leave tip end.

Reducing Flame
(Excess LP gas with oxygen) For heating and soft soldering or silver brazing

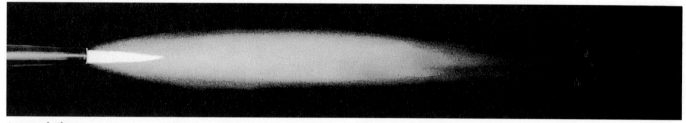

Neutral Flame
(LP gas with oxygen) Temperature 4579°F (2526°C). For brazing light material.

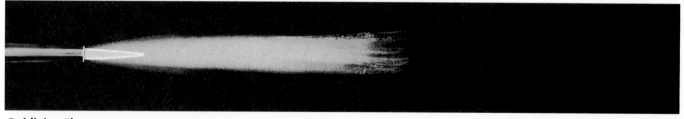

Oxidizing Flame
(LP gas with excess oxygen) Hottest flame about 5300°F (2927°C). For fusion welding and heavy braze welding.

Figure 6-4. *Color appearance of oxygen-LP gas flames. (Smith Equipment, Division of Tescom Corp.)*

Metal	Melting Temp.	
	°F	(°C)
Aluminum	1215	(657)
Brass (yellow)	1640	(893)
Bronze (cast)	1650	(899)
Copper	1920	(1049)
Iron, gray cast	2200	(1204)
Lead	620	(327)
Steel (.20%) SAE 1020	2800	(1538)
Solder (50-50)	420	(216)
Tin	450	(232)
Zinc	785	(418)

Figure 6-5. *Melting temperatures of some of the commonly used metals.*

oxyacetlyene welding with this type torch, the oxygen and fuel gas working pressures are usually the same.

The *injector-type torch* is usually used in connection with an acetylene generator. The torch operates with a relatively low acetylene pressure and a much higher oxygen pressure.

See Chapter 5 for further details concerning the construction and operation of each of these torches.

Study Chapter 1 and Heading 6.14 before attempting any welding practice.

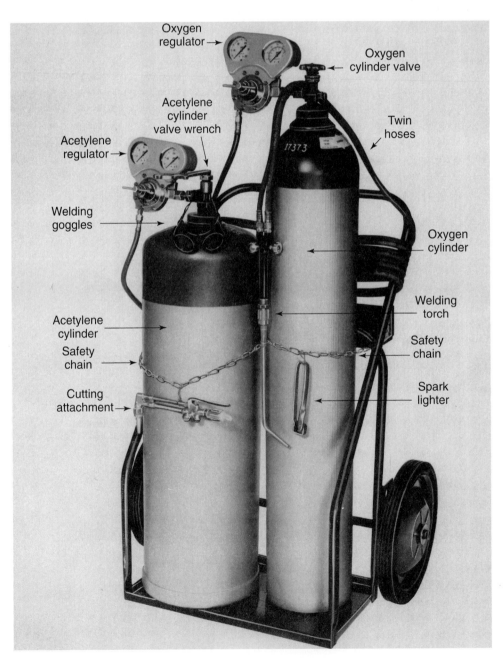

Figure 6-6. *A portable oxyacetylene gas welding and cutting station. (Modern Engineering Co., Inc.)*

6.5.1 Assembling the Oxyfuel Gas Welding Outfit

Proper handling of welding equipment is essential to safety, as well as being very important to obtaining good welds and a reasonable amount of service from the equipment. Oxygen cylinders and most fuel gas cylinders are usually owned by the companies that furnish these gases. A small demurrage (rental) charge is made for the use of the cylinders after a reasonable rent-free period. The rest of the welding equipment, however, is usually the property of the welder or of the company employing the welder.

Because of the high-pressure in the oxygen cylinder and the ease of combustion of fuel gases, especially acetylene, it is necessary that great care be used when handling cylinders. See Heading 6.5.2 for further information on handling cylinders. Always wear approved welding goggles when welding. See Figure 5-48 for suggested welding goggle shades.

Before welding equipment is used, it is very important to make sure the equipment is properly assembled. Check to see if the gas cylinders are in good condition. The cylinders should be fastened securely to a wall, post, or column by means of steel straps or chains. This will eliminate the possibility of them falling or being knocked over.

If the outfit is portable, the cylinders should be fastened to an approved cylinder truck by means of steel straps or chains. The cylinder truck should be so designed that it is almost impossible to knock it over accidentally. Whatever method is used to secure the cylinders, it should permit the cylinders to be changed easily.

Before attaching a regulator to a cylinder, the cylinder opening should be cleaned out. The cylinder opening is cleaned by quickly opening the cylinder valve slightly, then closing it. This is called *cracking* the valve. It allows some of the high-pressure gas to blow through the valve opening to remove dirt and other particles. **CAUTION: Be certain that there is no exposed flame in the area and that the cylinder opening is pointed away from any possible flame source when a fuel gas cylinder valve is blown clean.** Inspect the sealing surfaces and the fittings. Avoid using damaged or worn parts. The regulators may then be attached to the cylinder.

Only fixed-end wrenches, having wide jaws provided for the purpose, should be used on any cylinder and regulator fittings. Be sure that the regulator nut fits the cylinder valve fitting properly. **As a safety precaution, fuel gas and oxygen cylinder valves have different threads. Fuel cylinder valves are usually fitted with left-hand threads. Oxygen cylinder valves are fitted with right-hand threads. The thread diameters on the two cylinder valves are also different.** By using different threads, it is impossible to change the regulators from one type gas cylinder to another. This prevents the accidental mixing of gases. Many different varieties of cylinder and regulator fittings are in common use.

The hose connecting the regulators to the torch should be fastened firmly to the fittings at each end of the hose. The hose should be installed so that it is not twisted when the torch is held in the welding position. If the hose is twisted, it will strain the welder's hand and cause fatigue.

Before attaching the hose to the torch, the hose should be blown out or purged. With the regulators attached, open the cylinder valves, then gently open and close the regulator valves—first the fuel gas regulator, then the oxygen regulator. This brief purging will clear the hose lines. Where pipe threads are used, they should be sealed with pipe-thread sealing compound (such as a glycerine and litharge paste) or Teflon® tape.

Following the purging, the torch should be attached to the hose. Note that on oxyfuel gas welding equipment, the fuel gas hose nuts have left-hand threads, and the oxygen hose fittings have right-hand threads. After the welding equipment is assembled, test for leaks using a soap solution. Leak-testing must be done when installing any new cylinder or any new part of the apparatus.

To test for leaks, the recommended procedure is to put a soap and water *(soapsuds)* solution on the outside of the joints suspected of leaking. **A flame or oil of any kind should never be used to test for leaks.** Before testing for leaks, read Headings 6.5.4 and 6.5.6 on turning on the oxyacetylene welding outfit. To test for leaks, turn the regulator screws out all the way, open the cylinder valves, and build up from 5 psig to 14 psig (34.5kPa to 96.5kPa) of pressure in the regulators and hoses by slowly turning the regulator screws in (clockwise). Then, apply the soap solution to all joints. Leaks, if any, will be indicated by bubbles.

When first using welding equipment, remember that the procedure to follow is:

1. Learn the proper ways to prepare the station for use.
2. Learn the proper method of igniting the torch.
3. Learn how to adjust gas flows to produce a proper flame.
4. Learn the proper method of shutting down the equipment.

The proper procedures for handling the torch and selecting the proper size tip to use depend on several factors. These factors are the type of weld desired, the kind of metal used, the thickness of the metal, and the shape and position of the metal. Instructions on tip selection are found in Heading 6.5.3. Headings 6.5.4 and 6.5.6 cover the turning on and lighting of an oxyacetylene torch. Heading 6.14 reviews safety in oxyfuel gas welding.

6.5.2 Handling Fuel and Oxygen Cylinders Safely

Welding cylinders, when properly handled, are quite safe. Improperly handled, they may be very dangerous. **Cylinders should never be dropped or allowed to tip over. The cylinder safety cap, which encloses and protects the cylinder valve, should always be screwed into place when the cylinder is not in use, or when it is being moved.**

To move a cylinder, it is advisable to use a cylinder truck. The cylinder must be properly secured so that it cannot tip or fall off the truck.

Avoid the storage or use of cylinders in extremely hot locations. High temperatures may cause the cylinder

pressure to reach dangerously high levels. Check on local community building and fire codes and be certain that the cylinders are used and stored according to those codes.

6.5.3 Selecting a Welding Torch Tip of the Correct Size

The size of a welding torch tip is designated by a number stamped on the tip. The tip size is determined by the size of the orifice (opening). There is no standard system of numbering welding torch tip sizes. Manufacturers have their own numbering systems. For this reason, tip size instructions in this text are given in orifice "number drill size." *Number drills* form a series of drills in graduated diameters, numbered 1 through 80. The diameter of a Number 1 drill is 0.2280" (5.79mm). The diameter of a Number 80 drill is 0.0135" (0.34mm). *Note that the larger the number, the smaller the drill diameter.* See Chapter 33 for a table of number drill sizes. See Figure 6-7 for a list of manufacturers' welding tip numbers corresponding to number drill sizes. Once a welder becomes familiar with the operation of a certain manufacturer's torches and numbering system, it is seldom necessary to refer to orifice number drill sizes.

The orifice size determines the amount of fuel gas and oxygen fed to the flame. The orifice therefore determines the amount of heat produced by the torch: the larger the orifice, the greater the amount of heat generated. With a positive-pressure type torch, the tip sizes shown in Figure 6-8 should provide satisfactory results for oxyacetylene welding.

If the torch tip orifice is too small, not enough heat will be available to bring the metal to its melting and flowing temperature. If the torch tip is too large, poor welds will result for the following reasons:

- The weld will have to be made too fast.
- The welding rod will melt too quickly and the weld pool will be hard to control.
- The appearance and quality of the weld bead will be generally poor.

6.5.4 Turning on the Oxyacetylene Outfit and Lighting the Positive-Pressure Type Torch

To light the torch, the gases must be turned on and adjusted to the proper pressures. Proceed as follows:

1. Visually check the equipment for condition.
2. Inspect regulators.
3. Turn the regulator adjusting screws all the way out (counterclockwise) before opening the cylinder valves. This prevents damage to the regulator diaphragm.
4. Stand to one side of the regulator when opening the cylinder valves. A burst regulator or gauge could cause severe injury.
5. Slowly open the acetylene cylinder valve 1/4 turn to 1/2 turn counterclockwise. This will usually provide adequate flow and will permit rapid closing in an emergency. Use the proper-size wrench; leave it on the valve in case an emergency shutoff is needed.
6. Open the oxygen cylinder valve very slowly (turning counterclockwise) to prevent damage to the regulator diaphragm from the pressure and heat of the moving gas. When the regulator high-pressure gauge reaches its highest reading, turn the cylinder valve all the way open. This is necessary because the oxygen cylinder valve is one with a double seat (a back-seating valve). In the fully open position, this seat closes any possible opening along the valve stem through which the high-pressure oxygen could escape.
7. Open the acetylene torch valve one turn. Turn the acetylene regulator adjusting screw in slowly (clockwise) until the low-pressure acetylene gauge indicates a pressure that is correct for the tip size. (Never use acetylene gas at a gauge pressure above 15 psig (103.4kPa). An approximate setting may be arrived at as shown in the table, Figure 6-8. Turn off the acetylene torch valve using finger-tip force only. Important. Check the low-pressure gauge for an indication of possible regulator defects. See Heading 6.5.5.
8. Adjust the oxygen torch pressure. Open the torch oxygen valve one turn. Turn the oxygen regulator adjusting screw in (clockwise) until the low-pressure oxygen gauge indicates the pressure which is correct for the tip orifice. Then turn off the oxygen torch valve. The regulator pressures have now been adjusted to approximately the proper levels.

    Important. Check the low-pressure gauges after a few seconds to see if the pressure reading has increased from the working pressure set. If so, close the station down immediately. This increase on the low-pressure gauge is an indication that the pressure regulator valve is leaking. To prevent damage to the needle valves and seats, use only fingertip force to open and close the torch valves.
9. Purge the system before lighting the torch. To ensure that the proper gases are in the respective hoses (no air or oxygen in the acetylene hose; no fuel gas in the oxygen hose), the system must be purged. This is done by allowing the acetylene to flow through the acetylene hose and oxygen to flow through the oxygen hose for a short time before lighting. Note: If Steps 1 through 8 are followed, the system will have been purged and the torch is ready for lighting.
10. After purging, crack (open) the acetylene torch valve no more than 1/16 turn. Use a spark lighter to ignite the acetylene gas coming out of the tip.
11. Continue to open the acetylene torch valve slowly until the acetylene flame jumps away from the end of the tip slightly. This indicates that the proper amount of acetylene is being fed to the tip. A quick flip of the torch should make the flame leap away from the tip 1/16" (1.6mm) and come back again. If the flame will not move back to the tip, too much acetylene has been turned on. (If the tip is worn, it may be difficult to make the flame jump away from it.)

Trade name	Series	71	70	69	68	67	66	56	55	54	53	52	51	50	49	48	45	44	43	42	41	40	36	35	34	1/8	30	29	28	27	26	
Airco	All				1			3		4			5			6		7				8	9				10⁵					
Canadian Liquid Air	All				1			3		4			5							7					9		10					
Craftsman	AA			1						3										5									8⁶			
Dockson 4EC, 4SC, 7SC; 3EC, 5EC, 6EC, & 7EC	All						3		5			6				7				4			5							7	8	
Gasweld	G25, G35			1					4		5		6			7		8			9	10	12	13			16					
Gasweld	G55			2					4		5		6					8														
Gasweld	AVG										6		7		8																	
Harris	2890-F	00							1				2																			
Harris	6290				00				1			2																				
Harris	7490-A								1					2																		
Harris	23,13-F, 23A swedged)															3				9		10										
Harris	17F swedged)																									15⁷				19		
K-G	AP, APM, APL				1				3	4			5¹			6		7			8⁸			9			10⁹					
Liquidweld	90, 70, 72		00						2		3						4		5				6					7	8	9	10	11
Liquidweld	80, 82		00						2		3						4		5				6					7	8			
Marquette	A			0				2		4			5							7			8					11		12		
Marquette	B				2						5		6					7				8	9					11				
Marquette	F					1				4	5		6					6				8	9							12		
Marquette	G	0			2					3				5		6		7				8	9							12		
Marquette	H & J	0			2		1			4				5																		
Meco	All			1						4			5							7			8					9	10			
National	B	2		3					7		8		9			10		11		12			13		14							
National	G, P					2				3			4						6				8				10					
National	R		00					2			3							5										7	8	9	10	11
Oxweld	W-15					2		4			5		6	7		6																
Oxweld	W-29		2					6			9	12		15																		
Oxweld	W-17, W-22, W-26							6			9	12		15			20					30		30			55	70		85		
Oxweld	W-45, W-47									9		12	15	20								40	40				70					100
Powr-Craft	84-5881 (Montgomery Ward)						1		5					7						9												
Prest-O-Lite	420										15			20						30												
Prest-O-Weld	W-109	2							4				6²		5		8					10	11					29				
Prest-O-Weld	W-110, W-111								4				7				9															
Prest-O-Weld	W-120								9		12	15		20			30					40					70					
Prest-O-Weld	W-121, W-122								9		12	15		20			30													85		
Purox	33									4			5			6						8										
Purox	34				2					4			6			6												13				
Purox	35							2	2		5		5			8							10									
Purox	W-200								9		12	15		20			30													70		
Purox	W-201, W-202								9		12	15		20			30					40					70			85		
Purox	00-D									2		4	5																			
Rego	GX, GXU, SX				68				55		53		50¹⁰							42			36¹¹									
Smith	Pipeliner MW³	101		102		103		106		107		108		109		110				111				112								
Smith	Airline AW³	101		102		103		106		107		108		109		110				111				112								
Smith	Silver Star LW³	101		102		103		106		107		108		109		110				111		708		112			113					
Smith	LW⁴	700								703		704		705		706				707											710	711
Torchweld	GP 570, 870				68				55		53		50							42			36¹²									
Torchweld	71, 370, 170				68				55		53		50							42			36									
Victor	All			00		00½		2	2½		3		3½			4				5			6					7	8	9	10	11
Weldit	All	1								4			5			6		7	8		9	10	12				16					

¹APL to No. 5 only. ²W-110 to No. 6 Tip only. ³Soft-flame tip. ⁴Heavy-duty tip. ⁵Airco tip. No. 11-drill size No. 25; No. 12-drill size No. 20;
No. 13-drill size No. 10; No. 14-drill size No. 2; No. 15-1/4" drill. ⁶Tips No. 7 and 8 require special gooseneck. ⁷13-F and 17-F Swedged to No. 15 tip only.
⁸APM to No. 8 tip only. ⁹AP to No. 10 only. ¹⁰SX to No. 46 tip only. ¹¹GXU to No. 31 only. ¹²GP 570 to No. 31 only. (Welding Engineer)

Figure 6-7. *A table of manufacturers' gas welding tips with tip numbers and orifice drill sizes.*

Another method for determining the correct amount of acetylene is to increase the flow until the flame becomes *turbulent* (rough) at a distance of 3/4" to 1" (19mm to 25.4mm) from the torch tip. With the correct amount of acetylene, the flame will no longer smoke or release soot. Look at Figure 6-9 and compare the flames. Refer to Figure 6-3 for a color illustration of the flames.

12. After the acetylene is regulated, slowly open the oxygen valve on the torch. As the oxygen is fed into the flame, the brilliant acetylene flame turns purple and a small inner cone starts to form. This inner cone is white and becomes an even more brilliant white as more oxygen is added. When first formed, the extremity of this inner cone will have a blurred and irregular contour. As the oxygen flow increases, the inner cone loses its blurred edge and becomes a round, smooth cone. Stop the adjustment at this point. Any increase in oxygen will result in an oxidizing flame. (Too much oxygen will burn or oxidize the metal being heated.) The tip of this inner cone is the hottest part of the flame. See Figure 6-3 for a color illustration. The correct quantities of gases for the smaller tip sizes may also be detected by listening to

Metal Thickness	Size* Welding Tip Orifice	Welding Rod Diameter	Oxygen		Acetylene		Welding Speed Ft./hr.
			Psig Pressure	Cu. ft./hr.	Psig Pressure	Cu. ft./hr.	
1/32"	74	1/16"	1	1.1	1	1	
1/16"	69	1/16"	1	2.2	1	2	
3/32"	64	1/16" or 3/32"	2	5.5	2	5	20
1/8"	57	3/32" or 1/8"	3	9.9	3	9	16
3/16"	55	1/8"	4	17.6	4	16	14
1/4"	52	1/8" or 3/16"	5	27.5	5	25	12
5/16"	49	1/8" or 3/16"	6	33.	6	30	10
3/8"	45	3/16"	7	44.	7	40	9
1/2"	42	3/16"	7	66.	7	60	8

*Note the tip orifice size as shown is the number drill size. These recommendations are approximate. The torch manufacturer's recommendations should be carefully followed.

Figure 6-8. *A table showing the relationships between welding tip size, gas pressures, welding rod diameters, and metal thickness. This table is for oxyacetylene welding with a positive-pressure type torch.*

the torch flame. When correctly adjusted, it should emit a soft purr, not a sharp irritating hiss.

13. If the torch burns with an irregular contour (feather) to the cone, the flame is called a *carburizing flame.* There is an excess of acetylene. See Figure 6-3. If the inner cone has a very sharp point and if it hisses excessively, it usually means that too much oxygen is being used. If the flame has a smooth inner cone, the flame is called "neutral." See Figure 6-3.

There is another method of adjusting the welding torch:

1. Turn on the cylinder gases, as described previously.
2. Open the acetylene torch valve one turn. Slowly turn in the adjusting screw on the acetylene regulator. When the acetylene starts flowing, light the acetylene. Resume turning the acetylene regulator screw in until the acetylene is made to jump away from the torch, or the turbulence is correct (as in the first method). Refer to Figure 6-9.
3. Open the oxygen torch valve one turn. Turn the oxygen regulator adjusting screw in slowly until enough oxygen is being fed to the torch to completely consume all the acetylene and a neutral flame is obtained. (The neutral flame is described earlier.) Also see the color illustration in Figure 6-3.
4. This method may be used in place of the first method. In cases where the welder uses a long hose, it will compensate for the pressure drop in the hoses. The welder may choose either of the two methods. Results of either are generally satisfactory.

6.5.5 Regulator Gauge Readings

After adjusting the desired pressure on the regulator low-pressure gauge, the torch valve is closed. You will see

a slight rise in the pressure on the low-pressure gauge. The pressure reading will rise and stop. This is normal. The flowing pressure (*dynamic pressure*) is always lower than the pressure that exists when the torch valve is closed (*static pressure*).

If the low-pressure gauge reading continues to rise (creeps) after the torch valve is turned off, this indicates a leaking regulator nozzle and seat. Turn off the cylinder valve immediately or the low-pressure gauge may receive the full cylinder pressure. This will cause the gauge to rupture. If such a continuous rise (creep) occurs, replace the regulator or have it repaired before using the station again.

Review Heading 6.14.2 on regulator safety before attempting to weld.

6.5.6 Turning on the Oxyacetylene Outfit and Lighting the Injector-Type Welding Torch

The steps for lighting an injector-type torch are:

1. Inspect the equipment to make sure all parts are in good operating condition.
2. Inspect the regulators. The adjusting screws of the regulators should be turned all the way out (counterclockwise), before the cylinder valves are opened. This prevents damage to the regulator diaphragm when the cylinder valve is opened. Never stand in front of the regulator when opening the cylinder valves. Injury may result if the regulator or gauges rupture.
3. Open the oxygen cylinder valve very slowly until the regulator high-pressure gauge reaches its maximum reading, then turn the valve all the way open. The oxygen cylinder valve is turned all the way out

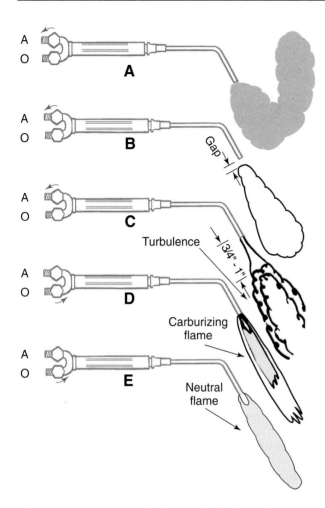

Figure 6-9. *Recommended steps for lighting the oxyacetylene welding torch. A—Open the acetylene torch valve slightly and light the acetylene with a spark lighter. B—The correct amount of acetylene is flowing if the flame jumps away from the tip with the torch is shaken, or C—As shown here, a turbulence is created in the acetylene flame and the sooty smoke is eliminated. D—Begin turning on the oxygen by opening the torch oxygen valve. E—Continue to turn the oxygen torch valve until the middle flame is eliminated and a rounded inner cone is seen.*

because this valve has a double (back) seat. In the full-open position, this seat closes any possible opening through which the high-pressure oxygen might escape along the valve stem.

4. With a cylinder wrench, slowly open the acetylene cylinder valve a quarter to a half turn. Leave the wrench on the acetylene cylinder valve stem so the cylinder may be shut off quickly in an emergency.

5. To adjust the oxygen torch pressure, open the torch oxygen valve fully (about 1 1/2 turns). Turn the pressure adjusting screw in on the oxygen regulator until the low-pressure (delivery) gauge registers the approximate pressure shown in Figure 6-10. Then, close the oxygen valve. **Important: Check the low-pressure gauge for an indication of regulator failure. See Heading 6.5.4.** Use only fingertip force to close the torch valves. Too much force will damage the needle valves.

6. Open the torch acetylene valve 1/2 turn. Turn in the pressure adjusting screw on the acetylene regulator until the low-pressure (delivery) gauge shows a pressure of 1 psig (7kPa). Close the torch acetylene valve. By following this procedure, the system will have been purged.

7. To light the torch after purging, open the torch oxygen valve 1/4 turn. Open the torch acetylene valve 1/2 turn. Light the gases at the tip with a lighter. Open the torch oxygen valve fully (1 1/2 turns) and adjust the torch acetylene valve to the desired flame.

8. Should the acetylene delivery pressure become so low that a delivery pressure of 1 psig (7kPa) can no longer be obtained, proceed as follows:
    a. Open the torch acetylene valve all the way.
    b. Turn the oxygen regulator pressure adjusting screw to the left until the flame shows an excess acetylene feather about four times as long as the inner cone.
    c. Adjust the torch acetylene valve to give the desired flame.

Metal thickness	Torch tip number drill size	Oxygen regulator pressure psi	Acetylene regulator pressure psi
1/16" (1.59mm)	56	8-20 (55.2-137.9 kPa)	1 (7 kPa)
1/8" (3.18mm)	53	11-25 (75.8-172.4 kPa)	1 (7 kPa)
1/4" (6.35mm)	48	12-23 (82.7-158.6 kPa)	1 (7 kPa)

Figure 6-10. *This table shows correct oxygen and acetylene pressures for welding different thicknesses of metal using an injector-type torch.*

6.5.7 Torch Adjustments

The torch may be adjusted to produce the following flame characteristics:

- Neutral flame.
- Carburizing flame.
- Oxidizing flame.

In general, the neutral flame is the one desired. However, in welding aluminum, in brazing, and in some other operations where oxidizing of the metals would interfere with welding, a slightly carburizing flame is often used. Figures 6-2 and 6-3 illustrate the appearance of each of these flames. While a slightly carburizing flame may be recommended for certain work, a *neutral* flame usually will do just as well. However, because of slight fluctuations in gas pressures, it is difficult to maintain a perfectly neutral flame. The flame may vary from neutral to slightly oxidizing or carburizing. To avoid the possibility of an oxidizing flame, a slightly carburizing flame is usually safer.

The torch may occasionally backfire (pop). A *backfire* is a small explosion of the flame at the torch tip, which may result from any one of several avoidable conditions. The most frequent cause is preignition of the gases. Some causes of backfiring are:

- The gas is flowing out too slowly and the pressures are too low for the size tip (orifice) used. The gases are therefore burning faster (flame propagation) than they can flow out of the tip. This trouble may be corrected by adjusting to a slightly higher pressure for both the oxygen and the acetylene.
- The tip may become overheated from overuse, from operating in a hot corner, or from being too close to the weld. Cool the tip.
- The inside of the tip may have carbon deposits or a hot metal particle may be lodged inside the orifice. These particles become overheated and act as ignitors. Correct this by cleaning the tip. See Headings 5.6.4 and 6.14. Backfiring happens rarely, but could occur when the inner cone of the flame is submerged in the weld pool.

A *flashback* is the burning of the gases in the mixing chamber or beyond toward the regulator. **Flashback is an extremely dangerous occurrence! Action must be taken immediately to extinguish the flame.**

Generally, when flashback takes place, a squealing or sharp hissing noise is heard. If flashback occurs, the torch oxygen valve should be turned off first, then the fuel gas torch valve. This is done in an attempt to keep the flashback from going beyond the torch. Close the fuel gas and oxygen cylinder valves to cut off the supply of the gases.

Inspect for damage. If flashback occurs, the hose, torch, and regulators are usually damaged and must be replaced or overhauled.

Flashback can occur for the following reasons:

- Failure to purge the system before lighting the torch. Purging insures that the proper gas is flowing in each hose and torch passage. Any oxygen in the fuel gas passages is cleared out and the fuel gas is cleared from the oxygen passages prior to lighting the flame.
- An overheated tip.

Review Heading 6.14.3 on torch safety before attempting to weld.

6.5.8 Turning Off the Torch and Shutting Down the Oxyacetylene Outfit

If the welder is to leave the welding station for just a few minutes, he or she can simply turn off the torch by closing the torch valves, then lay the torch aside. To properly extinguish the flame, first turn off the acetylene (fuel gas) torch valve, then the oxygen torch valve.

If the equipment is not to be used for some time, however, the outfit should be completely shut down. These instructions may be followed for both the positive-pressure torch and the injector-type torch. To shut down the oxyacetylene outfit:

1. Close the acetylene hand valve on the torch, then the oxygen valve. This extinguishes the flame and eliminates the soot.
2. Tightly close the cylinder valves.
3. Open the hand valves on the torch.
4. Wait until the high-pressure and the low-pressure gauges on *both* the acetylene and the oxygen regulators read zero.
5. Lightly close both hand valves on the torch, then hang up the torch.
6. Turn the adjusting screws on both the acetylene and oxygen regulators all the way out. **Caution: If turning out the regulator screws is not the last step, pressure will remain in the high-pressure gauge and regulator.**

6.6 TORCH POSITIONS AND MOVEMENTS

Moving the torch in the direction that the tip is pointing is called *forehand welding*. The torch is held at an angle of 15° to 75° to the work. The angle will depend on the tip size used, metal thickness, and other welding conditions. A 30° to 45° angle is typical. See Figure 6-11. In forehand welding, the flame spreads over the work ahead of the weld. This preheats the metal before it comes under the high-temperature flame.

Use either the *oscillating torch motion* or the *circular torch motion*. Figure 6-12 shows different torch motions used by welders. In either case, the cone of the flame should never go outside of the weld pool. The tip of the inner flame cone should be about 1/16" (1.6mm) to 1/8" (3.2mm) above the metal. Figure 6-13 shows a butt weld in progress.

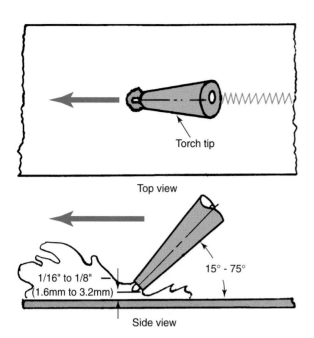

Figure 6-11. Recommended torch angle, direction, and flame distance from the metal when running a continuous weld pool in the flat welding position. Angles between 15° and 75° are used to compensate for metal thickness and tip size.

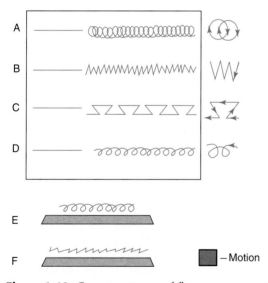

Figure 6-12. Common types of flame movement patterns used when oxyfuel gas welding. All motion must remain within the size of the weld pool.

6.7 RUNNING A CONTINUOUS WELD POOL

Before attempting a weld of any kind, it is recommended that the beginning welder practice running a *continuous weld pool*. Once described by the term *"puddling,"* this is the creation and control of the molten pool of metal which is carried along the seam of the parts to be welded together.

Figure 6-13. Oxyacetylene welding a butt joint. These steel plates are 5/16" (7.94mm) thick. Note the position and angle of the torch flame and filler rod.

An experienced welder can, by watching the appearance of the weld pool carefully, tell the following about the weld:

- The amount of weld penetration.
- The torch adjustment and heat provided.
- How and when to move the torch.
- When and how often to add filler metal.

The size (diameter) of the weld pool will be in proportion to its depth. The welder may judge the depth, or penetration, of a weld by watching and controlling the size of the pool of molten metal. On very thin metal, the penetration or depth of the weld pool will be greater, in proportion to the width, than with thicker metal.

The appearance of the surface of the pool will indicate the condition of adjustment of the torch. A neutral flame, when melting a good grade of metal, will give a smooth, glossy appearance to the weld pool. The edge of the pool away from the torch will have a small bright incandescent spot. This spot will move actively around the edge of the pool. If this spot is oversize, the flame is *not neutral*. If the weld pool bubbles and sparks excessively, an oxidizing flame is in use. A poor quality and/or dirty metal will also spark as it is being welded. If the flame is carburizing to any great extent, the weld pool will have a dull and dirty (sooty) appearance.

The tip of the inner cone of the torch flame must be held within the boundary of the weld pool at all times. A correctly adjusted flame prevents the oxygen in the atmosphere from coming in contact with the surface of the pool and causing an oxidizing condition. The hottest area of the flame is 1/16" to 1/8" (1.6mm to 3.2mm) from the end of the inner flame cone. See Figure 6-14. If the weld pool sinks or sags too far, indicating too much penetration, lower the angle of the torch or increase the speed of movement rather than draw the torch away from the surface.

Figure 6-15 shows a continuous weld pool in progress. Note the width of the weld pool and the torch movement which controls the width of the pool. Note also the *eye of the weld pool,* a bright flake of oxide that indicates the movement of the molten metal. A partially completed weld pool practice piece is shown in Figure 6-16. Note the side view showing the penetration.

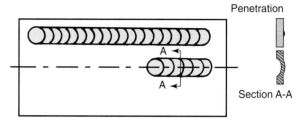

Figure 6-16. *Weld pool procedure. This illustrates a partly completed exercise piece. Note the penetration. The welder has obtained uniform width and has carried the weld pool in a straight line. The penetration is uniform.*

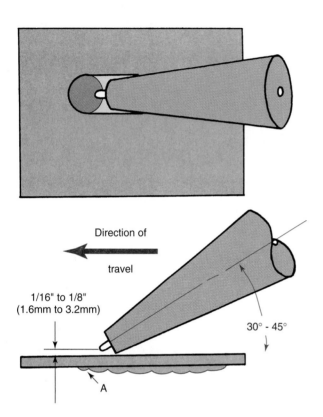

Figure 6-14. *Continuous weld pool procedure. This illustration shows the correct position of the torch in relation to the base metal. Detail A shows the penetration (sag) below the undersurface of the base metal.*

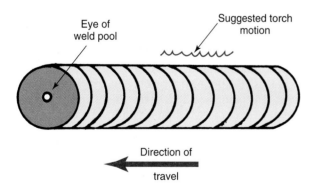

Figure 6-15. *This illustration shows the continuous weld pool in progress.*

Before starting to practice with welding rod, the beginning welder should be able to produce five consecutive passes by running a continuous weld pool. Each should be at least 5" (130mm) long, without any holes melted through the metal and with good penetration. The passes must also be straight (in-line) and uniform in width. Beginners who can run a good continuous weld pool have proved that they are familiar with the proper use of the torch to control a molten pool of metal. Such a beginner is ready to learn how to manipulate the welding rod.

6.7.1 Type of Welds Made without the Use of a Welding Rod

The overlapped square-groove outside corner joint can be welded without the use of a welding rod. It therefore makes an excellent welding exercise for beginners. This exercise teaches how to weld by using some of the base metal as the filler metal. The pieces are placed one against the other at right angles, so that the vertical piece extends beyond the surface of the horizontal sheet approximately 1/32" to 1/16" (0.8mm to 1.6mm) as shown in Figure 6-17. The two pieces are then tacked together at their ends. A *tack weld* is a small weld used to hold parts together prior to making the completed weld. The extended metal serves as the filler metal.

The weld must have good penetration, but the penetration should not show on the inside corner. The welder will find that very little torch motion is needed for this exercise. The torch tip should be slightly tilted, making the flame point inward toward the flat or horizontal surface. The weld should be all on the horizontal surface. None of it should run over onto the vertical edge. This is necessary, because in many cases of metal finishing, a weld is made into a right-angle corner by grinding the excess metal from the one side. After checking the weld for appearance, the penetration may be tested by bending the two pieces of metal open like the pages of a book. Any cracking or breaking of the metal pieces at the joint will indicate a lack of penetration or fusion.

Another good exercise to help in learning the use of a welding torch when welding without welding rod is an edge weld on a flanged butt joint.

To prepare the metal for this weld, bend a 90° flange about 1/4" to 1/2" (6.4mm to 12.7mm) long on two pieces

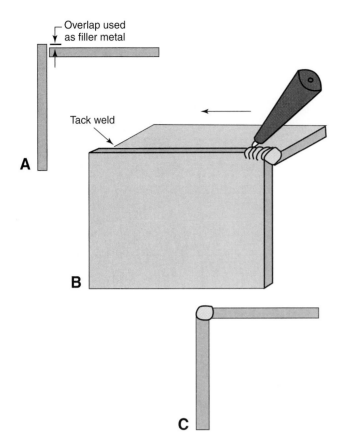

Figure 6-17. Steps in performing the overlapped square-groove outside corner joint without a welding rod. A—Metal in position to weld. B—Weld in progress. C—Appearance of the finished weld.

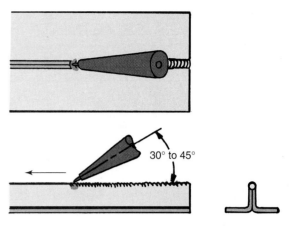

Figure 6-18. A weld on a flanged butt joint in progress. Note that the base metal is used as the filler metal, so a welding rod is not needed. The joint should be tack welded several times before starting the main weld.

of sheet steel about 1/32" to 1/16" (0.8mm to 1.6mm) thick. Be sure the lengths of the two flanges are equal. Place the flanges together along their lengths and tack weld several times to hold the metals in position. Fuse the two flanges with the welding torch using the flanges as the filler metal, in the same manner as you did when welding the outside corner joint. Figure 6-18 shows an edge weld on a flanged butt joint in progress.

6.7.2 Types of Welds Made with a Welding Rod

To become proficient in the art of gas welding, certain fundamental exercises must be planned and practiced until satisfactory welds can be performed consistently. The different fundamental gas welding operations may be classified according to the type of joint and position of the weld.

The basic joints (See Figure 3-1) are:
- Butt joint.
- Lap joint.
- Outside corner joint.
- Inside corner joint
- T-joint.

The types of welds made on these basic joint designs are the *square-groove weld*, the *groove weld*, and the *fillet weld*. See Figures 3-4 and 3-5.

The basic welding positions are:
- *Flat welding position*, a horizontal weld on a horizontal surface.
- *Horizontal welding position*, a horizontal weld on a vertical surface.
- *Vertical welding position*, a vertical weld on a vertical surface.
- *Overhead welding position*, a horizontal weld on a horizontal surface over the welder's head.

The welding of each of the joints should be practiced in each of the positions. Figure 6-19 illustrates some typical types of joints. Welding of these joints should be performed first on thin sheet steel and later on steel plate of at least 1/8" (3.2mm) thickness. After obtaining the necessary skill on steel plate with these exercises, the welder may then proceed to study special welding applications such as pipe welding, aluminum welding, or cast iron welding.

6.7.3 Use of Welding Rod in Running a Bead

Running a continuous weld pool without the use of welding rod is done mainly on outside corner or flange joints. *Welding rod* (filler metal) is added when extra metal is needed to create the correct weld shape and strength. Welding thicker metal joints using only the weld pool method will cause thinning of the metal in the weld area. To obtain a strong weld, metal from a welding rod is fed into the weld pool. This will decrease its depth and increase the thickness of the metal in the weld area. See Heading 5.10.3 for specifications of welding rods. The bead in a correctly made weld should be slightly convex (crowned) for extra thickness and strength of the weld. Fillet welds are often made with a concave bead. The fillet must be made large enough to provide sufficient strength to the joint. Figure 6-20 illustrates how welding rod material is added to make a slightly convex bead.

To weld using welding rod, bring the torch to that part of the joint where the weld is to start. Melt a small weld pool on the surfaces of the two pieces and allow the pools to flow together. At the same time, with the other

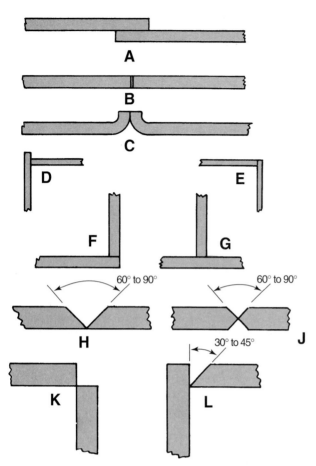

Figure 6-19. *Some typical welding joint designs: A—Sheet steel lap joint in the flat welding position. B—Sheet steel butt joint in the flat welding position. C—Flanged butt joint in the flat welding position. D and E—Outside corner joints. F—Inside corner joints. G—A T-joint. H, J, K, and L—Joints designed for metal plate. Note that when welding joints A, B, and E through L, welding rod is used as the filler metal. When welding the joints at C and D on sheet metal, no welding rod is required as a filler metal because the metal pieces themselves are melted to form the bead and to join the pieces together.*

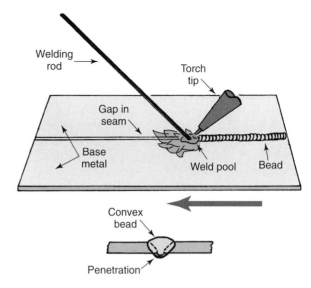

Figure 6-20. *Recommended torch and welding rod positions for welding a butt joint in the flat welding position.*

hand, bring the welding rod to within 3/8″ (9.5mm) of the torch flame and 1/16″ to 1/8″ (1.6mm to 3.2mm) from the surface of the weld pool. In this position, the rod will become preheated, and will melt sooner when dipped into the weld pool. When the welder judges that the weld pool needs additional metal, the end of the welding rod is inserted into the leading edge of the weld pool, and some of the rod is melted off and mixed with the molten base metal. The welder continues to dip the rod into the weld pool until enough welding rod metal is added to raise the weld pool to a slight crown. At the same time, he or she continues the torch forward motion without interruption. *Torch control at this time is very important.* Control over the

weld pool condition and welding rod melting can be done with small changes of the torch position. As soon as enough welding rod metal has been added to make the correctly shaped bead, stop adding the welding rod. Keep the welding rod close to the weld pool and flame to maintain the end of the welding rod in a preheated condition.

If the welding rod is withdrawn too far from the torch, it will become too cool. This cool rod, in turn, will chill the weld pool when it is again inserted . On the other hand, if the welding rod is held too close to the flame of the torch, it will become too hot. If it should become molten, drops of molten welding rod will be blown by the flame onto the cooler parts of the metal being welded. Such a condition will result in a very uneven bead, poor fusion, and probably poor penetration. As the welding rod melts, it becomes necessary to change the position of your hand on the rod. To change the hand position, lay the welding rod down with the hot end away from your body. Then pick it up at a new position. **Do not place the rod against the body to slide the hand to a new position. Many welders have been injured when the tip of the welding rod burned through clothing and touched the body.**

The beginner is sometimes tempted to use different size welding rods for the same weld. This change should be avoided because:

- If a 3/32″ (2.38mm) welding rod is correct, and a change is made to 1/16″ (1.59mm) welding rod, it will be extremely difficult to add enough welding rod to obtain a good weld.
- The smaller rod will make it more difficult to control the weld pool.
- There will be a tendency to burn (oxidize) the smaller size welding rod.

If the welder changes to 1/8″ (3.18mm) welding rod when 3/32″ (2.38mm) welding rod is correct, the following troubles might result:

- The larger welding rod will cool the weld pool too much while it is being added, preventing consistent welding.
- There will be a tendency to add too much welding rod, destroying penetration and building the weld higher than it should be on the top surface.

Whether to use a 1/16" (1.59mm), 3/32" (2.38mm), or 1/8" (3.18mm) welding rod is not so important as adhering to one size after becoming used to it for a certain thickness of metal and a certain size tip, See Figure 6-8 for a table of recommended welding rod sizes and Chapter 5 for more detailed information on welding rod choices.

A good weld—with good fusion, good bead, and good penetration—is obtainable only by attaining skill in the handling of the welding torch and the welding rod in harmony with each other. The torch motion should be constant in forward speed and in the width of the motion. The proper distance between the flame cone and the metal must also be maintained. The slant (angle) of the torch with respect to the surface should always be the same. The filler metal additions should be made at regular intervals to keep the bead a uniform size and shape.

6.8 BUTT JOINT WELDING

The butt joint is one of the most common welds made with the oxyfuel gas torch. The instructions which follow will aid the beginner in making this weld on thin steel.

Procure two pieces of mild steel approximately 1" (25mm) wide and 5" (127mm) long. The pieces should be clean and flat, and the edges should be straight. Place the two pieces of metal across two firebricks, so that the bricks support the ends of the metal. Place the edges of the two pieces of metal together at the end where the weld is to start. As the weld proceeds along the joint, the metal will shrink. As it cools from a molten state, it tends to pull the two pieces of metal together. This shrinkage may warp the metal or cause the edges to lap one over the other. The welder may prepare the metal for this expansion and contraction by:

- Tack welding (fusing) the two pieces of the metal together before proceeding with the welding as in Figure 6-21A. This method will produce some internal strain, but will keep the ends sufficiently in line to enable the welder to make a good weld. Tack weld every two inches or so along the joint.
- Tapering the gap between the two pieces of metal to allow for contraction as in Figure 6-21B. The approximate contraction is from 1/8" to 1/4" per foot, or from 10mm to 20mm per meter of length. As the weld pool gets wider, the contraction increases.
- Using especially prepared wedges placed between the two pieces of the joint to prevent the contraction of the metals as the weld cools, Figure 6-21C. This method is more generally used with long joints.

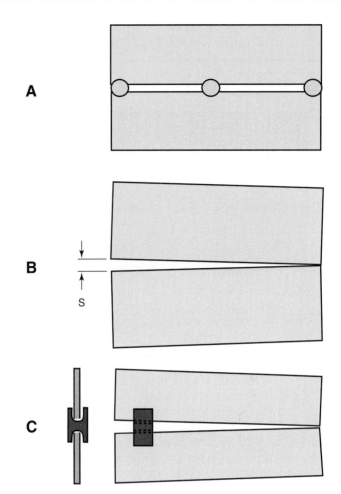

Figure 6-21. *Some methods used to maintain correct position of welded pieces, since the weld metal shrinks as it cools and solidifies. A—"Tacking" pieces together before welding. B—Allowing for shrinkage (S). C—Use of special wedges.*

- Clamping the metal in a heavy fixture to minimize movement.

After the metal is prepared, light the torch, adjust it to a neutral flame, and proceed as follows: Bring the torch to the point where the weld is to start, holding it at a 30° to 45° angle to the metal with the tip pointing in the direction of travel. Hold the inner cone approximately 1/16" to 1/8" (1.6mm to 3.2mm) away from the metal. With the other hand, bring the end of the welding rod to approximately 3/8" (10mm) away from the welding torch and about 1/8" (3mm) above the metal. The torch flame will melt a pool on the edge of each of the pieces of metal. The weld pool should spread equally over the two pieces of metal. Apply the welding rod (filler metal) as directed in Heading 6.7.2.

Advance the torch a very short distance until the weld pool again reaches the size of the previous pool. The tip motion should continue as the torch moves forward. The welding rod is again dipped into the pool and the weld bead is built up to a crown. Continue this procedure throughout the length of the weld joint, using a continuous torch motion. See Figure 6-12 for suggested torch motions.

The tip should be kept at a uniform distance from the weld, and the torch angle with the metal should remain unchanged. The welding rod should be added in uniform amounts, and at regular intervals. After the weld has been finished, allow it to cool, and then inspect it. Figure 6-22 is a magnified photograph of a butt weld in mild steel.

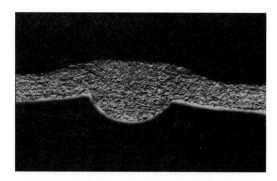

Figure 6-22. *Macrograph (4X) of a butt joint weld in mild steel. The metal has been etched to show the grain and fusion. Note that the penetration is excessive in relation to the thickness of the base metal.*

There is a procedure in butt joint welding in which the heat source penetrates completely through the workpiece before the welding rod is added. This process is called *keyhole welding*. A hole is formed at the leading edge of the weld metal. As the heat source progresses, the molten filler metal and base metal fill in behind the hole to form the weld bead and provide excellent penetration. The keyhole procedure is often used on thicker sections of metal that have been beveled (ground at an angle). See Figure 6-20. As long as the keyhole continues to appear, penetration is certain. See Figure 6-23 for an example of the keyhole procedure.

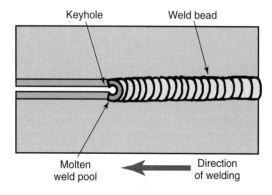

Figure 6-23. *The keyhole welding procedure. Note the hole (keyhole) at the leading edge of the molten weld pool. The continual appearance of the keyhole will ensure good, uniform penetration.*

6.9 LAP JOINT WELDING

The lap joint is one of the five basic joint designs. The joint consists of lapping one piece of sheet metal over the one to which it is to be welded, as shown in Figure 6-24.

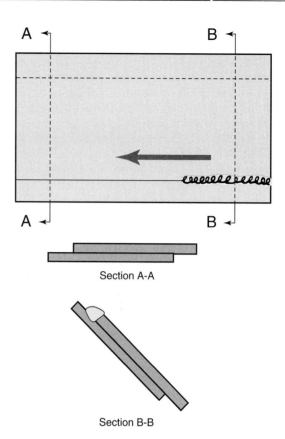

Figure 6-24. *Lap joint welding. Section A–A shows the position of the two pieces in relation to each other. The two pieces should be placed as close together as possible (no gaps). Section B–B shows the finished weld.*

This weld should be first performed in the flat welding position. Although the welding technique is typical, several things must be kept in mind in order to obtain a satisfactory lap weld:

- It will be found difficult to heat the bottom piece of metal to a molten state before the top metal edge melts too much, making the weld very ragged. The way to prevent this is to concentrate the torch flame on the lower surface. The bottom piece of metal requires two-thirds of the total heat, as shown in Figure 6-25.
- The welded portion (the weld nugget) must be at least as thick as the original metal. To provide this thickness, add enough welding rod metal to make a *convex* (curved outward) bead.

The beginner usually has a tendency to perform this weld without heating the metal surface to obtain fusion between the edge and the surface. The destructive separation test will quickly show the lack of fusion on such a lap joint. To perform this test, the lap joint is placed in a vise and the piece containing the edge is bent toward the surface piece until it is flat. A poorly made lap weld will break near the toe of the weld where it contacts the surface piece. A good weld will tear some of the surface metal away as it separates from the surface piece.

A special form of lap welding is known as plug welding. A *plug weld* on a lap joint may be defined as: "A

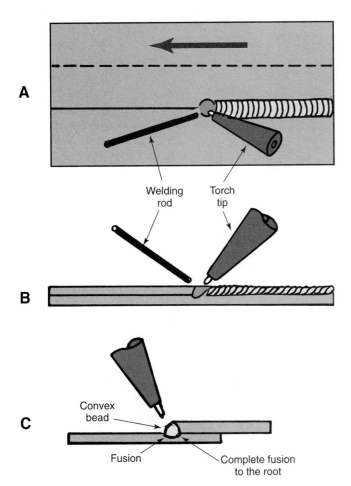

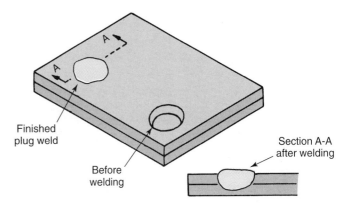

Figure 6-26. *A plug weld. This method of welding is used to join two or more pieces, but not at their edges.*

Figure 6-25. *Recommended position of metal, torch, and welding rod when lap welding in the horizontal welding position. Note the position of the torch and welding rod in relation to the base metal. In C, the torch points more directly at the bottom surface than at the edge.*

circular weld made through a hole in one member of a lap joint, fusing that member to the other. The walls of the hole may or may not be parallel and they may be partially or completely filled." See Figure 6-26 for an example. The torch flame should be concentrated on the base plate in order to bring it to its melting temperature at the same time that the edges of the hole melt. In aircraft tubing repairs, the plug weld is known as a *rosette weld*. A *slot weld* is the same in all respects as a plug weld, except that instead of a round hole being drilled, an elongated hole or slot is cut.

6.10 OUTSIDE CORNER JOINT WELDING

Outside corner joints may be made without using welding rod; see Heading 6.7.1. An alternate method of performing an outside corner for practice welds is shown in Figure 6-27. This joint design offers an opportunity to make a V-groove weld without having to bevel the edges.

The keyhole method of welding can be practiced on this joint. In this joint, the two pieces do not overlap. While welding the outside corner joint, watch the outside edges of the metal. When the weld pool touches the outside edges of the metal, the welding rod should be placed in the pool. Enough filler metal should be melted to form a convex bead. The torch is moved forward and the welding rod is again placed in the weld pool when it touches the outside edges of the joint. This operation is repeated until the joint is completely welded. Figure 6-28 shows a macrograph of a completed square-groove outside corner joint made with welding rod.

6.11 INSIDE CORNER AND T-JOINT WELDING

This is a fairly easy exercise to perform, but one in which the welder may find it difficult to obtain sufficient penetration. Two pieces of stock are placed with their surfaces at a right angle to each other, either in an L-formation (corner joint) or in an inverted "T" shape (T-joint). The inside corner joint or T-joint is welded as shown in Figure 6-29. When welding in this position, the torch flame is placed close into the corner. It is difficult to obtain additional oxygen from the air to complete the combustion of acetylene when welding in a corner. Therefore, it is sometimes necessary to open the torch oxygen valve slightly to provide enough oxygen for complete combustion. This adjustment would result in an oxidizing flame under normal circumstances, but in this special case, it produces a neutral flame. The exercise may be set up in two different ways.

The preferred method is to have the two pieces set at an angle of 45° from the horizontal, forming a trough as shown in Figure 6-30. This is the flat (downhand) welding position. The welder will find that the exercise is easier with the specimen set up in this manner. This position means that the joint is in a 45° position, and that the face (top) of the weld is in the flat welding position. The two pieces should be tacked at points along the joint before making the weld.

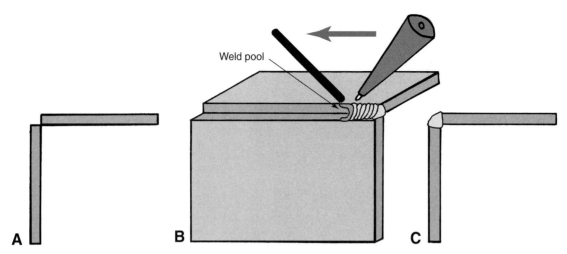

Figure 6-27. *Recommended steps when welding an outside corner joint using a welding rod: This arrangement allows the welder to make a V-groove weld and practice using the keyhole method. A—Metal is in position. B—Weld is in progress. C—Weld is completed.*

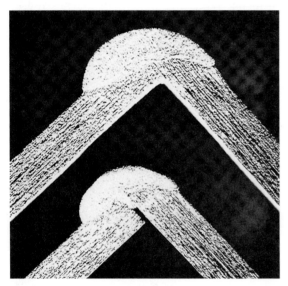

Figure 6-28. *Cross sections of welded square-groove outside corner joints (a 4X macrograph). The weld at the top was tightly fitted prior to welding. The weld at the bottom has a space between the metals which creates a weaker joint. The metal has been etched to emphasize grain lines and fusion.*

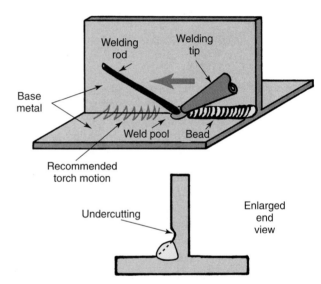

Figure 6-29. *Fillet welding on a T-joint in the horizontal welding position. Note the lack of filler metal on the vertical piece. This is a welding defect called undercutting. The recommended inclined half-moon torch motion will help eliminate undercutting.*

A second method is to have one piece horizontal and the other vertical. The weld in this joint will be made in the horizontal welding position. The angle of the weld face in this position is considered the horizontal welding position by definition. Before welding the pieces in this position, the welder should tack weld the joint to keep it in line while welding. It will not be necessary to provide for expansion and contraction since the metals are pulled one against the other as the weld cools and solidifies.

The technique of handling the torch and welding rod will be almost the same in this exercise as in the lap weld. The welder will find that very little torch motion is necessary. It is very important to produce a weld pool before attempting to add the welding rod; otherwise, insufficient penetration will result. To secure fusion at the *root* (bottom) of the weld, the torch should be held as close as possible without allowing the inner cone of the flame to touch the metal. The torch is then drawn back slightly when adding the welding rod. Before adding the filler metal in an inside corner joint, both surfaces must be melted. The surfaces are molten when the edges of the weld pool appear to run ahead, forming a "C" shape. The welding rod should be placed into the molten metal at the time that the C-shaped pool forms. See Figure 6-31. After the weld is completed, inspection should show a good bead, good fusion, consistent width, and a clean weld appearance. The weld must also have equal distribution, meaning that half of the weld is on one piece, and half on the other. In addition, neither piece of base metal should show any signs of

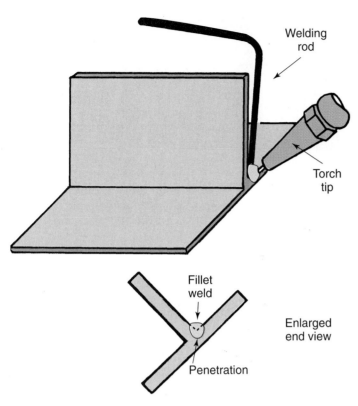

Figure 6-30. *Fillet welding of a T-joint in the flat welding position.*

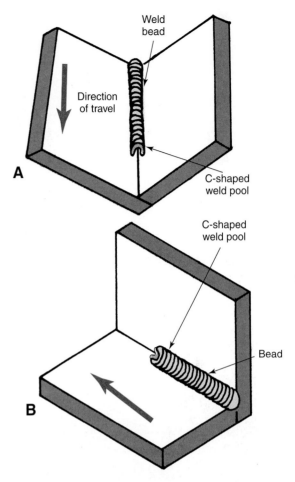

Figure 6-31. *Welding an inside corner joint. At A, the joint is set up in the flat welding position. At B, the joint is set up in the horizontal welding position. The C-shaped weld pool must be seen to ensure that both surfaces are melted before the filler metal is added.*

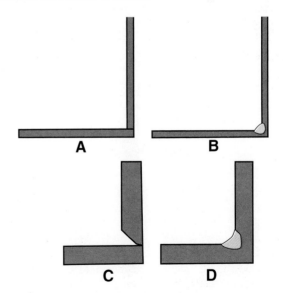

Figure 6-32. *End views of some popular corner joint preparations for fillet welds. A—Thin metal corner joint in position for horizontal fillet welding. B—Completed fillet weld on a thin metal corner joint. C—A single-bevel-groove inside corner joint on thick metal. D—A completed fillet weld on a single-bevel-groove outside corner joint.*

being *undercut.* Undercut means that the base metal is depressed at the toe of the weld bead, which leaves the metal thinner and weaker.

Use an inclined, half-moon torch motion when making an inside corner or T-joint in the horizontal welding position. This half-moon motion will help to eliminate undercutting on the vertical surface, as shown in Figure 6-29. Stop momentarily at each end of the weld pool. The inclined, half-moon torch motion will help to push the molten metal into the possibly undercut areas. In addition, adding the welding rod into the weld pool on the vertical surface will help prevent undercutting.

To test the weld for penetration, the two pieces of metal are closed together like the pages of a book. This may be done by striking the weldment with a hammer while holding it in a vise. **Caution: A metal screen or heavy material barrier should be installed around the vise whenever an attempt is made to break a weld. The screen will prevent broken metal pieces from flying into the shop area.** Lack of penetration, or fusion, is indicated by the metal peeling away from the base metal. Figure 6-32 shows some alternate setups for inside corner joint designs. Figure 6-33 illustrates some typical T-joint designs.

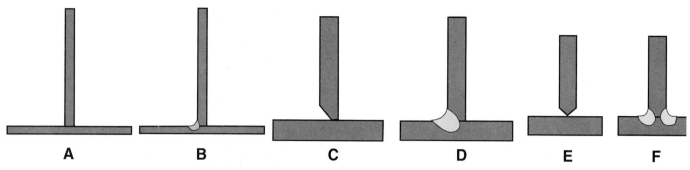

Figure 6-33. *End views of some popular T-joint preparations for fillet welds. A—A thin metal T-joint. B—A completed fillet weld on a thin metal T-joint. C—A prepared single-bevel-groove T-joint on thick metal. D—A completed fillet weld on a single-bevel-groove T-joint. E—A prepared double-bevel-groove T-joint on thick metal. F—Completed fillet welds on a double-bevel-groove T-joint.*

6.12 WELDING POSITIONS

Welds made in the flat welding position are the easiest to make. However, it is often necessary to make welds in various other positions. The recognized positions are shown in Figure 6-34.

- *Flat welding position.* The position used to weld from the upper side of the joint at a point where the weld axis is approximately horizontal and the weld face lies in an approximately horizontal plane.

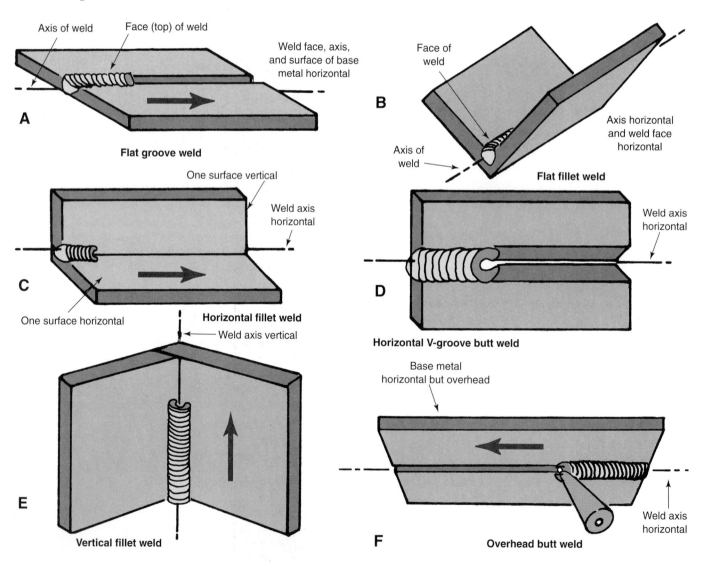

Figure 6-34. *Welding positions. Notice the position of the weld axis, weld face, and base metal surfaces in each example.*

- *Horizontal welding position* (Fillet weld). The position in which the weld is made on the upper side of an approximately horizontal surface and against an approximately vertical surface.
- *Horizontal welding position* (Groove weld). The position in which the weld face lies in an approximately vertical plane and the weld axis at the point of welding is approximately horizontal.
- *Vertical welding position.* The position in which the weld axis, at the point of welding, is approximately vertical, and the weld face lies in an approximately vertical plane.
- *Overhead welding position.* The position in which welding is performed from the underside of the joint.

6.12.1 Oxyfuel Gas Welding in the Horizontal Welding Position

This position consists of welding a horizontal joint on a vertical surface, as shown in Figure 6-35. Welding in this position should be practiced on butt joints, inside corner joints, and lap joints. A precaution to be noted: to produce an excellent weld, the torch tip should point at a slightly upward angle, instead of pointing directly along the weld joint. This tip position will permit the force from the velocity of the gases to keep the molten metal from sagging. The welding rod should be placed into the upper edge of the weld pool. This will result in a more evenly distributed bead.

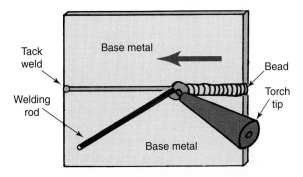

Figure 6-35. *Welding a horizontal butt joint weld. The base metal surfaces are vertical. Note that the torch flame is directed upward to help hold the molten metal in place.*

It will be found that, with practice, it is relatively easy to obtain consistent penetration in the horizontal and vertical welding welding positions. The welder will have little difficulty in obtaining a good-looking and acceptable bead. However, beads made in the horizontal and vertical welding positions generally will not have as good an appearance as welds made in the flat welding position.

6.12.2 Oxyfuel Gas Welding in the Vertical Welding Position

Vertical welding consists of welding in a vertical direction on a vertical surface. Exercises should be per-

formed on various weld joints including the butt joint, Figure 6-36. After the first few attempts, vertical welding will not be difficult if the following precautions are observed.

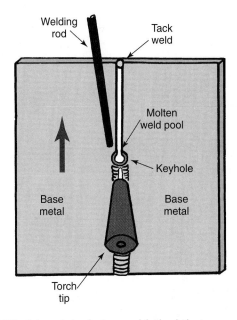

Figure 6-36. *A butt joint being welded while in a vertical position. Note the position and angle of the welding rod and torch tip with respect to the weld.*

The weld should generally proceed upward. The torch is inclined from the surface of the metal at an angle of approximately 15° to 30°. The torch tip is pointed upward. This tip position enables the force from the stream of gas to keep the molten metal from falling or sagging downward due to gravitational pull.

A very small torch motion should be used, so that the gas stream will keep the molten metal continually in position and not allow it to sag.

The weld should be made in the same manner as a flat weld. The torch and the welding rod angle from the base metal surface is the same as when flat welding. The welding rod should be inserted into the upper edge of the molten weld pool. As the weld proceeds, the bead width, fusion, and penetration should be continually checked, just as it is when making a flat weld.

6.12.3 Oxyfuel Gas Welding in the Overhead Welding Position

Overhead welding is considered, by many welders, to be the most difficult welding position. The welder should be skilled in welding in all other positions before starting overhead welding as shown in Figure 6-37. This requires consistent, diligent practice to reach a satisfactory degree of skill. **The welder's body, hands, and head should be protected with fire-resistant work clothes, a cap, and gauntlet gloves. He or she should wear high-top shoes or**

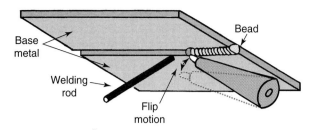

Figure 6-37. *A butt joint being welded while in the overhead welding position. Note the flip motion used to control the temperature of the weld pool. When welding overhead, wear gloves, cap, and long sleeves. Be sure all clothing is fire-resistant.*

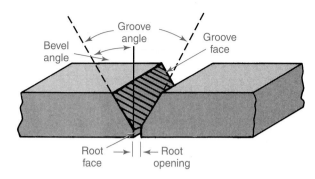

Figure 6-38. *Standard terms for the various parts of a V-groove butt joint.*

welding spats when performing this exercise. Shirts, trousers, and coveralls should have covered pockets. There should be no cuffs on the legs of trousers or coveralls. Shirts or coveralls should be buttoned at the collar to prevent hot metal from entering.

Overhead welding should be practiced on all the standard joints. There should be at least two excellent samples of each type of joint before practice welding is concluded. The exercise should be mounted approximately 6" to 12" (150mm to 300mm) above the welder's head to be most comfortable, and the welder should stand to one side of the seam, welding parallel to his or her shoulders.

The position of the welding rod and the torch is almost the same in reference to the weld pool as when welding in other positions. The force of the gas velocity in the torch flame, plus the *surface tension* (the attraction of the molecules for each other) of the metal in the pool, helps to overcome the pull of gravity on the molten metal. However, be careful to keep the metal as close to its lowest flow temperature as possible. Any superheating of the metal produces a more fluid condition and may cause the molten metal to fall.

One method of keeping the molten weld pool from becoming too hot is to use a flip motion of the torch tip. The flip motion is a quick movement of the tip from the weld pool and back again. The momentary removal of the flame allows the pool and bead to cool slightly. Control of the weld pool and good timing when inserting the welding rod into the leading edge of the molten pool will result in welds of quality comparable to flat-position welds.

6.12.4 Joint Preparation for Plate Welding

Welders should be able to successfully weld both thin metal (sheet) and thick metal (plate). The American Welding Society has standardized the names of the joints and the names of those parts of the joint that are in the weld zone. Figure 6-38 shows standard terms for various parts of a groove weld.

Several edge preparation designs have been developed to make the edges of thick metal sections ready for welding. The edges are prepared in these ways to ensure complete penetration. Figure 6-39 shows the edge preparation for various types of butt joints. Square, bevel, and V-groove joint preparations are the most common. The bevel or V-groove may be ground, machined, or flame-cut.

J-groove or U-groove preparation may be machined or flame-cut (gouged) with special cutting tips. The bevel and V-grooves are more economical in respect to quantity of welding rod metal used, gas or electricity cost, and in time saved. Figure 6-40 shows some suggested angles and measurements used when preparing metal edges for welding a groove-type butt joint.

Welding rod motion and torch motion must be used in a coordinated fashion when oxyfuel gas-welding thick plate joints. Figure 6-41 shows how the welding rod is moved to one side of the weld pool as the torch is brought to the other side. The oscillating motion is then reversed. This permits excellent control of the large weld pool and provides adequate fusion and buildup.

6.12.5 Backhand Welding

Backhand welding, as shown in Figures 6-42 and 6-43, requires that the torch be held at an angle of 30° to 45° with the work, and the flame directed back over the portion of the work that has been welded. Directing the flame in this manner tends to anneal the completed weld, relieving the welding stresses to a great extent. In addition, the backhand direction of the flame tends to help the welder form a higher bead and attain good weld penetration. Backhand welding is commonly used in welding cast iron and when welding thick sections of other metals. Relieving the stresses created by welding is necessary in these heavier sections. In backhand welding, the continued spread of the torch flame on the portion just welded tends to maintain a rather large pool of molten metal. So that the edge of the pool may solidify into a bead, it is necessary to move the flame upward (flip motion), at frequent regular intervals. This action allows the edge of the weld pool to cool slightly and solidify.

6.12.6 Multiple-Pass Welding

Welding a number of thin *passes* (layers of weld), one on top of another, is known as *multiple-pass welding.* Multiple-pass welding is well-suited to making welds on thick metal. It also may be used on thin sections where necessary.

The determining factor of whether to use a single pass or multiple passes is the thickness and width of the weld

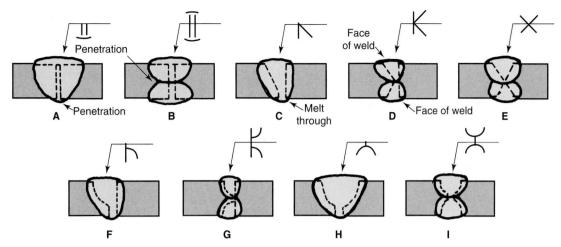

Figure 6-39. *Methods used to prepare metal edges before butt welding. The method used depends on metal thickness and whether it can be welded from both sides. A—Square-groove joint welded from one side. B—A double-square-groove joint welded from both sides. C—A welded single-bevel-groove joint. D—A double-bevel-groove joint welded from both sides. E—A double-V-groove joint welded form both sides. F—A single-J-groove joint welded from one side. G—A double-J-groove joint welded from both sides. H—A single-U-groove joint welded from one side. I—A double-U-groove joint welded from both sides. See Chapter 3 for an explanation of the weld symbols shown.*

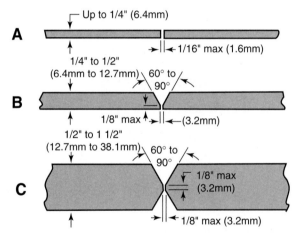

Figure 6-40. *Suggested angles and dimensions for butt joints.*

pool that a welder can handle. Also, if a pass is too thick, impurities may not be able to rise to the surface before the weld pool cools.

The layers of the weld may be made continuously for the full length of the joint, or may be done by the "step" method illustrated in Figure 6-44. The continuous-pass, multiple-layer weld is generally preferred.

6.13 APPEARANCE OF A GOOD WELD

The weld is usually inspected by careful visual examination. Some inspectors use a magnifying glass of two- to ten-power magnification.

The weld should be of consistent width throughout its length. It should be straight, so the edges form two parallel lines.

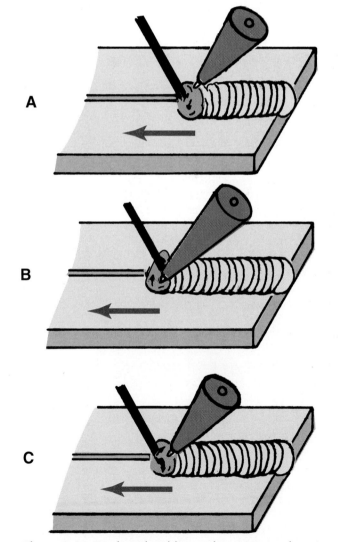

Figure 6-41. *Torch and welding rod positions and motions used when welding thicker metals where wide weld pools are required. The flame tip and welding rod are on opposite sides of the weld pool. Their positions are continually changed as the weld proceeds.*

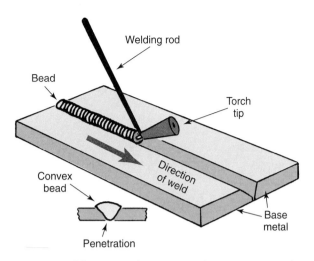

Figure 6-42. *Welding a single-V-groove butt joint using the backhand method. Note that the torch tip points away from the direction of the weld progress.*

The weld should be slightly *crowned* or convex (built up above the surface of the parent metal). This crown should be consistent (even). A gauge is sometimes used to check the weld contour.

The weld should have the appearance of being fused into the base metal, have a blended appearance, and not have a distinct edge between it and the base metal. The surface of the weld should have a ripple throughout its length. The ripples should be evenly spaced.

The weld should have a clean appearance. There should be no color spots, no scale on the weld, and no rough pitted appearance to the weld. Lap welds and corner welds should normally show no visible penetration on the side opposite the bead. On a butt weld, turn the specimen over and check the penetration. The degree of penetration will be indicated by the sag of the lower surface of the weld. The amount of sag depends on the thickness of the metal and the width of the molten pool maintained while welding. Complete penetration and fusion is hard to

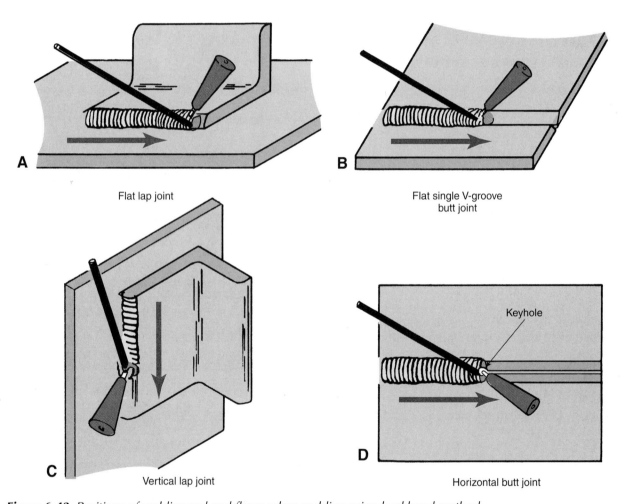

Figure 6-43. *Positions of welding rod and flame when welding using backhand method.*

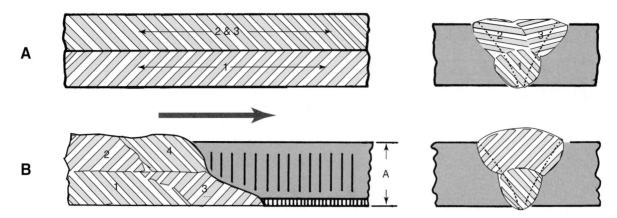

Figure 6-44. *The multiple pass and "step pass" welding methods. A—Several continuous passes can be used to weld a thick piece. B—A step weld. The weld progress is in small steps from bottom to top, bottom to top, and so on.*

determine; one method of testing for penetration is to bend the weld along its length. This is done by clamping the weld in a vise so that the weld line is in line with the top of the vise jaws. The upper piece is then bent to a 90° angle or greater until it breaks. Another method to use is the guided bend test described in Heading 30.10. If the weld has not penetrated or fused properly, it will break at the edge of the joint without tearing away the base metal.

Caution: When striking metal in a vise, always place a screen around the vise to prevent pieces of metal from flying across the shop.

The welder should obtain consistent and efficient penetration and fusion. It is possible to secure a very good weld with excessive penetration, and still build the upper surfaces of the weld up to the correct height. The weld will be of sufficient strength, but a weld of this nature will *cost more* than the weld described previously. It is only necessary to produce a weld that is as strong, or a little stronger, than the original metal. It is not possible to produce sufficiently strong butt joints if full penetration is not obtained.

Figure 6-45 shows the cross section of both butt joint and lap joint welds. Note that B and D show poor penetration.

Figure 6-46 shows cross-sectional views of both good and poor lap and butt joints. In a lap joint, the main fault is lack of fusion along the toe of the bead, as shown at B. To correct this problem, direct more heat toward the bottom metal. The overlap at C is caused when too much welding rod is added to the weld pool.

Figure 6-47 shows severe undercutting in cross section. On an inside corner weld, this condition should be improved by stopping momentarily at the edges of the weld. Undercutting may also be decreased by placing the welding rod in the upper edge of the weld pool. Another method used to eliminate undercutting is to use a torch motion that pushes the molten pool up toward the undercutting area. See Figure 6-29. In the lap weld, undercutting is caused by poor positioning of the torch that overheats the weld metal. It is also caused by improper torch movement, too small a tip with too high a gas velocity, or by using an undersized welding rod.

A specially designed welding bench may be used to practice welding under the ideal conditions. See Figures 6-48 and 6-49.

6.14 SAFETY IN OXYFUEL GAS WELDING

Oxyfuel gas welding equipment is safe if it is properly used. It does, however, possess great potential destructive power if carelessly used. Therefore, it is important that the welder be familiar with all of the potential dangers in the welding processes. Most of the safety hazards of oxyfuel gas welding were pointed out as the operation of the equipment was explained in this chapter. **Welders should know all these precautions, and follow**

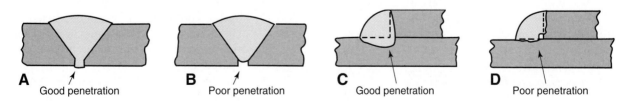

Figure 6-45. *Cross sections of welds. A and B—Examples of butt welds. C and D—Examples of lap joint welds. Note in B and D that the fusion has not been completed to the root (bottom) of the joint.*

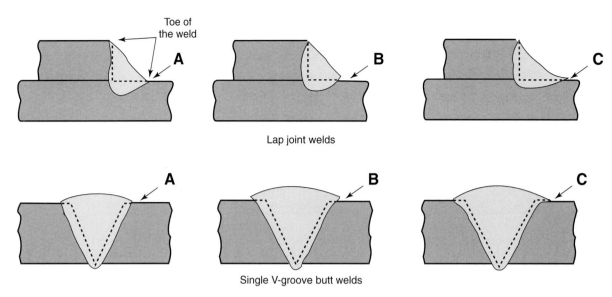

Figure 6-46. *Cross sections of good and poor butt and lap welds. A—Properly made welds with proper fusion at the toe (edge) of the weld. B—Poor fusion at the toe of the weld. C—Poor fusion and overlapping of filler metal at the toe of the weld.*

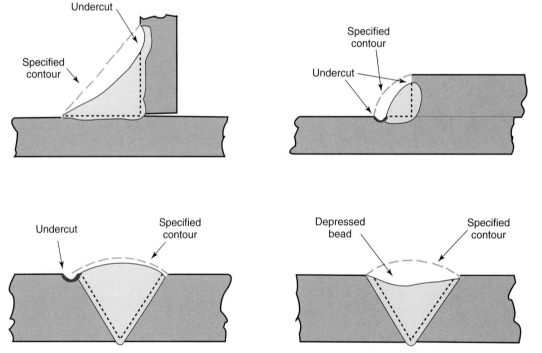

Figure 6-47.. *Cross sections of finished welds, showing some typical faults that may occur. The shape of the bead contour is usually specified. When undercutting occurs, or if the face of the weld is not built up to specifications, the welded joint is weakened.*

them for their own safety, for the safety of fellow workers, and to protect the equipment. The following precautions are reviewed, and elaborated on, in the interest of safety:

6.14.1 Protective clothing and shields

- Wear goggles. Various shades are required for various welding applications. When welding or cutting heavier metals, a darker (higher-numbered) shade of lens is required.
- Wear a pair of leather welding gloves. Gauntlet type gloves are recommended.

- For overhead welding, wear a cap and a leather apron and cape or a leather jacket. Leathers should also be worn whenever there are greater than normal hazards from molten metal and sparks.
- Remove combustibles, especially matches, from pockets.
- Avoid wearing trousers with cuffs and open pockets into which sparks may fall. Avoid wearing oily articles of clothing.

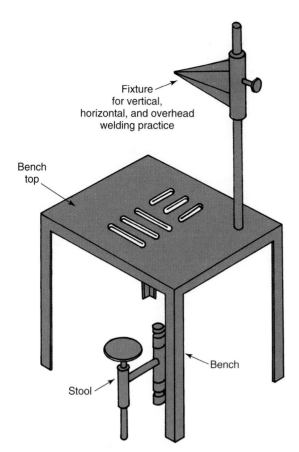

Figure 6-48. *A well-planned welding bench or table. Note the built-in stool and the facilities for holding work when welding in the overhead welding position.*

Figure 6-49. *A well-designed bench enables a person to practice welding in any position and at various heights. A—A clamp used for holding welding exercises flat, horizontal, or vertical. B—Height adjustment that enables positioning of a welding exercise at various heights above the bench. C—This fixture may be removed from the bench when necessary. (Greene Mfg., Inc.)*

- Use only fireproof materials to support articles being welded or cut.

6.14.2 Welding or cutting of tanks and containers

- Do not weld or cut a tank or container, unless the device has been processed to be safe for such operations. See the American Welding Society recommendations for procedures to be followed in preparing for welding and/or cutting certain types of containers that have held combustibles.
- Remember, many substances that are not usually considered flammable or explosive become vaporized (and therefore, explosive) when heated to a high temperature.

6.14.3 Handling oxygen and fuel gas cylinders and equipment

- All equipment used for oxygen, such as cylinders, regulators, valves, and torches, must be kept free of oil.
- Check for leaks, using soap and water. Never use gas welding equipment with leaks.

- Open oxygen and fuel gas cylinder valves slowly. Open oxygen cylinders fully when in use, to eliminate possible leakage around the cylinder valve stem.
- Purge oxygen and fuel gas valves, regulators, lines, and torch passages before use.
- Most fabrics that have been saturated with oxygen will burn easily, rapidly, and fiercely. Such clothing should not be worn for at least 30 minutes after the oxygen source has been removed.
- Support pressurized cylinders so they cannot tip over. If a valve is broken off an oxygen cylinder, the cylinder can become a rocket propelled with tremendous force. A fuel gas cylinder can behave like a flame thrower if the cylinder valve is broken off and the gas is ignited.
- When opening the cylinder valve, stand to one side of oxygen fuel gas regulators.
- Always call oxygen "oxygen." It should never be called "air." Call acetylene "acetylene," not "gas." Identify the content of cylinders by the name marked on the cylinder. If a cylinder is unnamed, do not use it.
- Keep all cylinders away from exposure to high temperatures. Remember that the pressure in a cylinder increases with the temperature.

- Store full and empty cylinders separately in a well-ventilated space.
- Mark empty cylinders "MT" with chalk as soon as they are taken out of service.
- If a cylinder leaks around a valve or a fuse plug, tag it to indicate the fault. Move the cylinder to a safe area, and immediately notify the supplier to pick it up.
- Keep cylinder caps screwed on all cylinders that are not in use, particularly while they are being moved.
- Keep the regulators in good repair. Do not use a leaking (creeping) regulator.

6.14.4 The flame

- If the torch is to be laid down, the flame must be turned off.
- Be sure that no combustible material is in the area where an oxyacetylene torch is to be lighted.

6.14.5 Handling the oxyfuel gas welding station

- Purge (blow out) the cylinder valves before attaching the regulators. **Caution: When purging fuel gas valves, regulator, and hoses, be sure no open flames are near, because fuel gases are flammable.**
- Use a well-fitting wrench to attach regulators and hoses.
- Purge the hose before attaching a torch.
- Fuel gas hose fittings generally use left-hand threads. A groove on the *periphery* (around the sides) of the acetylene hose nut indicates that it has a left-hand thread. Oxygen hose fittings have right-hand threads. Do not interchange fittings on hoses.
- Use a spark lighter to light a torch.
- Keep all hoses away from oil and grease. Examine the hoses regularly for leaks. Leaking fuel gas may cause a severe explosion or fire. If a hose is burned or damaged by a flashback, replace it. A flashback usually burns the inner wall of the hose and makes it unsafe to use. Never repair a hose by binding it with tape.
- The adjusting screw on the regulator must be turned out (counterclockwise, regulator closed) before opening the cylinder valve. Open the cylinder valve slowly. Generally, the fuel gas cylinder needs to be opened only 1/4 to 1/2 turn. The wrench should be left on the valve stem to permit quick closing of the valve in an emergency. Adjust oxygen and fuel gas regulator pressures as recommended by the manufacturer of the torch.
- **Never use acetylene pressure in excess of 15 psig (103.4kPa).**

- Backfires are generally caused by a dirty tip, an overheated tip, or insufficient working pressures. After correcting the condition, the torch may be relighted.
- A flashback is a burning back of the flame into the tip, torch, hose, or even into the regulator. It is characterized by a squealing or sharp hissing sound and by a smoky or sharp-pointed flame. **In case of a flashback, immediately extinguish the flame by first closing the torch oxygen valve, then the torch fuel gas valve. Also close the cylinders and regulators. A flashback indicates something radically wrong with the equipment. Before the outfit is used again, it must be thoroughly examined for damage and the cause of the flashback identified and corrected.**
- It is recommended that flashback arrestors and check valves be installed between the hoses and the torch valves or on the regulators or in both locations.

The *Occupational Safety and Health Administration* (OSHA) has established many compulsory safety requirements for the welding industry. The safety precautions in this text will conform to OSHA requirements.

6.14.6 Metal Fume Hazard

When heated, many metals release irritating and toxic fumes. Metals that release the most dangerous fumes are: cadmium, zinc, lead, and beryllium.

Cadmium is widely used as a rust-protective plating over steel. It is white in color. **The fumes generated by welding, brazing, or cutting cadmium-coated metals are very dangerous. If it is necessary to perform welding operations on cadmium-coated metals, very thorough ventilation must be provided.** Cadmium coatings may usually be identified by the fact that the metal, when gently heated with a torch, will turn to a yellow-gold color.

Zinc is widely used in die-cast metal parts. Galvanized sheet metal is an example of a zinc-coated metal. Zinc is white in color and melts at a rather low temperature, about 787°F (419°C). When zinc is alloyed with other metals, this melting temperature may be either raised or lowered. **When heated by a welding torch or an electric arc, zinc gives off a white vapor that is very irritating to the respiratory system. Very thorough ventilation must be used when welding or heating zinc parts or zinc-coated materials.**

Lead is not as widely used as cadmium or zinc. However, some plating tanks are lead-lined. Lead is also used for tanks and pipes for handling certain liquids and gases. Most electric storage batteries (like those used in automobiles) have lead plates, posts and cell connectors. Some paints may contain lead oxide pigments. Many solders contain lead in varying amounts.

When heating any substances containing lead, thorough ventilation is required. Lead poisoning may occur without the welder being warned in any way of the dan-

ger, since its fumes are not irritating. The human body does not appear to be able to throw off lead taken into the body through either the lungs or the digestive system. Therefore, a person exposed to lead fumes or lead in any form may slowly build up an accumulation of the metal until acute lead poisoning develops.

Beryllium has a high modulus of elasticity. It is used in aerospace when metals which are light but stiff are needed. It is also used in the atomic energy field. Its vapors are toxic. Excellent ventilation must be provided when beryllium is welded.

6.14.7 Welding Regulator Safety

The regulator is perhaps the single most important device contributing to safety in oxyfuel gas welding. If handled properly, its safety qualities will be preserved. See Heading 5.4 for the operation and use of regulators. If the regulator is abused, it may fail to perform its safety function. For maximum useful life and safety from a regulator, observe the following rules:

- Clean the cylinder passages prior to fitting a regulator to a replacement cylinder. This is done by slightly and quickly opening and closing the cylinder valve (cracking the cylinder valve). **Aim the cylinder valve opening away from other workers and any open flames during this operation**

- Examine the condition of the threads on both the regulator and the cylinder fittings. The regulator fitting should screw on the cylinder valve easily. Have the fittings repaired rather than use great force to assemble them. **Note: Fuel gas cylinders have left-hand threads. All oxygen cylinders have right-hand threads. Regulators, hoses, torches, etc., have left-hand threads for fuel gas and right-hand threads for oxygen.**

- Use the proper wrench to tighten the regulator fitting to the cylinder. Be sure that the wrench fits the regulator nut properly. **Never use common pliers or pipe wrenches**

- Be sure that the regulator adjustment is turned all the way out before opening the cylinder valve. In this position, no gas will flow through the regulator into the low-pressure side.

- Open the cylinder valve slowly. A full cylinder, opened quickly, may cause a high-enough pressure and a resulting heat of compression to damage the regulator. The temperature at the regulator seat may approach 1000°F (538°C) and the seat could fail. If the full cylinder pressure is allowed to enter the regulator too rapidly, the gauge may be damaged or burst. This may cause the gauge glass to blow outward.

All safety-approved pressure regulators incorporate a safety disc on the low-pressure side. This disc is designed to burst at a pressure below the safe pressure limits of the regulator. This is generally below the pressure at which the diaphragm would burst. If the nozzle- or stem-type valve should leak severely, this safety device will keep the regu-

lator diaphragm from bursting. However, even with all these safety devices, regulators have exploded!

If the regulator valve seat is in poor condition or leaking, the heat of friction along with the heat of compression may cause the regulator seat to melt or decompose. **Do not stand in front of a regulator as the cylinder valve is opened, even when opening it very slowly.**

- **After setting the working pressures, check the gauges to ensure that the pressure does not continue to increase. If it does, this indicates that the regulator seat is leaking. The outfit should be shut down immediately or the gauge may rupture.**

- The pressure to be used in a gauge should never be greater than one-half to two-thirds of the maximum calibration of the dial. This means that if the gauge is calibrated up to 300 psig, a 200 psig reading should be the maximum used.

- Teflon® tape or a paste made of glycerine and litharge should be used to seal the threads that connect the gauge to the regulator. Do not use pipe compounds containing oil.

- **Never use acetylene at a gauge pressure above 15 psig (103.4kPa).**

- Never use oil or petroleum-based products on oxyfuel gas equipment. Do not allow oil or grease to come in contact with any part of oxyfuel gas welding equipment. Never work with welding equipment when wearing greasy or oily gloves or clothing.

- Test for leaks *only* with a soap and water solution.

- Never interchange oxygen and fuel gas regulators or gauges.

- Around electrical equipment, insulate the cylinders with wood or rubber to prevent grounding. A grounded cylinder may allow an electrical arc to form and the resulting heating or burning of the cylinder may cause an explosion.

6.14.8 Safety When Handling Torches

The following are some pointers concerning safety when handling torches. These recommendations concern safety for both the welder and the equipment:

- Do not put a cold tip in a hot tip tube; the hot tip tube will produce a shrinking action on the tip as it cools.

- Use only a clean wood surface or a leather surface to clean the end of a tip. Keep the oxygen flowing during this operation to prevent plugging the orifice.

- Be careful when cleaning a tip with a tip cleaner. Do not increase the size of the orifice or to cause it to become out-of-round or tapered.

- **Always extinguish a torch before allowing it out of your hands.**

- Torch hand valves should be turned only with the fingertips.

- If a torch backfires, find the trouble and remedy it before continuing to use the torch.
- Each welding station should be provided with a hook upon which to hang the unlighted torch.
- Be careful that a torch is not directed toward another person while it is being lighted.
- Be sure no flammable material is near the welding station.
- If a flashback occurs, quickly close the oxygen torch valve and then the fuel gas torch valve. The oxygen cylinder and fuel gas cylinder and regulators should then be closed. The welding or cutting outfit should not be used again until the cause for the flashback has been corrected.

TEST YOUR KNOWLEDGE OF SAFETY

Write your answers on a separate sheet of paper. Do not write in this book.

1. Why must the regulator adjusting screws be turned out until loose before opening the cylinder valves?
2. Name at least two things that must be done with cylinders to protect them from accidents or harm.
3. The welder should stand to one side of the regulator and gauges when the cylinder valve is turned on. Why?
4. What is the maximum gauge pressure that may be set for acetylene?
5. Why is it necessary to use a spark lighter or gas economizer flame and *not* a match to light the oxyacetylene flame?
6. Why must the oxygen and acetylene regulator, hose, and torch passages be purged before lighting the torch?
7. What is flashback? What must be done immediately if a hissing or squealing sound is heard?
8. Good ventilation is always required when welding. Name four metals that require especially good ventilation when being welded.
9. What personal precautions must a welder take when welding in the overhead welding position?
10. When the welding rod melts down, what procedure should be followed in order to change the position of your hand on the rod?

TEST YOUR KNOWLEDGE

Write your answers on a separate sheet of paper. Do not write in this book.

1. Give a definition of the term "welding."
2. What size torch tip and welding rod is suggested for welding 1/8" (3.18mm) steel?
3. What is meant by undercutting a welded joint?
4. What gases are produced by the combustion of the oxygen-acetylene flame?
5. What is the approximate temperature of the oxyacetylene flame?
6. Does oxygen burn?
7. Name the four main parts of an oxyfuel gas welding station.
8. What should be used when checking an oxyfuel gas station for leaks?
9. What is a number drill?
10. How are the sizes of welding tip orifices designated?
11. What are two popular types of welding torches?
12. List the steps required for starting an oxyacetylene station and lighting the torch.
13. What is meant by purging the system?
14. List, in proper order, the six steps to be followed when you are shutting down an oxyacetylene welding station.
15. How does the weld pool indicate the penetration being obtained?
16. What is the approximate angle between the welding tip and the weld when welding flat stock in the flat welding position?
17. How is welding rod metal added to a weld?
18. Name three things that may cause a torch to "pop" or backfire.
19. What is the position of the torch in forehand welding? Backhand welding?
20. Name four basic types of welded joints.
21. Name or illustrate four methods used to compensate for shrinkage when welding a butt joint.
22. When welding an inside corner joint, a flame adjustment of a slightly oxidizing nature is used. Why?
23. What are the four recognized welding positions?
24. How does the welding position affect torch position?
25. How may undercutting be prevented on an inside corner joint that is welded in the horizontal welding position?

Chapter 7

OXYFUEL GAS CUTTING EQUIPMENT AND SUPPLIES

LEARNING OBJECTIVES

After studying this chapter, you will be able to:
* Identify the various components of an oxyfuel gas cutting outfit.
* Cite the maximum safe working pressure used for acetylene gas.
* Describe the equipment used to semiautomate or fully automate the cutting process.

Oxyfuel gas cutting deals with high pressures, flammable gases, flying sparks, possible rough handling, and other severe services. Therefore, the equipment must be reliable and rugged, and yet quick to respond to changes in demand.

The design and construction of an oxyfuel gas cutting outfit is similar to an oxyfuel gas welding outfit. There are major differences, however, in the design of the torch and in the oxygen regulator for heavy-duty cutting.

For safety, quality of results and economical operation, it is very important that the welder become thoroughly familiar with the design and construction of this equipment and its correct use.

7.1 COMPLETE PORTABLE OXYFUEL GAS CUTTING OUTFIT

The complete portable oxyfuel gas cutting outfit consists of these components:
• Cylinder truck.
• Oxygen cylinder.
• Oxygen regulator.
• Oxygen hose and fittings.
• Fuel gas cylinder.
• Fuel gas regulator.
• Fuel gas hose and fittings.
• Cutting torch.

This equipment is shown in Figure 7-1. Each part of the cutting outfit, with the exception of the cylinders, hoses and fittings, will be described in this chapter. See Chapter 5 for information on those parts that are the same as found in an oxyfuel gas welding outfit. The parts that are unique to a cutting outfit will be described in the order in which they are assembled on the complete oxyfuel gas cutting outfit.

7.2 THE CYLINDER TRUCK

The cylinders must be securely fastened in an upright position with a chain or clamping device. If the cutting outfit is portable, the cylinder truck must be designed in such a way that it is resistant to tipping. The cylinders must be securely fastened to the cylinder truck as shown in Figure 7-1. The possibility of damage to the cylinders must be reduced to a minimum.

7.3 REGULATORS FOR OXYFUEL GAS CUTTING

A standard fuel gas regulator is usable for cutting stations. However, the oxygen regulator is quite often a heavy-duty type designed for cutting applications. Since the oxygen cutting pressures sometimes reach 100 psig to 150 psig (689.5kPa to 1034kPa) and because the gas volume required may be quite high, the oxygen regulators usually have heavy-duty springs and a high capacity regulator orifice. The two-stage regulator is preferred for cutting applications. The regulator fittings used to attach the regulator to the cylinders are standard and so are the hose fittings. If heavy cutting is to be done, it is advisable to check the oxygen regulator by model number and manufacturer's specifications to be sure the correct regulator is being used. See Heading 5.4 for more information regarding regulators.

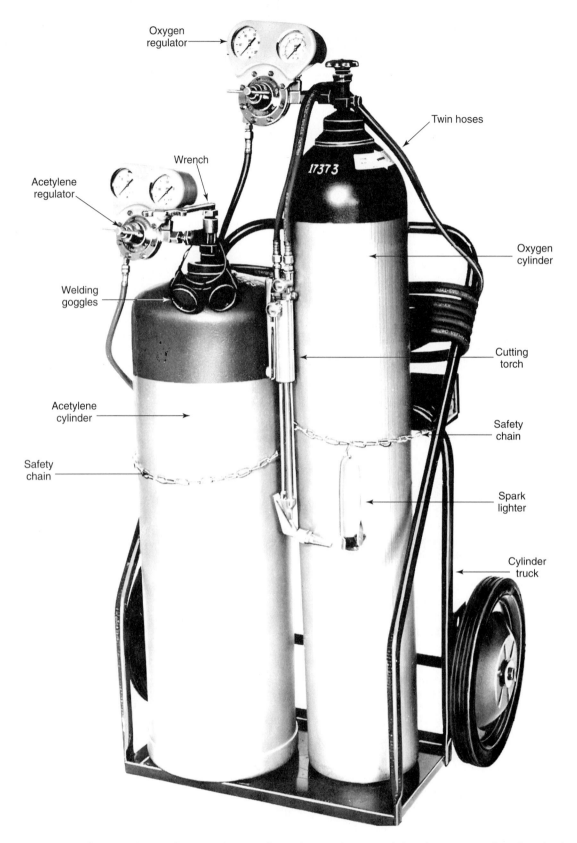

Figure 7-1. *Oxyacetylene cutting outfit. Note the guards on the regulators and the chains around each cylinder holding them securely to the cylinder truck. (Thermadyne Industries, Inc.)*

7.4 THE CUTTING TORCH

The cutting torch carries the fuel gas and oxygen in separate tubes to the mixing chamber within the torch head or torch body. In the mixing chamber, the two gases combine. The combined gases then travel to the torch preheat orifices (holes) in the tip to produce the preheat flames. The torch must also carry the cutting oxygen to a separate orifice in the tip. As it emerges from the tip, it oxidizes the metal and blows it away to form a clean *kerf* (cut).

The torch is usually constructed from a yellow brass body, stainless steel tubes that carry the gases, and a brass head or tip holder. All these parts are silver-soldered together. The copper tip is held into the torch head by a threaded nut. The torch is equipped with three valves:

- A fuel gas valve similar to the welding torch.
- An oxygen valve similar to the welding torch.
- A cutting oxygen valve, button- or lever-operated, with an automatic spring closing device.

A cutting torch may consist of an attachment that may be fastened to a welding torch body as shown in Figure 7-2, or it may be a torch designed for cutting operations only, as shown in Figures 4-4 and 7-1

The combination welding and cutting torch is most popular in small shops where welding and cutting are auxiliary operations to the work in the shop.

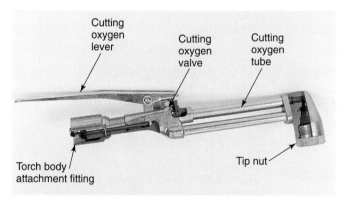

Figure 7-2. *This cutting torch attachment has been cut away to show some of the gas passages. (CONCOA).*

7.4.1 Cutting Torch Tips

Just as in welding, the proper size cutting tip is very important if quality work is to be done. The preheat flames must furnish just the right amount of heat, and the oxygen jet orifice must deliver the correct amount of oxygen at just the right pressure and velocity to produce a clean kerf. All of this must be done with a minimum consumption of oxygen and fuel gases. Careless workers, or those not acquainted with the correct procedure, may waste both oxygen and fuel gas.

Each manufacturer has cutting tips of different designs. The orifice arrangements and the copper alloy tip

material are much the same among various manufacturers. The sealing surface of the tip which fits into the torch head, however, often differs in design among manufacturers. Figure 7-3 shows a number of different tip-to-torch head sealing designs. It also illustrates a number of different orifice sizes and designs. Figure 7-4 is a table showing several different orifice arrangements and their suggested uses.

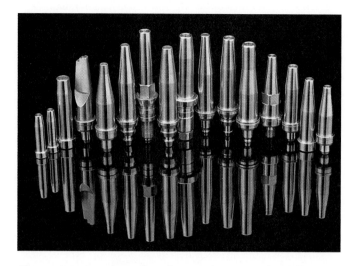

Figure 7-3. *A variety of torch tip styles. Note the different designs of the tip-to-torch-head sealing surfaces for various torch manufacturers. (The American Torch Tip Company)*

The internal construction of a cutting tip that is designed to mix the preheat gases at the tip is shown in Figure 7-5. Some cutting torches are designed to mix the preheat flame gases in the handle (body). The tips and seats are designed and constructed to produce a good flow of gases, to keep the tips as cool as possible, and to produce leakproof joints. The sealing areas are generally *metal-to-metal* between the tip and the torch head. These surfaces must be kept clean and free from damage. If these joints leak, preheat gases may mix with the cutting oxygen or may escape to the atmosphere. Figures 7-3 and 7-6 shows the details of several cutting tip seating designs. A unique torch arrangement is shown in Figure 7-7. This tip design permits various sizes and shapes of tips to be screwed onto a single torch head adapter.

It is very important that the orifices and passages be kept clean and free of burrs, to permit free gas flow, and to form a well-shaped flame. Figure 7-8 shows a cutting tip orifice being cleaned.

Cleaning a cutting tip is done in the same manner as cleaning a welding tip. With a cutting tip there may be two or more orifice sizes. Move the tip cleaner straight in and out of the hole, or a *bell-mouthed orifice* may result. Study Figure 7-9, which illustrates several tips. Tip A is in need of repair. Tips B and D are beyond repair and must be replaced. Tip C is in good condition. Figure 7-10 shows a tool used to recondition the flame end of a cutting tip. Since it is extremely important that the sealing surfaces be

In the cutaway illustration (Figure 7-2), the following labels appear: Cutting oxygen lever, Cutting oxygen valve, Cutting oxygen tube, Torch body attachment fitting, Tip nut.

Number of Preheat Orifices		Degree of Preheat	Application
	2	Medium	For straight line or circular cutting of clean plate.
	2	Light	For splitting angle iron, trimming plate and sheet metal cutting.
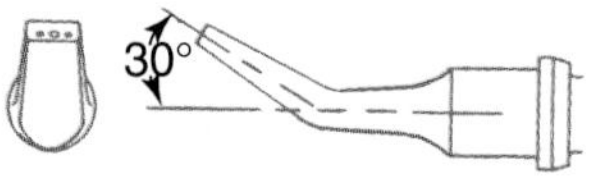 30°	2	Light	For hand cutting rivet heads and machine cutting 30° bevels.
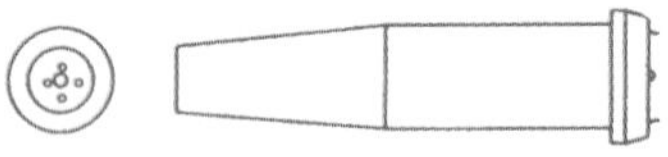	4	Light	For straight line and shape cutting clean plate.
	4, 6, 8	Medium	For rusty or painted surfaces.
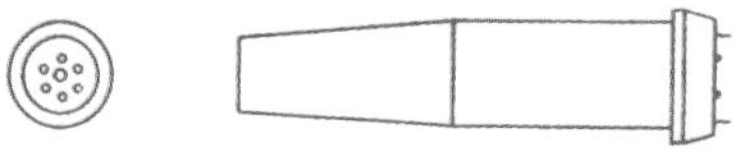	6	Heavy	For cast iron cutting and preparing welding Vs.
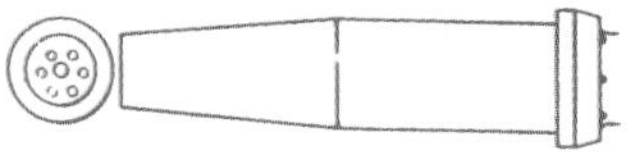	6	Very heavy	For general cutting, also for cutting cast iron and stainless steel.
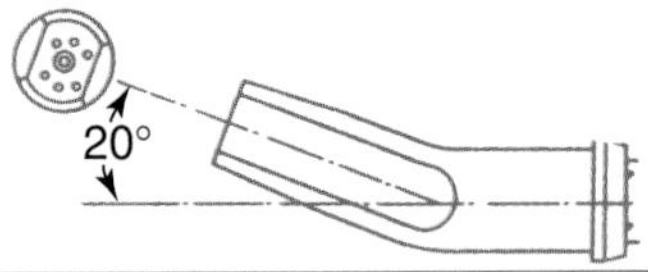 20°	6	Medium	For grooving, flame machining, gouging, and removing imperfect welds.
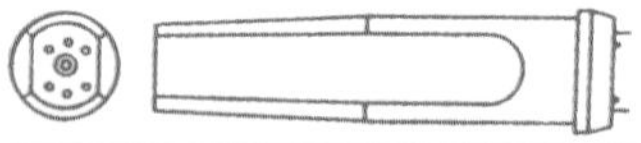	6	Medium	For grooving, gouging, or removing imperfect welds.
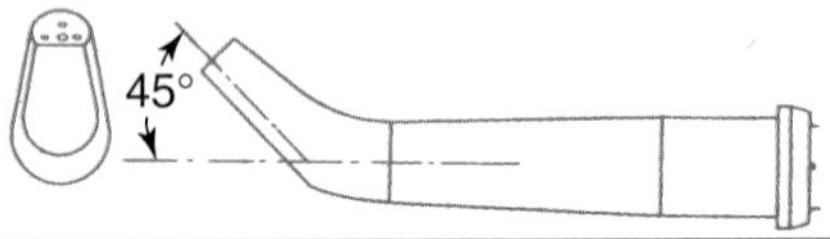 45°	3	Medium	For machine cutting 45° bevel or hand cutting rivet heads.
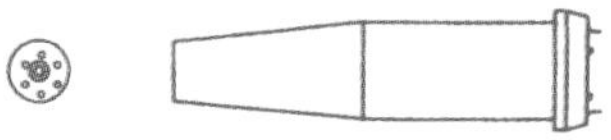	6	Heavy	Flared cutting orifices provide a large oxygen stream of low velocity for rivet head removal (washing).

Figure 7-4. *A table of some common cutting torch tip designs and their uses.*

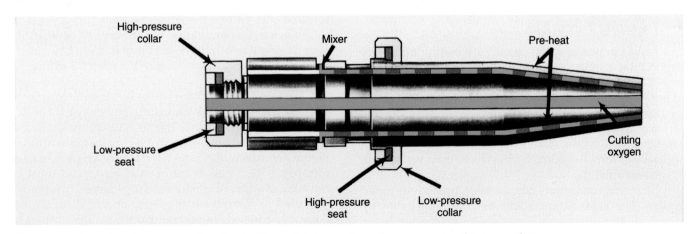

Figure 7-5. *A cutting tip cross section. Study Figure 4-4 to see how the gases enter this type of tip.*

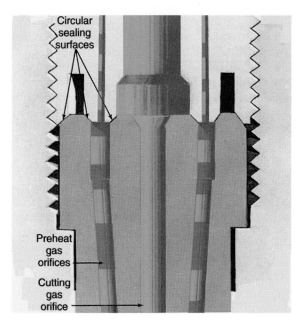

Figure 7-6. *Cutting tip cross section which shows details of the cutting tip seating design. These metal-to-metal sealing surfaces are machined.*

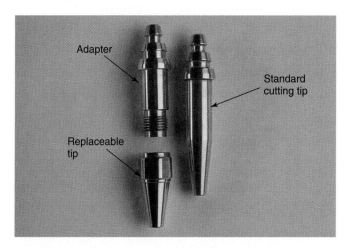

Figure 7-7. *A unique cutting tip design. A variety of tip styles can be attached to a single tip-to-head sealing adapter. (American Torch Tip Company)*

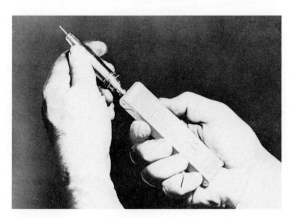

Figure 7-8. *Cutting oxygen orifice being cleaned. A different-sized wire broach may be required to clean the preheat orifices. (Thermacote-Welco Co.)*

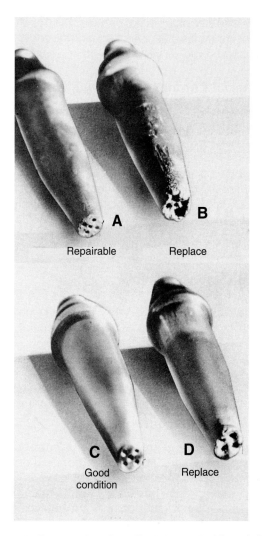

Figure 7-9. *Four cutting tips. Tip A is repairable, while tip C is in good condition. Tips B and D must be discarded.*

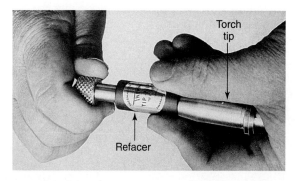

Figure 7-10. *A special tool used to reface the flame end of a cutting tip. This could be used on the tip in Figure 7-9A to make it flat again. (Thermacote-Welco Co.)*

kept clean and free of scratches or burrs, the tips should be stored in a container that cannot scratch the seats. An aluminum or wood rack is best.

7.4.2 Special Cutting and Gouging Tips

Many special cutting tips have been designed for specific applications. Special tips are made for use with various fuel gases such as acetylene, liquefied petroleum (LP), MAPP, and propane. Tips for cutting rivets or for gouging metal are other examples of special tips. See Figure 7-4 for a table of special cutting and gouging tips and their suggested uses. Local welding supply companies can provide information regarding sources for special cutting tips.

7.5 TORCH GUIDES

The welder should always try to cut a smooth kerf , and to cut accurately to a dimension. *Freehand cutting* (holding the torch in your hands) makes both of these objectives very difficult. Many mechanical, electrical, and electronic devices have been developed to help produce clean cuts, accurate size cuts, and exact duplicate pieces.

7.5.1 Mechanical Guides

Mechanical guides are used to help control the position of the torch. They do not control the speed of the cutting operation. Therefore, the welder must be very skilled; otherwise rough, ragged cuts may result. To cut straight edges, a simple torch guide may be assembled. A band-type clamp may be attached to the torch tip and a length of angle iron clamped to the base metal to ensure a straight cut, as shown in Figure 7-11.

Another popular type guide uses one or two small steel wheels mounted on the torch tip to reduce the friction. This attachment allows the welder to follow a line and maintain a constant flame distance along the cut line.

A metal template may be used when cutting several parts with an irregular shape. These templates, which may

be magnetic, are attached to the plate to be cut. The torch tip is kept in contact with the template and moved around the outside of the pattern to cut the correct shape.

To cut arcs or circles, a circle guide may be used. This device consists of an adjustable-length rod with a center pivoting point and a means of holding the tip. Figure 7-12 illustrates a circle-cutting guide. Figure 7-13 shows a circle guide in use.

7.5.2 Motorized Carriages

An *electric motor-driven carriage* and track may be used when making long straight cuts. A motorized carriage has a variable-speed motor that allows the carriage to move along its track at any desired speed. The track may be rigid, or it may be flexible to permit a curve to be followed.

The carriage typically has four wheels; one is driven by the variable-speed electric motor, two pivot to follow

Figure 7-11. *An aid to cutting straight lines with a torch. Put a 3/4" band type clamp on the cutting torch tip, as shown. Fasten the clamp so the end of the band will ride on top of a piece of angle iron clamped onto the stock to be cut. Locate the clamp so that the tip of the torch is the right distance above the metal. To cut in a straight line, hold the torch tip against the angle iron as you cut.*

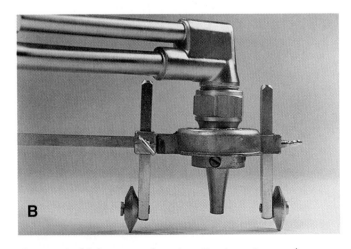

Figure 7-12. *A circle-cutting guide. A—This guide has a magnetic base to hold the center location. B—A cutting torch installed on the circle guide. Two small wheels support the torch as it moves around the circle. (Lenco-NLC, Inc.)*

Figure 7-13. A circle-cutting guide in use. Note the safety clothing worn by this welder. (Smith Equipment, Division of Tescom Corp.)

the track, and one is freewheeling. The track may be attached to a weldment by means of strong magnets. See Figure 7-14. Some carriages are designed so that they remain on their track in any position. If a welding torch is mounted on it, this type of carriage may then follow horizontally around large weldments to either cut them or weld them. Tracks also may be mounted so that the carriage and its torch can climb vertically.

The torch, either oxyfuel gas or arc, is mounted off the side of the carriage. It is mounted so that it can be adjusted inward and outward. This adjustment may be made by means of a clamped sliding bar, or a rack-and-pinion gear.

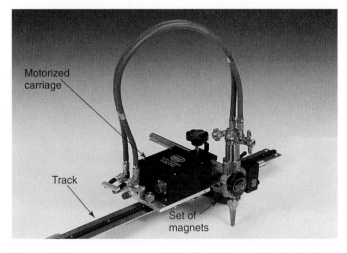

Figure 7-14. The track associated with this motorized carriage is held to the base metal with strong magnets. The track is somewhat flexible and could be attached to a large-diameter object. The carriage may carry a cutting torch or a welding torch. (Bug-O Systems, Inc.)

The torch is held at the end of the moving bar. See Figure 7-15. The tip is also adjustable by means of a rack-and-pinion gear arrangement to move it toward or away from the base metal. See Figures 7-15 and 7-16. The motorized carriage has an on/off switch, a motor speed control, a for-

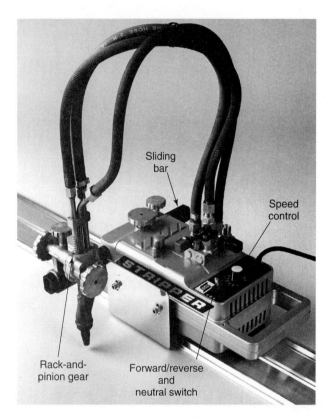

Figure 7-15. A motorized carriage. The torch is moved up and down by means of a rack-and-pinion gear. It is moved in and out with a sliding bar. (ESAB Welding and Cutting Products)

ward and reversing switch, and an engaging switch or clutch.

Figure 7-16 shows an electrically driven carriage being used on a straight track. Before using a motorized carriage, the welder should disengage the clutch and run the carriage, by hand, the full length of the cut. This is done to ensure that the hoses and electric cord to the carriage will not get caught on anything during the cut.

A unique motorized cutting torch is shown in Figure 7-17. This torch has a variable speed motor built into the torch body. Oxygen, fuel gas, and cutting oxygen valves are in standard locations. A wheel, driven by the motor in the torch body, is mounted at the torch head. The torch head may be steered by means

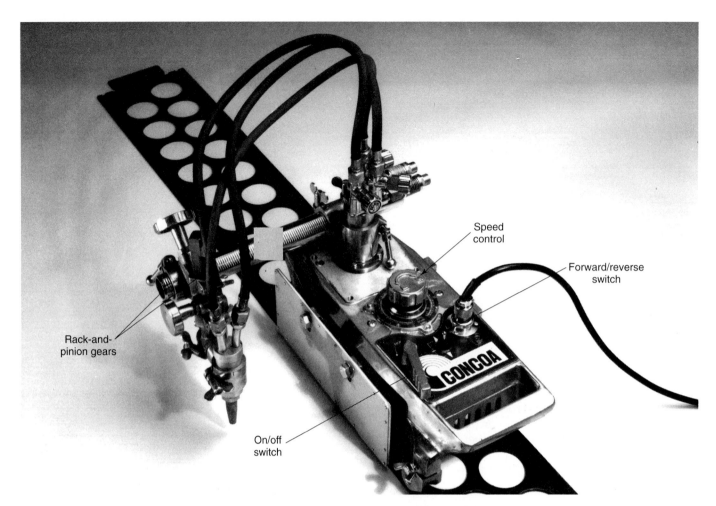

Figure 7-16. *This carriage has rack-and-pinion gears to control the movement of the torch and the bar that holds the torch. (CONCOA)*

of a knob on its top. This torch may be set to move along at a desired speed and steered manually along a straight or curved path as required.

7.5.3 Tracer Devices

In an *electronically guided pattern tracer,* a light source is used to follow a trace line on a pattern or drawing. The trace line may be as thin as .040 of an inch.

A pinpoint of light is directed onto the trace line on a template drawing. Some of this light is reflected into a *photoelectric cell* which produces a small current. This small current is made larger through an amplifier and signals two *servomotors* (steering motors) in the tracer body. The tracer movements on the **X** (longitudinal) and **Y** (lateral) axis are controlled by these servomotors and cause

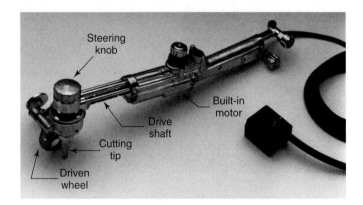

Figure 7-17. *A built-in motor in this hand-held torch drives the wheel at the torch head. This wheel may be steered by turning the knob above the torch head. (CONCOA)*

the tracer to constantly follow the trace lines on the pattern. See Figure 7-18. The electronic pattern tracer is normally used with one or more cutting torches mounted on a cantilevered beam over a cutting table. The beam holding the torch(es) also has two servomotors to control the movement of the beam and torches on the X and Y axes. As the electronic pattern tracer moves, it sends signals to the servomotors on the beam. These servomotors then move the beam and torches to cut the same pattern on the metal under the torches as the tracer is following on the pattern. See Figure 7-19. The pattern used may be a drawing or it may be cut from black-colored material and mounted on a white background.

The electronic circuit is so designed that the direction the servomotor (steering motor) turns is determined by the intensity of the light reflected from the template surface as shown in Figure 7-20. When the light source reflects off a

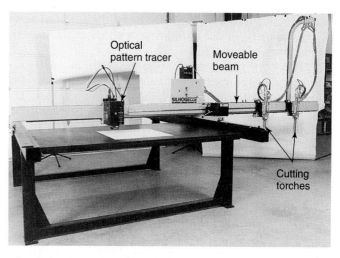

Figure 7-19. *An optical pattern tracer and two cutting torches. The torches move to cut the metal under them as the tracer moves around the pattern. (ESAB Welding and Cutting Products)*

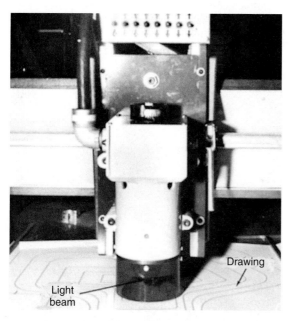

Figure 7-18. *An electronic pattern tracer uses a light beam to sense changes in the line direction on a black and white pattern. The unit electronically controls the movements of the cutting torch to cut the base metal to the shape of the pattern.*

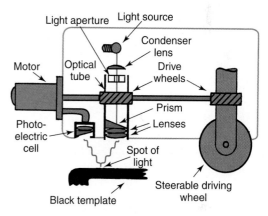

Figure 7-20. *A diagrammatic sketch of an electronic pattern tracer that uses an optical system in the scanning head for controlling torch movements.*

dark surface to the photoelectric cell, the servomotor turns in one direction, When the light is reflected from a white surface, the servomotor turns in the opposite direction. The servomotor does not change direction when the light source reflects equally from a black surface and a white surface (as when the light source is right on the edge of the black template or line).

The light spot which follows the template is aimed to slightly lead the position of the carriage. This gives the servomotor time to correct the position of the steering wheel as changes in the pattern occur.

If the light is following a straight line on the template, the reflected light will be equal from the black surface of the template and the white background surface. The servomotor is not being called to change direction; it will be

neutral. When the light source reaches a curve on the template, it leaves the black pattern. The photoelectric cell now picks up a more intense reflected light from the white background. This causes the servomotor to energize and change the position of the steerable wheel to bring the light source back to the edge of the black pattern. As the curve continues or changes, the steerable wheel is constantly bringing the carriage and light source back to the neutral position at the edge of the pattern. A complex or irregular pattern can be followed to a very precise degree with this type of electronic carriage.

Another method used to control one or more cutting torches automatically is the use of a ***magnetic tracer.*** This system uses a steel pattern and a magnetized pattern tracer (follower). The tracer is normally a knurled, cylindrical steel roller that is rotated by a small motor. The tracer mechanism and the torch or torches are mechanically connected. The tracer follows the pattern as it rotates and is kept in contact with the pattern by its magnetism. As the tracer moves, it follows the pattern and the torch(es) move to cut the metal under it to the same shape as the pattern. See Figures 7-21 and 7-22.

A control unit is used to set the speed of rotation of the tracer. As the speed of rotation increases, the speed of movement of the tracer over the pattern increases. The speed of the cutting torch also increases, since it moves at the same speed as the tracer.

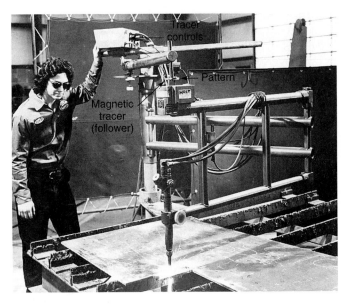

Figure 7-22. *A magnetic tracer which mounts the magnetic pattern follower above the cutting tip. As the tracer rotates and follows the pattern, the cutting torch travels a duplicate path. (ESAB Welding and Cutting Products)*

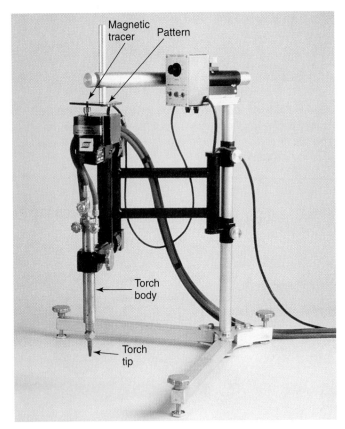

Figure 7-21. *A motorized magnetic pattern tracer. As the magnetic tracer moves around the metal pattern, the torch moves with it to cut the metal into the shape of the pattern. (ESAB Welding and Cutting Products)*

7.6 MULTIPLE TORCHES

There are many cases when it is necessary to make more than one piece of a specially shaped metal part. To do this, two or more torches may be mounted on a cutting machine, as shown in Figure 7-19. The torches have to be carefully mounted to ensure that the cuts are accurate. Figure 7-23 illustrates a device that allows two cuts to be made at the same time using either a hand or mechanically operated cutting torch.

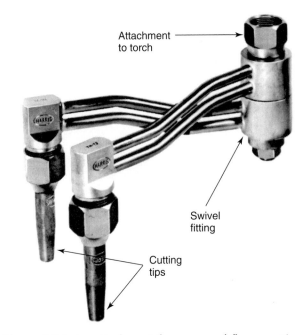

Figure 7-23. *An attachment for a manual flame-cutting torch that enables two cuts to be made at one time. The swivel fitting permits placing the two tips in various positions. (Harris Calorific Div. of The Lincoln Electric Co.)*

TEST YOUR KNOWLEDGE

Write your answers on a separate sheet of paper. Do not write in this book.

1. How are the fuel gas and oxygen cylinders fastened to the cutting outfit or portable truck?
2. What type of fuel gas regulator is used for a cutting station?
3. The oxygen pressure required in oxyfuel gas cutting can reach _____ to _____ psig.
4. Does the oxygen regulator for a cutting outfit differ from the one used on a welding outfit? How?
5. The fuel gas and oxygen are combined in the _____ _____ .
6. What is another name for the slot that is made in the base metal while cutting?
7. The cutting torch is usually constructed from the following types of metals: a _____ _____ body, _____ _____ tubes that carry the gases, and a _____ head or tip holder.
8. Of what material is a cutting tip made?
9. How is the cutting tip connected to the torch?
10. What are the three valves on a cutting torch?
11. What is a cutting attachment?
12. Are all cutting tips that are manufactured in the United States constructed the same?
13. If a cutting torch tip has three orifices, what are the names of the orifices?
14. Why is it so important that the orifices and passages of the cutting tip be free of dirt, scratches, or dents?
15. What is the best way to store cutting tips when they are not in use?
16. How may a welder cut a straight kerf without using a cutting machine?
17. Two types of guides are available to make cutting easier and better. The two types are _____ guides and _____ guides.
18. How does an electronic tracer follow a pattern?
19. When using a multiple-torch cutting machine with an electronic tracer, how many patterns must be used to operate all the cutting torches?
20. List three fuel gases that can be used for cutting.

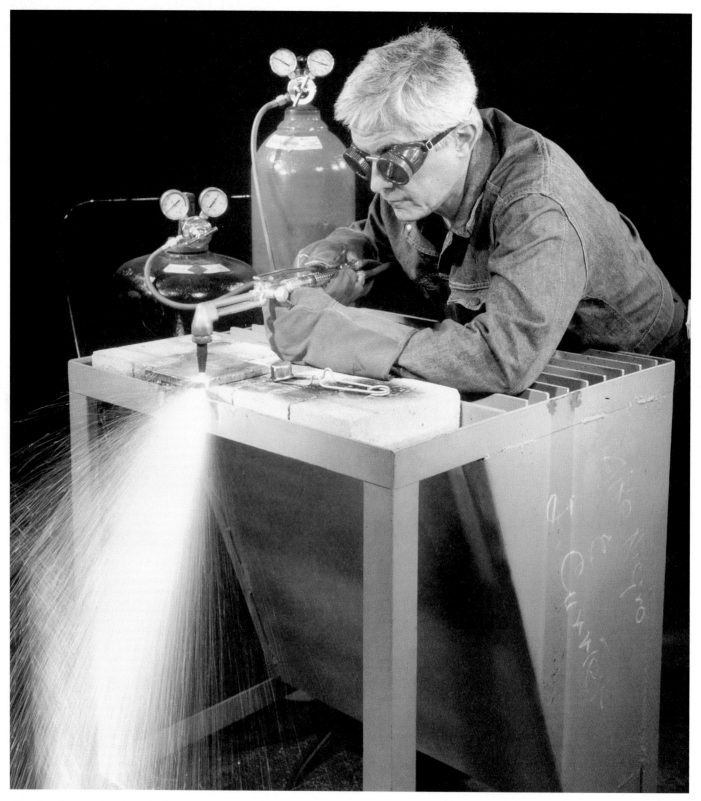

The fundamentals of oxyfuel gas cutting and oxyfuel gas welding are similar. Note the use of firebricks to support the piece being cut. (CONCOA)

Chapter 8

OXYFUEL GAS CUTTING

LEARNING OBJECTIVES

After studying this chapter, you will be able to:

* Describe the function of each component of an oxyfuel gas cutting outfit.
* Correctly and safely assemble an oxyfuel gas cutting outfit.
* Correctly and safely test the complete cutting outfit for leaks.
* Describe the difference between a positive-pressure and injector type cutting torch.
* Correctly select the proper cutting tip and working pressures for cutting a particular thickness of steel.
* Distinguish between a well cut surface and a poorly cut surface, and be able to adjust where necessary to perform a good cut.
* Describe the gouging process and state when it is used.
* Pass a test on safe practices and procedures while oxyfuel gas cutting or gouging.
* List all the safety equipment that should be worn and used when oxyfuel gas cutting in various positions.

In oxyfuel gas cutting of metal, an oxyfuel gas flame is used to heat the metal and an oxygen jet is used to perform the cutting. Oxyfuel gas cutting and oxygen cutting are both approved AWS terms.

The art of oxyfuel gas cutting has progressed rapidly. It is now possible to accurately flame cut both very thin and very thick steel sections. For production work, many layers of metal may be cut at the same time (*stack cutting*). This process greatly reduces both time and costs.

Oxyfuel gas cutting is particularly useful when shape-cutting metal parts. Oxyfuel gas cutting can be done with great accuracy. It leaves the edges smooth enough to satisfy most finished job requirements. Oxygen cutting can

be done much faster than machine saw cutting. Standard rolled plates or sections may be used for many fabrications in shipbuilding, machine frames, and building structures. They are cut to size, then welded together to form a solid steel structure. Such structures, known as *weldments,* are strong, economical to build, and present an attractive appearance.

8.1 THE HEAT OF COMBUSTION OF STEEL

Burning is the rapid oxidation of a material. Virtually all materials will burn, if they are first heated to their *ignition temperature* in the presence of oxygen. Steel is a combustible material, since it will burn or oxidize. During the process of burning, steel releases a considerable amount of heat that is measured in British thermal units (Btu) or joules. This is called the *heat of combustion*. The heat of combustion helps to maintain the metal in the cutting area at the ignition temperature, so that cutting may progress smoothly.

8.2 OXYFUEL GAS CUTTING PROCESS

A special torch and tip is used in the oxyfuel gas cutting process. See Figure 8-1. This cutting tip has one or more preheating orifices (holes) to provide the flames used to heat the metal being cut to its ignition temperature. Cutting oxygen exits from a central orifice when the welder presses the cutting oxygen lever. The ignition temperature of steel occurs when its color is cherry red and its temperature is about 1300°F to 1400°F (704°C to 760°C).

The oxyfuel gas preheating flames are adjusted and used in the same manner as when this flame is used for welding. The preheating flames are used to heat a spot on the base metal to its ignition temperature. When the ignition temperature is reached, the oxygen lever is depressed and a stream of pressurized oxygen burns the metal and

171

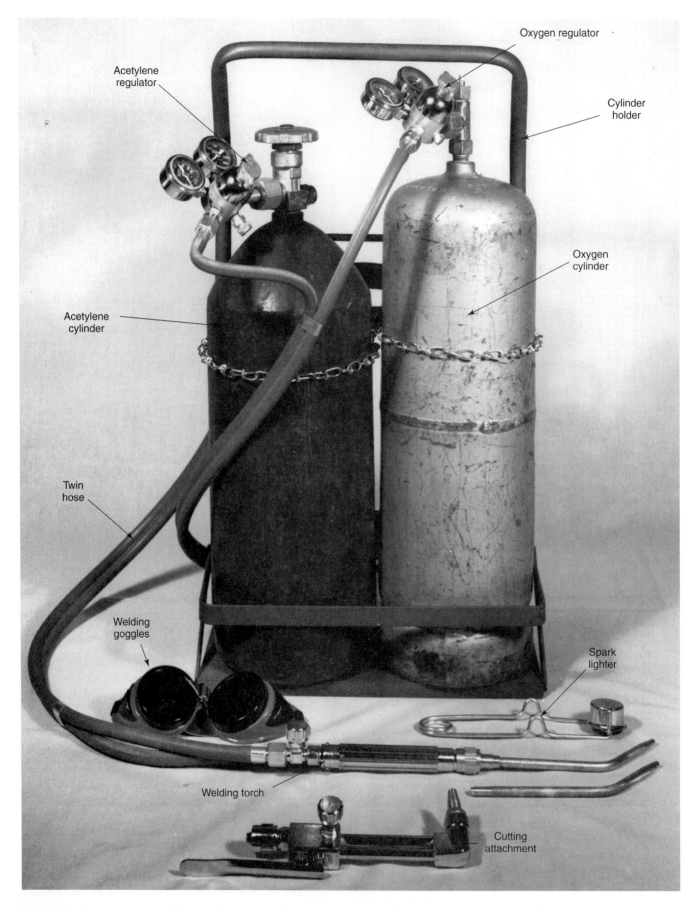

Figure 8-1. *Oxyacetylene welding and cutting outfit. Note the small cylinders used on this portable unit. (Goss, Inc.)*

blows it away from the base metal. As the torch is moved along a predetermined line, the metal is cut, forming the cut or *kerf*. The preheating flames are kept burning throughout the entire cutting process. These flames are required to ensure that the base metal never drops below its ignition temperature during cutting.

8.3 CUTTING OUTFIT

An outfit used for manual oxyfuel gas cutting is similar to the oxyacetylene welding outfit shown in Figure 8-1. The outfits differ only in the torch and possibly the oxygen regulator used. The term *cutting outfit* is used to include all equipment required to perform a cut. A cutting *station* would include the outfit, lighting, ventilation, a cutting table, and possibly a booth.

Since the cutting torch must provide an oxygen cutting jet, it is quite different from a welding torch. Because the oxygen pressures for cutting are usually higher than the pressures employed when welding, an oxygen regulator with a higher working pressure should be used. Also, the inside diameter of the oxygen hose should be larger.

Carefully review Chapters 5 and 7 before connecting and operating the oxyfuel gas cutting outfit. Information in those chapters concerning cylinders, manifolds, regulators, hoses, torches, and tips also applies to the oxyfuel gas cutting outfit.

8.4 CUTTING TORCH

A cutting torch is similar to a welding torch, but has a separate passageway for the oxygen jet. Review Chapter 5 for a more complete description of the oxyfuel gas cutting torch.

This chapter will deal with only the procedure for oxyfuel gas cutting. See Chapter 24 for other methods of cutting.

In an oxyfuel gas cutting torch, the preheating flame(s) comes from one or more orifices arranged around a central oxygen orifice. Figures 8-2 and 8-3 show two different designs of cutting torches. The welder controls the cutting operation through the use of a *cutting oxygen*

lever. In operation, a preheating flame is maintained at the tip through small orifices arranged around the cutting oxygen orifice. Cuts are made by depressing the cutting oxygen lever on the torch, which controls the flow of oxygen from the center orifice, as shown in Figure 8-4. As in welding, the cutting torch is connected to oxygen and fuel gas cylinders. See Heading 6.5.

Two different types of oxyfuel gas cutting torches are in use:
- Positive-pressure torch.
- Injector-type torch.

Before the cutting torch is turned on, the desired cutting tip must be inserted into the torch and aligned. Most cutting tips have an even number (2,4,6,8) of preheating orifices. For the best cutting quality, two of the preheating orifices are aligned with the cutting line and the others are set on each side of the cutting line. See Figure 8-5.

8.4.1 Lighting the Positive-Pressure Type Oxyacetylene Cutting Torch

To turn on the outfit, purge the system, and light the oxyacetylene positive-pressure type hand cutting torch, proceed as follows:
1. Check the equipment for condition.
2. Inspect the regulators. Turn adjusting screws all the way out (regulator closed).

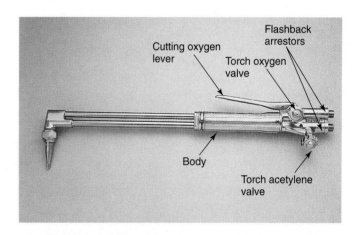

Figure 8-2. *A cutting torch with flashback arrestors built into the torch body. (Victor Equipment Co.)*

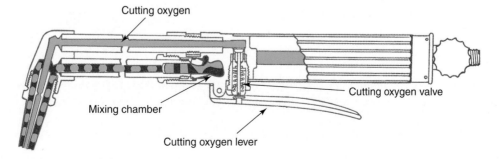

Figure 8-3. *Schematic cross-sectional drawing of cutting torch with preheat mixing chamber located in the torch body. (ESAB Welding and Cutting Products)*

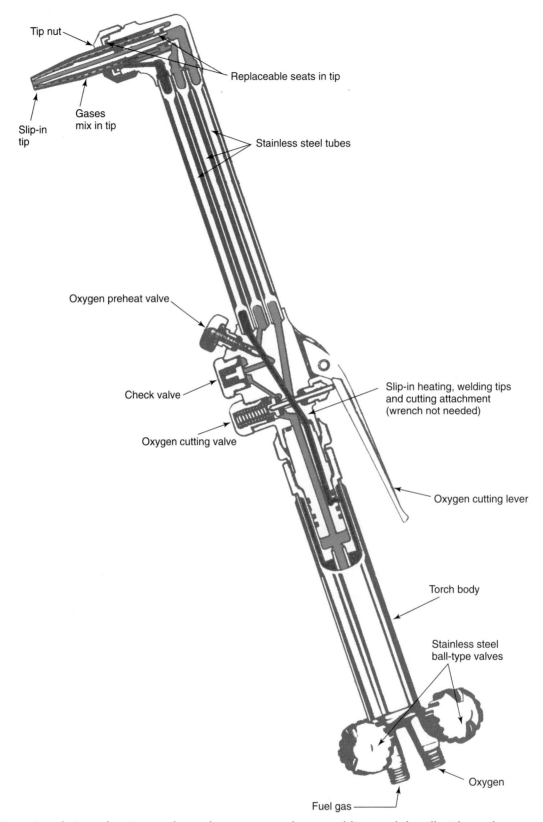

Tip nut

Replaceable seats in tip

Gases
mix in tip

Slip-in
tip

Stainless steel tubes

Oxygen preheat valve

Check valve

Slip-in heating, welding tips
and cutting attachment
(wrench not needed)

Oxygen cutting valve

Oxygen cutting lever

Torch body

Stainless steel
ball-type valves

Oxygen

Fuel gas

Figure 8-4. *A sectioned view of cutting torch attachment mounted on a welding torch handle. The preheat gases are mixed in the tip of the torch where it connects into the torch head. (Smith Equipment, Division of Tescom Corp.)*

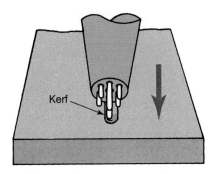

Figure 8-5. *Correct alignment of the preheating orifices on cutting torch. Two preheating orifices should line up with the cutting line.*

3. Open the oxygen cylinder valve very slowly until the regulator high-pressure gauge reaches its maximum reading. Then, turn the cylinder valve all the way open to close the double seating (back-seating) valve. **While doing this, the welder should stand to one side of the gauges.**
4. Open the acetylene cylinder valve slowly 1/4 to 1/2 turn. **Leave the acetylene cylinder valve wrench in place so the cylinder valve may be shut off quickly if necessary.**
5. Open the torch oxygen valve one full turn. Next, open the oxygen *cutting* valve and adjust the oxygen regulator to give the desired oxygen working pressure. Figure 8-6 shows a single-stage oxygen pressure regulator. If the oxygen cutting valve is not opened while adjusting the oxygen pressure, oxygen pressure will drop when this valve is opened during the cutting operation. This will result in a reduced preheat flame, and possibly, a poor cut. Close the torch oxygen valve and the oxygen cutting valve. See the table in Figure 8-7 for oxygen and acetylene working pressure.
6. Open the acetylene torch valve one turn. Slowly turn in the acetylene regulator adjusting screw until the

Figure 8-6. *A single-stage oxygen regulator. The low-pressure gauge reads up to 200 psig (1400kPa), indicating that it may be used with a cutting torch. (Victor Equipment Company, Division of Thermadyne Industries)*

low-pressure acetylene gauge indicates working pressure corresponding to tip size. Refer to Figure 8-7. Close the torch acetylene valve. **Caution: Never use acetylene at a pressure above 15 psig (103kPa).**
7. **After setting the working pressures, check the low-pressure gauge readings to make sure that the pressures are not rising. A rising gauge pressure indicates that the regulator pressure valve is leaking. The outfit should be shut down immediately or the gauge may rupture.**

Metal Thickness		Cutting Orifice Dia.		Cutting Speed		Gas Flow					
						Ft³/hr. L/min.					
in.	mm	in.	mm	in./min.	mm/sec.	Cutting Oxygen		Acetylene		Mps	
1/8	3.2	0.020-0.040	0.51-1.02	16-32	6.8-13.5	15-45	7.2-21.2	3-9	2-4	2-10	2-4
1/4	6.4	0.030-0.060	0.76-1.52	16-26	6.8-11.0	30-55	14.2-26	3-9	2-4	4-10	2-5
3/8	9.5	0.030-0.060	0.76-1.52	15-24	6.4-10.1	40-70	18.9-33	6-12	3-5	4-10	2-5
1/2	12.7	0.040-0.060	1.02-1.52	12-23	5.1-9.7	55-85	26-40	6-12	3-5	6-10	2-5
3/4	19.1	0.045-0.060	1.14-1.52	12-21	5.1-8.9	100-150	47-71	7-14	3-6	8-15	3-5
1	25.4	0.045-0.060	1.14-1.52	9-18	3.8-7.6	110-160	52-76	7-14	4-7	8-15	4-7
1 1/2	38.1	0.060-0.080	1.52-2.03	6-14	2.5-5.9	110-175	52-83	8-16	4-8	8-15	4-8
2	50.8	0.060-0.080	1.52-2.03	6-13	2.5-5.5	130-190	61-90	8-16	4-8	8-20	4-8

Figure 8-7. *Table showing approximate oxygen and acetylene pressures used when cutting steel sheet and plate with the positive-pressure cutting torch. Most cutting tip manufacturers will recommend a range of at least three sizes for any cutting condition. One end of the range gives the greatest economy; other end, maximum speed. For the learner, the middle-sized tip is usually the best. The table is for a torch using a tip with four preheat orifices. For heavier or faster cutting, a tip with 6, 8, 10, or 12 preheat orifices may be used.*

8. To light the torch, open the torch acetylene valve approximately 1/16 turn. **Then, use a flint lighter to ignite the acetylene.**

9. Open the torch acetylene valve until the acetylene flame jumps away from the end of the tip slightly and back again when the torch is given a shake or whipping action. An alternate method of adjusting the acetylene, after the torch is lighted, is to turn on the acetylene until most of the smoke clears from the flame. See Figures 4-5 and 4-6 for illustrations of various conditions of the flame adjustment.

10. Now open the torch oxygen valve, and adjust it to obtain a neutral flame. Open the cutting oxygen valve, and readjust the preheat flame if necessary. The neutral flame may be altered when the cutting oxygen valve is opened. The torch is now adjusted and is ready to be used as a cutting torch. See Figure 8-8.

Figure 8-8. *An instructor showing a welder how to adjust the cutting torch flames. (American Welding Society)*

8.4.2 Lighting the Injector-Type Oxyacetylene Cutting Torch

Figure 8-9 shows typical gas flow through an injector-type oxyacetylene cutting torch. The following procedure is used to turn on the oxyacetylene cutting outfit, purge the system, and light the torch:

1. Check the equipment to make sure all parts are in good operating condition.

2. Inspect the regulators. The adjusting screws of the regulators should be turned all the way out (closing the regulator).

3. Very slowly open the oxygen cylinder valve until the regulator high-pressure gauge reaches its maximum reading. Then, turn the cylinder valve all the way open.

4. Using an acetylene cylinder wrench, slowly open the acetylene cylinder valve 1/4 to 1/2 turn. **Leave the wrench in place on the acetylene cylinder valve.**

5. Open the torch oxygen valve 1/4 turn. Open the torch oxygen cutting orifice lever wide open. Adjust the oxygen regulator screw to give the correct working pressure, as shown in Figure 8-10. Close the torch oxygen valves.

6. Open the torch acetylene valve fully. Adjust the acetylene working pressure on the regulator to the pressure shown in Figure 8-10. Close the torch acetylene valve.

7. **Check the low-pressure gauges for any gradual increase in pressures. If they are increasing, the regulator seat may be leaking. Shut the station down immediately.**

8. The pressures are now adjusted, and the cutting torch is ready to be lighted.

9. To light the torch, open the torch oxygen valve 1/4 turn. Open the torch acetylene valve fully. **Use a spark lighter to ignite the fuel gas.**

10. Press down the oxygen cutting lever and adjust the torch acetylene valve until the preheating flames are neutral.

11. The torch is now ready for oxyacetylene cutting operations.

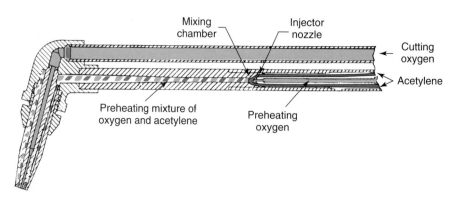

Figure 8-9. *Cross section of an injector-type cutting torch. The acetylene is drawn into the mixing chambers by the pulling action (suction) of the oxygen jet. This injector-type cutting torch is particularly adaptable for use with acetylene generators that operate under low pressure.*

Metal thickness		Preheat orifice drill size	Cutting orifice drill size	Oxygen regulator pressure		Acetylene regulator pressure	
in.	mm			psig	kPa	psig	kPa
1/8 -1/4	3.2 - 6.4	75	67	15-20	103-138	1	7
1/4 -3/8	6.4 - 9.5	74	62	20-25	138-172	1	7
3/8 -1/2	9.5 - 12.7	72	59	25-30	172-207	1	7
1/2 -3/4	12.7 - 19.1	71	55	30-35	207-241	1	7
3/4 -1	19.1 - 25.4	70	54	35-40	241-276	1	7
1 1/2 -2	38.1 - 50.8	68	51	45-50	310-345	1	7

Figure 8-10. *Table of the approximate oxygen and acetylene pressures used when cutting steel with injector-type cutting torch.*

8.5 USING A CUTTING TORCH

The cutting torch must be carefully used if cuts are to be clean and accurate. The tip must be in excellent condition and set to align properly with the cutting line. See Heading 8.4. The preheating flames must be correctly adjusted, and the cutting oxygen pressure must be correct.

To cut, bring the tip of the inner cone of the preheating flames to the edge of the metal to be cut. The cutting torch should be held so that the inner cone of the preheat flames is about 1/16" to 1/8" (1.8mm to 3.2mm) from the surface of the metal being cut, as shown in Figure 8-11.

As soon as a spot on the cutting line has been heated to a bright cherry red color, open the cutting oxygen valve all the way. The jet of oxygen coming through the center of the tip (*oxygen jet*) will cause the heated metal to burn (oxidize) away, forming the kerf (cut).

One of the best indications of a good cutting operation is the appearance of the slag stream at the bottom of the cut. The following may be noted:
- The ideal slag stream passes directly through a plate which is less than 1" (25.4mm) thick. For economic use of oxygen, some drag is desirable when cutting thicker steel sections. (Drag is a measurement, made in the direction of travel, between the entry and exit points of the cutting jet.)
- If the slag stream lags excessively behind the torch tip travel:
  - The flame adjustment may be incorrect.
  - The cutting oxygen pressure adjustment may be too low.
  - The tip travel is too fast and the metal is not being preheated enough.

8.5.1 Cutting Attachments

Most manufacturers of oxyfuel gas welding and cutting torches market a cutting attachment. The attachment is connected to the welding torch body to change from a welding torch to a cutting torch. The cost of such an attachment and the welding torch is usually less than the cost of a separate welding torch and a cutting torch. For portable kits, such an attachment saves space. To connect a cutting attachment, it is only necessary to remove the welding tip tube and screw on the cutting attachment.

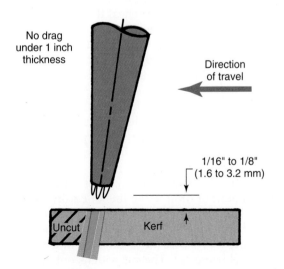
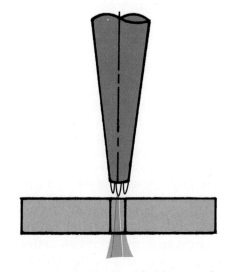

Figure 8-11. *The cutting torch position used for cutting with the oxyfuel gas cutting torch. The tip is held perpendicular or inclined slightly in the direction of travel.*

The operation of the torch with the cutting attachment is the same as the operation of a regular cutting torch. Figures 7-2 and 8-12 illustrate a cutting attachment. There are usually two oxygen torch valves, one acetylene torch valve, and a cutting oxygen lever involved when a cutting attachment is used. The welding torch body has an oxygen valve and a fuel gas valve. The cutting attachment has an oxygen torch valve and a cutting oxygen lever.

To make flame adjustments, the oxygen torch valve on the cutting attachment should be opened one or more turns. The torch valves on the torch body are then used to adjust the flames.

Caution: The torch valves on the welding torch body must be turned off before the cutting attachment is disconnected.

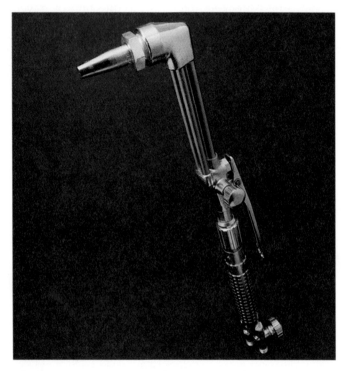

Figure 8-12. *A cutting attachment screwed into place on a welding torch body. (CONCOA)*

8.5.2 Cutting Tips

The cutting tip will normally have at least two orifices. One orifice, which is usually in the center of the tip, is for the cutting oxygen: one or more smaller orifices are for preheating the metal to be cut, as shown in Figure 8-11.

Cutting tips may be of one-piece or two-piece construction. One-piece tips are used for oxyacetylene cutting only. Two-piece tips are used for all other gas cutting: natural gas, propane, MAPP, etc.

For satisfactory service, tips must be kept in good condition. The orifice end must be clean. The tip surface must be at a right angle to the orifices, so that the preheat flame is shaped and aimed properly. The sealing faces of the tip, (which contacts the torch body) must be clean and free from scratches, burrs, nicks, etc., or these joints may

free from scratches, burrs, nicks, etc., or these joints may leak.

Clean the tips as described in Heading 7.4.1. Always store the extra tips in soft holders. A wooden block with holes drilled in it works well. See Chapter 7 for more detailed information.

8.6 CUTTING STEEL WITH OXYFUEL GAS CUTTING TORCH

Metals that may be cut with the oxyfuel gas cutting torch may be divided into two classes:
- Metals whose *oxides* (compound of oxygen and the metal) have a lower melting temperature than the metal itself.
- Metals whose oxides have a higher melting temperature than the metal itself.

Practically all steels fall under the first classification and, therefore, cutting presents little difficulty. When the cutting jet is turned on, the iron oxides which form will melt at a lower temperature than the base metal. These oxides are easily blown away by the cutting oxygen jet, leaving a clean and straight cut. Figure 8-13 shows a cut in progress. When an expert handles the cutting torch, or if an automatic cutting machine is used, the kerf formed will have a smoothness of machine-like quality.

The second group includes cast iron, some alloy steels, stainless steel, and nonferrous metals. These metals present a complication in cutting because the oxide has a higher melting temperature than the metal. It is almost impossible to cut an even kerf. It is very important that these oxides, called *refractory oxides*, be reduced by chemical action or be prevented from forming. See Chapter 24 for special cutting processes used with refractory oxides.

Items of importance to be watched in cutting are:
- Pressure of the oxygen fed to the cut.
- Size of the oxygen jet orifice.
- Speed of the cutting torch across the metal.
- Distance of the preheat flame from the metal.
- Size of the preheat flames or the amount of heat delivered to the base metal.
- Torch tip position (angle) relative to the metal.
- The alignment of the torch tip orifices with the kerf.

It should be noted that the oxygen pressure will determine the velocity of the oxygen jet. The orifice size will determine the amount of oxygen delivered in *cfh* (cubic feet per hour) or *L/min* (liters per minute) at any particular pressure.

The cut should proceed just fast enough to provide a slight amount of drag at the line of cutting. If the drag is too small, the oxygen consumption is too great. If the drag is large, the cutting tip orifices may be too small for the job.

Figure 8-14 shows the result if too much oxygen is fed to the steel being cut. The cut widens out as the jet penetrates the thickness of the metal. This leaves a ***bell-mouthed kerf*** on the side of the metal away from the torch.

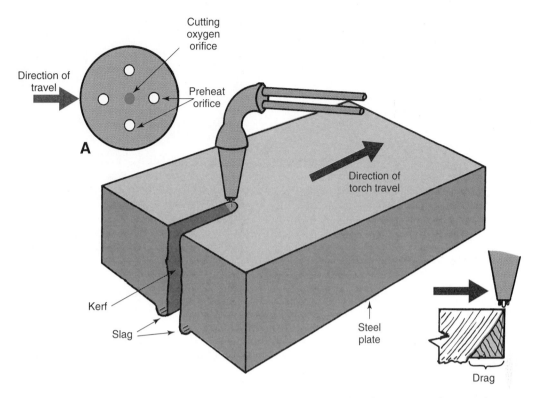

Figure 8-13. *An oxyfuel gas cutting torch being used to cut a steel plate. Detail A shows an end view of a typical cutting tip with a center oxygen orifice and four preheat orifices.*

Figure 8-15 shows the result if the torch is moved too rapidly across the work. When the torch is moved too rapidly, the metal at the bottom (far side) of the cut will not be burned away. This is because it does not receive enough heating and oxygen to complete the cut. The large drag results in a turbulent action of the torch gases which will leave a kerf that is very rough and irregular in shape.

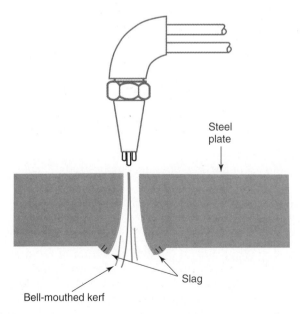

Figure 8-14. *The effect of using too much oxygen when cutting steel. Note how the kerf widens at the bottom of the plate to create a bell-mouthed kerf.*

If the torch is moved too slowly across the work, the preheated metal will be completely burned away, and the preheating flame wasted.

If the metal preheat temperature is lost, the oxygen to the cutting orifice should be closed off. The metal should again be preheated to the proper temperature by the preheating flames. When a cherry red color is obtained, the cutting oxygen should be turned on and the cut continued. These starting and stopping actions may also cause an irregular cut.

Metal that is very dirty and rusty should be cleaned before starting the cutting operation. The impurities on the metal will slow the cutting speed, and may cause a rough and irregular kerf. Figure 8-16 shows a number of completed cuts. The cause of each poor cut is explained in the captions.

The torch motion to be used in cutting is a matter of the welder's own experience. Usually, no motion is used. In some cases, the thickness of the metal requires an *oscillating* (side-to-side) motion in order to obtain the necessary width of cut.

The welder, when cutting, should stand in a comfortable position that permits looking into the cut as it is being formed. The torch movement should be away from the welder, rather than toward him or her, in order to see into the kerf. The cut may be made from right to left with a good view of the kerf. The torch is usually held with both hands for best control, as shown in Figure 8-17.

Normally, the tip is perpendicular to the surface being cut. The end of the inner cone of the preheating flame should be held just above the metal. If the cutting tip has

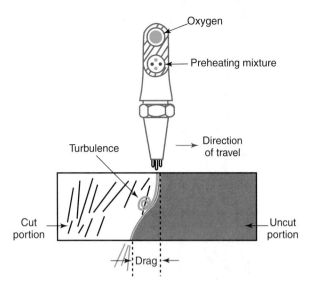

Figure 8-15. *The effect of moving a cutting torch too rapidly across the work. The drag becomes too large.*

4, 6, or more preheating orifices, one orifice should precede (lead) the cutting orifice. One orifice should follow the cut. The other orifices should be aligned to heat each side of the kerf equally. See Figure 8-18.

The welder should wear leggings, safety boots with high tops, and trousers without cuffs. (If trousers do have cuffs, they should be covered to keep them from catching the white-hot metal slag as it drops from the cut.) Place a container under the cut to catch the very hot liquid slag. This container should be lined with a heat-resistant (refractory) material.

8.6.1 Cutting Thin Steel

Cutting steel that is 1/8" (3.18mm) or less in thickness requires the use of the smallest cutting tip available. A tip with few preheat holes is often used. In addition, the tip is usually pointed in the direction the torch is traveling. If even a small tip size seems too large, change the tip angle to 15°-20°, as shown in Figure 8-19. This lowered angle effectively increases the thickness of the metal being cut. On very thin metal, holding the tip near vertical will produce too much preheating. The resulting cut will be very poor. Many welders actually rest the edge of the tip on the metal during this process. Be careful to keep the end of the preheating inner cone *just above* the metal.

8.6.2 Cutting Thick Steel

Steel over 1/2" (13mm) thick should be cut by holding the torch so the tip is perpendicular to the surface of the base metal being cut. Figure 8-20 shows the position of the cutting torch tip orifice when cutting thick steel.

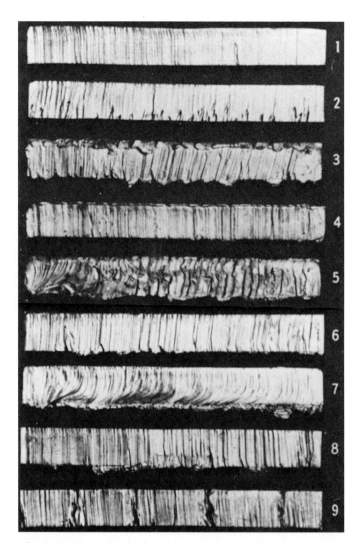

Figure 8-16. *Typical edge conditions resulting from oxyfuel gas cutting operations: (1) good cut in 1" (25mm) plate–the edge is square, and the drag lines are essentially vertical and not too pronounced; (2) preheat flames were too small for this cut, and the cutting speed was too slow, causing bad gouging at the bottom; (3) preheating flames were too long, with the result that the top surface melted over, the cut edge is irregular, and there is an excessive amount of adhering slag; (4) oxygen pressure was too low, with the result that the top edge melted over because of the slow cutting speed; (5) oxygen pressure was too high and the nozzle size too small, so that control of the cut was lost; (6) cutting speed was too slow, with the result that the irregularities of the drag lines are emphasized; (7) cutting speed was too fast, resulting in a pronounced break in the dragline, and an irregular cut edge; (8) torch travel was unsteady, with the result that the cut edge is wavy and irregular; (9) cut was lost and not carefully restarted, causing bad gouges at the restarting point. (American Welding Society)*

Figure 8-17. *This welder is using a cutting torch to cut a 1/4 in. (6.4mm) steel plate. (American Welding Society)*

The cut is normally started at the edge of the stock. The torch may be moved from left to right, or from right to left. Either direction is good if it permits the welder to look into the kerf and check cutting progress. If a line of travel is chalked on the metal, torch movement from right to left enables a right-handed welder to most easily follow the guidelines on the metal. Figure 8-21 shows the progress of a cut in thick steel.

After the edge has been heated to a dull cherry red, the oxygen jet should be opened all the way by pressing on the cutting lever. As soon as the cutting action starts, move the torch tip at a steady rate. Avoid an unsteady movement of the torch, or the cut will be irregular and the cutting action may stop.

To start a cut faster in thick plate, the welder may begin at the corner of the metal by slanting the torch in the direction opposite the direction of travel, as shown in Figure 8-22. As the corner is cut, the welder moves the

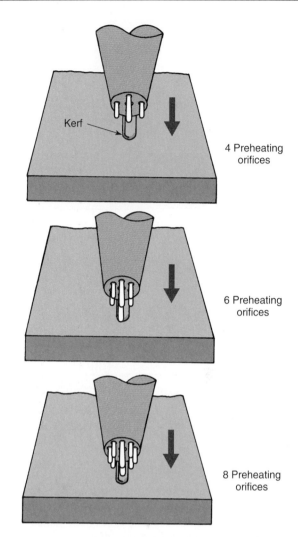

Figure 8-18. *Proper alignment of the preheating orifices on the cutting tip. Notice that, in each case, one orifice leads the oxygen jet and one follows.*

torch to a vertical position, until finally the total thickness is cut, also shown in Figure 8-22. The cut may then proceed.

Two other methods are used to start cuts. One is to *nick* the edge of the metal, where the cut is to start, with a cold chisel. The sharp edges of the metal upset by the chisel will preheat and oxidize rapidly under the cutting torch. This makes it possible to start without preheating the entire edge of a thick plate.

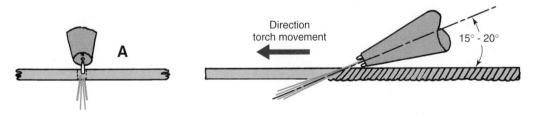

Figure 8-19. *A recommended procedure for cutting thin steel. Notice that the two preheat flames are in line with the cut (kerf).*

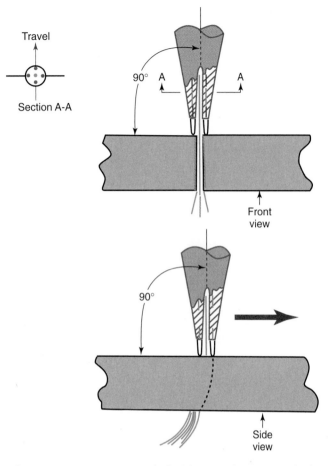

Travel

Section A-A

90° A A

Front view

90°

Side view

Figure 8-20. *A recommended technique for cutting thick steel. Note the position of the torch tip preheat orifices in relation to the line of the cut. Two preheat flames are in the line of the torch progress. This position enables one preheat flame to be ahead of the cut, two flames to heat the sides of the cut, and one flame to heat down in the kerf.*

The second method is to place an iron filler rod under the preheating flames at the edge of a thick plate. The filler rod will reach the cherry red temperature quickly. When the cutting oxygen is turned on, the rod will oxidize and cause the thicker plate to start oxidizing. Figure 8-23 shows a welder cutting a thick section.

8.6.3 Cutting Chamfers (Bevels)

Another important torch cutting operation is the cutting of chamfers for bevel and V-groove joints on the edges of steel plate prior to welding, as shown in Figure 8-24. Thicker pieces of steel must have the chamfered edge preparation so the weld will penetrate through the thickness of the metal.

Bevel angles may be cut at the same time the metal is being cut to size and shape. The chamfer may be cut as a separate operation, however. Instructions relative to cutting thick metal, given previously, are usable when cutting bevel angles. It is important to obtain a high-quality cut when making the bevel. This will minimize any further plate preparation and ensure a good fit-up of the joint to be welded.

8.6.4 Cutting Pipe or Tubing

One of the most popular uses of the oxyfuel gas cutting torch is to prepare pipe for joining by welding. See Chapter 23 for more pipe-welding instructions. The cutting torch is especially useful for preparing odd-shaped joints on the job. Also, it may be used extensively to chamfer the edges of thick pipe to provide a bevel or V-groove joint for the welder.

The exact procedure to follow when cutting pipe depends on the pipe's diameter. For small-diameter pipe, it is best to keep the tip almost tangent to the circumfer-

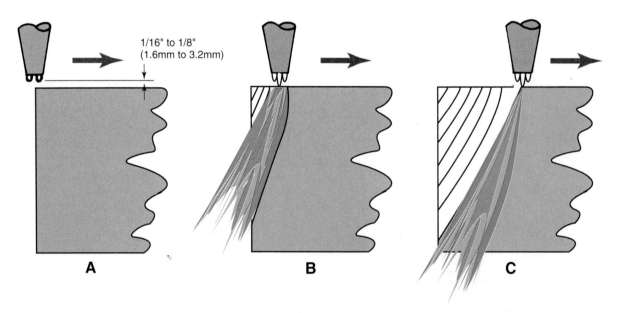

1/16" to 1/8" (1.6mm to 3.2mm)

A **B** **C**

Figure 8-21. *Progress of a cut in thick steel. A—Preheat flames are 1/16" to 1/8" (1.6 to 3.2 mm) from the metal surface. The torch is held in this spot until the metal becomes cherry red. B—Note that the torch is moved slowly to maintain the rapid oxidation, even though the cut is only part way through the metal. C—Note that as the cut is made through the entire thickness, the bottom of the kerf lags slightly behind the top edge.*

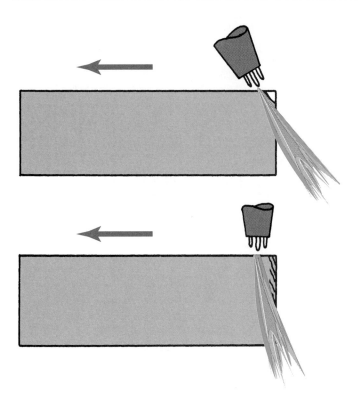

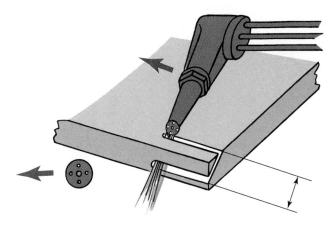

Figure 8-24. *A method of obtaining a beveled edge on thick metal plate using a cutting torch. Note the alignment of the cutting tip holes. Two are in line with the kerf and one on each side.*

ence of the pipe. (Note a similar approach to cutting very thin metal in Figure 8-19.) This will prevent cutting through both sides of the pipe at once. Attempting to cut through the two thicknesses of the pipe simultaneously usually results in a poor cut.

With a pipe diameter of approximately 4" (102mm) or larger, it is possible to keep the torch tip perpendicular to the pipe surfaces while cutting, without burning through the other side. Of the two methods, the perpendicular position permits a cleaner and straighter cut. If the welder's helper rotates the pipe as it is being cut, a very clean cut can be obtained. Most welders start the cut at the extreme edge of the pipe and then cut back to the marked chamfer ring.

When chamfering pipe by hand, it is best to point the torch toward the end of the pipe. This procedure provides a clean chamfer, permits more accurate cutting, and should not leave any excessive oxide clinging to the pipe when the cut is completed. Cutting machines which use a mechanism to revolve the torch around the pipe produce excellent chamfered edges.

Figure 8-25 shows how to use a cutting torch to bevel or chamfer the end of the pipe. It must be remembered that when cutting pipe, it is not the diameter of the pipe that determines the size of the cutting torch tip. The *thickness of the pipe wall* is the controlling factor. The proper welding codes and procedures must be followed whenever pipe is cut for use in structural or pressure vessel applications.

8.6.5 Piercing and Cutting Holes

Holes may be pierced in steel plates rapidly and with accurate results. To *pierce* is to produce a relatively small hole (in comparison to the size of the metal surface) through a steel plate. The process consists of holding the cutting torch with the nozzle perpendicular to the surface of the metal, and preheating the spot to be cut until it is a bright cherry red. After the metal is brought up to the proper temperature (1400°F, or 760°C), the oxygen jet may be turned on very slowly. At the same time, the nozzle

Figure 8-22. *The correct oxyfuel gas cutting torch position for the starting cut; also the required change in the torch position as the cut progresses.*

Figure 8-23. *A cut in progress on thick steel.*

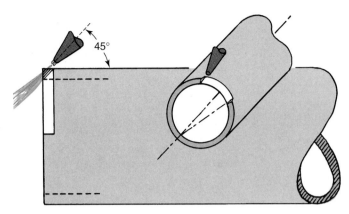

Figure 8-25. *A cutting torch being used to cut a bevel on a steel pipe.*

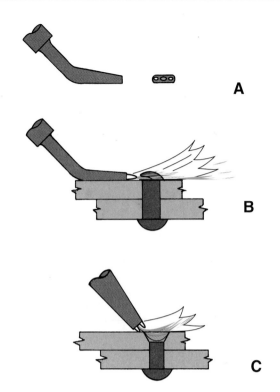

Figure 8-26. *Recommended methods of removing rivet heads using a cutting torch. A—Special rivet-cutting tip. B—Special tip used to cut a round-head rivet. C—Regular cutting tip used in cutting out a countersunk rivet.*

should be raised enough to eliminate the slag being blown back into the nozzle orifices. A greater amount of heat is required to preheat the surface for piercing than when starting on an edge. It is, therefore, recommended that the welder use at least the next larger tip in relation to the thickness of the metal than recommended in Figures 8-7 and 8-10. See Heading 24.1.2 for use of the oxygen lance for cutting holes in thick plate. **Caution: Until the steel plate is melted through the bottom surface, molten metal will be blown upward by the pressure of the cutting oxygen.**

To cut larger holes in steel plate, the typical steel cutting method described in this chapter is recommended. It is good practice to outline the hole first, using special chalk. This outline is used as a guide to permit the welder to cut an accurate hole. If the size of the hole warrants, it is best to do the cutting with an automatic machine, or with a radius bar attachment clamped to the torch head.

8.6.6 Cutting and Removing Rivet Heads

The cutting torch is frequently used in salvage operations when dismantling large fabricated structures that have been assembled with rivets. The torch is used to remove the rivet heads.

Two typical rivet shapes are:
- Round-head.
- Countersunk-head.

The procedure for removing these heads is fundamentally the same as for any cutting operation, but one additional precaution should be observed. If possible, the welder should do the cutting without damaging the steel plate. To do this, it is very important that the size of the tip is carefully chosen. If too large a tip is used, the steel plate will be damaged as the rivet head is being removed. If too small a tip is used, the method becomes too slow. Practically all welding equipment companies recommend special-shaped cutting tips for rivet cutting.

The procedure for cutting the round-head rivet is to preheat the head of the rivet to a bright cherry red. The steel plate is usually adequately protected from the preheating flame by a coating of *scale* (oxide). The special cutting tip is placed on the base metal and the rivet head cut

in the usual manner. See Figure 8-26. The appearance of an accurately performed job shows a clean removal of the head without any score marks from the cutting jet on the steel plate.

Removal of a countersunk-head rivet is more difficult than the round-head rivet. The rivet is usually tightly embedded in the plate. Countersunk rivet heads may be removed, however, with very little (if any) damage to the steel plate. This is done by carefully selecting the tip size and by carefully cutting around the countersunk angle.

8.6.7 Gouging with the Cutting Torch

Gouging is a process that will remove metal from the surface of a part to a desired depth. When properly done, gouging will cut a U-shaped groove into the surface of the base metal. Gouging may be done using the oxyfuel gas process or one of several arc cutting processes.

Gouging may be used to open up a defect (crack) that may occur in a part and provide a well-shaped U-groove in preparation for a welded repair. It can be used to cut out areas of a completed weld that are judged to be defective, preparing the weld for rewelding.

Oxyfuel gas *gouging* differs from oxyfuel gas *cutting*. When gouging, a lower oxygen cutting pressure and a special larger diameter cutting tip orifice are used. The resulting lower-pressure oxygen cutting jet allows the welder to move more slowly along the gouge line. The lower pressure also helps to prevent cutting through the base metal. See Figure 8-27. The less forceful oxygen stream oxidizes

Figure 8-27. A typical oxygen gouging operation. A low-velocity cutting jet is used to maintain better control of the gouge width and depth.

stream oxidizes the surface metal only and penetrates more slowly. This enables the welder to gouge or groove the base metal with greater accuracy.

In a gouging tip, there are five or six preheat orifices to provide for even distribution of the preheat flames. Gouging tips are sometimes produced with a small bump on the underside of the tip. This small bump helps the welder to achieve a more even depth of gouge by keeping the tip at a uniform distance above the metal surface. In an automatic cutting machine, a gouging tip is capable of creating a gouge with a very accurate depth and width.

If the gouging cut is not started properly, it is possible to cut too deeply or actually cut through the entire thickness of the base metal. The speed at which the torch is moved along the gouging line is important. Too-rapid movement of the torch will create a groove that is too shallow and narrow. Moving the torch too slowly will create a gouge that is too deep and wide.

8.7 CUTTING FERROUS ALLOY METALS

The introduction to industry of many alloy steels has made it necessary to develop new cutting techniques so that these metals can be successfully and economically cut. Of the alloy steels, stainless steel is perhaps the most widely used. Stainless steel consists of chromium, nickel, and other elements added to iron. See Chapters 21, 27, and 28 for more information about alloy metals.

Many of these alloy metals have melting temperatures below that of steel. The oxides formed when cutting have a melting temperature higher than that of the original metal. These high-melting-point oxides must be reduced and/or removed from the cut as it proceeds, or the cutting action will stop. It has been found that for the same relative thickness of metal, stainless steels need approximately 20% more preheating flame and 20% more oxygen for the cutting. It has also been found that it is a good

practice to use a slightly carburizing flame when preheating stainless steel.

The metal to be cut should be placed so that the cutting tip and flame are in a horizontal position. The cut should start at the top of the metal and proceed downward in a vertical line. A slight, but quick, up and down motion of the torch facilitates the removal of the slag. Figure 8-28 shows how the torch should be moved up and down to facilitate the slag removal. Using a regular cutting torch, it is difficult to obtain as clean and narrow a kerf when cutting alloy metals as when cutting straight carbon steels.

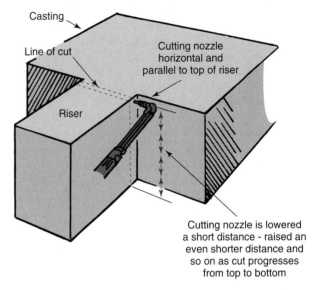

Figure 8-28. A technique used for the oxyacetylene cutting of chromium steel. Note how the torch is raised and lowered to assist in removing the slag from the cut.

As in the case of steel, the alloy metals must be preheated before the cutting operation is started. Stainless steels, especially, must be preheated to a white heat before the cutting oxygen is turned on. The cutting action is much more violent with stainless steels than with straight carbon steels. Cutting takes place with considerable sparking and blowing of the slag.

In situations where the progress of the cutting is frequently interrupted by the presence of unmeltable slag, the welder may find it advisable to hold a mild steel welding rod in the kerf of the metal. The mild steel, mixed with the alloy steel, dilutes or reduces the percentage of the alloys in the area of the cut. The cutting properties of the alloy metal in the area of the cut thus become more like those of mild steel. The cut proceeds more smoothly.

Adding welding rod to the cut is also useful when cutting poor grade steels, cast irons, and old, oxidized steel castings. Powder cutting, plasma arc cutting, and inert gas arc cutting have proven much more practical for cutting steel alloys and nonferrous metals. For full details on these processes, refer to Chapters 16, 17, and 24.

8.8 CUTTING CAST IRON

As mentioned previously, it is more difficult to cut cast iron than steel. This is because iron oxides of cast iron melt at a higher temperature than the cast iron itself. However, cutting has been performed successfully on cast iron in many salvage shops and foundries. When cutting cast iron, it is important to preheat the whole casting before the cutting is started. The metal should not be heated to a temperature that is too high, since this will oxidize the surface and make cutting difficult. A preheat temperature of about 500°F (260°C) is usually satisfactory.

When cutting cast iron, a carburizing flame is recommended to prevent oxides from forming on the surface before the cutting starts. The cast iron kerf is always wider than a steel kerf, because of the oxidation difficulties. After completing a cut on cast iron, the casting should be cooled very slowly if a gray cast iron is desired. Rapid cooling will create a white cast iron grain structure.

It is difficult to cut cast iron with the usual oxyfuel gas cutting torch. Other more satisfactory methods of cutting have been developed. These include oxygen arc, plasma arc, chemical flux, metal powder, gas tungsten arc, and exothermic cutting. See Chapters 16, 17, and 24 for descriptions of these cutting processes.

8.9 AUTOMATIC CUTTING

Automatic cutting machines are constantly being improved. Cutting processes such as metal powder cutting (POC), plasma arc cutting (PAC), flux cutting (FOC), air carbon arc (CAC-A), exothermic cutting, and water jet cutting are being done using automatic machines.

Automatic cutting equipment requires a controller or a computer to control the operation. The cutting process is completely controlled and does not require adjustment while cutting. The controller or computer is programmed with the path and speed of the torch or torches. To start and stop the flow of gases, solenoid valves are used. Electric motors, actuated by the controller or computer, are used to very precisely control the movement of the torch or torches. Automatic cutting equipment will produce almost identical parts every time.

Feedback controls are used to monitor the automatic cutting process and make necessary corrections. Artificial vision devices may be used to keep the preheating flames at a specified height above the base metal at all times.

Semiautomatic cutting, or mechanized cutting, is similar to automatic cutting except that some changes may be required to the process while cutting is taking place. These processes often use electronic and magnetic tracers to follow a pattern. As the pattern is followed, electric motors move the torch or torches to cut the metal in the same shape as the pattern. Electronic tracers permit a pattern to be followed so that extremely accurate shapes can be produced. Electronic tracers are described in Heading 7.5.3. Chapter 25 describes semiautomatic and automatic processes.

There are many automatic and semiautomatic mechanisms available to perform cutting operations. Multiple-torch cutting machines are used extensively in industry. They will cut several exact copies of a desired part at one time. Figure 8-29 shows automatic cutting equipment. Inexpensive metal templates or line drawings can be used to guide the mechanical or electronic tracers.

Figure 8-29. A gantry-type shape cutter. One plasma arc and five oxyfuel gas cutting torches are mounted on this computer-operated cutting machine. The required torch movements are programmed into the computer at the left. (ESAB Welding and Cutting Products)

Practically all pattern tracers and automatic cutting machines and their torches are moved by variable speed electric motors known as *servomotors*.

Special cutting torches are mounted on a light, rigid rail. These cutting torches may be adjusted to vary the flame tip-to-metal surface distance. See Figure 8-30. This rail, with all its mechanisms, is called a *gantry*. Both rail and torches are moved on the X-Y (lateral and longitudinal) axes by electric servomotors. As the tracer moves around the pattern, the servomotors move the rail and torches to duplicate the pattern. A servomotor gets feedback information from the pattern tracer. This feedback causes the servomotor to move the rail and torches in the proper direction to duplicate the pattern.

The operation of these machines necessitates four important adjustments:

- Adjustment for the number of in./min. (cm/min.) that will be cut.
- Gas pressure must be carefully adjusted to ensure a clean cut through the thickness of the metal without wasting fuel.
- Flame adjustments on each torch must be carefully made.
- Distance of the torch tip from the metal being cut must be carefully adjusted to obtain the best results. This adjusting is done by means of a graduated scale on the torch body, and a *vernier gear mechanism* (measuring device that permits fine adjustment), to raise and/or lower the torch.

Once set up, these cutting machines are self-operating. However, the initial adjustments must be very carefully made by the operator.

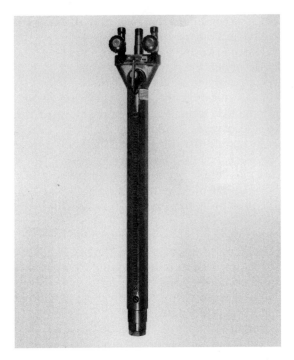

Figure 8-30. *An oxyfuel gas cutting torch designed for use on a motorized carriage or gantry-type cutting machine. Note the rack-type gear attached to the side of this torch. This permits the distance between the tips of the flames and the metal to be adjusted. (CONCOA)*

A magnetic tracer with a steel pattern can be used as a cutting torch guiding device. The tracer, driven by a small, adjustable motor, rotates and follows the pattern. A cutting torch, or torches, are normally mechanically attached to the tracer mechanism. As the tracer moves, the torches move in the same pattern to cut the metal. The speed of cutting is determined by the speed of rotation of the pattern tracer. The variable-speed motor on the tracer must be manually set. It allows the tracer and the cutting torches to move at the correct speed for the thickness of metal being cut.

8.10 SAFETY IN OXYFUEL GAS CUTTING

Oxyfuel gas cutting may be safely performed. However, as in all oxyfuel gas welding and cutting, proper procedures must be followed in order to eliminate dangers which might be caused by carelessness or incorrect handling of oxygen and fuel gas equipment. Certain potential hazards exist and the welder should always observe approved procedures to eliminate these dangers. See Heading 6.14.

Cutting operations are accompanied by considerable sparking and flying of sparks, which are *globules* (ball-shaped particles) of molten metal. To avoid accidents from this hazard:

- Floors on which cutting is done should be concrete or other fireproof material.
- Workbenches and other necessary shop furniture should be of metal or other fire-resistant material.
- Oil, paper, wood shavings, gasoline, lint, or other flammable materials, should not be in the room in which flame cutting is performed.
- Leather or other slow-burning fabrics should be worn. Trousers should be without cuffs. Pockets and clothing should be inspected for possible flammable materials such as buttons, combs, plastic rules, matches, pencils, and other items.
- Objects being flame-cut may present additional hazards. Tanks and containers may be welded or cut only by an experienced welder. **It is generally required to pass an inert gas or steam through the tank as it is being cut. The flow of steam or inert gas is intended to displace any combustible gases in the tank. This process is often required for hours prior to any attempted cutting operation. The tank may also be filled with water, except in the area of the work. In all cases, the tank must be vented to prevent the entrapment (holding) of potentially explosive gases. This work should never be done except under the supervision of a qualified safety engineer. A small amount of flammable material in such a container or tank may cause a powerful explosion.**
- Certain metals, such as magnesium, may burn with an explosive force if flame-cut. **Be certain of what is being cut.**
- Face and hands must be protected from metal splatter.
- A fire extinguisher should always be at hand while flame cutting for use in emergencies. The posting of a fire watch while cutting is recommended.
- In Chapter 6, the correct procedures for setting up and handling cylinders, regulators, hoses, and torches were explained. Be sure to know these procedures. It is a good idea to review Chapter 6.
- Review Heading 6.14.1 on metal fume hazards. Be sure to know what metal is being worked on before performing any oxyfuel gas cutting operation.
- The normal atmosphere contains about 21% oxygen by volume. As the oxygen content in an enclosed space is increased above this percentage, there is an increasing danger of a spark or flame causing a fire or explosion.
- **Never use acetylene at a pressure greater than 15 psig (103kPa). Acetylene becomes unstable above that pressure and may explode.**

TEST YOUR KNOWLEDGE

Write your answers on a separate sheet of paper. Do not write in this book.

1. Which is the correct term to describe the oxyfuel gas cutting process: "flame cutting," "burning," or "oxygen cutting"?
2. Is an oxyfuel gas cut smooth or rough when complete?
3. Are the preheat flames used after the cutting has begun? Why?
4. What are the two types of flame-cutting torches called?
5. When adjusting pressures before lighting a cutting torch, why is it necessary to check the pressures with the torch cutting lever in the fully open position?
6. What oxygen and acetylene pressures are used when cutting 1/2" (12.7mm) sheet steel with a positive-pressure torch?
7. Why is it necessary to provide a higher oxygen pressure with a cutting torch than with a welding torch?
8. What is the purpose of the preheating jets or orifices in a cutting torch tip? When is the cutting oxygen valve opened?
9. List two conditions that may cause the slag stream to lag behind the torch tip (drag).
10. When are two-piece cutting tips used?
11. If too much oxygen is used when cutting, the cut widens out and leaves a _____ kerf.
12. What is meant by the terms "kerf" and "drag"?
13. Should the cutting torch be moved toward, from side to side, or away from the welder?
14. How should the cutting torch be held, in relation to the work, when cutting thin sheet steel?
15. Is it possible to cut a piece to shape and to place a bevel on the metal at the same time?
16. What controls the size tip to be used when cutting pipe?
17. When piercing a hole in a plate, how is the torch held in relation to the base metal? What are the requirements for preheat when piercing, compared to starting a cut on the edge of a plate?
18. How is it possible to cut off rivet heads without injuring metal plates that are riveted together?
19. What procedure is recommended when it is necessary to cut stainless steel or cast iron with an oxyfuel gas cutting torch?
20. What safety precautions must be considered before starting a flame-cutting operation?

Chapter 9

SOLDERING

LEARNING OBJECTIVES

After studying this chapter, you will be able to:
* Define the process of soldering.
* Select the proper filler metal and flux for soldering a particular joint.
* Describe the terms "solidus" and "liquidus."
* Identify the three types of fluxes and make judgments about when to use each.
* List the steps that must be completed prior to soldering to ensure a strong finished joint.
* List, in the proper order, the steps that must be completed to ensure a satisfactory soldered joint.
* Describe at least three possible health hazards that can be present when soldering.

Soldering is a term applied to the *coalescence* (growing together) of metals at or near their surfaces. The base metals being soldered may be two *like* metals or two *unlike* metals. Soldering causes the coalescence of two or more base metals without melting any of them. The theory of soldering is that a molten binding or joining metal, called *solder,* adheres to the clean surfaces of the base metals. This adhering is accomplished by means of molecular attraction at or near the mating surfaces. The molecules of solder *entwine* (wrap around) with the base metal's molecules and form a very strong bond. In some cases, the metals in the solder may form a surface alloy with one or both of the base metals. Refer to Heading 4.3 and Figure 4-7.

In soldering, the joining metal melts and flows at temperatures less than 840°F (450°C). The biggest advantage of soldering is minimum warpage and minimum disturbance of the heat treatment of the parent metals being joined. If the joining metal melts and flows above 840°F (450°C), the process is called *brazing*. Brazing is described in Chapter 10. This method of identifying soldering and brazing was established by the American Welding Society.

9.1 SOLDERING PRINCIPLES

Soldering is used where a leakproof joint, neatness, a low-resistance electrical joint, and sanitation are desired. The joint produced by means of soldering is not as strong as a brazed or welded joint, and in many cases, a mechanical joint is used together with the solder seam. The most commonly used solders contain alloys of tin, antimony, silver, cadmium, indium, aluminum, or lead. The proportions of the metals in a soldering alloy are varied, producing a variety of properties. Different solders are produced to meet various needs. There are different solders for joining tin, copper, brass, aluminum, bronze, sheet iron, sheet steel, and even glass. Soldered joints have excellent heat conductivity and electrical conductivity. Because of this and the localized and limited heat inputs while soldering, this process is widely used in the electronics industry. See Figure 9-1. The soldered joint usually has less strength than the metals being joined. Soldered assemblies must be kept at a low operating temperature to prevent the soldered joint from failing.

9.1.1 Cleaning Methods Prior to Soldering

Metal surfaces should be cleaned prior to soldering to ensure a good, strong soldered joint. *Mechanical cleaning* may be done by machining, sanding, or wire-brushing. The wire brush should be clean. On most metals, it is advisable to use a stainless steel wire brush. *Chemical cleaning* is also done. A different chemical solution is generally required for each metal or alloy to be soldered.

Some of the recommended cleaning solutions and the metals they are used on are listed below:
* For copper and copper alloys, *solvents* such as toluene, mixed acetates, and trichloroethylene may be used. Alkaline cleaning solutions may also be used. **CAUTION: Vapors from trichloroethylene and other chlorinated hydrocarbons are toxic and may create phosgene gas when heated. They should only be used when the ventilation is adequate.**

Figure 9-1. *This television repair technician is making a soldered repair with a small electric soldering iron.*

- *Pickling* is a process for removing scale and oxides from metals using acids or other chemical solutions. To pickle copper or copper alloys, a 5% -10% solution of sulfuric acid is used at a temperature of 125°F to 150°F (51°C to 65°C).
- Aluminum may be degreased with a *degreasing solvent*. Mechanical cleaning is also advised.
- Stainless steel should be mechanically cleaned. Pickling is recommended using 15% to 20% nitric acid and 2% to 5% hydrofluoric acid at 140°F to 180°F (60°C to 82°C).
- For most surfaces, degreasing solvents or alkaline solutions can be used to remove grease and oil. Hydrochloric, nitric, phosphoric, sulfuric, or hydrofluoric acids can be used for pickling.
- Parts that have been pickled should be dried quickly and soldered immediately while the surfaces are clean.

Always wear chemical-type eye goggles, rubber gloves, and long sleeves when using cleaning solutions, pickling solutions, or acids. Chemical-type eye goggles do not have holes in the shield area above the eyes.

If any of the chemical solutions, acids, or fluxes come in contact with your skin, wash the area thoroughly under running water. If such chemicals contact your eyes, wash thoroughly under running water and have a doctor check them immediately.

9.2 SOLDER ALLOYS

An *alloy* is a mixture of two or more metals or nonmetals. A metal alloy has characteristics different from any of the pure metals in the alloy. Metal characteristics such as hardness, melting temperature, and strength are changed when an alloy is created.

For example, alloying tin and lead creates a lower-melting-point metal. Lead melts at 621°F (327°C) and tin melts at 450°F (232°C). If they are combined in equal parts (a "50/50" alloy), the melting temperature will be 361°F to 421°F (183°C to 217°C). The alloy melts at a temperature *lower* than pure tin or pure lead.

Soldering alloys must melt below 840°F (450°C) and below the melting point of the base metals on which they are used.

Tin-lead solder alloys are still used, but to a lesser degree than previously. Lead has been found to be *toxic* (very harmful) when used on water lines and food containers. Tin-lead may be alloyed in percentages from 5%

tin and 95% lead to 70% tin and 30% lead. The tin content is always given first: 5/95, or 70/30.

Figure 9-2 is a graph that shows the solidus and liquidus lines for tin-lead solders. The *solidus line* (A, B, C, D, E) indicates the temperature at which the metal or alloy begins to melt. The *liquidus line* (A, C, E) indicates the temperature at which the metal or alloy is completely liquid. The graph covers the range from 0% tin and 100% lead to 100% tin and 0% lead. This graph is called the *tin-lead phase diagram.*

For a 40/60 tin-lead solder, the solidus point is at 361°F (183°C). The liquidus point is at 455°F (235°C). This alloy, therefore, is gradually melting between these two temperatures.

An alloy of 62/38 tin-lead falls at the eutectic point for tin-lead alloys. The *eutectic point* is the point on the graph where the alloy begins to melt and completely melts at *one* temperature. All other alloys melt over a *range* of temperatures from solidus to liquidus. See Figure 9-3 for a table that shows the solidus and liquidus temperatures for tin-lead alloys.

Tin-lead solders may be used to join most metals. The 35/65, 40/60, and 50/50 alloys have the best wetting (flowing) properties, strength, and economy. They are therefore, widely used. High-lead content solders like 10/90, 15/85, and 20/80 have the best mechanical (strength) properties of all the alloys.

Other soldering alloys include tin-antimony, tin-antimony-lead, tin-silver, tin-silver-lead, tin-zinc, cadmium-silver, cadmium-zinc, zinc-aluminum, and tin-lead-indium.

A 95% tin and 5% antimony solder has a melting range of 450°F to 464°F (232°C to 240°C). For use on drinking water lines, a 95% tin and 5% antimony solder is recommended, since it is not toxic like the tin-lead solders. It is also used for soldering in refrigeration and air conditioning. See Figure 9-4.

Antimony may be added to tin-lead solder. The mechanical properties of a tin-lead solder are improved by adding up to 6% antimony. The antimony percentage replaces the tin in the alloy. Antimony alloy solders are not advised for use on galvanized iron or steel. The antimony may mix with the zinc coating and produce a gritty, brittle solder mixture.

Tin-silver alloys, which contain no lead, are used to join stainless steel on food handling equipment. This solder alloy is 96% tin and 4% silver. See Figure 9-4. When soldering silver surfaces in electronic equipment, a tin-lead-silver alloy is used. This alloy is 62% tin, 36% lead, and 2% silver.

For high strength and *cryogenic* (extremely cold) applications, a high-lead-content alloy is used. Such an alloy contains 5% tin, 94.5% lead, and 0.5% silver. See Figure 9-4 for other tin-lead-silver alloy combinations.

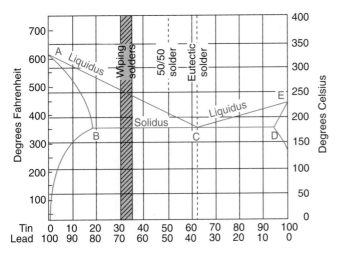

Figure 9-2. *A graph of the solidus and liquidus temperatures for alloys of tin and lead. The solidus temperature is the point at which the metal begins to melt. At the liquidus temperature, all the metal is a liquid.*

ASTM Solder Classification	Composition (% by weight)		Solidus Temperature		Liquidus Temperature	
	Tin	Lead	°F	°C	°F	°C
5	5	95	572	300	596	314
10	10	90	514	268	573	301
15	15	85	437	225	553	290
20	20	80	361	183	535	280
25	25	75	361	183	511	267
30	30	70	361	183	491	255
35	35	65	361	183	477	247
40	40	60	361	183	455	235
45	45	55	361	183	441	228
50	50	50	361	183	421	217
60	60	40	361	183	374	190
*	62	38	361	183	361	183
70	70	30	361	183	378	192

Figure 9-3. *The solidus and liquidus temperatures for various tin-lead solders. *This is the eutectic alloy. Its liquidus and solidus temperatures are the same at 361°F (183°C).*

Solder Alloy Compositions and Melting Temperatures

Alloy	Composition (% by Weight)								Solidus Temperature		Liquidus Temperature	
	Tin	Lead	Silver	Antimony	Cadmium	Zinc	Aluminum	Indium	°F	°C	°F	°C
Tin-Antimony	95			5					450	232	464	240
Lead-Tin-Silver	96		4						430	221	430	221
	62	36	2						354	180	372	190
	5	94.5	0.5						561	294	574	301
	2.5	97	0.5						577	303	590	310
	1.0	97.5	1.5						588	309	588	309
Tin-Zinc	91					9			390	199	390	199
	80					20			390	199	518	269
	70					30			390	199	592	311
	60					40			390	199	645	340
	30					70			390	199	708	375
Silver-Cadmium			5		95				640	338	740	393
Cadmium-Zinc					82.5	17.5			509	265	509	265
					40	60			509	265	635	335
					10	90			509	265	750	399
Zinc-Aluminum						95	5		720	382	720	382
Tin-Lead-Indium	50							50	243	117	257	125
	37.5	37.5						25	230	138	230	138
		50						50	356	180	408	209

Figure 9-4. Solder alloy compositions and melting temperatures. Alloys containing cadmium are indicated in red type, since they are a health hazard. Note that the melting temperature of an alloy occurs in a range between the solidus and liquidus temperatures.

Only inorganic fluxes are recommended for use with these solders. This is due to the high melting temperatures involved. Fluxes will be studied in Heading 9-3.

Tin-zinc solders are used to solder aluminum. Zinc content as high as 40%, or the addition of 1% to 2% aluminum, improves corrosion resistance.

A 95% cadmium and 5% silver solder is used to join parts that must operate at high temperatures. This alloy melts between 640°F and 740°F (338°C and 393°C). See Figure 9-4. Aluminum-to-aluminum and aluminum-to-other-metal joints can be accomplished with this solder. High-strength joints are also possible with this alloy.

Cadmium and zinc alloy solders are used to solder aluminum. They provide good corrosion resistance when used with the proper flux. See Figure 9-4.

A 95% zinc and 5% aluminum alloy was developed for use in soldering aluminum. Its melting point is very high at 720°F (382°C).

Soldering alloys and fluxes containing certain chemicals are toxic or dangerous to your health. The American Welding Society booklet Z49.1-94, *Safety in Welding and Cutting,* lists the following as having very low Permissible Exposure Levels (PEL) or Threshold Limit Values (TLV):

antimony	chromium	mercury
arsenic	cobalt	nickel
barium	copper	ozone
beryllium	lead	selenium
cadmium	manganese	silver
		vanadium

Soldering alloys and fluxes containing any of these materals pose a health hazard. Extremely good ventilation is necessary when they are present.

Indium solder, containing 50% indium and 50% tin, adheres to glass. This solder may be used to solder glass to glass or glass to metal. Indium solder does not require any special techniques in its use. See Figure 9-4.

Fusible alloys are used in soldering operations where the temperature should remain below 361°F (183°C). These alloys contain bismuth as about 50% of the alloying metal. See Figure 9-5.

The low melting temperatures of these alloys make them useful when soldering on metals or in areas where higher temperatures cannot be tolerated. Such applications are:

Fusible Alloy Solders									
Composition (% by Weight)						Solidus Temperature		Liquidus Temperature	
Alloy	Bismuth	Lead	Tin	Cadmium	Antimony	°F	°C	°F	°C
Eutectic	52	40		8		197	91	197	91
Eutectic	52.5	32	15.5			203	95	203	95
Bending	50	25	12.5	12.5		158	70	165	74
Rose's	50	28	22			204	96	225	107
Lipowitz	50	26.7	13.3	10		158	70	158	70
Matrix	48	28.5	14.5		9	217	102	440	227
Mold or Pattern	55.5	44.5				255	124	255	124

Figure 9-5. *Fusible alloy solders. Cadmium alloy numbers are in red type. For safety, cadmium soldering must be done with excellent ventilation.*

- Soldering heat-treated areas.
- Soldering on fire sprinkler links where the solder must melt at a low temperature to permit sprinklers to actuate in a fire.
- Soldering near heat-sensitive devices.

When using bismuth alloy solders, a corrosive flux must be used. If the surfaces are plated with tin or tin-lead, a rosin flux can be used. More information about fluxes is found in Heading 9.3.

Solder comes in a variety of solid shapes including ingots, ribbons, foils, solid and flux cored wires, and in paste forms. Any size, alloy, weight, or form is available on special order.

For further information regarding solder specifications refer to the following:

American Society for Testing and Materials publications:

ASTM B32, *Solder Metal;* ASTM B284, *Rosin Flux Core Solder;* ASTM B486, *Paste Solder.*

United States Government publication: QQ-S-571, *Solders.*

9.3 SOLDERING FLUXES

Surfaces to be soldered must be very clean. The base metal surfaces are first cleaned with solvents. They are cleaned again during the soldering process by *soldering fluxes*. A good flux will keep the solder and base metal from oxidizing while they are being soldered.

Wetting is an action in which a heated solder or flux becomes liquid and flows and spreads evenly, adhering to the base metal in thin continuous layers. Soldering flux should have a good wetting action.

There are three classifications of fluxes: *organic, inorganic,* and *rosin-based.*

In chemistry, organic compounds are mixtures that contain carbon. **Organic fluxes** are moderately active and have a medium level of cleaning ability. Organic fluxes are corrosive during the soldering operation; after soldering, they generally become noncorrosive or inert. Water-soluble organic fluxes are ideal for use on electronic assemblies. The flux residue can be removed by washing with water. These fluxes do char or burn easily. The temperatures at which an organic flux is most efficient range between 200°F to 600°F (93°C to 316°C). These fluxes consist of organic acids and bases.

The fumes produced during soldering can be corrosive. Care must be taken to prevent these corrosive fumes from damaging parts in the area. Organic flux residues are easily cleaned and removed after soldering is completed.

Inorganic fluxes do not contain carbon compounds. They are considered the most active because they clean surfaces better than other types of flux. Torch, oven, resistance, or induction soldering can be done with these fluxes, because they do not char or burn easily. Inorganic fluxes are highly corrosive. All areas that have had contact with an inorganic flux must be cleaned after soldering to stop any possible corrosive action. This flux is not recommended for soldering electrical joints.

Rosin-based fluxes are the least active of the flux types. They are the least effective in cleaning off metal oxides or tarnishes. All types of electrical and electronic soldering are best done with a rosin-based flux. Rosin fluxes are classified by their activity as:

- Nonactive.
- Mildly active.
- Fully active.

Federal specifications designate rosin fluxes as R, RMA, and RA.

Nonactive (R) fluxes are rosins dissolved in alcohol or turpentine. These fluxes are only used on highly solderable surfaces.

Mildly active (RMA) fluxes are used on highly solderable surfaces. They clean better and faster than nonactive fluxes.

Fully active (RA) fluxes are most commonly used. They are most active and clean best. They may be corrosive, but only during the soldering operation. After the soldering is completed, this flux is generally noncorrosive. Flux residues that are known to be corrosive must be removed from the joint after soldering. Figure 9-6 shows an automated soldering operation. Note the chemical corrosion on the torch heads and holders.

Figure 9-6. *Two automated soldering torches heating a part that has a piece of solder prepositioned in the joint. (Handy & Harman/Lucas-Milhaupt, Inc.)*

Rosin fluxes are easily cleaned from parts after soldering, and they generally leave no corrosive residue. See Figure 9-7 for suggested fluxes for use in soldering electric and electronic connections.

Fluxes may be purchased as dry powder, cores within soldering wires, pastes, and liquids. See Figure 9-8 for a table that shows the flux requirements for soldering various metals.

9.3.1 Flux Residue Removal

Inorganic fluxes contain inorganic acids and salts. These are highly corrosive and must be removed completely. Organic fluxes which contain mild organic acids or urea compounds should also be removed to prevent continuing corrosion.

Rosin fluxes may generally be left on the joints. However, active rosin fluxes that contain organic com-

pounds must be removed. Rosin fluxes should be removed if the joint is to be painted or otherwise finished.

Inorganic fluxes are chloride salts, such as sodium chloride and zinc-aluminum chloride. Sodium- and zinc chloride-based fluxes are corrosive. Their residue will absorb water and cause rust to form. Sodium- and zinc chloride-based fluxes are best removed with hot water containing 2% hydrochloric acid. Follow all safety procedures listed in Heading 9.1.1. This application should be followed with a hot water and mild detergent rinse. If necessary, the initial cleaning should be followed by washing with hot water and sodium carbonate (washing soda), then a hot water rinse.

Aluminum joints soldered with active-type fluxes may generally be cleaned with hot water. If this is not effective, then the joint must be scrubbed with hot water, then dipped in a 2% sulfuric acid solution. This is followed by immersion into a 1% solution of nitric acid. Follow all safety recommendations.

Organic fluxes are generally easily removed with hot water. An organic solvent may be required to remove greasy or oily flux residues.

If rosin flux residues must be removed, alcohol or chlorinated hydrocarbons may be used. Certain rosin residues must be removed with an organic solvent followed with a hot water rinse.

9.4 SOLDERING PROCEDURES

Many methods may be used to solder single or multiple parts, using manual methods or automatic soldering equipment. Some of these methods are:
- Electrically heated or air-fuel gas heated soldering irons. See Figures 9-1 and 9-15.
- Torch soldering. The metal in Figure 9-9 is being torch-soldered.
- Dip soldering.
- Wave soldering.
- Oven soldering.
- Resistance soldering.
- Induction soldering.
- Infrared soldering.

Before describing soldering procedures in detail, it should be pointed out that several things must be done in order to produce successful soldering. These are:
- Metals to be soldered together must be chemically and/or mechanically cleaned. All the oxides, grease, and dirt must be removed.
- Metals to be soldered together must be heated.
- Metals to be soldered must be firmly supported during the soldering operation, and until they cool.
- The proper flux must be used. This flux must be fresh and it must be as chemically pure (CP) as possible.
- The solder should be melted only by the heat in the metals to be soldered together.

		Rosin Fluxes			Organic fluxes (Water soluble)
Metals	**Solderability**	**Non-activated**	**Mildly activated**	**Activated**	
Platinum Gold Copper Silver Cadmium plate Tin (Hot-dipped) Tin plate Solder plate	Easy to solder	X	X	X	X
Lead Nickel plate Brass Bronze Rhodium Beryllium copper	Less easy to solder	Not suitable		X	X
Galvanized iron Tin-nickel Nickel-iron Mild steel	Difficult to solder	Not suitable			X
Chromium Nickel-chromium Nickel-copper Stainless steel	Very difficult to solder	Requires precoating for electronic applications			Not suitable
Aluminum Aluminum-bronze	Most difficult to solder				Not suitable
Beryllium Titanium	Not solderable				

Table title: Fluxes for Electronic Soldering

Figure 9-7. Suggested fluxes for use in electronic soldering on a variety of metals. Some metals are easier to solder than others. Beryllium and titanium are not solderable.

- The soldering operation should be done as quickly as possible.
- An excess of solder is useless and unsightly.
- Most solder fluxes should be thoroughly removed from the joint as soon as possible after the soldering operation is completed to prevent continuing corrosion.

The metals may be cleaned chemically if they are then thoroughly rinsed and dried. The metals may also be cleaned by filing or wire brushing. Use only clean tools, clean steel wool, and/or clean stainless steel wool. The metals should be heated to just above the flow temperature of the solder. Clean heat should be used.

If either piece of base metal moves while the metal is cooling from its liquid temperature to its solidification temperature, the solder will probably contain cracks and will fail. It is sometimes necessary to firmly support the metals with a fixture, clamps, or other means, to make sure they do not move while the soldering operation is in progress.

In order to keep the flux clean, be sure the main flux container is sealed when not in use. Remove only the quantity of flux needed for the particular job. Apply the flux with a clean brush, paddle, injector, or automatic applicator. Brushes and paddles should be thoroughly washed each day. Water-based fluxes should be used immediately. With most paste-type fluxes, you may wait as long as an hour before using the flux. The soldered joint should be completed as soon as possible once the flux is heated, since any delay will cause flux salts to form.

One method used to properly proportion the flux to the joint to be soldered is to use a paste made of the flux and filler metal in a powder form. Such a paste is shown in Figure 9-10. This is obtainable with a variety of solders and brazing alloys mixed with the proper fluxes. These fluxes are applied by using dispensers as shown in Figures 9-11 and 9-12.

Recommended Fluxes for Various Metals

Base metal, alloy or applied finish	Flux Recommendations		
	Corrosive	Non-corrosive	Special flux and/or solder
Aluminum			X
Aluminum-bronze			X
*Beryllium			
Beryllium copper	X		
Brass	X	X	
Cadmium	X	X	
Cast iron			X
*Chromium			
Copper	X	X	
Copper-chromium	X		
Copper-nickel	X		
Copper-silicon	X		
Gold		X	
Inconel			X
Lead	X	X	
Magnesium			X
*Manganese-bronze			
Monel	X		
Nickel	X		
Nichrome			X
Palladium		X	
Platinum		X	
Rhodium	X		
Silver	X	X	
Stainless steel			X
Steel	X		
Tin	X	X	
Tin-bronze	X	X	
Tin-lead	X	X	
Tin-nickel	X	X	
Tin-Zinc	X	X	
*Titanium			
Zinc	X		
Zinc die castings			X

Figure 9-8. Flux recommendations for soldering various metals. *Soldering not recommended.

Figure 9-9. An assembly being completed using the torch soldering process. (Handy & Harman/Lucas-Milhaupt, Inc.)

Figure 9-10. A kit of various special paste solders. These compounds contain both a paste flux and a filler metal. A syringe is provided for applying the mixture to the soldered joint. (Fusion, Inc.)

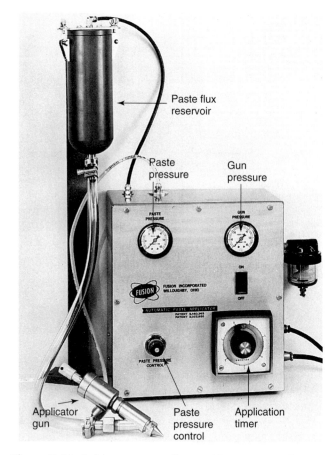

Paste flux reservoir

Paste pressure

Gun pressure

Applicator gun

Paste pressure control

Application timer

Figure 9-11. Solder paste applicator. The amount of paste applied is controlled by the paste pressure and by the timer to 1/20 sec. The quantity applied by weight can vary from 0.001 oz. (0.128 g) to 1 oz. (137.8 g).

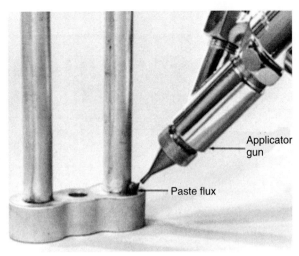

Figure 9-12. A close-up of soldering flux paste being applied to a part. The amount applied is very precise.

9.4.1 Joint Design for Soldering or Brazing

The design for joints to be soldered or brazed is considerably different from joint design for welding. Figure 9-13 shows a variety of soldered items.

Solder and brazing alloys are weaker than the base metals they are used to join. Therefore, joints must be designed so that they do not depend on the strength of the filler metal used.

There are many joint designs used for soldering and brazing. Examples of these joints are shown in Figure 9-14. Clearance between the parts of a soldered or brazed joint must be small enough to allow capillary action to take place. **Capillary action** causes a liquid to be drawn into the space between tightly fitted parts. This desired clearance is about 0.003" (0.076mm). Too large a clearance may stop the capillary drawing of solder into the joint. Large clearances, if filled with solder or brazing alloy, will be weak. Forces on the joint may cause the joint to fail in the weaker filler metal.

Figure 9-13. A variety of soldered and brazed products, including a frying pan, an ice cream scoop, and a partially complete welding torch. (Handy & Harman/Lucas-Milhaupt, Inc.)

9.4.2 Tinning

Adhering a very thin layer or film of solder to a metal surface is called *tinning*. The word "tinning" is from an old sheet metal practice in which thin layers of tin were applied to steel. This tinning was done with heat to prevent corrosion. "Tin cans" are not actually *tin*, but are made of steel with an extremely thin coating of tin or a tin alloy on the surface of the steel. Copper wire also is frequently tinned as it is manufactured. In all soldering operations, the solder tins the surfaces as the process travels along the joint. Some assemblies are easier to solder if each part is first tinned. The parts are assembled after tinning. They are then reheated to cause the tinning solder to flow and complete the joint. This process for soldering is called *sweating* a joint. It can be done with a torch or soldering iron. Tinning normally requires an active corrosive flux. Noncorrosive flux can be used where the final assembly is difficult to clean.

9.5 SOLDERING IRON METHOD

The electrically heated soldering iron method is popular because it is convenient and easy to use. The tip of the soldering iron is heated electrically. The tip, usually made of copper or a copper alloy, heats the seam to be soldered. The end of the copper tip may be plated. The efficiency of heat transfer from the copper tip to the base metal makes copper the ideal metal for the tip. Also, copper is easily tinned so that the molten solder will adhere more readily to the soldering tip.

The process of cleaning and coating the soldering iron tip with solder is called *tinning*. To tin a soldering tip, a small amount of the solder alloy being used is touched to the heated tip. The solder will flow onto the tip. Excess solder is immediately wiped off with a towel or wet sponge.

Small electrically heated soldering irons are used for electrical and electronic applications. See Figure 9-1. Soldering on heavy metal roofing is often done using a large air-fuel gas soldering iron, as shown in Figure 9-15.

The advantage of the soldering iron is that its heat is concentrated on a small area. It normally does not overheat surrounding parts. The temperature of the tip on electrical soldering irons is regulated by a controller that uses a temperature feedback circuit. See Figure 9-16.

9.6 TORCH SOLDERING METHOD

Soldering torches provide a fast and flexible method for providing heat for soldering. Several types of soldering torches are the:
- Oxyfuel gas torch.
- Compressed air-natural gas torch.
- Air-fuel gas torch. Gases such as propane or MAPP may be used. See Figures 9-17 and 9-18.

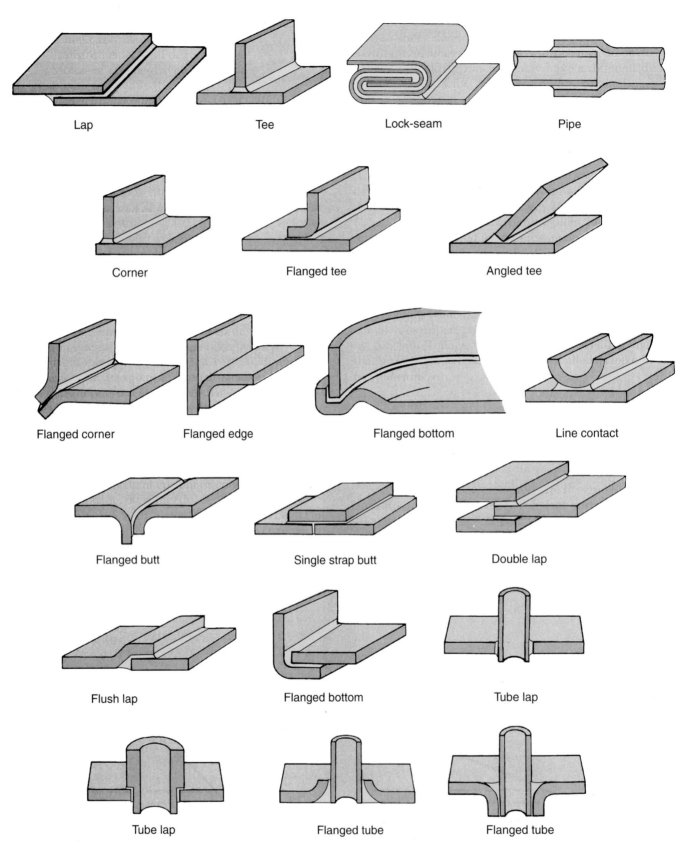

Figure 9-14. *Suggested joint designs for soldered and brazed joints. The spacing is exaggerated for purposes of illustration. Normal spacing and filler metal thickness is approximately 0.003" (0.076mm). (The Aluminum Association)*

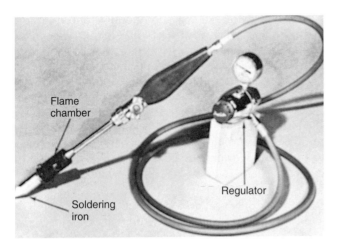

Figure 9-15. A closed flame air-fuel gas torch equipped with a soldering iron attachment. The regulator is attached to the fuel gas cylinder. (ESAB Welding and Cutting Products)

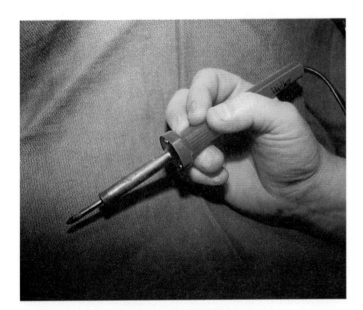

Figure 9-16. A small electric soldering iron.

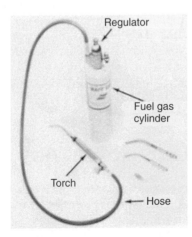

Figure 9-17. An air-fuel gas soldering and brazing outfit. The regulator is attached to the fuel gas cylinder.

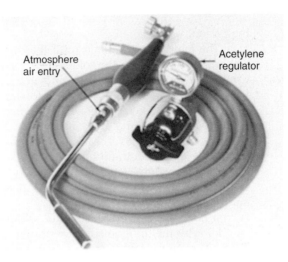

Figure 9-18. Air-fuel gas torch and regulator. This torch is equipped with a "swirljet" tip. The air and fuel gas is swirled through a set of propeller-like vanes at the rear of the flame tube. This produces a high heat output. (ESAB Welding and Cutting Products)

In order to solder satisfactorily with a torch, the following conditions must be met:

- The flame must heat the metals to be soldered.
- The flame must be clean, so that the surfaces will not be oxidized or dirtied by the flame gases.
- Heat from the flame must be concentrated.
- The amount of heat must be easily adjusted. Figure 9-19 shows a variety of tip sizes and shapes for soldering with the air-fuel gas flame.

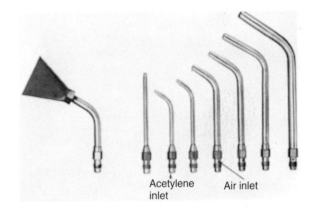

Figure 9-19. Variety of tips used with an air-fuel gas torch. More heat may be obtained by using a larger tip size; however, each torch tip flame operates at the same temperature. (ESAB Welding and Cutting Products)

The general procedure for torch soldering is as follows:

- Clean the surfaces to be soldered. Refer to Heading 9.1.1.
- Apply the correct flux to the joint surfaces. Refer to Heading 9.3.

- Support the joint through the soldering and cooling period.
- Heat the joint with the torch until the metal is just hot enough to melt the solder alloy used.
- Continually move the torch toward and away from the base metal. Move the flame toward the joint to keep the metal hot enough to melt the solder. Move the flame away from the joint to keep the metal from overheating. Touch the solder to the joint. Continue to add solder to the joint until it is completely filled.
- If the solder does not adhere, stop the operation and clean and reflux the joint.
- A common soldering error is to add solder to a cool joint by melting the solder with the torch. The base metal should melt the solder.

Another error is to use too much solder. Only the properly adhered solder at the surface is effective. If too much solder is used, a neat-looking joint may still be obtained by carefully wiping off the excess solder while it is still molten, using a clean, thick cloth. It is very important to avoid overheating the base metal.

Torch soldering is often used to restore the contour of damaged parts, especially automobile sheet metal. Figure 9-20 shows the edge of a replacement panel being soldered onto the existing panel. A thick coat of solder is applied. The solder is then "dressed" after it has cooled to create a smooth blended surface.

The refrigeration, air conditioning, and plumbing industries use a great number of soldered fittings. See Figure 9-21. Before soldering is begun, the pipe and fitting must be mechanically cleaned and fluxed, and the tubing firmly supported. Soldered pipe and tubing fittings use the principle of capillary action to draw the solder into the fitting. A small clearance is provided between the pipe or tube and the fitting. This allows solder to be drawn into the joint. Tube and pipe fittings may be soldered or brazed. See Chapter 10 for brazing procedures.

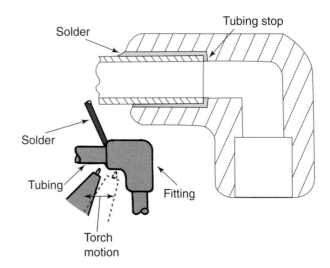

Figure 9-21. *Soldered tubing or pipe joint. Solder is drawn into the joint by capillary action. Note that the solder is applied at a point away from the torch flame.*

Before cast iron can be soldered, it must first be filed, machine-blasted, or shot-blasted to remove the oxide skin. The mechanically cleaned surface must then be degreased and cleaned in a special molten flux bath that will remove the surface graphite.

Many solder joints are made automatically. Parts to be soldered are mounted in fixtures. Flux and solder are often added to the joint automatically. Heating is automatic and the soldered joint is cooled before the parts are ejected from the fixtures. See Figure 9-22. Some installations use formed and preplaced solder rings or solder foil instead of automatic feeding of a solder-flux mixture or a solder filler wire. Solder for automatic operations comes in many designed forms such as sheets, rings, or ribbons, as shown in Figure 9-23. Precleaning and postcleaning are sometimes a part of the automatic process, as well.

9.7 DIP SOLDERING METHOD

This method consists of melting a quantity of solder in a tank or pot. The solder is protected by means of a hood or chemical covering, such as powdered charcoal, to prevent oxidation. The articles to be soldered are dipped in a flux bath, and then in the solder bath. This method is a labor-saving device used to either solder-coat surfaces to

Figure 9-20. *This auto body repair technician is applying body solder to the seam of a replacement panel during the repair of an automobile.*

Figure 9-22. *A twelve-station rotary soldering machine. The flux which contains powdered solder is automatically applied. Part assemblies are progressively heated until the flux and solder flow. The parts are then cooled and automatically ejected from the machine.*

make them rustproof, or to fasten the various parts of an assembly together.

The articles to be soldered together are usually assembled and then acid-cleaned (pickled). They are then thoroughly washed and dried before being dipped into the molten solder. The articles are lifted out of the bath and any excess solder is allowed to drain from them.

Parts to be dipped into molten solder must be dry, since a small amount of moisture will produce instant high-temperature steam. This may cause an eruption of the molten solder, with possible injury to workers and destruction of property.

9.8 WAVE SOLDERING

Molten solder in large tanks may be made to create small waves, as in the ocean. The wave height is precisely controlled by means of venturis, baffles, and screens to contact only the *bottom* of the part to be soldered.

Parts such as electronic circuit boards may be soldered by passing them over the molten solder so that they are touched by the waves. See Figure 9-24. A typical solder wave tank may be up to 24" (0.61m) wide. Parts may be moved over the wave crest at a rate of 12' (3.66m) per minute. With this speed, up to 1200 electronic joints could be soldered per minute per foot (0.3m) of wave width. See Figure 9-25.

9.9 OVEN AND INFRARED SOLDERING

Ovens heated by fuel gas or electricity are used to solder a large number of joints at one time. To *oven solder*, parts are assembled with preformed and preplaced solder forms in each joint. See Figure 9-23. A precise amount of

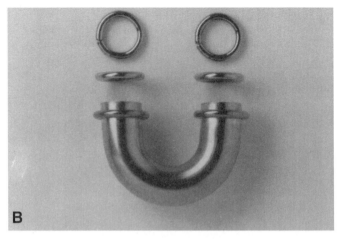

Figure 9-23. Solder is produced in a large variety of forms. Many of these forms are prepositioned in the joint. The joint is then heated in some manner, completing the soldering process. A—A variety of solder forms. Note how the ring forms are produced in long tubular shapes for high-production applications. B—A tubing connector with prepositioned solder rings. (Handy & Harman/Lucas-Milhaupt, Inc.; J.W.Harris Co., Inc.)

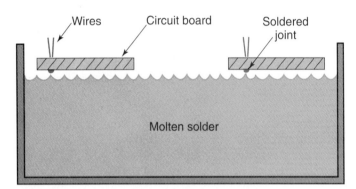

Figure 9-24. Wave soldering schematic. Electronic circuit boards are passed over the molten solder wave. The crests of the waves just touch the joints. Several thousand joints per minute can be made with this process.

flux is placed on each joint. The parts are then placed in the soldering oven and heated until the solder flows. In large production facilities, the parts are passed through the oven on a conveyor assembly. After the joints are made, the parts are cooled and cleaned.

Infrared soldering is similar to oven soldering. The difference is that the parts are heated by infrared lamps. This heating usually occurs in an enclosed space, so that workers are not exposed to the infrared light rays.

9.10 RESISTANCE AND INDUCTION SOLDERING

In *resistance soldering,* parts are assembled with pre-placed and preformed solder forms, and flux is applied to each joint. Electricity is passed through the base metal,

Figure 9-25. Wave soldering is often used to connect electronic components mounted on printed circuit boards. A—The printed circuit board at top, with its multicolored components, is about to pass over the solder wave. B—A bottom view of the completed circuit board, showing the soldered connections. (Electrovert)

causing it to heat because of its resistance to electrical flow. The metal is heated until the solder begins to flow. The electric current is then stopped. After soldering, the parts are cooled and cleaned.

In *induction soldering,* the parts are prepared as above. The heating of the base metal is done by passing a current through an insulated copper wire coil. The part to be soldered is placed inside the coil of copper. As the current passes through the coil, it causes *(induces)* a current to flow in the part also. This current in the part occurs without actual electric contact. The part is heated by resistance to this induced current.

When the part is hot enough to cause the solder to flow, the current to the coil is turned off. There is no longer any induction heating, so the part begins to cool.

9.11 STAINLESS STEEL SOLDERING

Many kinds of stainless steels are now being used. Some of the more common stainless steels are:
- 200 series (approximately 16% to 19% chromium, 3.5% to 6% nickel).
- 301 to 308 series (approximately 16% to 24% chromium, 6% to 12% nickel).
- 309 to 314 series (approximately 22% to 26% chromium, 12% to 22% nickel).
- 316 to 347 series (approximately 16% to 19% chromium, 10% to 13% nickel).

Properties of stainless steels are described in Chapter 21.

Stainless steel joints can be soldered, brazed, or welded. Soldering is used often to make low-strength, leakproof joints, and to eliminate cracks and crevices.

A corrosive flux is needed to promote adhesion of the solders, since the chromium surface film on the stainless steel resists ordinary fluxes. A highly active inorganic flux is generally used. It is very important that these corrosive fluxes be completely removed after soldering. A solution of warm water with 2% hydrochloric acid is used to clean the parts after soldering. This cleaning is followed by a washing with detergent and warm water. **Follow all safety precautions carefully.**

A 50/50 solder makes a good joint, but its color does not match the color of stainless steel. Special solders (usually with no lead content) are available to match stainless steel's color. Solder for stainless steel comes in rods 1/8" (3.2mm) in diameter by 15" (.38m) long.

9.11.1 Stainless Steel Soldering Fluxes

The surface of stainless steel must be cleaned prior to soldering. This can be done by mechanical means such as rubbing with a stainless steel wire brush. Stainless steel may also be precleaned by chemical pickling. Pickling is accomplished with a nitric-hydrofluoric acid mixture. This is a mixture of 15% to 20% nitric acid (HNO_3) and 2% to 5% hydrofluoric acid (HF) at 140°F to 180°F (60°C to 82°C).

Follow all safety precautions for working with acids. Only experienced persons should mix the acid (always add the acid slowly to the water—not water to acid). Wear goggles, face guards, rubber-lined clothing and rubber gloves when preparing this cleaning mixture.

After the acid has been on the stainless steel surface for approximately four minutes, rinse with warm water. After the surface dries, apply a zinc chloride flux, then carefully heat the metal and apply solder as usual. Great care is needed, since stainless steel is difficult to solder. You will find that the solder will adhere more easily and better if the stainless steel surface is scratched, using a small scrubbing motion with the solder filler rod. This action seems to mechanically remove any residual film that prevents the solder from adhering to the surface.

9.12 SOLDERING ALUMINUM ALLOYS

Aluminum and many of its alloys can be soldered in the same manner as other metals. The most easily soldered alloys are those that contain no more than 5% silicon and 1% magnesium. Other alloys are more difficult to solder because of a poorer flux wetting action. Wrought alloy products are easiest to solder. Castings are harder to solder due to their varied composition and surface irregularities.

Precleaning is necessary. Solvent degreasing alone may be sufficient. Occasionally, mechanical cleaning with a stainless steel wire brush is necessary. A modified chemical flux such as zinc chloride and ammonia compounds may be used. In the reaction flux used on aluminum, zinc chloride is the main ingredient.

When heated, the flux mixes with the aluminum oxides and cleans the surface. Both the chemical and reactive types of fluxes are highly corrosive and should be removed after soldering.

There are three soldering alloys generally recommended for aluminum. These are:
- Tin-lead plus zinc and cadmium.
- Zinc-cadmium or zinc-tin based.
- Zinc-based with aluminum and copper.

The zinc-based alloys with aluminum and copper added have the best wetting action, strength, and corrosion resistance. However, they are the hardest to apply.

The same basic procedure is necessary for soldering aluminum as for any other metal. See Heading 9.3.1. The metal, the flux, and the filler rod must be clean. Good support for the parts while they are being joined is important.

Usually, the joint is heated with a carburizing flame. A rather large torch movement is used. The solder should be melted only by the heat from the metal to be joined. The surface of the metal is often rubbed with the soldering filler rod as it is applied. This action helps to remove any remaining aluminum oxides from the surface.

9.13 SOLDERING DIE CASTINGS

Some zinc die castings can be soldered. However, it is a difficult operation because of the varying composition of zinc die casting metal. The solder used is usually an alloy of 82.5% cadmium and 17.5% zinc. This alloy melts (flows) at 508°F (264°C). Any surface coatings or plating must be removed prior to soldering.

Die castings can sometimes be soldered without flux. Die castings may be successfully soldered by first coating the die casting with nickel and then soldering with a tin-lead solder.

Cadmium fumes are toxic and deadly. Soldering on, or with, cadmium or cadmium alloys should only be done under extremely good ventilation conditions.

9.14 TESTING AND INSPECTING SOLDERED JOINTS

Soldered joints are usually tested in two ways.
- They are tested for leaks, when applicable.
- They are tested for humidity and chemical corrosion resistance, when applicable.

A joint can be tested for leaks *hydrostatically* using water under pressure.

The joint can be tested for humidity corrosion resistance by placing it in a cabinet under 100% humidity and at about 100°F (38°C) for 72 hours. Chemical resistance is checked by exposing the joint to an atmosphere containing the chemical that it will be exposed to in actual service. A good joint will reveal no evidence of corrosion under low-magnification inspection.

Visual inspection of the joint will usually reveal such defects as poor adhesion, incomplete soldering, too much solder, overheating, or dirt inclusion.

9.15 REVIEW OF SOLDERING SAFETY PRACTICE

Soldering, brazing or welding with or on alloys containing any of the materials listed in Heading 9.2, especially cadmium or beryllium, can be extremely hazardous.

- **Fumes from cadmium or beryllium compounds are extremely toxic. Several deaths have been reported from inhaling cadmium oxide fumes. Skin contact with cadmium and beryllium should also be avoided.** An expert in industrial hygiene should be consulted whenever cadmium or beryllium compounds are to be used, or when repairs are to be made on parts containing these metals.
- **Fluxes containing fluoride compounds are also toxic.**
- Good ventilation is essential when soldering or brazing and the welder should always observe good safety practices.
- A common hazard when soldering is exposure of the skin, eyes, and clothing to acid fluxes. **Always work in a way that flux will not be spilled on the skin or clothing. Always wear**

chemical-type eye goggles, rubber gloves, and long sleeves when using cleaning solutions, pickling solutions, or acids. Chemical-type eye goggles do not have holes in the shield area above the eyes.
- **If any of these chemical solutions, acids, or fluxes come in contact with your skin, wash the area thoroughly under running water. If such chemicals contact your eyes, wash them thoroughly under running water and have a doctor check them immediately.**
- Be sure that there are no flammable fumes from gasoline, acetylene, or other flammable gases present where soldering is being performed.
- Perhaps the greatest hazard met in soldering is attempting to solder gasoline or other fuel tanks. The job should never be attempted before all safety references have been read and a fire marshal consulted. The fire marshal will normally advise on the safe and proper procedures for soldering or welding on containers which have held combustible liquids or gases.

TEST YOUR KNOWLEDGE

Write your answers on a separate sheet of paper. Do not write in this book.
1. Can soldering be used to join two different base metals together?
2. How does soldering differ from welding?
3. How does soldering differ from brazing?
4. *True or False?* A soldered joint typically has the same strength as the base metal.
5. What does pickling do to a metal?
6. List three common solder alloys.
7. What element, used in some solders, is a health hazard?
8. There are three classifications of fluxes. List all three.
9. What is the purpose of a soldering flux?
10. Which classification of fluxes will best clean a surface?
11. What type soldering flux is recommended in Figure 9-7 for use on brass?
12. After soldering, what method is used to clean a solder joint when an inorganic flux has been used?
13. Name five (5) soldering methods.
14. Should the clearance between two metals being soldered be large or small?
15. When torch soldering, where does the heat required to melt the solder come from?
16. When the dip soldering method is used, the molten solder is in a tank or pot. What is done to prevent the solder from oxidizing?
17. What type flux should be used when soldering stainless steel?
18. How should the surface of aluminum be prepared for soldering?
19. What type solder is used on aluminum?
20. What is the chemical composition of the eutectic tin-lead solder?

Chapter 10

BRAZING AND BRAZE WELDING

LEARNING OBJECTIVES

After studying this chapter, you will be able to:

* Differentiate between brazing and braze welding.
* Select the proper filler metal and flux for brazing or braze welding a particular joint.
* List the things which must be done prior to brazing or braze welding to ensure a strong finished joint.
* List, in the proper order, the procedures for making an acceptable brazed joint.
* List, in the proper order, the procedures for making an acceptable braze welded joint.
* Describe at least one possible health hazard involved when brazing or braze welding.

Both brazing and braze welding are metal-joining processes that are performed at temperatures *above* 840°F (450°C), as compared to soldering, which is performed at temperatures *below* 840°F (450°C). Refer to Heading 4.4 and to Figure 4-8.

The American Welding Society defines these processes as follows:

Brazing: "A group of welding processes which produces coalescence of materials by heating them to a suitable temperature and by using a filler metal having a liquidus above 840°F (450°C) and below the solidus of the base metal. The filler metal is distributed between the closely fitted surfaces of the joint by capillary action." (*Coalescence* means the joining or uniting of materials.)

Braze welding: "A welding process variation in which a filler metal, having a liquidus above 840°F (450°C) and below the solidus of the base metal, is used. Unlike brazing, in braze welding the filler metal is not distributed in the joint by capillary action."

Brazing has been used for centuries. Blacksmiths, jewelers, armorers, and other crafters used the process on

large and small articles before recorded history. This joining method has grown steadily both in volume and popularity. It is an important industrial process, as well as a jewelry making and repair process. The art of brazing has become more of a science as the knowledge of chemistry, physics, and metallurgy has increased. Refer to Figure 4-2, which shows the oxyacetylene flames used for brazing and braze welding. Also refer to Figure 4-9, which shows the oxygen-LP gas flames used for brazing and braze welding.

The terms *brazing* and *braze welding* usually imply the use of a *nonferrous alloy* (one that does not contain iron). Nonferrous alloys consist of alloys of copper, tin, zinc, aluminum, beryllium, magnesium, silver, gold, and other metals.

Brass is an alloy consisting chiefly of copper and zinc. *Bronze* is an alloy consisting chiefly of copper and tin. Most rods used in both brazing and braze welding on ferrous metals are brass alloys, rather than bronze. The brands which are called bronze usually contain a small percentage (about 1%) of tin.

10.1 BRAZING AND BRAZE WELDING PRINCIPLES

Brazing is an adhesion process in which the metals being *joined* are heated but not melted. The brazing filler metal melts and flows at temperatures above 840°F (450°C). *Adhesion* is the molecular attraction exerted between surfaces.

A brazed joint is stronger than a soldered joint, because of the strength of the alloys used. In some instances, it is as strong as a welded joint. Brazing is used where mechanical strength and leakproof joints are desired. Brazing and braze welding are superior to welding in some applications, since they do not affect the heat treatment of the original metals as much as welding.

Brazing and braze welding also warp the original metals less than welding, and make it possible to join dissimilar metals. For example, steel tubing may be brazed to cast iron, copper tubing brazed to steel, and tool steel brazed to low-carbon steel.

Brazing is done on metals that fit together tightly. The molten filler metal enters the joint by *capillary action* (a physical process in which a liquid is drawn into the space between two tightly fitted surfaces). Very thin layers of filler metal are used when brazing. The joints and the material being brazed must be specially designed for the purpose. When brazing, poor fit and alignment will result in poor joints and an inefficient use of brazing filler metal.

In braze welding, joint designs used for oxyfuel gas or arc welding are satisfactory. Such joint designs are discussed in Chapters 6 and 12. When braze welding, thick layers of the brazing filler metal are used.

When brazing or braze welding, the metals must be cleaned thoroughly prior to adding the filler metal. The proper flux and filler metal must be used and post-cleaning is sometimes required.

It is important to provide excellent ventilation when brazing or braze welding. Fumes from some heated fluxes and vaporized filler metals are toxic and may affect the respiratory system, eyes, or skin. A list of many of the dangerous chemicals that might be contained in fluxes and brazing filler metals is found in Heading 9.2.

10.2 JOINT DESIGNS FOR BRAZING AND BRAZE WELDING

The brazing process is similar to the soldering process, since it uses a thin layer of filler metal. It also requires specially designed, tightly fitting joints. Brazing also requires that the joint be well supported until it has cooled. Examples of well-designed brazing joints may be seen in Figure 9-14. A few examples of joints designed for brazing, both good and bad, are shown in Figures 10-1 and 10-2.

Braze welding is a process in which a thick layer of filler metal is used. A convex or concave bead is usually employed. The fit of the joint is not as critical as that required for brazing. Figures 10-3, 10-4, and 10-5 illustrate several well-designed braze welding joints.

The proper clearance between the parts in a brazing joint is critical. If the spacing is too large, capillary action may not take place. If the space is too small, an insufficient amount of filler metal may be deposited in the joint. See Figure 10-6 for recommended brazing joint clearances.

Brazing is often done on dissimilar metals. Dissimilar metals have different expansion rates, so special attention

Figure 10-1. *Joints designed to produce good brazing results. (Handy & Harman /Lucas-Milhaupt, Inc.)*

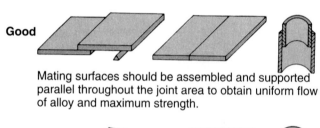

Good

Mating surfaces should be assembled and supported parallel throughout the joint area to obtain uniform flow of alloy and maximum strength.

Poor

Mismated lap joints, V-type joints, and flared tubular joints waste brazing alloy and may reduce the joint strength.

Good

Proper fit and alignment in tubular joints is an assurance of high-strength joints.

Poor

Poor fit can interrupt the pull of capillary action, reduce strength, and make it difficult to get leak-tight joints.

Figure 10-2. *Well-designed joints which have been prepared for brazing. Some poorly-designed joints are shown for comparison.*

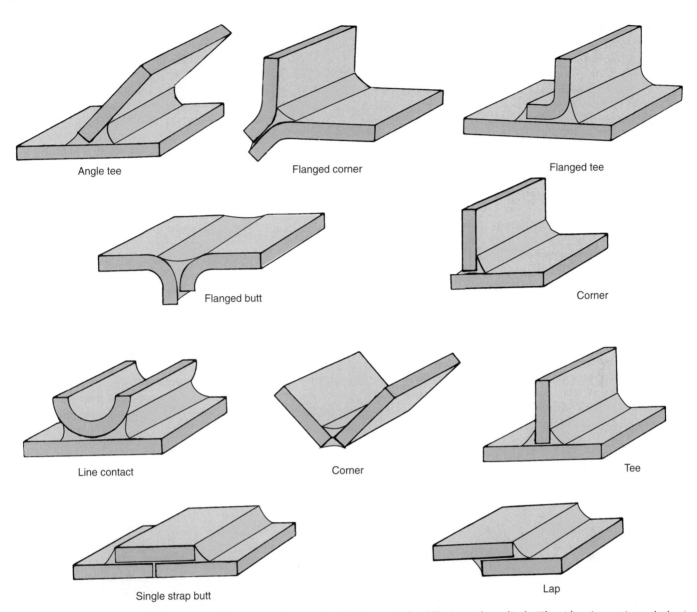

Figure 10-3. Well-designed braze welding joints. Note the thickness of the filler metal applied. (The Aluminum Association)

to the clearance in these joints is required. These varying expansion rates must be considered if the parts are clamped, restrained in jigs, or fit into one another.

Brass expands more than steel at a given temperature. If a brass part is fitted into a steel part, the clearance between them, at room temperature, must be larger than normal. If a steel part fits inside a brass part, the clearance should be closer than normal. See Figure 10-7.

10.3 CLEANING BASE METALS PRIOR TO BRAZING OR BRAZE WELDING

In order to braze successfully, capillary action must occur. It will occur properly only when the parts to be joined are clean. Clean parts are also required when braze welding. Part cleaning prior to brazing or braze welding

may be accomplished by chemical and/or mechanical means.

Oil and grease must be removed, using a proper solvent (*chemical cleaning*). Attempting to mechanically clean oily or greasy surfaces will usually result in scrubbing the oil and grease more deeply into the surfaces of the parts.

Following the removal of oil and grease, a chemical pickling solution is used. This *pickling solution* removes any scale and rust from the surfaces of the parts. The chemicals used on various metals are the same as those described in Heading 9.1.1. **Be sure to follow all safety procedures.**

Mechanical cleaning is done with a grinding wheel, grit blasts, emery cloth, and/or a wire brush. Mechanical cleaning should be followed with a rinsing and drying operation. This rinsing will wash away small particles removed during the mechanical cleaning.

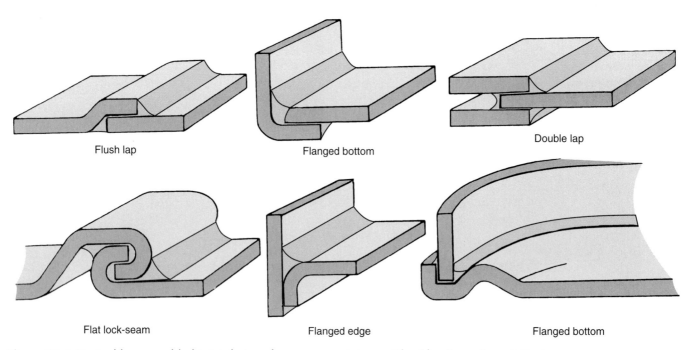

Figure 10-4. *Typical braze-welded joint designs for use on containers. (The Aluminum Association)*

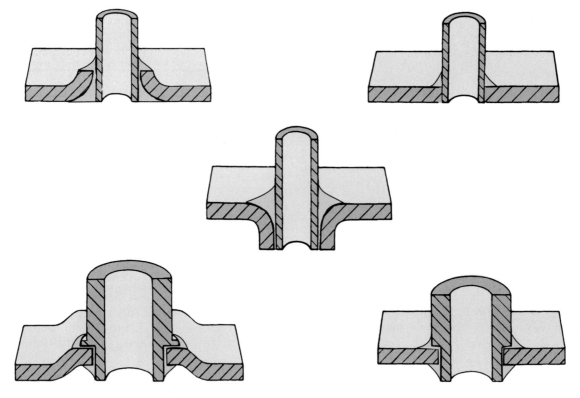

Figure 10-5. *Designs for braze welding sheet and tube joints. (The Aluminum Association)*

Filler metal AWS classification	Joint clearance		Notes
	in.	**mm**	
BAlSi group	0.006-0.010 0.010-0.025	0.15-0.25 0.25-0.61	For length of overlap less than 1/4" (6.35mm) For length of overlap greater than 1/4" (6.35mm)
BCuP group	0.001-0.005	0.03-0.12	} Flux brazing (mineral fluxes) Atmosphere brazing (gas-phase fluxes)
BAg group	0.002-0.005 0.001-0.002	0.05-0.12 0.03-0.05	
BAu group	0.002-0.005 0.000-0.002	0.05-0.12 0.00-0.05	Flux brazing (mineral fluxes) Atmosphere brazing (gas-phase fluxes)
BCu group	0.000-0.002	0.00-0.05	Atmosphere brazing (gas-phase fluxes)
BCuZn group	0.002-0.005	0.05-0.12	Flux brazing (mineral fluxes)
BMg group	0.004-0.010	0.10-0.25	Flux brazing (mineral fluxes)
BNi group	0.002-0.005 0.000-0.002	0.05-0.12 0.00-0.05	General applications (flux or atmosphere) Free-flowing types, atmosphere brazing

Chemical abbreviations

BAlSi = Aluminum-silicon alloy
BCuP = Copper-phosphorus alloy
BAg = Silver-based alloy
BAu = Gold-based alloy
BCu = Copper-based alloy
BCuZn = Copper-zinc alloy
BMg = Manganese-based alloy
BNi = Nickel-based alloy

Figure 10-6. Recommended brazing joint clearances at brazing temperatures for various filler metal groups. The "B" in the filler metal classification stands for brazing. Chemical abbreviations are also given.

10.4 BRAZING AND BRAZE WELDING FLUXES

The American Welding Society defines a *flux* as "a material used to hinder or prevent the formation of oxides and other undesirable substances in molten metal and on solid metal surfaces, and to dissolve or otherwise facilitate the removal of such substances."

A brazing or braze welding flux must be of a composition that will keep clean both the brazing filler metal and the metals being joined during the joining operation. The fluxes must be chemically pure. Generally, the manufacturers mark their fluxes *C.P.* (chemically pure).

The ingredients of a brazing flux are typically chlorides, fluorides, borax, borates, fluoborates, boric acid, wetting agents, and water.

These ingredients are mixed in varying amounts by manufacturers of brazing fluxes. The American Welding Society (AWS) has classified brazing fluxes into six categories. These fluxes are designated for use on various metals. Figure 10-8 lists these AWS flux categories and the suggested uses for each type of flux. The base metal combinations used in brazements often vary greatly. It is often necessary to experiment with fluxes to determine the best one to use for a given combination of metals.

Fluxes used for braze welding must withstand higher temperatures for longer periods than do brazing fluxes.

Borax or boric acid (borax plus water) is a common base for brazing and braze welding fluxes for use with steel and iron. A popular mixture is 75% borax (powdered form) and 25% boric acid (liquid form) mixed to form a paste. Other ratios of these two chemicals are also used to form more solid or more fluid fluxes as needed. These mixtures range from 75% down to 25% powdered borax, with the remainder being boric acid. Some of the commercial fluxes also contain small amounts of phosphorus and halogen salts (a *halogen* is any of the iodine, bromine, fluorine, chlorine, and astatine chemical elements).

Alkaline bifluoride is used as a flux for brazing or braze welding stainless steel, silicon bronze, aluminum, or beryllium copper alloys. **The fumes from these fluxes are generally harmful to health, so good ventilation is very necessary.**

A special flux made of sodium cyanide salts is excellent when silver brazing tungsten to copper. **The fumes from this flux are very dangerous. Avoid breathing the fumes and do not let the flux contact the skin. This flux should be handled only by people specially trained in its use. It must be kept dry, and away from acids, nitrates, and other oxidizing agents.**

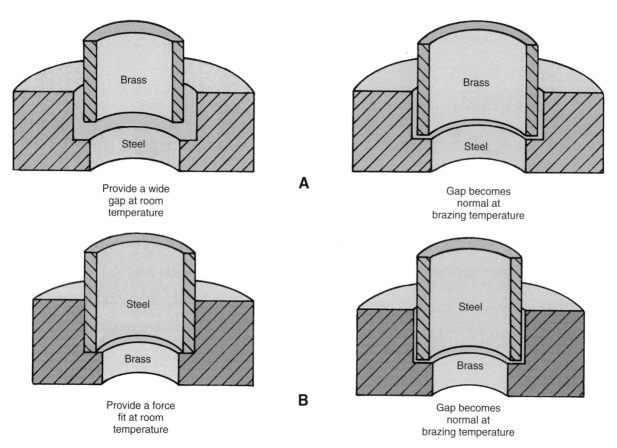

Figure 10-7. *The fitting of dissimilar metals must allow for the different rates of expansion that occur as they are heated. Brass expands at a faster rate than steel. A—Loosely fitted brass in steel fits properly at the brazing temperature. B—Steel force-fitted into brass at room temperature has the correct clearance at the brazing temperature.*

Brazable metal	Recommended filler metal	Brazing temperature range		Flux type	Flux ingredients
		°F	°C		
Aluminum alloys	BAlSi BAlSi BAlSi	1000-1140	538-615	FB1-A FB1-B FB1-C	Chlorides, Fluorides
Magnesium alloys	BMg	900-1150	482-621	FB2-A	Chlorides, Fluorides
Aluminum bronze, aluminum brass, iron or nickel containing Al or Ti	BAg & BCuP	1050-1700	565-927	FB3-A FB3-C	Chlorides, Fluorides, Borates, Wetting agents
All metals not listed above	BCu, BAg, BNi, BAu, RBCuZn	1350-2200	732-1203	FB3-D FB3-H FB3-I FB3-J FB3-K	Boric acid, Fluoborates, Borates, Wetting agents
	BCuP & BAg	1050-1600	565-886	FB3-E FB3-F FB3-G FB4-A	Boric acid, Fluoborates, Borates, Wetting agents
Note: These fluxes may come in either a liquid, powder, paste, or slurry form.					

Figure 10-8.. *Brazable metals and the filler metal and flux types suggested for use when brazing them.*

The following criteria should be considered when choosing a brazing flux:

- The base metal or metals used.
- The filler metal used.
- The heat source (oxyfuel gas, resistance, carbon arc, etc.)
- Ease of flux residue removal.
- Possible corrosive action on the metals involved.
- Health hazards, when using toxic fluxes.
- Electrical conductivity (in resistance brazing, the flux should conduct electricity).
- Fluxes containing water must not be used when dip brazing. The water turns to steam and causes the liquid metal to erupt from the dip bath.

Fluxes may be purchased in powder, paste, or liquid form, depending upon how they are to be applied to the metals being joined.

10.4.1 Methods of Applying Brazing and Braze Welding Flux

Fluxes are applied in whatever way is most efficient. The material may be sprayed, brushed, or applied with a pressurized applicator. The amount used should be sufficient to last throughout the brazing or braze welding operation.

When brazing or braze welding, the brazing rod is usually coated with flux. To apply the flux to the brazing rod on the job, heat the end of the rod for about 3" to 4" (75mm to 100mm) and dip or roll it in the proper flux. This process must be repeated as the rod and flux are consumed. See Figure 10-9. Brazing rods also may be purchased with the flux commercially applied to the entire rod surface.

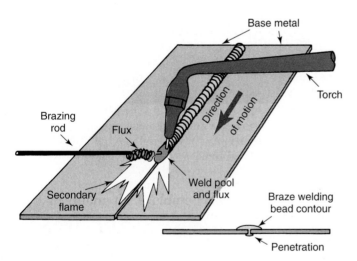

Figure 10-9. *A butt joint being braze-welded using the oxyfuel gas flame as the heat source. Flux has been applied to the brazing rod. The torch is being moved in a forehand direction.*

A method that eliminates the handling of a separate flux is to feed the flux into the fuel gas. This brings the flux to the joint being brazed or braze welded along with the gases, as shown in Figure 10-10. The flux is usually dissolved in alcohol. The installation requires two separate lengths of fuel gas hose with fittings. Several models of gas fluxing equipment, complete with reserve flux tanks, are shown in Figure 10-11.

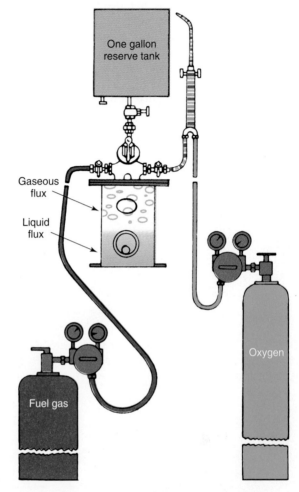

Figure 10-10. *An oxyfuel gas brazing station with a gas fluxing unit in the fuel gas supply line. (The Gasflux Company)*

Figure 10-11. *Four models of gas fluxing units. (The Gasflux Company)*

The fuel gas is fed through a container of liquid flux, where a controlled amount of the flux mixes with the fuel gas and is fed to the operation through the torch tip. This method not only eliminates the separate operation of adding flux, but assures a continuous flow of flux of the correct amount. It results in an excellent and clean joint.

Powdered braze metals can also be injected into the welding flame. Using this method, the heated powder particles are protected from oxidation as they are transferred to the surface of the base metal. Figure 10-12 shows a hopper-type powder brazing torch.

Another method is to thoroughly mix powdered brazing filler metal with the flux in the proper proportions to form a paste. This combination is then added to the joint. This mixture may be hand-fed using flexible plastic bottles, or it may be gun-fed automatically or by hand. See Figure 9-12.

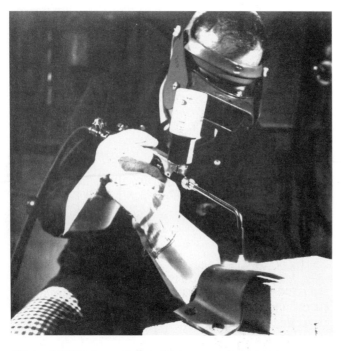

Figure 10-12. *A special oxyfuel gas torch with a hopper feed for ultrafine powder fluxes and/or metals. The torch can be used for brazing, braze welding, metal surfacing, and welding. (Wear Technology Div., Cabot Corp.)*

10.5 BRAZING FILLER METAL ALLOYS

For every brazable metal, there are one or more filler metal brazing rods manufactured. Some of the factors which must be considered when selecting a brazing filler metal are:
- The brazing and service temperatures required.
- Compatibility with the base metal(s).
- Method of heating.

Figure 10-13 lists a number of brazable metals and suggested metal combinations. The composition of a number of brazing filler metal alloys is shown in Figure 10-14. *Aluminum-silicon* filler metals are used on the following aluminum and aluminum alloys: 1060, 1100, 1350, 3003, 3004, 5005, 5050, 6053, 6061, 6062, 6063, 6951, and A712.0 and C712.0 cast alloys. These brazed assemblies will generally withstand a constant temperature of 300°F (149°C). See Figure 28-16 for the different aluminum alloy designations.

There are two *magnesium* filler metals in general use. The BMg-1 is used to join AZ10A, K1A, or M1A magnesium alloys. Filler metal BMg-2a has a lower melting temperature and is used with the magnesium alloys AZ31B and ZE10A. Service temperature for these alloys is generally 250°F (121°C).

Copper and *copper-zinc* filler metals are generally used on ferrous metals. They may also be used on nonferrous metals. The continuous operating temperature for these alloys is 400°F (204°C).

Copper-phosphorus alloys are used to join copper and copper alloys. They are also used for silver, tungsten, and molybdenum. They should not be used on ferrous, nickel-based, or copper-nickel (with more than 10% nickel) alloys. The service temperature is 300°F (149°C).

Silver-based filler metal alloys may be used to join ferrous or nonferrous metals. Silver alloy filler metals are not recommended for use on aluminum or magnesium. Cadmium added to the silver-zinc-copper alloy makes the metal more fluid. It also improves the metal's wetting qualities. Cadmium also lowers the melting and flow temperatures of silver-based filler metals. The service temperature is 400°F (204°C).

Many alloys of silver have been developed. Each has a different melting point, strength, color, and flow characteristics. Some silver alloys melt at relatively low temperatures; others at fairly high temperatures.

The metals which form the alloys of silver are gold, copper, cadmium, and zinc. The best grades of silver alloys are the ones that are formed partly of gold. This type of silver alloy is used mostly in jewelry work and precision instrument work. The silver alloy compositions are shown in Figure 10-14.

Silver brazing is often called "silver soldering." However, the correct term is silver *brazing* because the temperature of the process is above 840°F (450°C).

Silver alloy brazing rods are used in the fabrication of jewelry and precision instruments which require strong joints. The original use of silver brazing alloys was in manufacture of jewelry. At present, most industries have some applications for silver brazing. The joints on most high-quality welding and cutting torches are silver-brazed, as shown in Figure 10-15.

Because of their cost, silver alloys are most often used for brazing rather than braze welding. The application or the method of performing silver brazing varies little from the brazing previously described.

Each silver brazing alloy has its own particular fields of application. These alloys vary in melting (solidus) temperature and in flow (liquidus) temperature. The term

	Al & Al alloys	Mg & Mg alloys	Cu & Cu alloys	Carbon & low alloy steels	Cast iron	Stainless steels	Ni & Ni alloys
Al & Al alloys	BAlSi						
Mg & Mg alloys	X	BMg					
Cu & Cu alloys	X	X	BAg, BAu, BCuP, RBCuZn				
Carbon & low alloy steels	BAlSi	X	BAg, BAu, RBCuZn	BAg, BAu, BCu, RBCuZn, BNi			
Cast iron	X	X	BAg, BAu, RBCuZn	BAg, RBCuZn	BAg, RBCuZn, BNi		
Stainless steel	BAlSi	X	BAg, BAu,	BAg, BAu, BCu, BNi	BAg, BAu, BCu, BNi	BAg, BAu, BCu, BNi	
Ni & Ni alloys	X	X	BAg, BAu, RBCuZn	BAg, BAu, BCu, RBCuZn, BNi	BAg, BCu, RBCuZn	BAg, BAu, BCu, BNi	BAg, BAu, BCu, BNi
Ti & Ti alloys	BAlSi	X	BAg	BAg	BAg	BAg	
Be, Zr & alloys (reactive metals)	X BAlSi (Be)	X	BAg	BAg, BNi	BAg, BNi	BAg, BNi	BAg, BNi
W, Mo, Ta, Cb & alloys (refractory metals)	X	X	BAg	BAg, BCu, BNi	BAg, BCu, BNi	BAg, BCu, BNi	BAg, BCu, BNi
Tool steels	X	X	BAg, BAu, RBCuZn, BNi	BAg, BAu, BCu, RBCuZn, BNi	BAg, BAu, RBCuZn, BNi	BAg, BAu, BCu, BNi	BAg, BAu, BCu, RBCuZn, BNi

*Filler metals:	BAlSi - Aluminum silicon	BCuP	- Copper phosphorous
	BAg - Silver base	RBCuZn	- Copper zinc
	BAu - Gold base	BMg	- Magnesium base
	BCu - Copper	BNi	- Nickel base

Figure 10-13. *Brazable metal combinations and suggested filler metals. This table may be used if you are brazing two different metals. Find one metal in the left column and the other on the top line of the chart. Where the vertical and horizontal lines for these two metals intersect, the suggested filler metal may be found. "X" indicates that brazing this combination is not recommended.*

melting temperature (solidus) means the temperature at which the alloy begins to melt. *Flow temperature* (liquidus) is the temperature at which all the brazing alloy is liquid. The flow temperatures vary with the alloy content, from about 1145°F (618°C) to 1640°F (893°C).

The wide difference between the melting temperatures and the flow temperatures of the various metals in a brazing alloy can cause some difficulties when brazing. While applying one of these alloys, the lower-temperature metals may flow into the joint and leave the higher-melting-temperature alloy metal behind on the surface.

This action causes a change in color and strength. It also causes difficulty in flowing the remaining alloy onto

the base metal. It is best, therefore, to heat the alloy quickly. This is done, first, to minimize oxidation, and second, to prevent alloy separation. The separating nature of the wide temperature range alloys actually is an advantage in joints that are poorly fit-up. The higher-temperature metals in the alloy will help to bridge the gaps.

Cadmium and zinc are used as alloying metals in silver brazing alloys because they have the peculiar ability to "wet" (flow) and to alloy with iron. Cadmium and zinc also lower the alloy melting and flowing temperatures.

There is danger of producing harmful fumes if alloys containing zinc and cadmium are overheated and the metals vaporize. The work area should be very well-ventilated.

AWS filler metal classification	Ag	Al	Au	Cd	Cr	Cu	Ni	Si	Zn	B	C	Fe	Li	Mg	Mn	Sn	Ti	P	Pd	Zr	Others	Solidus¹ °F	Solidus¹ °C	Liquidus² °F	Liquidus² °C
BAlSi-2		91.0				.25		7.5	.20			.8									.15	1070	577	1135	613
BAlSi-3		84.4			.15	4.0		10.0	.20			.8		.15							.15	970	521	1085	585
BAlSi-4		86.3				.30		12.0	.20			.8		.10							.15	1070	577	1080	582
BAlSi-5		88.4				.30		10.0	.10			.8		.05			.20				.15	1070	577	1095	591
BAlSi-6		88.5				.25		7.5	.20			.8		2.5							.15	1038	559	1125	607
BAlSi-7		87.0				.25		10.0	.20			.8		1.5								1038	559	1105	596
BAlSi-8		85.9				.25		12.0	.20			.8		1.5							.15	1038	559	1075	579
BCuP-1						94.9												5.0				1310	710	1695	924
BCuP-2						92.6												7.3				1310	710	1460	793
BCuP-3	5.0					88.9												6.0				1190	643	1495	813
BCuP-4	6.0					86.6												7.3				1190	643	1325	718
BCuP-5	15.0					79.9												5.0				1190	643	1475	802
BCuP-6	2.0					90.9												7.0				1190	643	1450	788
BCuP-7	5.0					88.1												6.8				1190	643	1420	771
BAu-1			37.5			62.4															.15	1815	991	1860	1016
BAu-2			80.0			19.9																1635	891	1635	891
BAu-3			35.0			64.9																1785	974	1885	1029
BAu-4			82.0			17.9																1740	949	1740	949
BAu-5			30.0				36.0												34.0			2075	1135	2130	1166
BMg-2a		9.0							2.0			.005		88.4	.15						.30	830	443	1110	599
		12.0				.05		.05	5.0					82.7							.30	770	410	1050	566
BCu-1						99.9*															.1	1980	1082	1980	1082
BCu-1a						99.0*	.005														.3	1980	1082	1980	1082
BCu-2						86.5*															.5	1980	1082	1980	1082
BAg-1	45.0			24.0		15.0			16.0												.15	1125	607	1145	618
BAg-1a	50.0			18.0		15.5			16.5													1160	627	1175	635
BAg-2	35.0			18.0		26.0			21.0													1125	607	1295	702
BAg-2a	30.0			20.0		27.0			23.0													1125	607	1310	710
BAg-3	50.0			16.0		15.5	3.0		15.5													1170	632	1270	688
BAg-4	40.0					30.0	2.0		28.0													1240	671	1435	779
BAg-5	45.0					30.0			25.0													1250	677	1370	743
BAg-6	50.0					34.0			16.0													1270	688	1425	774
BAg-7	56.0					22.0			17.0							5.0						1145	618	1205	652
BAg-8	72.0					28.0																1435	779	1435	779
BAg-8a	72.0					27.6							.38									1410	766	1435	779
BAg-13	54.0					40.0	1.0		5.0													1325	718	1575	857
BAg-13a	56.0					42.0	2.0															1420	771	1640	893
BAg-18	60.0					29.8										10.0		.025				1115	602	1325	718
BAg-19	92.5					7.2							.23									1435	779	1635	891
BAg-20	30.0					38.0			32.0													1250	677	1410	766
BAg-21	63.0					28.5	2.5									6.0						1275	691	1475	802
BNi-1		.05			14.0		72.4	4.5		3.2	.75	4.5					.05	.02		.05	.50	1790	977	1900	1038
BNi-1a					14.0		73.1	4.5		3.2	.06	4.5						.02				1790	977	1970	1077
BNi-2					7.0		81.3	4.5		3.2	.06	3.0						.02				1780	971	1970	1077
BNi-3							91.1	4.5		3.2	.06	.5						.02				1780	971	1830	999
BNi-4							92.5	3.5		1.8	.06	1.5										1800	982	1900	1038
BNi-5					19.0		70.0	10.2		.03	.10							.02				1975	1079	2075	1135
BNi-6							88.3				.10							11.0				1610	877	1610	877
BNi-7					14.0		74.9	.10		.01	.08							10.1				1630	888	1630	888
BNi-8						4.5	64.7	7.0			.10	.2			23.0			.02				1800	982	1850	1010

1-Solidus = melting temperature   2-Liquidus = flow temperature   * - Minimum

Figure 10-14. *A table of brazing filler metal alloys. The nominal chemical composition and the melting (solidus) and flowing (liquidus) temperatures are shown. See Figure 10-6 or Figure 34-4 for chemical abbreviations. Note that the cadmium (Cd) alloys are printed in red type. Cadmium fumes are extremely toxic. (American Welding Society)*

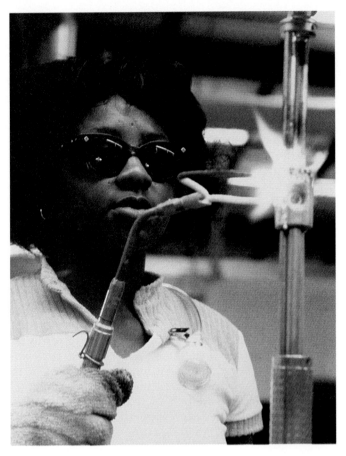

Figure 10-15. A cutting torch is being silver-brazed. Note the attachment that allows two flames to be used for better heating. (CONCOA)

Silver-brazed joints are very strong if properly made. The strength of any brazed joint varies with the thickness of the filler metal. When joining stainless steel butt joints by silver brazing, the tensile strength varies, as shown in Figure 10-16.

Silver brazing is one of the best methods for connecting parts in a leakproof manner and for providing maximum strength. The resulting joints are strong and will

stand up under severe conditions. The excellent adhesion qualities of a silver brazing alloy is are shown in Figure 10-17. When joining copper, brass, or bronze parts, a copper-silver-phosphorus alloy may be used. This alloy is less expensive than most silver brazing alloys. No flux is necessary on copper, but brass (copper-zinc) is usually brazed using a flux. This type of alloy is not used on steel or iron alloys. The silver content is approximately 15%. The alloy flows at 1300°F (704°C). A microphotograph of this joint is shown in Figure 10-18. The absence of flux when brazing copper improves the visibility of the joint during the brazing operation.

Figure 10-17. A microphotograph of a silver alloy-brazed steel joint. The brazing filler metal in the joint is Easy-Flo, which is an alloy consisting of 45% Ag, 15% Cu, 16% Zn, and 24% Cd.

Strength of Silver-Brazed Butt Joints			
Thickness of the brazing filler metal		**Tensile strength**	
inches	millimeters	psi	MPa
0.002	0.05	133,000	917
0.003	0.08	115,000	793
0.006	0.15	90,000	621
0.009	0.23	83,000	572
0.012	0.30	76,000	524
0.015 app. 1/64	0.38	70,000	483

Figure 10-16. A table showing how the strength of a silver-brazed joint is affected by varying thicknesses of silver brazing filler metal in the joint.

Figure 10-18. A microphotograph of a copper and brass part joined by a 15% silver plus copper and phosphorus brazing alloy.

Gold filler metals are used in electronics and many space technology applications. They are used to braze ferrous, nickel, and cobalt base metals. Gold alloys are used when volatile filler metal components are undesirable. They are also used when oxidation- and corrosion-resistance are required. Gold filler metal alloys have a service temperature of 800°F (427°C).

Nickel-based filler metals work well in cryogenic (very low temperature) applications. They also have a high service temperature of 1800°F (982°C). They are used on carbon and stainless steels, and cobalt- or copper-based alloys.

Brazing filler metals may be applied by means of an applicator. These applicators are generally pneumatic (air-powered). Filler metals are applied in this manner for automatic multistation brazing machines or for oven-heated processes. Such a filler metal applicator is shown in Figure 10-19.

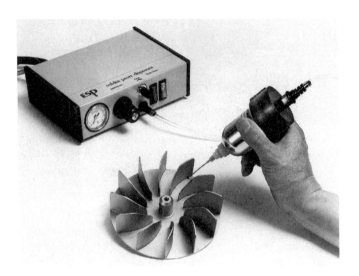

Figure 10-19. *Solder paste or soldering flux can be applied to a joint using a pneumatically powered dispenser. The pressure on the paste or flux and the time that it flows are set on the control box. These two variables will control the amount of material deposited in the joint, from 0.001 oz. to 1.0 oz. (Wall Colmonoy Corp.)*

Cobalt-based filler metals have the highest service temperature at 1900°F (1038°C). They are generally used with cobalt alloys.

10.5.1 Braze Welding Filler Metals

There are six AWS classifications for copper- or copper-zinc based brazing and braze welding filler metals. Their chemical composition and solidus and liquidus temperatures may be seen in Figure 10-20. Tin, iron, silicon, and manganese are added to some filler rods. These alloying elements improve the flow characteristics, scavenge oxygen, and increase strength and hardness. They also decrease the tendency of zinc to vaporize. The addition of 10% nickel also improves the strength of brass rods. The service temperature is 500°F (260°C). Brass brazing rods have a tensile strength of 40,000 psi to 60,000 psi (276MPa to 414MPa).

It should be remembered that a braze-welded joint generally involves two dissimilar metals. When dissimilar metals are in contact, a small electrical flow is created under certain conditions. This causes a form of corrosion called *galvanic corrosion.*

Braze welding filler metal may be less resistant to chemical solutions than the base metal. This lessened corrosion resistance and the possibility of galvanic corrosion must be considered in design applications.

10.6 BRAZING AND BRAZE WELDING PROCESSES

Heat sources for brazing may be:
- A molten bath of brazing metal alloy, similar to the dip solder method. See Heading 9.7.
- Torch heating with an oxyfuel gas or air-fuel gas torch. Figure 10-21 illustrates a multistation torch brazing machine.
- Controlled atmosphere furnaces.
- Electric resistance heating.
- Induction heating.

| AWS classification | Nominal chemical composition (percentage) | | | | | | | | | | | Temperature | | | |
| | Cu | Zn | Sn | Fe | Mn | Ni | P | Pb | Al | Si | Others (max.) | Solidus | | Liquidus | |
												°F	°C	°F	°C
BCu-1	99.9 min.						.08	.02	.01		.10	1980	1082	1980	1082
BCu-1a	99.0 min.										.30	1980	1082	1980	1082
BCu-2	86.5 min.										.50	1980	1082	1980	1082
RBCuZn-A	59.0	39.8	.63					.05	.01		.50	1630	888	1650	899
RBCuZn-C	58.0	39.4	.95	.72	.26			.05	.01	.10	.50	1590	866	1630	888
RBCuZn-D	48.0	41.0				10.0	.25	.05	.01	.15	.50	1690	921	1715	935
	B - brazing			RB - welding or brazing rod											

Figure 10-20. *Copper- and copper-zinc-based brazing filler metals. For more information, see AWS A5.8-92,"Specification for Filler Metals for Brazing and Braze Welding."*

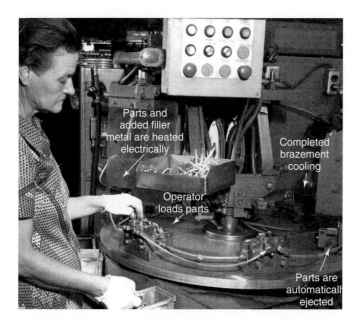

Figure 10-21. *A multistation torch brazing machine. Parts are manually loaded and unloaded. The flux is applied and the part is heated and cooled automatically as the table rotates. (Handy & Harman /Lucas-Milhaupt, Inc.)*

Braze welding usually requires more heat than brazing. Since braze welding is similar to welding, the operation is usually done with an oxyfuel gas torch or gas tungsten arc.

Brazing and braze welding procedures are quite similar. The main differences are in the joint design and in the quantity of brazing rod applied to the joint.

Some useful pointers applicable to both brazing and braze welding include:

- Metals to be joined must be mechanically and chemically clean.
- Two pieces to be brazed together must be fitted properly; that is, the metals should not be spaced too far apart or forced together. The braze -welded joint is prepared in the same manner as any oxyfuel gas- or arc-welded joint.
- The two pieces must be firmly supported during the brazing or braze welding and cooling operations. Any movement of either part while the joining metal is molten or plastic will weaken the joint.
- The metals to be joined must be heated to a temperature slightly above the melting temperature of the brazing filler metal, but below the melting temperature of the metal being brazed or braze welded.
- Clean, fresh flux must be used to reduce oxidation and to float the oxides to the surface as the operation proceeds.
- A heat source, as listed previously, is required to obtain a high enough temperature to obtain a good joint.
- The oxyfuel gas torch flame is usually adjusted to a neutral flame. Refer to Figure 4-2. A carbur-

izing (reducing) flame will produce an exceptionally neat-looking joint, but strength will be sacrificed. An oxidizing flame will produce a strong joint, but with a rougher-looking surface. A neutral flame will give the best results under ordinary conditions.

- Automobile body panels are often soldered or brazed after a repair. The solder or brazing material smooths out any flaws in the panel prior to painting. See Figure 10-22.
- As in steel welding, the brazing filler metal must penetrate to the other side of the joint. This penetration must be such that the joining metal covers 100% of the joint surfaces and adheres to these two surfaces. A minimum amount of joining material should be used.
- The metal must be cooled properly to obtain the desired properties in both the original metal and the joining metal. Heat treatment of metals is covered in Chapter 29.

10.6.1 Brazing or Braze Welding Steel with Copper and Zinc Alloys

Clean the surfaces as described in Heading 10.3. An emery wheel or emery cloth is *not* recommended, due to the possibility of embedding abrasive particles and oil in the metal. When braze welding, provide for metal contraction and expansion, as in the welding process. If the metal is thicker than 8 gage (0.128″ or 3.25mm), it must be grooved or chamfered to permit adequate penetration. The joint to be brazed must be designed and fitted so that a close fit of the two base metals is obtained.

Adjust the torch for the flame desired, using the same size tip as would be used for welding the same thickness of the base metal. When brazing, the flux is usually applied to the joint, while in braze welding the flux is often

Figure 10-22. *This autobody repair technician is soldering or "leading" the deck lid of an automobile. A propane torch provides the heat and a wooden stick is used to smooth the molten filler material.*

applied to the brazing rod. Apply the torch to the metals to be brazed or braze welded, heating them equally at the joint to a dull cherry red. The width of the bead when braze welding will be determined by how wide a portion of the metal is heated to cherry red. The brazing filler metal will not flow over the surface unless the surface is at the brazing filler metal flow temperature.

The width of the braze weld bead should be a little wider than a steel weld on the same thickness of metal. While heating the metal, the brazing filler metal rod should be kept near the torch flame to maintain a fairly high rod temperature, as shown in Figure 10-9. After the metals have been heated, bring the flux-coated portion of the brazing rod into contact with the cherry red metals, meanwhile maintaining the torch motion. The brazing rod will quickly melt and the molten metal will flow over or between the base metals. Do not overheat the brazing rod. Keep it away from the inner cone of the torch flame.

When braze welding, the bead should proceed along the joint, just as in the welding process (except the procedure should be faster). The torch flame is not held as close to the metal as in steel welding—it should be approximately double the distance. The width of the bead may be controlled by raising and lowering the torch flame. When the flame is held close to the metal, a wide bead will result. By drawing the torch away, the metal cools slightly and the bead will not be so wide.

The finished braze or braze weld should have the appearance of adequate fusion with the base metal. The brazing filler metal should penetrate through the joint and appear on the other side. A *white deposit* on the outside of the brazed or braze welded joint indicates an overheated joint. The color of the braze filler metal in the joint will indicate if it has been overheated. The best-looking brazed or braze welded joint will show a color very similar to the brazing filler metal used. If the filler metal is heated to an excessive temperature, some of the zinc will be burned out, leaving a coppery appearance. If an oxidizing flame is used, the brazing filler metal will have a red color due to the oxidation of the copper.

10.6.2 Brazing with Silver Alloys

Silver brazing is similar in all ways to other soldering and brazing processes. The steps to follow are:
1. Clean the joints mechanically and chemically (use clean materials and tools).
2. Fit the joint closely and support the joint.
3. Apply the proper flux.
4. Heat to the correct temperature.
5. Apply the silver brazing material.
6. Cool the joint.
7. Clean the joint thoroughly.

An oxyacetylene torch is an excellent heat source for silver brazing, as shown in Figure 10-23. Liquefied petroleum gas in conjunction with oxygen is acceptable.

Fluxes used for silver brazing must be clean, chemically pure, and fresh. Chlorides are popularly used as silver brazing fluxes. Borax made into a paste with water

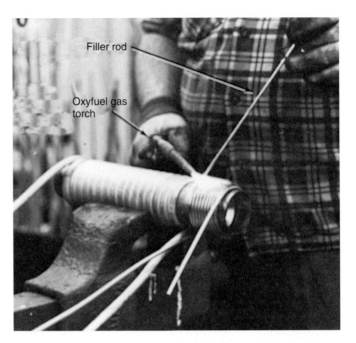

Figure 10-23. *Silver brazing small tubing to large pipe. (Handy & Harman/Lucas-Milhaupt, Inc.)*

may also be used successfully as a silver brazing flux. See Figure 10-8 for suggested fluxes to use.

The behavior of the flux is most helpful in silver brazing. It will indicate the temperature changes in the joint as follows:
1. The flux will dry out as moisture (water) boils away at 212°F (100°C).
2. Then the flux will turn milky and start to bubble at about 600°F (316°C).
3. Finally, it will turn into a clear liquid at about 1100°F (593°C). This is just short of the brazing temperature. *The clear appearance of the flux will indicate the time to start adding the filler metal.*

Figure 10-24 shows a silver brazing operation. The BAg-5 classification alloy melts at 1230°F (662°C) and flows at 1370°F (743°C). See Figure 10-14 for other silver brazing filler metal alloy melting and flow temperatures.

A large tip is recommended for heating the joint. The extra heat provided by the large tip permits a shorter brazing time, thus reducing the time for oxides to form. The basics of a brazing operation when using a flame as the source of heat are shown in Figure 10-25. The joint should be kept covered with the flame during the whole operation to prevent air from getting to it. The flame should not blow either the flux or the molten metal. Use a slight feather (reducing flame) on the inner cone for best joint appearance. Figure 10-26 shows a large assembly being silver brazed using an oxyacetylene flame as the source of heat.

It is necessary to heat both pieces evenly. If a thick piece is being joined to a thin piece, more heat must be applied to the thick piece. Keep the torch in motion while the alloy is being added, or local "hot spots" may develop, resulting in a poor joint.

A popular way to apply silver brazing filler metal when making a silver braze on a tubing joint is to use sil-

Figure 10-24. *Silver brazing copper pipe joints, using an oxyfuel gas torch as the source of heat. (J.W. Harris Co., Inc.)*

Figure 10-26. *A large assembly being silver-brazed using the oxyacetylene torch as a source of heat. (Handy & Harman/Lucas-Milhaupt, Inc.)*

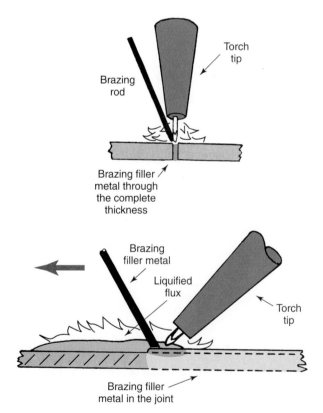

Figure 10-25. *Basic oxyfuel gas method of brazing a butt joint. (Handy & Harman /Lucas-Milhaupt, Inc.)*

ver alloy rings like the one shown in Figure 10-27. This is a practical way to add silver alloy. Preplaced brazing and soldering forms are economical when used on a production basis. See Figure 10-28. A method of brazing using preplaced brazing shims is shown in Figure 10-29.

It is necessary to thoroughly clean the completed silver brazed joint by washing in water or scrubbing. Any corrosive flux left on the metals will corrode them and also hinder any painting or plating operations.

The joint may be cooled quickly or slowly. Cooling with water is permissible. Water quenching may also serve to wash the flux from the joint.

Visual inspection of the joint will quickly reveal any places where the braze metal did not adhere. It is best to watch for this adherence and make any corrections during the brazing operation.

10.6.3 Brazing Stainless Steel

Stainless steel may be silver-brazed as easily as it is welded. A silver-brazed joint is strong if properly designed

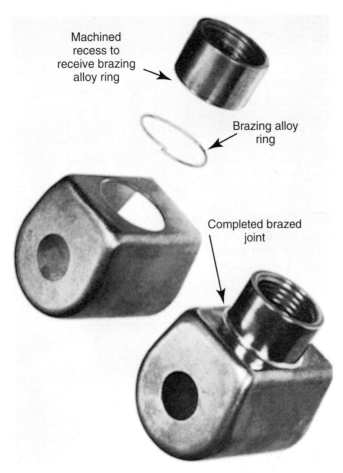

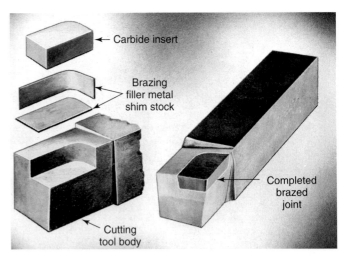

Figure 10-29. *A machine-tool bit, showing how the carbide insert is brazed to the tool bit body using preplaced brazing filler metal shims.*

and made. The metal does not warp to any great extent, because of the relatively low temperatures used for silver brazing. A good color match can usually be obtained. With stainless steel, a 45% silver alloy filler rod produces a good silver-brazed joint. Figure 10-30 shows adhesion properties of a silver brazing alloy between stainless steel and copper.

Figure 10-27. *A silver-brazed joint designed to use a preplaced silver alloy rings. The alloy forms almost perfect fillets and no further finishing is necessary. (Handy & Harman/Lucas-Milhaupt, Inc.)*

Figure 10-28. *Brazing filler metal may be obtained in sheet form, as wire, or in preshaped forms as shown. The preshaped forms are placed on or in the brazement before it is heated. (Handy & Harman/Lucas-Milhaupt, Inc.).*

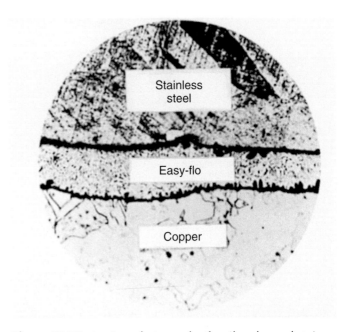

Figure 10-30. *A microphotograph of a silver-brazed stainless steel and copper joint.*

10.6.4 Brazing and Braze Welding Aluminum Alloys

Aluminum alloys can be brazed or braze welded using a number of the methods of heating. Oxyfuel gas is a popular means of heating aluminum for brazing. Dip brazing and furnace brazing are used when enough production is done to justify the cost. The carbon arc and electric resistance heating are also used.

Practically all of the common aluminum joints have been successfully brazed or braze welded. See Figures 10-3, 10-4, and 10-5.

In the torch method of aluminum brazing and braze welding, either the oxyhydrogen or the oxyacetylene flame may be used. In either case, a reducing flame with a 1" to 2" (25mm to 50mm) intermediate cone should be used. The following procedure produces good results:

1. Clean and degrease the surface. Braze or braze weld the joint as quickly as possible after cleaning the metal, while the surface is free of oxides.
2. Apply a fresh, chemically clean flux along the joint and on the brazing rod. The flux should be specially compounded for aluminum brazing or braze welding, and may be in either powder or paste form.
3. Heat the joint. As the flux chemicals turn liquid, start to apply the brazing rod. The rod will melt and flow into the joint and through the crevices to produce a heat brazed or braze welded joint. Hold the torch so the inner flame cone is 1" to 2" (25mm to 50mm) away from the joint.
4. Cool the finished braze or braze welded joint. Do not disturb the parts until the temperature has dropped to below 900°F (482°C).
5. Clean the flux from the joint. Use hot water, then a concentrated nitric acid solution. Follow all safety procedures. After this, wash again in boiling water. Any flux left on the metal will seriously corrode it.

Danger: The handling and use of concentrated acids should only be undertaken by an experienced person who has been given thorough training in the use of dangerous acids.

The finished weld should show 100% adhesion and the surface should be smooth and clean. The brazing filler metal should penetrate, with good adhesion, the full thickness of the joint. See Figure 10-31.

10.6.5 Brazing and Braze Welding Magnesium Alloys

Magnesium and magnesium alloys may be brazed or braze welded using the same procedures as used on other metals.

Cleaning methods vary with the magnesium alloy being brazed. See Figure 10-32. Mechanical cleaning with steel wool or abrasive cloth works very well. Chemical cleaning includes degreasing, bright pickling, chrome pickling, and modified chrome pickling. Degreasing is normally done with an alkaline solution. Refer to Chapter 34 for the chemical compositions of the various pickling solutions used on magnesium.

Figure 10-31. *A 15x microphotograph of a brazed inside corner joint on aluminum. Note that little or no base metal has melted.*

Magnesium alloy	Filler metal used		
	AZ125Xa	GA432	AZ92A
AZ10A	1, 2	1	1
AZ31B	1, 2	1, 2	*
AZ61A	*	1	*
K1A	1	*	1
M1A	1	1	1
ZE10A	1, 2, 3	1, 3	*
ZK21A	1	*	1
ZK60A	*	1, 4	*

1 - Mechanical cleaning
2 - Bright pickle (2 min.)
3 - Chrome pickle (2 min.)
4 - Modified chrome pickle followed by a 20% hydrofluoric acid dip (2 min.)

Figure 10-32. *Magnesium cleaning methods.*

The fluxes used in magnesium brazing and braze welding are shown in Figure 10-33. The brazable magnesium alloys and the suggested filler metals are shown in Figure 10-34. The composition of the magnesium brazing filler metal is shown in Figure 10-35.

Postcleaning of magnesium brazed and braze welded joints is extremely important. The cleaning should be done with hot flowing water. Mechanical scrubbing may also be required. After this initial cleaning, the parts are dipped into a chrome pickle for about two minutes.

Magnesium may be brazed using the torch, dip, or furnace methods. Heating must be very carefully controlled. The brazing temperature varies between 1080°F and 1160°F (582°C and 627°C).

Brazing process	Flux composition (percentage)	Approximate melting point	
		°F	°C
Torch	KCI - 45 NaCI - 26 LiCI - 23 NaF - 6	1000	538
Torch, dip or furnace	KCI - 42.5 NaCI - 10 LiCI - 37 NaF - 10 AlF$_3$·3NaF - 0.5	730	388

Figure 10-33. Magnesium brazing flux composition and use.

The procedure will not vary from that used with other metals, as indicated below:

1. Ensure that the joint is well-designed and, if brazing, clearances are proper to permit capillary action to occur.
2. Clean the metal.
3. Apply flux.
4. Use the proper flame (if using oxyfuel gas).
5. Heat the joint evenly to the correct temperature.
6. Apply the braze filler rod.
7. Support the metal during the heating and cooling process to avoid stress cracks.
8. Postclean the metal.

A good quality brazed magnesium joint is shown in Figures 10-36 and 10-37.

ASTM alloy designation	Base metal				Brazing range		Suitable filler metal	
	Solidus		Liquidus				BMg-1	BMg-2a
	°F	°C	°F	°C	°F	°C		
AZ10A	1170	632	1190	643	1080-1140	582-616	X	X
AZ31B	1050	566	1160	627	1080-1100	582-593		X
KIA	1200	649	1202	650	1080-1140	582-616	X	X
MIA	1198	648	1202	650	1080-1140	582-616	X	X
ZE10A	1100	593	1195	646	1080-1100	582-593		X
ZK21A	1159	626	1187	642	1080-1140	582-616	X	X

Figure 10-34. Magnesium brazing alloys.

AWS A5.8 classification	ASTM alloy designation	Filler metal				Brazing range	
		Solidus		Liquidus			
		°F	°C	°F	°C	°F	°C
BMg-1	AZ92A	830	443	1110	599	1120-1140	604-616
BMg-2a	AZ125A	770	410	1050	566	1080-1130	582-610

Figure 10-35. Magnesium brazing and braze welding filler metals.

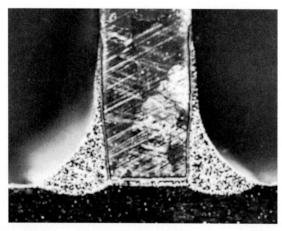

Figure 10-36. A magnesium brazed joint. The sample in this 8x macrophotograph has been etched. (Dow Chemical Co.)

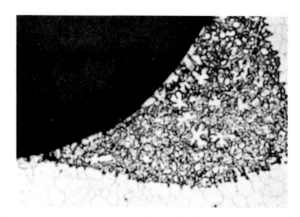

Figure 10-37. A 65x microphotograph of a magnesium-brazed inside corner joint. Note how the brazing alloy has penetrated the grain boundaries of the base metal.

10.6.6 Brazing and Braze Welding Cast Irons

There are various types of cast iron. These include the white, gray, malleable, and ductile cast irons. White cast iron is difficult to braze and is seldom attempted.

Cast iron is cleaned by mechanical and chemical means. Grinding, filing, grit blasting, and chipping are utilized to remove the cast iron surface crust and oxides. Electrochemical cleaning is done to remove graphite and silicon oxides from the surfaces. Graphite and silicon oxides prevent the filler metal from wetting and smoothly flowing over the surface. Methods of supporting and reinforcing a cast iron joint are shown in Chapter 21.

Fluxes recommended for cast irons contain the following ingredients: boric acid, borates, fluorides, fluoborates, and agents to improve the wetting action. Common borax makes a good cast iron brazing or braze welding flux. Recommended brazing and braze welding filler metal classifications are BCuZn, BAg, and BNi. The BCuZn (brass) filler metals adhere better when some nickel is added. See Figure 10-13 for filler metals recommended when joining various metal combinations.

For best results, cast irons should be preheated to 400°F to 600°F (204°C to 316°C). When brazing or braze welding malleable or ductile cast iron, the temperature must be kept below 1400°F (760°C). Above this temperature, the metallurgical structures may be damaged. The procedure for brazing cast irons are the same as for all metals.

Braze welding joints for cast irons are the same design as those used with other metals. See Figure 10-3. To produce a good braze weld on cast iron, the surfaces must be tinned. Tinning is done by heating the cast iron to a dull cherry red. A thin layer of filler metal is then applied to each surface. If the tinning operation is not done properly, the strength of the joint will be low.

When brazing cast iron, it is advisable to use a tip that will provide a high heat output with low gas pressures. This will provide a "soft" flame. A high-velocity, high-pressure flame tends to blow the flux away from the joint.

Braze welding is the most satisfactory way of repairing malleable iron castings. Welding malleable iron is not recommended. Figure 10-38 shows cast iron being braze welded. Since cast iron sections are generally quite thick, the backhand method is recommended. Multiple passes also may be required.

After brazing or braze welding, the metals should be cooled very slowly. Slow cooling will prevent white cast iron from forming. Figure 10-39 shows a cast iron joint which has been brazed with a silver-based filler metal.

10.6.7 Controlled-Atmosphere Brazing Furnace

A method of fabricating small articles is the controlled-atmosphere brazing furnace. It consists of a typical furnace, usually heated on the inside by electrical resistance coils or partly combusted fuel gases. The inside of the furnace is usually kept as nearly gas-tight as possible; little or no air is allowed to enter. Instead, the gas fed into the furnace is a low-pressure mixture of air and fuel gas.

Sufficient temperature is maintained inside the furnace to melt copper, brass, bronze, and silver alloys without melting the base metals.

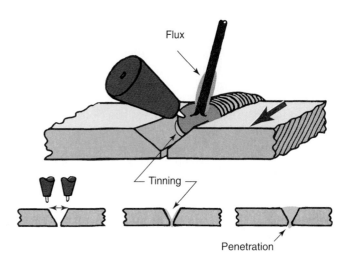

Figure 10-38. *A schematic showing the braze welding of cast iron. The backhand method is being used on this joint.*

Figure 10-39. *A microphotograph of steel being joined to cast iron by the silver brazing process. The cast iron was cleaned by a special process prior to the brazing operation. (Kolene Corp.)*

Developments in aluminum alloys enable the brazing of aluminum and its alloys in the reducing-atmosphere furnace. The operation of the furnace is described below.

Two articles, such as the ends of two small steel cylinders, are to be fastened together. They are first cleaned. Then, they are assembled in their final form . A small wire of the alloy being used to join them is placed inside the assembled joint. The assembly is then placed in the controlled-atmosphere furnace.

In the high temperature of the reducing-atmosphere furnace, the brazing filler metal wire melts and joins the two surfaces. The article is then removed from the furnace and cooled slowly in a neutral atmosphere. When finished, the joint is approximately as strong as a weld. The method is well-adapted to production work, since it ensures a high-quality joint. The reducing atmosphere in the furnace also simultaneously cleans the complete article, eliminating one manufacturing step. The reducing atmosphere of partly burned fuel gases eliminates any oxidation of the metals during the brazing operation.

Furnaces that are filled with an inert gas have been used for brazing operations. Brazing has also been done in vacuum furnaces.

10.7 HEAT-RESISTANT BRAZED JOINTS

There are several brazing applications where the brazed joint must retain most of its physical properties when exposed to high temperatures and/or highly corrosive conditions. Brazing filler metal alloys have been developed to withstand temperatures in the range of 2000°F (1093°C). These special brazing filler metal alloys must also have good corrosion resistance.

These alloys are usually made of either a nickel-chromium alloy or a silver-manganese alloy, as shown in Figure 10-40. The nickel-chromium alloy has good corrosion resistance. It is used on jet engine blades, stainless steel, and on low-carbon steel joints. The brazing is usually done in a furnace. The silver-manganese brazing filler metal is used mainly for joining stainless steel and high-nickel alloys. The silver-manganese alloy is also usually applied in a furnace.

10.8 SAFETY PRACTICES FOR BRAZING AND BRAZE WELDING

Brazing and braze welding operations are safe to perform if reasonable safety precautions are carefully followed. The normal precautions for the safe use of gas welding and arc welding equipment (as explained in previous chapters) must also be followed when brazing or braze welding. Refer to Headings 6.14 and 9.15. Also see the list of hazardous materials in Heading 9.2.

The special precautions to be taken when brazing or braze welding are:

- Ventilation must be excellent to eliminate the health hazards which are presented by toxic metal and flux fumes that are often present when brazing or braze welding.
- Many of the fluxes used are harmful to the skin; care should be taken in handling them so that no direct contact is made. If fluxes *do* come into contact with the skin, the skin should be washed thoroughly with soap and water.
- When acids are used, they should be handled only by persons thoroughly trained for this type of work.
- Some brazing filler metal alloys contain cadmium. When molten, and especially if overheated, these alloys emit cadmium oxide fumes to the atmosphere. Cadmium oxide fumes are very dangerous if inhaled. The limit value for cadmium oxide fumes is 0.1 milligrams per cubic meter of air for daily eight-hour exposures. This value represents the maximum tolerance under which workers may be exposed without adverse effects.

Cadmium fumes have no odor, and a lethal dose need not be sufficiently irritating to cause discomfort until after the worker has absorbed sufficient quantities to be in immediate danger of his or her life. Symptoms of headache, fever, irritation of the throat, vomiting, nausea, chills, weakness, and diarrhea generally may not appear until some hours after exposure. The primary injury is to the respiratory passages.

Filler metal classification	Composition (percentage)								Temperatures					
	Ni	Cr	Fe	Si	C	B	Ag	Mn	Solidus		Liquidus		Brazing	
									°F	°C	°F	°C	°F	°C
BNiCr	70	16	*	*	*	4			1850	1010	1950	1066	2000-2150	1093-1177
BAgMn							85	15	1760	960	1780	971	1780-2100	971-1149
*The total of these three elements is approximately 10% maximum.														

***Figure 10-40.** A table of brazing filler metals intended for use in high-temperature and/or corrosive conditions. In the American Welding Society classifications shown (BNiCr and BAgMn), the B means brazing; Ni, nickel; Cr, chromium; Ag, silver, and Mn, manganese. See Figure 34-4 for other chemical abbreviations.*

An increase in temperature over the molten state accelerates the quantity of the fumes produced. The use of an oxyacetylene flame as the heating source may produce higher concentrations of cadmium oxide fumes due to its higher temperatures. Flames produced by air-acetylene, air-natural gas, or oxygen-LP gas are therefore recommended.

Most states have industrial hygiene personnel, usually in the health department or in the department of labor. If there is any question concerning the amount of cadmium oxide fumes in the work area, the industrial hygiene department should be asked to take air samples. They will evaluate and report on the concentrations of cadmium oxide fumes present. Recommendations may then be made to correct situations that are potentially hazardous.

Eating or storing of food should not be permitted in the work area. Workers should wash both hands and faces before eating or leaving from work. Workers brazing with cadmium-bearing alloys should be made aware of the hazard involved and trained to take precautionary measures relative to the environment of the particular job.

Alloys free of cadmium are available and should be used wherever possible. Workers should be trained to recognize the brazing alloys which contain cadmium.

- Another brazing operation which must be handled carefully is the brazing of beryllium. Oxides from this metal are also very dangerous if inhaled. Thorough ventilation is a must when heating and working with this metal.

TEST YOUR KNOWLEDGE

Write your answers on a separate sheet of paper. Do not write in this book.

1. Explain the difference between the properties of brazing and soldering.
2. Explain the difference between brazing and braze welding.
3. Brass is an alloy of copper and _____. Bronze is an alloy of copper and _____.
4. Is the base metal melted when brazing? When braze welding?
5. Can brazing or braze welding be used to join two different base metals? If so, give an example.
6. What is meant by capillary action?
7. Why is some metal "pickled" before it is brazed or braze welded?
8. List four methods of mechanically cleaning a metal prior to brazing.
9. What filler metal and flux ingredients are recommended for brazable aluminum alloys? (See Figure 10-8.)
10. List at least five criteria that must be considered when selecting a flux for brazing or braze welding.
11. Name three methods for applying a brazing flux.
12. A chloride flux is often used when silver brazing. How can you tell when to add the filler metal? (See Heading 10.6.2)
13. What brazing filler metal classifications are recommended for brazing low-alloy steel to copper or copper alloy? (Use Figure 10-13.)
14. What service temperature can most brazing alloys withstand?
15. What is meant by the term "melting temperature"?
16. What is meant by the term "flow temperature"?
17. List two metals or elements that present a health hazard when in a filler or base metal.
18. Which process, brazing or braze welding, requires more heat?
19. When oxyfuel gas braze welding, what type of flame adjustment produces the strongest joint?
20. What type of flame adjustment should be used when torch brazing aluminum alloys?

Part 3

SHIELDED METAL ARC WELDING

Shielded metal arc welding uses a flux-covered electrode to generate a shielding gas that protects the weld pool from atmospheric contamination.
(American Welding Society)

Shielded metal arc welding (SMAW) being used in the field to construct a storage tank. (Miller Electric Mfg. Co.)

Chapter 11

SHIELDED METAL ARC WELDING EQUIPMENT AND SUPPLIES

Shielded metal arc welding (SMAW), sometimes called "stick welding," is the most popular form of electric arc welding. Refer back to Heading 4.5 for an overview of the principles and processes involved in SMAW. Shielded metal arc welding equipment is used in many places. Some uses of shielded metal arc welding are on farms, in service stations, in home shops, on large construction sites, and on pipelines.

The arc welding power source may produce alternating current (ac) or direct current (dc) for welding. This chapter will describe the various machines, accessories, and supplies required for ac and dc shielded metal arc welding. The next chapter will cover the techniques and skills of shielded metal arc welding.

A complete shielded metal arc welding (SMAW) station consists of a welding machine, an electrode lead, an electrode holder, a workpiece lead, and a work (ground) clamp. The station also includes a welding booth, work bench, stool, and the ventilation system.

Technology continues to change and improve arc welding machines. Figure 11-1 illustrates the changes that have occurred over the years. Arc welding power sources, other components of the SMAW station, accessories, and personal safety items will be discussed in the following headings.

Figure 11-1. A motor-generator type arc welding machine from 1925, contrasted with a modern inverter-type welding power source. (ESAB Group)

11.1 ARC WELDING POWER SOURCE CLASSIFICATIONS

Arc welding power sources, also called arc welding machines, are either ac, dc, or combination ac/dc machines. They are constructed to produce either a constant current measured in amperes, or a constant voltage (potential) measured in volts.

Alternating current (ac) power sources (see Heading 11.2.1 for details) may be of the following types:

- Transformer.
- Motor- or engine-driven alternator.
- Inverter.

Direct current (dc) power sources (see Heading 11.2.2 for details) are of the following types:

- Transformer with a dc rectifier.
- Motor- or engine-driven generator.
- Motor- or engine-driven alternator with a dc rectifier.
- Inverter.

Combination ac and dc arc welding power sources are of the following types:

- Transformer with a dc rectifier.
- Motor- or engine-driven alternator with a dc rectifier.
- Inverter.

An inverter welding power source is a transformer-type machine, but is usually referred to as an **inverter.** Inverters change incoming ac current to dc, then back to very-high-frequency ac current. This high-frequency alternating current then passes through a very small and efficient transformer. Inverters are much smaller, lighter in weight, and more efficient than regular transformer-type machines. Inverter power supplies can perform like either a constant current or a constant voltage machine.

To properly describe an arc welding power source, the following minimum information should be given:

- Type of power source.
- Whether constant current or constant voltage.
- Whether ac, dc, or ac/dc.

An example of such a description might be: "The welding machine is a transformer, constant current, ac machine." Other information that helps classify welding machines will be covered later, under Heading 11.3.1. This information consists of:

- Duty cycle.
- Welding current rating.

- Power requirement.

To meet the variety of demands on different jobs, welding machines are designed to produce ac or dc current. Each welding machine may have a different design and output capacity. When selecting a welding machine, the decision will also depend upon such additional factors as cost, portability, and the personal preferences of the user.

Arc welding machines also may also be classified according to a graph of the electrical output of the particular machine. Such a visual representation is called the **static volt-ampere curve.** A typical volt-ampere curve will be studied in detail under Heading 11.2 and as shown in Figure 11-2. The volt-ampere curve plots the electrical current output, measured in amperes, against the voltage output. Remember that current relates to the *welding heat* and that voltage relates to the *length of the arc.*

The general slope of the volt-ampere curve is called the **output slope** of the power source. Some output slopes are very flat, while others are very steep. The almost horizontal volt-ampere curves, or output slopes, are produced by constant voltage machines. An arc welding machine that produces a steep output slope is called a **constant current machine.**

11.2 CONSTANT CURRENT POWER SOURCES

Arc welding power sources are designed to produce an output that has a nearly constant current or a nearly constant voltage. See Chapter 13 for more information on constant voltage power sources. Figure 11-2 shows a typical volt-ampere curve for a **constant current power source.** Notice the relatively steep slope or "droop" of this curve. These welding machines are known as **droop curve machines** or **droopers.**

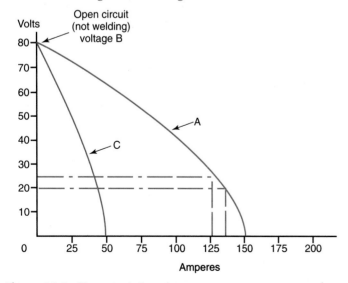

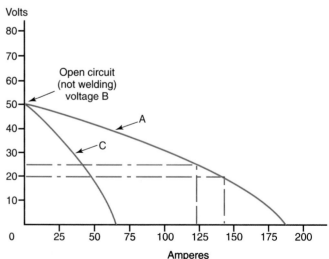

Figure 11-2. Characteristic volt-ampere curves, or output slopes, for a constant current arc welder. Left view shows the steep slope of a "drooper" machine. Right view shows output slopes of the same machine with a lower voltage setting. On the curves A and C, notice the relatively small percentage change in current for a large percentage change in voltage.

The arc voltage varies with the size or length of the arc gap. Examining Figure 11-2, you see a steeply sloping volt-ampere curve A in the left view. The open circuit voltage, or the voltage when not welding, has been selected to be 80V. The closed circuit or welding voltage is 20V. An increase in arc length causes an increase in the closed circuit voltage.

On the output slope of volt-ampere curve A, a change from 20V to 25V will result in a decrease in amperage from 135A to 126A. With an increase of 25% in voltage, only a 6.7% change occurs in the delivered current in curve A. Thus, if the welder varies the length of the arc, causing a change in voltage, the output slope shows that there will be very little change in the current and the weld quality will be maintained. The current in this machine, even though it varies somewhat, is considered constant.

The constant current power source is the type of machine desired when doing manual arc welding. The two manual arc welding processes are shielded metal arc welding (SMAW) and gas tungsten arc welding (GTAW). Additional information will be provided in Chapter 13.

The open circuit voltage curve for a setting of 50V on the machine is shown as curve B in the right view of Figure 11-2. The same 20V to 25V change in the welding voltage will result in a drop in current from 143A to 124A or 13.3%. This slower-sloping volt-ampere curve output causes a larger change in amperage with the same small change in voltage. A welder may wish to have this slower-sloping (flatter) volt-ampere output curve.

With a flatter output slope, the welder can control the molten weld pool and electrode melt rate by making small changes in the arc length. Control of the molten weld pool and electrode melt rate are most important when welding in the horizontal, vertical, and overhead welding positions.

11.2.1 Alternating Current Power Sources

Alternating current (ac) power sources are of either the transformer or alternator type, as classified in Heading 11.1.

The purpose of a transformer used in a welding machine is to change high-voltage, low-current electricity into the lower voltages and higher currents required for welding. Transformers are constructed of three principal electrical components. They are a primary winding, a secondary winding, and an iron core. See Figure 11-3.

The primary winding uses thinner wire than the secondary winding, because the primary winding carries less current. There are many more windings in the primary. When a transformer has more turns in the primary than in the secondary, it will decrease the voltage and increase the amperage from the primary to the secondary circuit. This type transformer is called a **step-down transformer**. A step-down transformer *reduces* (steps down) the voltage and *increases* the current.

When current flows in a wire, a magnetic field builds up around the wire. When the current stops flowing in the wire, the magnetic field collapses.

In a step-down welding transformer, this current-carrying wire is wrapped into a coil with many turns. The more turns there are, the stronger the magnetic field. This coil of wire is called a **primary winding**. A second coil of larger wire with fewer turns is placed next to the primary coil. It is called a **secondary winding**. The primary winding and secondary winding are not connected in any way. They are independent coils of wire, but are placed close together.

To cause a current to flow in the secondary, an alternating current is made to flow in the primary winding. As the current flows, a magnetic field builds up. The current momentarily stops when the alternating current changes direction in the primary circuit. When the current stops, the magnet field collapses and passes across the secondary winding. This collapse of the field *induces* (creates) a current in the secondary winding in one direction. Current in the primary begins to flow in the opposite direction. It builds a magnetic field and collapses when the current stops to change direction again. The collapsing magnetic field of the primary winding again passes across the secondary winding. A current is induced in the opposite direction. The process of inducing a current in the secondary takes place at the rate of 120 times per second, creating an alternating current.

A *laminated iron core* is used as the center for both the primary and secondary windings. Its purpose is to keep the magnetic field from wandering too far from the windings. If the primary and secondary windings are moved away from one another, the amount of current induced into the secondary decreases, because less of the magnetic field crosses the secondary windings.

An ac welding machine can also be an alternator-type machine. An *alternator* creates an ac electrical current

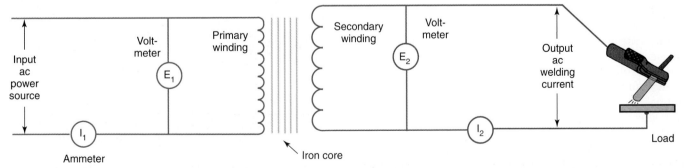

Figure 11-3. *The principal electrical components of an ac transformer. E_1 is the input or primary voltage; I_1 is the primary amperage. E_2 and I_2 are the secondary voltage and amperage.*

from a mechanical source. An electric motor or a gasoline or diesel engine is used to turn a rotor inside a number of electromagnets. The *rotor* is wrapped with multiple turns of wire. As this wire rotates in a magnetic field, an *alternating current* is created in the rotor windings. This created (generated) current is used for welding. Figure 11-4 shows an engine-driven ac alternator-type arc welding machine.

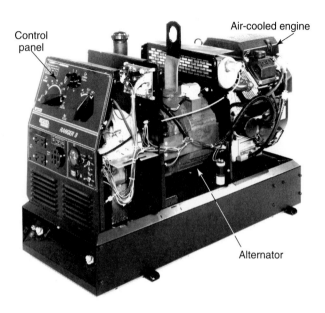

Figure 11-4. An engine-driven alternator-type power source. The outer cover has been removed to view the inner works. (The Lincoln Electric Co.)

Methods of controlling transformer output

Constant current ac transformers are controlled by a variety of methods. These control designs are:

- Solid-state.
- Movable coil.
- Movable shunt.
- Movable core reactor.
- Tapped secondary.
- Saturable reactor.
- Magnetic amplifier.

Current output from a transformer traditionally has been (and still is) accomplished in many welding power sources by moving coils and magnetic fields to control the amount of secondary current coming out of the transformer.

Newer arc welding power sources control the transformer output electronically, using solid-state devices. The most common *solid-state devices* are the diode and the silicon-controlled rectifier (SCR). A *diode* is a device that allows current to flow in only one direction. No current can pass in the opposite direction. Diodes are used to convert alternating current into direct current. This use of diodes is discussed in Heading 11.2.2.

Silicon-controlled rectifiers (SCRs) are like diodes. The difference is that a diode is like a one-way street. Any current going the proper direction, from the anode to the cathode, is allowed to flow. SCRs also allow current to flow

in only one direction, but do so *only after they are turned on*. Drawings of a diode and an SCR are shown in Figure 11-5. The gate on the SCR is the switch that, when turned on, allows current to flow in the proper direction. Once an SCR is turned on, the only way it can be turned off is by reversing the current direction or by stopping the current flow. By reversing the current direction, the SCR stops the current from flowing the wrong direction. The result is that no current flows through the SCR in either direction. Current will flow again once it tries to go the correct direction and the gate receives a signal to allow the current to flow.

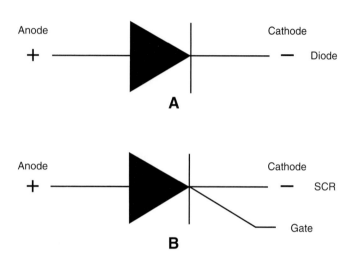

Figure 11-5. Comparison of a diode and an SCR. A—A diode, showing the anode and cathode terminals. B—An SCR, showing terminals and the gate connection.

SCRs are used in the secondary circuit of arc welding power sources. When a lot of current is required, the SCRs are turned on early in each half-cycle. If less current is required, the SCRs are turned on later in each half-cycle. See Figure 11-6. Controlling the current by turning SCRs on at different times controls the secondary current. This method of using SCRs eliminates the need to move primary or secondary coils. This method of controlling current is also used in resistance welding machines.

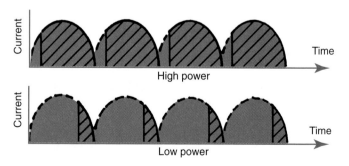

Figure 11-6. Phase control using a silicon-controlled rectifier (SCR).

Figure 11-7 illustrates how SCRs are used to control the secondary transformer output. The control circuitry provides precise control over the SCRs, thus accurately controlling the welding current. SCRs also provide the control necessary for pulse welding, programmed increases or decreases in current, and features like high initial currents to establish an arc followed by a lower steady flow of current. Silicon-controlled rectifiers can be used on constant voltage and constant current welding machines. They are also used in some inverter machines.

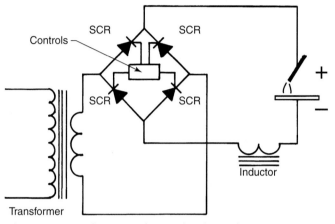

Figure 11-7. *A diagram of a single-phase dc power source that uses an SCR bridge for control.*

Another solid-state device is the transistor. A *transistor* is like an SCR. Current can only flow in one direction through a transistor. A transistor can turn the current on at different times like an SCR. The big difference is that a transistor can also turn the current off without the need to reverse current direction. Also, the transistor can allow different *amounts* of current to flow. The greater the signal applied to the transistor, the more current will flow. A small signal will allow a small amount of current to flow; a larger signal will allow more current to flow. If there is no signal, no current flows. Transistors are more expensive than SCRs and cannot carry as much current. Transistors are used in very-high-frequency inverters (over 10,000Hz), and in some other solid-state-controlled power sources.

Power sources that do not use solid-state devices to control the welding current employ mechanical or electric methods to adjust how the primary and secondary transformer coils interact with each other. An easy method is to move the primary and secondary coils together or apart. Either the primary or secondary winding coil may be moved to adjust the secondary output current. Normally, the secondary coil is fixed and the primary coil is moved. See Figure 11-8 for examples of current output with changing coil positions. A screw arrangement is often used to move the primary coil. The adjustment may be on the top or front of the machine. See Figure 11-9.

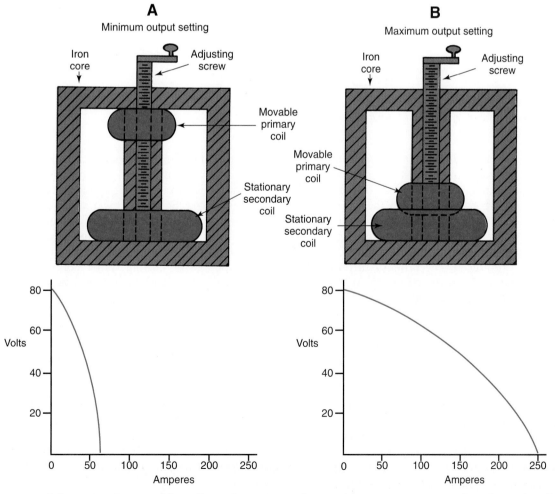

Figure 11-8. *Schematic of a movable coil transformer. A—The minimum setting occurs when the coils are farthest apart. B—The maximum output occurs when the coils are closest.*

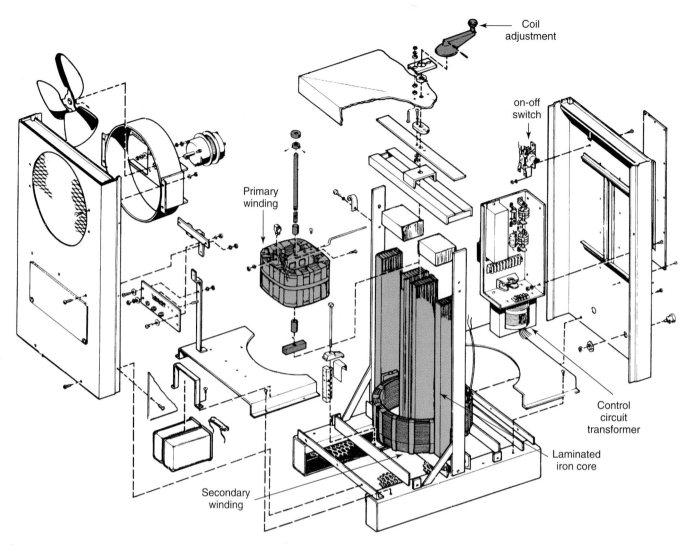

Figure 11-9. Exploded view of a movable coil transformer-type ac arc welder.

A schematic of the **movable shunt transformer control** is shown in Figure 11-10. The movable shunt is made of the same material as the main iron core. Both the primary and secondary coils are stationary. With the shunt moved away from the coils, the maximum number of magnetic lines of flux will cut the secondary windings when the primary current stops. In this position, the maximum welding current is produced. The primary current stops each time the current changes direction. When the shunt is adjusted all the way in, the output current is lowest. See Figure 11-10.

The **movable core reactor** type of transformer uses a primary and a secondary coil. In addition, a *third* coil with a movable core is used to control the current output. See Figure 11-11. The transformer and reactor produce a constant voltage. The reactor is wound to create a *bucking voltage* (counter voltage). This bucking voltage or **inductive reactance** increases the resistance of the secondary circuit. The varying inductive reactance controls the output current. The inductive reactance is least when the movable core is turned out. In this position, the output current from the transformer is greatest. When the movable core in the reactor is turned in, the transformer output is lowered.

Some welding transformers use a **tapped secondary winding,** as shown in Figure 11-12. In this design, the number of secondary windings used depends on where the tap or selector is placed. Selecting fewer secondary windings *increases* the secondary amperage. Selecting more secondary windings *decreases* the amperage used for welding. Changing the tap position makes large changes in the welding amperage or current. Finer control within the selected range is provided by another means, such as a movable core reactor. Figure 11-13 shows a welding machine that uses a tapped secondary winding.

Transformers may use a **saturable reactor control.** This system of control uses a low-amperage, low-voltage dc current to create a magnetic field. This magnetic field affects the magnetic field strength of the reactor coils in the secondary circuit. Large changes in the ac output current will occur with small changes in the dc current to the control coils. See Figure 11-14. This type of control requires no mechanical operation. It therefore lends itself well to remote control of the welding machine output current.

The **magnetic amplifier control** uses the welding current coils and diodes in series with the control coils. The load coil is used to assist the control coil to increase the

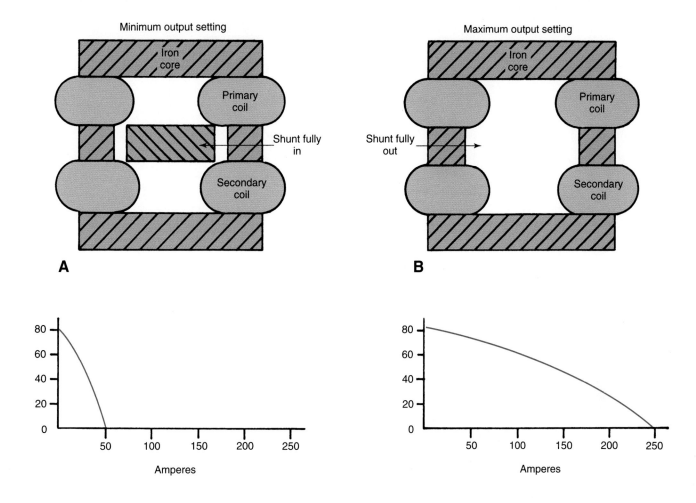

Minimum output setting

Maximum output setting

A

B

Amperes

Amperes

Figure 11-10. *Schematic of a movable shunt transformer control. A—-Shunt adjusted to provide the minimum output current. B—Shunt withdrawn to provide the maximum output current*

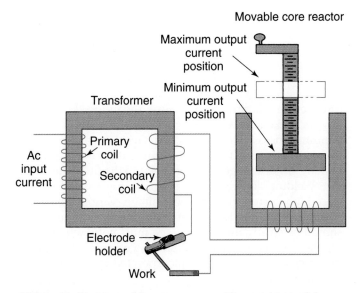

Figure 11-11. *Moveable core reactor. The position of the reactor core will cause the inductive reactance (resistance) of the secondary output coil to vary. This will cause the current output to change.*

magnetic field of the cores. The high magnetic field in the cores cause an inductive reactance in the secondary welding current. This increased reactance decreases the welding current from the transformer. See Figure 11-15 for the location of the cores and diodes.

11.2.2 Direct Current Power Sources

Direct current (dc) arc welding power sources are either the transformer-rectifier or the generator type, as classified in Heading 11.1.

In the transformer section of the transformer-rectifier welding machine, ac is changed from line voltage and current to an ac welding voltage and current. Line voltage supplied to the transformer is generally 220V or 440V at 60 Hz (cycles). The transformer changes the high voltage to an open circuit (no-load) welding voltage of 60V-80V. The current (amperage) welding is usually several hundred amps, depending on the construction of the welding machine. After leaving the transformer section, the welding current enters the rectifier. The rectifier changes the ac to dc.

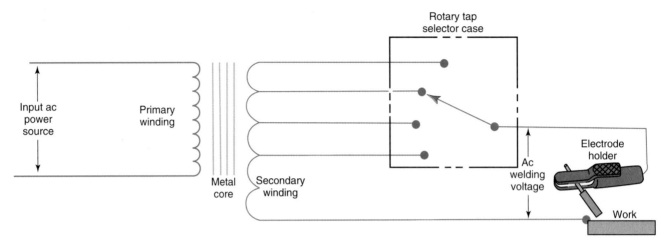

Figure 11-12. *An ac welding transformer with tapped secondary winding. The taps may be of the plug-unplug or the rotary contact type.*

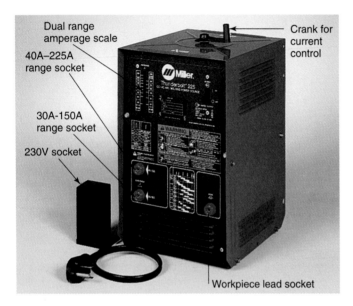

Figure 11-13. *A welding power source that uses two taps to select a high and low range of amperage.*
(Miller Electric Mfg. Co.)

A dc **constant current transformer-rectifier** may be either a single-phase or three-phase power source. Figure 11-16 illustrates both a single-phase and a three-phase rectifier schematic. These schematics illustrate the function of a diode. Alternating current leaving the transformer can only flow through the diode in one direction. The alternating current flowing into the rectifier is changing direction 120 times per second. The current flows from the rectifier in one direction only and has, therefore, been converted to direct current (dc). Figure 11-17 shows a welding machine that uses an ac transformer and a rectifier to convert ac to dc. The control of the current output of a dc transformer-rectifier is accomplished in the same manner as in the ac transformer.

Direct current may also be produced through the use of the dc generator or ac alternator with a rectifier. The dc *generator* is similar in construction to the ac alternator. The

generator uses a rotating armature, similar to the rotor in an alternator. The armature is wound with many separately wound coils. Each end of the *armature coil winding* is soldered to a copper terminal called a *commutator*. At least two carbon or copper contacts called *brushes* touch the commutator. One brush is a positive (+) terminal and one is a negative (–) terminal. See Figure 11-18 for a schematic of a generator. Two or more stationary, wire wound magnets are used. These are called *field windings*. The armature windings are rotated by some mechanical means, such as a motor or engine. As the armature windings rotate, they pass through (cut) the magnetic fields created by the field windings. This cutting of the magnetic field by the armature windings causes a current to be induced (created) in the armature coil. In Figure 11-19A, the armature wire is horizontal and not cutting through the magnetic field, so no current is induced. This position occurs twice in each revolution. In Figure 11-19B, the armature is vertical. It cuts through the magnetic field as it moves up from horizontal to a maximum current position at the top and bottom vertical positions. The induced current is picked up from the commutator by the brushes. A direct current is produced because the induced current is always picked up in one direction. The more armature windings, field windings, and brushes there are, the more evenly the dc will flow. The output voltage is changed by making small changes in the current to the field windings.

Welding generators are specially constructed to produce high current flow at low voltage. Various generator sizes are available. The current produced by the generator should be steady, and the voltage must not fluctuate during the welding procedure. A steady current is maintained by special devices that are incorporated in the design of the generator.

Some machines use a separate exciter to maintain good voltage and current flow characteristics. An *exciter* is a small generator electrically connected to the field windings of the large generator. The exciter keeps a constant voltage on the main fields and also prevents them from reversing their polarity. Trace the current flow in Figure

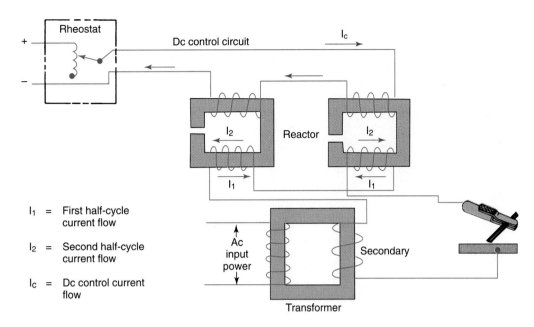

I₁ = First half-cycle current flow

I₂ = Second half-cycle current flow

I_c = Dc control current flow

Figure 11-14. *Saturable reactor-type transformer control. A low-voltage, low-amperage dc current is used in the control circuit. When the dc current is changed, the magnetic field in the reactor is changed. This, in turn, changes the secondary output current.*

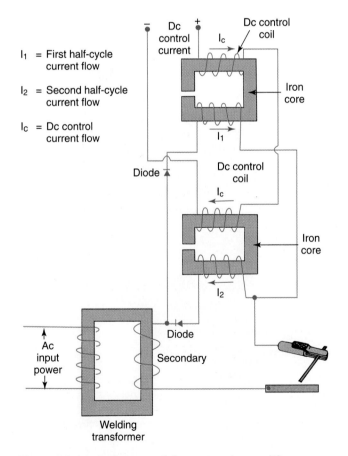

I₁ = First half-cycle current flow

I₂ = Second half-cycle current flow

I_c = Dc control current flow

Figure 11-15. *A diagram of the magnetic amplifier transformer output control. A diode allows current flow in only one direction. This control can be operated from remote locations.*

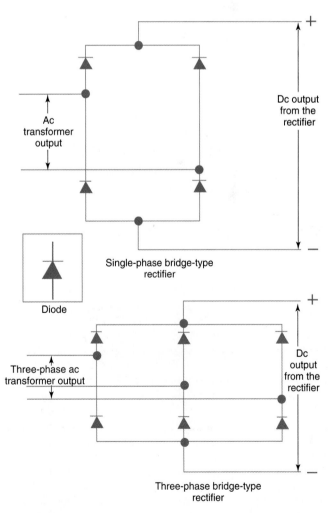

Figure 11-16. *Schematic drawings of single-phase and three-phase bridge-type rectifiers. Note that diodes allow current to travel in the direction of the arrow only. Ac is therefore changed to dc in the rectifier.*

Figure 11-17. *This dc welding machine uses a rectifier to change the ac input to dc output. The polarity is changed by means of a polarity switch. (Miller Electric Mfg. Co.)*

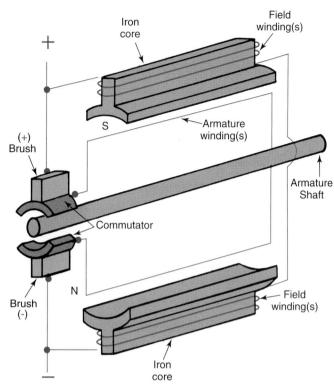

Figure 11-18. *A diagram of a generator. Notice that the field windings are in parallel with the armature windings. The commutator spins with the armature windings, while the brushes remain stationary.*

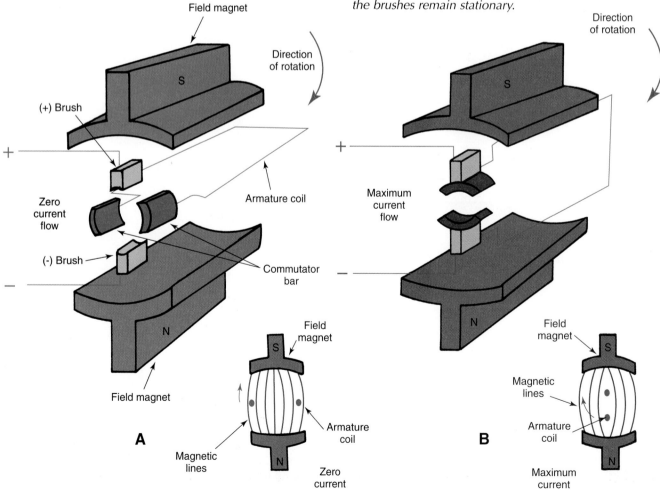

Figure 11-19. *Generator diagram. At A, the armature coil is at the zero current position. At B, the armature is at the maximum current position.*

11-20. Figure 11-21 illustrates a dc generator-type welding machine.

An alternator produces alternating current (ac), as discussed in Heading 11.2.1. An alternator used along with a rectifier will produce direct current. See Figure 11-16, which shows a single-phase rectifier. Figure 11-22 shows a small engine-driven ac/dc welding machine.

Direct current may flow in one of two directions. It may flow from the welding machine to the electrode holder, through the electrode, across the arc gap, and return to the machine through the workpiece lead. Current flowing in this direction is called *direct current electrode negative (DCEN)*. The electrode is negative in this case. Direct current electrode negative is also known as *direct current straight polarity (DCSP)*.

By reversing the position of the electrode and workpiece lead at the welding machine, the dc will first flow to the work. From the work, the current crosses the arc to the electrode, then to the electrode holder, and returns to the machine through the electrode lead. Current flowing in this direction is called *direct current electrode positive (DCEP)* or *direct current reverse polarity (DCRP)*.

Often there is a switch on a dc welding machine that allows the current to be changed from DCEN to DCEP. If there is not such a switch, the current is changed by reversing the electrode and workpiece leads at the machine. Direct current electrode negative is when the electrode is connected to the negative terminal and the work is connected to the positive terminal. Direct current electrode positive is when the electrode is positive and the work is negative.

Combination ac/dc arc welding power sources are also produced. An ac transformer is generally used to produce ac welding current. Rectifiers in series with the transformer are switched in when dc is required. An alternator may also be used to generate ac and rectifiers used to produce dc. Figure 11-23 illustrates an ac/dc arc welding machine. A switch on the machine allows the welder to select the desired current: ac, DCEP, or DCEN.

11.2.3 Inverter Arc Welding Power Sources

Inverter-type power sources are relatively new compared to regular transformer-rectifier, alternator, or generator-type machines. Inverters of equivalent output are much smaller and lighter in weight. See Figure 11-24. These power sources tend to use less input power with a higher efficiency level. They are often designed to provide multiple process capabilities.

An inverter power source is a variation of a conventional transformer-rectifier machine. The biggest difference is that the input power in an inverter is modified to produce high-frequency ac in the range of 1,000Hz to 50,000Hz (1kHz to 50kHz) across the main transformer, instead of the normal line frequency of 50Hz or 60Hz ac. This frequency increase allows the transformer to be reduced in size by up to 70%.

Figure 11-25 illustrates the steps that occur in an inverter. Refer to this figure to understand the inverter process. The first step in normal inverter operation is to

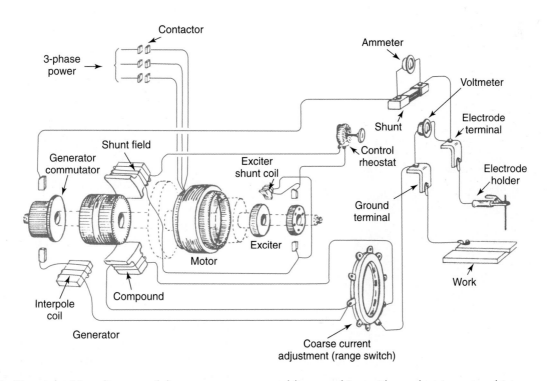

Figure 11-20. *Pictorial wiring diagram of dc motor-generator welding machine with an electric motor drive.*

Figure 11-23. *A combination ac and dc arc welding machine. (Hobart Brothers Co.)*

Figure 11-21. *An engine-driven dc generator type welding machine. This type of power source is often installed on the bed of a pickup truck for mobile use. (The Lincoln Electric Co.)*

Figure 11-22. *An engine-driven ac/dc power source. This power source has several electrical sockets for plugging in 115V and 230V equipment. (The Lincoln Electric Co.)*

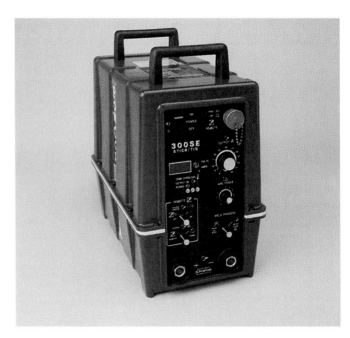

Figure 11-24. *An inverter power supply with multi-process capabilities. This unit measures only 17"H x 10"W x 18"D (43cm x 26cm x 49cm) (PowCon, Incorporated)*

use an input bridge rectifier to convert the ac to dc. The input is 50Hz or 60Hz ac and can be single-phase or three-phase.

The direct current is passed through an *inverter switcher.* The inverter switcher consists of a series of silicon-controlled rectifiers (SCRs) or transistors that are turned on and off very rapidly. They *invert* or *chop* this dc into very high-frequency, square wave ac. One or more

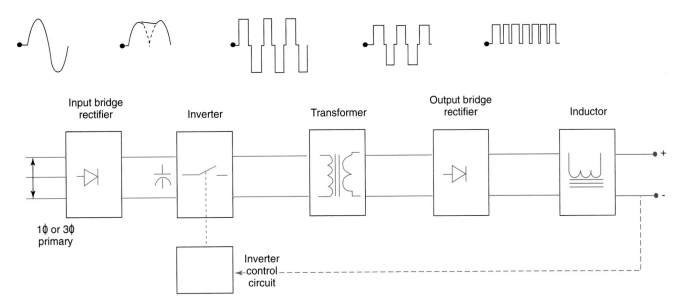

Figure 11-25. *The power supply sections, stages of conversion, and the resulting wave forms that take place in an inverter-type welding machine.*

SCRs or transistors are turned on to allow dc current to flow in one direction. These are turned off and another set is turned on to allow current to flow in the opposite direction. This process of turning SCRs or transistors on and off to chop or invert dc into ac repeats very rapidly to create high-frequency ac current. The frequency is 1kHz to 50 kHz. SCRs are used to create frequencies up to 10kHz. Transistors are used to create frequencies of 10kHz and above.

High-frequency square wave ac is used as the input to the primary of a step-down transformer. The input to the step-down transformer is high-voltage, low-current ac power. The output is low-voltage, high-current ac power. An output bridge rectifier is used to change this ac into dc for welding. The final step is an inductor used to smooth out the dc current.

An important part of the inverter process is the feedback controls used. This feedback can be used to allow one inverter power source to perform multiple welding or cutting processes (GTAW, GMAW, SMAW, PAC). Two separate conventional transformer designs, one for constant current and one for constant voltage, are not required. Instead, a solid-state control circuit is used to sample and control the welding output to produce the desired constant current or constant voltage output. See Figure 11-25.

One of the advantages of inverters is their light weight and small size. Using high-frequency power across the transformer is the factor that enables the transformer to be much smaller. As the frequency applied to a transformer increases, the area and thus the size of the transformer is reduced. This can be seen in the following equation:

$$NA = V/fB$$

Where:

N = number of turns on the transformer coil
A = cross-sectional area of the transformer core
V = inverter voltage
f = inverter operating frequency
B = flux density of the core material

If the number of turns on the transformer coil (N), the inverter voltage (V) and the flux density of the core material (B) are constant, then the cross-sectional area of the transformer core (A) can be reduced as the inverter operating frequency (f) increases.

Because the main transformer is normally the largest and heaviest component in a conventional transformer-rectifier power source, the reduced size of an inverter main transformer permits significant reductions in overall power source size and weight.

If ac output is desired, additional electronics are required. Another inverter switcher can be used to invert (chop) the output dc to create ac. The process of turning SCRs or transistors on and off to create ac was discussed earlier. The output alternating current can be at any frequency, not just 50Hz or 60Hz. Also the DCEN and DCEP portions of an ac cycle do not have to be the same length of time. Figure 11-26 shows an inverter circuit used to create an ac welding current.

Another method of obtaining ac is to use two power supplies in one unit. This method is called *dual source with inverter switching*. The dual source with inverter switching method uses an inverter to create ac current. This is the main welding current. A second power supply provides only dc current, and is used to produce additional current when the electrode is positive. Current flow in the electrode positive part of an ac cycle is often less than the current flow during the electrode negative part. This is especially true when gas tungsten arc welding. The second power supply adds its current to the main current to

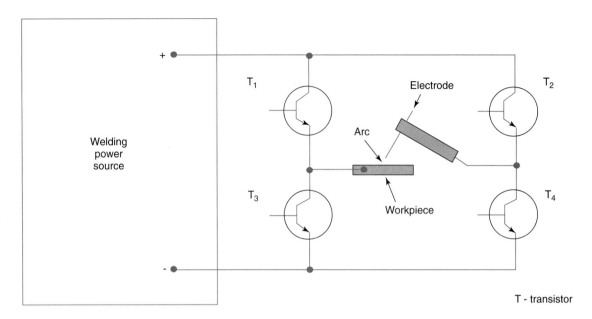

Figure 11-26. *An electrical diagram for an ac inverter.*

obtain desired current flow in the DCEP part of the ac cycle.

An added benefit of using a dual source with inverter switching is that the current in each part of the ac cycle can be different. The second power supply can add more current to the DCEP part of the cycle, as seen in Figure 11-27. A typical application will use a much longer DCEN time and a shorter DCEP time. However, the DCEP current can be greater than the DCEN current. This has application when gas tungsten arc welding aluminum. Figure 11-28 shows the inverter circuit used to create the ac and the additional DCEP current.

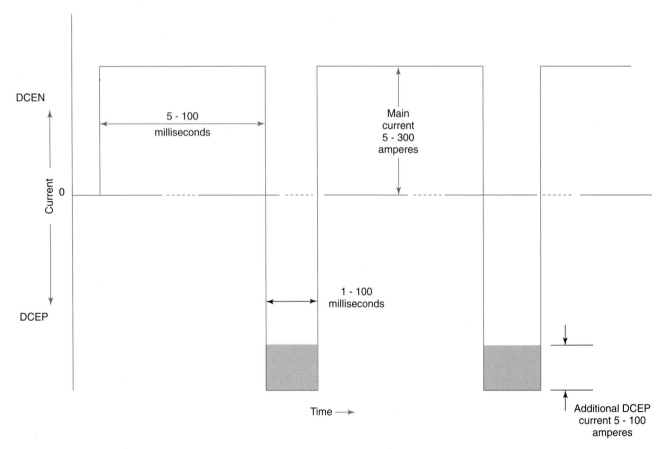

Figure 11-27. *This typical dual power supply waveform shows how the second power supply can add more current to the DCEP part of the cycle.*

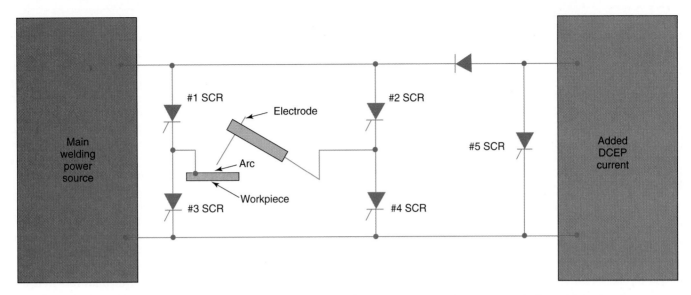

Figure 11-28. *A typical inverter electrical circuit used on a dual power source with inverter switching. This circuit uses two dc power supplies for controlling the heat during GTAW.*

Figure 11-29 shows a number of inverter power sources. Inverter power sources can be used for shielded metal arc, gas tungsten arc, and gas metal arc welding. They are also used for plasma arc cutting.

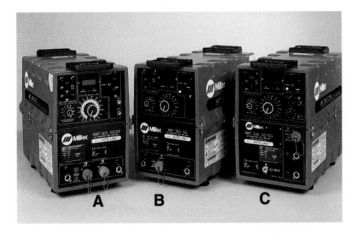

Figure 11-29. *Three different inverter power sources. A—A cc/cv inverter; B—A cc inverter; C—A cc/GTAW inverter with a gas supply connection. (Miller Electric Mfg. Co.)*

11.2.4 Cooling for Welding Power Sources

Most ac welding machines are air-cooled. Some of the smaller units have natural or gravity air-flow through the mechanism.

Some machines, particularly the larger ones, use forced-air circulation. An electric-motor-driven fan is connected into the primary circuit and provides forced-air circulation automatically when the unit is turned on, and while it is in operation. Refer to Figure 11-9.

Air must get in easily, pass around the mechanism, and exit easily. Air passageways must be kept open at all times. The machine should be kept clear of obstructions and should have its ventilation openings uncovered. Periodically, once or twice each year, the power should be disconnected and the casing should be removed for cleaning. Dust should be removed from the internal air passages and from other areas. **Wear goggles if compressed air is used.** A vacuum cleaner, in combination with a brush, produces excellent results.

11.2.5 Machine Installation

Since arc welding machines are single-phase and three-phase machines, they tend to disturb the electrical power circuit. Because of the power factor that will be imposed on the power circuit used, an arc welder will noticeably affect a single-phase or a three-phase circuit to which it may be connected. Power factor correction capacitors are used to improve the power factor of other electrical machinery on the same line. It is best to consult with the electrical utility company and with an electrical contractor before purchasing and installing an arc welding machine.

When using an alternating current machine, the arc welding leads should be as short as possible. The electrical resistance of long cables decreases the current available for welding. The cables should also be kept close together to minimize reactance, since this reactance will reduce the current output of the machine.

11.3 NEMA ARC WELDING POWER SOURCE CLASSIFICATIONS

The *National Electrical Manufacturers Association (NEMA)* has classified electric arc welding machines primarily by duty cycle. The classifications are as follows:

NEMA CLASS I. Machines manufactured to Class I standards that deliver rated outputs at a 60%, 80%, or 100% duty cycle.

NEMA CLASS II. Machines that deliver rated outputs at a 30%, 40%, 50% duty cycle.

NEMA CLASS III. Machines that deliver rated outputs at a 20% duty cycle.

11.3.1 Arc Welding Power Source Specifications

An arc welding machine is described and specified by the following:
- Rated output current rating.
- Power requirements.
- Duty cycle.

The ***rated output current rating*** is a term used to describe the amount of current (amperage) that a welding machine is rated to supply at a given voltage. These ratings are described in the NEMA publication EW-1. Figure 11-30 shows the NEMA rated output current for various classes of welding machines. See Heading 11.3, above, for a description of NEMA classifications.

Rated Output Current in Amperes		
Class I	Class II	Class III
200	150	180-230
250	175	235-295
300	200	
400	225	
500	250	
600	300	
800	350	
1000		
1200		
1500		

Figure 11-30. *National Electrical Manufacturer's Association (NEMA) rated output current for various classifications of arc welding machines.*

The power requirements are called the ***rated load voltage*** or ***welding voltage***. For Class I and II power sources under a 500A rating, the rated load voltage is calculated as follows: Voltage (E) = 20 + 0.04 x rated amperes. As an example, using a Class I machine with an output current rating of 300A, the rated load voltage would be:

Voltage = 20 + 0.04 x 300
= 20 + 12
= 32V

The welding rated output current rating for this welding machine is 300A at 32V. For welding machines with output currents above 600A, the rated load voltage used is 44 volts.

Electrical power input requirements for NEMA Class I and II transformer arc welders are as follows:
50 Hz (cycles)—220V, 380V, and 440V
60 Hz (cycles)—200V, 230V, 460V, and 575V

The electrical power input requirements for a NEMA Class III transformer is at 60Hz, 230V.

Duty cycle is: "the length of time that a welding machine can be used continually at its rated output in any ten (10) minute period."

Most welding machines are not required to operate 100% of the time. If they are used to weld one hundred percent of the time, a machine with a 100% duty cycle would be required. Work must be loaded and unloaded, and electrodes must be changed. Metal must be chipped, cleaned, and inspected. The duty cycle normally recommended for manual welding is 60%. Automatic and semi-automatic welding operations usually require 100% duty cycles. Light duty work, such as that done in the home hobby shop, could possibly use a duty cycle of 20%.

When purchasing a welding machine, the maximum duty cycle requirement must be considered. See Heading 11.3 for the NEMA arc welding machine classifications by duty cycle.

If an arc machine is used at lower than rated current outputs, the duty cycle may be increased under certain conditions. A graph supplied by the manufacturer will indicate how the duty cycle can vary with a given current setting. Figure 11-31 is an example of a graph showing welding amperage and duty cycle.

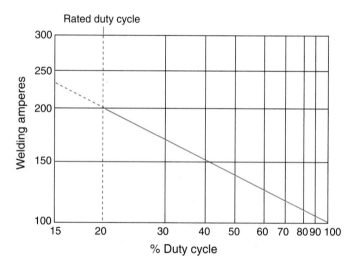

Figure 11-31. *A duty cycle-welding amperage chart. Such charts are developed for specific machines—this one is for the Miller Thunderbolt 225V machine. (Miller Electric Mfg. Co.)*

11.4 WELDING LEADS

Large diameter, superflexible leads (cables) are used to carry current from the welding machine to the work and back. The lead from the machine to the electrode holder is known as the ***electrode lead***. The lead from the work to the machine is known as the ***workpiece lead*** or ***ground lead***. See Figure 11-32.

Leads are well-insulated. They may be insulated with rubber and a woven fabric reinforcing layer, as shown in Figure 11-33A. Leads may be covered with neoprene, as shown in Figure 11-33B. The leads are usually subjected to considerable wear and should be checked periodically for

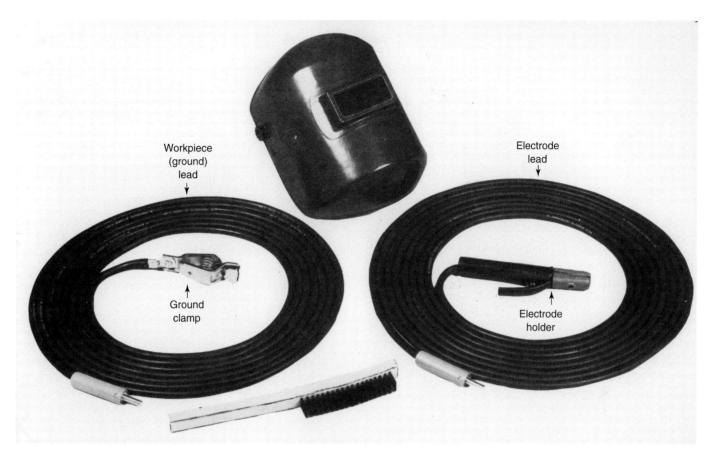

Figure 11-32. *Arc welding leads. (Miller Electric Mfg. Co.)*

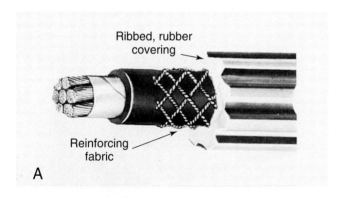

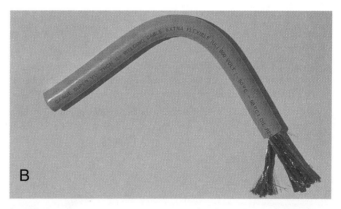

Figure 11-33. *Welding leads. A—Some leads have a heavy ribbed rubber outer covering over a mesh of reinforcing fabric. B—This piece of AWG #1/0 welding lead shows the large number of fine copper wires used. These small wires, woven into larger bundles, give the lead its flexibility. The bundle is wrapped in plastic and several strands of strong cord. Neoprene is then used as an insulator and a wear covering. (Carol Cable Co.)*

breaks in the insulation. The voltage carried by the leads is not excessive, varying between approximately 14V and 80V.

Leads are produced in several sizes. The smaller the number, the larger the diameter of the lead. Figure 11-34 is a list of sizes and current capacities for copper leads. The lead must be flexible to permit easy installation of the cable, and to reduce the strain on the arc welder's hand when welding. To produce this flexibility, as many as 800

to 2500 fine wires are used in each cable. The same diameter electric cable must be used on both the electrode and workpiece leads. The length of the lead has considerable effect on the size to be used for certain capacity machines.

The wiring to the motor, in case a motor-driven generator is used, must conform to local, state, and national electrical codes. Copper leads are preferred; however, aluminum leads have been used. Aluminum has about 61% the current-carrying capacity of copper. For a given

Current Capacity in Amperes Length in Feet and Meters					
Lead no.	Lead dia.		Length 0-50 ft. 0-15.2 m	Length 50-100 ft. 15.4-30.5 m	Length 100-250 ft. 30.5-76.2 m
	in.	mm	Amperage	Amperage	Amperage
4/0	.959	24.4	600	600	400
3/0	.827	21.0	500	400	300
2/0	.754	19.2	400	350	300
1/0	.720	18.3	300	300	200
1	.644	16.4	250	200	175
2	.604	15.3	200	195	150
3	.568	14.4	150	150	100
4	.531	13.5	125	100	75
Note: Lengths given are for the total combined length of the electrode and work leads.					

Figure 11-34. Arc welding lead recommendations. The voltage drop in these copper leads will be approximately 4V with all connections clean and tight.

current capacity, an aluminum lead will be larger in diameter, but only about one-half as heavy. Aluminum leads are pure, semi-annealed, electrolytic aluminum.

11.4.1 Connections for Leads

To consistently carry the large currents used in welding, all parts of the welding circuit (including all terminals) must be of heavy-duty design and construction.

The copper or aluminum leads are fastened to the welding machine and workpiece by means of insulated or uninsulated terminals. The uninsulated terminals are called *lugs* and are shown in Figure 11-35. These lugs are mechanically crimped to the leads. The lugs provide a firm means of attaching the electrode lead and the workpiece lead to the machine or worktable. These connections must be durable and must have low resistance, or the joint will overheat during welding. Less current will flow if the connection is loose.

Insulated lead terminals, also called *connectors*, are available in different styles. Figure 11-36 illustrates a type of quick-disconnecting terminal used in the United States. European-style terminals are called DIN connectors. DIN connections also use a quick disconnect method and can be seen in Figure 11-37. Adapters are available to connect the U.S. twist-style and DIN terminals.

Terminals are connected to the leads by the following methods:
- Mechanical
- Soldering
- Brazing

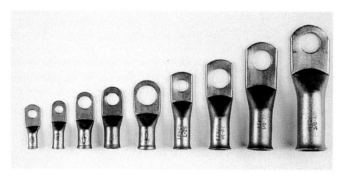

Figure 11-35. A variety of welding lead lugs. (Lenco – NLC, Inc.)

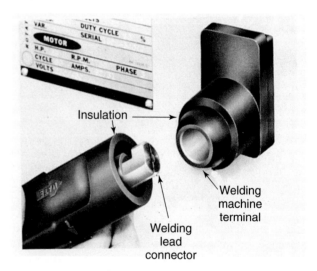

Figure 11-36. An insulated quick-connect and disconnect welding machine terminal.

Figure 11-37. *American and European quick connectors for welding machine leads. The European connector is called a DIN connector. The connectors marked with * are DIN connectors. (Weldcraft Products, Inc.)*

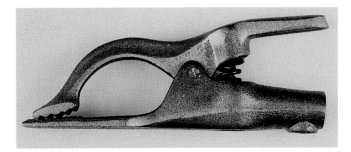

Figure 11-38. *A spring-loaded ground clamp for the workpiece lead. (Lenco – NLC, Inc.)*

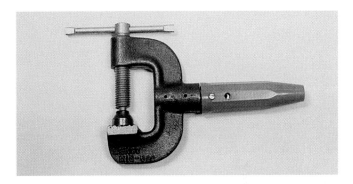

Figure 11-39. *A C-clamp-type ground clamp for the workpiece lead. (Lenco – NLC, Inc.)*

In the case of aluminum cables, it is claimed that it is best to clamp the aluminum to the electrode holder and to the other terminals. However, the lead can be successfully aluminum-brazed to either aluminum connections or copper connections. It is recommended that twisting of the lead, especially at the connections, be avoided. The lead may tend to separate from its terminals if this is done. Mechanical connections must be tight and clean.

Soldering and silver-brazing of connections must be done properly. It is necessary to make complete connection of the lead and connector to allow the electrical current to flow over the entire soldered or brazed surface area.

Good workpiece (ground) lead and electrode lead connections are important when welding. The workpiece lead can be fastened to the welding bench or table by means of a lug or insulated terminal. This method is practical when welding can be done on a welding table.

Frequently, the ground cable must be moved on a regular basis, or must be fastened to the article being welded, due to its size or location. The spring-loaded ground (workpiece) clamp, shown in Figure 11-38, is often used for on-site welding. A C-clamp type of ground clamp is shown in Figure 11-39. It is sometimes difficult to use a clamping device on a metal fabrication. Clamps must be used carefully on finished surfaces to avoid marring. A magnetic workpiece lead terminal is available which permits quick and secure fastening of the workpiece lead to the weldment. This makes it easy to change the position of the ground to obtain better arc characteristics, and does not injure or mar the article to be welded. The workpiece lead is either soldered or mechanically fastened to a permanent magnet or an electromagnet ground device. The operator may easily position the magnetic grounding device on most ferrous (iron) surfaces.

11.4.2 Electrode Holders

The *electrode holder* is the part of the arc welding equipment held by the operator when welding. The electrode holder holds the electrode. Many different styles and models have been produced, but they all have similar characteristics. The electrode lead is usually fastened to the electrode holder by means of a mechanical connection inside the electrode handle. Solid brass or copper is used inside the handle to conduct the electrical current from the lead connection to the lower jaw. The outer handle is made of an insulating material that has high heat and electrical resistance. Electrode holders are built to produce a balanced feeling when held in the operator's hand. There should be a good balance with the cable draped over the operator's arm, and with an average length of metallic electrode in the holder.

There are a number of methods used to clamp the electrode in the holder. The most common styles of electrode holder use a pinch-type construction and include a spring to produce the necessary pressure. Figure 11-40 shows an electrode holder, while Figure 11-41 shows the parts of the holder and how the workpiece lead is connected. Electrode holders come in different sizes. They are designed to hold different diameter electrodes and to carry different amounts of welding current. Select the smallest size that will handle the electrode diameter and current that will be used. A small size permits better welder comfort.

The electrode should be clean at the point where it is to contact the electrode holder. The electrode contact area may be cleaned with a wire brush or steel wool. The electrode holder jaws should also be kept clean by using a file, sandpaper, or other suitable means.

Figure 11-40. *A SMAW electrode holder that is rated for 250 amperes. (Lenco – NLC, Inc.)*

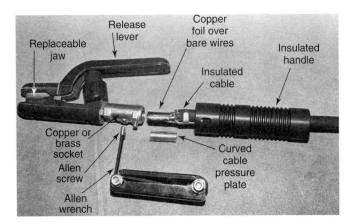

Figure 11-41. *An exploded view of a SMAW electrode holder. Notice the insulated handle and the parts required to connect the welding lead to the electrode holder.*

11.5 SMAW ELECTRODES

Shielded metal arc welding (SMAW) electrodes have a solid metal wire core and a thick flux covering (coating). These electrodes are identified by the wire diameter and by a series of letters and numbers. These letters and numbers identify the metal alloy and the intended use of the electrode.

The common SMAW electrode wire diameters are 1/16" (1.6mm), 3/32" (2.3mm), 1/8" (3.2mm), 5/32" (4.0mm), 3/16" (4.8mm), 7/32" (5.6mm), 1/4" (6.4mm), 5/16" (7.9mm), and 3/8" (9.5mm). They may be obtained in lengths from 11" to 18" (28cm to 46cm). The most frequently used length is 14" (36cm).

They are usually purchased in 50 lb. (22.7 kg) packages. Electrodes may be packaged in cardboard cartons or in hermetically sealed (airtight) metal cans. Figure 11-42 shows one method of storing electrodes.

SMAW electrodes are produced for welding on many metals and alloys, including the following:

- Carbon steels (ANSI/AWS A5.1).
- Low-alloy steels (ANSI/AWS A5.5).
- Corrosion-resistant steels (ANSI/AWS A5.4).
- Cast irons (ANSI/AWS A5.15).
- Aluminum and aluminum alloys (ANSI/AWS A5.3).
- Copper and copper alloys (ANSI/AWS A5.6).
- Nickel and nickel alloys (ANSI/AWS A5.11).
- Surfacing (ANSI/AWS A5.13 and A5.21)

Figure 11-42. *Storing electrodes. Each bin is marked with AWS number and electrode diameter.*

Covered electrodes serve many purposes in addition to adding filler metal to the molten weld pool. These additional functions are provided by the covering on the electrode. Heading 11.5.1 discusses the functions of the covering.

Dampness, usually caused by absorbing moisture from the air, destroys the effectiveness of most electrode coverings. The dampness will introduce hydrogen into the weld. This will cause cracking or brittleness when welding some metals. Many welding procedures require that electrodes be thoroughly dried prior to use. This is done in specially built drying ovens. See Figure 11-43.

Figure 11-43. *Two types of portable electrode drying ovens. The large gray oven is designed for shop use, and will hold up to 350 lb. of 18" electrodes, while the two yellow units are portable and can be easily moved to the job site for field welding. The square portable oven will hold 50 lb. of 18" electrodes, and the round unit will hold 13 lb. (Phoenix Products Co., Inc.)*

The time it takes for an electrode to pick up moisture from the air will vary from thirty minutes to four hours, depending on the electrode. This time period is known for each electrode in use. Welders, therefore, take only enough electrodes from the oven to last for this *moisture-pickup time period*. Electrodes must be handled carefully to prevent breaking the flux coating. Many welding procedures will not permit an electrode to be used if the flux coating is chipped. Figure 11-44 illustrates an electrode dispenser that is carried by the welder. This dispenser is sealed to keep the electrodes dry. The dispenser is also tough enough to prevent physical damage to the electrodes.

11.5.1 Electrode Covering Fundamentals

The covering on a shielded metal arc welding electrode is called *flux*. The flux on the electrode performs many different functions. These include:

- Producing a protective gas around the weld area.
- Providing fluxing elements and deoxidizers.
- Creating a solid coating over the weld as it cools.
- Establishing electrical characteristics.
- Adding alloying elements.

During the arc process, some of the flux covering changes to **neutral gases** or **reducing gases,** such as carbon monoxide (CO) or hydrogen (H_2). These gases, as they surround the arc proper, prevent air from coming in contact with the molten metal. They prevent oxygen in the air from combining with the molten metal. However, the gases usually do not protect the hot metal after the arc leaves that area of the weld.

The covering also contains special fluxing ingredients that work to remove impurities from the molten metal. Impurities are floated to the top of the molten weld pool.

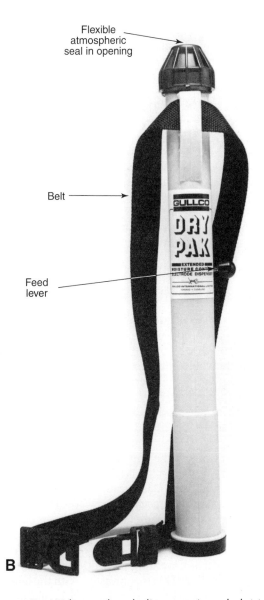

Figure 11-44. *Electrode dispenser. A—Welder wearing a loaded dispenser. B—With top closed, dispenser is sealed. Moving the lever feeds a new electrode through the top seal. (Gullco)*

As the electrode flux coating residue cools, it forms a coating of material over the weld called *slag*. This slag coating prevents the air from contacting the hot metal. The slag covering also allows the weld to cool more slowly and helps prevent a hard, brittle weld.

When welding with ac, the current changes direction and actually stops 120 times per second. To maintain an arc as the current changes direction, ingredients are added to the covering of the electrode to create an ionized gas. This ionized gas allows good arc stability when welding with ac.

The flux covering on a shielded metal arc electrode can also contain alloying elements. These alloying elements are added to the weld pool as the covering is melted. Iron powder and iron oxide also can be added to the coverings of steel electrodes. They increase the amount of metal deposited in a given period of time.

Certain coverings on steel electrodes are designed to be low in hydrogen. Hydrogen causes low ductility and cracking in certain cases. The use of low-hydrogen electrodes helps to eliminate these problems.

A good flux-covered electrode can produce a weld that has excellent physical and chemical properties.

11.6 CARBON AND LOW-ALLOY STEEL COVERED ELECTRODE CLASSIFICATION

There is a large variety of *carbon and low-alloy steel electrodes* on the market. The chemical composition of the steel wire used in the E60XX and E70XX electrodes is exactly the same. The composition of the coverings on the electrodes varies considerably, according to the planned use of the electrode.

The American Welding Society has developed a series of *identifying number classifications*. See Figures 11-45 and 11-46.

The *carbon and low-alloy steel electrode classification number* uses four or five digits. For carbon steels, the electrodes are either in the E60XX or E70XX series. The minimum allowable tensile strength for a weld made with an electrode in the 60 series is 62,000 psi (427MPa). Additional elongation may allow the tensile strength of some of these electrodes to go as low as 60,000 psi (414MPa). For the 70 series, the minimum tensile strength as welded is 72,000 psi (496MPa). However, additional elongation may allow the tensile strength of some of these to go down to 70,000 psi (483MPa). The metals are similar in composition for each classification number. Each manufacturer has its own compounds for the coverings. Therefore, very few of the electrodes behave exactly the same, even though the classification number may be identical. The letter E preceding the four or five digit number (EXXXX) indicates a welding electrode used in arc welding. See Figure 11-46. This is contrasted with the letters RG, which indicate a welding rod used for gas welding.

AWS classifi- cation	Type of covering	Capable of producing satisfactory welds in position shown[a]	Type of current[b]
E60 series electrodes			
E6010	High cellulose sodium	F, V, OH, H	DCEP
E6011	High cellulose potassium	F, V, OH, H	ac or DCEP
E6012	High titania sodium	F, V, OH, H	ac or DCEN
E6013	High titania potassium	F, V, OH, H	ac or dc, either polarity
E6020	High iron oxide	H-fillets	ac or DCEN
E6022[c]	High iron oxide	F	ac or dc, either polarity
E6027	High iron oxide, iron powder	H-fillets, F	ac or DCEN
E70 series electrodes			
E7014	Iron powder, titania	F, V, OH, H	ac or dc, either polarity
E7015	Low hydrogen sodium	F, V, OH, H	DCEP
E7016	Low hydrogen potassium	F, V, OH, H	ac or DCEP
E7018	Low hydrogen potassium, iron powder	F, V, OH, H	ac or DCEP
E7024	Iron powder, titania	H-fillets, F	ac or dc, either polarity
E7027	High iron oxide, iron powder	H-fillets, F	ac or DCEN
E7028	Low hydrogen potassium, iron powder	H-fillets, F	ac or DCEP
E7048	Low hydrogen potassium, iron powder	F, OH, H, V-down	ac or DCEP

a. The abbreviations, F, V, V-down, OH, H, and H-fillets indicate the welding positions as follows:
     F   =   Flat
     H   =   Horizontal
     H-fillets   =   Horizontal fillets
     V-down   =   Vertical down
     V   =   Vertical   } For electrodes 3/16" (4.8mm) and under,
     H   =   Overhead   } except 5/32" (4.0mm) and under for classifications E7014, E7015, E7016, and E7018.
b. The term DCEP refers to direct current, electrode positive (dc reverse polarity). The term DCEN refers to direct current electrode negative (dc straight polarity).
c. Electrodes of the E6022 classification are for single-pass welds.

Figure 11-45. *AWS Carbon Steel Covered Arc Welding Electrodes. This information is from AWS A5.1-94. (American Welding Society.)*

The meaning of the digits in the AWS classification number follows. The first two or three digits of the four or five digit number (E60XX or E100XX) represent the *tensile strength.* That is, **60** means 60,000 psi (414MPa) and **100** means 100,000 psi (689MPa). The value 60,000 psi (pounds per square inch) also may be shown as 60 ksi. The letter "k" represents 1000 lbs.(a "kilopound"), so 60 ksi (kilopounds per square inch) is the same as 60,000 psi (pounds per square inch). The tensile strength may be given in the "as-welded" or the "stress-relieved" condition. See the electrode manufacturer's specification to determine under

AWS classification[a]	Type of covering	Capable of producing satisfactory welds in positions shown[b]	Type of current[c]
E70 series—Minimum tensile strength of deposited metal, 70,000 psi (480 MPa)			
E7010-X	High cellulose sodium	F, V, OH, H	DCEP
E7011-X	High cellulose potassium	F, V, OH, H	ac or DCEP
E7015-X	Low hydrogen sodium	F, V, OH, H	DCEP
E7016-X	Low hydrogen potassium	F, V, OH, H	ac or DCEP
E7018-X	Iron powder, low hydrogen	F, V, OH, H	ac or DCEP
E7020-X	High iron oxide	{ H-fillets / F	ac or DCEN / ac or DC, either polarity
E7027-X	Iron powder, iron oxide	{ H-fillets / F	ac or DCEN / ac or dc, either polarity
E80 series—Minimum tensile strength of deposited metal, 80,000 psi (550 MPa)			
E8010-X	High cellulose sodium	F, V, OH, H	DCEP
E8011-X	High cellulose potassium	F, V, OH, H	ac or DCEP
E8013-X	High titania potassium	F, V, OH, H	ac or dc, either polarity
E8015-X	Low hydrogen sodium	F, V, OH, H	DCEP
E8016-X	Low hydrogen potassium	F, V, OH, H	ac or DCEP
E8018-X	Iron powder, low hydrogen	F, V, OH, H	ac or DCEP
E90 series—Minimum tensile strength of deposited metal, 90,000 psi (620 MPa)			
E9010-X	High cellulose sodium	F, V, OH, H	DCEP
E9011-X	High cellulose potassium	F, V, OH, H	ac or DCEP
E9013-X	High titania potassium	F, V, OH, H	ac or dc, either polarity
E9015-X	Low hydrogen sodium	F, V, OH, H	DCEP
E9016-X	Low hydrogen potassium	F, V, OH, H	ac or DCEP
E9018-X	Iron powder, low hydrogen	F, V, OH, H	ac or DCEP
E100 series—Minimum tensile strength of deposited metal, 100,000 psi (690 MPa)			
E10010-X	High cellulose sodium	F, V, OH, H	DCEP
E10011-X	High cellulose potassium	F, V, OH, H	ac or DCEP
E10013-X	High titania potassium	F, V, OH, H	ac or dc, either polarity
E10015-X	Low hydrogen sodium	F, V, OH, H	DCEP
E10016-X	Low hydrogen potassium	F, V, OH, H	ac or DCEP
E10018-X	Iron powder, low hydrogen	F, V, OH, H	ac or DCEP
E110 series—Minimum tensile strength of deposited metal, 110,000 psi (760 MPa)			
E11015-X	Low hydrogen sodium	F, V, OH, H	DCEP
E11016-X	Low hydrogen potassium	F, V, OH, H	ac or DCEP
E11018-X	Iron powder, low hydrogen	F, V, OH, H	ac or DCEP
E120 series—Minimum tensile strength of deposited metal, 120,000 psi (830 MPa)			
E12015-X	Low hydrogen sodium	F, V, OH, H	DCEP
E12016-X	Low hydrogen potassium	F, V, OH, H	ac or DCEP
E12018-X	Iron powder, low hydrogen	F, V, OH, H	ac or DCEP

a. The letter suffix "X" as used in this table stands for the suffices A1, B1, B2, etc. (see Figure 11-47) and designates the chemical composition of the deposited weld metal.

b. The abbreviations F, V, OH, H, and H-fillets indicate welding positions as follows: F = Flat; H = Horizontal; H-fillets = Horizontal fillets. V = Vertical } For electrodes 3/16" (4.8mm) and under, except 5/32" (4.0mm) and under for classifications OH = Overhead } EXX15-X, EXX16-X, and EXX18-X.

c. DCEP means electrode positive (reverse polarity). DCEN means electrode negative (straight polarity).

Figure 11-46. *AWS Low-Alloy Steel Covered Arc Welding Electrodes. This information is from AWS A5.5. See Figure 11-47 for the letter suffixes used in place of the "X" with these electrode numbers. (American Welding Society)*

what condition the indicated tensile strength occurs. "As-welded" means without post heating. "Stress-relieved" means the weld has been given a heat treatment after welding to relieve stress caused by the welding process. See Chapter 29 for an explanation of stress caused by welding.

The second digit from the right indicates the *recommended position* of the joint that the electrode is designed to weld. For example, EXX**1**X indicates that this electrode will weld in all positions; EXX**2**X electrodes are used for welds in the flat or horizontal welding position. The EXX**4**X is recommended for flat, horizontal, overhead, and downhill welding.

The right hand digit indicates the *power supply (ac, DCEN,* or *DCEP)*, the *type of covering,* and the *presence of iron powder* or *low-hydrogen characteristic* (or both).

The last two digits should be looked at together to determine the proper application and the covering composition for an electrode. For example:

Electrode Number	Covering Composition
EXX10	High-cellulose, sodium
EXX11	High-cellulose, potassium
EXX12	High-titania, sodium
EXX13	High-titania, potassium
EXX14	Iron powder, titania
EXX15	Low-hydrogen, sodium
EXX16	Low-hydrogen, potassium
EXX18	Iron powder, low-hydrogen, potassium
EXX20	High-iron oxide
EXX22	High-iron oxide
EXX24	Iron powder, titania
EXX27	Iron powder, high-iron oxide
EXX28	Iron powder, low-hydrogen, potassium
EXX48	Iron powder, low-hydrogen, potassium

See Figures 11-47 and 11-48, which show the recommended position and polarity for various carbon and low-alloy steel electrodes.

Occasionally an electrode number may have a letter and number *after* the normal four or five digits, such as E7010-A1 or E8016-B2. The letter and number combination or *suffix* is used with low-alloy steel electrodes. The suffix indicates the chemical composition of the deposited weld metal. See Figure 11-47. The letter **A** indicates a carbon-molybdenum steel electrode. The letter **B** stands for chromium-molybdenum steel electrode. The letter **C** is a nickel steel electrode, and the letter **D**, a manganese-molybdenum steel electrode. The final digit in the suffix indicates the chemical composition under one of these broad chemical classifications. The exact chemical composition may be obtained from the electrode manufacturer.

The letter **G** is used for all other low-alloy electrodes with minimum values of molybdenum (0.20% minimum); chromium (0.30% minimum); manganese (1% minimum); silicon (0.80% minimum); nickel (0.50% minimum); and vanadium (0.10% minimum) specified. Only one of these elements is required to meet the alloy requirements of the **G** classification.

-A1	1/2% Mo
-B1	1/2% Cr, 1/2% Mo
-B2	1 1/4% Cr, 1/2% Mo
-B3	2 1/4% Cr, 1% Mo
-C1	2 1/2% Ni
-C2	3 1/4% Ni
-C3	1% Ni, .35% Mo, .15% Cr
-D1 and D2	.25-.45% Mo, 1.25-2.00% Mn
-G	.50 min Ni, .30 min Cr, .20 min Mo, .10 min V, 1.00 min Mn, .80 min Si (Only one of the listed elements is required for the G classification).

Figure 11-47. *Approximate chemical composition of suffix numbers of the AWS Electrode Numbering System.*

An example of a complete electrode classification is E8016-B2:

- **E** indicates electrode.
- **80** indicates tensile strength—80,000 psi or 80 ksi.
- **16** indicates a low-hydrogen, potassium covering used with ac or DCEP (reverse polarity). DCEP means *direct current electrode positive.* See Heading 12.2 for more information on electrode polarity.
- The **1** indicates it is an all-position electrode.
- The suffix **B2** indicates that the deposited metal chemical composition is a low-alloy chromium-molybdenum steel with 1.25% chromium and 0.50% molybdenum.

Manufacturers imprint AWS numbers on the covering material near the grip end for identification. Figure 11-48 shows how shielded metal arc welding electrodes are marked with AWS numbers.

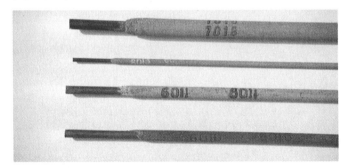

Figure 11-48. *Several electrodes that show how the identifying numbers are placed on the covering.*

11.6.1 Low-Hydrogen Electrodes

Hydrogen has harmful effects on alloy steels. It causes a low ductility weld and underbead cracking. This is called **hydrogen embrittlement** or **hydrogen cracking**.

Low-hydrogen electrodes deposit a minimum of hydrogen in the weld. The low hydrogen condition is obtained by using a special covering in the EXXX5, 6, or 8 category. See Figure 11-45. Sodium, potassium, and iron powder coverings are used. These electrodes conform to AWS E7015, E7016, E7018, E7028, and E7048 specifications. They are used on hard-to-weld free machining steels, low-carbon, low-alloy, and hardenable steels. The slag is very fluid, but good flat or convex beads are easily obtained.

Right hand Digit	Covering compositions	Application (use)
5 E-7015	Low-hydrogen sodium type.	This is a low-hydrogen electrode for welding low-carbon alloy steels. Power shovels and other earthmoving machinery require this rod. The weld files or machines easily. Use DCEP (DCRP) only.
6 E-7016	Same as "5" but with potassium salts used for arc stabilizing.	It has the same general application as (5) above, except it can be used on either DCEP (DCRP) or ac.
8 E-7028 E-8018	Iron powder (Low-hydrogen), flat position only. Iron powder plus low-hydrogen sodium covering.	For low-carbon alloy steels, use dc or ac. Similar to (5) and (6) DCEP (DCRP) or ac. Heavy covering allows the use of high-speed drag welding. Ac or DCEP (DCRP) may be used.

Figure 11-49. *Low-hydrogen electrode compositions and applications. These coverings will withstand a high temperature, so high currents (amperages) may be used.*

Figure 11-49 shows the composition and application of some low-hydrogen electrode coverings. These special coverings contain practically no organic material. Recommended arc welding machine settings for these electrodes are shown in Figure 11-50.

The deposited metal has excellent tensile and ductile qualities, and is exceptionally clean as may be seen by X-ray inspection. Figure 11-51 illustrates the physical properties of a weld made with a low-hydrogen E7016 electrode.

The E7016 electrodes may be used with either ac or DCEP (DCRP). Low-hydrogen electrodes should be dried by baking at 250°F (121°C) before using. If they have been exposed to the atmosphere for an appreciable period, they should be re-baked at 500°F to 700°F (260°C to 371°C) for one hour. This baking will remove any moisture that may be in the coating. The electrodes are then stored at 250°F (121°C) until they are used.

11.6.2 Carbon and Low-Alloy Steel Iron-Powder Electrodes

The addition of iron powder to the covering of shielded arc electrodes changes the arc behavior. It also greatly increases the amount of filler metal deposited. Much higher currents can be used to produce welds faster. The addition of iron powder also produces more easily cleaned welds, less spatter, and better bead shapes. Figure 11-52 shows a change in filler metal deposited from 4.5 lb./hr. (2.0 kg/hr.) with the E6012 electrode to 9 lb./hr. (4.1 kg/hr.) with the E7028 electrode.

Current Settings for Low-Hydrogen Electrodes

Electrode diameter	Amps (flat)	Amps (vertical and overhead)	Volts
1/8"	140-150	120-140	22-26
5/32"	170-190	160-180	22-26
3/16"	190-250	200-220	22-26
7/32"	260-320		24-27
1/4"	280-350		24-27
5/16"	360-450		26-29

Figure 11-50. *Recommended arc machine settings when using low-hydrogen electrodes in flat, vertical, and overhead positions.*

	As welded
Yield point	60,000 psi (414 MPa)
Tensile (minimum)	70,000 psi (480 MPa)
Elongation, % in 2" (50.8mm)	22%
Charpy V notch at -20°F (-29°C)	20 ft. lb. (27.1 Nm)

Figure 11-51. *AWS specifications for welds made with E7016 low-hydrogen electrodes.*

	Electrode Classification			
	E7028	E7024	E7018	E6012
Iron powder content	45%	39%	24%	0%
Amount deposited	9 lb./hr.	8 lb./hr.	5.5 lb./hr.	4.5 lb./hr.
Deposit efficiency increase over E6012	100%	78%	22%	0%

Figure 11-52. *Effect of using iron powder in electrode coatings. Note the change in the amount of filler metal deposited per hour. The deposit efficiency increase is based on the E6012 electrode.*

The arc obtained is smooth and steady. You may use about 25% more current with the EXX18 iron-powder electrode, and as much as 50% more current with the EXX24 and EXX27 iron-powder electrodes, because of their heavy coverings.

When electrodes with a high content of iron powder (39% and above) in their covering are used, the weld pool is so fluid that only flat position welding or horizontal fillet welds are practical. Electrodes with high iron-powder content have a number 2 before the last digit: EXX**2**X.

11.6.3 Corrosion-Resistant Steel Electrodes

Corrosion-resistant steel electrodes are often called *stainless steel electrodes*. The specification that covers these electrodes is titled ANSI/AWS A5.4, *Specification for Covered Corrosion-Resisting Chromium and Chromium-Nickel Steel Welding Electrodes*. These electrodes are used to weld the different stainless steels, which are corrosion-resistant.

Below is a listing of all the AWS classification numbers for corrosion-resistant steel electrodes:

E209	E310	E330
E219	E310H	E330H
E240	E310Cb	E347
E307	E310Mo	E349
E308	E312	E410
E308H	E316	E410NiMo
E308L	E316H	E430
E308Mo	E316L	E502
E308MoL	E317	E505
E309	E317L	E630
E309L	E318	E7Cr
E309Cb	E320	E16-8-2
E309Mo	E320LR	

Occasionally these electrode numbers may be followed by a two-digit suffix.

Examples: E320-15
              E310-16

The -15 electrode is used with direct current electrode positive (DCEP). The suffix -16 is used with ac or direct current electrode positive (DCEP).

Electrodes up to 5/32" (3.97mm) in size may be used in all positions. Electrodes 3/16" (4.76mm) and larger are used only in the flat or horizontal-fillet welding positions.

The numbering system in this classification calls out the chemical composition of the electrode. Each of the electrode numbers on the above list are a type of chromium or chromium-nickel alloy. The **L** that follows some electrode designations means the electrode has a low-carbon content. The letter **E** means electrode. To find the chemical composition of these corrosion-resistant electrodes, see the ANSI/AWS 5.4 publication.

11.6.4 Covered Electrodes for Cast Iron

Cast iron may be arc welded with cast iron, copper-based, nickel-based, or mild steel arc welding electrodes. The *cast iron electrode* is designated ECI. It contains 3.25% to 3.50% carbon and 2.75% to 3% silicon. It also contains less than 1% each of manganese, phosphorus, and 0.10%

sulfur. The remainder of the metal composition (approximately 92%) is pure iron.

Copper-based electrodes for cast iron are designated: ECuSn-A, ECuSn-C, and ECuAl-A2. The ECuSn-A electrode contains between 4% and 6% tin (Sn), 0.1% to 0.35% phosphorus, and traces of aluminum and lead. The remainder, or about 94% of the metal, is copper (Cu). The ECuSn-C electrode contains 7% to 9% tin (Sn), and traces of aluminum and lead. It also contains between 0.05% and 0.35% phosphorus. The remainder, or about 91%, is copper (Cu). The ECuAl-A2 classification contains 9% to 11% aluminum and 1.5% iron. It also contains a small amount of silicon, zinc, and lead. The remainder, or about 87%, is copper (cu). The four *nickel-based electrode* classifications are: ENi-Cl, ENiFe-Cl, ENiCu-A, ENiCu-B. The nickel-based electrodes contain from 45% to 85% nickel, depending on their designation. They also contain small percentages of carbon, silicon, manganese, sulfur, and copper. ECuFe-Cl may contain as much as 44% iron. The single mild steel electrode for cast iron is classified ESt. ESt contains 0.15% carbon, 0.30% to 0.60% manganese, 0.03% silicon, and 0.04% phosphorus and sulfur. The remainder, or about 99% of the electrode, is pure iron.

Exact chemical compositions for the electrodes may be found in the AWS 5.15 publication.

11.7 NONFERROUS ELECTRODE CLASSIFICATIONS

The American Welding Society has produced the following booklets on nonferrous electrodes:

AWS A5.3. *Specification for Aluminum and Aluminum Alloy Electrodes for Shielded Arc Welding.*

AWS A5.6. *Specification for Covered Copper and Copper Alloy Arc Welding Electrode.*

AWS A5.11. *Specification for Nickel and Nickel Alloy Welding Electrodes for Shielded Metal Arc Welding.*

All of the nonferrous electrode designations convey, in the letters which make up the designation, an idea of the major alloying ingredients. The letters are abbreviations for chemical elements as follows:

Al – aluminum	Fe – iron	Ni – nickel
Cr – chromium	Mn – manganese	Si – silicon
Cu – copper	Mo – molybdenum	Sn – tin

There are three *aluminum electrodes* in the AWS A5.3 specification. They are the E1100, with an aluminum content of 99%; the E3003, with an aluminum content of about 96.7%, and the E4043, with approximately a 92.3% aluminum content.

The *copper-based electrodes* range in copper content from about 68.9% to 99%. ECuNi is the exception. It contains up to 33% nickel. The copper content is about 62%, with the remainder made up of other elements. The AWS classifications are shown below:

ECu	ECuSn-C	ECuAl-B
ECuSi	ECuNi	ECuNiAl
ECuSn-A	ECuAl-A2	ECuMnNiAl

The *nickel-based electrodes* contain from 59% to 92% nickel. The exact chemical composition may be found in AWS A5.11. The eight AWS classifications are:

ENi-1	ENiCrFe-3
ENiCu-7	ENiCrFe-4
ENiCrFe-1	ENiMo-1
ENiCrFe-2	ENiMo-3

11.8 ELECTRODE CARE

It is important that the user follow the electrode manufacturer's recommendations as to ampere settings, base metal preparation, welding technique, welding position, and the like. Electrodes must not be used after being exposed to dampness because the steam generated by the heat of the arc may cause the covering to be "blown away," and also cause hydrogen inclusions. Questionable electrodes should be baked at 250ºF (121ºC) for several hours to drive off any moisture.

Because of the similarity in appearance of electrodes that may be very different in welding properties, it is important to label and store all electrodes in carefully marked bins. Use masking tape to bind electrodes into bundles, and label carefully when putting them in storage. Electrodes are costly, and loss of identification can lead to losses in time and money.

If an electrode is used beyond its ampere rating, it will overheat, and the covering will crack. This will spoil the electrode. The excess current will also cause considerable spattering of the molten metal.

11.9 POWER SOURCE REMOTE CONTROLS

Welders work on a variety of joints and metal thickness. They use different electrodes and weld in various positions. All these variables require changes in polarity and welding machine adjustments.

To eliminate the time required to travel back and forth to the welding machine, several manufacturers provide remote control devices that may be kept near the operator for convenient control of the machine.

The small portable remote control panel shown in Figure 11-53 provides for remote current adjustments. Using a remote current setting device of this type, the welder may turn on the welding machine, then climb into a restricted place. The welder can then adjust the welding machine to any current setting without changing any controls on the front of the welding machine itself. The saving in time is important, and better quality welds are produced because the machine is more accurately adjusted to the job requirements at all times.

Figure 11-53. *A remote amperage adjuster. With this device, a percentage of the amperage range set on the power supply can be adjusted remotely. (Miller Electric Mfg. Co.)*

11.10 WELD-CLEANING EQUIPMENT

It is very important that the base metals in weldments be cleaned prior to welding. It is difficult to weld dirty or corroded surfaces. If this is attempted, the resulting welds will normally be of poor quality. Many types of equipment and tools have been developed for the purpose of cleaning joints and welds. Cleaning may be done by using sandblasting machinery, rotary wire wheels, and tools such as chipping chisels, hammers, and wire brushes. Nonferrous metals may be chemically cleaned, especially in production welding situations. The amount and size of the welding done usually determines the kind of cleaning apparatus needed.

The slag that covers each weld bead must be removed before the next weld bead is laid, to prevent inclusions in the finished weld. The slag on the final bead must also be removed before the weld can be inspected or painted. This slag coating may be removed by a rotary wire wheel, or by tapping the slag with a *chipping hammer*. In either case, suitable eye protection must be worn.

The chipping hammer often is double-ended, as shown in Figure 11-54. One end is shaped like a chisel for general chipping. The other end is shaped like a pick, for reaching into corners and narrow spaces. Another type of chipping hammer is shown in Figure 11-55.

11.11 SHIELDS AND HELMETS

Electric arc welding makes necessary the use of special protective devices for skin surfaces, such as the hands, face, and eyes. The *arc welding helmet* is used to protect the face and eyes. It is mounted and supported on the head, Figure 11-56. Headbands on the helmets are

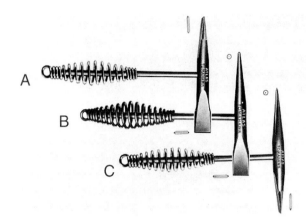

Figure 11-54. *Variety of chipping hammers. Note that in A, the blades are turned 90° to each other. In B and C, the chipping hammers have a blade at one end, and a pick at the other end. (Atlas Welding Accessories, Inc.)*

Figure 11-55. *Combination wire brush and chipping hammer. (Atlas Welding Accessories, Inc.)*

adjustable. There is a tension adjustment that keeps the helmet up. Adjust the tension so a slight nod of the head will allow the helmet to rotate down over your face. A hand-held welding shield is sometimes used by observers like inspectors, foremen, or instructors. See Figure 11-57.

The Occupational Safety and Health Act (OSHA) requires the use of a ***hard hat*** with the arc welding helmet on construction sites. A combination helmet and hard hat is shown in Figure 11-58. The face shield is usually made of fiber, plastic, or fiberglass, and is formed in a shape that covers the front half of the head. An aperture or opening at the level of the eyes provides visibility. This aperture is approximately 4 1/4" x 2" (108mm x 50.8mm) and is provided with at least two glass lenses. Larger square lenses are available in a 4 1/2" x 5 1/4" (114mm x 133mm) size. The larger lens is good for use by welders who wear bifocal eyeglasses. An outer clear lens, which is of double strength glass or plastic, is used to protect the inner and more expensive welding lens from metal spatter and abuse. Many operators also put another clear lens on the face side of the helmet to protect the colored lens from that side. Spring clips are generally used to provide a snug fitting of the lenses in the helmet.

A good grade of colored arc welding filter lens will remove approximately 99.5% of the infrared rays and

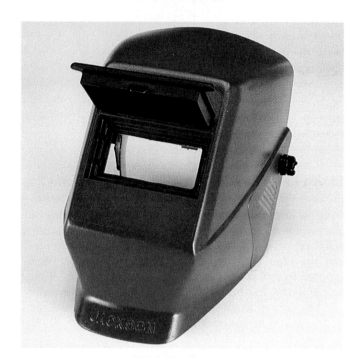

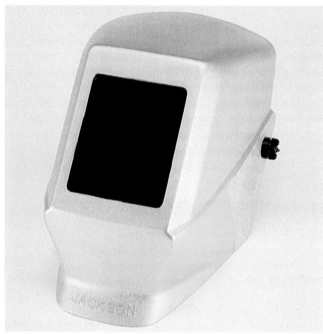

Figure 11-56. *Molded arc welding helmets. The model at top has a flip-up lens, while the one below has an extra-large lens for increased visibility. (Jackson Products, Div. Thermadyne Industries, Inc.)*

99.75% of the ultraviolet rays from the light emitted by the arc. These figures were developed by U.S. Bureau of Standards. Figure 11-59 shows recommended shades for various arc welding applications. Shade numbers 10, 12, and 14 are the common ones used for shielded metal arc welding (SMAW). Numbers 11 or 12 are used for gas metal arc welding (GMAW). When gas tungsten arc welding (GTAW), shade numbers 10-14 should be used. However, other shade numbers may be obtained and used, following the manufacturer's recommendations. The higher the

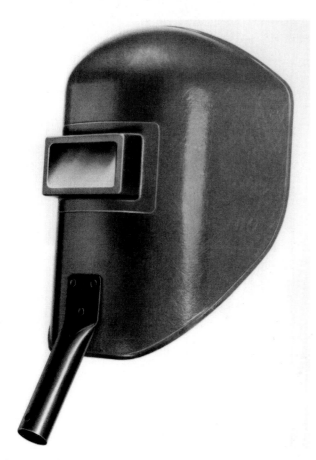

Figure 11-57. A hand-held welding helmet. (Fibre-Metal Products Co.)

	Lens shade number
SMAW (Shielded metal arc welding)	
Up to 5/32" electrodes	10
3/16" - 1/4" electrodes	12
5/16" - 3/8" electrodes	14
GMAW (Gas metal arc welding) (nonferrous)	
Up to 5/32" electrodes	11
GMAW (Gas metal arc welding) (ferrous)	
1/16" - 5/32" electrodes	12
GTAW (Gas tungsten arc welding)	10 to 14

Figure 11-59. Suggested lens shade numbers for various arc welding applications.

shade number, the lower the transmission of infrared or ultraviolet rays. A gold plating is being applied to some filter lenses. This gold plating reflects most of the harmful rays. The plating greatly improves the filtering capabilities of the regular filter lens. The same lens numbering system is used for the gold plated lenses.

Excess ultraviolet rays may cause eye pain for 8 to 18 hours after exposure. **Infrared light rays tend to injure eyesight; every precaution should be taken to shield the eyes from these rays.**

Filter lenses are of such density or shade that the operator cannot see through them until the arc is struck. Helmets have been developed which have a battery powered photoelectric cell built in. The lens is clear until an arc is struck. The circuitry of the photoelectric cell then instantly darkens the lens. This feature allows the electrode to be located and the arc started with great accuracy. These electronic quick change lenses will also protect the welder from nearby arc flashes. Figure 11-60 shows a helmet with an electronic quick change lens.

The helmet, or head shield, has a swing mounting that permits the forward part of the helmet to be lifted above the operator's face, without removing the headband from the head. Some welders who work continuously find that wearing a pair of ordinary welding goggles under the helmet helps to reduce eyestrain. These goggles eliminate reflected glare around the back of the helmet. Also available are lightweight goggles with a #1 or #2 filter lens. These goggles are called *flash goggles*. They enable the operator to set up work, chip welds (if the lens is tempered), peen, etc., and still have eye protection from flying particles and adjacent arc rays. **A welder must wear either a pair of welding goggles or a pair of safety glasses at all times.**

Some helmets are available into which fresh air is fed by means of a hose to increase the comfort of the operator. Helmets are also available which filter the air that the welder breathes. Refer to Figure 11-61. Figure 11-62 shows a pickup hood used to remove contaminated air from the welding area.

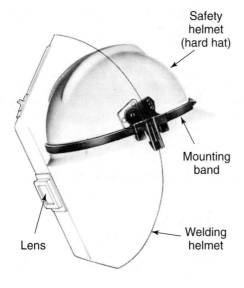

Safety helmet (hard hat)

Mounting band

Lens

Welding helmet

Figure 11-58. An arc welder's helmet used in connection with a safety helmet or "hard hat." (Kedman Co.)

Figure 11-60. An electronic quick-change arc welding helmet. (Jackson Products, Div. Thermadyne Industries, Inc.)

Figure 11-61. An air-supplied purifier similar to this one should be worn when welding on toxic materials or when welding in gas-filled areas.

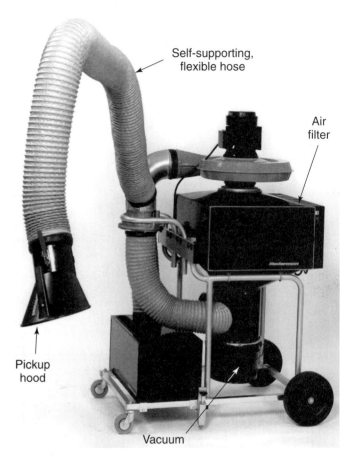

Figure 11-62. A portable fume extractor and filter unit. The pickup hood can be positioned anywhere the welder chooses. The large-diameter pickup hose is self-supporting. (Nederman, Inc.)

11.12 SPECIAL ARC WELDER CLOTHING

While an arc weld is in progress, the molten flux and the metal itself sometimes spatter for a considerable distance around the joint being welded. The operator must, therefore, be protected from the danger of being burned by these hot particles. Such clothing as gloves, gauntlet sleeves, aprons, and leggings are sometimes necessary, depending upon the type of welding being performed. **When performing welds in the overhead welding position, it is recommended that you wear a jacket or cape to protect the shoulders and arms. A cap or a special hooded arc helmet is needed to protect the head and hair.** All these clothing articles should be made of leather. They are referred to as *leathers.*

It is further recommended that the operator wear high-top shoes or boots. Trousers worn by the welder should not have cuffs. Cuffs may catch burning particles as they fall. Gloves should be worn to cover the hands and wrists and to prevent "sunburn."

All clothing should be carefully inspected to eliminate any place where the metal may catch and burn. Open pockets are especially dangerous.

Clothing worn while welding, other than the "leathers" mentioned, should be of heavy material. Thin clothing will permit infrared and ultraviolet rays to penetrate to the skin. If the skin is not properly protected, the operator will become "sunburned." Such burns, if they do occur, should be treated as a severe sunburn. If the burns are severe, a physician should be consulted. Easily ignited material, such as flammable combs, pens, or butane cigarette lighters, should not be carried by the operator while welding.

TEST YOUR KNOWLEDGE

Write your answers on a separate sheet of paper. Do not write in this book.
1. List the equipment required to outfit a SMAW station.
2. What welding processes use constant current arc welding machines?
3. What minimum information should be provided when describing an arc welding machine?
4. What is a typical open circuit (no load) voltage for an arc welding machine?
5. What are the advantages of using a constant current arc welding machine?
6. Explain how SCRs are used to control the welding current from a transformer.
7. How is dc created in an ac alternator or transformer-type welding machine?
8. What direct current polarity flows from the electrode, across the arc, to the work?
9. Briefly explain the steps taken in an inverter machine to turn ac input into dc welding current output.
10. Why is an inverter transformer so much smaller in size than a regular transformer?
11. When a welding machine is used at a higher current setting, does the duty cycle increase or decrease?
12. What is used to grip the electrode while welding?
13. What size welding leads are required to carry 300A a distance of 90' (27.4 m)?
14. What are five functions of the flux covering on the SMAW electrode?
15. What information is obtained from the first and second (possibly third) number in the AWS electrode classification?
16. The last two digits together tell the composition of the electrode covering. What is in an EXX24 electrode covering?
17. List four electrode numbers which contain low hydrogen. Which three last digits designate a low-hydrogen electrode?
18. List three AWS electrodes that contain iron powder.
19. What does the electrode classification E316-16 tell the welder?
20. What number filter lens is recommended for SMAW?

A welder uses the shielded metal arc process to join metal inserts that were cast into precast concrete panels. The panels form the corner of a retaining wall. Note the chain used to hold the panels in alignment while they are being welded. (Jack Klasey)

Chapter 12

SHIELDED METAL ARC WELDING

LEARNING OBJECTIVES

After studying this chapter, you will be able to:
* Describe how and why electricity flows in an electrical circuit and list the three variables in Ohm's law.
* Differentiate between DCEP and DCEN, and describe the difference in current flow.
* Demonstrate your ability to correctly complete a safety inspection on an arc welding station.
* Plan and list all the hand tools, personal safety equipment, and clothing required for a given welding task.
* Demonstrate your ability to select the correct electrode, current, and polarity to use while SMAW in a given welding position.
* Demonstrate your ability to strike a welding arc and produce acceptable stringer and weaving beads.
* Demonstrate your ability to correctly restart the arc and blend the new and old beads.
* Demonstrate your ability to perform acceptable SMA welds in all positions and on all five basic joints.
* Describe all types of weld defects and demonstrate your ability to inspect and differentiate between acceptable and unacceptable welds.
* Successfully pass a test on safety as related to arc welding work areas, equipment, clothing, tools, and correct welding procedures in all positions.

The most frequently used welding process is shielded metal arc welding (SMAW). Weldments in the shop and in the field are made with this process. Industrial machines, farm equipment, and vehicles are most frequently repaired using SMAW.

Both alternating current (ac) arc welding and direct current (dc) arc welding are used? Each has its own advantages and disadvantages. The welding technique is similar for both ac and dc. Electricity, as it arcs across the gap between the metal electrode and the work, creates a temperature of approximately 6500°F to 7000°F (3600°C to 3900°C). With the correct size electrode, this temperature is sufficient to melt any weldable metal.

12.1 DIRECT CURRENT (dc) ARC WELDING FUNDAMENTALS

Electric arc welding is defined by the American Welding Society as: "A group of welding processes which produce coalescence of metals by heating them with an arc, with or without the application of pressure, and with or without filler metal."

Shielded metal arc welding (SMAW) is done by producing an arc between the base metal and a consumable, flux-covered metal electrode. The electrode acts as an electrical conductor and filler metal. Figure 12-1 shows the electrical circuit for shielded metal arc welding.

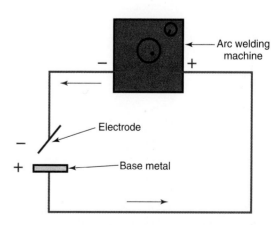

Figure 12-1. A dc shielded metal arc welding circuit.

Dc arc welding power sources, also called "welding machines," are of the following types:

- Ac transformer with dc rectifiers.
- Inverter.
- Motor- or engine-driven generator.
- Motor- or engine-driven alternator with dc rectifiers.

See Heading 11.2.2 for more information on dc arc welding power sources.

It is important to understand the voltage-current (ampere) characteristics of the dc arc welding machine. Under a no-load (open circuit) condition, when not welding, the voltage of the machine is about 60V to 80V. The current, measured in amperes, is zero. When the welding arc is struck, the current will increase and the voltage will decrease to between 15V and 40V. See Figure 12-2 for typical voltage and amperage meter readings on an arc welding machine.

Ohm's law for electricity states that voltage in a closed circuit has a constant relationship to the current and the resistance of the circuit. The Ohm's Law formula is:

$$V = I \times R$$

V is the voltage. I is the current. R is the resistance in the circuit.

On a constant current arc welding power source, the only adjustment on the welding machine is for current. If the electrode gap or the resistance in the circuit is held constant, the amperage increases when the voltage is

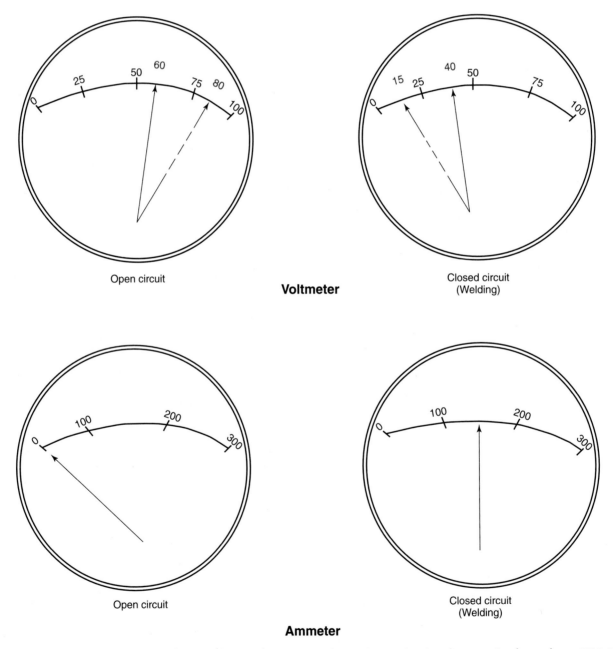

Figure 12-2. *Typical arc welding machine voltage and current readings. Open circuit voltage varies from about 60V–80V. Closed circuit amperage will correspond to the machine setting.*

increased. This may be seen in the formula V = I × R (constant). The current output is determined by the voltage from the welding machine.

Constant current power sources are drooper-type power sources. The volt-amp curve has a slope to it. Typical volt-amp curves are shown in Figure 11-20. Headings 11.2 and 13.2 discuss these power sources.

A welder who is welding with any drooper-type constant current welding machine must keep the arc length constant to maintain a constant current. *Arc length* is the distance from the end of the electrode to the base metal. When using a true constant current-type welding machine, the current will not change if the arc length is changed; however, a constant arc length is still required.

When welding with a drooper-type constant current welding machine, the current will change slightly as the arc length is changed. An increase in the arc length will cause the resistance in the circuit to increase. According to Ohm's law (V = I × R), to keep the current constant the voltage will increase as the resistance increases. However, the current does not remain constant. The current decreases as the arc length increases. Heading 11.2 and Figure 11-2 show that when the arc length is changed, both current and voltage will change.

Because a change in arc length will cause a change in welding current, it is important for the welder to keep the arc length constant while welding. However, the ability to slightly change the current by changing the arc length can be used by an experienced welder. Instead of stopping the arc and making an adjustment on the welding machine, the welder can vary the arc length to adjust the current and the heat input to the weld. Increasing the arc length will cause the current and the melting rate of the electrode to decrease. If a welder decreases the arc length, the current and the electrode melting rate will increase. In both cases, the voltage will change quite a bit, but the current will change only slightly (about 10%–15%).

To make a good quality weld, the arc must be maintained at a constant length while welding. This is the most important control of the heat generated in the arc. Since the electrode is constantly being melted, the welder must feed the electrode at the same rate it is melting to maintain a constant arc length. This is a skill that comes with practice.

To make a good weld, the welder must consider the following:
- The current (amperage) output of the welding machine.
- The diameter, polarity, and type of electrode.
- The arc and its manipulation.
- The preparation of the base metal.
- The type of base metal.

The *electrode* wire may be a ferrous or nonferrous metal. See Heading 11.15 for more information on electrodes.

The *welding arc*, when viewed through the helmet lens, is seen to be divided into two separate parts: the *stream* and the *flame*, as shown in Figure 12-3. The *arc flame* consists of neutral gases that appear to be pale red. The *vaporized metal* in the stream appears yellow. *Liquid metal* in the stream appears green. If the arc is longer than

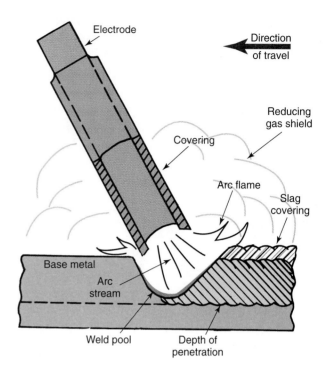

Figure 12-3. A covered electrode arc weld in progress.

normal, the flame gases can no longer protect the arc stream from oxidation. With a long welding arc, the weld will form oxides and nitrides, resulting in a very weak and brittle weld.

If the correct current flow and arc length are maintained, a good weld should result with direct current. The voltage and amperage required for any particular weld may be easily obtained from established tables. The correct arc length is entirely the operator's responsibility.

12.2 DCEN AND DCEP FUNDAMENTALS

The welding circuit shown in Figure 12-4 is known as a *direct current electrode negative (DCEN)* circuit. This circuit was formerly defined by the American Welding Society as direct current straight polarity (*DCSP*). In this circuit, the electrons are flowing from the negative terminal (or pole) of the machine to the electrode. The electrons continue to travel across the arc into the base metal and to the positive terminal or pole of the machine.

It is possible, and sometimes desirable, to reverse the direction of electron flow or *polarity* in the arc welding circuit. This may be done by disconnecting the electrode and workpiece leads and reversing their positions. Many machines have a switch which will change the circuit polarity. When electrons flow from the negative terminal or pole of the arc welding machine to the base metal, this circuit is known as *direct current electrode positive (DCEP)*. It was previously called a direct current reverse polarity (*DCRP*) circuit. In this circuit, the electrons flow from the negative pole of the welding machine to the work. Electrons travel across the arc to the electrode and then return to the positive terminal (pole) of the machine, as shown in Figure 12-5.

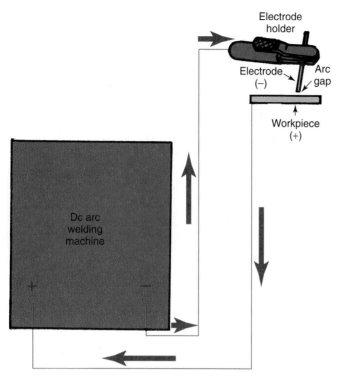

Figure 12-4. *A wiring diagram for a direct current electrode negative (DCEN) arc welding circuit. This circuit is also known as a direct current straight polarity (DCSP) circuit.*

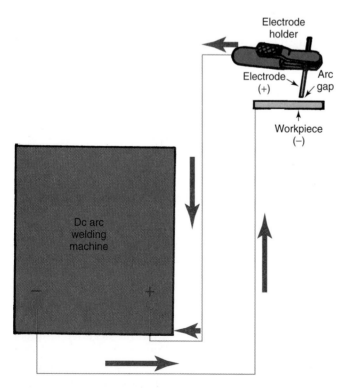

Figure 12-5. *A wiring diagram for a direct current electrode positive (DCEP) arc welding circuit. This circuit is also known as direct current reverse polarity (DCRP) circuit.*

The choice of when to use direct current electrode negative (DCEN) or direct current electrode positive (DCEP) is primarily determined by the electrode being used. Some SMAW electrodes are designed to use only DCEN or only DCEP. Other electrodes can be used with either DCEN or DCEP. See Heading 12.5.1 for information on electrodes.

In industrial practice, choice of the electrode may be made by the welder or may be specified by the welding procedure specification (WPS) and/or codes used. Specifications and codes are covered in Chapter 31. Once the electrode to be used is known, the polarity for that electrode may be determined by referring to the manufacturer's recommendations. See Figures 12-17 and 12-18.

The decision to use DCEN (DCSP) or DCEP (DCRP) often depends on such variables as:
- The depth of penetration desired.
- The rate at which filler metal is deposited.
- The position of the joint.
- The thickness of the base metal.
- The type of base metal.

DCEP (DCRP) produces better penetration than DCEN (DCSP). The SMAW electrodes that have the best penetrating abilities are E6010, E6011, and E7010. These electrodes use DCEP. There is a theory that with a DCEP covered electrode, there is a jet action and/or expansion of gases in the arc at the electrode tip. This expansion causes the molten metal to be propelled with great speed across the arc. The molten metal impacts the base metal with greater force. This heavy impact on the base metal helps to produce deep, penetrating welds.

When a high rate of filler metal deposit is required, an EXX2X electrode is recommended. DCEN (DCSP) is usually recommended for the EXX2X electrodes. Examples of the EXX2X electrodes that deposit a high rate of filler metal are E6020, E6027, E7027, or E7028. See Figure 12-17.

To weld out of position, an electrode intended for all positions must be used. Either DCEN or DCEP can be used. The electrode used will determine which polarity is correct.

Base metal thickness will affect which polarity is required. On thick material, a welder must obtain good penetration. However, on thin material, excessive penetration should be avoided.

Shielded metal arc welding (SMAW) can be used to weld nickel, aluminum, and copper. Electrodes designed to weld these metals are generally used with DCEP (DCRP).

These considerations and others discussed in Heading 12.5.1 will determine what electrode and which current, DCEN or DCEP, should be used on a particular welding job.

12.3 ALTERNATING CURRENT (ac) ARC WELDING FUNDAMENTALS

Several types of ac arc welding machines are used. The different types are:

- Transformer-type.
- Inverter.
- Motor- and engine-driven alternator.

A transformer-type machine is shown in Figure 12-6. An inverter is shown in Figure 12-7 and an engine-driven alternator is shown in Figure 12-8. Also see Heading 11.2.1.

In alternating current (ac), the current reverses its direction of flow 120 times per second. As shown in Figure 12-9, it requires 1/60 of a second to complete a cycle or

Figure 12-8. A portable engine-driven alternator. This power source provides ac, DCEN, or DCEP. (Lincoln Electric Co.)

hertz (Hz). The current flow completes 60 Hz (cycles) per second and is called 60 Hz (cycle) current. In most parts of the world, 50 Hz current is used.

Figure 12-10 shows what happens at the arc in one cycle of a typical ac transformer-type arc welder. The voltage at both points A and B is zero. Beginning at the left side

Figure 12-6. A 230A ac arc welding machine. There are high and low amperage taps and a workpiece lead receptacle. (Century Mfg. Co.)

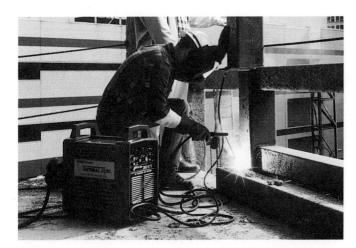

Figure 12-7. An inverter-type power source being used in structural welding. (Thermadyne Industries, Inc.)

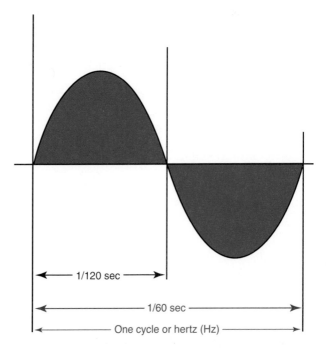

Figure 12-9. Sine wave form of single-phase alternating current.

of the graph, the voltage builds up to a maximum in one direction to point C, and then back to zero at point A. The voltage then builds up to maximum in the other direction to point D, then back to zero again at point B. This action is repeated at the rate of 60 Hz (cycles) per second.

When ac welding with 60 Hz current, the voltage and current are at zero 120 times each second, as shown in Figure 12-10. Each time the current crosses the zero point, the welding arc stops. To reestablish the arc, the voltage must increase enough to enable the current to jump the arc gap and maintain the arc. It is important that the voltage leads the current as each passes through zero. This will help make the ac arc stable. The ac arc welding machine must be designed to have the voltage lead the current.

Another method used to stabilize the ac arc is to increase the ionization of the material in the arc. *Ionization* is a physical phenomenon in which a particle obtains an electrical charge. These ionized or charged particles in the space between the electrode and the workpiece make it easy for the arc to jump the gap. Electrodes that are intended for ac welding have ionizing agents in the electrode covering. These agents help to ionize the materials in the arc gap and help to stabilize the ac arc.

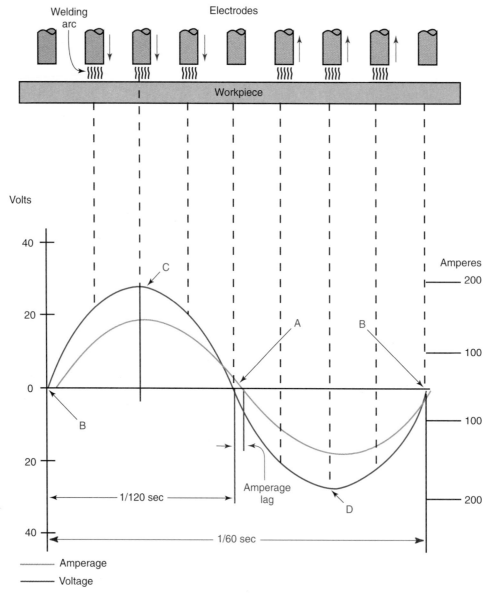

Figure 12-10. *The sine wave curve of alternating current at 60 Hz (cycles). At points A and B, the voltage value is zero. At points C and D, voltage is at maximum. The two zero values that occur in each cycle (every 1/60 of a second) may make it difficult to strike and maintain an ac arc at small current values. Because a certain voltage is required to overcome electron inertia in a circuit, there is usually a small lag or lead between the voltage and amperage (current). The voltage usually leads the amperage.*

Ac welding can be done only with electrodes that are designed for use with alternating current. If an electrode is not intended for use with ac, the arc will be very unstable.

Welds performed with ac electrodes show good penetration. Larger diameter electrodes are used to increase the metal deposition rates. Higher travel speeds can be obtained when large ac currents and large electrodes are used. There is no arc blow when ac welding. See Heading 12.6 for a description of arc blow.

Figure 12-10 shows the current flow measured in amperes (the yellow curve), as related to the voltage or potential (the blue curve). It should be noted that the voltage leads the current, or the current lags the voltage slightly in time. This slight difference between the voltage and current must be designed into the ac arc welding machine.

12.4 SELECTING AN ARC WELDING MACHINE

Deciding whether an ac or dc arc welding machine is best to buy or use depends on several factors. Selecting the type of current to use should be done after considering their individual advantages and disadvantages.

Advantages of the dc constant current-type arc welding machine include:
- The ability to choose direct current electrode positive (DCEP) or DCRP.
  - DCEP or DCRP produces deeper penetrating welds than DCEN.
  - DCEP or DCRP can be used in positions other than flat position or downhand welding.
  - Electrodes designed to weld nickel, aluminum, and copper generally use DCEP.
- The ability to choose direct current electrode negative (DCEN) or DCSP.
  - DCEN or DCSP is recommended for EXX2X electrodes that have high metal deposition rates.
  - DCEN or DCSP can also be used in welding positions other than flat.

The disadvantage of the direct current (dc), constant current arc welding machine is that a dc arc welder is generally more expensive than an ac arc welding machine of the same quality, current output, NEMA classification, and duty cycle.

Advantages of the alternating current (ac), constant current, arc welding machine include:
- Welds made with ac arc welding machines and electrodes have moderate penetration.
- Large-diameter electrodes can be used with high ac currents to obtain greater filler metal deposition rates and faster welding speeds.
- Ac arc welding machines are generally less expensive than dc arc welding machines of equal quality, current output, NEMA classification, and duty cycle rating.

The major disadvantage of ac arc welding machines is that not all SMAW electrodes can be used with alternating current.

The choice of which arc welding machine to use or buy must be made on the basis of what type welds are to be made, the economics of the welding machine purchase, and personal preference. Welding machines may also be purchased with the capability of selecting ac or dc operation. Such combination ac/dc welding machines are more expensive than single- current output machines. However, they offer the welder the opportunity to better match the current output to the welding requirements of the job.

12.4.1 Inspecting an Arc Welding Station

The *arc welding station*, as discussed in the beginning of Chapter 11, includes:
- Arc welding power source .
- Electrode lead and terminals.
- Workpiece lead and terminals.
- Electrode holder.
- Workbench.
- Ventilation.
- Stool.
- Booth.

Before beginning to weld, check all parts of the arc welding station. This should be done to ensure the efficiency of the station and your safety.

Before the inspection is made, the arc welding machine should be turned *off*. The arc welding machine should be as close as possible to the booth or the parts to be welded. This is necessary to eliminate long leads that increase electrical resistance. Check that the electrode and workpiece leads are tightly attached to the machine. Inspect each lead, checking for any damage that may have occurred to the covering. Damage to the leads, particularly the workpiece lead, can occur from rolling over them with lift trucks, pallet movers, and other wheeled vehicles. If the electrode and workpiece lead must temporarily run across an aisle, it is advisable to cover them with channel iron or similar material for protection.

Inspect the electrode holder to make certain that the handle insulation is not cracked. Check also that the electrode lead is tightly fastened into the holder. The electrode holder jaws should be clean for good electrode contact. Make certain that the workpiece lead is making good contact with a cleaned area on the workbench. Any loose connections will cause an increase in resistance in the welding circuit.

Each booth must have an insulated hook or hanger on which the electrode holder is hung when not in use. The booth curtains or walls should not have holes in them. Holes in the booth could present arc flash dangers to persons outside the booth. If a portable arc welding machine is used, it is advisable to set up a portable booth to protect others from arc flash.

Check the ventilation pickup hose, or duct, for holes that would lower its efficiency. Turn on the ventilation system and check to see that it is working. If the ventilation system efficiency seems to be inadequate, it should be checked and repaired. The inlet to the ventilation pickup duct should be placed so that fumes are removed before they can reach the welder's face.

12.4.2 Safety, Protective Clothing, and Shielding

Before you proceed further with the study of shielded metal arc welding, you should study safety. Refer to Chapter 1 on safety and to the following specifics about shielded metal arc welding.

Arc welding performed with proper safety equipment presents no great safety hazards. Welders should learn the correct procedures for arc welding so that the hazards which do exist may be properly recognized and injury avoided.

The chief hazards to be avoided in arc welding are:
- Radiation from the arc; ultraviolet and infrared rays.
- Flying sparks and globules of molten metal.
- Electric shock.
- Fumes.
- Burns.

Radiation from the arc presents some dangers. **Eyes must be protected from radiation from the arc by use of a helmet or hand-held face shield with approved lenses. Never look at an arc from any distance, unless your eyes are protected by approved filter lenses**. Figure 12-11 shows a helmet with filter lenses installed. Filter lenses for SMAW can be Number 10, 12, or 14, depending on the electrode diameter. See Figure 11-59 for a chart of filter lenses.

Face, hands, arms, and other skin surfaces must be covered. Wear gloves and keep other parts of the body covered by clothing of sufficient weight to shut out the rays from the arc. Without proper clothing, burns comparable to sunburn will result.

The arc welding operation should be shielded so that no one may accidentally look directly at the arc or have it shine or reflect into their eyes. Such an arc "flash" may cause a person to be temporarily blinded. The effect is to see a white spot similar to the effect of a photographer's flash. The severity of an arc flash and the time it will take to recover varies with the length of time the person was exposed to the arc. A long exposure has been known to cause permanent damage to the retina of the eye. If someone is severely "flashed," special treatment should be administered at once by a physician.

Arc welding is usually accompanied by *flying sparks*. These present a hazard if they strike unprotected skin, lodge on flammable clothing, or hit other flammable material. It is advisable to wear suitable weight clothing and cuffless trousers. See Figure 12-12.

Pockets should be covered so they will not collect sparks. **Remove flammable materials, such as matches, lighters, plastic combs, or pens. High shoes with safety toes should be worn.**

The possibility of dangerous *electric shock* can be avoided by working on a dry floor, using insulated electrode holders, and wearing dry welding gloves. **Avoid using arc welding equipment in wet or damp areas.**

Figure 12-11. *An arc welding helmet with a flip-up lens. Arc welding helmets are used to protect the eyes and face from harmful rays and flying molten metal. (Fibre-Metal Products Co.)*

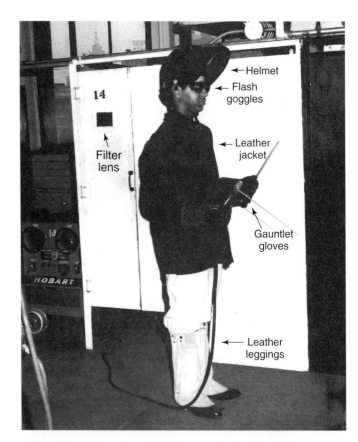

Figure 12-12. *A welder equipped with flash goggles, leather jacket, gauntlet gloves, and leather leggings. Note that the arc welding booth is fitted with a filter lens so that the instructor may observe the arc from outside the booth.*

The health hazard of *fumes* developed by the electrode covering and molten metal, may be avoided by the use of proper ventilating equipment. Certain special jobs require forced airflow (ventilation) into the welder's helmet. The fumes generated in the welding arc may contain poisonous metal oxides. **Arc welding should never be done in an area which is not well-ventilated.**

Hot metal will cause severe burns. Use leather gloves with tight-fitting cuffs that overlap the sleeves of the jacket. Many welders wear an apron of leather or other heavy material for protection. **Hot metal should be handled with tongs or pliers.** In a welding shop, all metal should be treated as if it is hot.

12.5 STARTING, STOPPING, AND ADJUSTING THE ARC WELDING POWER SOURCE FOR SMAW

Before beginning to weld, inspect the complete arc welding station to make certain it is safe for use. See Heading 12.4.1. An arc welding machine should never be started or stopped under load (with a closed circuit). Make certain that the electrode holder is hung on an insulated hanger before turning the machine on or off. The electrode holder should never be left on the workbench.

Arc welding machines powered by ac are easy to start and stop. An on-off switch or buttons are provided on the machine. An engine-driven arc welding machine must have the engine started. Once the engine is running and up to its operating speed, the welding current may be turned on using an on-off switch.

Constant current-type arc welding machines are used for manual arc welding processes. The desired current is set on the machine. The voltage on a constant current machine is not set. It varies as the welding circuit resistance changes to maintain a constant or relatively constant current output. All electrical connections must be tight. The length of the arc gap will then control the welding circuit resistance, and therefore, the circuit voltage.

Amperage (current) controls vary in appearance, location, and operation on various manufacturer's machines. See illustrations of arc welding machines in Chapters 11 and 12.

Figure 12-13 shows an ac transformer with a hand crank on the top of the machine. As this hand crank is rotated, the primary coil is moved to vary the current output. A pointer moves with the primary coil, indicating the current setting on a scale on the outside of the cabinet. Figure 12-14 illustrates a machine that uses a tap-type control for coarse current adjustment and a crank on top of the machine for fine adjustment.

Figure 12-15 shows a shielded metal arc welding machine that uses a coarse current adjustment and a fine current adjustment. The fine current adjustment is made using a rotary knob. Fine adjustment controls are often

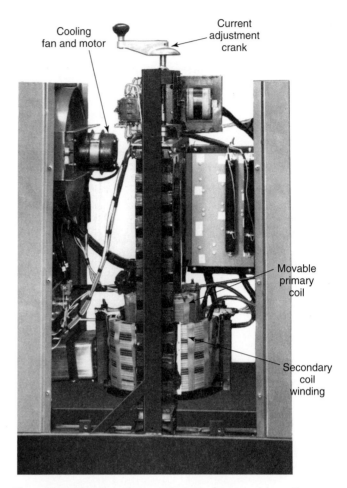

Figure 12-13. *Internal construction of a movable coil ac arc welder. (Applied Power, Inc.)*

Figure 12-14. *A welding power source with high- and low-range amperage taps. Fine current control is achieved by rotating the crank on the top of the machine. (Miller Electric Mfg. Co.)*

marked with numbers from 0 to 10, or from 0 to 100. These numbers do not represent current, but rather *percentages*. Once a coarse current range is selected, the fine adjustment will adjust the current within percentages of that coarse range. The machine will deliver the amperage set on the low end of the coarse setting, plus the percentage of the range set on the fine adjustment.

Figure 12-15. *A constant current ac/dc arc welding machine. An amperage range is first set on the large dial. A fine adjustment within that amperage range can then be made on the small dial. (Hobart Brothers Co.)*

Example 1:
If the coarse range setting is 90-110A, the current
    delivered = 90A:
        Plus
If the fine adjustment setting is 50%, the current
    delivered is:
        50% × current range or
        0.50 × (110-90) = 0.50 × 20A = 10A.
Then
    total current delivered = (90A + 10A) = 100A.
Example 2:
If the coarse range setting is 200-250A, the current
    delivered = 200A:
        Plus

If the fine adjustment is set on 70%, the current
    delivered is:
        70% × current range or
        0.70 × (250-200)
        =0.70 × 50A = 35A.
Then
    total current delivered = (200A + 35A) = 235A.
    Figure 12-16 is a photo of an inverter power supply. There is only one knob used to set the current.

Figure 12-16. *An inverter-type power source. (Miller Electric Mfg. Co.)*

12.5.1 Selecting the Proper Electrode

When attempting to select an electrode for shielded metal arc welding (SMAW), the following must be considered:
- The weld groove design.
- Tensile strength of the required weld.
- The base metal composition.
- The position of the weld joint.
- The rate at which you want to deposit the weld metal.
- The type of arc welding current used.
- The penetration required.
- The metal thickness.

- The experience of the welder.
- The specifications for the weld to be made.

When a *groove weld* is made, the electrode must be small enough in diameter to easily manipulate it in the root of the weld. A small-diameter electrode is used for the root pass to ensure full penetration. After the root pass is made and cleaned, larger electrodes may be used to finish the weld. E60XX, 70XX, 80XX, 90XX, 100XX, and higher may be used depending on the weld strength required. Refer to Figures 12-17 and 12-18 as you study this material.

AWS classifi-cation	Type of covering	Capable of producing satisfactory welds in position shown[a]	Type of current[b]
E60 series electrodes			
E6010	High cellulose sodium	F, V, OH, H	DCEP
E6011	High cellulose potassium	F, V, OH, H	ac or DCEP
E6012	High titania sodium	F, V, OH, H	ac or DCEN
E6013	High titania potassium	F, V, OH, H	ac or dc, either polarity
E6020	High iron oxide	H-fillets	ac or DCEN
E6022[c]	High iron oxide	F	ac or dc, either polarity
E6027	High iron oxide, iron powder	H-fillets, F	ac or DCEN
E70 series electrodes			
E7014	Iron powder, titania	F, V, OH, H	ac or dc, either polarity
E7015	Low hydrogen sodium	F, V, OH, H	DCEP
E7016	Low hydrogen potassium	F, V, OH, H	ac or DCEP
E7018	Low hydrogen potassium, iron powder	F, V, OH, H	ac or DCEP
E7024	Iron powder, titania	H-fillets, F	ac or dc, either polarity
E7027	High iron oxide, iron powder	H-fillets, F	ac or DCEN
E7028	Low hydrogen potassium, iron powder	H-fillets, F	ac or DCEP
E7048	Low hydrogen potassium, iron powder	F, OH, H, V-down	ac or DCEP

a. The abbreviations F, V, V-down, OH, H, and H-fillets indicate the welding positions as follows:
   F = Flat
   H = Horizontal
   H-fillets = Horizontal fillets
   V-down = Vertical down
   V = Vertical } For electrodes 3/16" (4.8mm) and under, except 5/32" (4.0mm) and under for classifications E7014, E7015, E7016, and E7018.
   H = Overhead }
b. The term DCEP refers to direct current, electrode positive (dc reverse polarity). The term DCEN refers to direct current, electrode negative (dc straight polarity).
c. Electrodes of the E6022 classification are for single-pass welds.

Figure 12-17. AWS Carbon Steel Covered Arc Welding Electrodes. This information is from AWS A5.1. (American Welding Society)

The *metal composition* of the base metal will determine the metal composition of the electrode used. Refer to Headings 11.5 through 11.7.

The *weld joint position* will determine the electrode used. If the joint is in the flat welding position or is a horizontal fillet weld, a larger-diameter electrode may be used. An iron powder electrode with high metal deposition rates may be chosen. If the weld is in the vertical, horizontal, or overhead welding positions, a smaller diameter electrode may be selected. A smaller diameter electrode will form a smaller weld pool and will be easier to control. The metal in a larger weld pool tends to run out due to gravitational force.

Electrodes are made to be most effective with one type of welding current: ac, DCEN (DCSP), or DCEP (DCRP). When selecting an electrode, the type of current produced by the arc welding machine must be known. Ac and some DCEP (DCRP) electrodes produce deeper penetrating welds than other electrodes. Refer to Heading 11.6 and Figures 12-17 and 12-18.

When welding on *thin metal*, DCEN (DCSP) may be used. Small DCEN (DCSP) electrodes used with a low-current setting produce a soft arc action with small penetrating abilities. See electrodes E6012 and E6013 in Figure 12-19. The largest electrode diameter possible should be used. However, it must not create too large a weld or overheat the base metal. The experience of the welder is a large factor in choosing an electrode. An experienced welder can produce a sound weld with a much larger diameter electrode than a beginner can.

Very often, no choice is allowed in the selection of an electrode. Whenever a qualified welding procedure specification is used, the diameter and type of electrode are specified. The amount of current to be used is also specified. Chapter 31 has more information on welding procedure specifications. Figures 12-19 and 12-20 provide information regarding electrode diameters, current ranges used, and suggested metal thickness applications for E60XX and E70XX series electrodes.

12.5.2 Striking the Arc

One of the first lessons to be mastered, when learning to arc weld, is to produce an arc between the metal electrode and the base metal. To strike a welding arc, the electrode must first touch the base metal. The end of the electrode must then be withdrawn to the correct arc distance or length.

First attempts to strike a welding arc may cause the electrode to stick. That is to say, the electrode may weld itself to the base metal. When this occurs, follow these steps to separate the electrode from the base metal.

- If welding small pieces of material, lift the electrode holder without releasing the electrode. This will pick up the electrode and the base metal. Some arcing will occur between the base metal and the table you are welding on. If you release the electrode from the electrode holder, arcing will occur between the electrode and the holder.

AWS classification[a]	Type of covering	Capable of producing satisfactory welds in positions shown[b]	Type of current[c]
E70 series – Minimum tensile strength of deposited metal, 70,000 psi (480 MPa)			
E7010-X	High cellulose sodium	F, V, OH, H	DCEP
E7011-X	High cellulose potassium	F, V, OH, H	ac or DCEP
E7015-X	Low hydrogen sodium	F, V, OH, H	DCEP
E7016-X	Low hydrogen potassium	F, V, OH, H	ac or DCEP
E7018-X	Iron powder, low hydrogen	F, V, OH, H	ac or DCEP
E7020-X	High iron oxide	{ H-fillets { F	ac or DCEN ac or dc, either polarity
E7027-X	Iron powder, iron oxide	{ H-fillets { F	ac or DCEN ac or dc, either polarity
E80 series – Minimum tensile strength of deposited metal, 80,000 psi (550 MPa)			
E8010-X	High cellulose sodium	F, V, OH, H	DCEP
E8011-X	High cellulose potassium	F, V, OH, H	ac or DCEP
E8013-X	High titania potassium	F, V, OH, H	ac or dc, either polarity
E8015-X	Low hydrogen sodium	F, V, OH, H	DCEP
E8016-X	Low hydrogen potassium	F, V, OH, H	ac or DCEP
E8018-X	Iron powder, low hydrogen	F, V, OH, H	ac or DCEP
E90 series – Minimum tensile strength of deposited metal, 90,000 psi (620 MPa)			
E9010-X	High cellulose sodium	F, V, OH, H	DCEP
E9011-X	High cellulose potassium	F, V, OH, H	ac or DCEP
E9013-X	High titania potassium	F, V, OH, H	ac or dc, either polarity
E9015-X	Low hydrogen sodium	F, V, OH, H	DCEP
E9016-X	Low hydrogen potassium	F, V, OH, H	ac or DCEP
E9018-X	Iron powder, low hydrogen	F, V, OH, H	ac or DCEP
E100 series – Minimum tensile strength of deposited metal, 100,000 psi (690 MPa)			
E10010-X	High cellulose sodium	F, V, OH, H	DCEP
E10011-X	High cellulose potassium	F, V, OH, H	ac or DCEP
E10013-X	High titania potassium	F, V, OH, H	ac or dc, either polarity
E10015-X	Low hydrogen sodium	F, V, OH, H	DCEP
E10016-X	Low hydrogen potassium	F, V, OH, H	ac or DCEP
E10018-X	Iron powder, low hydrogen	F, V, OH, H	ac or DCEP
E110 series – Minimum tensile strength of deposited metal, 110,000 psi (760 MPa)			
E11015-X	Low hydrogen sodium	F, V, OH, H	DCEP
E11016-X	Low hydrogen potassium	F, V, OH, H	ac or DCEP
E11018-X	Iron powder, low hydrogen	F, V, OH, H	ac or DCEP
E120 series – Minimum tensile strength of deposited metal, 120,000 psi (830 MPa)			
E12015-X	Low hydrogen sodium	F, V, OH, H	DCEP
E12016-X	Low hydrogen potassium	F, V, OH, H	ac or DCEP
E12018-X	Iron powder, low hydrogen	F, V, OH, H	ac or DCEP

a. The letter suffix "X" as used in this table stands for the suffixes A1, B1, B2, etc., and designates the chemical composition of the deposited weld metal.
b. The abbreviations F, V, OH, H, and H-fillets indicate welding positions as follows: F = Flat; H = Horizontal; H-fillets = Horizontal fillets.
  V   =   Vertical   } For electrodes 3/16 in. (4.8mm) and under, except 5/32 in. (4.0mm) and under for classifications
  OH =   Overhead   } EXX15-X, EXX16-X, and EXX18-X.
c. DCEP means electrode positive (reverse polarity). DCEN means electrode negative (straight polarity).

Figure 12-18. *AWS Low-Alloy Steel Covered Arc Welding Electrodes. This information is from AWS 5.5. See Figure11-47 for the letter suffixes used in place of the "X" with these electrode numbers. (American Welding Society)*

Suggested metal thickness		Electrode size		E6010 and E6011	E6012	E6013	E6020	E6022	E6027
in.	mm	in.	mm						
1/16 & less	1.6 & less	1/16	1.6		20-40	20-40			
1/16-5/64	1.6-2.0	5/64	2.0		25-60	25-60			
5/64-1/8	2.0-3.2	3/32	2.4	40-80	35-85	45-90			125-185
1/8-1/4	3.2-6.4	1/8	3.2	75-125	80-140	80-130	100-150	110-160	169-240
1/4-3/8	6.4-9.5	5/32	4.0	110-170	110-190	105-180	130-190	140-190	210-300
3/8-1/2	9.5-12.7	3/16	4.8	140-215	140-240	150-230	175-250	170-400	250-350
1/2-3/4	12.7-19.1	7/32	5.6	170-250	200-320	210-300	225-310	370-520	300-420
3/4-1	19.1-25.4	1/4	6.4	210-320	250-400	250-350	275-375		375-475
1 - up	25.4 - up	5/16	8.0	275-425	300-500	320-430	340-450		

Figure 12-19. A table of E60XX series electrodes with suggested metal thickness applications and amperage ranges. These values are suggested and may be varied as required.

Suggested metal thickness		Electrode size		E7014	E7015 and E7016	E7018	E7024 and E7028	E7027	E7048
in.	mm	in.	mm						
5/64-1/8	2.0-3.2	3/32*	2.4*	80-125	65-110	70-100	100-145		
1/8-1/4	3.2-6.4	1/8	3.2	110-160	100-150	115-165	140-190	125-185	80-140
1/4-3/8	6.4-9.5	5/32	4.0	150-210	140-200	150-220	180-250	160-240	150-220
3/8-1/2	9.5-12.7	3/16	4.8	200-275	180-255	200-275	230-305	210-300	210-270
1/2-3/4	12.7-19.1	7/32	5.6	260-340	240-320	260-340	275-365	250-350	
3/4-1	19.1-25.4	1/4	6.4	330-415	300-390	315-400	335-430	300-420	
1 - up	25.4 - up	5/16*	8.0*	390-500	375-475	375-470	400-525	375-475	

Note: When welding vertically up, currents near the lower limit of the range are generally used.
*: These diameters are not manufactured in the E7028 classification.

Figure 12-20. A table of E70XX series electrodes with suggested metal thickness applications and amperage ranges. These values are suggested and may be varied as required.

- When welding on a large weldment it is not possible to lift the weldment, so you must release the electrode from the electrode holder. Inspect the jaws on the electrode holder to make sure no significant damage was done. If large pits or arc damage has occured, replace the jaws on the electrode holder.

Another problem that a beginning welder may experience is withdrawing the electrode too far after touching the metal. This will cause the voltage requirement to be too great to maintain the arc. The arc will, therefore, break and go out (extinguish). Only experience and practice will overcome these difficulties.

There are two common methods of striking (producing) the arc. The welder may use a *glancing* or *scratching motion* with the end of the electrode or a *straight down-and-up motion* with the electrode. Figure 12-21 illustrates both methods of striking an arc.

If the arc breaks continually, regardless of how careful the welder may be, it is probably due to a too-low current setting on the machine. If the electrode spatters excessively, and if it becomes overheated while welding, the current setting is too high.

As soon as the welding arc is struck and becomes stabilized, the base metal begins melting and the filler metal deposit begins. It is, therefore, important to strike the arc in

exactly the right spot or the metal may be marred. Most welders position the arc end of the electrode just above the exact spot where the weld is to start. They then lower the helmet in front of their eyes before actually contacting the metal with the electrode. Figure 12-22 shows a welder preparing to strike an arc. Another method to ensure the arc will be started in the right place is to use a photoelectric lens in the helmet. A photoelectric lens remains clear until the arc is struck, then instantly darkens. This helps the welder to strike the arc in precisely the right spot without raising or lowering the helmet. See Figure 11-60.

12.5.3 Running a Bead

Once the welding arc is struck and stabilizes, the *weld pool* will begin to form. As the welder moves the electrode forward, the *weld bead* forms. To make a good weld on any joint in any position, the first skill which must be mastered is to run (form) a bead.

To run a good weld bead in SMAW, the following factors must be controlled manually by the welder:
- Arc gap distance or arc length.
- Speed of forward motion.
- Bead width.
- Electrode angle or position.

The *arc length* must be varied slightly as different electrode diameters are used. However, for covered elec-

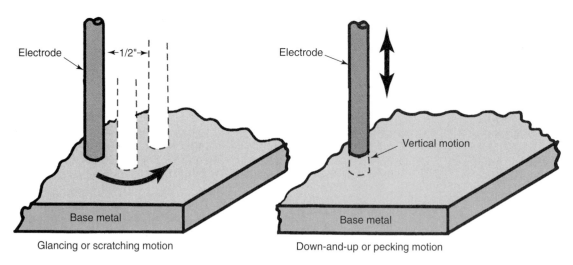

Figure 12-21. *Two methods of striking an arc. The vertical motion method is most often used; however, it takes some practice to be skillful in its use.*

Figure 12-22. *A welding student practices SMAW in the flat welding position.*

trodes, the arc length will be about 3/16" (about 5mm) to 1/4" (about 6mm). Arc welding should be done with one hand. A welder must use the other hand at times to hold a part while tacking it in place.

One way of checking if the arc length is proper is to listen to the *sound* of the arc. A proper arc length will produce a crackling or hissing sound. Too short an arc length may short out while welding. When small solidified metal drops are seen on the base metal surface, *spattering* is occurring. Too long an arc length will cause a great deal of filler metal spattering. To create a bead which has a uniform width and height and uniformly spaced ripples, a *uniform forward speed* must be maintained.

There are two types of beads used in arc welding. They are called *stringer beads* and *weaving beads*.

To make a stringer bead, the only motion of the electrode is forward. The bead width should be 2–3 times the diameter of the electrode. For example, with a 1/8" (3.2mm) electrode, the normal stringer bead should be 1/4" (6.4mm) to 3/8" (9.6mm). in width (1/8" × 2 = 1/4"

1/4" (6.4mm) to 3/8" (9.6mm). in width (1/8" × 2 = 1/4" or 6.4mm; 1/8" × 3 = 3/8" or 9.6mm). See Figure 12-23A.

When a weaving bead is made, the electrode is moved uniformly back and forth across the weld line while also moving forward. With such a motion, the bead may be made as wide as desired. However, for best bead control, it is recommended that a weaving bead be no wider than six times the electrode diameter. For example, using a 1/8" (3.2mm) electrode, the weaving bead should be no wider than 1/8" × 6 = 3/4" wide (19.2mm). See Figure 12-23B.

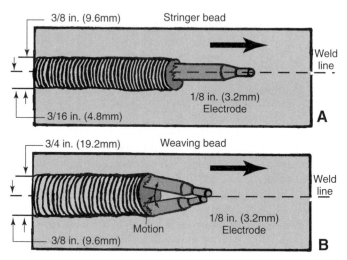

Figure 12-23. *Suggested dimensions for a stringer bead and a weaving bead. The stringer bead is the electrode diameter ×2 to × 3 wide. The weaving bead is the electrode diameter up to × 6 wide.*

The *electrode angle,* or position, is tipped forward 20° in the direction of travel. It is kept in line with the weld line for a stringer bead. See Figure 12-24.

To practice the stringer bead, use a piece of mild steel about 1/4" (6.4mm) thick, 2" (50mm) wide, and 6"

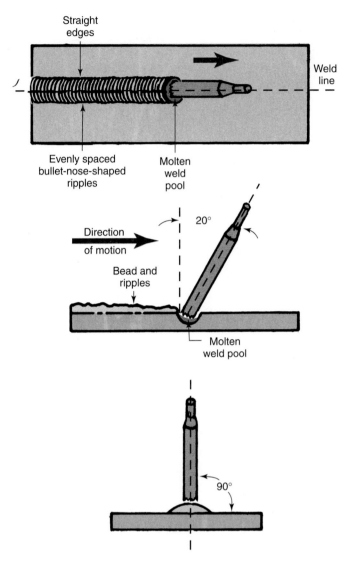

Figure 12-24. Three views of an arc bead in progress. Note that the electrode is inclined 20° in the direction of travel. The completed arc bead should have straight edges, evenly spaced ripples, and uniform height.

(152mm) long. Select an ac or dc electrode 1/8" (3.2mm) in diameter. See Figures 12-19 and 12-20.

Clean the bare end of the electrode if necessary and place it in the electrode holder. The arc should always be struck about 3/8" (9.6mm) ahead of where the bead should begin. The electrode and arc must then be moved rapidly to the spot where the weld is to begin. During this brief period of time, the arc has a chance to stabilize. The bead can then begin with a steady arc. While holding a uniform arc length, move slowly in the direction of motion. Right-handers usually weld best from left to right. Left-handers usually weld best from right to left. With the electrode tipped 20° in the direction of travel, a good view of the weld pool is possible. This slight angle also permits the force of the arc to push the molten metal to the rear of the weld pool to form the bead ripples.

When the width of the stringer bead reaches the desired size, move slightly forward. Watch the weld pool

grow in size and move slightly forward again. This action of watching and moving is continued until it appears to be a uniform forward motion.

The *speed of forward motion* is judged by two factors while welding. These are:

- The bead width.
- The bullet-nose-shaped appearance of the ripples at the rear of the molten weld pool.

If the forward speed is proper, the back of the weld pool will have a bullet-nose-shaped ripple. This will indicate that the weld bead height is correct. **Weld bead height** is the amount of buildup above the surface of the base metal. If the rear of the weld pool is less curved or straight, then the travel speed is too low. Too slow a travel speed will cause the weld bead height to be too high. If the weld pool shape becomes more pointed, the speed is too fast and the buildup is too low. Watch the weld pool shape while welding and adjust the travel speed to maintain a round, bullet-nose-shaped weld pool.

The speed of forward motion can also be judged *after* welding by examining:

- The amount of buildup of the weld metal.
- The shaped appearance of the ripples of the weld bead.

The completed stringer bead will be even in width, with evenly spaced, bullet-nose-shaped ripples. It will have the proper width and bead height. The weld height is normally about one-quarter the bead width.

The proper current setting is very important to make a quality weld or bead. A bead that is made with the correct current, arc length, and forward speed will have a cross section like the one shown in Figure 12-25A. If the current is too low, the bead will be high with poor penetration, as seen in Figure 12-25B. If the current is too high, the electrode will overheat and spatter excessively. The bead also will be low. See Figure 12-25C. A very porous bead or weld will result. Gas pockets and impurities will be trapped within the bead.

With a short arc, the bead will be high, with poor penetration and overlap, Figure 12-25D. If the arc length is too long, the bead will be too low with poor penetration and undercut. See Figure 12-25E. Moving too slowly will create a bead which is wide and high, as shown in Figure 12-25F. A rapid forward motion will result in a low, narrow bead, Figure 12-25G. Figure 12-26 labels the effects of weld beads created with less than ideal conditions.

A practical application of the weld bead is the rebuilding of worn surfaces in welding maintenance work. Shafting, excavation implements, gear teeth, wheels of various kinds, journals, etc., frequently become worn to the extent that they must either be discarded or rebuilt. Rebuilding these surfaces by laying arc beads side by side over the worn areas in one or more layers, then refinishing, has become an important arc welding maintenance operation.

Another application of bead work is **hard-surfacing** or wear-resistant surfacing. Laying beads of special metallic alloys side by side on a soft steel surface provides a surface that is extremely hard and resistant to abrasion. See Chapter 26.

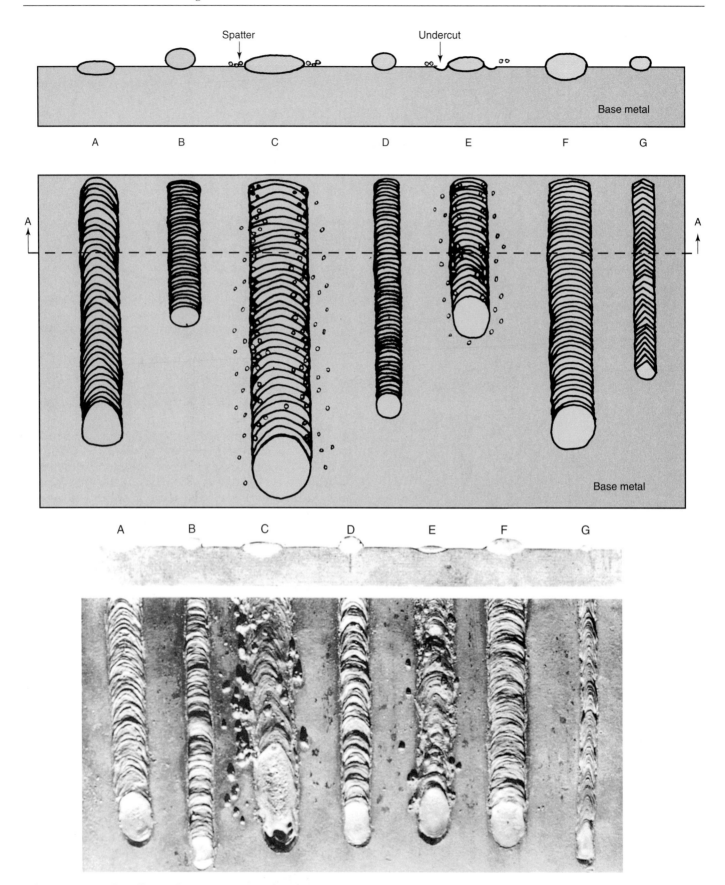

Figure 12-25. *The effects of current, arc length, and travel speed on covered electrode beads. A—Correct current, arc length, and travel speed; B—Amperage too low; C—Amperage too high; D—Too short an arc length; E—Arc length too long; F—Travel speed too slow; G—Travel speed too fast. (American Welding Society)*

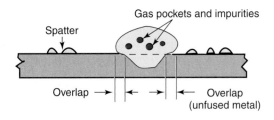

Figure 12-26. *Results of improperly made arc bead.*

12.5.4 Restarting and Finishing an Arc Welding Bead

When a SMAW bead is stopped prior to completion, a deep crater is left in the base metal. Restarting the arc and completing the bead must be done with care. If the restart is done correctly, the bead ripples will be uniform. It will be difficult to see where the bead was stopped and restarted.

Before restriking the welding arc, the previous bead must be cleaned as described in Heading 12.5.5. Restrike the arc about 3/8″ (about 10mm) ahead of the forward edge of the crater. The arc is then moved backward rapidly until the new molten weld pool just touches the rear edge of the previous crater. As soon as the two edges touch, the electrode is moved forward to continue the weld. If this is done correctly, the ripples of the old and the new bead will match. See Figure 12-27.

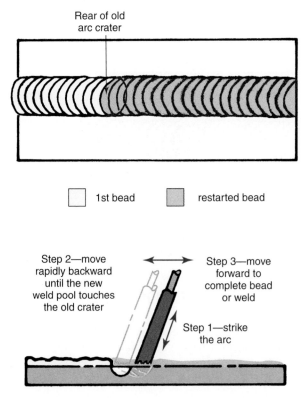

Figure 12-27. *Steps to restarting an arc bead.*

There are two ways to finish a bead or weld without leaving a crater. One method is to use a *run-off tab*. A piece of metal of the same type and thickness is tack welded to the end of the base metal being welded. The arc bead or weld is completed on the base metal and continued on the run-off tab. When the weld is stopped, the crater is on the run-off tab. The run-off tab is cut off, leaving a full-thickness bead at the end of the base metal.

A similar procedure is used to start a weld. A *run-on tab* is tacked to the base metal at the starting end. The welding arc is struck on the run-on tab. When the run-on tab is cut off, it leaves a well-defined, full-thickness bead at the beginning of the base metal.

Another method used to finish a weld without leaving a crater is to *reverse electrode motion* as the end of the weld is reached. The electrode is moved to the trailing edge of the weld pool. When the weld pool is filled, the electrode is then lifted until the arc is broken.

12.5.5 Cleaning the Bead

When shielded metal electrodes are used, a brittle *slag* coating is left, covering the weld bead. This slag covering must be removed prior to restarting a bead. It must also be removed after completing the bead, prior to welding over a bead, and before painting.

If the slag is not removed before restarting or welding over a bead, many slag inclusions will result. *Slag inclusions* are pieces of slag trapped or included in the weld. Slag is generally removed manually with a chipping hammer and a wire brush. The slag may also be removed mechanically by shot peening, wire brushing, or chipping.

12.6 DC ARC BLOW

As explained earlier, once started, the ac arc is quite stable. The *dc arc*, however, may have a tendency at times to wander from the weld line. This wandering, called *arc blow*, is usually caused by the forces of the magnetic field around the dc electrode. All electrical conductors are surrounded by a magnetic field when current is flowing. If the current travels continually in one direction, the magnetism can become quite strong. Ac electrodes are not affected because of the constantly changing direction of the current. These reversals virtually cancel the magnetic blow effects in the ac circuit.

The magnetic field or lines of flux travel easily in metal. They move through air with greater difficulty. Figure 12-28 shows a butt weld with the workpiece lead connected near the beginning of the weld. When the arc is struck, a magnetic field is created around the dc electrode. The magnetic field prefers to travel in the base metal, not in the air. Therefore, the magnetic field forces the molten filler metal to blow inward from the end of the weld joint toward the center of the work. This is called *forward arc blow*. See Figures 12-28 and 12-29. In the center area of the weld joint, the arc and molten filler metal act normally.

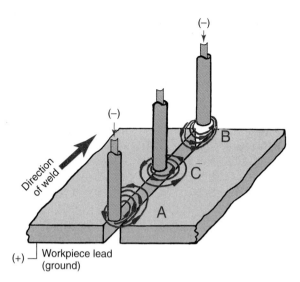

Figure 12-28. *The magnetic field around an electrode is deflected at the ends of the joint (A and B). It attempts to flow through the metal and not through the air. This concentration of the magnetic flux at ends of the joint forces the arc toward the center of the base metal. Thus, the arc is seen to "blow" away from the area directly under the electrode. Notice that the magnetic field is not distorted in the center areas of the weld at C.*

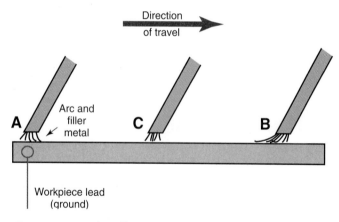

Figure 12-29. *The effects of dc arc blow on the arc and filler metal. As the arc is started at A, the arc is blown toward the right. In the center, C, the arc travels straight down. As the arc approaches the end of the weld at B, the arc and filler metal are blown toward the workpiece lead end of the weld.*

As the welder nears the end of the joint, the magnetic flux intensifies ahead of the electrode. This happens as the magnetic flux tries to stay in the metal, rather than travel out into the air. The arc and molten metal are now blown back toward the beginning of the weld. This action is known as **backward arc blow**. See Figure 12-28 and 12-29. Very seldom does arc blow occur across the weld axis (sideways).

If the arc blow is extremely strong, certain preventive or corrective measures can be taken. One or more of the following may be used to correct magnetic arc blow:

- Place the ground connections as far from the weld joint as possible.
- If forward arc blow is a problem, connect the workpiece lead (ground) near the end of the weld joint.
- If backward arc blow is a problem, place the workpiece lead (ground) near the start of the weld. It will also help to weld toward a large tack weld. The large tack weld will give the magnetic field a place to flow. This will prevent a crowding of the magnetic field which causes arc blow.
- Reduce the welding current. This will reduce the strength of the magnetic field.
- Position the electrode so that the arc force counteracts the arc blow force.
- Use the shortest arc that will produce a good bead. The short arc will permit the filler metal to enter the arc pool before it is blown away. A short arc will also permit the arc force to overcome the arc blow force.
- Weld toward a run-off tab or heavy tack weld.
- Wrap the electrode lead around the base metal in the direction which will counteract the arc blow force.
- Change to an ac welding machine and electrodes.
- Use the backstep method of welding.

The **backstep method** makes use of a number of short welds. The weld bead is divided into several sections. The first segment is started away from the beginning of the joint. The weld is made toward the beginning. Each section is welded back toward the previous section. See Figure 12-30.

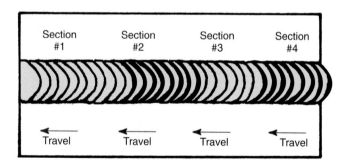

Figure 12-30. *An example of the backstep method of welding. Each section is welded back toward the previous section.*

12.7 ARC WELDED JOINT DESIGNS

Shielded metal arc welding can be done on any of the basic joint designs. Chapter 3 discusses these joint designs. The five basic joint designs are:

- Butt.
- Lap.
- Corner.
- T-joint.
- Edge.

See Figure 12-31. Refer also to Heading 3.1 for a complete discussion of joint designs and edge preparation.

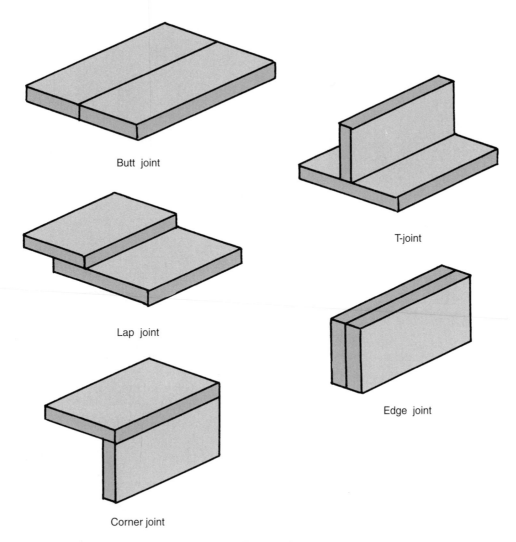

Butt joint

Lap joint

Corner joint

T-joint

Edge joint

Figure 12-31. *Common arc welding joint designs. See also Heading 3.1.*

The lap joint is made with square edges on the base metal. The butt joint, corner joint, T-joint, and edge joint weld may require edge preparation. The edges may be flame cut, ground, or machined to the required shape or angle. Edge preparation is done to ensure complete penetration of the weld.

It is necessary to know the proper names for the various parts of a weld in order to discuss welds and their quality. Refer to Figure 12-32 and to Chapter 3 for the proper names used in joints and welds.

Weld joints may be in any position. A weld may be made on any of the five basic joints and in every conceivable welding position. These are:

- *Flat welding position* (downhand). When the weld axis and weld face are horizontal. See Figure 12-33.
- *Horizontal welding position.* When the weld axis is horizontal and the weld face is vertical or near vertical. See Figure 12-34A.
- *Vertical welding position.* When the weld axis is vertical. See Figure 12-34B.
- *Overhead welding position.* When the weld axis is horizontal, but the weld is made from the underside. See Figure 12-34C.

12.7.1 Weld Flaws and Defects

Completed welds may have a variety of flaws. A *flaw* is anything about a weld that is an imperfection. If a flaw is large, it is called a *defect*. Many flaws and defects can be seen with the naked eye. Others may be found only through the use of destructive testing or nondestructive evaluation. Refer to Chapter 30 for information on testing of welds.

By means of a visual inspection, a welder may find the following weld flaws and defects:

- Poor weld proportions.
- Undercutting.
- Lack of penetration.
- Surface flaws and defects.

The weld face should have relatively small, evenly spaced ripples. The weld face on a groove joint should be wide enough to span the complete groove. A groove weld should have complete penetration.

On the lap joint and the unprepared corner joint, the weld normally does not penetrate to the other side. On a groove-type corner joint or T-joint, the weld may penetrate to the metal face opposite the bevel. On thicker metals,

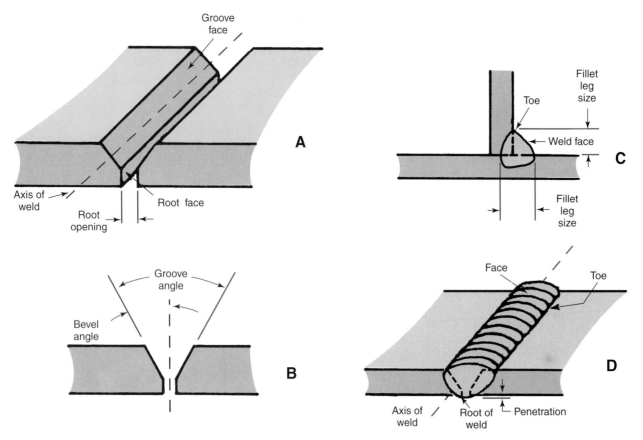

Figure 12-32. *Proper names for parts of a weld joint and weld. A and B illustrate parts of the joint. C and D illustrate parts of the weld.*

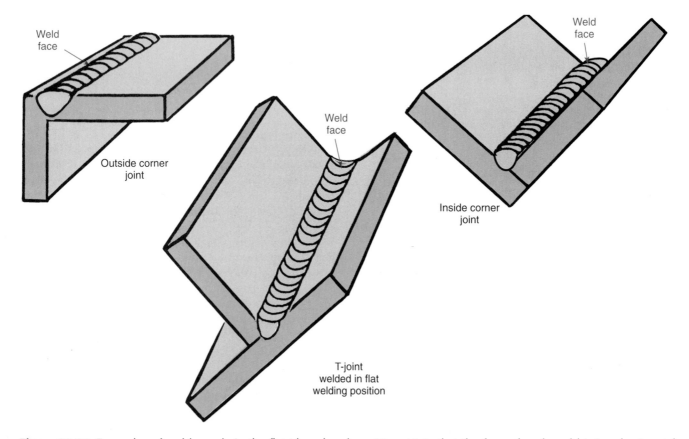

Figure 12-33. *Examples of welds made in the flat (downhand) position. Note that the face of each weld is in a horizontal plane.*

The weld illustrated in Figure 12-36B shows an **undercut** condition. Note that the toes of the weld are cut deep into the base metal. This weakens the base metal and is a defect that must be repaired. Undercutting is caused by improper welding technique. Often a long arc or too high a current is the cause. In Figure 12-36B, the cause was high current. The excessive penetration and spatter indicate a high current setting.

In Figure 12-36C, the weld has been made with a low current setting or a short arc length. The weld is overlapped at the toe of the weld. The weld also has poor penetration.

Figure 12-37 illustrates the cross section of three well-formed fillet welds. The weld at A has a flat face contour. At B, the weld face is convex (curved outward) and at C, the weld face is concave (curved inward). Note that there is no undercut at the toe of the weld. Figure 12-38 is a magnified cross section of a fillet weld. Note the undercut at the toe of the weld at the vertical metal surface.

The three fillet welds shown in cross section in Figure 12-39 are unacceptable welds. At A, the weld contour is poorly shaped; at B, the weld is undercut; and at C, the weld is overlapped.

The metal surfaces of the finished weld should be free from excessive spatter. *High and low spots* in the weld bead are not acceptable. This generally indicates a weld speed which was not uniform. Highs and lows may also indicate a poor restarting technique. If the bead has a number of small pit holes, this indicates porosity. **Porosity** is an indication of gases trapped inside the weld. The cause may be a welding speed which is too fast, or a welding current that is too low. The solution may be to slow down the welding speed and/or increase the welding current. Both of these will allow the weld pool to remain liquid long enough to allow all the gases to reach the surface before the metal becomes solid. Porosity can be eliminated by using low hydrogen electrodes. Always dry low hydrogen electrodes before use to remove any moisture that may cause porosity.

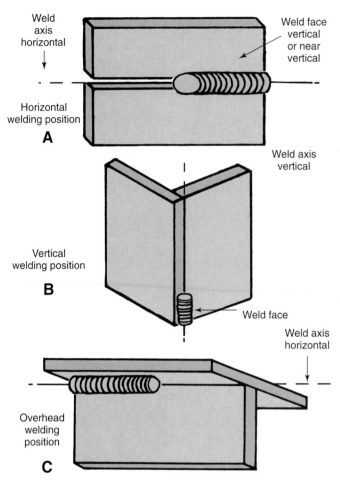

Figure 12-34. *The horizontal, vertical, and overhead welding positions.*

welding may be done from both sides. Refer to Figure 12-35.

A weld with a properly **contoured** (shaped) face is shown in Figure 12-36A.

Figure 12-35. *Fillet welds compared. A—Conventional fillet weld. B—Deep penetration fillet weld.*

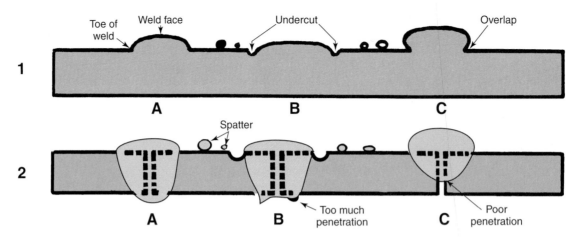

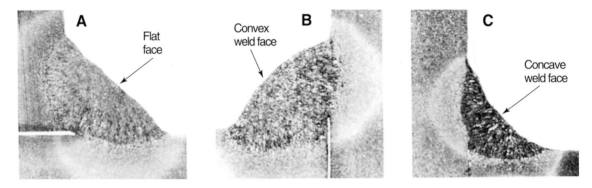

Figure 12-36. A properly made bead is shown at A; it has good contour and penetration. At B, the bead is undercut and the base metal is thus weakened. At C, the bead was made with insufficient heat; the bead is overlapped with poor fusion and penetration.

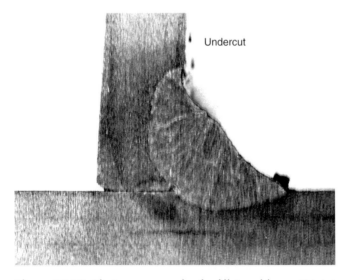

Figure 12-37. Cross sections of acceptable fillet welds (etched and magnified four times). (USX Corp.)

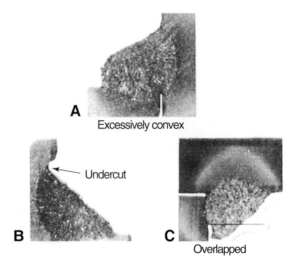

Figure 12-38. Photomacrograph of a fillet weld on a T-joint using dc. Note the undercut on the vertical piece of metal.

Figure 12-39. Cross sections of unacceptable fillet welds (etched and magnified four times). These are unacceptable because of excessive contour in A, undercutting in B, and overlapping in C. This weld was made in the overhead position.

Additional surface flaws can be seen during a visual inspection. These include spatter, slag inclusions, and cracks in the weld bead or weld crater.

Spatter can be controlled by using the correct current and arc length. Slag inclusions can be caused by not cleaning between the weld passes. Preheating and higher welding current help to eliminate slag inclusions. Cracks are caused by various factors. It is important to use the correct welding electrode type and diameter for a welding job. Preheat is required when welding some metals. Also, good welding technique is important to minimize (and often eliminate) most cracking.

If a defect is found, the weld must be repaired. Often, the poor section is ground down or cut out by a gouging process and rewelded.

12.7.2 Edge Joint SMAW

The *edge weld* is defined by AWS as: "A joint between the edges of two or more parallel or nearly parallel members." See Figure 12-40. The edge weld may be made in any position. This type of joint is the easiest of the various joints to arc weld.

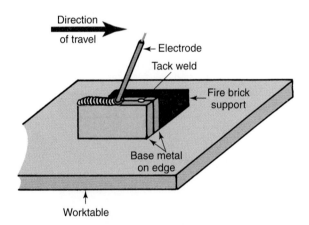

Figure 12-40. *A suggested setup for arc welding an edge joint in the flat welding position.*

On thin metal, no edge preparation is necessary. On thicker pieces of base metal, the edge should be ground, gouged, or machined to provide a bevel-, V-, U-, or J-groove. See Figure 3-7.

To gain experience in joining two pieces together by arc welding, the welder should obtain two pieces of metal approximately 1/4" (about 6.4mm) thick. A 1/8" (3.2 mm) E60XX electrode should be used. The current setting on the machine should be about 100A. Before beginning this weld (or any weld), it is advisable to run a few practice beads. A piece of the same metal being welded should be used for practice. Using the suggested E60XX electrode and about 100A, run a bead. If the bead is narrower than 3/8" or about 10mm (3 x electrode diameter), the current is probably too low. If the bead is wider than 3/8" (about 10mm), the current may be set too high. Run test beads and continue to reset the arc welding machine current until a bead three times the electrode diameter is achieved. The bead

should be run at a forward speed that is comfortable and easy to control. If the bead width and forward speed are proper, the amperage is right for the electrode and the welder (operator). Using test beads may ensure a better bead on the finished part.

This thickness of metal (about 1/4" or 6.4mm) should require no edge preparation. The pieces should be clamped together and grounded to the table. Firebricks may be used to prop the pieces up so that a flat welding position is possible.

A *tack weld* is a small, well-fused weld in one spot that is used to hold parts in proper alignment while welding. A weak tack weld may break during the welding operation and allow the metal to shift its position. Before making the edge joint or any welded joint, the metal should be tack welded as often as required to hold it in proper alignment. The number of tack welds used is optional. However, it is suggested that a tack weld be made at intervals of about 3" (about 75 mm).

After the welding arc is struck, the welder must watch the weld pool as it increases in diameter. When the edge of the weld pool touches the outside edge of the metal, the electrode is moved forward This process is repeated in an almost continuous operation until the weld is completed. On extremely thick pieces, the groove in the joint may not touch the outside edge of the metal. In this case, the electrode is moved forward when the weld pool just touches the edge of the groove. Refer to Figure 12-41.

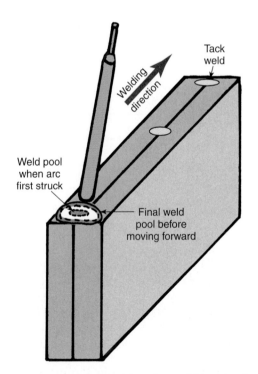

Figure 12-41. *An edge weld as the weld bead is begun. Note that the weld pool must increase in size until it touches the edges of the metal before the electrode is moved forward.*

12.7.3 Lap-Joint SMAW in the Flat Welding Position

The *lap joint* in the flat welding position is common, although it is not the strongest joint design. The weld normally used on a lap joint is the fillet weld. A fillet weld is basically triangular in shape. The weld face may have a flat, convex, or concave shape. The desired shape may be given on the weld symbol. A convex shape is slightly stronger and normally preferred. See Figure 12-42. The electrode should be aimed into the bottom of the joint. The electrode should point more toward the metal surface than the edge. It takes more heat to melt the surface. On thin metal, no sideways motion is needed to form a good fillet. When larger fillets are required, a weaving motion may be necessary.

12.7.4 Corner or T-Joint SMAW in the Flat Welding Position

The *corner joint* may be made as an *inside* or an *outside* corner joint. The inside corner joint may be made as a bevel- or J-groove, or without edge preparation as a square-groove. A fillet weld is often used on the inside corner joint. See Figure 12-44. The outside corner joint is prepared as a butt joint. The joint may be cut as a bevel-, V-, J-, or U-groove, or without edge preparation. Refer to Figure 12-45.

The weld on either the inside or outside corner should penetrate through the root. The weld should not penetrate through the base metal. Both edges of the bead should blend smoothly with the metal surfaces. There should be no overlapping or undercutting. The bead must be straight and have a uniform width and contour. See Figure 12-46.

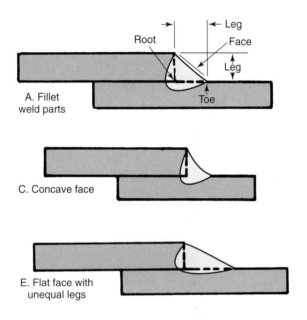

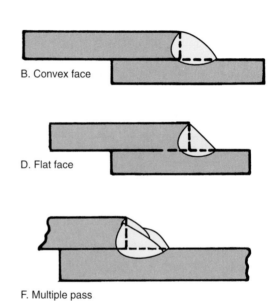

Figure 12-42. Fillet weld shapes. A—Fillet weld parts. The leg size is given on the weld symbol. B—Convex face. C—Concave face. D—Flat face. E—Fillet with unequal legs. F—Multiple-pass weld lap joint.

Figure 12-43 illustrates the position of the base metal while making a fillet weld on a lap joint in the flat welding position. The fillet weld on a lap joint is made on the edge of one piece and the surface of the other piece. More heat is required to melt the surface than is required to melt the edge. To distribute the heat, a weaving motion must be used with most of the motion taking place on the surface, not the edge.

The finished bead should have the proper contour, and must be straight, even in width, smooth, and clean. It should show good fusion between the bead and the base metal. The bead should blend smoothly with the base metal. If a distinct line is seen, the fusion is probably poor. There should be no overlap or undercutting in the fillet. The weld must penetrate through the root of the joint. It should not penetrate through the base metal.

The *T-joint* is formed by placing one piece of base metal on the other to form a T-shape. This joint may be welded from one or both sides. It may be in the form of a square-, V-, or J-groove. If the T-joint is to be welded from both sides, it is usually easier to complete the welds if the entire workpiece is repositioned after one weld is completed. See Figure 12-47. The T-joint fillet weld is made in the same manner as the weld on an inside corner joint.

These welds, as all welds, are easier to make in the downhand, or flat, welding position. A weaving motion is required only on large, wide joints. The electrode angles are the same as those used when running a bead. However, the welder should be certain to melt the surfaces as the inside corner or T-joint is welded. Too often, the beginner makes an inside corner or T-joint without proper fusion of the metal surfaces.

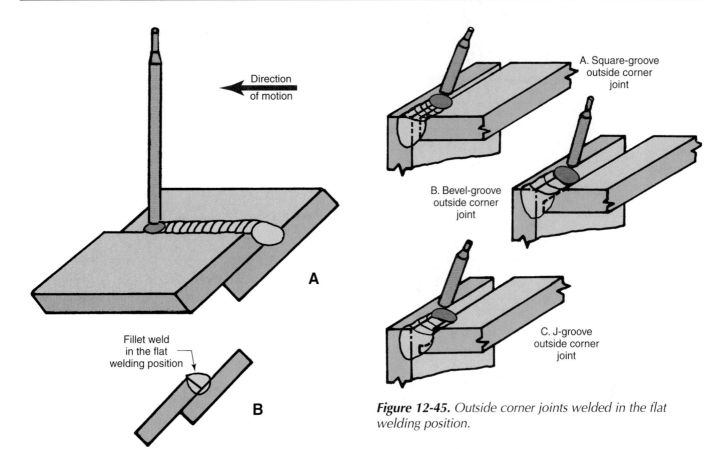

Figure 12-43. *Fillet weld on a lap joint done in the flat welding position. A—The weld in progress. B—The base metal position for welding a lap joint in the flat welding position.*

Figure 12-45. *Outside corner joints welded in the flat welding position.*

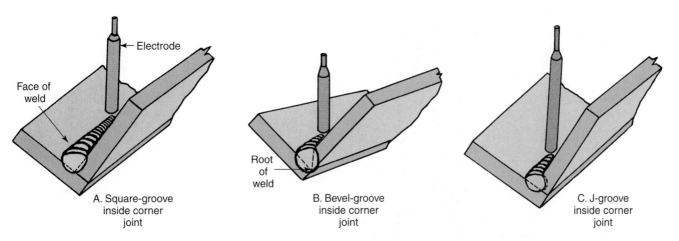

Figure 12-44. *Inside corner joints welded in the flat welding position. Note that the weld face is horizontal in the flat welding position.*

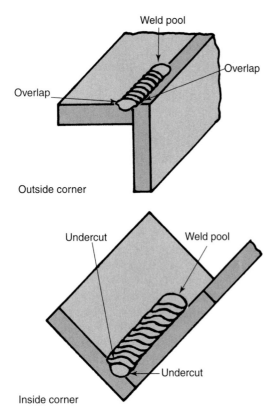

Figure 12-46. *Examples of overlap and undercut on an outside and inside corner joint.*

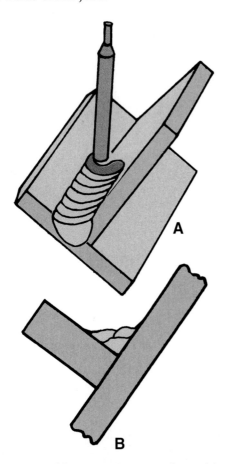

Figure 12-47. *Welding a T-joint. A—Fillet weld in progress on a T-joint in the flat welding position. B—T-joint welded with multiple passes.*

12.7.5 Butt-Joint SMAW in the Flat Welding Position

It is suggested that the butt weld be practiced on low carbon steel 1/4" to 3/8" (6.4mm to 9.6mm) thick. Steel 1/4" (6.4mm) thick and thicker should have the edge prepared prior to welding. This will ensure complete penetration of the base metal. A V-groove joint is suggested. Both metal edges should be ground to about a 45° angle. The angle should be cut to within 1/16" (1.6mm) of the bottom, as shown in Figure 12-48. The base metal joint should be tacked to hold the parts while welding.

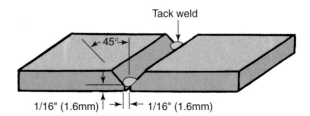

Figure 12-48. *Preparing a V-groove butt joint for welding. The edges have been beveled and the metal tack welded to hold it in proper alignment. Note that both the root opening and root face are 1/16" (1.6mm).*

The first pass in a butt joint must melt both pieces of metal and obtain complete penetration. When welding thinner base metal, only one weld pass is required. Thicker metal will require additional passes to fill the weld joint.

To ensure complete penetration in a groove weld, the **keyhole method** should be used. The keyhole is seen only on the first, or root, pass. As the weld pool forms and melts through the thinner metal at the root face, a small hole is formed. The hole looks like the old-fashioned keyhole. See Figure 12-49. The keyhole is continually filled in as the weld moves ahead. If the diameter of the keyhole is kept constant, the amount of penetration will be uniform as well.

When welding a long joint, more than one electrode is often required to complete the weld. When an electrode is down to a very short length, lift it from the joint. Do not fill the keyhole. The weld joint should look like Figure 12-49. Place a new electrode in the electrode holder. Strike the arc about 3/8" (about 10mm) ahead of the crater. The electrode is moved back over the rear edge of the keyhole crater to remelt it and join the two weld beads together. Starting the arc in this way allows time for the arc to stabilize. It also ensures proper preheating of the base metal.

The same technique is used on any weld bead that must be stopped and restarted. The only difference is there is no keyhole, except in the first pass of a butt joint. Other weld beads require the welder to remelt the existing crater and join the two weld beads together.

When the end of the weld joint is reached, slowly move the arc to the back of the weld pool. This motion will fill the weld pool. Then lift the electrode to break the arc.

Base metals 1/4" (6.4mm) and thicker will require more than one weld bead to fill the joint. If two or more beads are used to complete the weld, the welder must

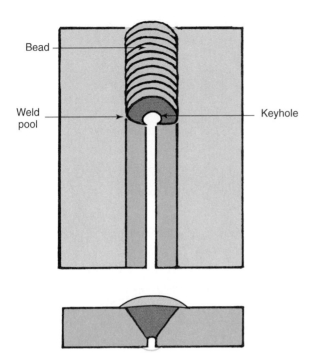

Figure 12-49. *Two views of a V-groove butt weld in progress. Note the keyhole near the leading edge of the weld pool.*

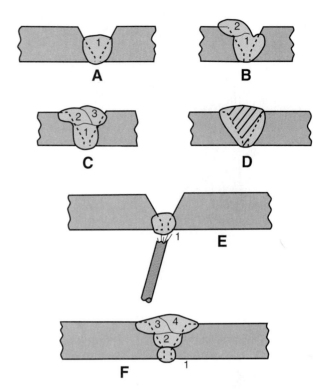

Figure 12-50. *Arc welding joints welded with multiple passes. A—The first, or root, pass in a butt joint. B—Second pass. C—Third pass. D—Finished weld. E and F—A butt joint in which the root pass is made from the penetration side of the joint.*

clean the previous bead before attempting to make the next. This cleaning operation prevents trapping slag in the next bead or pass. The previous weld bead should be penetrated by the new bead. The final bead or beads used to fill the weld joint should be built above the original top surface of the metal. See Figure 12-50.

When the welding of the joint is complete, allow the weld and slag to cool. Then remove the slag and wire brush the weld surface. Guard your eyes against flying particles. **Always wear safety goggles when cleaning metal.**

Inspect the weld for straightness, constant width, smoothness, penetration, gas bubbles, fusion, spatter, and buildup. It should have a clean-looking bead with straight edges and constant width of bead. The height of the bead should be constant. The ripples should be evenly spaced. The penetration should just show through the under part of the weld joint. The weld should have no small cavities that would indicate too long an arc. It should have good fusion. *Fusion* means a good bond between the filler metal and the base metal. Fusion is indicated by a smooth blending of the filler and base metals at the edges of the weld. There should not be a distinct edge or line between the filler metal and the base metal. There should be little or no spatter. Spattering is the result of too long an arc or too high a current. If the metal is loosely connected to the table, the resulting wandering arc may give poor fusion and weld appearance.

12.7.6 SMAW in the Horizontal Welding Position

When welding a butt, edge, or outside corner joint in the *horizontal welding position*, the electrode should be pointed upward at an angle of about 20°. This is to counteract the sag of the molten metal in the weld pool. The electrode is also inclined about 20° in the direction of travel of the weld, as shown in Figure 12-51. Gravity tends to work against the welder in all positions other than the flat welding position. Therefore, a short arc gap should be used to ensure the filler metal will travel across the horizontally positioned arc.

Be sure to eliminate undercutting at the edge of the bead. Undercutting is usually the result of excess current for the size of the electrode used, or poor electrode motion. Stop momentarily at the end of each motion or swing of the electrode. This will deposit additional metal and will help to eliminate undercutting.

When welding a T- or inside-corner joint, with one piece of the base metal in a near vertical position and the other in a near horizontal position, the electrode is inclined 20° in the direction of travel. The electrode is also positioned at about a 40°–50° angle to the horizontal piece of metal as shown in Figure 12-52. The motion used is usually some type of backward slanting weave motion. The forward motion may take place on the vertical piece. Figure 12-53 shows a T-joint being welded.

The suggested electrode angles for making a fillet weld on a horizontal lap joint are shown in Figure 12-54.

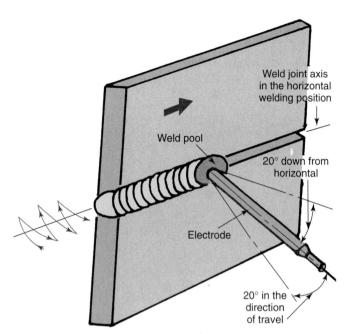

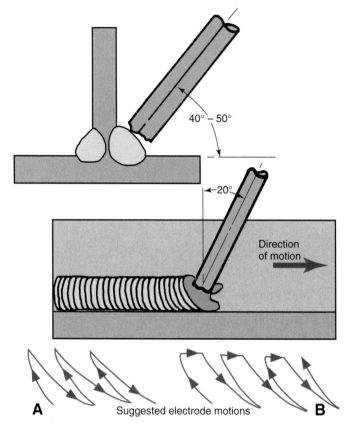

Figure 12-51. *Suggested electrode position for welding a butt joint in the horizontal welding position. The electrode is tipped 20° forward and 20° downward. Use the suggested electrode motion or one similar to it. The upward and backward motion uses the force of the arc to push the filler metal into position. Stopping momentarily at the end of each motion will help to eliminate undercutting.*

Figure 12-52. *Suggested electrode position for fillet welding on an inside corner or T-joint. The two suggested electrode motions have a backward slant. In B, the forward motion takes place on the vertical piece. Both motions use the arc force to overcome gravity.*

The electrode should point more toward the surface than toward the edge. It takes more heat to melt the surface of the metal than the edge of the metal.

Whatever electrode motion is used, remember that it should have a slight backward slant. This backward slant lets the arc force push the filler metal up as it attempts to sag down with gravity.

12.7.7 SMAW in the Vertical Position

Whenever possible, welds should be made with the seams in the flat welding position. In some industries, special turntables are used to rotate the work so that this position may be obtained. However, many welds have to be done in the vertical, horizontal, or overhead welding positions because of the project size. Typical structures that cannot be rotated are large construction projects like buildings, bridges, and pipelines. Welds made in various positions must be of the same quality and strength as welds done in the flat welding position. Figure 12-55 illustrates the metal and electrode positions for welding a *vertical* butt joint.

Welding in the vertical welding position may be done in either of two directions. The welder may weld *uphill* (sometimes referred to as "vertically up") or *downhill* ("vertically down"). Working in either direction, the weld must be made so that the electrode flux or slag is not entrapped or included in the weld metal. The weld must also be made so that it does not run or drip. This can result from allowing the filler metal in the weld pool to stay molten too long.

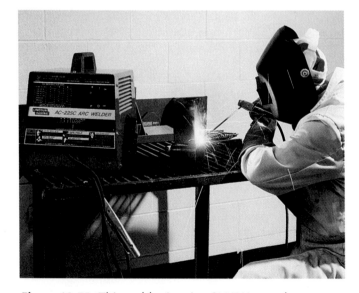

Figure 12-53. *This welder is using SMAW to make an inside corner joint on two square tubes. (The Lincoln Electric Co.)*

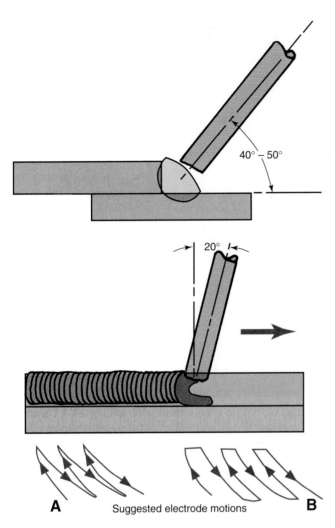

Figure 12-54. *Suggested electrode position for fillet welding on a horizontal lap joint. Note the two suggested backward slanting electrode motions.*

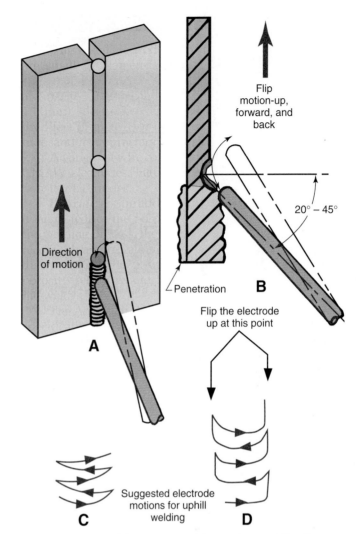

Figure 12-55. *Welding a vertical butt joint. A—The flip motion used to control and cool the weld pool. This motion may be used on any type of joint. B—The electrode should point upward at 20°–45°. The angle may vary as required. Note the motions shown in C and D.*

The following actions may be taken to prevent or control filler metal sagging:

- Hold a short arc. This will permit more filler metal to transfer from the electrode to the weld pool.
- Use a current setting that is as low as possible, but will still produce well-fused welds.
- Make multiple pass welds. Molten metal in a narrow weld pool will cool more rapidly and tend to sag less.
- Use a flip motion to allow the metal in the weld pool to cool.
- Use an electrode motion which will allow the metal in the weld pool to cool. See Figure 12-55 for suggested electrode motions.

The uphill method of welding is generally preferred. When welding downhill, the electrode slag or flux has a tendency to run into the molten weld pool. This is prevented by welding at a speed fast enough to stay ahead of the molten slag. Welding at such a rapid speed generally does not produce adequate penetration. The downhill method may work best on thin metal. See Figure 12-56.

When welding uphill, the greatest problem is controlling the metal in the weld pool. If this metal becomes too hot, or the weld pool becomes too wide, the molten metal may drip down and out of the weld. To control the weld pool heat, and allow the metal time to cool, a flip motion is used. With a flip motion, the electrode end is momentarily moved forward and raised slightly. It cannot be raised too far or the arc will stop. The electrode is then brought back to the rear of the weld pool to continue the weld. During the time that the electrode and arc are moved up, forward, and back again, the weld pool cools slightly. This flip action is continued throughout the weld. If more time is required for cooling, the electrode is moved farther forward before returning to the rear of the weld pool. Figure 12-57 shows a vertical weld in progress.

The vertical butt, edge, and outside corner joints may be a square-, V-, bevel-, J-, or U-groove. Practice pieces may be positioned and tack welded in the flat welding position. These pieces must then be held in the vertical welding

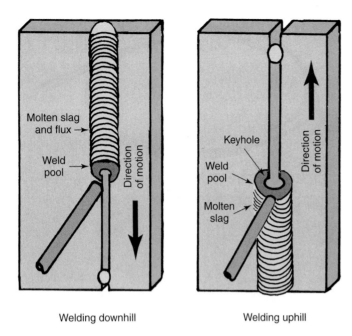

Figure 12-56. *The downhill and uphill welding methods. Uphill is generally preferred. When welding downhill, there is a danger of having the falling slag or flux mix with the molten metal.*

Figure 12-57. *A welder shown making a fillet weld in the vertical welding position on a large field welded structure. (Hobart Brothers Company)*

position. This may be done by clamping them in a welding fixture, as shown in Figures 6-48 and 6-49.

The motion used on the vertical lap joint is shown in Figure 12-58. A small weaving motion is often used. Forward movement and any flip motion required is done on the surface, not on the edge. The weld pool should just touch the edge of the one piece momentarily. The electrode and weld pool are then moved away to the surface.

The electrode angles used for the vertical inside corner and T-joint are shown in Figure 12-59. This figure also shows the electrode motion suggested for the filler weld used on this joint.

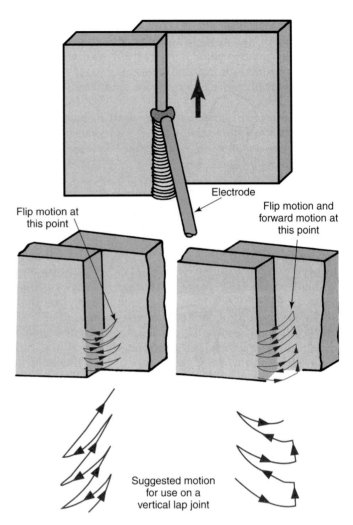

Figure 12-58. *A view of a vertical fillet weld in progress on a lap joint. Note the suggested motions. Both use a flip motion on the surface piece.*

All vertical welds should have the same appearance as welds done in the flat welding position. The bead must be straight with a uniform width. The weld bead must be properly fused, with no overlap or undercutting. The ripples of the bead should be uniform. The face of the weld should have the required contour.

12.7.8 Arc Welding in the Overhead Welding Position

Overhead arc welding is generally the most difficult. It can also be dangerous for a welder who is not wearing the correct protective clothing. The beginner should practice making beads in the overhead welding position before attempting to weld seams. Figure 12-60 illustrates a V-groove butt joint being welded in the overhead welding position.

On practice pieces, the joint is tack welded in the flat welding position. It is then placed in a fixture to hold it during overhead welding. See Figures 6-48 and 6-49. The angles of the electrode are much the same as in the flat or horizontal welding positions. To keep the metal in the weld pool from overheating, a flip motion is used.

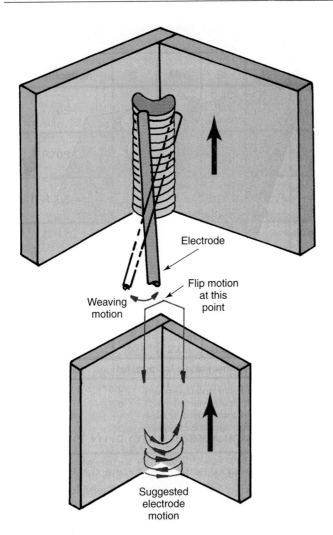

Figure 12-59. A fillet weld on a vertical inside corner joint. A weaving motion is often used with a flip motion to keep the filler metal from sagging.

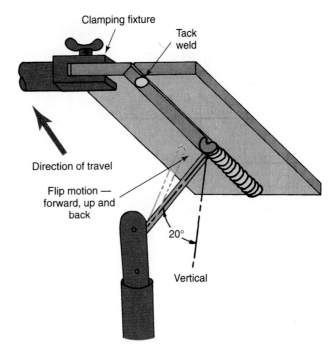

Figure 12-60. A V-groove butt weld in progress. This is the overhead welding position. Note the position of the electrode in the electrode holder.

Most electrode holders have jaws designed to allow the electrode to be held in a variety of positions. The electrode should be placed into the electrode holder at an angle that is comfortable for the welder to use. See Figure 12-61. A covered electrode should never be bent to change its angle for welding. Bending a covered electrode will crack the covering and cause it to fall off. Lack of a flux covering on the electrode will result in a poor weld. It is most important to keep metal in the molten weld pool from falling due to gravity. To keep it cool and to control the weld pool, a flip motion is used. Heading 12.7.7 discusses the flip motion.

A cap, leather cape or coat, and good-quality leather gloves should be worn when overhead welding. Coveralls must be buttoned at the collar and all pockets should have closed flaps on them. Figure 12-61 shows a welder wearing proper protective clothing, practicing an overhead T-joint.

12.8 REVIEW OF SAFETY IN SMAW

The following are some safety rules that must be carefully observed if welding accidents are to be prevented:

- Protect your eyes and face from harmful rays and sparks by using an approved type of helmet. See Figures 11-56, 11-57, and 11-60.
- Recommended clothing and shoes must always be worn. Leathers should be worn, especially during overhead welding.
- Avoid open pockets and cuffs, because they may catch hot sparks and the clothing may ignite.
- The floor on which the welder stands should be kept dry to eliminate the chance of an electrical shock. The welder may choose to stand on a wooden platform for comfort and insulation
- Only an experienced electrician should work on electrical power connections used in the electric arc welding machine.
- The welder should wear heavy, gauntlet-type gloves.
- When arc welding, all skin should be covered to prevent burns from the arc rays.
- The welder should have adequate ventilation or a source of filtered air to protect the nose, throat, and lungs from harmful and irritating fumes generated in the electric arc. Figure 12-62 shows a welder using a source of filtered air and a ventilation system.

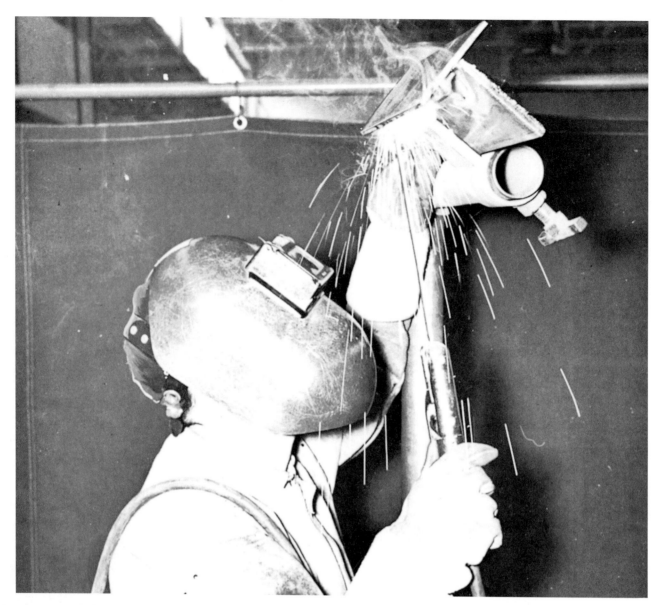

Figure 12-61. *Making a practice overhead arc weld. Note the position of the electrode in the electrode holder. This welder is wearing a cape to protect arms and shoulders, a cap, and gauntlet gloves.*

Figure 12-62. *This welder is using a welding fume extractor pickup and a powered air purifier as he welds a large assembly. (Nederman, Inc.)*

- Always use equipment approved by the National Electrical Manufacturers Association (NEMA).
- **Never use homemade or unapproved transformer equipment. It may be dangerous, particularly if the primary and secondary winding should become electrically shorted.**
- Never operate an ac welding machine with the welding cables wrapped around the welding machine. The magnetic field produced by the welding cables may interfere with the magnetic circuit inside the welder. The primary circuit should always be installed according to the prevailing electrical code.

TEST YOUR KNOWLEDGE

Write your answers on a separate sheet of paper. Do not write in this book.

1. What is the approximate temperature of the arc in SMAW?
2. List three types of dc arc welding machines.
3. List two types of ac arc welding machines.
4. What is the normal no-load voltage range in dc arc welding?
5. How is the ac arc stabilized?
6. What is arc blow?
7. In what direction do the electrons flow in a DCEN (DCSP) circuit?
8. List three variables that may determine the choice of which dc polarity to use.
9. Why is it important that the electrode and workpiece lead connections are tight?
10. Using Ohm's law, what is the total resistance in a welding circuit that has an arc voltage of 18V and a current of 120A? Show your work.
11. What is the danger of arc flash?
12. Give two reasons why a small diameter electrode should be used.
13. To run a good bead, what factors must the welder control in manual SMAW?
14. How wide is a stringer bead? A weaving bead?
15. What does a bead that was made with too long an arc gap look like?
16. How is an arc restarted on a previous, unfinished bead?
17. How is a weld bead properly ended?
18. What is undercut?
19. What is the best position in which to weld any weld joint?
20. What is the flip motion and why is it used?

The shielded metal arc process being used in industry to weld a fillet joint on a circular casting. SMAW is also widely used in construction work and repairs to equipment made in the field. (Hornell Speedglass, Inc.)

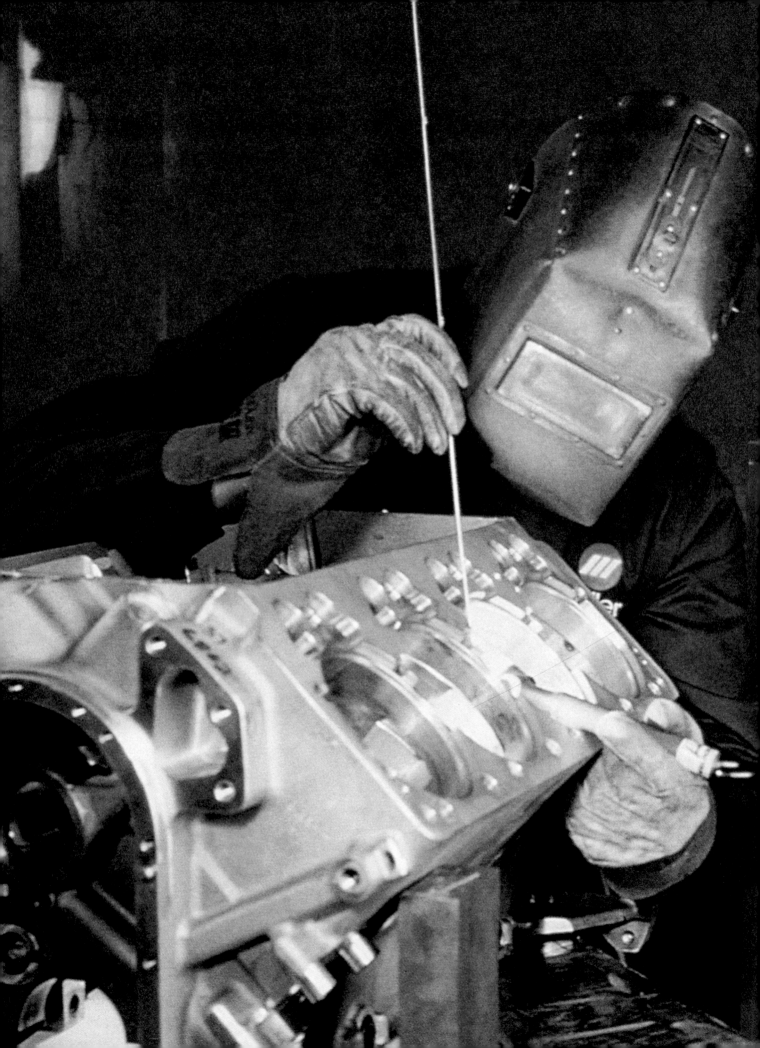

Part 4

GAS TUNGSTEN AND GAS METAL ARC WELDING

Welding an aluminum engine block with gas tungsten arc welding, using an inverter power supply. (Miller Electric Mfg. Co.)

This welder is using GMAW to fabricate a base ring for an industrial robot. (Hornell Speedglas, Inc.)

Chapter 13

GTAW AND GMAW EQUIPMENT AND SUPPLIES

LEARNING OBJECTIVES

After studying this chapter, you will be able to:

* Identify and describe the function of each component of a GTAW, GMAW, and FCAW station.
* Name the various types of shielding gases used in GTAW, describe their characteristics, and evaluate their effectiveness.
* Name the various types of shielding gases used in GMAW, describe their characteristics, and evaluate their effectiveness.
* Identify and specify the type of electrode used for GTAW, referring to the tables provided in the book and using the AWS electrode classification system.
* Identify and specify the various electrode wires used for GMAW and FCAW, using the tables provided in the book and the AWS electrode classification system.

Gas tungsten arc welding (GTAW) is a welding process in which an arc is struck between a nonconsumable tungsten electrode and the metal workpiece. The weld area is shielded by inert (chemically inactive) gas to prevent contamination. Filler metal may or may not be added to the weld. It is possible to join most weldable metals with GTAW. This process is excellent for welding root passes in heavy metal sections. GTAW is also known, in shop terms, as tungsten inert gas or *TIG welding*. Heading 4.6 describes the basics of GTAW.

Gas metal arc welding (GMAW) is a welding process in which an arc is struck between a consumable metal electrode and the metal workpiece. The consumable electrode wire is fed to the welding gun from a large spool that holds several hundred feet (meters) of wire. The consum-

able electrode is the filler metal. The weld area is protected by a shielding gas. This process is used in production, in welding shops, and in automobile body repair shops. It is capable of making excellent welds almost continuously. The welding skills required for this process are not as great as those required for some manual welding processes. In shop terms, this process is also known as metal inert gas or *MIG welding*. Refer to Heading 4.7 for a basic description of the GMAW process.

Flux cored arc welding (FCAW) is similar in most respects to the gas metal arc welding (GMAW) process. The difference is in the electrode wire used. FCAW uses a hollow-core electrode that contains flux and alloying materials. The flux core provides a gaseous shield around the arc. FCAW may also use a shielding gas provided through the gun, similar to the GMAW process. Refer to Heading 4.8 for a basic description of the FCAW process.

The equipment and supplies used for each process will be explained in this chapter. Chapter 14 will cover gas tungsten arc welding techniques and principles, while Chapter 15 will cover gas metal arc welding and flux cored arc welding principles and techniques. Studying these chapters, coupled with actual welding practice, will build the skills and techniques you need to master these welding processes.

13.1 THE GAS TUNGSTEN ARC WELDING STATION

The typical gas tungsten arc welding (GTAW) outfit will contain the following equipment and supplies:

* An ac or a dc or an ac/dc arc welding power source.
* Shielding gas cylinders or facilities to handle liquid gases.
* A shielding gas regulator.
* A gas flowmeter.
* Shielding gas hoses and fittings.
* Electrode lead, workpiece lead, and hoses.
* A welding torch (electrode holder).

- Tungsten electrodes.
- Welding rods.
- Optional accessories.
    - A water cooling system with hoses for heavy duty welding operations.
    - Foot rheostat.
    - Arc timers.

Figure 13-1 shows a schematic drawing of a gas tungsten arc welding outfit. The booth and exhaust system are not shown in this illustration. Refer also to Figure 4-11.

13.2 ARC WELDING POWER SOURCES FOR GTAW

Welding power sources for gas tungsten arc welding (GTAW) can be alternating current (ac), direct current (dc), or both alternating and direct current (ac/dc) in one machine. The most popular arc welding power source is the transformer-rectifier type. This machine produces both ac and dc current. Other arc welding power sources include the transformer and generator. The transformer type produces only ac; generator-type machines produce ac, dc, or both types of current.

Gas tungsten arc welding power sources must supply a *constant current*. In constant current machines, the volt-ampere curve is either drooping or a true constant current.

A drooping volt-ampere curve has a steep slope. The amperage will change slightly as the voltage changes. This allows a welder to vary the amperage by changing the arc length. Newer designs of power supply have nearly a vertical line on the volt-ampere curve. This is a true constant current power source, because the amperage is constant even though the voltage changes. This is made possible by using electronics to control the power supply. These electronically controlled power sources are more efficient, more responsive, and more repeatable. They also produce an excellent pulsed welding current.

Figure 13-2 shows volt-ampere curves for drooping and constant current welding power sources. Two drooping curves are shown in this illustration. Curve 1 is for an 80V open circuit voltage set for a 150A maximum welding current. A 5V change in the closed circuit (welding) voltage from 20V to 25V, represents a 25% change in voltage with only a 5% change in the current. Curve 2 represents a curve for an 80V open circuit voltage set for 50A maximum welding current. A 25% change in voltage (from 20V to 25V) will only change the amperage about 2A, or 4%. The amperage is therefore considered to be relatively constant with large changes in voltage.

Figure 13-2 also shows two curves (3 and 4), for an electronically controlled constant current power source. The same voltage change from 20V to 25V produces almost zero change in amperage.

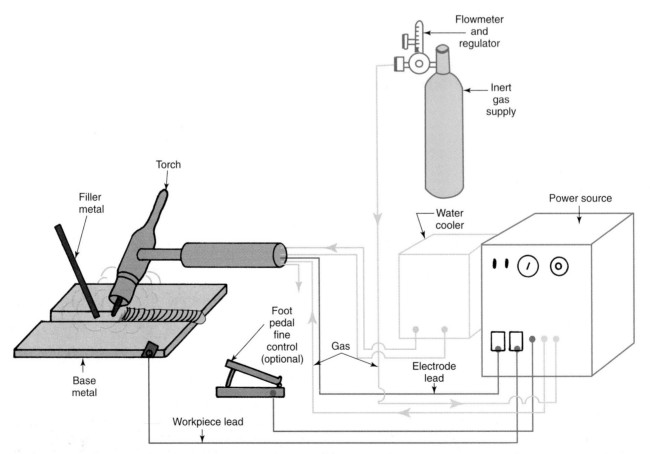

Figure 13-1. A diagram of a complete gas tungsten arc welding station.

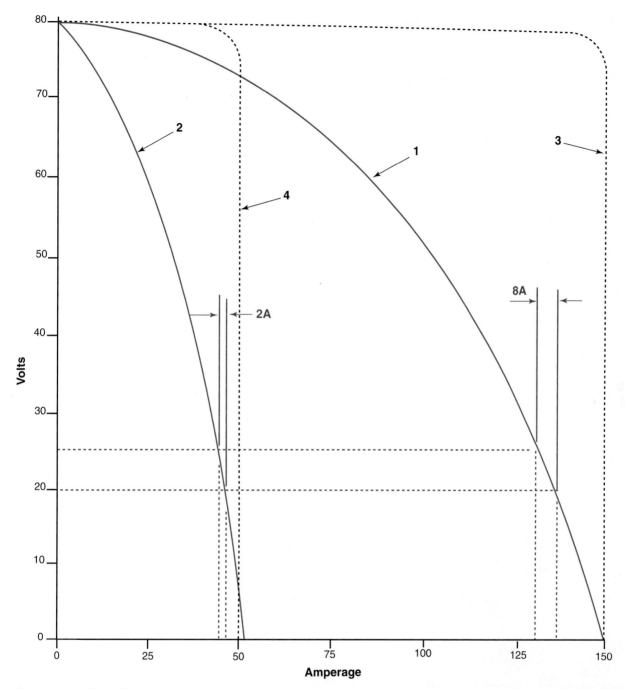

Figure 13-2. *The volt-ampere curves for a drooper type and constant current power source. Curves 1 and 2 are drooper curves which show a 2A and an 8A change with a change of 5V. Curves 3 and 4 are true constant current curves. They show a nearly constant amperage with voltage changes.*

The power source for GTAW may be an ac or a dc welding machine. Most often, the machine will produce both ac and dc constant currents. Figures 13-3 and 13-4 show GTAW power sources.

A direct current power source may be equipped with a device that produces a high-frequency voltage and feeds it into the welding circuit. High-frequency voltage, when used in a dc circuit, is used only for starting the arc. Once the dc welding arc is stabilized, the superimposed high-frequency starting voltage is automatically stopped. High-frequency voltage is produced by an oscillator or a similar

device. Several thousand volts are produced at a frequency of several million cycles per second or megahertz. The current in this high-frequency circuit is only a fraction of an ampere.

When ac welding, the current and voltage reverse directions many times each second. The electrode is positive, then the voltage switches so the electrode is negative. This repeats continuously when ac welding.

An ac power source must have a way to keep the current flowing while the current switches from electrode-negative to electrode-positive. Without a method to keep

Figure 13-3. An ac/dc, single-phase power source for GTAW. (Hobart Brothers Co.)

Figure 13-4. A constant current ac/dc power source for GTAW. It has preflow and postflow gas controls, spot welding controls, crater fill, and starting amperage controls. (Miller Electric Mfg. Co.)

the current flowing, there would be no arc during the electrode positive part of the cycle. When no current flows, the arc is said to be *rectified*. This is not good for the welding machine or the weld.

To reduce or eliminate such rectification, the following methods are suggested:

- Use a welding machine with a higher open circuit voltage of about 100V (rms). Welding machines for manual GTAW have a maximum open circuit voltage of 80V.
- Discharge capacitors at the start of the electrode positive half of each cycle.
- Put a high-frequency voltage supply, which produces several thousand volts, in series with the main transformer.
- Use a power source that produces a square-wave output.

Almost all GTAW machines currently produced use high-frequency or have a square-wave output.

The construction, operation, and control of ac, dc and inverter power sources are covered in Headings 11.1, 11.2, and 11.2.2. Review this basic information for a more complete understanding of welding power sources.

13.3 BALANCED AND UNBALANCED AC WAVES

Each alternating current cycle, when plotted against time on a graph, has a shape like a wave, as shown in Figure 13-5C. This wave shape is called a *sinusoidal wave*. Alternating current has equal current flowing during both halves of each wave, or cycle. However, when sinusoidal alternating current is used for welding, the current is less during the half of the ac cycle when the electrode is positive. This may be explained as follows:

During one half of an ac cycle, the electrode is negative and the workpiece is positive. After going through the zero current point, the cycle reverses. The electrode then becomes positive and the workpiece negative.

Electrons flow from negative to positive. During the electrode-negative half of the cycle, the hotter electrode gives off electrons readily. Since the electrode tip has a small surface, the electrons are concentrated and flow easily toward the large surface of the workpiece.

When the workpiece is negative, the electrons must flow from a large surface to a small point (the electrode). The electrons on the surface are not concentrated at any one point. Electrons, therefore, cannot easily travel from the surface to the electrode. This resistance of electron flow from a surface to a point is called *rectification*. Rectification may be so severe that the entire flow of current in the electrode-positive half of the cycle is stopped, as shown in Figure 13-5A.

The question may be asked by the welder: "What happens to the current during the electrode-positive half cycle?" No current flows during the electrode-positive half cycle. This lack of current flow makes the ac arc very unstable. Such rectification of the current is undesirable for both the welding operation and the welding power source.

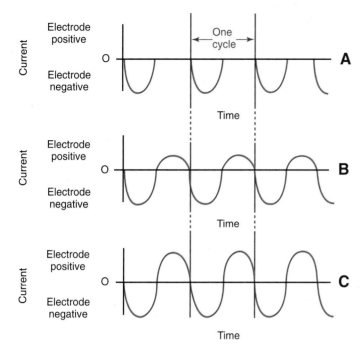

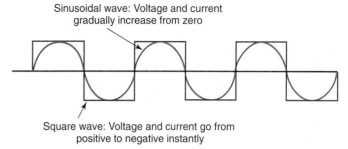

Sinusoidal wave: Voltage and current
gradually increase from zero

Square wave: Voltage and current go from
positive to negative instantly

Figure 13-6. *Comparing voltage and current in a sinusoidal ac wave and a square wave. The square wave goes from a positive value to a negative value instantly. The voltage and current allow the arc to reinitiate (restart). The traditional ac wave has voltage and current near zero for some period of time.*

Figure 13-5. *Alternating current plotted against time. A—Completely rectified ac wave with no current flow during electrode positive. B—Unbalanced wave. C—Balanced wave.*

To overcome this lack of flow during one half of the alternating current cycle, a high-frequency voltage is generated and fed into the welding circuit. Several thousand volts, with only a fraction of an ampere, are created. This high voltage has a frequency of several megahertz (million cycles per second).

A high-frequency unit is normally built into the gas tungsten arc welding machine. It consists of a step-up transformer, capacitors, a control rheostat, a set of spark gap points, and a coil to couple the high-frequency unit to the welding circuit.

High-frequency voltage flows continually in the welding circuit. This high-frequency voltage keeps the shielding gas in the arc area in an ionized state. When the gas is ionized, the arc is maintained during the half of the cycle when the electrode is positive. While the arc is maintained, some current will flow across to the electrode. However, because the work surface does not emit (give off) electrons easily, less current flows during this half of the ac cycle, or wave, as shown in, in Figure 13-5B. The wave shown is an unbalanced wave.

A completely balanced wave is shown in Figure 13-5C. In older machines, a balanced current output wave was achieved by using a large number of capacitors in series or a battery in the welding circuit. Newer GTAW power sources use electronics to create and maintain a balanced wave.

Most new GTAW power sources produce a square-wave current output. A square-wave power supply can change the current from electrode-positive to electrode-negative very quickly. See Figure 13-6. This provides a high voltage as the current crosses zero and allows the arc

to restart easily. The arc will *reinitiate* or restart even without the use of high-frequency or another arc stabilizing method.

Because the output current and voltage are controlled electronically, the amount of direct current electrode positive and direct current electrode negative can be adjusted. This allows the welder to adjust the amount of cleaning and the amount of penetration. See Figure 14-14, which shows how the amount of current in the electrode-positive and electrode-negative portion of the weld cycle can be adjusted to get a balanced wave, one with more penetration or one with more cleaning.

Balanced current is not necessary for most manual welding. It becomes more necessary for automatic welding as the duty cycles increase, often to near 100%. The advantages of a balanced current output wave are:
- Better oxide cleaning is provided because more current flows during the electrode-positive half cycle.
- A more stable arc.
- Reliable restarting of the arc during the electrode-positive half cycle.
- The output rating of the welding transformer does not need to be reduced because of output imbalance.

The advantages of the *unbalanced wave*, Figure 13-5B (or a square wave with greater electrode-negative current flow), are:
- Higher currents can be used with a given electrode, since the electrode-positive cycle has less current flowing than with a balanced wave.
- The power source is less expensive than for a balanced wave machine.
- Better penetration is possible with an unbalanced wave because penetration takes place during the electrode-negative half cycle.

The advantages of a square-wave power source that allows the ac balance to be adjusted are:
- The square wave provides high voltage and currents while the electrode switches polarities. This eliminates rectification.

- The welding power source can be adjusted to obtain a balanced wave and obtain the advantages of the balanced current wave listed above.
- Power source can be adjusted for greater penetration and still have the electrode-negative half cycle be present (not rectified).
- Power source can be adjusted for greater cleaning than can be obtained with a balanced wave.

13.4 SHIELDING GASES USED WITH GTAW

Inert gases are gases that will not react with metals or other gases. Inert gases are used to create a protective bubble around the arc, molten metals, and the tungsten electrode while welding is being done. Argon, helium, argon-helium, and argon-hydrogen mixtures are used as the inert gases when doing gas tungsten arc welding. Because this process uses a tungsten electrode and inert gases, it is known in the shop as *TIG*. TIG is an older term that stands for "tungsten inert gas." The correct name for the process, however, is gas tungsten arc welding. The abbreviation is GTAW.

Argon (Ar) has an atomic weight of 40 and a density of 1.78 gm/l. It is, therefore, a relatively heavy gas. Argon is usually furnished in heavy steel cylinders similar in construction to those used for oxygen. The usual cylinder size is 330 ft^3 (9.34 m^3). Argon as used for welding is a minimum 99.995% pure. It is capable of being shipped in the liquid form at –300ºF (–184ºC). Liquid argon can be shipped more cheaply than gaseous argon. However, equipment must be purchased by the user to vaporize the argon for use in welding.

Argon is used more often than helium for the following reasons:
- It provides easier arc starting.
- It provides a smoother, quieter arc action.
- It requires a lower arc voltage for a given arc length and current. This allows for better control of the weld pool.
- It has reduced penetration.
- It has a lower overall cost and is more readily available.
- It provides a greater metal cleaning action on aluminum and magnesium when used with alternating current.
- It is heavier than helium, and requires a lower flow rate for good shielding.
- Because it is heavier than helium, it provides for better cross-draft shielding.
- Because it requires lower arc voltages to maintain the arc, argon has great advantages for use with thin metal and in out-of-position welding.

Helium (He) is the lightest inert gas. It has an atomic weight of 4 and a density of 0.178 gm/l. Argon therefore is ten times heavier than helium. Helium used for welding is a minimum 99.95% pure. It is normally shipped and used as a gas. The cylinders are similar to those used for oxygen.

Helium is usually used in cylinders of the 330 ft^3 (9.34 m^3) size. Because helium is so light, it requires about two to three times more helium than argon to shield a given weld area.

The chief advantage of helium over argon is that helium can be used with greater arc voltages. Helium also yields a much higher available heat on the metal than is possible with argon. Helium is therefore used to weld thick sections of metal or metals with a high heat conductivity, such as copper or aluminum. Helium also produces greater weld penetration.

To produce the same available heat at the metal, higher currents must be used with argon than with helium. Studies have shown that undercutting will occur at the same current levels with either gas. Helium, therefore, will produce better weld results at higher speeds without undercutting than will argon.

Both helium and argon provide good cleaning action with direct current electrode positive (DCEP). With ac, which is used on aluminum and magnesium, argon provides a better cleaning action. Argon also provides better arc stability with ac than helium.

Argon and helium can be mixed to obtain desired results. Mixtures provide a combination of the results from each gas. Good arc starting and a stable arc can be combined with increased penetration and increased welding speeds.

Argon-hydrogen mixtures are used to produce higher welding speeds. The welding speed possible is in direct proportion to the amount of hydrogen added to argon gas. The addition of hydrogen to argon permits the mixture to carry higher arc voltages. Too much hydrogen will cause *porosity* (gas pockets) in the weld metal. Porosity weakens a weld. Mixtures of 65% argon and 35% hydrogen have been used on stainless steel with a 0.010" to 0.020" (approximately 0.250mm to 0.500mm) root opening. The most common argon-hydrogen ratio is 85% argon and 15% hydrogen. Argon-hydrogen mixtures are used with stainless steels, nickel-copper, and nickel-based alloys. Hydrogen produces negative welding effects with most other metals.

Argon, helium, or argon-helium mixtures can be used for most GTAW welding jobs. Manual welding on thin metals is done best with argon because of the low arc voltages and welding currents required. Whenever there is a cross draft, it is advisable to construct a wind break. This will reduce the possibility of the shielding gas being blown away from the weld area.

13.4.1 Shielding Gas Cylinders

The gases described in the previous heading can be obtained in cylinders of various sizes. These cylinders are similar to the oxygen cylinders described in Chapter 5. They are manufactured to Interstate Commerce Commission (ICC) specifications.

Some gases are stored in cylinders as a gas. Some gases may also be stored as liquids in thermos-like tanks, similar to the Dewar flask shown in Figure 5-3. The quantity of gas in the cylinder is determined by:

- The high pressure gauge or volume scale, if stored as a gas.
- The weight, if stored as a liquid.

Shielding gases, with the exception of hydrogen, are not flammable. Inert gases will not burn, nor will they support combustion. The gases in cylinders are stored under high pressure. **The high pressures in the full cylinders make it necessary to handle the cylinders with care.** Refer to Chapter 1 and to Heading 6.5.2 for more information on handling cylinders.

The cap should be securely threaded over the cylinder valve whenever the cylinders are being moved, or when they are in storage. The cylinder should be fastened to a wall or very stable object when in use. Cylinders should be placed where it is virtually impossible to accidentally damage them with an arc or cutting torch. Gas cylinders should always be stored and used in an upright position.

13.4.2 Inert Gas Regulators

Shielding gas regulators are designed to perform in the same way as the oxygen, acetylene, and hydrogen regulators described in Chapter 5. Review that material for additional understanding.

Inert gas regulators have either a gauge or a pressure indicator to show the cylinder pressure. Some of them have only a flowmeter gauge on the gas delivery side. Figure 13-7 shows a regulator with a high-pressure gauge on the inlet side. On the outlet side, a gauge indicates the volume of gas flowing. This regulator has a constant outlet pressure to the flowmeter of approximately 50 psig (344.7 kPa).

Figure 13-8 shows a regulator and flowmeter. The fitting used to connect the regulator to the cylinder varies with the kind of gas. These regulators will deliver gas flows up to 60 ft^3/hr. (28.3 L/min). The flowmeter scales are accurate only if the gas entering them is at

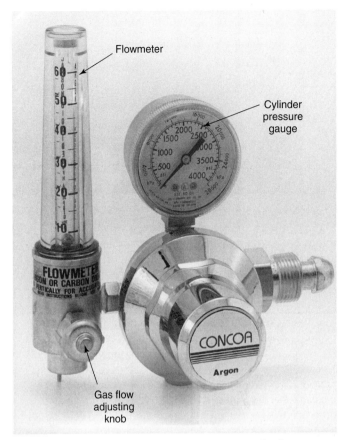

Figure 13-8. *A gas regulator and flowmeter. The low pressure to the flowmeter is preset. The flowmeter has two calibrations. One is for argon and the other for CO_2.* (CONCOA)

approximately 50 psig (345 kPa). If higher inlet pressures are used, the gas flow rate will be higher than the actual reading; the reverse is true if the inlet pressure is lower than 50 psig (345 kPa). It is therefore important to use accurately adjusted regulators. Figure 13-9 illustrates a two-stage regulator for argon gas. The gauge is a high-pressure gauge and is used to indicate the pressure in the cylinder.

A regulator is attached to an inert shielding gas cylinder. Start the threads by hand. Be careful not to cross the threads. Once the threads become snug, use a proper fitting wrench to finish tightening the nut on the regulator. Do not overtighten, just firmly secure the nut.

13.4.3 Flowmeters and Gas Mixers

The amount of gas around the arc can best be measured by the *volume* of gas coming out of the nozzle, rather than the pressure of the gas. Therefore the shielding gas system is usually equipped with a flowmeter that is calibrated in cubic feet per hour (ft^3/hr. or cfh) or in liters per minute (L/min).

In one type of flowmeter, a tapered plastic or glass tube contains a loosely fitted ball. As the gas flows up the tube, it passes around and lifts the ball. The more gas that moves up the tube, the higher the ball is lifted.

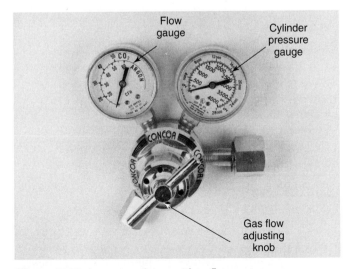

Figure 13-7. *A gas regulator with a flowmeter gauge instead of a low-pressure gauge. The flowmeter gauge may be used with CO_2 or argon. It is calibrated in cubic feet per hour. (CONCOA)*

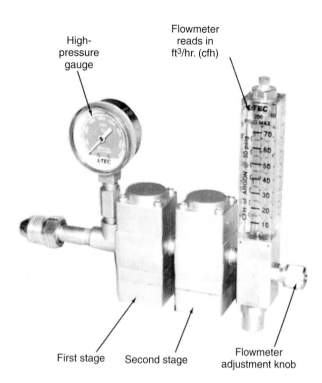

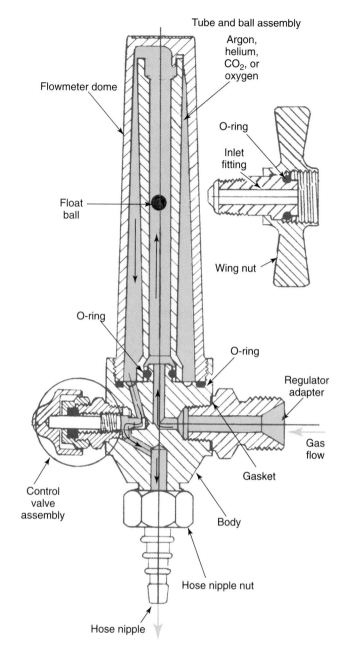

Figure 13-9. *A two-stage regulator with a flowmeter. The gas is fed to the flowmeter at 50 psig (345 kPa). This flowmeter has four separate calibrated gauges to accurately show the flow of argon, helium, carbon dioxide, or nitrogen. (ESAB Welding and Cutting Products).*

Figure 13-10. *A cross section of a floating ball-type flowmeter.*

Figure 13-10 shows a cross section of a tapered-tube flowmeter. For an accurate reading, it is important that this type instrument be mounted in a vertical position. Any slant will cause an inaccurate reading.

The tube and return gas housing are either clear plastic or glass. Some have a metal protecting cover. The joints between plastic tubes and the flowmeter body must be gas-tight. The scale on the inner tube is usually calibrated from 0 ft³/hr. (or cfh) to 60 ft³/hr. (0 L/min to 28.3 L/min). The flowmeter scale is usually read by aligning the top of the ball with the ft³/hr. or L/min reading desired. A different flowmeter or a different scale must be used for each gas. The reason for a different flowmeter or scale is that each shielding gas has a different density or weight. Because the weight of the gas is different, the height to which the ball will be lifted is different for helium, argon, and carbon dioxide, even if all have the same flow rate.

Some universal type flowmeters are available with a 0 to100 scale on the flow tube. A chart accompanying the flowmeter will indicate the flow in cfh (ft³/hr.) or L/min for gases of various densities.

Gas volume is related to pressure if the orifices and gas passages remain constant in size and shape. Therefore, a gas pressure gauge can be used as a volume gauge when the scale is calibrated as shown in Figure 13-6.

Most flowmeters have a needle valve to turn the gas flow on and off and to control the volume of the gas flow. Because this valve controls the volume, and because the pressure to the valve is constant, the valve orifice and the

needle must be accurate and in good condition at all times. Any abuse of the needle and/or seat will result in erratic volume feeds. Also, if the needle threads or packing around the needle have been damaged, the needle will not stay in adjustment. It is necessary that this valve be handled carefully. It should not be forced. *Fingertip handling only* is recommended.

It is sometimes necessary to use mixtures of inert gases in GTAW or GMAW. Most common mixtures are available from local gas distributors in cylinders. For some jobs and in some companies, individual gases are mixed. Such mixing must be done with great accuracy. Gases are mixed by percentages of volume. Each gas to be mixed is fed to a separate flowmeter, where the desired volume of each gas is set. These gases are then mixed by volume in a mixing chamber and fed to the welding torch through a final flowmeter.

13.5 ELECTRODE LEADS AND HOSES USED FOR GTAW

A water-cooled gas tungsten arc welding (GTAW) torch generally has two hoses and one combination hose and electrode lead connected to it. These hoses may be made of rubber or plastic materials. One hose carries the shielding gas to the torch. The second is a combination water hose and electrode lead. The electrode lead is a woven metal tube with excellent current carrying capabilities. This electrode lead is surrounded by a rubber or plastic tube. Welding current travels through the woven metal tube while water travels through the rubber or plastic tube. The third hose carries the water from the torch to the water drain. In a closed water system, the water is returned to the water cooling device and recirculated, Figure 13-11.

Figure 13-12 shows the hoses and fittings used in a typical water-cooled GTAW torch. Figure 13-13 shows a similar torch completely assembled for use.

Light-duty GTAW torches are gas-cooled. These torches have only one hose connected to them. This hose is a combination electrode lead and shielding gas hose. The

electrode lead may be a flexible cable or a woven tube. The shielding gas flows through a rubber or plastic tube which surrounds the electrode lead. Figure 13-14 shows an extremely light-duty GTAW torch which is gas-cooled. Figure 13-15 and 13-16 illustrate light-duty gas-cooled GTAW torches.

Figure 13-11. *A water circulating and cooling unit used with closed-circuit water-cooled systems. The coolant container is available in larger sizes, as well. This unit can be used with GTAW, GMAW, PAW, resistance, and electroslag welding. (Bernard Welding Equipment Co.)*

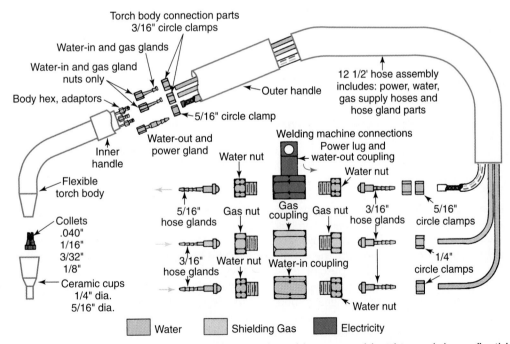

Figure 13-12. *Exploded view of a gas tungsten arc welding torch and hose assembly. This torch has a flexible body that can be bent to suit the job. (Falstrom Co.)*

Figure 13-13. *A water-cooled GTAW torch with its hoses and fittings. The water outlet hose also carries the current to the torch and electrode.*

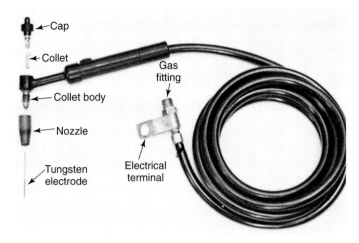

Figure 13-15. *An air-cooled, gas tungsten arc welding torch. A short cap and electrode permit use of this torch in difficult-to-reach places.*

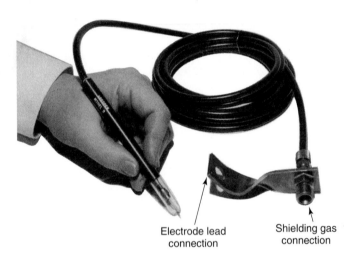

Electrode lead connection   Shielding gas connection

Figure 13-14. *A pen-sized gas tungsten arc welding torch. (Argopen)*

13.6 GTAW TORCHES

Gas tungsten arc welding torches come in many different designs, sizes, amperages, and shapes. Torches designed for lower amperages and light-duty use are cooled by the shielding gas flowing through the torch. Figures 13-14, 13-15, and 13-16 show examples of gas-cooled torches. Current capacity for the largest shielding gas-cooled torches is about 200A. Smaller gas-cooled torches carry significantly less current.

Torches used for higher amperages and those that are used to do a lot of welding at lower amperages are water-cooled. Most torches used in production and most torches used for continuous automatic welding are water-cooled. Figures 13-13 and 13-17 show water-cooled torches. Figure 13-18 shows a cross section of a water-cooled torch. Figure 13-19 shows how the different parts of a water-cooled GTAW torch are assembled. These parts are also shown in

Figure 13-20. Figure 13-15 shows some of the torch parts being assembled.

There are two basic torch designs. These are:
- Torches designed for general access. These torches usually use a long electrode and have a long cap. They also use standard collets, collet bodies, and nozzles. See Figure 13-17.
- Torches designed for work in areas where space is limited. These torches are called stubby or stubs. They use a very short electrode and have a very short cap. These also use stubby collets, stubby collet bodies, and stubby nozzles. See Figure 13-16.

There are variations of the general access design. These variations allow a welder to select different parts of the torch for a specific job.

13.6.1 Collets, Collet Bodies, and Gas Lenses

Tungsten electrodes are held in a *collet.* The function of the collet is to firmly hold the electrode and make an electrical connection to it. Collets are made with various inside diameters. A collet must be selected to match the diameter of electrode that is being used. Figure 13-21 shows a typical collet.

A collet is installed into the top of a torch assembly, as shown in Figure 13-22. To make a collet tighten around the electrode, the cap is tightened. Tightening the cap forces the collet against the collet body, squeezing the collet. The inside diameter of the collet reduces in size, firmly gripping the electrode.

The *collet body* locates the collet and transfers the electrical current to the collet. The collet in turn transfers the electrical current to the electrode. The collet body is threaded into the end of the torch. Examples of collet bodies are shown in Figures 13-19 and 13-20. Collet bodies are usually made for a single size of collet. When the electrode diameter is changed, both the collet and the collet body need to be changed.

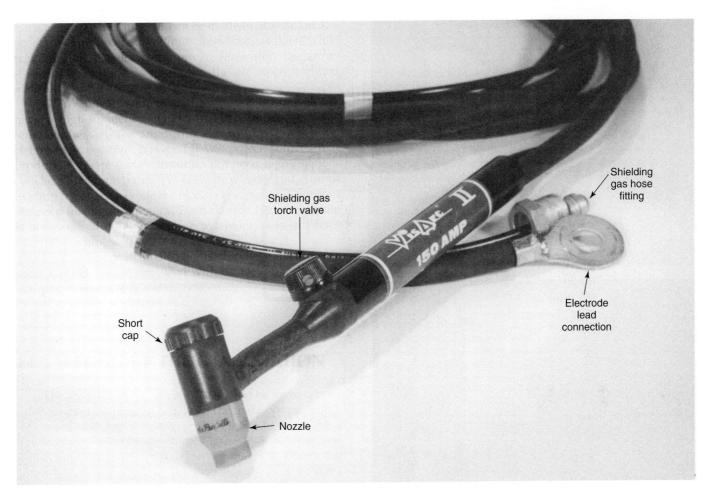

Figure 13-16. *An air-cooled GTAW torch that uses a short cap and stub length tungsten electrodes. This torch can be used in confined spaces. The valve allows the welder to control the shielding gas flow at the torch. Notice that the gas hose and electrode lead are separate. (Air Products & Chemicals, Inc.)*

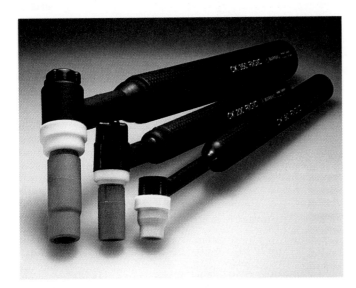

Figure 13-17. *Three GTAW torches. The two torches to the left are water-cooled; the torch at the right is gas-cooled. Note the difference in length from the nozzle to the cap. (CK Worldwide, Inc.)*

A collet body with a *gas lens* can be very useful to a welder. The purpose of a gas lens is to make the shielding gas exit the nozzle more as a *column* than as a *turbulent stream* of gas that begins to spread out after exiting. See Figure 13-23. The column of gas allows the electrode to stick out farther for visibility and for better access to the weld area.

A gas lens is a series of stainless steel wire mesh screens. The collet body with gas lens is installed in the torch in place of a standard collet body. See Figure 13-19. A different gas lens is required for each different electrode diameter. Different nozzles are also required because the gas lens has a larger diameter than a standard collet body.

13.6.2 Nozzles

Gas nozzles are used to direct the shielding gas over the tungsten electrode and to cover the weld area with shielding gas. Nozzles must be able to withstand very high temperatures because they are very close to the arc. Different designs are available to meet the requirements of different welding jobs.

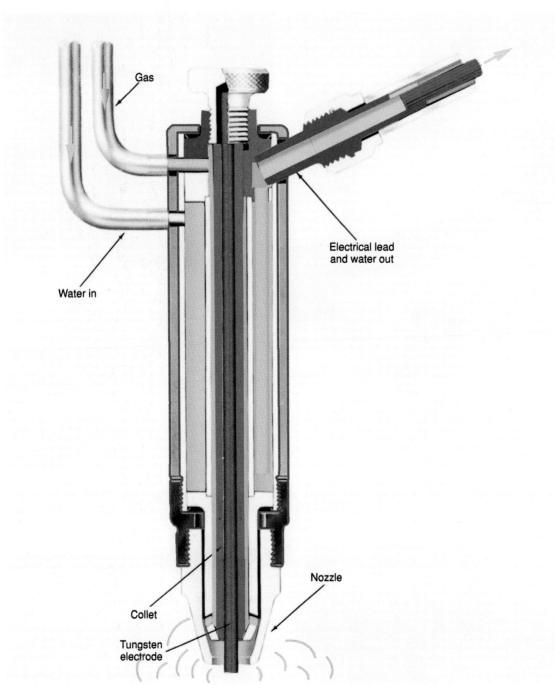

Figure 13-18. *A cross section of a water-cooled gas tungsten arc welding torch designed for automatic welding. This torch is designed with a current capacity of up to 500A. (Weldma Co.)*

Nozzles are made from different materials. Most commonly, nozzles are made from a ceramic material. Other materials used include metal-jacketed ceramics, metal, and fused quartz.

One end of a gas nozzle must attach to the end of the torch. Different designs are necessary to properly attach to different manufacturers' torches. The nozzle may be threaded onto the torch or held by friction.

The exit end of the GTAW nozzle is more standard. The exit diameter is identified with a number. The number is the exit diameter measured in 1/16" (1.6mm) incre-

ments. A number 6 nozzle thus has a diameter of 3/8" (9.6mm):

6 x 1/16" = 6/16" or 3/8" (6 x 1.6mm = 9.6mm)

A number 8 nozzle diameter is 1/2" (12.8mm):

8 x 1/16" = 8/16" or 1/2" (8 x 1.6mm = 12.8mm)

Nozzle exit diameters must be the correct size for the job. If they are too small, they will not allow the shielding gas to cover the weld area. They cannot be too large, or the velocity of the gas coming out will be too slow and will easily be blown away. A high velocity and a small diame-

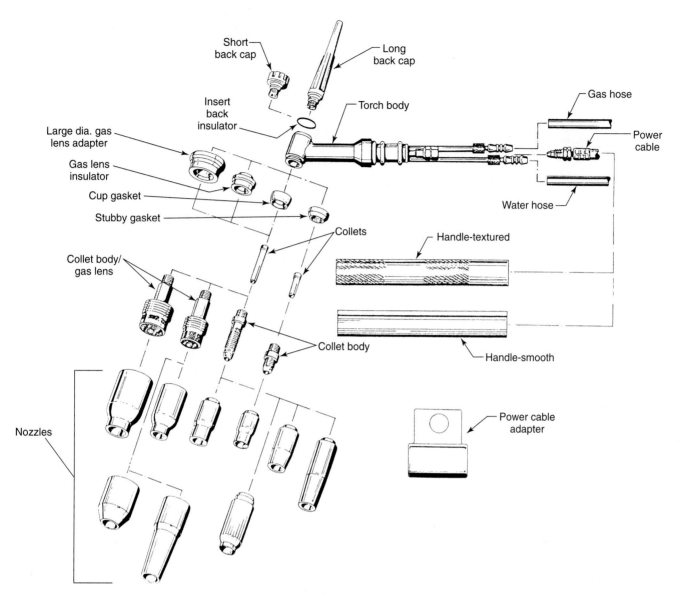

Figure 13-19. The various parts in a typical water-cooled GTAW torch, and how they fit together.

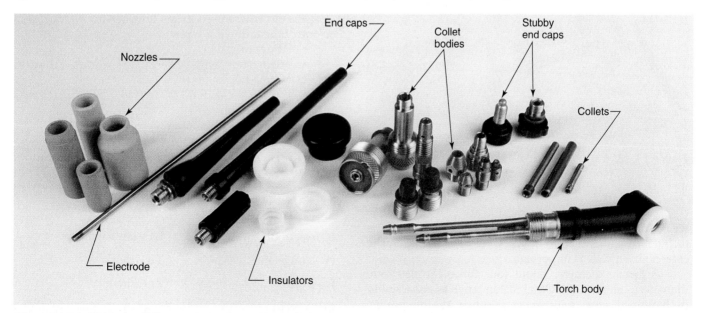

Figure 13-20. Note the different sizes of nozzles, collets, and collet bodies available for a typical water-cooled GTAW torch. (American Torch Tip Co.)

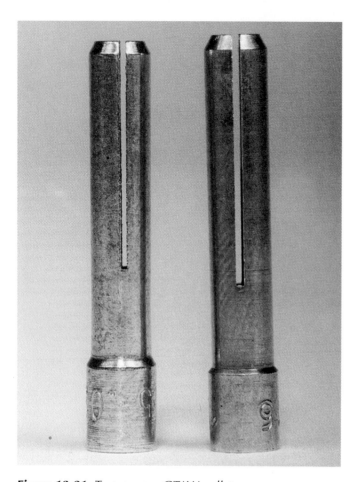

Figure 13-21. *Two copper GTAW collets.*

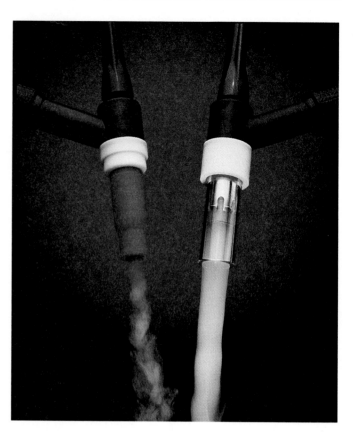

Figure 13-23. *The torch at the right has a gas lens installed. The gas flow from the conventional GTAW nozzle at the left is turbulent and dissipates more rapidly. (CK Worldwide, Inc.)*

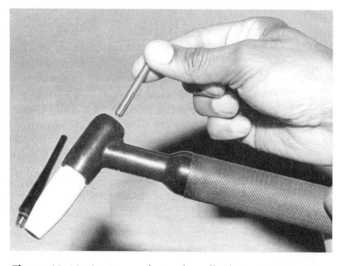

Figure 13-22. *A copper electrode collet being installed in a GTAW torch. Note the long electrode cap in the background.*

ter is also not good, because air may be caught up in the turbulence and contaminate the weld area. Selecting the correct size nozzle is important.

A nozzle diameter about four times the electrode diameter is about right. Other factors, such as accessibility of the weld area, may affect selection of the nozzle.

Different nozzle styles can be seen in Figure 13-20 and other figures in this chapter.

13.7 TUNGSTEN ELECTRODES

The *electrodes* used in gas tungsten arc welding may be one of the following types:
- Pure tungsten.
- Tungsten with 1% or 2% thoria (thorium dioxide).
- Tungsten with 0.15% to 0.40% zirconia (zirconium oxide).
- Tungsten with 2% ceria (cerium oxide).
- Tungsten with 1% lanthana (lanthanum oxide).

The chemical composition of tungsten electrodes, from the AWS A5.12 specification, is shown in Figure 13-24. Letters and numbers used in tungsten electrode classifications are interpreted as follows:

E – electrode	**La** – lanthana
W – tungsten	**Th** – thoria
P – pure	**Zr** – zirconia
Ce – ceria	

AWS classification	Tungsten min. percent (by difference)	Percent				
		Ceria (CeO)	Lanthana (LaO)	Thoria (ThO)	Zirconia (ZrO)	Total other elements or oxides, max.
EWP	99.5	–	–	–	–	0.5
EWCe-2	97.3	1.8 - 2.2	–	–	–	0.5
EWLa-1	98.3	–	0.9 - 1.2	–	–	0.5
EWTh-1	98.3	–	–	0.8 - 1.2	–	0.5
EWTh-2	97.3	–	–	1.7 - 2.2	–	0.5
EWZr-1	99.1	–	–	–	0.15 - 0.40	0.5
EWG	94.5	Not Specified				

Figure 13-24. The chemical composition of the AWS classifications for tungsten electrodes. (AWS A5.12)

Pure tungsten electrodes are the least expensive. However, they carry less current than alloyed electrodes. Pure tungsten electrodes are usually used only with ac welding. Pure tungsten electrodes may split or break down and cause inclusions (tungsten) in the weld if used with excessive current.

Electrodes with thoria added carry more current than pure tungsten or zirconia electrodes. It is easier to strike an arc and maintain a stable arc with thoria electrodes. These electrodes have a greater resistance to contamination of the weld. Electrodes with thoria added are usually used only with direct current.

Zirconia added to tungsten electrodes gives the electrode qualities which fall somewhere between pure tungsten and tungsten with thoria added. Tungsten electrodes with zirconia are the best electrode to use when ac welding aluminum or magnesium. They do not cause inclusions as pure tungsten electrodes can. For this reason, tungsten electrodes with zirconia are used for high quality applications.

Electrodes with ceria or lanthana are very similar to electrodes with thoria. They can be used with ac or dc. They allow for a stable arc; arc starting is easy.

Since these electrodes are identical in appearance, an approved color code is used to identify the type of electrode. The color is a painted band near one end of the electrode or a color painted on the end of the electrode. The AWS color codes are listed in Figure 13-25.

Tungsten electrode diameters are available in diameters of 0.010", 0.020", 0.040", 1/16", 3/32", 1/8", 5/32", 3/16", and 1/4" (0.25, 0.51, 1.02, 1.59, 2.38, 3.18, 3.97, 4.76, and 6.35mm). Electrodes come in lengths of 3", 6", 7", 12", 18", or 24" (76, 152, 178, 305, 457, or 610mm). The surface of a tungsten electrode, when purchased is either ground or chemically cleaned.

13.7.1 Care of Tungsten Electrodes

A tungsten electrode must be straight. If it is bent or off-center in the nozzle, the arc may wander to one side and produce uneven penetration. Contamination of the weld may also occur. Make sure the electrode is straight and is prepared correctly.

Sometimes the electrode comes in contact with the molten weld pool. This contaminates the tungsten electrode. The end of such an electrode must be removed to form a new clean end. A pair of pliers may be used to break off the dirty end of the electrode. Another method is to use a grinding wheel to notch the electrode just behind the contamination and then break the electrode using pliers. This method of notching the electrode is shown in Figure 13-26A.

Tungsten electrodes must be clean and must have good electrical contact with the collet. The tungsten should be adjusted to extend about 1/8" (1.6mm) beyond the end of the nozzle.

It is extremely important that shielding gas always protect the electrode and the weld area. Shielding gas hose connections must be tight to prevent air or moisture from mixing with the shielding gas, then coming in contact with the electrode. Such contamination of the shielding gas would be harmful to the weld and to the electrode.

Preparing the electrode end for welding is very important. Refer to Heading 14.3.5 on selecting and preparing a tungsten electrode for welding.

AWS classification	Defined	Color
EWP	Pure tungsten	Green
EWCe-2	2% Ceria	Orange
EWLa-1	1% Lanthana	Black
EWTh-1	1% Thoria	Yellow
EWTh-2	2% Thoria	Red
EWZr-1	0.15-0.40% Zirconia	Brown
EWG	Other	Gray

Figure 13-25. The AWS color code for various types of tungsten electrodes. (AWS A5.12)

13.8 FILLER METALS USED WITH GTAW

Gas tungsten arc welding is used to weld almost all weldable metals. The choice of filler metals, therefore, is very large.

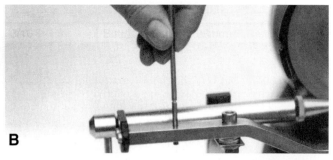

Figure 13-26. *Three steps are required to prepare a tungsten electrode for welding. A—Notching the electrode prior to breaking. B—Breaking the electrode in special fixture. C—Grinding the electrode in a special grinder. (Ind. Schweisstechnic E. Jankus)*

Figure 13-27 lists the American Welding Society (AWS) filler metal specification booklets for each type of base metal welded. Each "A5" specification booklet contains a list of filler metals, their suggested uses, and current settings in addition to their chemical compositions. Filler metal is available in various diameters. The filler metal can be purchased cut to length, usually 24" (610mm) or 36" (915mm), or as a continuous coil. Continuous coil is used

AWS specification number	Booklet Title
A5.2	Iron and steel gas-welding rods.
A5.7	Copper and copper alloy bare welding rods and electrodes.
A5.9	Corrosion-resisting chromium and chromium-nickel steel bare and composite metal cored and stranded arc welding electrodes and welding rods.
A5.10	Aluminum and aluminum alloy welding rods and bare electrodes.
A5.13	Surfacing welding rods and electrodes.
A5.14	Nickel and nickel alloy bare welding rods and electrodes.
A5.16	Titanium and titanium alloy bare welding rods and electrodes.
A5.18	Mild steel electrodes for gas shielded arc welding.
A5.19	Magnesium-alloy welding rods and bare electrodes.
A5.21	Composite surfacing welding rods and electrodes.
A5.24	Zirconium and zirconium alloy bare welding rods and electrodes.

Figure 13-27. *A list of AWS specifications and booklets for filler metals that may be used in GTAW and GMAW.*

in semiautomatic, mechanized, or automatic welding operations. GTAW with continuous coil filler metal requires a wire feeder and a method to guide the filler metal to the weld pool. Figure 13-28 shows equipment used to feed GTAW wire.

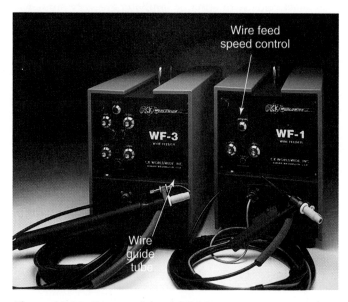

Figure 13-28. *Two models of GTAW power sources with an automatic wire drive to provide filler metal to the weld pool during the welding process. Note the wire guide tube. (CK Worldwide, Inc.)*

13.9 THE GAS METAL ARC WELDING STATION

Gas metal arc welding (GMAW) uses the equipment listed below to make a weld. This equipment is used to feed a continuous wire electrode to the weld. As the electrode melts, it also melts the base metal to form a weld.

The typical GMAW outfit contains the following equipment and supplies:
- A constant voltage dc arc welding power source.
- An electrode wire feed mechanism.
- Shielding gas cylinders. See Heading 13.4.1.
- A gas regulator. See Heading 13.4.2.
- A shielding gas flowmeter. See Heading 13.4.3.
- Shielding gas, coolant hoses and fittings, and the electrode and workpiece leads.
- A GMAW welding gun.
- Electrode wire.
- Optional equipment.
    - A coolant system. See Figure 13-11.
    - Remote controls.

Figure 13-29 is a schematic drawing of a GMAW outfit. A complete station would include the arc welding booth, a ventilation system, and a welding bench. Refer also to Figure 4-12.

13.10 ARC WELDING POWER SOURCES FOR GMAW

Gas metal arc welding (GMAW) uses a welding power source and a wire feeder that feeds the electrode wire through the welding gun to the weld. Welding power sources used for GMAW produce a constant voltage. The following types of power sources are used when gas metal arc welding:
- Transformer-rectifier type. See Figure 13-30.
- Inverter type. See Figure 13-31.
- Engine-driven generator type. See Figure 13-32.

The preferred power sources are the transformer-rectifier and inverter type machines. They provide better response and control over the welding process then an engine-driven generator machine. When welding where no source of electrical power is available, however, an engine-driven generator type machine must be used. See Chapter 11 for information on the construction, operation, and control of these machines.

Direct current electrode positive (DCEP), also called reverse polarity (DCRP), is used when GMAW. When GMAW, constant voltage is required. A constant voltage power source will maintain a constant arc voltage, which means it will maintain a constant arc length. To change the arc length, the set voltage must be changed. The output of a constant voltage power source has a very flat volt-ampere curve, as shown in Figure 13-33.

Constant voltage power sources have the ability to self adjust to maintain a constant arc length. When the welding gun is moved closer to the work, the arc length should decrease. However, the machine will deliver more current to burn off the electrode faster and maintain the arc length determined by the set voltage. Also if the welding gun is moved further from the work, the machine will deliver less current, so the electrode will burn off more slowly. By burning off the wire at a slower rate, the arc length and voltage remain constant.

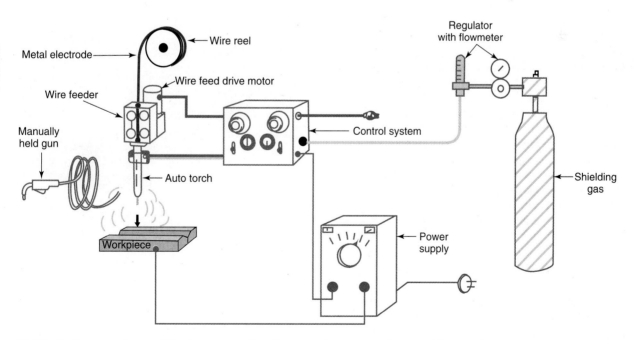

Figure 13-29. A diagram of a combination manual and automatic gas metal arc welding outfit.

Figure 13-30. *A transformer-rectifier type constant voltage (cv) power source for GMAW. (Hobart Brothers Co.)*

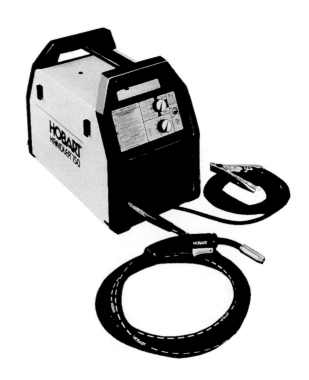

Figure 13-31. *An inverter type power source for GMAW. (Hobart Brothers Co.)*

Figure 13-33 shows how large a current change will occur when the voltage changes only slightly. A welder sets a welding machine and wire feeder to weld at 22V and 200A (Point B on Figure 13-33). If the gun is moved closer to the work, the voltage will attempt to decrease to 19V. To prevent this change in voltage, the machine will increase the current automatically to 300A. See Point C in Figure 13-33. This increase in current will quickly burn off the electrode wire and actually maintain the voltage at 22V. If the gun is moved away from the work and the voltage attempts to increase to 25V, the current will decrease. The electrode will burn off at a slower rate. This will allow the arc length and the voltage to remain very close to their set value of 22V.

The slope of the volt-ampere curve is important. Some welding machines allow the slope to be changed; others machines have only one preset slope. Electronically controlled machines have different slopes preset for common types of metal and methods of transfer. The slope of a machine is very important when using short circuiting transfer. Short circuiting transfer and other transfer methods are discussed in Headings 15.1.1 through 15.1.4.

Figure 13-34 shows various slopes for GMAW power sources. Nonferrous metals and large-diameter flux cored electrode wires use a slope of 1.5V to 2V per 100A. A medium slope of 2V to 3V per 100A is used for GMAW with carbon dioxide (CO_2) gas and small-diameter flux cored electrode wires, as shown in Figure 13-34B. A steeper slope, Figure 13-34C, has a slope of 3V to 4V per 100A. This steeper slope is recommended for short circuiting arc transfer.

A slope of 1.5V to 2V per 100A is used with large-diameter electrode wires to allow a large current change to burn off the large-diameter electrode wire. A steeper slope of 3V to 4V per 100A is used when using short circuiting

Figure 13-32. *An engine-driven generator type power source. (Miller Electric Mfg. Co.)*

transfer. This prevents excessive current when the short circuit occurs. If too-steep a slope is used with short circuiting transfer, the molten metal drop will not separate from the electrode and the arc will not restart.

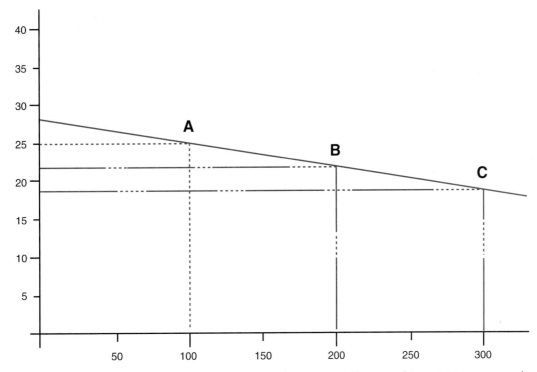

Figure 13-33. *Characteristic volt-ampere curve for a constant voltage arc welding machine. A 100-ampere change (from 200A to 300A) results from a voltage change of 3V from 22V to 19V.*

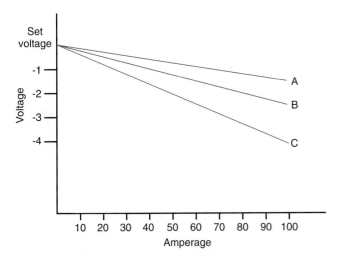

Figure 13-34. *Volt-ampere curves for various welding machine slope settings. Curve A—1.5V to 2V per 100A for GMAW of nonferrous metals. Curve B—2V to 3V per 100A for GMAW with CO_2. Curve C—3V to 4V per 100A for GMAW short circuiting arc transfer.*

13.11 WIRE FEEDERS USED WITH GMAW

The wire electrode used in gas metal arc welding comes in large coils (spools) or drums. These wire coils or drums contain hundreds or even thousands of feet (meters) of wire.

Small coils of wire are mounted on free-turning axles near the wire feed mechanism. If the larger drums of wire are used, the drum is placed near the wire feeder. The wire is smoothly drawn from either the coil or drum.

Wire feeders consist of a coil-mounting device, a set of drive wheels for the wire, and an adjustable, constant-speed motor to turn the wire drive wheels. Figure 13-35 shows a wire feeder with the drive wheels outside the wire

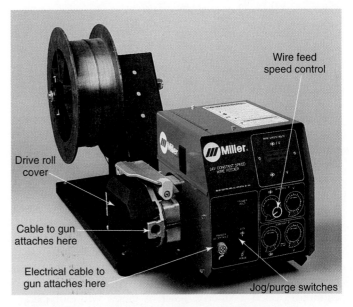

Figure 13-35. *A GMAW wire feeder and controls. The drive rolls are under the black cover at the left. (Miller Electric Mfg. Co.)*

feed control unit. Both drive wheels have gear teeth on the outside circumference. The driven wheel moves the second wheel by means of the gear teeth. One of the wheels is driven by the adjustable dc type constant-speed motor. The tension on the wire passing between these wheels is adjusted by moving the wheels closer together or farther apart. This is done by turning a tensioning device.

Sometimes, wire feeders are contained inside the GMAW machine. Figure 13-30 shows such a machine. Other welding outfits have the wire drive unit separate from the welding machine. See Figure 13-36.

A switch on the wire drive control panel is used to rotate the drive motor slowly. This is done to move the wire slowly through the electrode lead to the gun. The wire is moved slowly so that it does not bend or kink within the electrode cable. The switch on the wire drive control panel is called an *inch switch* or a *jog switch*. See the switch in Figure 13-35.

The shielding gas hose may have air in it before its first use after a long shutdown period. To clear or purge the hose and gun, the shielding gas is turned on for a short period prior to welding. A *purge switch* may be provided for this purpose on the wire feed control box.

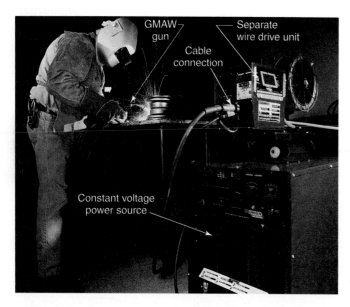

Figure 13-36. *A constant voltage (cv) power source with a separate wire feeder are being used to GMAW a small steel wheel. (The Lincoln Electric Co.)*

13.12 GMAW SHIELDING GASES

The shielding gases used with gas metal arc welding (GMAW) may be inert, reactive, or mixtures of both types of gases. Inert gases used are argon (Ar) and helium (He). These gases, being *inert*, will not react with other chemical elements. *Reactive gases* will react with other chemical elements. Used properly, reactive gases will not cause defects in welds.

The reactive gases used in GMAW are carbon dioxide (CO_2) and oxygen (O_2). Hydrogen (H) and nitrogen (N_2) are used, but only in highly specialized applications. Their use results in better control of penetration. Hydrogen and nitrogen will cause embrittlement and porosity in the welds on most metals.

Mixtures of argon and helium; argon and oxygen; argon and carbon dioxide; and helium, argon, and carbon dioxide are used. The gas or gas mixture used will be determined by the metal being welded and the type of arc transfer method desired. See Headings 15.1.1 through 15.1.4 for an explanation of metal transfer methods.

Pure argon (Ar) and helium (He) are excellent gases for protecting the arc, metal electrode, and weld metal from contamination. They are not, however, as suitable for some GMAW processes as mixtures of gases. Gas mixtures seem to create arc stability, reduce spatter, and improve the bead contour.

Reactive gases like carbon dioxide, oxygen, and nitrogen are not practical to use alone as shielding gases. Carbon dioxide is the exception. It is inexpensive and works well on carbon and low-alloy steels. Carbon dioxide generally costs about one-tenth as much as pure argon gas.

13.12.1 Argon and Helium Gases Used in GMAW

Argon and helium are the gases generally used with nonferrous metals, as seen in Figure 13-37. Helium conducts heat better in the arc than argon. Helium is used when high heat input is required in a welding application. Helium gas, therefore, is chosen for use on thick metals. It is also used on metals, like copper and aluminum, which conduct heat away from the weld area rapidly. When welding thin metal and metals that conduct heat poorly, argon is a good choice. Argon is often used for out-of-position welding because of its lower heat conductivity. Because argon is ten times heavier than helium, it shields better. Less gas is required to provide a good shield. Argon costs less than helium, and because less argon is required to obtain good shielding, the cost of using argon as a shielding gas is a lot less than using helium.

The weld bead contour and penetration are also affected by the gas used. Welds made with argon generally have deeper penetration. They also have a tendency to undercut at the edges. Welds made with helium generally have wider and thicker beads. Figure 13-38 shows the shape of welds made with various gases and gas mixtures.

Argon used with the gas metal arc spray transfer process tends to produce deeper penetration through the centerline of the bead. True spray transfer occurs only when an argon or very high percentage argon shielding gas is used. For spray transfer in steel, the amount of argon must be at least 90%. The GMAW globular and short circuiting arc metal transfer methods produce wider beads with shallower penetration.

13.12.2 Gas Mixtures Used in GMAW

Welds made with pure argon have deep penetration. On carbon and low-alloy steels, the filler metal tends to

Shielding gas or mixture	Chemical behavior	Metals and applications
Argon	Inert	Virtually all metals except steels.
Helium	Inert	Aluminum, magnesium, and copper alloys for greater heat input and to minimize porosity.
Ar + He (20-80% to 50-50%)	Inert	Aluminum, magnesium, and copper alloys for greater heat input and to minimize porosity (better arc action than 100% helium).
Nitrogen		Greater heat input on copper (Europe).
Ar + 25-30% N_2		Greater heat input on copper (Europe); better arc action than 100% nitrogen.
Ar + 1-2% O_2	Slightly oxidizing	Stainless and alloy steels; some deoxidized copper alloys.
Ar + 3-5% O_2	Oxidizing	Carbon and some low-alloy steels.
CO_2	Oxidizing	Carbon and some low-alloy steels.
Ar + 20-50% CO_2	Oxidizing	Various steels, chiefly short circuiting arc.
Ar + 10% CO_2 + 5% O_2	Oxidizing	Various steels (Europe).
CO_2 + 20% O_2	Oxidizing	Various steels (Japan).
90% He + 7.5% Ar + 2.5% CO_2	Slightly oxidizing	Stainless steels for good corrosion resistance, short circuiting arc.
60-70% He + 25-35% Ar + 4-5% CO_2	Oxidizing	Low-alloy steels for toughness, short circuiting arc.

Figure 13-37. *Shielding gases used with various metals. (American Welding Society)*

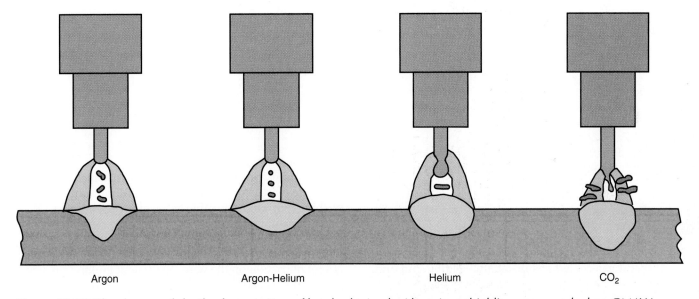

Argon   Argon-Helium   Helium CO_2

Figure 13-38. *The shape and depth of penetration of beads obtained with various shielding gases used when GMAW.*

draw in from the toe of the weld (weld edge) when argon is used as the shielding gas. This can cause undercutting. To overcome this tendency of the filler metal not to flow out to the edges of the weld, a gas mixture is used. For steels, oxygen (O_2) or carbon dioxide (CO_2) is added to argon. This addition of O_2 or CO_2 to argon tends to cause better metal transfer and flow. It also reduces metal spatter and stabilizes the arc.

As little as 0.5% oxygen will cause noticeable improvement. Additions of 1% to 5% oxygen or 3% to 10% carbon dioxide are usually used. Figure 13-37 shows some gas mixtures and their applications.

Adding oxygen or carbon dioxide to argon or helium causes the shielding gas to become slightly oxidizing. This may cause porosity in some ferrous metals. To offset this oxidizing tendency, a deoxidizer is added to the electrode wire.

13.12.3 Carbon Dioxide Gas

Carbon dioxide is used only on carbon steel and low-alloy steels. Carbon dioxide (CO_2) reduces to carbon monoxide (CO) and oxygen (O_2) in the arc. However, these gases return to CO_2 as they cool.

Carbon dioxide is not an inert gas. It will react with the base metal and oxidize the base metal. When using CO_2, an electrode with alloys to eliminate this oxidation must be chosen.

Carbon dioxide is 50% heavier than air and its ability to shield the arc is quite good. Moisture-free carbon dioxide must be used, or the hydrogen generated while welding will cause weld porosity and brittleness.

Carbon dioxide has a rather high electrical resistance and it, therefore, has a rather critical arc length. Even small changes in the arc length will produce spattering and a "wild arc." A very short and constant arc length must be maintained. Carbon dioxide is furnished in liquid form in 50 lb. (23kg) cylinders. These cylinders are approximately 9" (229mm) in diameter and 51" (1.30m) high and weigh 105 lb. (48kg), when empty. Each pound (0.45kg) of liquid will furnish 8.7 ft³ (0.25m³) of gas, which is equal to 435 ft³ (12.3m³) per cylinder. It must change from a liquid to a gas as it is being used. This change from liquid to gaseous CO_2 is dependent on the room temperature. When the delivery line from the liquid tank is opened, carbon dioxide gas "boils" or bubbles out of the liquid. The expansion of the gas as it leaves the liquid and passes through the regulator causes the CO_2 gas to cool. If moisture is present, it may condense and freeze in the regulator, blocking the gas passage. Excessive moisture may also be indicated by erratic flowmeter operation. It is recommended that CO_2 with a –20°F (–29°C) or lower dew point be used.

One cylinder can furnish only about 35 ft³/hr.(16.5 L/min) when the cylinder is in a 70°F (21°C) room. Sometimes, two or more cylinders must be connected in parallel to furnish enough gas for welding. The pressure in the cylinder when liquid is present is about 835 psig at 70°F (5760 kPa at 21°C).

Carbon dioxide is also available in the gaseous form in cylinders. These cylinders are similar to the oxygen cylinders described in Chapter 5.

13.13 THE GMAW WELDING GUN

A typical gas-cooled gas metal arc welding gun and its electrode lead and gas hose are shown in Figure 13-39. The gas hose, the electrode lead, and the electrode wire carrier, called a *liner*, are all built into one large cable. The gun has a fully insulated handle, a trigger switch, a current contact tube, and a nozzle. The trigger switch turns on the current and wire feeder and starts the shielding gas flowing. Figure 13-40 shows the main parts of a gas-cooled GMAW gun.

Gas metal arc welding guns are usually gas-cooled. Gas flowing through the gun and air on the outside of the welding gun keep the gun from overheating. Carbon dioxide GMAW guns may be used for intermittent duty up to 600A. Figure 13-41 shows a gas-cooled GMAW gun. Gas metal arc welding using a gas-cooled gun with argon or helium must be done at much lower currents than are possible with carbon dioxide.

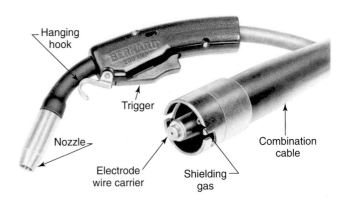

Figure 13-39. *A gas-cooled GMAW gun. Note how the electrode wire travels through the center of the combination cable. (Bernard Welding Equipment Co.)*

GMAW guns are water-cooled when used with argon or helium above 300A or for continuous duty. Torches used with CO_2 above 600A, or for continuous duty, are also water-cooled. Figure 13-42 shows a water-cooled GMAW gun. In addition to an electrical lead and a shielding gas hose, a water-cooled gun must have two additional hoses to carry water to the gun and back.

Gas metal arc welding guns that have self-contained wire drive units are available. See Figures 13-43 and 13-44. These guns have an electric motor to drive the electrode wire and a small coil of wire enclosed in the gun.

Another type of gun that is used is called a pull-type welding gun. When welding long distances from the wire feeder or when welding using aluminum wire (especially smaller diameter aluminum wire), a pull-type welding gun is used. See Figure 13-44.

A pull-type gun has a motor in it, which pulls the electrode wire. Attempting to feed aluminum wire by pushing it through the cable to the gun will cause it to kink or bend. Aluminum electrode wire is not rigid enough to be pushed. Kinking can also happen when feeding steel or stainless steel electrode wire a long distance, because there is a lot of friction on the wire. To prevent kinking the wire, a pull gun is used. The pull-type gun pulls the wire while the regular wire drive unit pushes the wire. This is a push-pull set-up. Figure 13-45 shows a push-pull system that includes a power source with a wire feeder and a pull-type welding gun. Welding can be done 50' (16m) from the wire feeder.

Welding guns used for automatic GMAW often have a straight body and nozzle. They do not have a handle. They are firmly attached to a carriage, robot, or other mechanism for welding, as shown in Figure 13-46. Since automatic welding is done almost continuously, the gun is almost always water-cooled. Gas-cooled guns are used when welding with lower amperages that will not cause them to overheat. Electrode wire and shielding gas is fed through the gun body to the arc area in the same manner as shown in Figure 13-40.

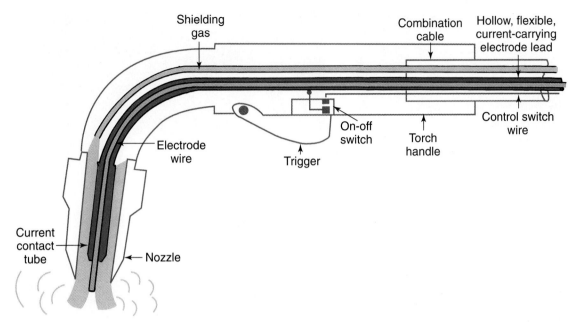

Figure 13-40. *A cross sectional drawing of an air-cooled GMAW gun. The shielding gas, electrode wire, and control switch wire are carried in a combination cable. The electrode wire runs through the hollow electrode lead.*

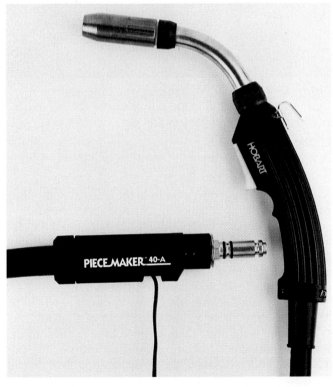

Figure 13-41. *A gas-cooled GMAW gun. The flow of shielding gas helps to cool the gun. (Hobart Brothers Co.)*

13.13.1 Contact Tubes, Nozzles, and Liners

Some of the parts of the welding gun need to be changed or replaced on occasion. These include the contact tube, nozzle, and liner.

The *contact tube* is the part that transfers the electrical current from the welding gun to the electrode wire.

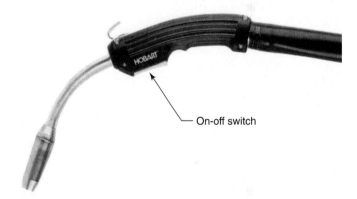

Figure 13-42. *A water-cooled GMAW gun. The gas and inlet water hoses and combination power and water outlet hose are under a plastic cover to protect them. (Hobart Brothers Co.)*

Contact tubes, also referred to as *contact tips*, are made from copper or copper alloys. The contact tube must have a hole diameter 0.005" to 0.010" (0.13mm to 0.25mm) larger than the diameter of the electrode wire being used. This is necessary to make good contact. Each contact tube is usually stamped with the electrode diameter it is designed for. Figure 13-47 shows different contact tubes, nozzles, and adaptors and how they are assembled.

The electrode wire continuously rubs against the contact tube, causing wear that enlarges the hole. As the hole increases in size, it makes less solid contact with the electrode wire and allows the electrode and arc to wander. This will cause problems when welding. The contact tube must be replaced when it gets worn.

Figure 13-43. *A self-contained wire feeder and GMAW gun. A small spool of wire, drive motor, and drive wheels are contained within this type of gun.*

Figure 13-44. *Three pull guns. These guns pull the wire while the wire drive at the power source pushes it. The gun at bottom may also be used as a self-contained wire drive gun. (M.K. Products, Inc.)*

Nozzles used for GMAW direct the shielding gas to the weld area. Nozzles are exposed to very high temperatures for long periods of time. Nozzles are usually made from copper or copper alloy. The exterior may be bright chrome-plated to reflect the heat and keep the nozzle

Figure 13-45. *A GMAW push-pull wire drive system. A push-type wire drive is in the power source cabinet and a pull drive is built into the welding gun. (M.K. Products, Inc.)*

cooler. The nozzle is insulated from the torch. Usually, they have an insulator permanently installed on the inside.

Nozzle designs vary. One end must fit onto the welding gun being used. Nozzles are attached to the welding gun using two different methods. One is by threading the nozzle onto the gun. The second way is to use an adaptor and slide-on nozzle. Both types are shown in Figure 13-47.

The exit end of the nozzle also varies. It needs to be large enough to allow a continuous flow of shielding gas to cover the weld area. It should not have spatter inside, since this will affect the flow of the shielding gas. The exit diameter is selected based on the application. Nozzle exit diameters are measured in 1/16" (1.6mm) increments. An example is a number 6 nozzle which has a 3/8" (6 × 1/16") or 9.5mm (6 × 1.6mm) exit diameter.

A *liner* is installed in the cable that carries the electrode from the wire feeder to the welding gun. The electrode wire is constantly moving through this cable and would wear out the cable. To protect the cable, a liner is installed. For most applications, a coiled steel wire liner is used. This coiled steel liner is sometimes called a conduit. When using aluminum electrode wire, a nylon liner is used. Figure 13-48 shows a coiled steel liner.

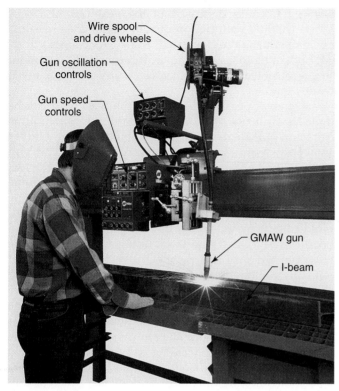

Figure 13-46. *A GMAW gun mounted on a rigid beam in order to make linear (straight line) welds on a structural "I" beam. This GMAW system has an oscillator which moves the gun from side to side along the weld line. (Miller Electric Mfg. Co.)*

13.14 GMAW ELECTRODES

Bare metal electrodes used for gas metal arc welding are small in diameter. Diameters range from 0.020" to 0.125" (0.51mm to 3.18mm). The most common diameters are 0.030", 0.035", and 0.045" (0.76mm, 0.89mm, and 1.14mm). Electrode wire is sold in spools of several hundred feet or meters, or in drums containing up to several thousand feet or meters of wire coiled inside.

A variety of wire compositions are available for use as GMAW electrodes. Figure 13-27 lists the AWS specifications for electrodes.

Deoxidizers are added to most electrode wires. This is done to reduce the porosity in the finished weld. During the welding process, the deoxidizer added to the wire will react with any oxygen, nitrogen, or hydrogen. The deoxidizers reduce the possibility of these gases producing porosity, which would lower the mechanical strength of the weld.

The American Welding Society chemical composition specifications for carbon steel filler metal used with GMAW are shown in Figure 13-49. Figure 13-50 lists the AWS filler wire specifications for GMAW low-alloy steels.

The numbers and letters used in the AWS classifications in Figures 13-49 and 13-50 have the following meanings:

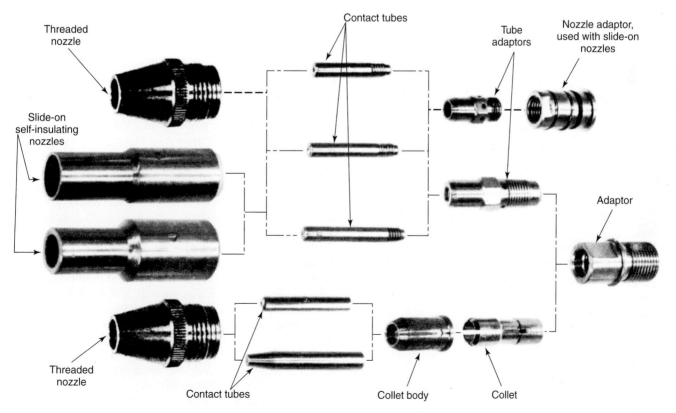

Figure 13-47. *Parts for a gas-cooled GMAW gun. (American Torch Tip Co.)*

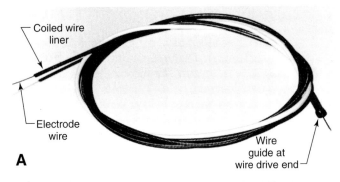

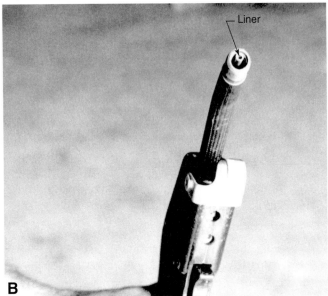

Figure 13-48. A cable liner is used to carry the electrode through the combination cable to the welding gun. A—The wire guide used in the wire drive. B—The liner within the welding gun.

Example: **ER70S-2**

> **E** – electrode
>
> **R** – rod
>
> **70** – tensile strength in ksi (1000 psi)
>
> **S** – solid rod
>
> **2** – variations in chemical compositions

Examples of dash numbers:

> **B2** – chrome-molybdenum steel
>
> **B3L** – chrome-molybdenum steel with a lower carbon content
>
> **Ni (1-3)** – Nickel steel
>
> **D2** – manganese-molybdenum steel

13.15 SMOKE EXTRACTING SYSTEMS

The GMAW process can generate some smoke. Flux cored arc welding (FCAW) often produces quite a bit of

smoke. FCAW is discussed in the next few headings. When welding with carbon dioxide (CO_2), there is some carbon monoxide (CO) created. Some ozone is also created. Both of these gases are toxic. It is especially important to prevent breathing these gases or any metal or metal oxide fumes.

Removal of these gases from the weld area can be done by means of a GMAW torch equipped with a fume extractor. Fume extractors may be built into the gun, as shown in Figure 13-51. Gas fumes, contaminated air, and air near the weld are drawn away by the torch, filtered, and released to the atmosphere. Figure 13-52 shows a fume extractor in use.

Another method to remove fumes is to use a large air removal system, as shown in Figure 13-53. This pickup hood can be located near the welding area to remove the contaminated air. This type of exhaust system can be used with any welding process or other process in the welding shop, such as grinding.

An inexpensive option to move fumes away from the welder is to use a portable fan. This does not eliminate the fumes, only moves them. The fan cannot blow too much air or it will blow the shielding gas away from the weld and cause contamination of the weld.

Sometimes, welding must be done in a closed or confined area, or on metals that are hazardous. In such cases, it is necessary for the welder to wear a purified air breathing apparatus to supply fresh, clean air. Refer to Heading 1.3.2. Figure 13-54 shows a welder wearing a purified air breathing apparatus.

Whenever it is necessary to use a purified air breathing apparatus, always make sure the system is operating properly before entering a closed, confined, or contaminated area and before doing any welding.

13.16 FLUX CORED ARC WELDING (FCAW)

The *flux cored arc welding* (FCAW) process is used to weld ferrous metals. Ferrous metals are those with iron as the main element. Flux cored arc welding is very similar to gas metal arc welding. The difference is in the metal electrode that is used. Flux cored electrode wire is hollow or tubular. Fluxing ingredients are contained within the tubular electrode. The shielding of the weld from atmospheric contamination is accomplished by two methods. The *self-shielded FCAW* method uses the gases formed when the fluxing agents within the hollow electrode vaporize in the heat of the arc. See Figure 13-55. A second method is called *gas-shielded FCAW*. In this method, a shielding gas, such as argon or carbon dioxide, is used in addition to the vaporizing flux in the electrode core. See Figure 13-56. Refer to Heading 4.8 for an additional description of FCAW.

Flux cored arc welding combines the benefits of shielded metal arc welding (SMAW), submerged arc welding (SAW), and gas metal arc welding (GMAW). These include:

AWS classification	C	Mn	Si	P	S	Ni[a]	Cr[a]	Mo[a]	V[a]	Cu[b]	Ti	Zr	Al
				Chemical Composition, Weight Percent									
ER70S-2	0.07	0.90 to 1.40	0.40 to 0.70	0.025	0.035					0.50	0.05 to 0.15	0.02 to 0.12	0.05 to 0.15
ER70S-3	0.06 to 0.15	0.90 to 1.40	0.45 to 0.70								—	—	—
ER70S-4	0.07 to 0.15	1.00 to 1.50	0.65 to 0.85								—	—	—
ER70S-5	0.07 to 0.19	0.90 to 1.40	0.30 to 0.60								—	—	0.50 to 0.90
ER70S-6	0.07 to 0.15	1.40 to 1.85	0.80 to 1.15								—	—	—
ER70S-7	0.07 to 0.15	1.50 to 2.00	0.50 to 0.80								—	—	—
ER70S-G	No chemical requirements[c]												

a. These elements may be present but are not intentionally added.
b. The maximum weight percent of copper in the rod or electrode due to any coating plus the residual copper content in the steel shall be 0.50.
c. For this classification, there are no chemical requirements for the elements listed, with the exception that there shall be no intentional addition of Ni, Cr, Mo, or V.

Figure 13-49. *Chemical composition specifications for carbon steel filler metal used with GMAW. For exact limitations and more information, see AWS specification A5.18.*

Chemical Composition

AWS classification	Carbon	Manganese	Silicon	Phosphorus	Sulfur	Nickel	Chromium	Molybdenum	Vanadium	Titanium	Zirconium	Aluminum	Copper[a]	Total other elements
Chromium-molybdenum steel electrodes and rods														
ER80S-B2	0.07-0.12	0.40-0.70	0.40-0.70	0.025	0.025	0.20	1.20-1.50	0.40-0.65	—	—	—	—	0.35	0.50
ER80S-B2L	0.05	0.40-0.70	0.40-0.70	0.025	0.025	0.20	1.20-1.50	0.40-0.65	—	—	—	—	0.35	0.50
ER90S-B3	0.07-0.12	0.40-0.70	0.40-0.70	0.025	0.025	0.20	2.30-2.70	0.90-1.20	—	—	—	—	0.35	0.50
ER90S-B3L	0.05	0.40-0.70	0.40-0.70	0.025	0.025	0.20	2.30-2.70	0.90-1.20	—	—	—	—	0.35	0.50
Nickel steel electrodes and rods														
ER80S-Ni1	0.12	1.25	0.40-0.80	0.025	0.025	0.80-1.10	0.15	0.35	0.05	—	—	—	0.35	0.50
ER80S-Ni2	0.12	1.25	0.40-0.80	0.025	0.025	2.00-2.75	—	—	—	—	—	—	0.35	0.50
ER80S-Ni3	0.12	1.25	0.40-0.80	0.025	0.025	3.00-3.75	—	—	—	—	—	—	0.35	0.50
Manganese-molybdenum steel electrodes and rods														
ER80S-D2	0.07-0.12	1.60-2.10	0.50-0.80	0.025	0.025	0.15	—	0.40-0.60	—	—	—	—	0.50	0.50
Other low-alloy steel electrodes and rods														
ER100S-1	0.08	1.25-1.80	0.20-0.50	0.010	0.010	1.40-2.10	0.30	0.25-0.55	0.05	0.10	0.10	0.10	0.25	0.50
ER100S-2	0.12	1.25-1.80	0.20-0.60	0.010	0.010	0.80-1.25	0.30	0.20-0.55	0.05	0.10	0.10	0.10	0.35-0.65	0.50
ER110S-1	0.09	1.40-1.80	0.20-0.55	0.010	0.010	1.90-2.60	0.50	0.25-0.55	0.04	0.10	0.10	0.10	0.25	0.50
ER120S-1	0.10	1.40-1.80	0.25-0.60	0.010	0.010	2.00-2.80	0.60	0.30-0.65	0.03	0.10	0.10	0.10	0.25	0.50
ERXXS-G	Subject to agreement between supplier and purchaser.[b]													

a. The maximum weight percent of copper in the rod or electrode due to any coating plus the residual copper content in the steel shall comply with the stated value.
b. In order to meet the requirements of the G classification, the electrode must have as a minimum either 0.50 percent nickel, 0.30 percent chromium, or 0.20 percent molybdenum.

Figure 13-50. *The chemical composition specifications for low-alloy steel filler metals used with GMAW. For exact limitations and more information, see AWS specification A5.28.*

- The use of fluxing agents to dissolve and remove oxides and undesirable substances.
- A thin slag layer to shield the hot weld bead.
- The ability to weld continuously for long periods.
- The metallurgical effects which can be controlled by using various fluxing and alloying elements.

Flux cored arc welding is generally done semiautomatically. The welding gun is held and manipulated by the welder when doing semiautomatic welding. Flux cored arc welding may be used on fully automatic machines with excellent results.

Self-shielded FCAW gives better results than gas-shielded FCAW when cross winds are present. The shielding gas for gas-shielded FCAW may be disturbed or removed by a cross wind. This would result in poor welds. The self-shielded FCAW process also is preferred for

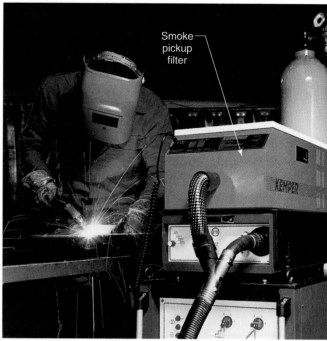

Figure 13-52. *A smoke pickup and filter set up on top of a GMAW power source. The pickup hose is attached to the GMAW gun. (Kemper Purification Systems, Inc.)*

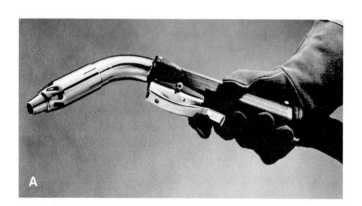

Figure 13-51. *A—A GMAW gun with a built-in smoke extractor. B—A portable smoke collector and filter. (Bernard Welding Equipment Co.)*

Figure 13-53. *A smoke extractor is attached to a pickup hood by a flexible tube and positioned over the weld area. The smoke is removed from the weld area and filtered. (Kemper Purification Systems, Inc.)*

welding in hard-to-reach or hard-to-see places. Since there is no shielding gas, there is no nozzle so visibility is improved. The electrode extension may be greater. Moving the arc out and away from the electrode contact tube permits the welder to see the joint more easily.

13.17 THE FLUX CORED ARC WELDING STATION

The equipment required in a flux cored arc welding station is similar in most respects to the equipment used in gas metal arc welding. The equipment required for FCAW is as follows:

- A constant voltage dc arc welding power source. See Heading 13.10.
- An electrode wire feed mechanism. See Heading 13.11.
- Shielding gas cylinders. See Heading 13.12.
- A gas regulator. See Heading 13.4.2.
- A shielding gas flowmeter. See Heading 13.4.3.
- Shielding gas and coolant hoses and fittings. See Heading 13.5. Shielding gas is used only for gas-shielded FCAW.
- Electrode lead and workpiece lead.
- A FCAW welding gun. See GMAW gun, Heading 13.13.
- Flux cored metal electrode wire.
- Optional equipment:
  - ◆ A coolant system. See Figure 13-11.
  - ◆ Remote controls.
- A booth, table, and a ventilation system.

The differences occur in the wire drive and shielding gas equipment.

Shielding gas is used with the gas-shielded FCAW process. Therefore, shielding gas cylinders, regulators, flowmeters, gauges, and hoses are required. When self-shielded FCAW wire electrodes are used, none of the shielding gas equipment listed above is necessary. Also when self-shielded FCAW is done, there is no need for a gas nozzle. See Figure 13-55.

The gun for FCAW may be gas-cooled or water-cooled. Gas-cooled guns are generally used up to 200A. Water-cooled guns are used when using currents of more than 200A. Water cooling is also used on guns which operate on a 100% duty cycle.

Flux cored arc welding electrodes are tubular and easily flattened. The wire drive wheels used for FCAW are usually knurled to firmly grip, but not crush, the tubular electrodes. The adjustment of these drive wheels must be carefully done to permit the wire to be driven, but not flattened.

13.18 FLUX CORED ARC WELDING ELECTRODE WIRE

The tubular, flux cored filler wire used in flux cored arc welding is made as follows:

1. Flat strip steel is formed by rolls into a "U" shape.
2. The U-shaped area is then filled with a carefully prepared granular flux and/or alloying material.
3. After filling, the metal is closed and rolled into a round shape. This closing and rolling compresses the flux material.
4. The tubular wire is then passed through drawing (forming) dies. This further compresses the granular flux and forms a perfectly round form of an exact diameter.
5. In a continuous operation, the completed flux cored wire is then wrapped on spools or into coil drums.

Flux cored wire has the advantage of varying the core ingredients to match any weld requirements. Such welding requirements may include:

- Adding deoxidizers, such as silicon, manganese, or aluminum, to reduce weld porosity.
- Adding denitrifiers, such as aluminum, to reduce nitrogen by forming stable nitrides.
- Providing mechanical, metallurgical, and corrosion-resistant properties to the weld metal by adding alloying elements.
- Forming gases to shield the weld area from the oxygen and nitrogen in the atmosphere.
- Creating a slag covering over the weld bead to shield it while it cools.
- Stabilizing the arc by providing for better ionization of the arc.
- Trapping the impurities in the molten weld metal and floating them to the top of the weld to form slag.

Figure 13-57 gives the chemical composition for flux cored arc welding electrode wire used on carbon steels. These electrodes may be used for self-shielded or gas-shielded FCAW. *Dash numbers* are used at the end of the carbon steel electrode classifications. They are used to indicate electrode polarity, the shielding gas used, or the number of weld passes recommended. See Figure 13-58 for the meaning of these dash numbers.

The electrode number and letter classifications are explained as follows:

Example: **EXXT-1**

E – electrode

XX – two-digit number that specifies minimum tensile strength of the electrode in thousands of pounds per square inch (ksi). The second digit represents the position in which the electrode is to be used. A 0 is for flat and horizontal fillet welds. A 1 is for all-position welding.

T – indicates the electrode is tubular

1 to **11** – a grouping of chemical composition, method of shielding, or its suitability to make single or multiple pass welds.

Example: **E70T-3**

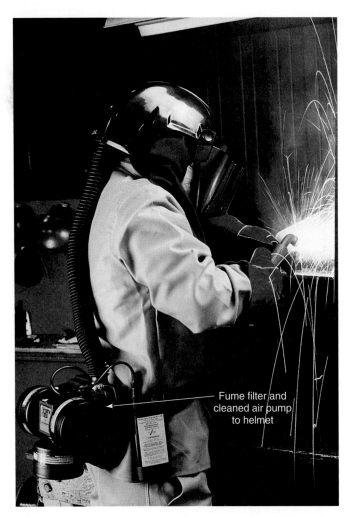

Figure 13-54. *This welder is wearing a powered air purifier around the waist. The purifier pumps cleaned air to the hood and helmet combination. (Racal Health and Safety, Inc.)*

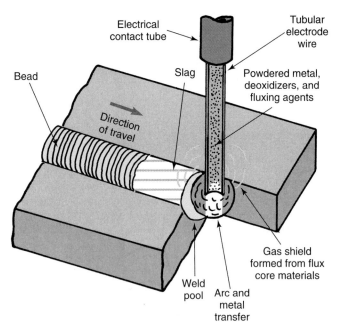

Figure 13-55. *This sketch shows a self-shielded flux cored arc weld in progress. Note that there is no nozzle or shielding gas required.*

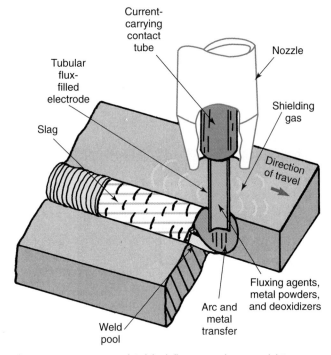

Figure 13-56. *A gas-shielded flux cored arc weld in progress is shown in this sketch.*

This is a tubular electrode with 70 ksi (483MPa) tensile strength to be used for flat or horizontal fillet welds. The dash three (-3) in this case means it is self-shielded and intended for single-pass welds. For a complete description of these FCAW electrodes, refer to the AWS 5.20 electrode specification manual.

Chemical compositions and AWS classification numbers for low-alloy FCAW electrodes are shown in Figure 13-59.

The meaning of electrode numbers and letters used in Figure 13-59 are shown below:

Example: **E80T1-B2H**

    E – electrode

    80 – 80 ksi (80,000 psi or 552MPa) tensile strength

    T1 – intended usage
    B – major alloying ingredients as follows:
      A – carbon-molybdenum

      B – chromium-molybdenum

    Ni – nickel

    D – manganese-molybdenum

    K – all other low-alloy electrodes

    2 – Chemical composition group. See brackets on Figure 13-57.
    H – comparative carbon content

      H – higher carbon
      L – lower carbon

Chemical Composition, Percent[a]										
AWS classification	Carbon	Phos-phorus	Sulfur	Vanadium	Silicon	Nickel	Chromium	Molyb-denum	Manganese	Alumi-num[b]
EXXT-1 EXXT-4 EXXT-5 EXXT-6 EXXT-7 EXXT-8 EXXT-11 EXXT-G	—	0.04	0.03	0.08	0.09	0.50	0.20	0.30	1.75	1.8
EXXT-GS EXXT-2 EXXT-3 EXXT-10	No chemical requirements									

a.   Values are maximums. Composition limits are intended to insure a plain carbon steel deposit.
b.   For self-shielded electrodes only.

Figure 13-57. *The chemical composition of various carbon steel electrodes for flux cored arc welding. Note that aluminum is used as a denitrifier only for self-shielded electrodes. See Figure 13-58 for the meanings of the dash numbers. (American Welding Society, A5.20.)*

AWS classification	External shielding medium	Current and polarity
EXXT-1 (Multiple-pass)	CO_2	dc, electrode positive
EXXT-2 (Single-pass)	CO_2	dc, electrode positive
EXXT-3 (Single-pass)	None	dc, electrode positive
EXXT-4 (Multiple-pass)	None	dc, electrode positive
EXXT-5 (Multiple-pass)	CO_2	dc, electrode positive
EXXT-6 (Multiple-pass)	None	dc, electrode positive
EXXT-7 (Multiple-pass)	None	dc, electrode negative
EXXT-8 (Multiple-pass)	None	dc, electrode negative
EXXT-10 (Single-pass)	None	dc, electrode negative
EXXT-11 (Multiple-pass)	None	dc, electrode negative
EXXT-G (Multiple-pass)	a	a
EXXT-GS (Single-pass)	a	a

a.   As agreed upon between supplier and user.

Figure 13-58. *The meanings and uses of the dash numbers at the end of carbon steel FCAW electrode designations. The letter "G" is used when a manufacturer and purchaser develop a special electrode.*

Figure 13-60 lists the chemical composition of the deposited weld metal when using chromium and chromium-nickel flux cored electrodes.

The electrode classification numbers are explained below:

Example: **E308T-X** and **E308LT-X**

E – electrode

308 – the stainless steel classification

T – tubular electrode

LT – low carbon, tubular electrode

X – numbers from 1 to 3 plus the letter "G"

1 – used with carbon dioxide (CO_2)

2 – used with argon plus 2% oxygen

3 – self-shielded. No external shielding gas used.

G – not specified. Used only for special compositions.

Example: **E316T-3**

This describes a tubular (**T**) stainless steel (**316**) electrode (**E**) used without any external shielding gas (**-3**).

For complete specifications and information on flux cored arc welding electrodes, refer to one of the following AWS electrode specifications:

A5.20, *Specifications for Carbon Steel Electrodes for Flux Cored Arc Welding.*

A5.22, *Specifications for Flux Cored Corrosion-Resisting Chromium and Chromium-Nickel Steel Electrodes.*

A5.29, *Specifications for Low-Alloy Steel Flux Cored Arc Welding Electrodes.*

13.19 ACCESSORY DEVICES

When water-cooled guns are used, it is important that water flow be maintained. If the water flow stops or slows, the torch or gun and the electrode lead may quickly overheat and be damaged. To protect the equipment, a safety switch is sometimes placed in the water circuit. If the water flow decreases below a set limit, a switch opens and shuts off the electrical power source. Figure 13-61 shows a flow safety device.

Chemical Composition, Percent											
AWS classification	C	Mn	P	S	Si	Ni	Cr	Mo	V	Al[a]	Cu
Carbon-molybdenum steel electrodes											
E70T5-A1 E80T1-A1 E81T1-A1	0.12	1.25	0.03	0.03	0.80	—	—	0.40/0.65	—	—	—
Chromium-molybdenum steel electrodes											
E81T1-B1	0.12	1.25	0.03	0.03	0.80	—	0.40/0.65	0.40/0.65	—	—	—
E80T5-B2L	0.05	1.25	0.03	0.03	0.80	—	1.00/1.50	0.40/0.65	—	—	—
E80T1-B2 E81T1-B2 E80T5-B2	0.12	1.25	0.03	0.03	0.80	—	1.00/1.50	0.40/0.65	—	—	—
E80T1-B2H	0.10/0.15	1.25	0.03	0.03	0.80	—	1.00/1.50	0.40/0.65	—	—	—
E90T1-B3L	0.05	1.25	0.03	0.03	0.80	—	2.00/2.50	0.90/1.20	—	—	—
E90T1-B3 E91T1-B3 E90T5-B3	0.12	1.25	0.03	0.03	0.80	—	2.00/2.50	0.90/1.20	—	—	—
E100T1-B3 E90T1-B3H	0.10/0.15	1.25	0.03	0.03	0.80	—	2.00/2.50	0.90/1.20	—	—	—
Nickel steel electrodes											
E71T8-Ni1 E80T1-Ni1 E81T1-Ni1 E80T5-Ni1	0.12	1.50	0.03	0.03	0.80	0.80/1.10	0.15	0.35	0.05	1.8	—
E71T8-Ni2 E80T1-Ni2 E81T1-Ni2 E80T5-Ni2 E90T1-Ni2 E91T1-Ni2	0.12	1.50	0.03	0.03	0.80	1.75/2.75	—	—	—	1.8	—
E80T5-Ni3 E90T5-Ni3	0.12	1.50	0.03	0.03	0.80	2.75/3.75	—	—	—	—	—
Manganese-molybdenum steel electrodes											
E91T1-D1	0.12	1.25/2.00	0.03	0.03	0.80	—	—	0.25/0.55	—	—	—
E90T5-D2 E100T5-D2	0.15	1.65/2.25	0.03	0.03	0.80	—	—	0.25/0.55	—	—	—
E90T1-D3	0.12	1.00/1.75	0.03	0.03	0.80	—	—	0.40/0.65	—	—	—
All other low-alloy steel electrodes											
E80T5-K1	0.15	0.80/1.40	0.03	0.03	0.80	0.80/1.10	0.15	0.20/0.65	0.05	—	—
E70T4-K2 E71T8-K2 E80T1-K2 E90T1-K2 E91T1-K2 E80T5-K2 E90T5-K2	0.15	0.50/1.75	0.03	0.03	0.80	1.00/2.00	0.15	0.35	0.05	1.8	—
E100T1-K3 E110T1-K3 E100T5-K3 E110T5-K3	0.15	0.75/2.25	0.03	0.03	0.80	1.25/2.60	0.15	0.25/0.65	0.05	—	—
E110T5-K4 E111T1-K4 E120T5-K4	0.15	1.20/2.25	0.03	0.03	0.80	1.75/2.60	0.20/0.60	0.30/0.65	0.05	—	—
E120T1-K5	0.10/0.25	0.60/1.60	0.03	0.03	0.80	0.75/2.00	0.20/0.70	0.15/0.55	0.05	—	—
E61T8-K6 E71T8-K6	0.15	0.50/1.50	0.03	0.03	0.80	0.40/1.10	0.15	0.15	0.05	1.8	—
E101T1-K7	0.15	1.00/1.75	0.03	0.03	0.80	2.00/2.75	—	—	—	—	—
EXXXTX-G	—	1.00 min*	0.03	0.03	0.80 min*	0.50 min*	0.30 min*	0.20 min*	0.10 min*	1.8	—
E80T1-W	0.12	0.50/1.30	0.03	0.03	0.35/0.80	0.40/0.80	0.45/0.70	—	—	—	0.30/0.75

a. For self-shielded electrodes only.
* All values are maximum except where min (minimum) is indicated.

Figure 13-59. *The chemical composition and AWS classification numbers for low-alloy FCAW electrodes. Numbers and letters are explained in Heading 13.18.*

AWS classification	C	Cr	Ni	Mo	Cb + Ta	Mn	Si	P	S	Fe	Cu
E307T-X	0.13	18.0-20.5	9.0-10.5	0.5-1.5	—	3.3-4.75	1.0	0.04	0.03	Rem	0.5
E308T-X	0.08	18.0-21.0	9.0-11.0	0.5	—	0.5-2.5					
E308LT-X	a	18.0-21.0	9.0-11.0	0.5	—						
E308MoT-X	0.08	18.0-21.0	9.0-12.0	2.0-3.0	—						
E308MoLT-X	a	18.0-21.0	9.0-12.0	2.0-3.0							
E309T-X	0.10	22.0-25.0	12.0-14.0	0.5	—						
E309CbLT-X	a	22.0-25.0	12.0-14.0		0.70-1.00						
E309LT-X	a	22.0-25.0	12.0-14.0		—						
E310T-X	0.20	25.0-28.0	20.0-22.5		—	1.0-2.5		0.03			
E312T-X	0.15	28.0-32.0	8.0-10.5		—	0.5-2.5		0.04			
E316T-X	0.08	17.0-20.0	11.0-14.0	2.0-3.0	—						
E316LT-X	a	17.0-20.0	11.0-14.0	2.0-3.0	—						
E317LT-X	a	18.0-21.0	12.0-14.0	3.0-4.0	—						
E347T-X	0.08	18.0-21.0	9.0-11.0	0.5	8 x C min to 1.0 max						
E409T-X	0.10	10.5-13.0	0.60	0.5	—	0.8					
E410T-X	0.12	11.0-13.5	0.60	0.5	—	1.2					
E410NiMoT-X	0.06	11.0-12.5	4.0-5.0	0.40-0.70	—	1.0					
E410NiTiT-X	a	11.0-12.0	3.6-4.5	0.05	—	0.70	0.60	0.03			
E430T-X	0.10	15.0-18.0	0.60	0.5	—	1.2	1.0	0.04			
E502T-X	0.10	4.0-6.0	0.40	0.45-0.65	—	1.2					
E505T-X	0.10	8.0-10.5	0.40	0.85-1.20	—	1.2					
E307T-3	0.13	19.5-22.0	9.0-10.5	0.5-1.5	—	3.3-4.75					
E308T-3	0.08	19.5-22.0	9.0-11.0	0.5	—	0.5-2.5					
E308LT-3	0.03	19.5-22.0	9.0-11.0	0.5	—						
E308MoT-3	0.08	18.0-21.0	9.0-12.0	2.0-3.0	—						
E308MoLT-3	0.03	18.0-21.0	9.0-12.0	2.0-3.0	—						
E309T-3	0.10	23.0-25.5	12.0-14.0	0.5	—						
E309LT-3	0.03	23.0-25.5	12.0-14.0		—						
E309CbLT-3	0.03	23.0-25.5	12.0-14.0		0.70-1.00						
E310T-3	0.20	25.0-28.0	20.0-22.5		—	1.0-2.5		0.03			
E312T-3	0.15	28.0-32.0	8.0-10.5		—	0.5-2.5		0.04			
E316T-3	0.08	18.0-20.5	11.0-14.0	2.0-3.0	—						
E316LT-3	0.03	18.0-20.5	11.0-14.0	2.0-3.0	—						
E317LT-3	0.03	18.5-21.0	13.0-15.0	3.0-4.0	—						
E347T-3	0.08	19.0-21.5	9.0-11.0	0.5	8 x C min to 1.0 max						
E409T-3	0.10	10.5-13.0	0.60	0.5	—	0.80					
E410T-3	0.12	11.0-13.5	0.60	0.5	—	1.0					
E410NiMoT-3	0.06	11.0-12.5	4.0-5.0	0.40-0.70	—	1.0					
E410NiTiT-3	0.04	11.0-12.0	2.6-4.5	0.5	—	0.70	0.50	0.03			
E430T-3	0.10	15.0-18.0	0.60	0.5	—	1.0	1.0	0.04			
EXXXT-G	As agreed upon between supplier and purchaser.										

a. The carbon content shall be 0.04% maximum when the suffix "X" is "1"; it shall be 0.03% when the suffix "X" is "2".

Figure 13-60. *The chemical composition of the deposited weld metal for flux cored chromium and chromium-nickel steel electrodes. (American Welding Society, A5.22)*

Some systems use a (thermal) heat fuse to protect the water-cooled welding gun. If the water flow stops or decreases to a dangerous minimum, the fuse link will overheat and open to interrupt the electrical current flow. See Figure 13-62.

Most arc welding machines are equipped with a remote control circuit switch and external plug. The

Figure 13-61. *Water flow safety control. If the water flow decreases below a set limit, the current is shut off until the flow of water resumes.*

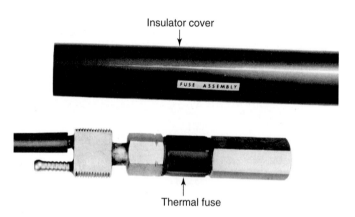

Figure 13-62. *Fuse and hose assembly with insulator cover removed. If the thermal fuse overheats, it will shut off the current flow to the torch. (ESAB Welding and Cutting Products)*

remote control is used to vary the voltage or amperage within a rough setting range. The welder at the job site can adjust the arc welding machine at a distance with this control. Figure 13-63 shows one type of remote control device. Several types of remote controls and their uses are described in Heading 11.9.

13.20 FILTER LENSES FOR USE WHEN GAS-SHIELDED ARC WELDING

Because of the clearer atmosphere around the arc, the operator must use arc welding lenses with a darker shade to reduce eye fatigue and possible eye damage. Most helmets for gas arc welding use a clear cover lens and a filter lens. Sometimes a clear cover lens is used on the

Figure 13-63. *A remote amperage adjuster. With this device, a percentage of the amperage range set on the power supply can be adjusted remotely. (Miller Electric Mfg. Co.)*

inside of the helmet, as well. It is very important that all these lenses be clean. Figure 13-64 shows a helmet with a lens that automatically darkens when an arc is struck. Figure 13-65 shows the effect of such an automatic-darkening lens. Figure 13-66 lists recommended shade numbers for gas tungsten, gas metal, and flux cored arc welding.

13.21 PROTECTIVE CLOTHING

Fire-resistant clothing and accessories of leather are recommended. Wool also is satisfactory. Cotton does not provide sufficient protection and deteriorates rapidly under infrared and ultraviolet rays. Always wear dark clothing to reduce reflection of light behind the helmet. **The clothing should be without cuffs or open pockets, since these can collect sparks.**

Leather or leather-palm gloves should be worn. See Chapters 1 and 11 for additional information on clothing recommended for arc welding.

13.22 SAFETY REVIEW

When using gas tungsten, gas metal, or flux cored arc welding equipment, follow all the safety precautions normally used with arc welding equipment.

Figure 13-64. An electronic quick change lens is installed in this arc welding helmet. The welder can see the weld joint clearly until the arc is struck. When the arc is struck, the lens darkens in about 0.00004 second. (Jackson Products, Division Thermadyne Industries, Inc.)

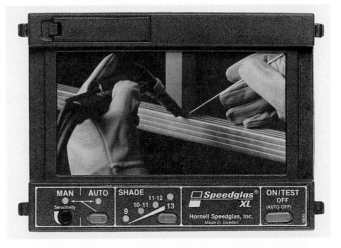

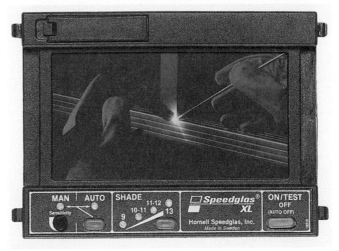

Figure 13-65. The action of an electronic quick-change lens is shown in this pair of photos. A—View before arc is struck. B—View after arc is struck. (Hornell Speedglas, Inc.)

- Electrical connection leads should be in good condition and tight. They should be protected from accidental damage from shop traffic.
- When arc welding, it is necessary to prevent fumes and contaminated air from reaching the welder's face. Adequate ventilation and filtration equipment must be used. In some welding conditions, it is necessary to use purified air breathing apparatus.
- Proper clothing should be worn to prevent burns from hot metal and ultraviolet and infrared rays.

- Flammable materials should not be carried in the pockets of clothing. Pockets should be closed and cuffs rolled down to prevent hot metal from going into them.
- A welding helmet with the proper number filtering lens for the type of welding being done must be used.
- Shielding curtains should be placed around all jobs so that workers in the area are protected from arc flashes.

Guide for Shade Numbers				
Operation	**Electrode size 1/32 in. (mm)**	**Arc current (A)**	**Minimum protective shade**	**Suggested* shade no. (comfort)**
Shielded metal arc welding	Less than 3 (2.5) 3-5 (2.5-4) 5-8 (4-6.4) More than 8 (6.4)	Less than 60 60-160 160-250 250-550	7 8 10 11	— 10 12 14
Gas metal arc welding and flux cored arc welding		Less than 60 60-160 160-250 250-500	7 10 10 10	— 11 12 14
Gas tungsten arc welding		Less than 50 50-150	8 8	10 12
Air carbon Arc cutting	(Light) (Heavy)	150-500 Less than 500 500-1000	10 10 11	14 12 14
Plasma arc welding		Less than 20 20-100 100-400 400-800	6 8 10 11	6 to 8 10 12 14
Plasma arc cutting	(Light)** (Medium)** (Heavy)**	Less than 300 300-400 400-800	8 9 10	9 12 14
Torch brazing	—	—	—	3 or 4
Torch soldering	—	—	—	2
Carbon arc welding	—	—	—	14

Plate thickness			
	in.	**mm**	

Gas welding
  Light
  Medium
  Heavy
	
	Under 1/8
	1/8 to 1/2
	Over 1/2
		
		Under 3.2
		3.2 to 12.7
		Over 12.7
			
			4 or 5
			5 or 6
			6 or 8

Oxygen cutting
  Light
  Medium
  Heavy
	
	Under 1
	1 to 6
	Over 6
		
		Under 25
		25 to 150
		Over 150
			
			3 or 4
			4 or 5
			5 or 6

*As a rule of thumb, start with a shade that is too dark to see the weld zone. Then go to a lighter shade which gives sufficient view of the weld zone without going below the minimum. In oxyfuel gas welding or cutting where the torch produces a high yellow light, it is desirable to use a filter lens that absorbs the yellow or sodium line in the visible light of the (spectrum) operation.

**These values apply where the actual arc is clearly seen. Experience has shown that lighter filters may be used when the arc is hidden by the workpiece.

Figure 13-66. *A guide to the correct welding lens shade for various welding processes and applications. (ANSI/AWS Z49.1)*

TEST YOUR KNOWLEDGE

Write your answers on a separate sheet of paper. Do not write in this book.

1. Name the welding processes covered in this chapter which may use a shielding gas to shield the weld area.
2. What type welding current may be used with GTAW?
3. A GTAW electrode is held in the torch by a _____ .
4. GTAW welding power sources are constant _____ machines.
5. What is the advantage of using a constant voltage power source?
6. Why is a high-frequency voltage used continually in an ac circuit that is being used for GTAW?
7. List five reasons why argon is more frequently used for GTAW than helium.
8. When GTAW, which shielding gas is best for use in cross drafts? Gives the best cleaning action? Provides the greatest heat conductivity?
9. A regulator controls and measures _____ . A flowmeter measures and controls the _____ of gas flow.
10. The maximum amperage used on a gas-cooled GTAW torch is generally _____ amperes.
11. What determines the nozzle diameter when selecting a nozzle for GTAW or GMAW?
12. What is the advantage of using thoriated tungsten electrodes?
13. The color coding of a 2% thoriated tungsten electrode is _____ .
14. The arc welding power source used for GMAW and FCAW produces a constant _____ .
15. GMAW power sources have a sloping voltage curve. How much slope should the curve have for use with carbon dioxide (CO_2) gas?
16. Increasing the rate of speed on the wire feed in GMAW or FCAW increases the _____ in the circuit.
17. The shielding gas mixture suggested in Figure 13-34 for use with low-alloy steel is a mixture of _____ and 2% _____ .
18. Which gas produces the deepest penetration when used in GMAW?
19. To produce better metal transfer through the arc, reduce spatter, and stabilize the arc, _____ and _____ are added to argon or helium.
20. Describe the following GMAW electrode completely: ER90S-B3L
21. What two methods are used in FCAW to shield the weld from atmospheric contamination?
22. When feeding a FCAW electrode through the wire feed mechanism, care must be taken not to _____ the flux cored wire.
23. List five ways that FCAW wires are altered to meet the requirements of a weld.
24. Describe the following FCAW electrode completely: E80T-2.
25. What number filter lens is recommended when GTAW steel with 80A? When GMAW steel, using 220A?

Manual GTAW requires greater skill than other manual welding processes, but is capable of producing very high quality welds. Careful addition of filler metal at the proper time is critical. (Hornell Speedglas, Inc.)

Chapter 14

GAS TUNGSTEN ARC WELDING

LEARNING OBJECTIVES

After studying this chapter, you will be able to:

* Contrast the effects of DCEN, DCEP, and ac on surface cleaning efficiency, electrode life, and weld characteristics.
* Select the proper welding power source, polarity, shielding gas, flow rate, tungsten electrode type and diameter, nozzle size, and filler metal required to produce an acceptable GTA weld.
* Properly assemble and adjust all the variables required to produce an acceptable GTA weld.
* Correctly prepare a tungsten electrode for welding with ac or dc.
* Correctly prepare metals for welding and perform acceptable welds on all types of joints in all positions.
* Identify the potential safety hazards involved when using the GTAW process in a working environment and describe ways of safely dealing with these hazards.
* Pass a safety test on the proper and safe use of the GTAW process.

Metal	Weldable
Aluminum and aluminum alloys	Yes
Bronze and brass	Yes
Copper and copper alloys	Yes
Cast iron	
Malleable	Possible, but not popular
Nodular	Possible, but not popular
Inconel	Yes
Lead	Possible, but not popular
Magnesium	Yes
Monel	Yes
Nickel and nickel alloys	Yes
Precious metals (gold, silver, platinum, etc.)	Yes
Steel, alloy	Yes
Steel, stainless	Yes
Titanium	Yes
Tungsten	Possible, but not popular

Figure 14-1. A list of metals showing their weldability.

Gas tungsten arc welding (GTAW) is a process used to produce high-quality welds in virtually all weldable metals. See Figure 14-1. It is done manually, semiautomatically, or automatically. Gas tungsten arc welding (GTAW) is also known as TIG (tungsten inert gas) welding. It was called "heliarc welding" when it was first developed.

Welding is done with a tungsten electrode. The tungsten electrode is not consumed—it does not melt and enter the weld. Gas tungsten arc welding may be done in any welding position. GTAW is used in many applications. One of these is to weld thin-walled pipe and tube. The process is also used almost exclusively to weld the root bead in heavy walled pipe in petroleum, chemical, and power generating applications. Filler metal may or may not be required with GTAW. Flange joints, lap joints, and outside corner joints on thin metal may be designed for welding without filler metal. Inert gases are used to shield the electrode and base metal from atmospheric contamination.

The welder must manipulate the gas tungsten arc welding torch to accurately control the arc length and the size of the weld pool. The welder must also carefully add filler metal at the proper times when doing manual GTAW. The desired result is a very high-quality weld. Manual gas tungsten arc welding requires more welder skill than other manual welding processes.

14.1 GAS TUNGSTEN ARC WELDING PRINCIPLES

Gas tungsten arc welding requires the use of a constant current power source, an inert gas, gas regulating equipment, a torch, and (when needed) filler metal. Refer to Heading 4.6 and Figure 4-11.

Direct current (dc) or alternating current (ac) power supplies may be used. Either direct current electrode negative (DCEN) or direct current electrode positive (DCEP) may be used. When doing GTAW with DCEN (also known as DCSP, or direct current straight polarity), approximately 70% of the heat is generated on the workpiece and 30% is generated at the electrode. In DCEP (also known as DCRP, or direct current reverse polarity), approximately 70% of the heat is generated at the electrode and 30% on the workpiece. Figure 14-2 shows how the electrons flow and how the heat is distributed and penetration obtained when DCEN, DCEP, or ac is used with GTAW. DCEN produces the best penetration. DCEP has less penetration, but provides a cleaning action. When DCEP is used for GTAW, a larger diameter electrode must be used than when DCEN is selected. The current-carrying capacity of an electrode using DCEP is only about 1/10 that of an electrode using DCEN.

Filler metal is added to the weld pool either manually or automatically. When doing manual GTAW using filler metal, the filler metal is held in one hand and the torch in the other. The end of the filler metal is held close to the arc for two reasons:
- To preheat the filler metal.
- To keep the filler metal in the protective shielding gas envelope.

Automatic wire feeders can be used with manual GTAW. Manual welding with an automatic wire feeder is called *semiautomatic welding*, because it is a combination of manual and automatic welding.

The filler metal used for semiautomatic or automatic GTAW is in the form of spooled wire. The wire is fed into the weld pool as shown in Figure 14-3. Metal thicknesses below 0.02" (0.51mm) are usually joined without adding filler metal. Thicknesses above 0.02" can be joined without filler metal, depending on the joint design. Edge joints, some outside corner joints, and some flange joints can be welded without adding filler metal.

Direct current electrode negative (DCEN) is used to weld carbon and alloy steels, stainless steel, nickel and nickel alloys, titanium, copper, and most copper alloys. As shown in Figure 14-2, the electrons flow from the negative electrode to the workpiece. Welds with good penetration are produced using DCEN.

Alternating current or direct current electrode positive (DCEP) is used when surface oxides must be removed. Surface oxides occur on aluminum, magnesium, and some other nonferrous metals. These metal oxides melt at a

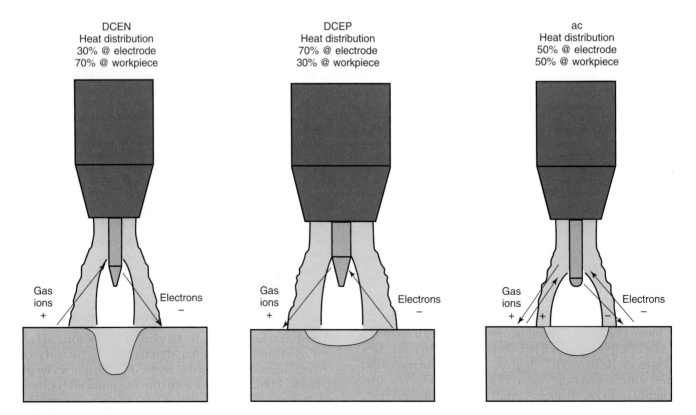

Figure 14-2. The approximate heat distribution for GTAW using DCEN, DCEP, and ac. Note the larger electrode diameter required for DCEP and the depth of penetration for DCEN.

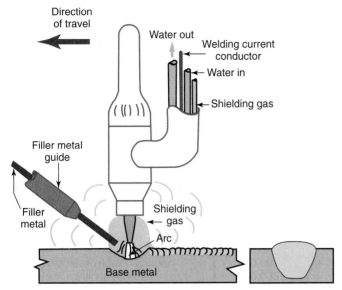

Figure 14-3. *Automatic GTAW. The filler metal guide feeds the filler metal to the desired position in the weld pool at a preset feed rate.*

higher temperature than the base metal. The oxides therefore make it hard to weld the base metal if they are not removed.

Electrons flow from the workpiece to the electrode when DCEP is used, and during one half of the ac cycle. Positively charged shielding gas ions travel from the nozzle to the negatively charged workpiece. See Figure 14-2. These shielding gas ions strike the workpiece surface with sufficient force to break up the oxides. DCEP and ac both

work well in breaking up the surface oxides on aluminum and magnesium. See Figure 14-4.

Ac gives better penetration than DCEP. This occurs because the electrode is negative during half the ac cycle, which provides penetration. Ac can also be used with a smaller electrode diameter for a given current flow than DCEP, because of the heat distribution at the electrode.

DCEP requires an electrode larger than one used when ac welding. DCEP requires a much larger electrode than DCEN when used with the same current. A larger electrode diameter is required when using DCEP because of the great amount of heat created at the electrode. See Figure 14-2. The current-carrying capacity of an electrode using DCEP is about 1/10 that of an electrode using DCEN.

Gas tungsten arc welding is done with ac, a steady flowing direct current, or direct current that is pulsing. Figure 14-5 graphically shows the current flow used in ***pulsed arc GTAW***. The amperage is changed between high and low currents. Electronics within the welding machine create the pulsed wave form shown in Figure 14-5. Welding is done during the high-amperage interval. This high amperage is usually referred to as the ***peak current***.

During the low-amperage period, the arc is maintained but the current output of the arc is reduced. This low-amperage period is called the ***background current***. During a welding operation, the weld pool cools slightly during the low-amperage period. This allows a lower overall heat input into the base metal. Pulsed arc GTAW is ideal for welding out of position. It allows for controlled heating and cooling periods. In shielded metal arc welding (SMAW), a flip motion is used to allow a period for cooling. With pulsed arc GTAW, the welder does not have to move the torch to accomplish a cooling period.

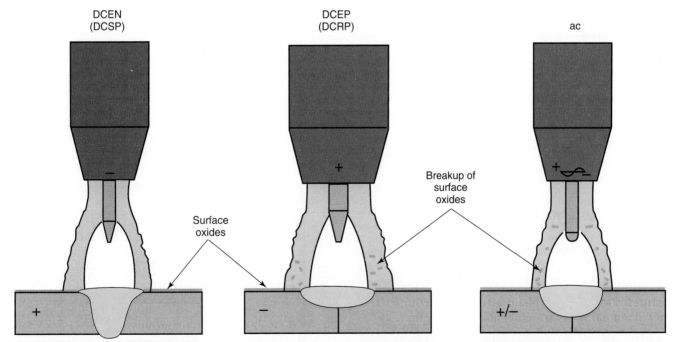

Figure 14-4. *Examples of how the surface oxides are broken up when DCEP and ac current is used. The heavy gas ions striking the surface cause the oxides to break up near the weld bead area. In DCEN, however, the gas ions travel from the workpiece to the electrode. The gas ions in DCEN, therefore, do not strike the surface. No surface oxides are removed when DCEN is used.*

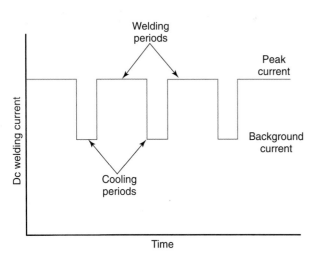

Figure 14-5. *The high and low amperages in a pulsed-arc GTAW sequence.*

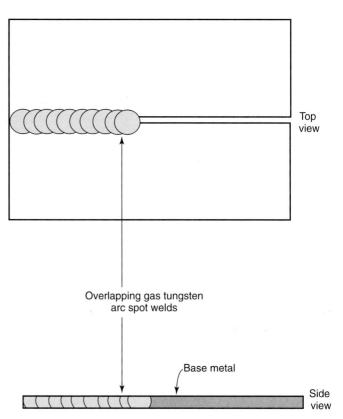

Figure 14-6. *Two views of a pulsed-current GTAW. A series of overlapping gas tungsten arc spot welds make up the welded seam.*

At slow pulse rates, a weld is a series of overlapping arc spot welds. The torch is moved from one arc spot location to the next during the cooling period. Each arc spot weld overlaps the previous one, as shown in Figure 14-6. The physical characteristics and appearance of the finished weld are excellent. The total heat input into the base metal is reduced. Not all gas tungsten arc welding machines can provide a pulsed current.

Some machines can provide upslope and downslope currents. **Upslope** is used at the beginning of a weld to gradually increase the current from zero to the welding current. **Downslope** is a gradual decrease in current at the end of a weld. Downslope allows the final weld pool to be completely filled and slowly cooled. Figure 14-7 shows upslope, downslope, and pulsed currents.

Pulsed current is used with excellent results where the metals are not well-fitted. It is also used when parts of unequal thicknesses are welded. Gas tungsten arc welding, like other welding processes, works best when accurately fitted metals of equal thickness are welded.

When automatic GTAW, the operator must set the power source to control the arc voltage and welding current. He or she must also set the wire feed speed and the torch travel speed.

14.2 GTAW POWER SOURCES

Welding power sources for GTAW must provide a constant current. The most common power sources are transformer-rectifier machines that provide ac and dc constant currents. These machines are relatively large and heavy.

Power sources using inverter technology also provide constant current. Inverter machines are much smaller and lighter than the more common transformer type machines. To inverter machines often have the ability to do multiple processes. See Chapter 11 for additional infor-

mation on inverter type machines and why they are so much smaller.

Transformer type power sources used for GTAW are designed to control the amount of current delivered to the torch. Usually, a knob on the control panel is used to set either the current used or the maximum current available.

Machines used for GTAW do not control the amount of shielding gas flow. They only control *when* the shield-

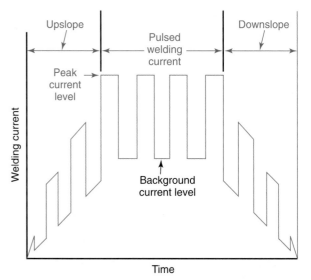

Figure 14-7. *A programmed upslope, pulsed welding current, and downslope for GTAW on tubing or piping.*

ing gas flows. Shielding gas can be made to flow before the arc starts. This is called *preflow*. Shielding gas flows while welding and usually continues to flow after the arc is stopped. Shielding gas flow after the arc stops is called *postflow*.

GTAW power sources that are used for ac welding have a high-frequency generator. High frequency helps reinitiate (restart) the arc during the half cycle when the electrode is positive. Newer GTAW power sources that produce a square-wave output do not need high frequency. The square wave delivers enough voltage and current to jump the arc gap while the electrode is positive, without the need for high frequency. Even though high frequency is not required on square-wave machines, it is still built into the machines.

High frequency can also be used to help start the arc when dc welding. When used with dc welding, the high frequency is automatically turned off by the machine once the arc is started and stabilized.

14.3 SETTING UP THE GTAW STATION

The arc welding power source should be given a brief safety inspection prior to use. With the main power switch off, check both the workpiece and electrode leads for a tight connection on the machine. Figure 14-8 shows where connections are made on one type of machine. Check the entire length of each lead for evidence of wear or cuts. Wear or cuts on the outside of the electrode and workpiece leads may indicate internal damage to these conducting leads.

The ground lead clamp or connection should be checked for a good connection. The contact area of the workpiece connection or clamp must be clean. A clean contact area provides the best current-flow conditions.

Check to make sure the shielding gas cylinder is properly secured. **A chain or other safety device should be used to prevent a gas cylinder from being knocked over.** Check that the regulator and gas flowmeter are attached properly. Connect a hose from the flowmeter to the welding machine. Check to make sure there are no leaks. Leaks allow expensive shielding gas to escape. Leaks may allow air to enter the gas lines and cause contamination of the electrode and the workpiece.

Connect the fittings from the torch to the proper locations on the welding machine. A gas-cooled torch will only have a fitting for shielding gas. A water-cooled torch will have connections for water in, water out, and for shielding gas in. Refer to Figure 14-8.

Often a water cooler is used. Connect the water cooler to the welding machine or to the torch, according to the manufacturer's recommendations. Usually, the water coming from the water cooler goes to the water-cooled welding torch. Water returning from the torch goes to the water cooler inlet. Figure 14-9 shows many components of a GTAW station.

A remote control foot switch, shown in Figure 14-10, performs many functions. When the pedal (foot switch) is

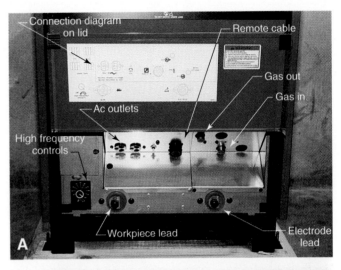

Figure 14-8. *The various lead and hose connections under the lower panel of a GTAW power source. A—The panel prior to any connections. B—The panel with leads and hoses connected.*

pressed, the shielding gas begins to flow. The pedal may be used as a current on-off switch. When the remote control is used as an on-off switch, the current set on the welding machine cannot be changed by the welder while welding. The current is either flowing at the set amount (*on*) or it is not flowing (*off*).

The foot switch also can be used to control and vary the welding current. When the foot pedal is used this way, the welding current can be controlled by the welder while welding. Pressing down on the pedal increases the welding current. The welder can vary the welding current from a very low level up to the current set on the welding machine.

A torch-mounted current control is shown in Figure 14-11. This current control method is used instead of a foot pedal control. Like the foot pedal, the torch-mounted control can be used as an on-off switch or to adjust the current while welding.

Whichever type of control is used, it is connected to the welding machine. A switch on the welding machine is used to select how the foot switch or torch-mounted cur-

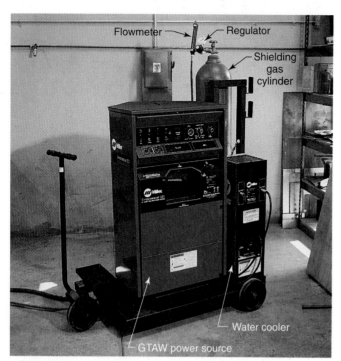

Figure 14-9. *A GTAW outfit showing the components and connections.*

Figure 14-10. *This foot switch may be used to change the amperage or to turn the power on and off. Its function is determined by the position of the panel/remote switch on the power source. (Miller Electric Mfg. Co.)*

rent control will perform. The switch on the welding machine has two positions, *panel* and *remote*. Placing the switch in the panel position will cause the foot pedal or torch-mounted control to function like an on-off switch. Placing the switch on the welding machine in the remote position will allow the welder to use the foot pedal or torch-mounted control to vary the current while welding. Heading 14.3.1 explains this switch and other controls on the GTAW machine.

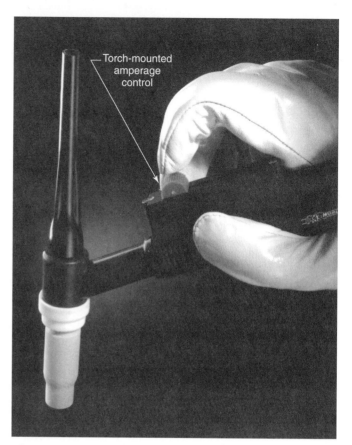

Figure 14-11. *A torch-mounted amperage control. (CK Worldwide, Inc.)*

While setting up the GTAW station, look for all potential safety problems. Make sure other workers will not be exposed to harmful arc rays while welding. It is recommended that all flammable materials be removed from the welding area. Also make sure gas cylinders are properly secured. Welding leads should not be laid across aisles where they can get damaged. If they must run across an aisle, cover them with a length of steel channel to protect them.

14.3.1 Setting Up the GTAW Power Source

Once the welding station has been checked, then the welding power source or machine must be properly set up. Figure 14-12 shows a gas tungsten arc welding machine. Figure 14-13 is a closeup of the control panel for this same machine. This control panel contains a number of controls, which are explained below. Not all welding machines have all these controls. Different manufacturers call functions by slightly different names, but the functions are the same.

1. The POWER (ON-OFF) pushbuttons are used to turn the arc welding machine primary circuit on or off. The machine should not be turned off while the arc is struck. The machine should only be turned on or off while the GTAW torch is hung on an insulated hook.
2. Selector switch for the type of current to be used. Select DCEN, DCEP, or ac.

Figure 14-12. A typical GTAW power source. (Miller Electric Mfg. Co.)

3. The AMPERAGE ADJUSTMENT knob used to set the welding current or the maximum welding current. The amperage switch (#5) selects how this amperage adjustment knob will function.

4. This welding machine can be used to weld using SMAW and GTAW. The MODE switch is used to select which process is being used.

5. If the current is to be set on the panel (using knob #3) and not changed while welding, the AMPERAGE switch is placed in the PANEL position. If the welder is allowed to vary the current while welding, using a foot pedal or thumb switch, the amperage switch is placed in the REMOTE position.

6. The secondary voltage (OUTPUT) switch may be kept on all the time. This is used when SMAW or when GTAW using the touch start method. Usually, when gas tungsten arc welding, the switch is in the remote position and the secondary voltage is turned on and off using a foot pedal or thumb switch.

7. The HF (high frequency) switch has three positions. In the START position, high frequency is applied to the welding circuit only until the arc is struck and has stabilized. This position may be selected when dc is used. In the CONTINUOUS position, high frequency is applied to the welding circuit constantly. This position is chosen when ac is used for GTAW. In the OFF position, high frequency is not used at all. The off

position is used when touch-starting the tungsten electrode. When using a square-wave ac current, high frequency is not required. Ac welding with a square wave can be done without any high frequency, so this switch can be in the OFF position. Place the switch in the OFF position when SMAW, as well.

8. The AMPERAGE display (ammeter) will indicate the current flow while welding.

9. The VOLTAGE display (voltmeter) will show open circuit voltage and it will show the closed circuit voltage while welding.

10. The ARC/BALANCE CONTROL (identified as AC WAVE BALANCE on some machines) provides the means for controlling the penetration and cleaning action. See Figure 14-14. Refer to Heading 13.3 for an explanation of balanced and unbalanced wave forms.

11. The POSTFLOW timer is used to control how long (in seconds) the shielding gas will flow after the arc is stopped. The gas should flow for a few seconds to keep the hot electrode and weld metal from becoming contaminated. A good rule to follow is to continue the gas flow 1 second for each 10A of current used.

12. PREFLOW TIME is used to cause the shielding gas to flow for some period prior to starting the arc.

13. SPOT TIME is used to make GTAW spot welds. The length of time the current will flow is set using this control.

14. The START CONTROL switch, amperage, and time knobs all work together. When this switch is ON, the arc starting current and time for the starting current to flow are set. When the starting current time runs out, the current changes to the preset welding amperage.

15. CRATER FILL is a gradual decrease in current to allow the weld pool to be filled with filler metal.

16. Some dc GTAW machines have the ability to pulse the welding current, as shown in Figure 14-15. The ON-OFF switch activates the pulse controls.

17. BACKGROUND AMPERAGE is the lower current value set when using a pulsed wave. Background amperage is set as a percentage of the peak or high current.

18. The duration of the pulse cycle, when used, may be varied. Typical pulse frequencies will vary from ten per second to one in two seconds. The number of pulses per second are set on the PULSES PER SECOND control.

19. Control of the duration of the peak current pulse is set using the % ON TIME control. The higher the percent on time, the shorter the background or low-current time.

Figure 14-16 shows another type of GTAW (TIG) arc welding machine. The control panel on this arc welding machine contains many of the same controls shown in Figure 14-13. Controls and switches may appear in a different form or location on various machines. These controls serve the same purposes, however, as those described for the machine shown in Figure 14-13. Figure 14-17 shows an inverter machine that is used for gas tungsten arc welding.

Figure 14-13. *A closeup of a GTAW power source control panel. The functions of the controls identified by number are discussed in the text. (Miller Electric Mfg. Co.)*

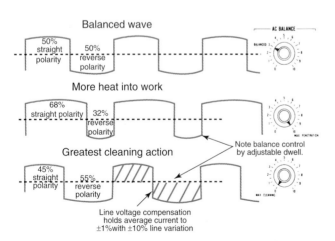

Figure 14-14. *Variations in the square-wave patterns that occur with the setting of the ac balance control. Note how the ac balance can be adjusted to obtain more cleaning, more penetration, or a balanced wave. (Miller Electric Mfg. Co.)*

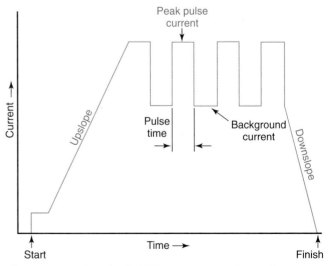

Figure 14-15. *A pulsed GTAW program. The peak current is for welding. The background current maintains the arc but allows the weld to cool.*

Figure 14-16. *A GTAW (TIG) welding machine and its control panel. The numbers on these controls correspond to those shown in Figure 14-13. Functions of the controls are discussed in the text. (The Lincoln Electric Co.)*

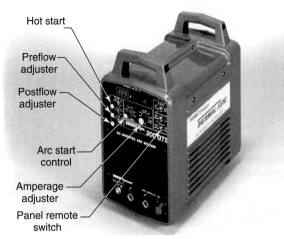

Figure 14-17. *A dc inverter welding power supply. (Thermal Dynamics)*

Base material	Direct current		Alternating current
	DCEN	DCEP	
	DCSP	DCRP	ac
Aluminum up to 3/32"	P	G	E
Aluminum over 3/32"	P	P	E
Aluminum bronze	P	G	E
Aluminum castings	P	P	E
Beryllium copper	P	G	E
Brass alloys	E	P	G
Copper-base alloys	E	P	G
Cast iron	E	P	G
Deoxidized copper	E	P	P
Dissimilar metals	E	P	G
Hard facing	G	P	E
High-alloy steels	E	P	G
High-carbon steels	E	P	G
Low-alloy steels	E	P	G
Low-carbon steels	E	P	G
Magnesium up to 1/8"	P	G	E
Magnesium over 1/8"	P	P	E
Magnesium castings	P	G	E
Nickel & Ni-alloys	E	P	G
Stainless steel	E	P	G
Silicon bronze	E	P	P
Titanium	E	P	G
E-Excellent G-Good P-Poor			

Figure 14-18. *Suggested current and polarity for use with GTAW on various base metals. (Welding & Fabrication Data Book)*

14.3.2 Selecting the Proper Current for GTAW

When selecting the type of current to use for a welding job, usually the most important factor is the type of base metal to be welded. Direct current electrode negative gives the best penetration. Direct current electrode positive gives the least penetration, but is excellent for removing surface oxides from metals, especially aluminum and magnesium. Alternating current combines the penetration and cleaning qualities of DCEN and DCEP. Figure 14-18 lists the most common base metals and the suggested current and polarity to use when welding each metal.

After selecting the type of current, the *amount* of current must be selected. Figures 14-19 through 14-23 are tables that list current ranges for different base metals and different metal thicknesses. These figures also identify the suggested shielding gas types, flow rates, electrode diameters, and filler metal diameters.

14.3.3 Selecting the Proper Shielding Gas for GTAW

High-quality welds using the gas tungsten arc are produced because the weld area is protected by an inert shielding gas. Argon is the most common shielding gas used for ac and dc GTAW. See Heading 13.4 for more information on shielding gases.

Metal thickness	Joint type	Tungsten electrode diameter	Filler rod diameter (if req'd.)	Amperage	Gas		
					Type	Flow (cfh)	L/min*
1/16" (1.59mm)	Butt	1/16" (1.59mm)	1/16" (1.59mm)	60-70	Argon	15	7.08
	Lap	1/16"	1/16"	70-90	Argon	15	
	Corner	1/16"	1/16"	60-70	Argon	15	
	Fillet	1/16"	1/16"	70-90	Argon	15	
1/8" (3.18mm)	Butt	1/16-3/32" (1.59-2.38mm)	3/32" (2.38mm)	80-100	Argon	15	7.08
	Lap	1/16-3/32"	3/32"	90-115	Argon	15	
	Corner	1/16-3/32"	3/32"	80-100	Argon	15	
	Fillet	1/16-3/32"	3/32"	90-115	Argon	15	
3/16" (4.76mm)	Butt	3/32" (2.38mm)	1/8" (3.18mm)	115-135	Argon	20	9.44
	Lap	3/32"	1/8"	140-165	Argon	20	
	Corner	3/32"	1/8"	115-135	Argon	20	
	Fillet	3/32"	1/8"	140-170	Argon	20	
1/4" (6.35mm)	Butt	1/8" (3.18mm)	5/32" (4.0mm)	160-175	Argon	20	9.44
	Lap	1/8"	5/32"	170-200	Argon	20	
	Corner	1/8"	5/32"	160-175	Argon	20	
	Fillet	1/8"	5/32"	175-210	Argon	20	

*Liters per minute

Figure 14-19. *Variables for manual gas tungsten arc welding mild steel using DCEN (DCSP).*

Metal thickness	Joint type	Tungsten electrode diameter	Filler rod diameter (if req'd.)	Amperage	Gas		
					Type	Flow (cfh)	L/min
1/16" (1.59mm)	Butt	1/16" (1.59mm)	1/16" (1.59mm)	60-85	Argon	15	7.08
	Lap	1/16"	1/16"	70-90	Argon	15	
	Corner	1/16"	1/16"	60-85	Argon	15	
	Fillet	1/16"	1/16"	75-100	Argon	15	
1/8" (3.18mm)	Butt	3/32-1/8" (2.38-3.18mm)	3/32" (2.38mm)	125-150	Argon	20	9.44
	Lap	3/32-1/8"	3/32"	130-160	Argon	20	
	Corner	3/32-1/8"	3/32"	120-140	Argon	20	
	Fillet	3/32-1/8"	3/32"	130-160	Argon	20	
3/16" (4.76mm)	Butt	1/8-5/32" (3.18-4.0mm)	1/8" (3.18mm)	180-225	Argon	20	11.80
	Lap	1/8-5/32"	1/8"	190-240	Argon	20	
	Corner	1/8-5/32"	1/8"	180-225	Argon	20	
	Fillet	1/8-5/32"	1/8"	190-240	Argon	20	
1/4" (6.35mm)	Butt	5/32-3/16" (4.0-4.76mm)	3/16" (4.76mm)	240-280	Argon	25	14.16
	Lap	5/32-3/16"	3/16"	250-320	Argon	25	
	Corner	5/32-3/16"	3/16"	240-280	Argon	25	
	Fillet	5/32-3/16"	3/16"	250-320	Argon	25	

Figure 14-20. *A table of the variables used when manually welding aluminum with the gas tungsten arc using ac and high frequency.*

Argon (Ar) provides a smoother, quieter arc. Argon requires a lower arc voltage and provides a better cleaning action than helium. Because of the lower required arc voltage, argon is used when ac welding. Argon is 10 times heavier than helium. Because argon is heavier than air, it provides better shielding than helium, so less gas is required.

Helium (He) provides a higher available heat at the workpiece than does argon. GTAW done with helium gas produces deeper penetration than GTAW done with argon. Helium is better for use on thicker metal sections than argon because of its higher available heat.

Argon and helium gas mixtures are used in some applications. These mixtures contain up to 75% helium. They produce a weld with deeper penetration and have a good cleaning action.

Hydrogen (H_2) may be added to argon when welding stainless steel, nickel-copper, or nickel-based alloys. The

Metal thickness	Joint type	Tungsten electrode diameter	Filler rod diameter (if req'd.)	Amperage	Gas		
					Type	Flow (cfh)	L/min
1/16" (1.59mm)	Butt	1/16" (1.59mm)	1/16" (1.59mm)	40-60	Argon	15	7.08
	Lap	1/16"	1/16"	50-70	Argon	15	
	Corner	1/16"	1/16"	40-60	Argon	15	
	Fillet	1/16"	1/16"	50-70	Argon	15	
1/8" (3.18mm)	Butt	3/32" (2.38mm)	3/32" (2.38mm)	65-85	Argon	15	7.08
	Lap	3/32"	3/32"	90-110	Argon	15	
	Corner	3/32"	3/32"	65-85	Argon	15	
	Fillet	3/32"	3/32"	90-110	Argon	15	
3/16" (4.76mm)	Butt	3/32" (2.38mm)	1/8" (3.18mm)	100-125	Argon	20	9.44
	Lap	3/32"	1/8"	125-150	Argon	20	
	Corner	3/32"	1/8"	100-125	Argon	20	
	Fillet	3/32"	1/8"	125-150	Argon	20	
1/4" (6.35mm)	Butt	1/8" (3.18mm)	5/32" (4.0mm)	135-160	Argon	20	9.44
	Lap	1/8"	5/32"	160-180	Argon	20	
	Corner	1/8"	5/32"	135-160	Argon	20	
	Fillet	1/8"	5/32"	160-180	Argon	20	

Figure 14-21. *A table of the variables used when manually welding stainless steel with the gas tungsten arc and DCEN (DCSP).*

Metal thickness	Joint type	Tungsten electrode diameter	Filler rod diameter (if req'd.)	Amperage[1] with backup	W/O backup	Gas		
						Type	Flow (cfh)	L/min
1/16" (1.59mm)	All	1/16" (1.59mm)	3/32" (2.38mm)	60	35	Argon	13	6.14
3/32" (2.38mm)	All	1/16" (1.59mm)	1/8" (3.18mm)	90	60	Argon	15	7.08
1/8" (3.18mm)	All	1/16" (1.59mm)	1/8" (3.18mm)	115	85	Argon	20	9.44
3/16" (4.76mm)	All	1/16" (1.59mm)	5/32" (4.0mm)	120	75	Argon	20	9.44
1/4" (6.35mm)	All	3/32" (2.38mm)	5/32" (4.0mm)	130	85	Argon	20	9.44
3/8" (9.53mm)	All	3/32" (2.38mm)	3/16" (4.76mm)	180	100	Argon	25	11.80
1/2" (12.7mm)	All	5/32" (4.0mm)	3/16" (4.76mm)	—	250	Argon	25	11.80
3/4" (19.05mm)	All	3/16" (4.76mm)	1/4" (6.35mm)	—	370	Argon	35	16.52

1 - Use alternating current with a constant high frequency (AC-HF)

Figure 14-22. *Variables for welding magnesium manually with the gas tungsten arc using ac and high frequency.*

Metal thickness	Joint type	Tungsten electrode diameter	Filler rod diameter (if req'd.)	Amperage[1]	Gas		
					Type	Flow (cfh)	L/min
1/16" (1.59mm)	All	1/16" (1.59mm)	1/16" (1.59mm)	110-150	Argon	15	7.08
1/8" (3.18mm)	All	3/32" (2.38mm)	3/32" (2.38mm)	175-250	Argon	15	7.08
3/16" (4.76mm)	All	1/8" (3.18mm)	1/8" (3.18mm)	250-325	Argon	18	9.50
1/4" (6.35mm)	All	1/8" (3.18mm)	1/8" (3.18mm)	300-375	Argon	22	10.38
3/8" (9.53mm)	All	3/16" (4.76mm)	3/16" (4.76mm)	375-450	Argon	25	11.80
1/2" (12.7mm)	All	3/16" (4.76mm)	1/4" (6.35mm)	525-700	Argon	30	14.16

1 - Use DCEN (DCSP)

Figure 14-23. *Variables for welding deoxidized copper using the gas tungsten arc and DCEN (DCSP).*

addition of hydrogen to argon permits increased welding speeds. Hydrogen is not recommended for use on other metals, because it produces hydrogen cracks in the welds.

Figure 14-24 lists the shielding gases and current types for various metals. For manual gas tungsten arc welding, argon is the most common shielding gas. Argon-helium mixtures are sometimes used. When doing automatic GTAW, argon, helium, argon-helium, and argon-hydrogen mixtures are used. Gas mixtures can be obtained in cylinders from local welding gas distributors. Shielding gas mixtures can also be mixed using a gas mixer.

14.3.4 Selecting the Correct Shielding Gas Flow Rate for GTAW

The *flow rate* is the volume of gas flowing. The rate is measured in cubic feet per hour (ft³/hr. or cfm) or liters per minute (L/min.). This flow rate varies with the base metal being welded, the thickness of the base metal, and the position of the welded joint. A higher gas flow rate is required when welding overhead. This is necessary because argon, which is heavier than air, tends to fall away from the overhead joint.

After determining the correct gas and desired flow rate, the rate must be properly set on the flowmeter. Refer to Heading 13.4.3 for an explanation of flowmeters and flow gauges. The vertical tube gas flowmeter is most common. See Figure 14-25 for a schematic of a gas flowmeter.

Before setting the flowmeter, the shielding gas cylinder must be opened. The procedure for opening the shielding gas cylinder is as follows:

1. Turn the regulator adjusting screw outward in a counterclockwise direction. This ensures that the regulator is closed. Note: When a preset pressure is set on the regulator, no adjusting handle is used. Once the pressure is set, an acorn nut is placed over the adjusting screw. In this case, the regulator is always open.
2. Open the cylinder valve *slowly*. Continue until it is fully opened. This is necessary because a back seating valve is used to seal the valve stem from leakage. A

back seating valve is also used in an oxygen cylinder. Refer to Heading 5.2.1 and Figure 5-5.

3. If a regulator adjusting screw is used, turn it in to the pressure at which the flowmeter is calibrated. Most flowmeters are calibrated at 50 psig (345.7 kPa). The calibrating pressure should be indicated somewhere on the flowmeter.

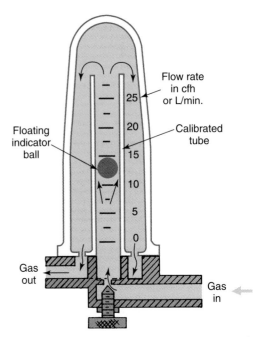

Figure 14-25. *A schematic of a gas flowmeter used on GTAW or GMAW outfits. When adjusting this gauge, the top of the ball indicates the gas flow through the meter. The rate of flow is measured in cubic feet per hour (ft³/hr. or cfh) or liters per minute (L/min). This flowmeter is set for 15 ft³/hr.*

Metal	Thickness	Manual	Automatic (machine)
Aluminum and aluminum alloys	Under 1/8" (3.2mm)	Ar[1] (AC-HF)	Ar (AC-HF) or He[2] (DCEN)
	Over 1/8" (3.2mm)	Ar (AC-HF)[3]	Ar-He (AC-HF) or He (DCEN)[4]
Copper	Under 1/8" (3.2mm)	Ar-He (DCEN)	Ar-He (DCEN)
	Over 1/8" (3.2mm)	He (DCEN)	He (DCEN)
Nickel alloys	Under 1/8" (3.2mm)	Ar (DCEN)	Ar-He (DCEN) or He (DCEN)
	Over 1/8" (3.2mm)	Ar-He (DCEN)	He (DCEN)
Steel, carbon	Under 1/8" (3.2mm)	Ar (DCEN)	Ar (DCEN)
	Over 1/8" (3.2mm)	Ar (DCEN)	Ar-He (DCEN) or He (DCEN)
Steel, stainless	Under 1/8" (3.2mm)	Ar (DCEN)	Ar-He (DCEN) or Ar-H₂[5] (DCEN)
	Over 1/8" (3.2mm)	Ar-He (DCEN)	He (DCEN)
Titanium and its alloys	Under 1/8" (3.2mm)	Ar (DCEN)	Ar (DCEN) or Ar-He (DCEN)
	Over 1/8" (3.2mm)	Ar-He (DCEN)	He (DCEN)

1 - Ar (argon)
2 - He (helium)
3 - AC-HF (alternating current, high frequency)
4 - DCEN (direct current electrode negative, also DCSP)
5 - H₂ (hydrogen)

Figure 14-24. *Suggested choices of shielding gases and current type for welding different metals and metal thicknesses.*

4. Press the foot pedal or thumb contactor switch to cause the shielding gas to flow to the torch.

5. Turn the adjusting knob out on the flowmeter until the top of the floating indicator ball is at the flow rate line desired.

6. Release the foot or thumb contactor switch to stop the gas flow.

14.3.5 Selecting and Preparing a Tungsten Electrode

The selection of the correct type and diameter of tungsten electrode is extremely important to performing a successful gas tungsten arc weld. Heading 13.7 describes the various types of tungsten electrodes used. It also lists the diameters and lengths of electrodes available.

The types of tungsten electrodes, with their color codes shown in parentheses, are:
- Pure tungsten (green).
- Tungsten with 1% or 2% thoria added (1%-yellow, 2%-red). These are called *thoriated tungsten electrodes.*
- Tungsten with from 0.15% to 0.40% zirconia added (brown). These are called *zirconiated electrodes.*
- Tungsten with 2% ceria (orange). These are called *ceriated electrodes.*
- Tungsten with 1% lanthana (black).
- Tungsten electrodes that do not fit into the above classifications (gray). These electrodes can have different amounts of the alloys listed above, or different alloys added, or a combination of alloys.

Figure 14-26 shows tungsten electrodes with color markings to identify their type.

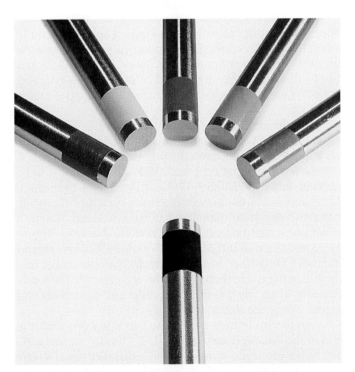

Figure 14-26. *The standard end colors for GTAW electrodes: brown, tungsten and zirconia; yellow, 1% thoria; red, 2% thoria; orange, 2% ceria; green, pure tungsten; black, 1% lanthana. (Osram Sylvania, Inc.)*

When ac welding, pure tungsten or zirconiated electrodes are preferred. They will form a ball or hemisphere on the end when they are heated. The ball shape is desirable because it reduces current rectification and allows the ac to flow more easily. See Figure 14-27A. Ceriated electrodes will also do fairly well when used for ac welding.

To form a ball or hemisphere, set the current to DCEP. Strike an arc on a clean piece of copper. Copper will not melt easily and will not contaminate the electrode readily. Increase the current until the end of the electrode forms a ball about the same diameter as the electrode.

To obtain full current capacity from a pure or zirconiated tungsten electrode when used with ac, the electrode is not ground to a point. Instead, as noted earlier, a ball the same size as the electrode diameter will form when the arc is struck. This ball may be up to one and one half times the size of the electrode diameter, but should never exceed that limit. A ball larger than one and one-half times the electrode diameter may melt off and fall into the weld. If the ball on the end of the tip is much larger than the electrode diameter, the current may be set too high. Electrodes used with ac can be ground to a taper, so that a smaller ball will form on the tip. A tapered and balled tip reduces the current capacity of the electrode when used with ac.

Dc welding is usually done with thoriated tungsten electrodes. Thoria increases the electron emissions from a tungsten electrode. The addition of 1% to 2% thoria also keeps the electrode tip from forming and maintaining a ball. Thoriated electrodes and other electrodes used for dc welding are ground to a point and the point is then blunted. Refer to Figure 14-27B.

If a thoriated electrode is used with ac, it will initially form a ball on the end. During use, the ball breaks down quickly and forms a number of small projections at the tip. These projections will cause a wild wandering of the arc. Therefore, thoriated electrodes should only be used for direct current GTAW. See Figure 14-27C.

Electrodes with ceria and lanthana added can be used for ac and dc welding. They do not form a ball as well as pure or zirconiated tungsten.

A pure tungsten electrode used with ac is shown in Figure 14-27D. The amperage in this case was too high. The ball on the end has begun to melt and bend to one side. With continued use, the ball might have fallen into the weld and contaminated it. Excessive current may also cause tungsten electrodes to split near the end.

A pure tungsten electrode that was ground to a point and used with direct current electrode negative (DCEN or DCSP) is illustrated in Figure 14-27E. Notice the small ball formed at the tip. Pointing of pure tungsten electrodes is not recommended. The small ball formed at the tip may melt off and fall into the weld.

Tungsten electrodes of any type must be protected from contamination. The electrode should not be touched to the base metal, weld metal, or filler metal while it is hot. Figure 14-27F illustrates a tungsten electrode that was touched by the filler rod. Such contamination prevents the electrode from emitting or receiving electrons effectively. The contamination must be removed. One way to remove the contamination is to break off the tip of the electrode and reshape it.

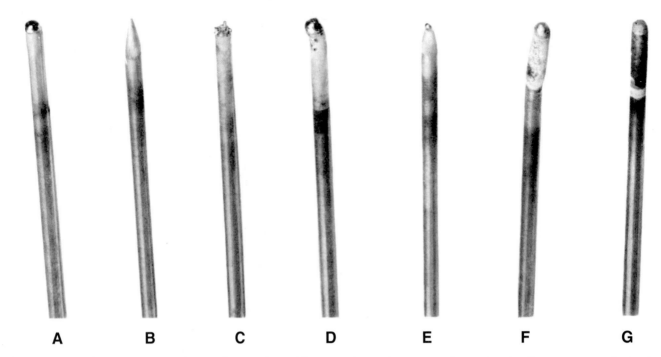

A   B   C   D   E   F   G

Figure 14-27. The appearance of several properly and improperly used tungsten electrodes. A—Pure tungsten electrode with ball or hemisphere. B—Thoriated electrode ground to point. C—Thoriated electrode used with ac. D—Pure tungsten electrode used with excessive amperage, ac. E—Pure tungsten electrode ground to a point. F—Tungsten electrode contaminated by filler rod. G—Tungsten electrode contaminated by air. (Miller Electric Mfg. Co.)

The hot tungsten electrode and metal in the weld area may be contaminated by oxygen and nitrogen in the air and by airborne dirt. To prevent this contamination, a shielding gas is used. The shielding gas is allowed to flow over the electrode and the weld area after the arc is broken. The timing of this shielding gas postflow is set on the arc welding machine panel. See Figure 14-13 and Heading 14.3.3.

In Figure 14-27G, this electrode was contaminated by the air. The black surface of the electrode indicates that the shielding gas did not have sufficient postflow time to protect it. An electrode contaminated by oxidation must not be reused. If it is used as is, the oxides will fall from the electrode into the weld. The electrode must be ground to remove the contamination or broken off behind the contamination. The altered electrode must then be retipped before use.

Tungsten electrodes come in sizes from 0.010" to 1/4" (0.25mm to 6.4mm). The most frequently used lengths are 6" and 7" (152mm and 178mm). Each diameter and type of electrode has a maximum amperage that it can carry. This amperage limit varies with the current, polarity, and shielding gas used. Figure 14-28 lists the operating current range for each type of tungsten electrode.

Special grinding wheels should be used for pointing tungsten electrodes. These wheels should be used only for grinding tungsten electrodes. Keeping the electrodes free from contamination is essential.

Silicon carbide or alumina oxide grinding wheels are preferred. Alumina oxide wheels cut more slowly, but last longer than silicon carbide wheels. Electrodes should be rough-ground on an 80-grit grinding wheel. The finish grinding should be done on a 120-grit wheel. A grinding wheel with an open structure is best because it will run cooler and pick up less contamination.

Electrodes should be ground in a lengthwise direction. The grinding marks on the tapered area must run in a lengthwise direction. This method of grinding ensures the best current carrying characteristics. Figures 14-29 and 14-30 illustrate the suggested method for grinding tungsten electrodes. **Always wear safety goggles when grinding**

Another method of accurately pointing a tungsten electrode is by using a flame, as shown in Figure 14-31. A very repeatable tip is created using this method.

Determine what type and thickness of metal is to be welded. Refer to Heading 13.7 and Figures 14-19 through 14-23 and select the correct electrode type and diameter. After the electrode diameter has been selected, obtain a collet and collet body for that electrode diameter. Figure 14-32 shows these and other torch parts for GTAW. Thread the collet body into the GTAW torch. Place the collet into the top of the torch, then thread the end cap onto the torch. Place the electrode into the torch collet and tighten the end cap to secure the electrode in place.

An electrode may be installed so it is even with the end of the nozzle. It may extend 1/2" (13mm) or more. As a rule, the electrode should extend a distance equal to one to two electrode diameters, but not more than the nozzle diameter. Some joints, such as corner or T-joints, require a longer electrode extension. As the electrode extension increases, higher gas flows must be used. Another option

Pure tungsten diameter (mm)	Current Range – Amperes			
	DCEN or DCSP argon	DCEP or DCRP argon	Ac-unbalanced wave-argon	Ac-balanced wave-argon
.010" (.254)	Up to 15	*	Up to 15	Up to 10
.020" (.508)	5-20	*	5-20	10-20
.040" (1.02)	15-80	*	10-60	20-30
1/16" (1.59)	70-150	10-20	50-100	30-80
3/32" (2.38)	125-225	15-30	100-160	60-130
1/8" (3.18)	225-360	25-40	150-210	100-180
5/32" (3.97)	360-450	40-55	200-275	160-240
3/16" (4.76)	450-720	55-80	250-350	190-300
1/4" (6.35)	720-950	80-125	325-450	250-400
Thorium (1% and 2%), cerium, and lanthanum alloyed tungsten				
.010" (.254)	Up to 25	*	Up to 20	Up to 15
.020" (.508)	15-40	*	15-35	5-20
.040" (1.02)	25-85	*	20-80	20-60
1/16" (1.59)	50-160	10-20	50-150	60-120
3/32" (2.38)	135-235	15-30	130-250	100-180
1/8" (3.18)	250-400	25-40	225-360	160-250
5/32" (3.97)	400-500	40-55	300-450	200-320
3/16" (4.76)	500-750	55-80	400-550	290-390
1/4" (6.35)	750-1000	80-125	600-800	340-525
Zirconium alloyed tungsten				
.010" (.254)	*	*	Up to 20	Up to 15
.020" (.508)	*	*	15-35	5-20
.040" (1.02)	*	*	20-80	20-60
1/16" (1.59)	*	*	50-150	60-120
3/32" (2.38)	*	*	130-250	100-180
1/8" (3.18)	*	*	225-360	160-250
5/32" (3.97)	*	*	300-450	200-320
3/16" (4.76)	*	*	400-550	290-390
1/4" (6.35)	*	*	600-800	340-525
Special alloyed tungsten	Refer to manufacturer for specific information about alloy content and recommended current ranges.			
* Not recommended				
The figures listed are intended as a guide, and are a composite of recommendations from American Welding Society and electrode manufactures.				

***Figure 14-28.** The suggested current range for various types and sizes of tungsten electrodes.*

is to use a gas lens. See Heading 13.6.2 for information on gas lenses. Also see Figure 13-23.

14.3.6 Selecting the Correct GTAW Torch Nozzle

Nozzles used on gas tungsten arc welding torches vary in size and method of attachment. Figures 14-32 and 14-33. The end of the nozzle that attaches to the torch varies in design. This design variation is necessary to permit attachment to different manufacturer's torches.

Most nozzles used for GTAW are manufactured from ceramic materials. The *exit diameter* (diameter of the nozzle closest to the arc) is manufactured in a variety of sizes; as shown in Figure 14-34. GTAW nozzles are also made in various lengths from short ("stubby") nozzles to extra-long.

Manufacturers have their own distinct part number for each nozzle. However, the exit diameter for any nozzle is specified with a number that represents the diameter in 1/16" (1.6mm) increments.

The diameter of the nozzle must be large enough to allow the entire weld area to be covered by the shielding gas. The exit diameter can be neither too large nor too small, or poor shielding gas coverage will result. Refer to Heading 13.6.2 for additional information on nozzles.

The choice of a nozzle size is often a compromise that is necessary to meet the requirements of the job. Small-diameter nozzles are often used to permit a constant arc length. An example is shown in Figure 14-35. In this case, the important root pass is to be welded. The nozzle is constantly touching the sides of the groove as it is rocked back and forth across the root opening. It is also kept in contact with the groove opening as the weld moves forward

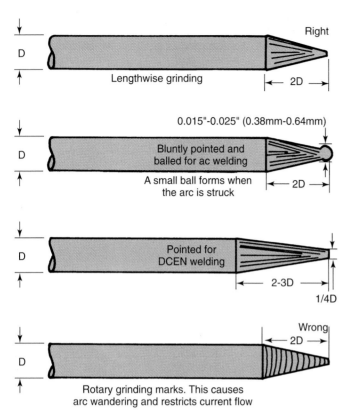

Figure 14-29. *Methods of grinding tungsten electrodes. Note the lengthwise direction of the grinding marks on the properly ground electrodes.*

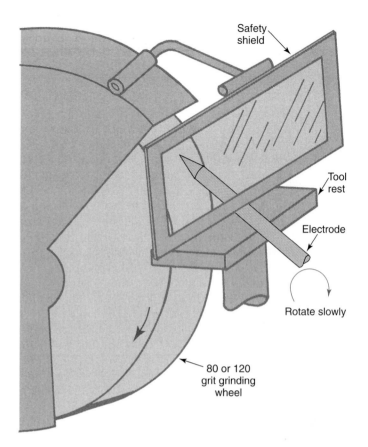

Figure 14-30. *The correct position for grinding a tungsten electrode. The grinding marks on the taper must run lengthwise on the electrode. Use an 80-grit wheel for rough grinding and a 120-grit wheel for finish grinding.*

slowly. This choice of a small nozzle diameter allows the welder to reach the bottom of the groove. It also allows the welder to keep a constant arc length.

14.3.7 Selecting the Correct Filler Metal for Use with GTAW

Filler metal used for gas tungsten arc welding (GTAW) is generally bare wire. The filler metal is purchased in coils of several hundred feet, or in precut lengths. Coiled wire is cut to any desired length by the welder when doing manual welding. If a wire feeder is used for semiautomatic or automatic welding, the wire is fed from the coil. Precut wire is usually purchased in lengths of 24" or 36" (610mm or 914mm).

Filler metal alloy must be selected for the specific job. See Figure 13-27 for a listing of the AWS specifications for filler metals used with GTAW. Steel welding rods are not copper-coated as they are in oxyfuel gas welding. The copper coating would cause spatter, which could contaminate the tungsten electrode.

Common diameters of filler wires vary from 1/16" to 1/4" (1.6mm to 6.4mm). Smaller-diameter wire is readily available in coils to 0.015" (0.4mm). Precut filler rods are readily available up to 1/4" (6.4mm). The proper filler metal diameters to use when welding various thicknesses of metals are shown in Figures 14-19 to 14-23.

Figure 14-31. *A patented flame device for sharpening accurately pointing tungsten electrodes. (Swanstrom Centerline)*

14.4 PREPARING METAL FOR WELDING

GTAW is generally chosen as a welding method because it produces very high-quality welds. Therefore, before welding is attempted, the surface of the base metal must be cleaned. All rust, dirt, oil, grease, and other materials that may be on the metal surface must be removed. Any such materials left on the base metal may cause

Figure 14-32. *Various GTAW nozzles, caps, collets, and collet bodies. (American Torch Tip Co.)*

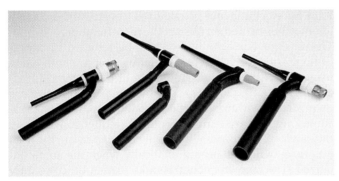

Figure 14-33. *A variety of GTAW torch designs. Note the different angles of the torch heads. (American Torch Tip Co.)*

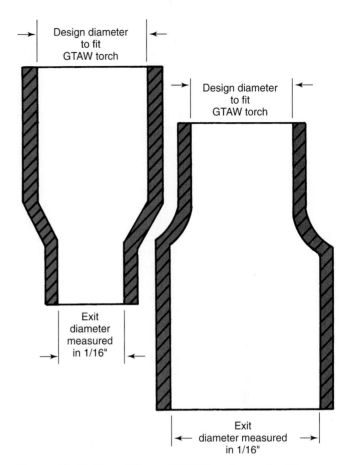

Figure 14-34. *Two different GTAW torch nozzles. These nozzles show the design diameter that permits a nozzle to fit a certain torch. The variation in exit diameter is also shown.*

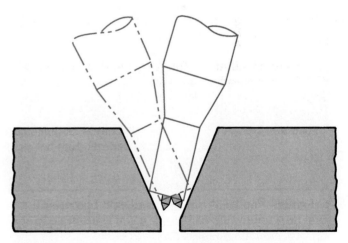

Figure 14-35. *Using a small exit diameter allows the welder to reach the bottom of a groove weld. It also allows the welder to maintain a constant arc length if the nozzle is held against the sides of the groove.*

contamination of the weld. This may result in weak and defective welds.

Aluminum and magnesium oxides on the surface of aluminum and magnesium are removed while welding with ac or DCEP (DCRP). No such surface cleaning occurs when DCEN (DCSP) is used. If it is necessary to use DCEN when welding aluminum or magnesium, the surface oxides must be removed prior to welding. This generally is done chemically, or mechanically using an abrasive cloth or a stainless steel brush. Even if ac or DCEP is used, cleaning prior to welding is a good practice. **Always wear goggles to protect your eyes when using chemical solutions.**

The joint designs used for GTAW are the same as those used for SMAW. On thin metal sections, the flange-type joint is used without filler metal. When gas tungsten arc welding metal over 3/16″ (4.8mm) thick, the edge is generally beveled or machined to the desired groove shape.

If parts must be beveled or cut to shape or size, they may be cut by thermal or mechanical means. Plasma arc cutting produces a clean, fairly smooth, and accurate cut. Machining, sawing, shearing, and grinding are the mechanical processes usually used.

Metal parts of a weldment must be held in alignment while welding. This may be done by using jigs and fixtures. It is often done by tack welding the parts into place.

To control the depth of penetration, a backing strip or ring may be used. The backing ring or strip may also be used when welding metals that may exhibit *hot shortness*. Hot shortness occurs on such metals as aluminum, magnesium, and copper. These metals, when they reach the melting temperature, become very weak. As metal in the weld area reaches the melting temperature, it may fall away. This would leave a large hole in the base metal. A backing strip or ring helps to prevent this falling away of the metal.

Shielding gas may be used on the back or root side of the weld when welding metals like magnesium, titanium, or zirconium. This backing gas is applied by various means. It may be applied through a metal channel attached to the root side of the weld. In pipes or closed containers, the inside of the weldment may be filled with the shielding gas. A vent must be provided when gas is used in a closed container. This vent allows the gas to escape, thus avoiding a pressure buildup that could prevent good penetration.

Another method used to keep contamination away from the weld is to weld the parts in a chamber. See Figure 14-36. The chamber may be filled with shielding gas or it may be a vacuum. Welding parts in a chamber is a way of making sure all parts of the weld and base metal are protected. Titanium is often welded in a chamber. More information about welding titanium is found in Chapter 22.

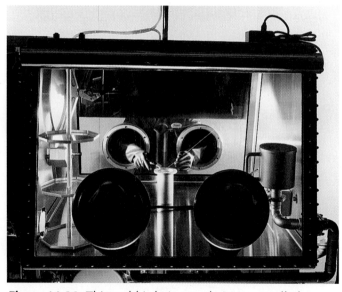

Figure 14-36. *This weld is being made in a controlled atmosphere chamber. The chamber is filled with an inert gas. (Vacuum/Atmospheres Co.)*

Caution: Never enter an area filled with a shielding gas unless you are wearing supplied air breathing equipment. A person can lose consciousness in as little as seven seconds in spaces filled with shielding gases like argon. Death from suffocation has resulted. This is because shielding gases displace oxygen.

14.5 METHODS OF STARTING THE ARC

The gas tungsten arc may be started in one of three ways. These are by touch starting, or by the application of a superimposed high frequency, or by high-voltage starting.

To start the arc, the remote finger- or foot-operated contactor switch must be depressed. See Heading 14.3 and Figures 14-10 and 14-11. This switch also causes the shielding gas to flow prior to starting the arc.

When *touch starting* is used, the electrode is touched to the base metal, then withdrawn about 1/8" (3mm). After a few seconds, the arc will stabilize (run smoothly). It may then be brought down to a short arc length of 1/32" to 3/32" (0.8mm to 2.4mm). By touching the tungsten electrode to the base metal, either may become contaminated because of some base metal or tungsten transfer.

To start the arc using *superimposed high frequency*, place the nozzle on the metal as shown in Figure 14-37. With the electrode and nozzle in this position, the contactor switch is turned on to start the high-frequency current. The machine contactor switch may be actuated by a foot- or finger-operated remote switch. Another method, when using high-frequency start, is to hold the electrode horizontally about 1" (25mm) above the metal. The electrode is then rotated toward a vertical position. As the electrode comes near the base metal, the high frequency will jump the gap to start the arc. When using direct current, the high frequency will turn off automatically when the arc is stabilized. The high frequency should remain on constantly when using alternating current.

High-voltage starting is done with a high-voltage surge. The electrode is brought close to the base metal, as in high-frequency starting. When the contactor switch is pressed, a high-voltage surge causes the arc to jump the gap and start. After the arc is stabilized, the voltage surge stops automatically.

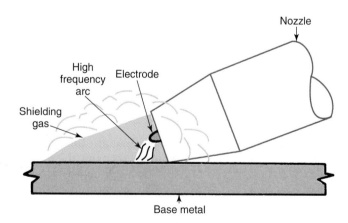

Figure 14-37. *The position of the electrode and nozzle for high-frequency arc starting.*

14.6 GAS TUNGSTEN ARC WELDING TECHNIQUES

One advantage of GTAW is that a weld may be made with a small heat-affected zone around it. Oxyfuel gas and SMAW heat a large area as the metal is raised to the melting temperature. This causes a large heat-affected zone and a potentially weaker metal area around the weld.

Another advantage of GTAW is that there is no metal transfer through the arc. There is no spattering of metal globules from the welding arc or weld pool. The arc action is very quiet and the completed weld is of high quality.

GTAW should be done with the lowest current necessary to melt the metal. Use the highest welding speed possible that will ensure a sound weld.

Once the arc is struck, it is directed to the area to be melted. A molten weld pool is formed under the arc and the filler rod is added to fill the pool. The width of the pool, when making stringer beads, should be about 2 to 3 times the diameter of the electrode used. If the bead must be wider, a weaving bead is used. Several stringer beads may also be used to fill a wide groove joint. Sufficient shielding gas must flow to protect the molten metal in the weld area from becoming contaminated.

The filler rod must not be withdrawn from the area protected by the shielding gas. If the filler rod is withdrawn while it is molten, it will become contaminated. If it is then melted into the weld, the weld will become contaminated.

After the arc is struck, heat a spot until a molten pool forms. The electrode should be held at about a 60°-75° angle from the workpiece. Hold the filler rod at about a 15°-20° angle to the workpiece. See Figure 14-38. When the molten weld pool reaches the desired size, add the filler rod to the pool. When the filler rod is to be added, the electrode should be moved to the back of the weld pool. The filler rod is then added to the forward part of the molten pool. Refer to Figure 14-39. This technique of adding the filler metal to the weld pool may be used for all weld joints in all positions.

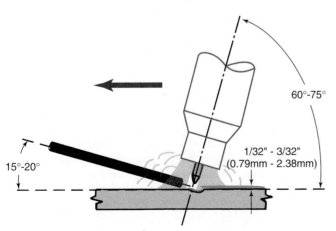

Figure 14-38. *The relative positions of the electrode and the filler rod for GTAW in the flat (downhand) position.*

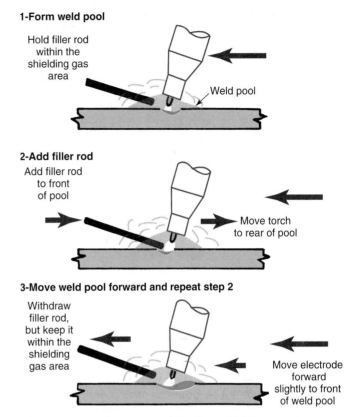

1-Form weld pool

Hold filler rod within the shielding gas area

Weld pool

2-Add filler rod

Add filler rod to front of pool

Move torch to rear of pool

3-Move weld pool forward and repeat step 2

Withdraw filler rod, but keep it within the shielding gas area

Move electrode forward slightly to front of weld pool

Figure 14-39. *The steps required to add filler metal during GTAW. This process is continuous until the weld is completed.*

14.7 SHUTTING DOWN THE GTAW STATION

Each time the arc is broken, shielding gas continues to flow for a few seconds. This is done to protect the weld metal, electrode, and filler metal from becoming contaminated by the surrounding atmosphere. The gas also continues to flow after the torch or foot-operated contactor (off-on) switch is turned off.

When welding is stopped for a short time, the GTAW torch should be hung on an insulated hook. If welding is to be stopped for a long period of time, the station should be shut down. After the gas postflow period, hang up the torch. Shut off the shielding gas cylinder. Turn on the torch or foot-operated contactor switch to start the gas flow. This is done to drain the complete shielding gas system of gas. Turn out the regulating screw on the regulator to turn it off. Screw in the flowmeter to shut it off. If the flowmeter is not turned off, the float ball will hit the top of the flowmeter very hard when the regulator is opened again. Turn off the arc welding power source switch.

14.8 WELDING JOINTS IN THE FLAT WELDING POSITION

Welding joint designs for GTAW are the same as those used for oxyfuel gas and shielded metal arc welding. The flat or downhand position is generally the easiest position in which to weld. Refer to Chapter 3 for more information on weld joints and weld symbols.

14.8.1 Square-Groove Weld on a Flange Joint

The metal should be bent up as shown in Figure 14-40. This bent-up edge is melted to form a weld. No filler rod is required. The electrode should be held at about 60°-75° from the base metal. The GTAW arc is applied to the flanged metal to form a weld pool on both edges. As the weld pool increases in size, it will touch the outer edges of the flange. When this occurs, the electrode must be moved ahead to the front edge of the weld pool. When the pool again touches the outer edges of the flange, the electrode is again moved ahead. With practice, this motion will become smooth to create a weld bead with uniformly spaced ripples. This is repeated until the weld is completed.

When the end of the weld joint is reached, the electrode should be moved backward slightly and the contactor switch used to decrease the welding current. If remote current control is not used, slowly raise the torch until the arc stops. Hold the electrode near the end of the weld until the shielding gas stops flowing. This postflow of shielding gas protects the weld and electrode as they cool. A general rule for postflow is to allow one second for every 10A of current.

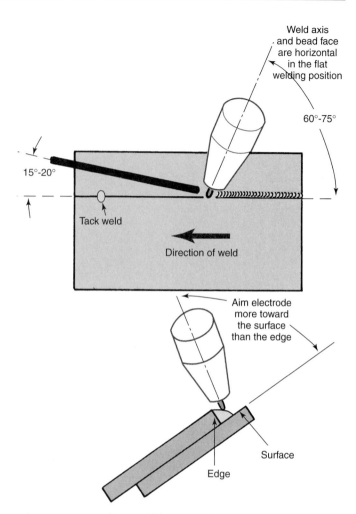

Figure 14-41. A lap weld being welded with GTAW in the flat (downhand) position. Notice the suggested angles for the electrode and filler metal.

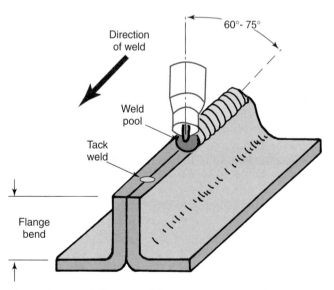

Figure 14-40. A flange weld in progress. Note the torch angle, weld pool size, and the tack weld.

14.8.2 Fillet Weld on a Lap Joint

The fillet weld on a lap joint should be set up as shown in Figure 14-41 and tacked about every 3" (75mm).

An edge and a surface are being heated on a lap joint. The electrode should be aimed more toward the surface, since it will take more heat to melt. The metal should be heated to the melting point. When the weld pool forms a C-shape, as shown in Figure 14-42, the electrode is moved to the rear of the pool. At the same time, the filler metal is added to the front of the weld pool. The C-shaped weld pool indicates that both pieces are molten and flowing together. Enough filler metal is melted into the weld pool to form the desired weld bead contour. A flat or convex bead is generally desired. The filler metal is then withdrawn slightly, but should remain within the shielding gas area. As the filler rod is withdrawn, the electrode is moved to the front of the weld pool. The C-shaped pool is again formed and the filler metal added. This procedure is continued to the end of the weld. When the end of the weld is reached, the electrode is slowly raised and moved to the rear of the weld pool. Filler metal is added to fill the weld pool. The electrode is then withdrawn to break the arc or the current is decreased until the arc stops. If a wide joint is welded, several passes may be required to fill the joint.

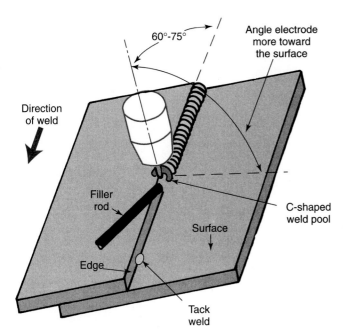

Figure 14-42. A fillet weld in progress on a lap joint. Notice that the electrode is aimed more toward the surface than toward the edge. The edge will melt more easily.

14.8.3 Fillet Weld on an Inside Corner Joint

The inside corner joint should be set up as shown in Figure 14-43. It should be tack welded about every 3" (75mm). The fillet weld is being made on two surfaces in this case. Each piece must be heated evenly. Therefore, the electrode should be kept at about a 45° angle to each piece, as shown in Figure 14-43. When both pieces are melted sufficiently to run together, the weld pool will appear C-

shaped. When the C-shaped weld pool forms, the electrode is moved to the rear of the pool. At the same time, the filler rod is added to the front of the weld pool. Continue to add the filler metal until a flat or convex bead forms. Withdraw the filler rod and move the electrode to the front of the weld pool. Re-form the C-shaped pool. Move the weld pool forward. When additional filler metal is required, add it to the front of the weld pool while moving the electrode to the rear of the pool. Repeat this action to the end of the weld. When the filler rod is withdrawn, always keep it within the area protected by the shielding gas. Finish off the weld by adding filler metal to fill the weld pool. Then, withdraw the torch or decrease the current until the arc stops. If a wide joint is welded, several passes will be required to fill the joint.

14.8.4 Square-Groove and V-Groove Welds on a Butt Joint

Metal under 3/16" (4.8mm) in thickness may not require edge preparation. If full penetration is difficult to obtain from one side, the edges should be machined or flame-cut. When metals are over 3/16" (4.8mm) in thickness, the edges must be machined or flame-cut to obtain full penetration.

A square-groove weld on a butt joint is shown in Figure 14-44. Note the electrode and filler rod position in relation to the weld line. The electrode should point straight down the weld line as shown in Figure 14-44B.

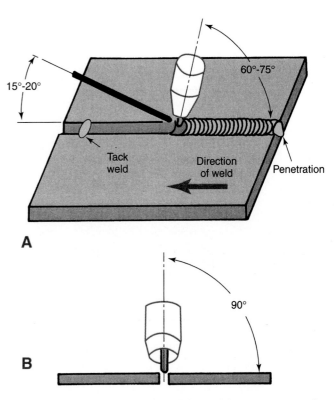

A

B

Figure 14-44. A square-groove butt weld in progress in the flat position. Note that the penetration is 100%. The torch angles are the same as for other weld joints. The electrode should point straight down the weld line, as shown in B.

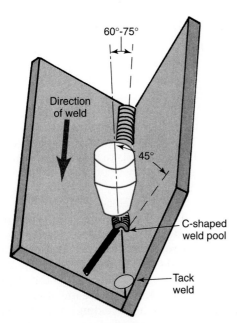

Figure 14-43. An inside corner joint. A fillet weld is in progress in the flat position using GTAW. In the flat position, the face of the weld should be near-horizontal.

If the electrode is pointed toward one piece or the other, that piece will melt more and the weld bead will pile up on that side. Notice also the suggested electrode and filler rod angles from the base metal. The weld joint designs used for GTAW are the same as those used for SMAW. See Figure 3-1 for additional information on basic welding designs.

GTAW is seldom used to weld metals over 1/4" (6.4mm) thick, except in very critical applications. On carbon steel, alloy steel, stainless steel, copper, and other metals over 1/4" (6.4mm), GTAW is used only for the root pass. After the root pass is laid, other types of welding processes are generally used. Other processes used may be SMAW, SAW, and GMAW.

The design shapes may be square-, U-, J-, V-, or bevel-groove. These design shapes may be cut on both sides of a thick metal butt weld. Figure 14-45 shows a single-sided, V-groove butt weld in progress. The torch and filler rod angles are the same as those used on other GTAW butt joints. Complete penetration is required in a butt weld. To obtain complete penetration, the *keyhole* method of welding is used. See Headings 6.8 and 12.7.5 for explanations of keyhole welding.

The width of a stringer bead, like the root pass, should only be 2 to 3 times as wide as the electrode diameter. To complete a V-groove or other wide-mouthed joint, a weaving bead might be used. A weaving bead done with the gas tungsten arc should only be about 6 times the electrode diameter in width. If a wider area must be welded, then several narrower passes must be used.

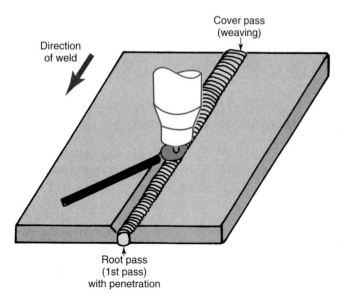

Figure 14-45. A V-groove butt weld, welded in the flat position with the gas tungsten arc. Two or more weld passes may be required to weld a thick joint.

14.8.5 Square- or V-Groove Outside Corner Joint

The square- or V-groove outside corner joints are set up as shown in Figure 14-46. They are welded in the same manner as a square-groove or V-groove butt weld.

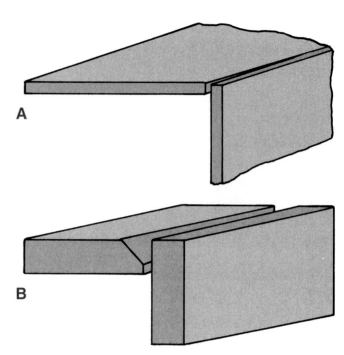

Figure 14-46. Outside corner joints. A—Square-groove butt joint. B—Bevel-groove butt joint. These joints are welded in the same manner as the square-groove and V-groove butt joints.

14.9 WELDING JOINTS IN THE HORIZONTAL WELDING POSITION

The welding procedures for GTAW explained in Heading 14.8 are used when welding in the horizontal position.

When horizontal welding is done, the weld line is horizontal and the face of the bead is in a near-vertical position. Molten weld metal tends to move downward. To prevent this downward movement of the weld metal, the following actions can be taken:
- Do not create a large-diameter weld pool.
- Add the filler rod to the upper edge of the weld pool.
- Point the electrode slightly upward. This upward angle will use the force of the arc to reduce sagging.

Figures 14-47 and 14-48 illustrate a lap joint being welded in the horizontal position. Notice that the angle of the electrode and filler rod from the base metal surface are similar to the angles used when welding in the flat position. The electrode may be pointed upward slightly to reduce sagging of the molten metal.

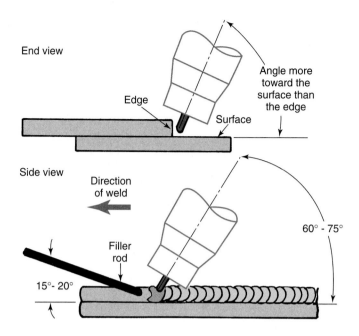

Figure 14-47. *A lap joint being welded with GTAW in the horizontal position. Notice the suggested angles for the electrode and filler rod.*

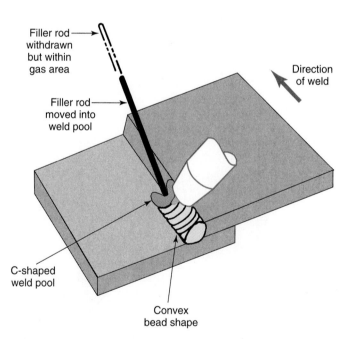

Figure 14-48. *A fillet weld on a lap joint in the horizontal position.*

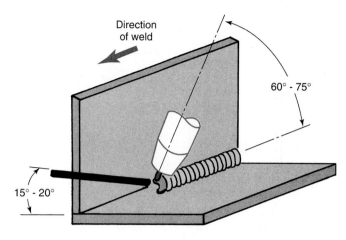

Figure 14-49. *GTAW used to make a fillet weld on an inside corner joint. This weld is being done in the horizontal position.*

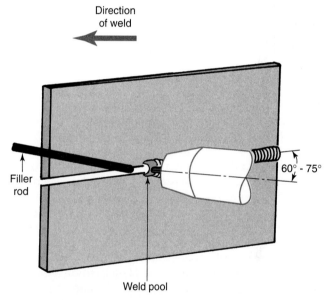

Figure 14-50. *Horizontal square-groove butt weld in the horizontal position. The keyhole method is used to obtain full penetration. The filler rod is held about 15°-20° from the base metal. The outer end of the filler rod is held slightly above the horizontal weld line.*

An inside corner joint is shown being welded in the horizontal position in Figure 14-49. A square-groove weld on a butt joint is shown in Figure 14-50. Figure 14-51 illustrates the suggested electrode and filler rod angles for use on a V-groove butt joint. The outside corner joints can be welded horizontally in the same manner as a square-groove or V-groove butt joint.

14.10 WELDING JOINTS IN THE VERTICAL WELDING POSITION

When welding vertically, it is important to use the lowest current possible to make a good weld. Lower current will help to keep the weld pool from becoming too large. If the weld pool is too large, the molten metal may flow away from it. A pulsed arc may be used to provide a good cooling period that will help control the molten pool. Refer to Heading 14.1 for an explanation of the pulsed arc.

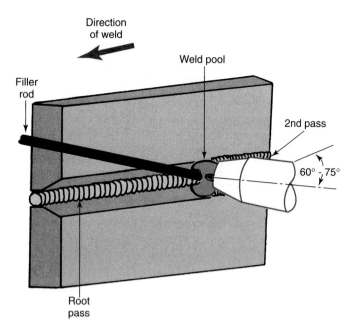

Figure 14-51. *Horizontal V-groove butt weld in the horizontal position. The filler rod is held about 15°-20° from the base metal. The outer end of the filler rod is held slightly above the horizontal weld line.*

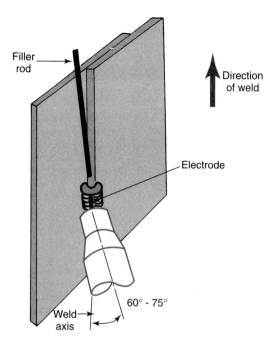

Figure 14-52. *A fillet weld being made on a lap joint in the vertical position.*

Electrode and filler rod angles are the same as for welding in the flat position. The electrode should be held about 60°-75° from the base metal surface. The filler rod is held about 15°-20° from the base metal.

Vertical welds may be made uphill (from the bottom up) or down hill (from the top down). Metal thicknesses over 1/2" (13mm) are seldom welded downhill. There is no flux to run into the molten weld pool when doing GTAW. Therefore, both uphill and downhill welding produce welds of excellent quality.

Figure 14-52 illustrates a fillet weld being made on a lap joint in the vertical position. The electrode should be aimed more toward the surface. This prevents the edge of the lapped metal from melting too quickly. Wait for the weld pool to form a C-shape before adding the filler metal to it. This crescent shape indicates that both pieces are molten and are running together. If the filler rod is added before the C-shaped weld pool forms, the filler metal may not fuse with both pieces of the joint. A fillet weld on an inside corner joint in the vertical position is shown in Figure 14-53. Like the fillet weld on a lap joint, do not add the filler metal until the C-shaped weld pool forms.

Figure 14-54 illustrates a square-groove weld being made on a butt joint. The outside corner joint shown in Figure 14-55 is being joined with a bevel-groove weld.

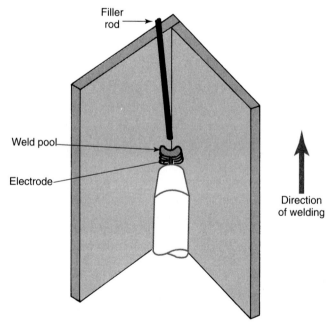

Figure 14-53. *A fillet weld being made on an inside corner joint in the vertical position. Note that the electrode is distributing the heat evenly to both surfaces. This weld is being made uphill.*

14.11 WELDING JOINTS IN THE OVERHEAD WELDING POSITION

Welds in the overhead position are relatively easy to make using the gas tungsten arc. The temperature and size of the weld pool must be controlled. This may be done by selecting the lowest possible effective current. Cooling of the weld pool is also possible by using a pulsed arc. Refer to Heading 14.1 for an explanation of the pulsed arc. Small beads are used to keep the weld pool small and controlled. Several passes will be necessary in this case to completely fill the weld groove. The angle of the electrode is kept at

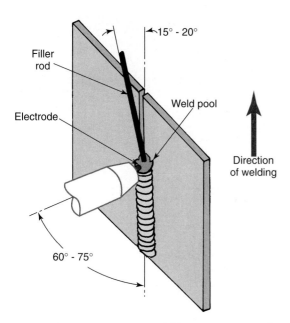

Figure 14-54. *A square-groove weld in progress on a butt joint in the vertical position. The weld is being made in the uphill direction, using the keyhole method to get full penetration.*

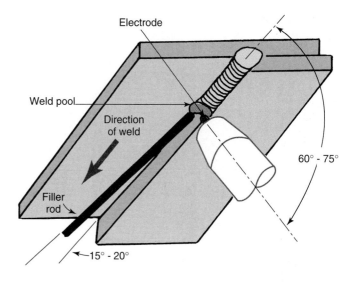

Figure 14-56. *A fillet weld on a lap joint in an overhead position. Note that the angles of the electrode and filler rod are the same as in the flat position.*

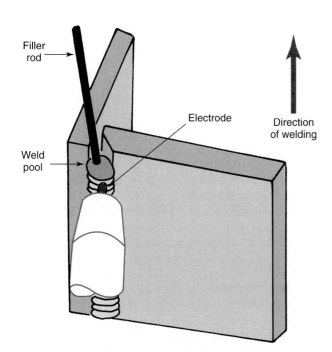

Figure 14-55. *A bevel-groove weld being welded uphill on an outside corner joint.*

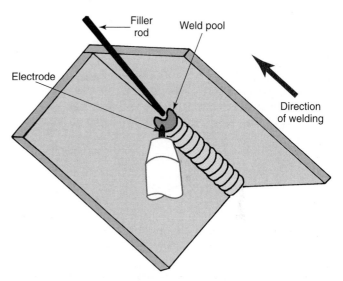

Figure 14-57. *A fillet weld on an inside corner joint in the overhead position.*

14.12 SEMIAUTOMATIC WELDING

Semiautomatic GTAW is very similar to manual GTAW, with the addition of a filler metal feed mechanism. The feed mechanism must have the following:

- A wire drive device.
- A speed control.
- An attachment to the torch that will guide the wire into the weld pool.

The welder must control the arc length and travel speed. Semiautomatic welding is limited to fairly straight weld joints because the filler metal only comes from one direction. Changing directions is not possible unless the entire torch changes direction so the filler metal continues

about 60°-75° from the surface of the base metal. An angle of 15°-20° is considered correct for the filler rod.

Figure 14-56 illustrates a fillet weld being made on a lap joint in the overhead position. An overhead fillet weld is being made on an inside corner joint in Figure 14-57. The overhead weld shown in Figure 14-58 is a U-groove on a butt joint. Figure 14-59 shows a square-groove weld being made overhead on an outside corner joint.

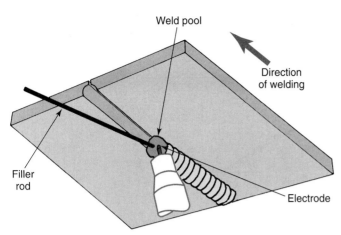

Figure 14-58. *A U-groove weld on a butt joint in the overhead position. The angle of the electrode and filler rod from the base metal is the same as for flat position welding.*

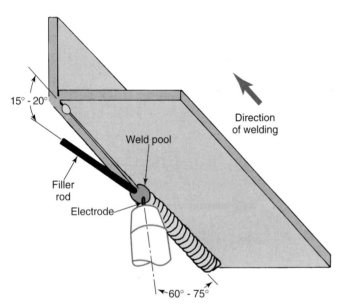

Figure 14-59. *A square-groove weld on an outside corner joint in the overhead position. Note the tack weld.*

to enter the weld pool from the same point. Figure 14-60 shows a cold wire feeder attachment on a GTAW torch.

14.13 AUTOMATIC AND MECHANIZED GTAW

Gas tungsten arc welding can be mechanized or done in a fully automatic manner. In mechanized and automatic GTAW, the torch is moved along the joint to be welded. The torch may be mounted on an electrically driven tracer, on a rigid frame (as when automatic flame cutting), or on a robot arm. See Figure 14-61. Filler metal is added, when required, by using a wire feed motor and drive

Figure 14-60. *Semiautomatic GTAW. The filler wire is automatically added to the weld pool as the welder controls the torch. (CK Worldwide)*

mechanism. The filler wire is fed into the weld pool at a constant rate by the feed mechanism.

An *automatic GTAW system* uses feedback signals to adjust the process as necessary to maintain a high-quality weld. Automatic GTAW often has a voltage feedback system to monitor the arc voltage. The system uses the arc voltage to automatically adjust the distance of the torch to the base metal, maintaining a constant arc length. Vision systems can be used to follow and track the weld seam, as seen in Figure 14-62. This is important if parts do not match up perfectly or if the joint is not a straight line. Permanent records can be kept on critical welds by videotaping the weld as it is made.

A robot using automatic GTAW can be programmed to weld a complete and complex assembly. Welding can be done in numerous positions and on different joint types with excellent results.

Mechanized GTAW does not have feedback systems like these described for automatic welding. In mechanized GTAW, the torch is moved along a weld seam at a set height above the base metal. There is no method for automatically adjusting the welding process if the metal is not flat or the seam is not straight.

Extremely accurate and sound welds are possible using automatic GTAW. Automatic GTAW can be used on metal as thin as 0.003" (0.076mm). Generally, metals up to

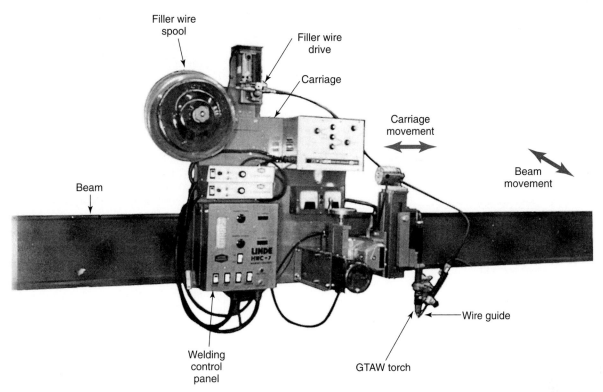

Figure 14-61. *A mechanized gas tungsten arc welding arrangement. The torch, wire feed, and welding control panel are mounted on a horizontal beam. The carriage moves along the beam. The beam can also be moved. (ESAB Welding and Cutting Products)*

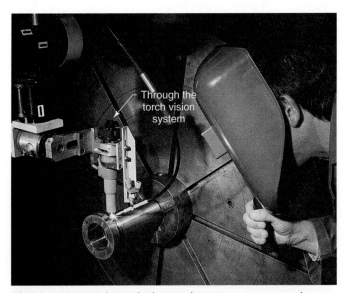

Figure 14-62. *A through-the-torch vision system used to make an automatic GTA weld on a part. This system allows a view straight down on the weld. The arc provides the light; a TV screen is used to monitor the weld pool. (Rocketdyne Division, Rockwell International Corp.)*

0.02″ (0.51mm) are welded without filler rod. This is made possible by using a flange joint design and melting the flange, which serves as the filler metal. Above 0.02″ (0.51mm), filler wire or rod is often required, depending upon the joint design.

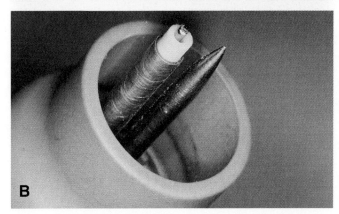

Figure 14-63. *Two methods used for feeding the filler wire into the weld pool through the GTAW nozzle. A—The electrode is grooved to accept the insulated wire feed guide. B—The wire feed guide is insulated and runs next to the electrode. (Rocketdyne Division, Rockwell International Corp.)*

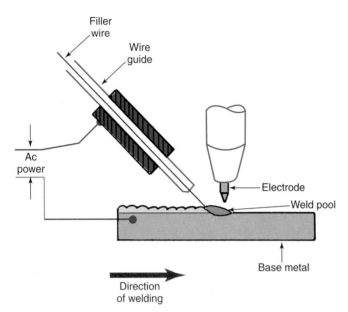

Figure 14-64. *A schematic of the hot filler wire process in use. The wire is heated and fed at a constant rate directly into the weld pool.*

The filler wire feed rate must be carefully set and the wire directed into the weld pool. The wire feed rate will determine the size of the bead buildup. Figure 14-63 shows specialized applications in which the wire is fed through the nozzle directly into the arc. This allows the torch to move in slightly different directions without the need for torch rotation to keep the wire feeder in the proper location.

Automatic GTAW may be done in all positions using pulsed arc. In high-current automatic welding applications, the wire guide may be water-cooled.

To speed the rate of welding, a hot filler wire process is sometimes used. The filler wire is fed, controlled, and directed in the same manner as cold filler metal. One important difference is that hot filler wire is added to the trailing edge of the weld pool, as opposed to cold filler wire, which is added to the front or leading edge of the weld pool.

The hot wire process heats the wire before it enters the weld pool. The filler metal is resistance-heated as it travels through the wire guide. See Figure 14-64. This saves much of the heat of the arc for melting the base metal. Since the filler metal is preheated, the welding speed is much faster.

14.14 GTAW TROUBLESHOOTING GUIDE

Occasionally, a problem such as those listed below will occur, which will affect the quality of the completed weld:
- An unstable arc.
- Rapid electrode consumption.

- Tungsten in the weld (inclusions).
- Porous welds.

Figure 14-65 lists problems, their possible causes, and how to correct them.

14.15 GTAW SAFETY

All safety precautions required for other electric arc welding processes apply to GTAW, as well. These include wearing proper protective clothing. Like other arc welding processes, GTAW creates ultraviolet and infrared radiation. This radiation can cause burns to the skin if exposed for long periods of time. Gloves are required to protect the hands. Gloves used in GTAW are often thinner than those used in shielded metal arc welding, because the welder needs to have a good feel of the torch and the filler metal.

The filler shade numbers recommended for GTAW are as follows:

Less than 50A	10
50A to150A	12
Over 150A	12-14

Wear supplied air breathing equipment when entering a space filled with an inert gas, CO, or CO_2.

Phosgene gas is formed when arc welding on parts that have been cleaned with chlorinated hydrogen solvents, if a film of the solvent is present. Adequate ventilation must be provided to remove this toxic gas.

TEST YOUR KNOWLEDGE

Write your answers on a separate sheet of paper. Do not write in this book.

1. Which dc electrical polarity, DCEN or DCEP, requires a larger electrode diameter? Why?
2. The best direct current polarity to use for cleaning surface oxides off aluminum or magnesium is DCE ____?
3. Why do the electrons not flow easily during the DCEP half of the ac cycle?
4. How many seconds of gas postflow should be selected if the machine is set to deliver 130A?
5. When GTAW 1/8" (3.2mm) mild steel, what amperage is suggested when welding a lap joint?
6. Which inert shielding gas generally provides the best surface cleaning action?
7. How many cubic feet per hour (cfh) and liters per minute (L/min) of argon are suggested for GTAW a 3/16" (4.8mm) aluminum butt joint? What diameter electrode is suggested for this weld?
8. When adjusting a flowmeter, what part of the floating ball is aligned with the desired flow rate line on the flowmeter?
9. What is the exit diameter of a number 4 size gas tungsten arc welding nozzle?
10. List two reasons why a small-diameter nozzle is often used.

Trouble	Possible Causes	How to Correct
Arc instability	1. Dirty, contaminated base material. 2. Joint is too narrow. 3. Contaminated electrode. 4. Electrode diameter too large. 5. Arc too long.	1. Use chemical cleaners, wire brush, abrasives as appropriate to clean base material. 2. Make groove wider; bring electrode closer to the work; decrease voltage. 3. Cut off end of electrode tip, dress tip. 4. Use small electrode, smallest diameter that will handle required current. 5. Bring electrode closer to work.
Rapid electrode consumption (use)	1. Inert shielding is inadequate, allowing oxidation of the electrode. 2. Operating on reverse polarity. 3. Electrode too small for required current. 4. Electrode holder is too hot. 5. Electrode contamination. 6. Oxidation of electrode during cooling.	1. Clean nozzle; bring nozzle closer to work; increase gas flow. 2. Change to straight polarity to use larger electrode. 3. Use larger electrode - See Figs. 14-19 to 14-23. 4. Change collet; use ground finish electrodes; check for proper collet contact. 5. Remove contaminated section of electrode. Electrode will continue to degrade as long as contaminants are present. 6. Continue gas flow for 10-15 seconds after arc stops. Rule: 1 second for each 10 amps.
Tungsten inclusions in work	1. Touch starting with electrode. 2. Electrode melts and alloys with base plate. 3. Fragmentation of electrode by thermal shock.	1. Use high-frequency starting device; use a copper striking plate. 2. Use lower current or larger electrode; use alloyed tungsten electrode (they run cooler) and reduce electrode extension. 3. Be sure electrode ends are not cracked, especially when using high currents. Remove cracked end. Use a small flat on the end of the electrode when dc welding.
Porosity	1. Gas impurities present; hydrogen, nitrogen, air, water vapor. 2. Use of old acetylene hose. 3. Gas and water hoses interchanged. 4. Oil film on base material.	1. Use welding grade inert gas (99.995% pure argon or 99.95% pure helium). Purge all lines before striking the arc and check for leaks. 2. Use new hose only. Acetylene impregnates a hose and makes it unsuitable with an inert gas. 3. Never interchange gas and water hoses. Use hoses of different colors. 4. Clean base material with a chemical cleaner that does not dissociate in the arc. Do not weld while material is wet.

Figure 14-65. A GTAW troubleshooting guide.

11. What kind of a tungsten electrode has a yellow band painted on it?
12. A ball is formed on the end of a pure tungsten electrode when welding with _____ current.
13. What will cause a tungsten electrode to split at the tip?
14. When grinding a point on a tungsten electrode, in what direction should the grinding marks run?
15. As a general rule, how far should the electrode extend beyond the nozzle?
16. Why is it extremely dangerous to enter a space filled with an inert gas without wearing proper breathing equipment?
17. List and describe three ways of starting the GTAW arc.
18. When gas tungsten arc welding, how is the electrode moved as the filler rod is added?
19. When GTAW, the electrode is held at _____°-_____° from the base metal and the filler rod is held at _____°-_____° from the base metal.
20. List two causes for rapid tungsten electrode consumption (use).

These two welders are using gas metal arc welding (GMAW) to fabricate a framework from square tubing. (Miller Electric Mfg. Co.)

Chapter 15

GAS METAL ARC WELDING

LEARNING OBJECTIVES

After studying this chapter, you will be able to:

* Contrast the various GMAW metal transfer methods, considering arc characteristics, weld characteristics, and the possibility of performing out-of-position welds.
* Select the proper arc welding machine, wire feeder, shielding gas, flow rate, contact tube, nozzle size, and electrode wire type to produce an acceptable GMA weld.
* Contrast the various types of shielding gases used when GMAW, and how they affect the shape and penetration of the completed welds.
* Properly assemble and adjust all the equipment required to produce an acceptable GMA and FCA weld.
* Correctly prepare metals for welding, and perform acceptable welds on all types of joints in all positions using GMAW and FCAW.
* Identify the potential safety hazards involved in the GMAW and FCAW process in a working environment; be able to describe ways of safely dealing with these hazards.
* Be able to pass a safety test on the proper use of the GMAW and FCAW process.

The *gas metal arc welding (GMAW)* process uses a solid wire electrode that is continuously fed into the weld pool. The wire electrode is consumed and becomes the filler metal. *Flux cored arc welding (FCAW)* is very similar to gas metal arc welding. One big difference is that FCAW uses an electrode wire with flux inside the wire. For an overview of these processes, refer to Headings 4.7 and 4.8 and also to Figures 4-12 and 4-13.

The growth in the use of GMAW is the result of several events. The continuous development and refinement of constant voltage arc power sources and wire feeders has

made GMAW more useful. Welding using GMAW is easy to learn, especially if a welder has already learned to weld using a different process. GMAW equipment is low in cost. Also, this process deposits more weld metal in lbs./hr. (kg/hr) than the shielded metal arc or gas tungsten arc welding processes. The low purchase cost, the ability to weld continuously, and the ability to deposit weld metal faster, make GMAW an attractive choice for welding.

GMAW can be used to produce high-quality welds on all commercially important metals such as aluminum, magnesium, stainless steels, carbon and alloy steels, copper, and others. GMAW may also be done easily in all welding positions.

15.1 GAS METAL ARC WELDING PRINCIPLES

Gas metal arc welding is generally used because of its high productivity. GMAW is done using solid wide electrodes. FCAW uses flux cored wire electrodes. See Figure 15-1. A shielding gas or gas mixture must be used with GMAW.

GMAW is done using DCEP (DCRP). Alternating current is never used. DCEN (DCSP) is rarely used for GMA welding, but has found very limited use for surfacing. DCEN (DCSP) is used with only one special electrode, called an *emissive electrode*. (AWS designation E70U-1).

For every pound of solid electrode wire used, 92%-98% becomes deposited weld metal. Flux cored arc welding wire is deposited with a wire efficiency of 82%-92%. As a comparison, shielded metal arc welding (SMAW) deposits 60%-70% of the electrode wire as weld metal. Some spatter does occur in the GMAW and FCAW processes. Very little stub loss occurs when continuously fed wire is used.

There is a very thin glass-like coating over the weld bead after GMA welding. No heavy slag is developed because the weld area is shielded by a gas. When FCAW, a slag covering is present. Some of the flux in the FCAW

365

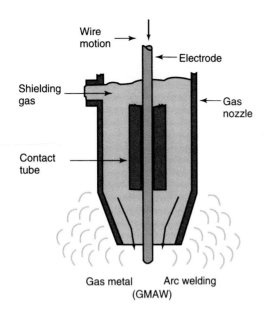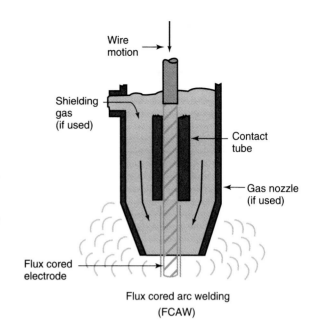

Figure 15-1. *Schematic views of GMAW and FCAW gas nozzles and electrodes. Shielding gas is not always used with FCAW. If shielding gas is not used, no nozzle is required.*

forms a gas around the weld area. Some of the flux forms a slag, covering the weld. Shielding gas may or may not be used when FCAW. More welder time can be spent on the welding task with a continuously fed wire process. This improves the cost efficiency of GMA and FCA welding.

The GMAW process can be adapted to a variety of job requirements by choosing the correct shielding gas, electrode size, and welding parameters. Welding parameters include the voltage, travel speed, and wire feed rate. The arc voltage and wire feed rate will determine the filler metal transfer method.

Metal transfer occurs in two ways. One is by the short circuiting method. The second is to transfer metal across the arc. Methods of transferring metal across the arc include:

- Globular transfer.
- Spray transfer.
- Pulsed spray transfer.

15.1.1 Short Circuit GMAW

Short circuit gas metal arc welding (GMAW-S) is used with relatively low welding currents. It also uses electrode wire sizes under 0.045" (1.1mm). This process is particularly useful on thin metal sections in all positions. All position welds are made easily because there is no metal transfer across the arc. The weld pool cools and solidifies rapidly using the short circuiting arc. Short circuiting transfer has a low heat input into the base metal.

Since short circuit gas metal arc welding has a low heat input, it is also used to weld thick sections in the overhead or vertical welding position. It is very effective in filling the large gaps of poorly fitted parts.

Refer to Figure 15-2 to see how the short circuiting arc method deposits metal. When the electrode touches the molten weld pool, the arc is no longer present. The surface

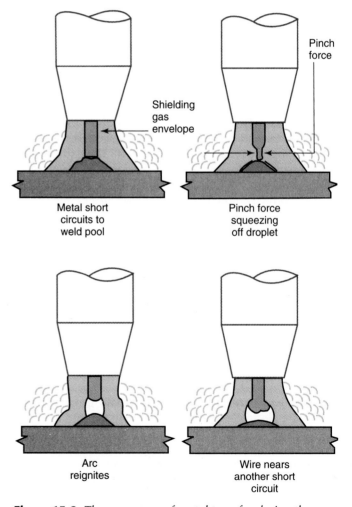

Figure 15-2. *The sequence of metal transfer during the short circuit GMAW method.*

tension of the pool pulls the molten metal from the end of the electrode into the pool. The ***pinch force*** around the electrode squeezes the molten end of the electrode. The combined effects of surface tension and the pinch force separate the molten metal and the electrode. The arc then reestablishes itself. The continuously fed electrode again touches the molten pool and the process repeats. The droplet transfer or short circuiting process repeats itself about 20 to 200 times per second. The strength of the pinch force depends on the arc voltage, the slope of the power source or welding machine, and the circuit resistance. These factors — voltage, slope, and resistance — affect the welding current. The frequency of the pinch force and the formation of droplets is controlled by the inductance of the power source.

If a 150A current is set on the arc welding machine, the amperage may rise rapidly to the maximum output of the machine when the electrode short-circuits. This could be 500A or more. To control and slow down this possible rapid rise in current, an inductance circuit is built into the arc welding machine.

Inductance is the property in an electric circuit that slows down the rate of the current change. Some arc welding machines have an electric coil built in near the welding current transformer coils. See Heading 11.2.1 for a discussion of inductance. The current traveling through an inductance coil creates a magnetic field. This magnetic field creates a current in the welding circuit that is in opposition to the welding current. Increasing inductance in a welding machine will slow down the increase of the welding current. Decreasing the inductance will increase the rate of change of the welding current.

When *too little* inductance is used, the current rises too rapidly. The pinch force is so great that the molten metal at the end of the electrode literally explodes. A great deal of spatter occurs in this case. When *too much* inductance is used, the current will not rise fast enough. The molten on the electrode is not heated sufficiently.

By properly balancing the inductance and slope, an ideal droplet transfer rate and pinch force can be obtained. See Figure 15-3 for the metal deposition rate for the short circuiting transfer method. Shielding gas also has an effect on short circuiting transfer. Inert gases must be used on all nonferrous base metals. Nonferrous base metals are those that do not contain iron as the main element. This grouping includes everything except steels, steel alloys, and cast

GMAW Method	Metal Deposited	
	lbs/hr	kg/hr
Short circuiting	2-6	0.9-2.7
Globular	4-7	1.8-3.2
Spray	6-12	2.7-5.4
Pulsed spray	2-6	0.9-2.7

Figure 15-3. *The approximate rate at which filler metal is deposited with various GMAW methods. (American Welding Society)*

irons. Adding helium to argon will increase the penetration. Argon and helium mixtures are used only on nonferrous base metals.

Carbon dioxide (CO_2) may be used as a shielding gas when GMA welding carbon and low-alloy steels. CO_2 will produce greater penetration, but will create more spatter than an inert gas used for shielding. Mixtures of argon and CO_2 are often used. They provide a good combination of improved penetration with minimal spatter. Stainless steel usually requires a mix of three gases. A typical mixture is 90% helium, 7 1/2% argon, and 2 1/2% CO_2.

15.1.2 Globular Transfer

Globular metal transfer gas metal arc welding occurs when the welding current is set slightly above the range used for short circuiting metal transfer. In the globular metal transfer process, the metal transfers across the arc as large, irregularly shaped drops. See Figure 15-4. The drops are usually larger than the electrode diameter.

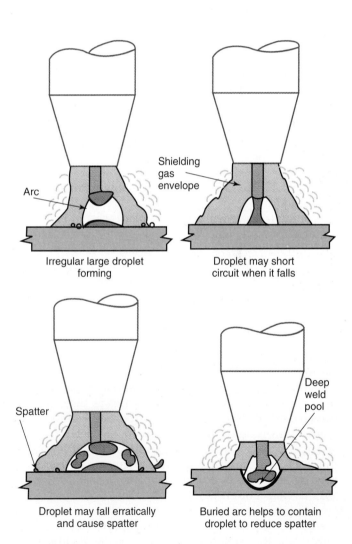

Figure 15-4. *GMAW globular metal transfer. Drops may fall erratically and cause spatter. Note that the buried arc may help contain the drops to reduce spatter.*

Drops form on the end of the electrode. Each drop grows so large that it falls from the electrode due to its own weight. When a high percentage of inert gas is used for shielding, the drops will fall into the weld pool. When a high percentage of carbon dioxide is used, the drops travel across the arc in random patterns, creating spatter. To minimize spatter, a shorter arc length can be used. However, a short arc will allow large drops to short to the work. The drops will explode, still creating a lot of spatter.

One way to minimize spatter when using CO_2 is to increase the current slightly. This will create a deep weld pool that is below the metal surface. This is referred to as a *buried arc* or *submerged arc*. Using a buried arc, much of the spatter is contained within the deep weld pool. With a buried arc, a combination of globular and short circuiting transfer occurs. Deeper penetration occurs when using a buried arc.

Welds of sufficient quality for many applications can be produced with this process. When using globular transfer, welding can be done only in the flat welding position, because the molten metal falls into the weld pool. Welds may be made faster with this process than with the short circuiting transfer method. See Figure 15-3 for the rate at which metal is deposited with this method.

15.1.3 Spray Transfer

Spray transfer gas metal arc welding will occur when the current and voltage settings are increased above those required for globular transfer. When spray transfer occurs, very fine droplets of metal form. These droplets travel at a high rate of speed directly through the arc stream to the weld pool. Figure 15-5 illustrates the spray arc metal transfer method.

Spray transfer will only take place when a high percentage of argon is used. When welding nonferrous metals and alloys, 100% argon shielding gas is used. When using spray transfer on carbon or low-alloy steels or stainless steels, a shielding gas mixture containing at least 90% argon is used

Before spray transfer can occur, a current setting above the *transition current* level must be made on the welding machine. The transition current varies with the electrode diameter, its composition, and the amount of electrode extension. A higher transition current is required for steel than aluminum. The transition current increases with the electrode diameter. It decreases as the electrode extends farther from the contact tube. See Figure 15-40. Until the transition current is exceeded, the metal transfers as large globules. Above the transition current level, the pinch force becomes great enough to squeeze the metal off the tip of the electrode as fine droplets. See Figure 15-6 for various transition current levels.

The droplets are squeezed off cleanly and transferred across the arc gap in a straight path. Spray transfer occurs only when at least 90% argon is used as the shielding gas. Common shielding gas mixtures for carbon and low-alloy steels are: 98% Ar plus 2% O_2; 95% Ar plus 5% O_2; 95% Ar plus 5% CO_2 ; and 90% Ar plus 10% CO_2. The spray transfer method produces deep penetration. The arc can be

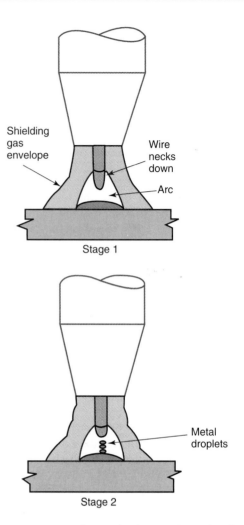

Figure 15-5. *Spray transfer method. Note how the droplets are concentrated in the center of the arc. Spray transfer will occur only when a high percentage of argon gas is used.*

directed easily by the welder. This is because the arc and metal spray pattern are stable and concentrated. Spray transfer is best done in the flat or horizontal welding position, and on metal over 1/8" (3mm) thick. See Figure 15-3 for the metal deposition rate.

15.1.4 Pulsed Spray Transfer

The *pulsed spray transfer* gas metal arc welding method is similar to the spray transfer method. See Figure 15-7. The current level for pulsed spray must be above the transition current level. Special circuits within the power source (welding machine) cause the current to pulse. A low-level current in the globular transfer range is used to maintain the arc. This current is called the *background current*. The current is increased at a regular frequency to the *peak current*. The peak current is above the transition current level. Since the background current is on for only a short time, no globular transfer actually occurs. During the peak current time period, spray transfer occurs. In this

Wire electrode type	Wire electrode diameter		Shielding gas	Minimum spray arc current, A
	in.	mm		
Mild steel	0.030	0.76	98% argon-2% oxygen	150
Mild steel	0.035	0.89	98% argon-2% oxygen	165
Mild steel	0.045	1.14	98% argon-2% oxygen	220
Mild steel	0.062	1.59	98% argon-2% oxygen	275
Stainless steel	0.035	0.89	99% argon-1% oxygen	170
Stainless steel	0.045	1.14	99% argon-1% oxygen	225
Stainless steel	0.062	1.59	99% argon-1% oxygen	285
Aluminum	0.030	0.76	argon	95
Aluminum	0.045	1.14	argon	135
Aluminum	0.062	1.59	argon	180
Deoxidized copper	0.035	0.89	argon	180
Deoxidized copper	0.045	1.14	argon	210
Deoxidized copper	0.062	1.59	argon	310
Silicon bronze	0.035	0.89	argon	165
Silicon bronze	0.045	1.14	argon	205
Silicon bronze	0.062	1.59	argon	270

Note: Spray transfer will only occur when high percentage of argon are used.

Figure 15-6. Approximate transition current levels to obtain spray transfer for various metals. (American Welding Society)

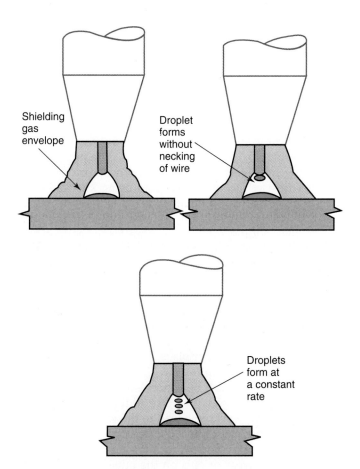

Shielding gas envelope

Droplet forms without necking of wire

Droplets form at a constant rate

Figure 15-7. Pulsed spray metal transfer method. Spray transfer only occurs during peak current.

method, no necking down of the wire occurs. The metal leaves the electrode in a spray of small droplets.

Spray transfer does not occur continually, therefore, the name pulsed spray transfer. The rate of metal transfer increases and the droplet size decreases as the pulse frequency increases. Basic welding machines with pulse capabilities allow the welder to select pulse frequencies of 60 or 120 pulses per second. Some machines allow the user to adjust the pulse frequencies. Pulse frequencies can go much higher than 120 pulses per second. The coolest spray transfer occurs at 60 pulses per second.

A lower average current level is used in pulsed spray than in spray transfer. This lower average current level makes it possible to weld out of position. Thin metal sections may also be welded more easily with the pulsed spray. This method creates very little metal spatter.

The pulsed spray transfer method can use larger-diameter electrode wire. This is an advantage. Larger-diameter electrodes are cheaper. Also, nonferrous wires of larger diameter can be fed through the wire drive unit more easily without kinking.

See Figure 15-3 for the metal deposition rate for the pulsed spray transfer method. Pulsed spray is also used to weld parts with silicon bronze filler wire. This process is sometimes called MIG brazing. Light steel parts in auto repair shops can be welded with very low heat inputs. This reduces the problems of distortion and melt-through.

15.2 GMAW POWER SOURCES

Welding power sources for GMAW provide a constant voltage. The most common types of power sources are transformer-rectifier machines. GMAW is done using DCEP. DCEN can be used in special applications. Ac is not used for gas metal arc welding.

Inverter power sources are much smaller and lighter than traditional transformer-rectifier machines. This type of power source is gaining in popularity. Quite often, an inverter arc welding machine will provide a choice of constant current or constant voltage from the same machine. The welder must select the constant voltage mode when GMAW. Performance of an inverter machine is very similar to a transformer-rectifier constant voltage machine.

Machines used for GMAW may have a wire feeder built into the power supply. The wire feeder may be an external unit, as seen in Figure 15-8.

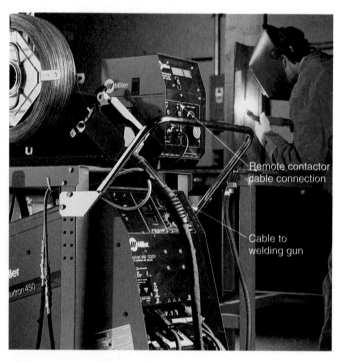

Figure 15-8. *An inverter power source with a separate wire feeder mounted on top of the welder.* *(Miller Electric Mfg. Co.)*

Transformer-rectifier machines are designed to control voltage. *Voltage* is one of the two important variables used to set the welding parameters for GMAW. On the front of the machine, there is a control that is used to set the voltage.

The second important variable is the *wire feed speed*. This control will be on the welding power source if the wire feeder is built into the power source. If the wire feeder is not in the power source, the wire feed speed is set on the external wire feeder. When the welder sets the wire feed speed, the wire feed speed and the appropriate welding current are being adjusted. A higher wire feed speed requires a higher current to melt the electrode wire faster. A slower wire feed speed requires less current to melt the electrode wire.

Inverter machine controls are very similar to those on a transformer-rectifier machine. If the inverter machine has the ability to do multiple processes, select GMAW or the constant voltage setting. After making this selection, the main control setting on the inverter machine is used to set the voltage. The wire feed speed adjustment sets the wire feed speed and also sets the appropriate current. The output and performance of an inverter are the same as those of a transformer-rectifier type machine. Refer to Heading 11.2.3 for more information on inverter machines.

15.3 SETTING UP THE GMAW STATION

Figure 15-9 illustrates a complete GMAW outfit. The same equipment may be used for flux cored arc welding. Remember, self-shielding FCAW does not require any shielding gas.

To prepare a GMAW or FCAW outfit for welding, the following steps should be taken:

1. Connect a separate wire feed unit to the welding power source, if required. The manufacturer's instructions should be followed to make these connections. Usually, a single cable assembly is enough to electrically connect the wire feeder to the welding machine. The welding lead from the positive terminal of the welding machine is usually connected to the wire feeder. Connecting the positive lead to the wire feeder will provide DCEP current. A shielding gas hose may also need to be connected.
2. Mount the desired electrode wire onto the wire feeder.
3. Determine what shielding gas is required. See Heading 15.3.3. No shielding gas is required when using self-shielded FCAW. Properly secure the cylinder (if used) to prevent it from being knocked over. Check that the regulator and gas flowmeter are attached properly. Connect a hose from the flowmeter to the welding machine or wire feeder as required. The hose and fittings should be checked to make sure there are no leaks.
4. Connect the welding gun to the correct place on the welding machine or wire feeder. Quite often there are two connections to be made. See Figure 15-8. One is the main cable, that is connected where the electrode wire exits the wire feeder. Attach this part of the cable assembly to its proper place. The second connection is for the electrical control circuit. This part of the cable assembly also must be connected to its proper place. A water-cooled gun will have additional connections.
5. Connect the workpiece lead to the welding machine. Both the welding gun cable assembly and the workpiece lead should be checked for any signs of wear or cuts. Such wear or cuts on the outside may indicate damage to the leads.
6. The workpiece clamp should be checked. The clamp should be clean so it can make a good electrical connection.

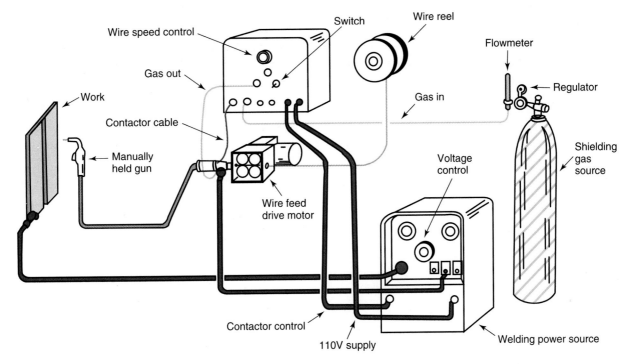

Figure 15-9. *Diagram of a complete gas metal arc welding (GMAW) outfit.*

7. If a water cooler is used, connect it to the welding machine or to the gun according to the manufactures' recommendations. Usually, the welding machine, wire feeder, or welding gun is connected to the outlet on the water cooler. This way, cool water flows from the cooler to the gun. Warm water returning from the gun is connected to the inlet on the water cooler.

When setting up the GMAW station, look for all potential safety problems. Spatter from GMAW or FCAW can cause a fire. All flammable materials must be removed from the welding area.

15.3.1 Setting Up the GMAW Power Source

Properly setting up of a GMAW power source is necessary to obtain the desired transfer method. Before setting up the power source, the following information needs to be known:

- The type of base metal to be welded.
- Base metal thickness.
- The type of transfer method to be used.
- The type of shielding gas to be used.
- The type and diameter of electrode wire.

Once these are known, the welding machine can be properly set up. Only a few controls must be set prior to welding. Figure 15-10 shows a welding machine and its controls.

Two switches must be set. One is to allow the voltage to be set on the panel or remote. The second switch allows a choice between a remote or panel contactor. Once these two switches are set, they are rarely changed. Once the remote or panel voltage and contactor switch are set, the

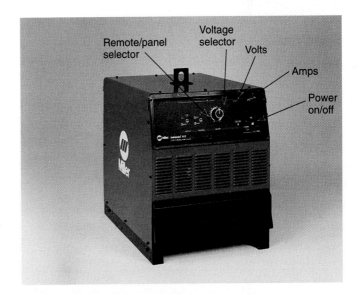

Figure 15-10. *A cc/cv welding power source. (Miller Electric Mfg. Co.)*

only adjustments that need to be made are the voltage and wire feed speed.

The next adjustment is the voltage setting. The voltage determines the arc length and helps determine the electrode transfer method. Other factors also affect the transfer method as discussed in Headings 15.1.1 through 15.1.4.

Welding machines with a wire feeder built in will have the wire feed speed adjustment on the machine itself. If the wire feeder is a separate unit, the wire feed speed adjustment is on the wire feeder. When adjusting the wire feed speed, the welder is adjusting the nominal amperage of the welding machine.

The following figures list the voltage and amperage settings for welding different base metals using both short circuiting transfer and spray transfer.

Metal	Metal Transfer Method	Figure No.
Mild and low-alloy steel	Short circuit	15-11
	Spray transfer	15-12
Stainless steel (300 series)	Short circuit	15-13
	Spray transfer	15-14
Aluminum and aluminum alloys	Short circuit	15-15
	Spray transfer	15-16

Electrode diameter		Arc voltage	Amperage range
in.	mm		
0.030	0.76	15-21	70-130
0.035	0.89	16-22	80-190
0.045	1.14	17-22	100-225

Note: The values shown are based on the use of CO_2 for mild steel and argon CO_2 for low-alloy steel.

Figure 15-11. Approximate machine settings for short circuiting metal transfer on mild and low-alloy steel.

Electrode diameter		Arc voltage	Amperage range
in.	mm		
0.030	0.76	24-28	150-265
0.035	0.89	24-28	175-290
0.045	1.14	24-30	200-315
1/16	1.59	24-32	275-500
3/32	2.38	24-33	350-600

Note: The values shown are based on the use of argon with 2%-5% oxygen for mild and low-alloy steel.

Figure 15-12. Approximate machine settings for spray arc transfer on mild or low-alloy steel.

Electrode diameter		Arc voltage	Amperage range
in.	mm		
0.030	0.76	17-22	50-145
0.035	0.89	17-22	65-175
0.045	1.14	17-22	100-210

Note: The values shown are based on a mixture of 90% helium; 7 1/2% argon; 2 1/2% CO_2. The flow rates were about 20 cfh (9.44 L/min.).

Figure 15-13. Approximate machine settings for short circuiting transfer on 300 series stainless steel.

Electrode diameter		Arc voltage	Amperage range
in.	mm		
0.030	0.76	24-28	160-210
0.035	0.89	24-29	180-255
0.045	1.14	24-30	200-300
1/16	1.59	24-32	215-325
3/32	2.38	24-32	225-375

Note: The values shown are based on the use of argon-oxygen shielding gas. The oxygen percentage varies from 1-5%.

Figure 15-14. Approximate machine settings for spray transfer on 300 series stainless steel.

Electrode diameter		Arc voltage	Amperage range
in.	mm		
0.030	0.76	15-18	45-120
0.035	0.89	17-19	50-150
0.047 (3/64)	1.19	16-20	60-175

Note: The values shown are based on the use of argon shielding gas.

Figure 15-15. Approximate machine settings for short circuiting transfer on aluminum and aluminum alloys.

Electrode diameter		Arc voltage	Amperage range
in.	mm		
0.030	0.76	22-28	90-150
0.035	0.89	22-28	100-175
0.047 (3/64)	1.19	22-28	120-210
1/16	1.59	24-30	160-300
3/32	2.38	24-32	220-450

Note: The values shown are based on the use of argon as the shielding gas.

Figure 15-16. Approximate machine settings for spray transfer on aluminum and aluminum alloys.

Globular transfer voltages and amperages will fall in the range between those shown for short circuiting and spray transfer. Pulsed spray background voltage settings will be slightly higher than the values shown for short circuiting transfer. The peak current must be above the transition current.

Figure 15-17 shows a welding machine on which welding variables are set and stored in electronic memory. The welding machine has a microprocessor inside. A *microprocessor* can be considered a small computer. The

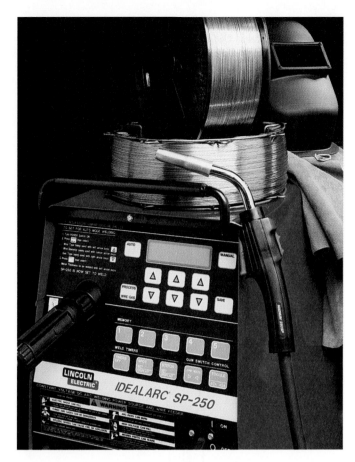

Figure 15-17. *A microprocessor-controlled power source and wire feeder with digital displays.*
(The Lincoln Electric Co.)

microprocessor is programmed by the manufacturer with a set of welding values. Based on a set of input data about a weld, the microprocessor determines the best settings or parameters for the welding application. Since the welding machine has determined the welding parameters, the type of transfer is also determined by the welding machine.

A welder using a microprocessor-equipped welding machine enters the following type information about the weld to be made: electrode wire type, wire diameter, type of shielding gas, and metal thickness. The microprocessor sets the voltage, wire feed speed, and possibly the slope, thus determining the type of metal transfer that will be used. Standard welding values are preprogrammed into the welding machine. Special welding parameters can be saved as a program in the machine. This program, or any preset values, can be recalled at any time in the future. Because the welding values are stored electronically, the welding machine will be set up exactly the same way each time.

Some power sources allow the slope to be changed. Many machines have a preset value for the slope. Heading 13.10 discusses slope.

Power sources that have the ability to pulse weld have additional controls to set up. These controls include an on-off switch, a background voltage adjustment, a peak amperage adjustment, and sometimes a "pulses per

second" adjustment. The background voltage is set relatively low, in the globular transfer range. The peak amperage is fairly high to cause spray transfer to occur. This peak amperage must be above the transition current. The pulses per second adjustment is used to set the number of times per second the current will pulse from the low value to the high value. Figure 15-18 shows a GMAW machine with these controls.

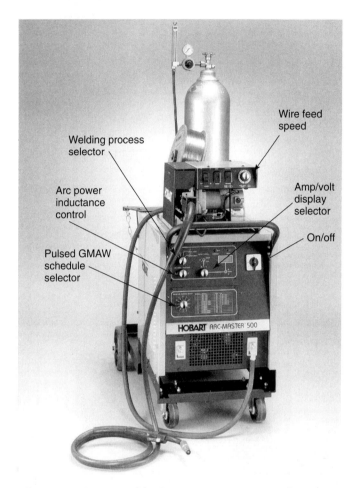

Figure 15-18. *A portable GMAW power source. The control panel includes a pulse schedule selector.*
(Hobart Brothers Co.)

15.3.2 Setting Up the Wire Feeder

Most wire feeders use a 115V ac motor; however, 24V dc motors are becoming very popular. Figure 15-19 shows a complete wire drive unit. Two mated gears are located in the wire drive unit. One gear is driven by an electric variable-speed motor. A roll is attached to each gear. Figure 15-20 illustrates a two-drive-roll wire drive system.

The lower roll on the wire drive unit shown in Figure 15-20 is adjustable in and out. The lower drive gear has spring washers behind it. By turning the adjustment bolt in the center of this gear, the gear and drive roll can be moved inward or outward. This adjustment is provided to align the groove in the wire drive roll with the center of the wire. Figure 15-21 illustrates the adjustment of the wire drive rolls.

Figure 15-19. *Hobart wire feeder. (Hobart Brothers Co.)*

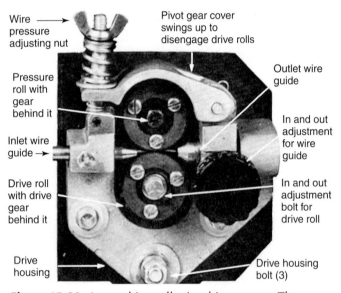

Figure 15-20. *A two-drive roll wire drive system. The upper pressure roll is pivoted out of the way when the wing nut is loosened and the gear cover lifted up. (Miller Electric Mfg. Co.)*

Figure 15-22 shows a wire drive unit with four drive rolls. This drive unit is similar to the two drive roll unit shown in Figure 15-20. The unit shown in Figure 15-22 has three wire guides. Wire guides must be in alignment with each other and with the center of the drive rolls. Figure 15-23 illustrates properly and improperly adjusted wire guides. The alignment of the wire guides is made at the factory. In time, an adjustment may be necessary. Each wire guide must be in perfect alignment with the other, as shown on the drive units in Figures 15-20 and 15-22. The end of each wire guide should be adjusted as close to the drive rolls as possible without touching them. After the wire guide is set, the securing bolt is tightened to hold the guide in place.

One problem that occurs occasionally during wire feeder operation is the wire getting jammed and forming a bird's nest. A ***bird's nest*** is a tangle of electrode wire that did not feed properly through the rolls and into the guide tube. Figure 15-24 shows such a bird's nest.

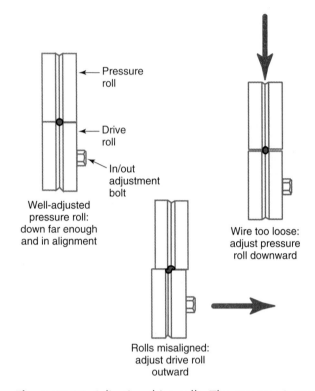

Figure 15-21. *Adjusting drive rolls. The pressure (upper) roll is adjusted up and down by means of the pressure-adjusting wing nut, as in Figure 15-20. The lower drive roll is adjusted in and out by means of an adjustment bolt.*

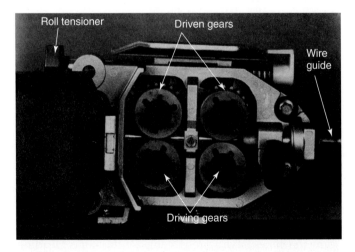

Figure 15-22. *A four-wheel wire drive system. The parts are similar to a two-wheel wire drive. The main drive gear is in the center behind the four visible gears. The main gear drives the gears behind the lower two rolls. (Miller Electric Mfg. Co.)*

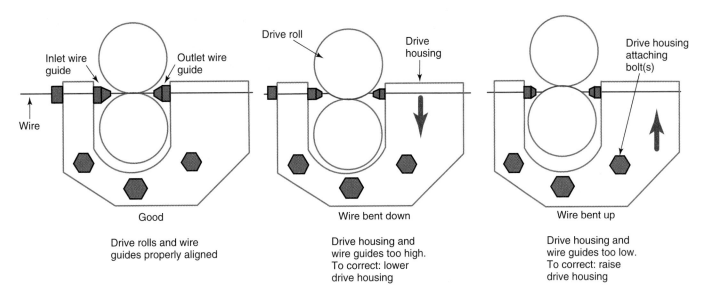

Figure 15-23. Properly and improperly aligned wire guides. If the wire bends going through the drive rolls, adjust the drive housing up or down. Loosen the drive housing bolts, align, and tighten the bolts.

A bird's nest can be caused by the following conditions:

- Stubbing the electrode onto the base metal. This is caused by holding the gun too close to the work, using too-low a voltage or by using too-high a wire feed speed.
- Misaligned guide tubes and rolls.
- A blockage in the cable or liner.

To correct the cause of a bird's nest, use the correct contact tube-to-work distance (see Figure 15-40) or adjust the settings on the welding machine or wire feeder. If this does not solve the problem, determine if there is a misalignment of the guide tubes and rolls or if there is a blockage in the cable or liner. Correcting these problems will eliminate the cause of the bird's nesting.

Figure 15-24. Removing a bird's nest by cutting the electrode wire behind the inlet wire guide and before the outlet wire guide.

To eliminate the bird's nest and continue welding, the following steps should be taken:

1. Raise the upper pressure roll.
2. Cut the electrode wire in two places to eliminate the bird's nest. See Figure 15-24.
3. Remove the electrode wire from the cable assembly going to the welding gun. This has eliminated the bird's nest.
4. Feed new wire into the cable assembly and lower the upper pressure roller.
5. Feed the electrode wire to the gun by pulling the trigger on the gun, or by pressing the inch switch.

To load a spool of electrode wire, place the spool onto the hub on the wire feeder. Secure the spool using the method available on the type of wire feeder being used.

Pressure rolls have one or two grooves cut in them. Select and install pressure rolls that have the same groove diameter as the diameter of the electrode wire being used. Remember to change the pressure rolls if the electrode diameter changes.

The final adjustment is to apply the proper force from the rolls to the electrode. Adjustment is made using a spring-loaded wing nut or knob. Tighten the knob to apply only enough force to drive the wire without slippage. Too much force on the rolls and wire may cause the solid wire to flatten (especially if the wire is aluminum). Flux cored electrodes may be crushed. If the wire is damaged, it will not feed through the wire cable and torch properly. If not enough force is applied to the wire, the rolls will slip and not drive the wire consistently.

Once the adjustments discussed so far have been made, the wire feeder is ready to feed wire continuously. Only the wire feed speed needs to be adjusted to meet the requirements for each welding job. Adjust the feed speed to obtain the amperage and transfer method desired.

Other features on wire feeders often include an *inch switch* or *jog switch*. This is used to feed wire to the gun at a relatively slow speed to prevent kinking the wire. Another switch is the *purge switch*. This is used to allow the shielding gas to flow, so that the shielding gas will fill the hose and remove (purge) all air.

Some wire feeders have a display that shows either the set or actual voltage or wire speed. Figure 15-25 shows such a wire feeder.

Figure 15-25. A wire drive unit. The visual display will show set or actual wire speed and voltage. (The Lincoln Electric Co.)

15.3.3 Inert Gases and Gas Mixtures Used for GMAW

The inert shielding gases and other gases used in shielding gas mixtures for GMAW are argon (Ar), helium (He), oxygen (O_2), carbon dioxide (CO_2), and nitrogen (N_2).

Inert gases used should be of a *welding grade*. Carbon dioxide gas is generally supplied 100% pure. Gas mixtures can be purchased from a welding gas distributor or can be mixed using a gas mixer like the one shown in Figure 15-26.

Each shielding gas and mixture of gases will have a different effect on the shape of the bead and the penetration. See Figure 15-27.

Factors that must be considered when choosing a shielding gas are:
- The type of metal transfer desired: short circuiting, globular, spray, or pulsed spray transfer.
- The desired bead shape, width, and weld penetration.
- The required welding speed.
- The undercutting tendencies of the gas.

Figure 15-26. A shielding gas mixer capable of mixing up to three different gases. (Thermco Instrument Corp.)

Figures 15-28 and 15-29 list shielding gases to be used with different metals and transfer methods. The shielding gases listed for short circuiting transfer are usually also used for globular transfer. Those gases listed for spray transfer are also used for pulsed spray transfer.

Inert gases, such as argon and helium, are chemically inactive and do not unite with other chemical elements. Nitrogen, oxygen, and carbon dioxide are *reactive gases*. They will mix or react with metals in a weld. With the exception of CO_2, reactive gases are not used alone as shielding gases. Nitrogen gas is used in Europe to weld copper.

As noted earlier, each gas and gas mixture has an effect on the type of metal transfer, and on the bead size, penetration, welding speed, and undercutting tendencies. Each of the important gases or gas mixtures is discussed in the following paragraphs. Also refer to Heading 13.12.

Argon

This gas causes a squeezing (constricting) of the arc. The results are a high current density (concentration) arc, deep penetration, a narrow bead, and almost no spatter. Argon ionizes more easily than helium and it conducts some electricity. Therefore, lower arc voltages are required for a given arc length. Argon conducts heat through the arc more slowly than helium. Argon has a lower thermal (heat) conductivity. It is an excellent choice for use on thin

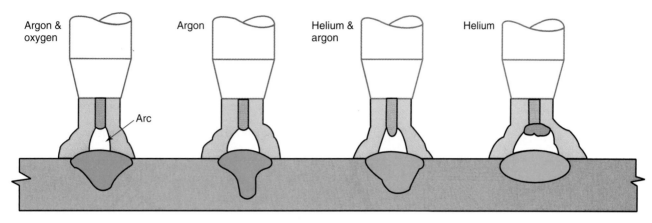

Figure 15-27. *Bead contours and penetration shapes that occur with various gases using DCEP polarity.*

Metal	Shielding gas	Advantages
Aluminum, copper, magnesium, nickel, and their alloys	argon and argon-helium	Argon satisfactory on sheet metal; argon-helium preferred on thicker sheet metal.
Steel, carbon	argon-20-25% CO_2	Less than 1/8" (3mm) thick; high welding speeds without melt-through; minimum distortion and spatter; good penetration.
	argon-50% CO_2	Greater than 1/8" (3mm) thick; minimum spatter; clean weld appearances; good weld pool control in vertical and overhead positions.
	CO_2[a]	Deeper penetration; faster welding speeds; minimum cost.
Steel, low-alloy	60-70% helium-25-35% argon-4-5% CO_2	Minimum reactivity; good toughness; excellent arc stability, wetting characteristics, and bead contour; little spatter.
	argon-20-25% CO_2	Fair toughness; excellent arc stability; wetting characteristics, and bead contour; little spatter.
Steel, stainless	90% helium-7.5% argon-2.5% CO_2	No effect on corrosion resistance; small heat-affected zone; no undercutting; minimum distortion; good arc stability.

a - CO_2 is used with globular transfer also.

Figure 15-28. *Suggested gases and gas mixtures for use in GMAW short circuiting transfer.*

metal. It is also good for out-of-position welds because of the low voltages required.

Argon is the most common inert gas used for welding nonferrous metals. It is used for all types of metal transfer. When welding steel and steel alloys using spray transfer, high percentages of argon, 90% or greater, must be used.

Pure argon used on carbon steel will cause undercutting using the spray transfer method. Because this undercutting is not acceptable, argon is usually mixed with small amounts of oxygen or carbon dioxide. Argon is heavier than helium, therefore, less gas is needed to protect a weld.

Metal	Shielding gas	Advantages
Aluminum	argon 75% helium-25% argon 90% helium-10% argon	0.1"-1" (0.25mm-25mm) thick; best metal transfer and arc stability; least spatter. 1-3" (25-76mm) thick; higher heat input than argon. 3" (76mm) thick; highest heat input; minimizes porosity.
Copper, nickel, & their alloys	argon helium-argon	Provides good wetting; good control of weld pool for thickness up to 1/8" (3mm). Higher heat inputs of 50% and 75% helium mixtures offset high heat conductivity of heavier gages.
Magnesium	argon	Excellent cleaning action.
Reactive metals (titanium, zirconium, tantalum)	argon	Good arc stability; minimum weld contamination. Inert gas backing is required to prevent air contamination on back of weld area.
Steel, carbon	argon-2-5% oxygen	Good arc stability; produces a more fluid and controllable weld pool; good coalescence and bead contour, minimizes undercutting; permits higher speeds, compared with argon.
Steel, low-alloy	argon-2% oxygen	Minimizes undercutting; provides good toughness.
Steel, stainless	argon-1% oxygen argon-2% oxygen	Good arc stability; produces a more fluid and controllable weld pool, good coalescence and bead contour, minimizes undercutting on heavier stainless steels. Provides better arc stability, coalescence, and welding speed than 1% oxygen mixture for thinner stainless steel materials.

Figure 15-29. Suggested gases and gas mixtures for use in GMAW spray transfer.

Helium

The inert gas helium (He) has a high heat-conducting ability. It transfers heat through the arc better than argon. Helium is used to weld thick metal sections. This gas is also used to weld metals that conduct heat well. Such metals as aluminum, magnesium, and copper will conduct heat away from the weld zone rapidly. More heat must be put into the metal, therefore, helium gas is the best choice. The arc voltages required for helium are higher and spatter is greater. Helium will allow filler metal to be deposited at a faster rate than is possible with argon. This gas is often used on nonferrous metals. It produces welds with wider bead reinforcements. Helium is lighter than argon and will require a greater gas flow to protect a weld as well as argon. In addition to requiring a greater flow rate that uses more shielding gas, helium is about 10% more expensive than argon. Even though the cost for helium may be greater than that for argon, the benefits of helium for the right welding application make helium an excellent choice.

Carbon dioxide

This gas has a higher thermal (heat) conductivity than argon. It requires a higher voltage than argon. Since carbon dioxide (CO_2) is heavy, it covers the weld well. Therefore, less gas is needed.

CO_2 costs about 80% less than argon. This price difference will vary from location to location. Beads made with CO_2 have a very good contour. The beads are wide and have deep penetration and no undercutting. The arc in a CO_2 atmosphere is unstable and a great deal of spattering occurs. This is reduced by holding a short arc. Deoxidizers like aluminum, manganese, or silicon are often added to the filler metal. The deoxidizers remove the oxygen from the weld metal. **Good ventilation is required when using pure CO_2. About 7%-12% of the CO_2 becomes dangerous CO (carbon monoxide) in the arc.** The amount of CO increases with the arc length.

Nitrogen

In Europe, nitrogen (N_2) is used where helium is not readily available. Mixtures containing nitrogen have been used to weld copper and copper alloys. One mixture used contains 70% argon and 30% nitrogen.

Argon-helium

Mixtures of argon and helium help to produce welds and welding conditions that are a balance between deep penetration and a stable arc. A mixture of 25% argon and 75% helium will give deeper penetration with the arc stability of a 100% argon gas. Spatter is almost zero when a 75% helium mixture is used. Argon-helium mixtures are used on thick nonferrous sections.

Argon-carbon dioxide

Mixing CO_2 in argon makes the molten metal in the weld pool more fluid. This helps to eliminate undercutting when GMA welding carbon steels using spray transfer. CO_2 also stabilizes the arc, reduces spatter, and promotes a straight-line (axial) metal transfer through the arc.

Argon-oxygen

Argon-oxygen gas mixtures are used on low-alloy, carbon, and stainless steels. A 1%-5% oxygen mixture will produce beads with penetration that is wider and less finger-shaped. Oxygen also improves the weld contour, makes the weld pool more fluid, and eliminates undercutting. Oxygen seems to stabilize the arc and reduce spatter. The use of oxygen will cause the metal surface to oxidize slightly. This oxidization will generally not reduce the strength or appearance of the weld to an unacceptable level. If more than 2% oxygen is used with low-alloy steel, a more expensive electrode wire with additional deoxidizers must be used.

Helium-argon-carbon dioxide

This shielding gas mixture is used to weld austentic stainless steel, using the short circuiting transfer method. The following mixture is often used and produces a low bead: 90% He; 7 1/2% Ar; 2 1/2% CO_2.

Metal transfer methods

The various GMAW metal transfer methods, and the gases suggested for use with them follow.

Short circuiting transfer

Pure argon or helium, or argon and helium mixtures, are used on aluminum and other nonferrous metals and their alloys. For carbon steels, pure CO_2 or a mixture of 75% argon and 25% CO_2 is often used. A mixture of helium, argon, and CO_2 is used to weld stainless steel.

Globular transfer

Argon with high percentages of CO_2, or pure CO_2, are used to weld low-carbon steels with globular transfer. With CO_2, the globules leave the wire in a random way and spatter is high. When argon or a high argon percentage gas mixture is used, the metal is squeezed off the wire and travels in a straighter line to the metal.

Spray and pulsed spray transfer

The spray transfer method will occur only in an atmosphere that has a high argon percentage. Pure argon or an argon-helium mixture is used on nonferrous metals. The following argon mixtures are used when welding low-carbon steels: argon with 2%-5% oxygen (O_2), and also argon with 5%-10% CO_2.

Small amounts of oxygen lower the transition current. Oxygen appears to decrease the surface tension of the molten metal on the wire. This allows the molten metal droplets to leave the electrode more easily. Oxygen makes the weld pool more fluid and reduces undercutting. It also acts to stabilize the arc.

Figure 15-30 lists shielding gas selections for GMAW on a number of metals.

15.3.4 Selecting the Proper Shielding Gas Flow Rate for GMAW

Enough gas must flow to create a straight line (laminar) flow. If too much gas comes out of the nozzle, the gas may become turbulent. See Figure 15-31. If it becomes turbulent, the shielding gas will mix with the atmosphere around the nozzle area. This will cause the weld to become contaminated.

When too little gas flows, the weld area is not properly protected. The weld will become contaminated and porosity will occur. The recommended rate of flow for a given nozzle is generally provided by the manufacturer. Once the correct flow rate is known, it can be used at all wire speeds. Too little gas will give a popping sound. Spatter will occur, the weld will have porosity showing, and the bead will be discolored. Refer to Figure 15-32 for some suggested gas flow rates for use with various metals and thicknesses. Set the proper flow rate, using the flowmeter.

The heavier shielding gases like CO_2 and argon will tend to drop away from the weld area when welding out of position. Therefore, the gas flow rates must be increased as the position moves from the flat to the horizontal, vertical, and overhead welding positions.

When a gas mixture is used, it may be necessary to use a double- or triple-unit gas mixer. Such units have a separate pressure regulator and flowmeter for each gas. See Figure 15-26. Premixed gas mixtures can be purchased from welding gas suppliers in cylinders, just like pure argon or oxygen.

Metals	Gases %	Uses and results
Aluminum	Ar	Good transfer, stable arc, little spatter. Removes oxides
	50Ar-50He 25Ar-75He He	Hot arc - 3/8" to 3/4" (10mm to 19mm) thickness. Remove oxides. Hot arc, less porosity, removes oxides - 1/2" to 1" (13mm to 25mm) Hotter, more gas; 1/2" (13mm) and up. Removes oxides.
Magnesium	Ar 75He-25Ar	Good cleaning. Hotter, less porosity, removes oxides.
Copper (deox.)	75He-25Ar Ar	Preferred. Good wetting, hot. For thinner materials.
Carbon steel	CO_2	Short circuiting arc: high quality, low current, out-of-position, medium spatter
	Ar-2 O_2	Globular arc: fast, cheap, spattery, deep penetration.
	Ar-5 O_2	Fast, stable, good bead shape, little undercut, fluid weld pool.
	75Ar-25CO_2	Short circuiting arc: fast, no melt-through, little distortion and spatter.
	50Ar-50CO_2	Short circuiting arc: deep penetration, low spatter.
Low-alloy steel	Ar-2 O_2	Removes oxides, eliminates undercut, good properties.
High-strength steels	60He-35Ar-5CO_2	Short circuiting arc: stable arc, good wetting and bead contour, little spatter. Good impacts.
	75Ar-25CO_2	Short circuiting arc: same except low impact.
Stainless steel	Ar-1 O_2	No undercutting. Stable arc, fluid weld, good shape.
	Ar-5 O_2	More stable arc .
	90He-7 1/2 Ar-2 1/2 CO_2	Short circuiting arc: small heat-affected zone, no undercut, little warping.
Nickel, monel	Ar	Good wetting - decreases fluidity.
Inconel	Ar-He	Stable arc on thinner material.

Figure 15-30. Some shielding gas selections for GMAW of various metals.

15.3.5 Selecting the Correct Gas Nozzles and Contact Tubes

The *gas nozzle* is located at the end of the GMAW gun. See Figures 15-33 and 15-34. It is designed to deliver the shielding gas to the weld area in a smooth, unrestricted manner. The gas nozzle is usually made of copper, which is a very good heat conductor. A copper nozzle will resist melting when exposed to the heat generated in the welding operation. GMAW nozzles and FCAW nozzles (if used) are the same. The construction of the nozzle end of a GMAW gun is shown in Figure 15-34.

Nozzles are made with different exit diameters. Gun manufacturers usually provide information on the correct nozzle to use for various applications. A general-purpose

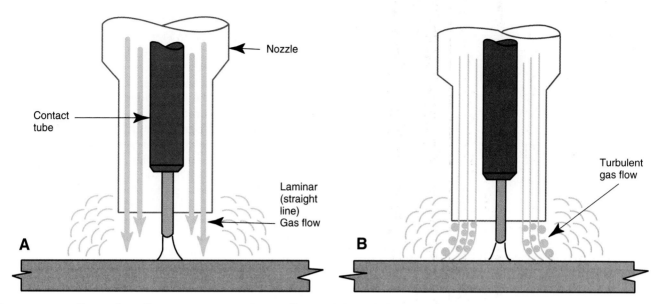

Figure 15-31. *Effects of gas flow rate. A—Laminar gas flow is the result of the proper gas flow rate. B—Turbulence occurs when too much gas is used.*

Metal	Type joint	Thickness		Weld position	Argon flow	
		in.	mm		ft³/hr.	L/min
Aluminum and aluminum alloys	All	1/16	1.59	F	25	11.80
		3/32	2.38	F,H,V,O	30	14.16
		1/8	3.18	F,H,V,O	30	14.16
		3/16	4.76	F,H,V,O	23-27	10.85-12.74
		1/4	6.35	F	40	18.88
				H,V	45	21.24
				O	60	28.32
		3/8	9.53	F	50	23.60
				H,V	55	25.96
				O	80	37.76
		3/4	19.05	F	60	28.32
				H,V,O	80	37.76
Stainless steel	Butt	1/16	1.59		30	14.16
	Butt	1/8-3/16	3.18-4.76		(98Ar-2O₂) 35	16.52
	60° Bevel	1/4-1/2	6.35-12.7		35	16.52
	60° Double Bevel	1/2-5/8	12.7-15.88		35	16.52
	Lap, 90° Fillet	1/8-5/16	3.18-7.94		35	16.52
Nickel and nickel alloys	All	Up to 3/8	Up to 9.53		25	11.80
Magnesium	Butt	0.025-0.190	0.64-4.83		40-60	18.88-28.32
		0.250-1.000	6.35-25.4		50-80	23.60-37.76

Figure 15-32. *Suggested gas flow rates for various metals and thicknesses.*

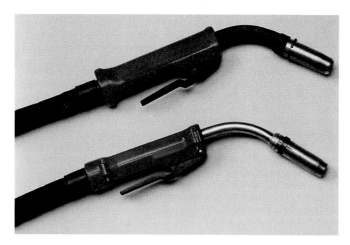

Figure 15-33. *Two GMAW guns. The nozzles on these two guns are held in place by tension. (Miller Electric Mfg. Co.)*

nozzle is often used and will work well for most applications. A variety of nozzles is shown in Figure 15-35. It can be seen that some nozzles thread onto the gun. Other nozzles are designed to slip onto or over a nozzle adaptor and

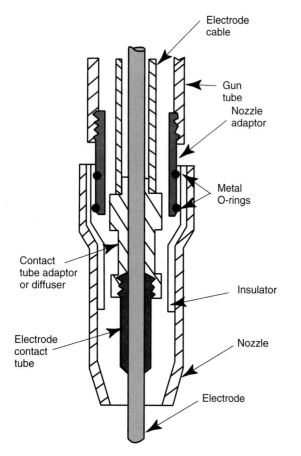

Figure 15-34. *A schematic drawing of the nozzle end of a GMAW or FCAW torch.*

are held by tension. Special nozzle shapes are also manufactured, as illustrated in Figure 15-36.

Under the nozzle lies the electrode *contact tube*. A contact tube makes the electrical connection between the welding gun and the electrode.

The contact tube is threaded into a part of the gun called a diffuser or an adaptor. See Figures 13-46 and 15-35. One end of the diffuser or adaptor threads into the welding gun. The other end has threads for installing the contact tube. Diffusers have holes around them to allow shielding gas to escape into the nozzle. Shielding gas exits the end of the nozzle to protect the weld area.

Contact tubes, also called *contact tips*, are made with a variety of inside diameters (ID) and lengths. The contact tube must be designed for the diameter of electrode wire being used. A good sliding electrical contact must be made with the electrode wire. Each time the wire diameter is changed, the contact tube must be changed so that the ID matches the diameter of the wire.

Most manufacturers of contact tubes make them in different lengths. Different lengths are used to help obtain different transfer methods. The longest tubes for a gun are usually used for short circuiting transfer. When using short circuiting transfer, the contact tube should be flush with the end of the nozzle or should stick out about 1/16" (1.6mm) beyond the end of the nozzle. With a long contact tube, minimal resistance heating of the wire takes place. See Figure 15-37.

Resistance heating of the electrode takes place after the electrode wire exits from the contact tube. The electrode extension distance, shown in Figure 15-40, is the distance over which the electrode is heated. The longer this distance, the more heating takes place. Figure 15-37 shows that a long contact tube minimizes the electrode extension and reduces the resistance heating of the electrode wire.

A medium-length contact tube is used for spray transfer. A medium-length contact tube usually keeps the end of the contact tube inside the end of the nozzle. This allows the welding current to preheat the wire more than when using a long contact tube.

Short contact tubes are used for flux cored arc welding. A flux cored electrode must be heated to a higher temperature than a solid electrode so that some of the flux will vaporize and create a shielding atmosphere around the weld. A shorter contact tube allows the electrode to be heated to a higher temperature. See Figure 15-37.

Contact tubes will wear and must be changed regularly. Eight hours of continuous welding with a steel electrode can excessively wear a contact tube. Regular replacement of the contact tube will ensure a continuous good electrical contact with the electrode wire.

Look at the contact tube occasionally. If the round hole is becoming elongated or if the arc appears to be fluctuating while welding, it is time to replace the tube. A fluctuating arc may be due to a worn contact tube not making consistent contact with the electrode.

While arc welding, the inside and outside of the nozzle and the outside of the contact tube can become spattered. This spatter can be kept from sticking by spraying or dipping the nozzle with a special proprietary anti-stick

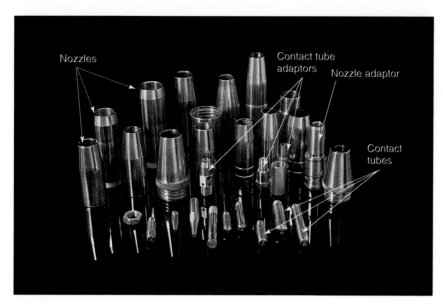

Figure 15-35. A number of different GMAW nozzles, contact tubes, and contact tube adaptors. (American Torch Tip Co.)

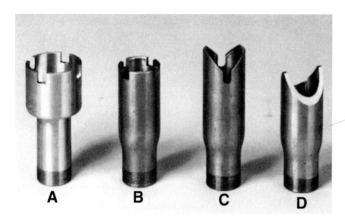

Figure 15-36. Special GMAW nozzles for spot or tack welding. A and B—Standard spot or tack welding nozzle. C—Nozzle for outside corner spot or tack welds. D—Nozzle for inside corner spot or tack welds. (Miller Electric Mfg. Co.)

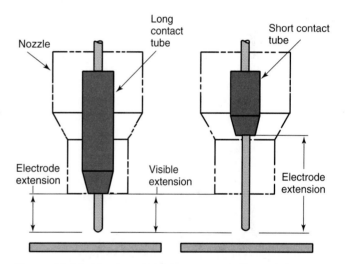

Figure 15-37. The length of the contact tube and the amount of electrode extension affects the amount of heating that the electrode wire receives. The heating takes place in the length of electrode wire that extends from the contact tube. With a long contact tube, there is less extension and thus, less heating.

compound. If the inside of the nozzle becomes spattered, the flow of shielding gas will become turbulent. Gas turbulence may cause weld contamination. To remove the spatter from the nozzle, a special cleaning reamer is used.

15.3.6 Selecting and Installing a Liner

The electrode wire travels from the wire feeder to the welding gun in a cable. Inside the cable a *liner*, also called a *conduit*, is installed. The liner protects the cable from the continuous wear of the electrode wire. The liner also prevents the electrode wire from getting tangled or stuck while traveling through the cable.

There are two types of liners. One is a hardened steel wire wound in a tight coil to form a tube. This wound steel liner is used for hard materials like steel and stainless steel wires. See Figure 15-38.

The second type liner is made of Teflon®. Teflon is a type of plastic. It is much softer than the wound steel liner material. Teflon liners are used with softer materials, especially aluminum wires.

Fine metal filings can accumulate in a coiled liner. It is a good idea to occasionally blow compressed air through this type of liner to remove these very fine particles. The electrode must not be in the liner when it is being blown out. **When blowing out the liner, always point the open end of the liner toward the floor or a trash can. Never allow the open end to point toward yourself or any other person.**

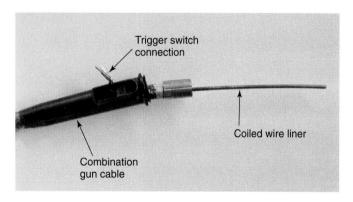

Trigger switch connection

Coiled wire liner

Combination gun cable

Figure 15-38. A coiled wire liner may be used in a GMAW cable to guide the electrode wire. The wire liner sticking out of this welding gun cable is replaced when worn.

Occasionally, liners get worn or become clogged with fine metal particles. A problem also can occur if the liner ever gets kinked. When any of these happen, the electrode wire will not feed smoothly. The liner must be replaced.

To replace a liner, or to change from one type to the other, disconnect the gun cable from the wire feeder. Remove the nozzle, contact tip, diffuser, and any setscrews used to keep the liner in place. Then, remove the liner from the gun and cable. Install the new liner. Push it firmly until it bottoms out against the far end of the cable. Secure the liner in place. Most liners are made slightly long and must be trimmed to a specific length. Each manufacturer has directions to follow. Reassemble the welding gun. Attach the cable to the wire feeder, then refeed the wire through the cable and liner to the gun.

15.3.7 Selecting the Correct GMAW or FCAW Electrode

Smaller-diameter wire usually costs more than larger-diameter wire. The rate at which filler metal is deposited when using small-diameter wire makes up for its added cost. Because of the small diameter and the high currents generally used in GMAW and FCAW, small-diameter electrode wire is melted more rapidly than larger-diameter wire. Small-diameter wire is thus deposited at a much higher rate.

The electrode wire used must match, or be compatible with, the base metal being welded. When CO_2 or O_2 is used on steel-based metals, it causes oxidation of the weld metal. Deoxidizer types of electrode wires must be used to neutralize this oxidation. Manganese, silicon, and aluminum are used as deoxidizers in steel electrode wires. Titanium, silicon, and phosphorus are the deoxidizers in copper electrodes.

For more information regarding GMAW electrodes, see Heading 13.14. Also refer to Figure 13-48 for carbon steel electrodes and Figure 13-49 for low-alloy electrodes. More information about FCAW electrodes may be found in Heading 13.18 and Figure 13-56.

Once the correct electrode is selected it should be loaded in the wire feeder as stated in Heading 15.3.2. The correct-size drive wheels must be used in the wire feeder. The wire should be fed through the electrode cable using the inch switch until about 2" to 3" (50mm to 75mm) extend beyond the nozzle.

15.4 PREPARING METAL FOR WELDING

Metal surfaces usually may be cleaned mechanically or chemically. Abrasive cloth or wire brushing may be used. On severely corroded areas, grinding may be used. Welding may be done on an oxidized (rusted) carbon steel or low-alloy steel surface without cleaning. However, if the surface is rusty, a deoxidizing electrode wire should be used. This reduces oxidation and weld porosity.

Joint designs for GMAW and FCAW are similar to those used for SMAW. The groove angle used when GMAW or FCAW may be smaller than the angle used when SMAW. See Figure 15-39. This narrower angle is possible for two reasons. The wire diameters used are smaller and GMAW penetrates better than SMAW. A 45° groove angle will take less filler metal to fill than a 75° groove angle. Welding time will also be less. Therefore, savings in filler metal and welder's time are possible.

15.5 ELECTRODE EXTENSION

Electrode extension is the amount that the end of the electrode wire sticks out beyond the end of the contact tube. See Figure 15-40. This distance is sometimes referred to as *stickout*.

A good extension for use with the short circuit GMAW transfer method is about 1/4" to 1/2" (6mm to 13mm). The correct electrode extension for all other transfer methods varies between 1/2" and 1" (13mm and 25mm). An electrode extension used for gas-shielded FCAW may vary from 1/2" to 1 1/2" (13mm to 38mm). The suggested electrode extension for use with self-shielding FCAW is 3/4" to 3 3/4" (19mm to 95mm).

Contact tips are made in different lengths. The different lengths help to establish the correct electrode extension. Longer tips are used for short circuiting transfer; shorter tips are used for spray transfer and FCAW. Heading 15.3.5 discusses different contact tip lengths.

As the electrode extension increases, the resistance heating of the electrode increases. Resistance causes the current to heat the wire along the electrode extension distance. A long extension may cause too much filler metal to be deposited with low heating by the arc. This may cause spatter, shallow penetration, and a low weld bead shape.

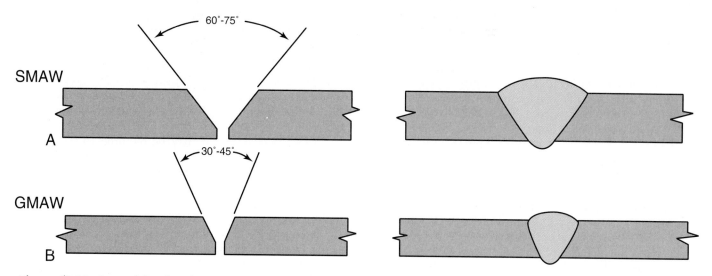

Figure 15-39. Arc welding beads compared. A—Typical groove angle and weld bead for SMAW. B—Typical groove angle and weld bead for GMAW and FCAW. Notice that less filler metal is required to fill the groove at B. Welding time will also be less.

15.6 WELDING PROCEDURES

Before beginning to weld, the welding station should be checked for safety. All electrical, gas, and water connections must be checked for tightness.

Weldments should be tack welded or placed into fixtures prior to welding. When complete joint penetration is required, backing is often recommended. *Backing* is used to control the penetration and may be in the form of a backing plate, strip, ring, or other design.

Most arc welding processes require the welder to control the arc length, welding speed, and torch or gun angle to obtain a good weld. In GMAW and FCAW, the arc length will remain constant and is determined by the arc voltage. The welder doing GMAW must watch and control the distance from the nozzle or contact tube to the work. See Figure 15-40. By controlling the nozzle-to-work distance, the welder will control the electrode extension distance. Heading 15.5 explains the importance of electrode extension.

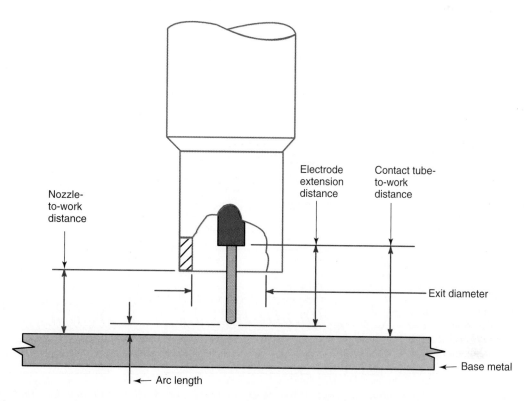

Figure 15-40. Electrode extension distance. Other distances important in GMAW and FCAW are also shown.

The welding speed will be determined by the appearance of the bead width and penetration. Torch angle will also affect the bead width and penetration. The terms forehand, backhand, and perpendicular welding are used.

In *forehand welding*, the tip of the electrode points in the direction of travel. When *backhand welding*, the electrode tip points away from the direction of travel. *Perpendicular welding* is done with the electrode at 90° to the base metal. Figure 15-41 shows the effects of these various methods.

The backhand method will give the best penetration. A 25° angle forward of perpendicular will give the best penetration in the flat welding position, as shown in Figure 15-41C. For the best control of the weld pool, an angle of 5°-15° forward of perpendicular is preferred for all positions.

To start welding, tip the top of the gun 5°-15° in the direction of travel and place the helmet down over your eyes. To start the arc, the wire feeder and the gas, squeeze the trigger on the gun. The wire will arc as soon as it feeds out far enough to touch the metal. No striking or up-and-down motion is required to start the arc as required with SMAW.

As the weld pool reaches the proper width, which occurs rapidly, the welder moves the welding gun forward. Continue to move the gun along the weld, watching the width of the weld pool to maintain a uniform size. Continue this procedure until the end of the weld is reached. A run-off tab may be used to ensure a full-width bead to the end of the weld. Without a run-off tab, the end of the weld may have a crater (depression). This depression can be reduced by moving the electrode to the end of the weld and then back over the completed bead about 1/2" (13mm). At the end of this reverse travel, the contactor switch is released. To shield the end of the weld, hold the gun in position to allow the gas postflow to protect the weld until it cools.

More than one pass may be required to fill a weld groove. Each pass should be cleaned before the next pass is laid. This is generally done with a wire brush or wheel. The glass-like coating on some gas metal arc welds is easily removed. The slag layer on a flux cored arc weld is heavier and requires more effort to remove.

Out-of-position welds require that leathers be worn. Molten base metal, filler metal, and spatter may fall on the welder. Therefore, a cap, coat, cape, and chaps should be worn to protect against burns.

15.7 SHUTTING DOWN THE STATION

When welding is stopped for an extended period, the station should be shut down. To shut down the station, proceed as follows:

1. Return the wire speed to zero.
2. Turn off the wire drive unit.
3. Turn off the shielding gas cylinder(s).

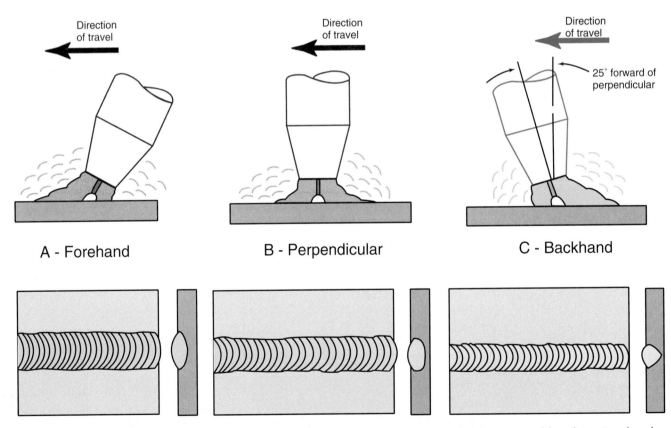

Figure 15-41. *Effects of the welding method on the bead. A—Forehand. B—Perpendicular. C—Backhand. Notice that the backhand method gives the deepest penetration.*

4. Squeeze the gun trigger and hold it in for a few seconds to bleed the gas lines.
5. Turn the flowmeter adjusting knob(s) in to close it.
6. Turn off the power switch on the arc welding power source.
7. Hang the gun on an insulated hook.
8. Turn out the pressure adjusting knob on the flowmeter regulator, if an adjustment knob is provided.

15.8 WELDING JOINTS IN THE FLAT WELDING POSITION

The face of a weld made in the flat welding position should be horizontal or nearly horizontal. The weld axis is also horizontal. See Figure 15-42. Any of the metal transfer methods may be used in the flat welding position. The method used will depend on the metal thickness and other factors. Figure 15-43 shows a welder practicing running a bead in the flat welding position.

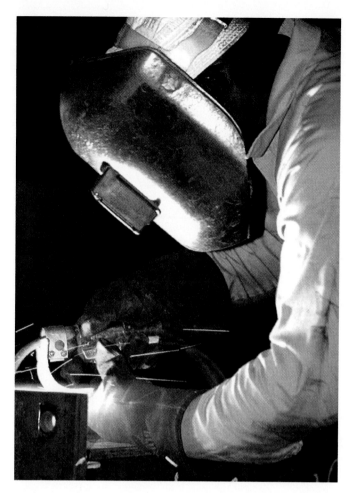

Figure 15-43. A welder running a bead in the flat welding position. (American Welding Society)

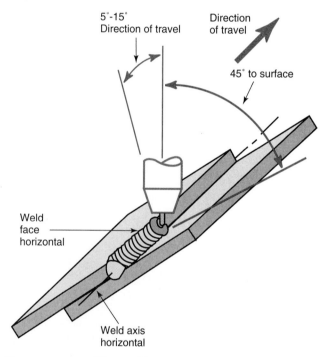

Figure 15-42. A fillet weld on a lap joint in the flat welding position. Note the angles used and the deep penetration of the weld. Also, notice that the weld face and axis are horizontal or near-horizontal.

15.8.1 Fillet Weld on a Lap Joint

The metal should be set up as shown in Figure 15-42. It should be tack welded about every 3" (75mm). This will hold it in position while the weld is made.

To make the fillet weld, the centerline of the electrode should be held at about 45° to the edge and metal surface.

The electrode should point more toward the surface if the edge begins to melt too quickly. The electrode and gun should be held between 5°-15° forward from a vertical line to the metal surface.

A C-shaped weld pool will form, as when GTAW. When the end of the weld is reached, reverse the direction for about 1/2" (13mm). This movement will help reduce the crater which occurs if the weld is stopped at the end of the joint. No matter what type weld is made, this same finish movement can be made. A run-off tab will totally eliminate the crater at the end of a weld.

15.8.2 Fillet Weld on an Inside Corner Joint

Fillet welds may be made on metal up to 3/8" (10mm) thick without edge groove preparation. This can be done because of the deep penetration possible with the spray transfer method. The centerline of the electrode should be held at 45° to each metal surface. If the backhand welding procedure is used, the electrode and gun are held between 5° and 15° forward of vertical. See Figure 15-44.

GMAW can generally weld 1/4" (6mm) beads on each pass. If the weld size is more than 1/4" (6mm) thick, two or more weld passes will be required.

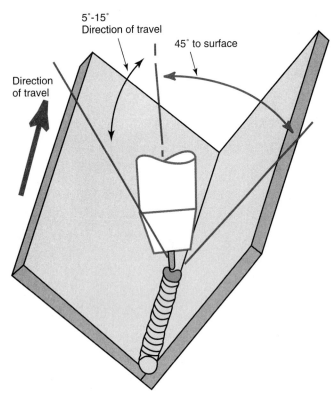

Figure 15-44. A fillet weld on an inside corner joint in the flat welding position. The electrode is 45° from each metal surface. It is also tipped 5°-15° forward in the direction of travel.

15.8.3 Groove Weld on a Butt Joint

Square-groove welds can be made on metal up to 3/8" (10mm) thick without edge shaping. Groove welds with shaped edges of any thickness can be made with the GMAW process. The groove angle on a V-groove butt weld can be narrower than is used with SMAW. Because of the penetration possible with spray transfer methods, the root face can be larger. The root opening can be smaller with GMAW than the opening used for SMAW.

The centerline of the electrode should be directly over the axis of the weld. An angle of between 5° and 15° forward of vertical is correct for the backhand welding method. See Figure 15-45.

A keyhole in the weld pool will indicate that complete penetration is occurring. One problem that may occur in a groove weld made with GMAW is whiskers. **Whiskers** are lengths of electrode wire that stick through the root side of a groove weld. Whiskers occur when the electrode wire is advanced ahead of the weld pool.

The wire goes through the weld root and burns off. The burned-off length is left stuck in the weld. Whiskers can be prevented by slowing the welding speed. They may also be prevented by reducing the wire feed speed. A small weaving motion may be used to keep the wire from getting ahead of the weld pool.

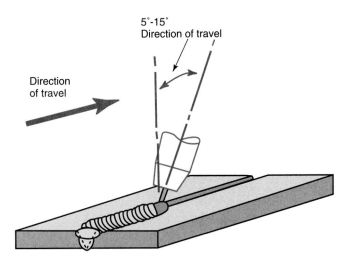

Figure 15-45. A V-groove weld on a butt joint in the flat welding position. Note the narrow (45°) groove possible with GMAW.

15.8.4 Groove Weld on an Outside Corner Joint

The outside corner joint is set up as shown in Figure 15-46. A square- or prepared-groove weld may be used. The electrode angles are the same as those used for welds made on a butt joint. Since groove welds are made on the outside corner joint, whiskers can occur.

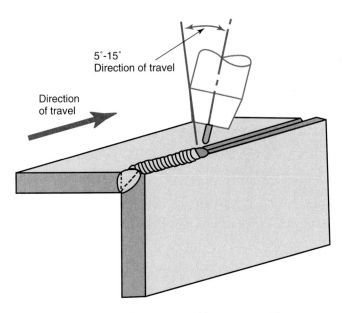

Figure 15-46. A bevel-groove weld on an outside corner joint in the flat welding position.

15.9 WELDING JOINTS IN THE HORIZONTAL WELDING POSITION

The face of a weld made in the horizontal welding position is in the vertical or near-vertical welding position. In the horizontal welding position, the centerline of weld axis runs in a horizontal or near-horizontal line. See Figure 15-47.

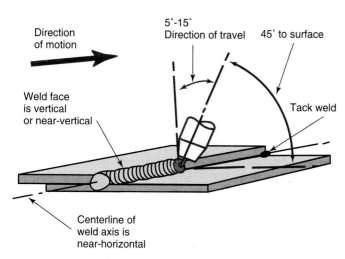

Figure 15-47. *A fillet weld on a lap joint in the horizontal welding position. In the horizontal welding position, the weld axis is near-horizontal and the face of the weld near-vertical.*

Short circuiting, globular, spray, or pulsed spray transfer methods may be used when welding horizontal fillet welds. Horizontal butt welds are limited to short circuiting and pulsed spray transfer. The weld pool is often too large and fluid when using globular or spray transfer. Also, metal transfer in globular transfer will not fall into the weld pool.

15.9.1 Fillet Weld on a Lap Joint

For practice welds, the metal should be set up and tack welded as shown in Figure 15-47. The centerline of the electrode should be about 45° to the edge and metal surface. It may point more toward the surface if the edge melts too quickly. The electrode or gun should tip about 5°-15° forward of vertical in the direction of travel. The typical C-shaped weld pool will indicate that both the edge and surface are melting properly.

15.9.2 Fillet Weld on an Inside Corner or T-Joint

Square or prepared-groove welds may be made in the horizontal welding position. The use of a V-, bevel-, U-, J-type prepared groove will depend on the metal thickness and joint design. The bead width used in GMAW does not have to be as wide for the same thickness as when doing SMAW. This is because the gas metal arc weld penetrates more. It does not need bead width and reinforcement to strengthen the weld.

The electrode should be held at 45° to each metal surface as seen in Figure 15-48. Aiming the wire more toward the vertical surface may improve the bead shape. This will help compensate for the molten metal sag. Incline the gun and the electrode about 5°-15° forward of vertical. See Figure 15-49.

Figure 15-48. *This welder is making a horizontal weld on a T-joint using the GMAW process. (American Welding Society)*

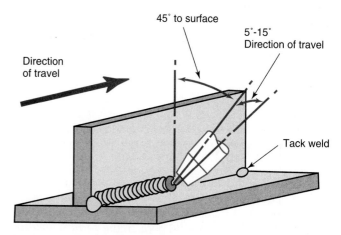

Figure 15-49. *A fillet weld on a T-joint in the horizontal welding position. Note the angles from the metal and in the direction of travel.*

15.9.3 Groove Weld on a Butt Joint or Outside Corner Joint

A square- or prepared-groove weld may be used. Figure 15-50 shows a U-groove weld in progress. The electrode centerline should be directly over the weld line. For best weld pool control, the electrode should tip 5°-15° in the direction of travel. The gun and electrode should also point upward slightly to keep the molten metal from sagging. Short circuiting transfer and pulsed spray transfer allow the molten weld pool to cool slightly.

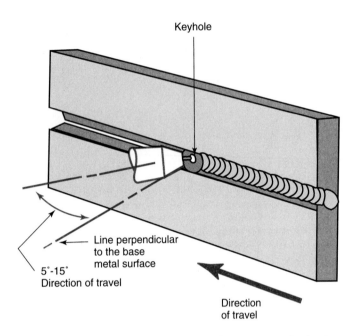

Figure 15-50. *A U-groove weld on a butt joint in the horizontal welding position. Note the keyhole at the root of the weld.*

To ensure complete penetration, watch for a continuous keyhole through the root pass. More than one pass is necessary on thicknesses above 3/16" (5mm). To completely fill the groove, an electrode weaving motion may be required. Figure 15-51 shows a horizontal butt joint being welded.

15.10 WELDING JOINTS IN THE VERTICAL WELDING POSITION

GMAW in the vertical welding position is done using the short circuiting or pulsed spray transfer method. Spray transfer may also be used, but only with small-diameter wire and a small molten weld pool. In the vertical welding position, the weld axis and the weld face are both vertical. Figure 15-52 shows a vertical weld.

GMAW may be made uphill (from the bottom up) or downhill (from the top down). Downhill welding is more difficult with FCAW. The flux material might flow into the

Figure 15-51. *A mechanized GMA welder mounted on a track. The track guides the GMAW gun along the circular butt joint on this large tank. (Bug-O Systems, Inc.)*

Figure 15-52. *A farmer using the FCAW process to repair a front loader bucket. (Hobart Brothers Co.)*

weld pool. This can be avoided if the welder can keep the weld pool ahead of the molten flux.

The centerline of the electrode should be tipped 5°-15° in the direction of travel, as in other position welds. This angle will permit the easiest weld pool control. The weld pool remains relatively cool when the short circuiting method of metal transfer is used. A properly adjusted pulsed spray arc will allow time between pulses for the weld pool to cool. Spray arc transfer can be used in some applications, but the weld pool must be kept small. To maintain a small weld pool, a higher travel speed must be used. The short circuiting method of metal transfer keeps the weld pool coolest.

15.10.1 Fillet Weld on a Lap Joint

Figure 15-53 illustrates a fillet weld being made in the vertical welding position. The angles of the electrode and gun are the same as for other positions. The electrode

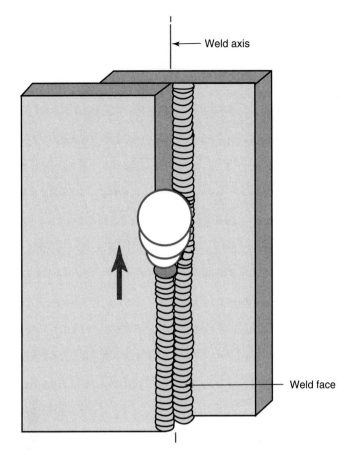

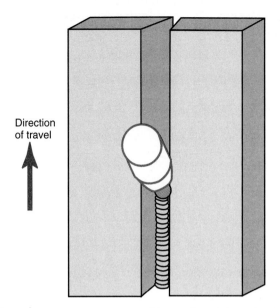

Figure 15-54. *A V-groove weld on a butt joint. The root pass is in progress.*

Figure 15-53. *A fillet weld on a lap joint in the vertical welding position. Two passes are being used on the weld. Notice that the weld axis and bead face are vertical.*

15.11 WELDING JOINTS IN THE OVERHEAD WELDING POSITION

The short circuiting and pulsed spray metal transfer methods are recommended for overhead welding. **When overhead welding, it is strongly suggested that a cap, coat, cape, and possibly chaps be worn. This is necessary to protect the welder from falling molten metal.**

The angle of the electrode to the joint surfaces is the same as for other welding positions. The electrode should be held more vertically when overhead welding. An angle of between 5° and 10° is suggested. The weld pool in short circuiting and pulsed arc transfer is relatively cool. A weaving motion is not required for the purpose of cooling the weld pool. As the weld pool increases in size, the possibility of the metal falling out or sagging increases. The use of several narrower beads, rather than a weaving motion, is recommended.

should tip about 5°-15° in the direction of motion. The centerline of the electrode should be at about 45° to the edge and the flat surface. If the edge of the metal melts too rapidly, point the electrode more toward the flat surface. Be certain that both the edge and surface are melting completely as the filler metal is added. The appearance of a C-shaped molten weld pool will indicate good fusion.

15.10.2 Fillet Weld on an Inside Corner Joint

The centerline of the electrode should be held at 45° to each surface. It should be tipped at 5°-15° in the direction of motion. A C-shaped weld pool will indicate good fusion is occurring. Short circuiting and pulsed spray transfer are best suited for vertical welding. Spray arc can be used with a weaving motion in some applications.

15.11.1 Fillet Welds

See Figures 15-55 and 15-56 for examples of the angles used. The centerline of the electrode should be 45° from each metal surface. It should be tipped about 5°-10° in the direction of travel.

15.11.2 Groove Welds

Figures 15-57 and 15-58 show the angles used to weld a butt joint and an outside corner joint in the overhead welding position.

15.10.3 Groove Weld on a Butt or Outside Corner Joint

A V-groove butt weld in progress is shown in Figure 15-54. The electrode centerline should be directly above the weld line. The electrode and torch should be inclined (tipped) 5°-15° in the direction of travel. A keyhole at the root of the weld will indicate complete penetration.

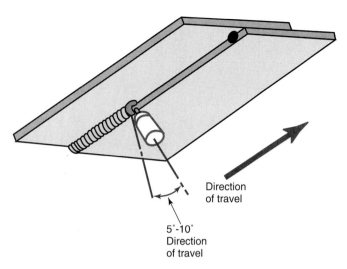

Figure 15-55. *A fillet weld on a lap joint in the overhead welding position. In this joint, two passes will be made. This is done to keep the weld pool size small and easy to manage.*

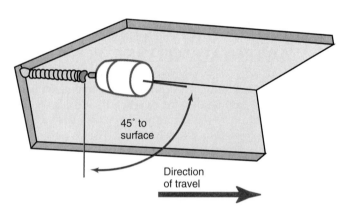

Figure 15-56. *A fillet weld on an inside corner joint. The electrode and gun are tipped 5°-10° in the direction of travel.*

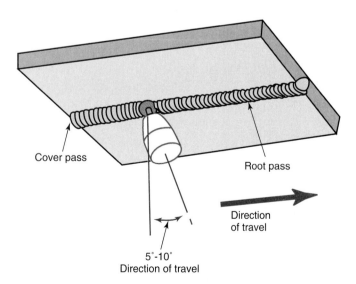

Figure 15-57. *Bevel-groove weld on a butt joint in an overhead welding position.*

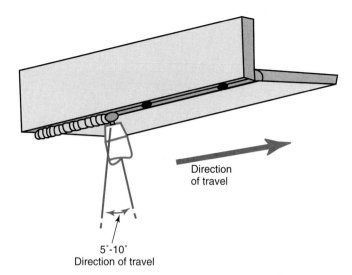

Figure 15-58. *A J-groove weld in an outside corner joint in the overhead welding position.*

15.12 AUTOMATIC GMAW AND FCAW

Gas metal arc welding and flux cored arc welding may be semiautomatic or fully automatic processes. In the semiautomatic process, the welder must direct and move the arc welding gun while the electrode wire feeds automatically into the weld pool.

Both GMAW and FCAW guns may be mounted on a motor-driven carriage or robot. Figure 15-59 shows a robot moving a GMAW gun. When the welding gun is controlled by a machine with feedback controls, the process becomes fully automatic. Refer to Chapter 25 for information on robots and other automatic welding equipment.

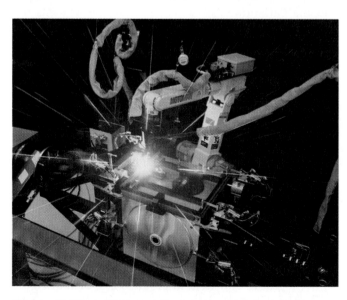

Figure 15-59. *A GMA welding gun mounted on a robot. The part is held in a fixture. (Motoman, Inc.)*

15.13 GAS METAL ARC SPOT WELDING

The gas metal arc welding power source and arc welding gun can be used to produce a weld in one small spot. Metals commonly welded with the gas metal arc spot welding process are low-carbon steel, stainless steel, aluminum, copper-based metals, and magnesium. Gas metal arc spot welding is generally done on metals under 1/16″ (1.6mm) thick. Metals up to 3/16″ (5mm) thick can be welded.

Small tack welds can be made on lap and corner joints. See Figure 15-60. A completed spot weld is shown in Figure 15-61. A spot weld is a weld made on overlapping pieces with the weld away from the edges. The gas metal arc welding power source must be equipped with special controls to do spot welding. The arc welding gun must be fitted with a special nozzle for spot welding. See Figure 15-36.

Several welding variables must be controlled to make gas metal arc spot welds. These variables are:

- Arc voltage.
- Welding current.
- Welding time.
- Electrode size and composition.
- Electrode extension.
- Shielding gas.

Voltage settings are made on the arc welding power source in the same way as when GMAW. If the voltage is increased, the arc length will increase. However, the pene-tration and the weld reinforcement (buildup) will decrease slightly.

Welding current is DCEP (DCRP). Current is controlled by varying the wire feed speed. Welding current greatly affects the spot weld penetration. Higher currents create greater penetration.

Welding time is controlled by timers in the arc welding equipment. Timers allow the same quality weld to be made each time. Spot welding times are usually about one second. Longer times are necessary when welding thicker metals. Penetration increases as the welding time is lengthened. The diameter of the weld area also increases as the welding time is increased.

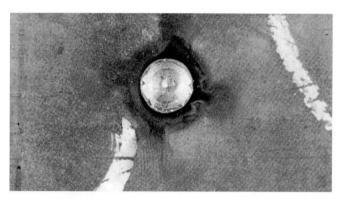

Figure 15-61. *A completed gas metal arc spot weld. (ESAB Welding and Cutting Products)*

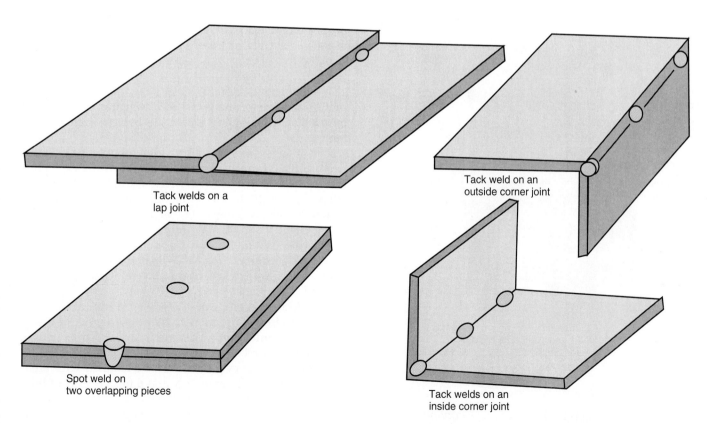

Figure 15-60. *Tack welds on lap, inside corner, and outside corner joints. Several spot welds are also shown. Notice the depth of penetration shown in section.*

The same size and type of solid wire used for welding may be used for spot welding a particular metal. Electrode extension, Figure 15-40, must remain constant during the GMA spot welding process. The extension distance is kept constant by using a special nozzle. Several GMA spot welding nozzle designs are shown in Figure 15-36. The end of the contact tube is set back from the end of the nozzle. This is done to keep the contact tube out of the weld. This setback will also reduce the possibility of the electrode melting up into the contact tube at the end of the weld cycle. The shielding gas used may be the same gas or gas mixture used for welding beads.

GMA spot welds on thin metals may be made in any position. As the metal thickness increases, GMA spot welding is limited to the flat welding position. Weld quality and uniformity is not as good as that possible with resistance spot welding. The big advantage of GMA spot welding over resistance spot welding is that access to only one side of the parts is required.

The GMA spot welding controls found on various gas metal arc welding machines differ. Figure 15-62 shows a GMAW power source. Some controls typically found on a GMA spot welding control panel are:

- *Control switch*. The switch used to change the gas metal arc welding machine from a regular welder to a spot welder.
- *Weld timer*. This control is for setting the welding time. The entire spot welding operation takes place in one or two seconds.
- *Burn-back adjustment*. Some machines have a burn-back adjustment. This control allows the

current to flow for a short time after the wire feed stops. The continued current flow prevents the wire from sticking in the weld pool. If the burn-back time is set too high, the electrode wire may burn back into the contact tube. If it is not set high enough, the wire will stick in the weld pool at the end of the welding time.

15.14 GMAW TROUBLESHOOTING GUIDE

Figure 15-63 is a chart that describes many typical troubles which may occur when making a gas metal arc weld. Steps to take to correct each problem are listed. The causes are shown along with methods for correcting each problem.

15.15 GMAW AND FCAW SAFETY

The safety precautions for arc welding covered in Chapter 1 and in other chapters of this book also apply to GMAW and FCAW.

Adequate eye protection must always be worn. If welding for long periods, flash goggles with a #2 lens shade should be worn under the arc helmet. A #11 lens is recommended for nonferrous GMAW and a #12 for ferrous GMAW. Lens shades up to #14 may be worn as required for comfort. Figure 15-64 shows a welder with proper hood and an electronic quick-changing lens installed. All welding should be done in booths or in areas shielded by curtains. This is done to protect others in the weld area from arc flashes.

Suitable dark clothing must be worn. This is done to protect all parts of the body from radiation or hot metal burns. Leather clothing offers the best protection from burns.

It is suggested that all welding should be done in well-ventilated areas. Ventilation and/or filtering equipment should be provided, as necessary, to keep the atmosphere around the welder clean. **Carbon monoxide gas is generated when using CO_2 as a shielding gas while doing GMAW and FCAW. Ozone is also produced when doing GMAW and FCAW. Ozone is a highly toxic gas. Metals still covered with chlorinated hydrocarbon solvents will form toxic (poisonous) phosgene gas when welded.**

Protect arc cables from damage. **Do not touch uninsulated electrode holders with bare skin or wet gloves. A fatal shock could result. Welding in wet or damp areas is not recommended.**

Shielding gas cylinders must be handled with great caution. Refer to Chapters 1, 5, and 6 for a review of how to handle high-pressure cylinders. Chapters 5 and 6 should also be referred to for instructions on how to attach regulators and other gas equipment.

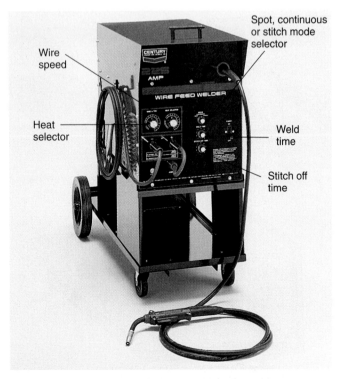

Figure 15-62. This GMAW outfit can make continuous, spot, or stitch-type welds. The wire feeder is under the top cover. (Century Mfg. Co.)

Trouble	Possible causes	How to correct
Difficult arc start	Polarity wrong	Check polarity, try reversing
	Insufficient shielding gas	Check valves, increase flow
	Poor ground	Check ground—return circuit
	Open circuit to start switch	Repair
Irregular wire feed, burn back	Insufficient drive roll pressure	Increase drive roll pressure
	Wire feed too slow	Check, adjust wire feed speed
	Contact tube plugged	Clean, replace contact tube
	Arcing in contact tube	Clean, replace contact tube
	Power circuit fluctuations	Check line voltage
	Polarity wrong	Check polarity, try reversing
	Torch overheating	Replace with higher amp gun
	Kinked electrode wire	Cut out, replace spool
	Conduit liner dirty or worn	Clean, replace
	Drive rolls jammed	Clean drive case, clean electrode wire
	Conduit too long	Shorten, install push-pull drive
Welding cables overheating	Cables too small	Check current requirements, replace
	Cable connections loose	Check, tighten
	Cables too long	Check current-carrying capacity
Unstable arc	Cable connections loose	Check, tighten
	Weld joint area dirty	Clean chemically or mechanically
Arc blow	Magnetic field in DC causes arc to wander	Rearrange or split ground connection
		Use brass or copper backing bars
		Counteract "blow" by direction of weld
		Replace magnetic work-bench
Undercut	Current too high	Use lower current setting
	Welding speed too high	Slow down
	Improper manipulation of gun	Change angle to fill undercut
	Arc length too long	Shorten arc length
Excessively wide bead	Current too high	Use lower current setting
	Welding speed too slow	Speed up
	Arc length too long	Shorten arc length
Incomplete penetration	Faulty joint design	Check root opening, root face dimensions, including angle
	Welding speed too rapid	Slow down welding speed
	Welding current too low	Increase welding current
	Arc length too long	Shorten arc length
	Improper welding angle	Correct faults, change gun angle
Incomplete fusion	Faulty joint preparation	Check root opening, root face dimensions, included angle
	Arc length too long	Shorten arc length
	Dirty joint	Clean chemically or mechanically
Dirty welds	Inadequate gas shielding	Hold nozzle closer to work
		Increase gas flow
		Decrease gun angle
		Check gun and cables for air and water leaks
		Shield arc from drafts
		Center contact tube in nozzle
		Replace damaged nozzle
	Dirty electrode wire	Keep wire spool on welder covered
		Keep unused wire in shipping containers
		Clean wire as it enters wire drive
	Dirty base metal	Clean chemically or mechanically
Porosity	Dirty electrode wire	See above, *Dirty welds*
	Dirty base metal	
See above, *Dirty welds*	Inadequate gas shielding	See above, *Dirty welds*
Cracked welds	Improper technique	Change angle of gun to improve shielding
	Faulty design	Check edge preparation and root spacing
	Faulty electrode	Check electrode wire for compatibility with base metal
	Shape of bead	Change travel speed or shielding gas to obtain more convex bead
	Travel speed too fast	Slow down
	Improper technique	Change angle of gun to improve deposition
	Rigidity of joint	Redesign joint, preheat and postheat, weave bead

Figure 15-63. *A troubleshooting guide for problems that might occur when GMAW. (Welding and Fabricating Data Book)*

Figure 15-64. *This welder is wearing a quick change filter lens in the welding helmet. The lens will darken to a protective shade in a fraction of a second after the arc is struck. (Jackson Products, Div. Thermadyne Industries, Inc.)*

TEST YOUR KNOWLEDGE

Write your answers on a separate sheet of paper. Do not write in this book.

1. Name three benefits of using GMAW.
2. The polarity used for almost all GMAW and FCAW is DCE _____ or DC _____ _____.
3. Name three metal transfer methods.
4. _____ is the property in an electric circuit that slows down the rate of current change.
5. Spray transfer will only occur when the current is set above the _____ current.
6. Spray transfer will only occur when at least _____ % argon is used.
7. What is MIG brazing?

8. How many pounds and kilograms of filler metal can be deposited per hour with the spray transfer method?
9. There are two main variables made on the welding machine prior to welding. Name the two variables.
10. A GMAW power source used for pulsed spray transfer must have what additional controls?
11. Using spray arc transfer, _____ volts and _____ amperes are used with a 0.045" (1.1mm) electrode to weld stainless steel.
12. _____ volts and _____ amps are used to weld mild steel using short circuiting transfer and 0.035" (0.9mm) diameter wire.
13. On the wire drive unit shown in Figure 15-20, wire guides and drive rolls are aligned by loosening the _____ _____ securing bolts and moving the _____ _____ up or down.
14. To feed the electrode wire through the electrode cable to the arc welding gun, the _____ switch is operated.
15. What factors must be considered when choosing a shielding gas?
16. Argon has a _____ thermal conductivity than helium, so _____ is used to weld thick aluminum or copper sections.
17. Why is good ventilation important when using CO_2 gas?
18. What effect does oxygen (O_2) have on the arc when mixed with argon?
19. Which gases are suggested for use with pulsed spray transfer?
20. What argon flow rate in ft^3/hr. and L/min should be used to weld 0.150" (3.8mm) thick magnesium in a butt joint?
21. What part of the gas metal arc welding gun contacts the electrode wire and passes electricity to the electrode?
22. How can metal spatter be kept from sticking to the nozzle?
23. Electrode extension is the distance from the end of the _____ _____ to the end of the _____ .
24. The suggested angle for the electrode and gun for best weld pool control with backhand welding in most positions is _____ ° to _____ ° forward of vertical.
25. Metals still covered with chlorinated hydrocarbon solvents will form a toxic _____ gas when welded.

This welder is using GMA to make a downhill weld on an outside corner joint. (Hornell Speedglas, Inc.)

Part 5

ARC CUTTING

This hand-held plasma arc cutting system can be used to cut steel, aluminum, or other metals up to 1" in thickness. (Hypertherm, Inc.)

A water-filled table is often used with plasma arc cutting, since cutting under water reduces the sound level. These two plasma arc torches are cutting duplicate patterns in steel plate. (Thermadyne Industries, Inc.)

Chapter 16

ARC AND OXYGEN ARC CUTTING EQUIPMENT AND SUPPLIES

Several of the electric arc processes are used in the welding industry to cut or gouge metal. The arc cutting processes permit a welder to make high-quality cuts. The cutting speeds used in the electric arc processes are usually much higher than those of the oxyfuel gas cutting processes. This makes the electric arc methods of cutting and gouging very economical. Improved equipment and supplies have contributed to the usefulness and efficiency of arc cutting.

Gouging is a process that is generally used to remove metal around a crack or a defective weld prior to rewelding. It employs an arc cutting process to cut a groove of a desired depth and width into the surface of a metal. Since gouging can be done using any arc process, the AWS does not give a specific abbreviation identification to gouging, but identifies only the cutting process.

The electric arc processes used for cutting and gouging include those listed below. Exothermic cutting and oxygen arc cutting are included in this chapter, even though they are classified as oxygen cutting processes by the American Welding Society.

• Plasma arc cutting (PAC).
• Air carbon arc cutting (CAC-A).
• Exothermic cutting.
• Oxygen arc cutting (AOC).
• Shielded metal arc cutting (SMAC).
• Gas metal arc cutting (GMAC).
• Gas tungsten arc cutting (GTAC).
• Carbon arc cutting (CAC).

When using any of the arc cutting or gouging processes, be certain to follow all safety precautions listed for both oxygen and arc cutting processes. Refer to Heading 8.10 for a review of oxyfuel gas cutting safety and Heading 17.8 for a review of arc cutting safety.

16.1 PLASMA ARC CUTTING (PAC) PRINCIPLES

Plasma arc cutting uses the principle of passing an electric arc through a quantity of gas traveling through a restricted outlet. The gas is heated by the electric arc to such a high temperature that it turns into what physicists call a *plasma*. Plasma is the *fourth* state of matter—not a gas, liquid, or solid. A great deal of heat energy is required to change a gas to a plasma. The heat energy required is furnished by the electric arc. The heat of the plasma is then released into the base metal as the plasma turns back to a gas. Refer to Heading 4.13.

Plasma arc cutting (PAC) was originally developed to cut nonferrous metals, using inert gases. Later developments made it possible to cut steel, using oxygen as the plasma gas. In the plasma process, temperatures as high as 25,200ºF (14 000ºC) are reached. A constant current (cc), direct current arc welding machine is used as the power source for PAC. Inverter-type power supplies are often used for plasma arc cutting. The polarity used on either the constant current or inverter power supply is DCEN (direct current electrode negative). A voltage of 50V to 200V may be required. Figure 16-1 is a drawing showing the equipment used for plasma arc cutting. Clean, narrow kerfs are possible at cutting speeds of 300 in/min (7.6m/min) on aluminum and over 200 in/min (5m/min) on stainless steel. Figure 16-2 shows the quality of a plasma cut surface.

Several ports are sometimes used in the torch. Only the center port has the arc plasma, while the surrounding ports provide a shielding gas protection. Gas flow may be as high as 250 ft^3/hr (118 L/min). The *dual-flow cutting process* uses one gas for the plasma and one for a

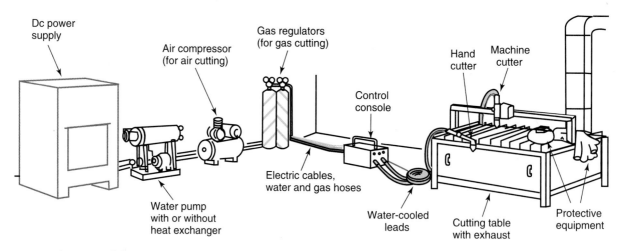

Figure 16-1. *A drawing of the equipment required for a plasma arc cutting station. (Thermadyne Industries, Inc.)*

Figure 16-2. *This 1" (25mm) mild steel plate was cut with a plasma arc cutting torch using compressed air as the gas.*

shielding gas. See Figure 16-3. Nitrogen is often used as the plasma gas. The shielding gas for mild steel is CO_2 or air. An argon and helium mixture is used as the shielding gas for aluminum.

Water may be used in specially built plasma torches in a similar manner as the shielding gas. See Figure 16-3. This process is known as *water shield plasma cutting*.

A water spray may be used to constrict the plasma gas flow, as shown in Figure 16-4. Narrow cuts can be produced using this method. Water-injection plasma cutting may be done on a cutting table covered with a shallow depth of water. The plasma arc cutting torch is started and the cutting is done under water. The water acts as a shielding medium and also quiets the sound of the plasma cutting operation. Water from the table may be filtered and recirculated for reuse in the torch. See Figure 16-5.

When cutting with the plasma arc without a water table, the gas is heated in the nozzle. The gas tries to expand, but is restricted by the nozzle. The nozzle acts like a venturi, causing the plasma to increase in speed. As the plasma passes through the nozzle, its speed increases to several thousand feet per second. It is this high gas velocity that removes the molten metal from the work and creates the kerf. The plasma (formed from a shielding gas) and the shielding gas or auxiliary shielding gas shields the sides of the kerf from contamination and oxidation.

There are two plasma arc processes. They are:
- Transferred arc.
- Nontransferred arc.

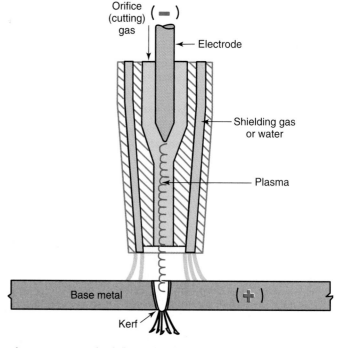

Figure 16-3. *A dual-flow plasma arc cutting torch nozzle. The orifice (cutting) gas becomes a plasma in the arc stream. The shielding gas or water flows out around the plasma orifice through a number of holes.*

In the *transferred arc process,* the work is electrically connected to the plasma arc torch, so that it becomes a part of the electrical circuit. An arc is created between the electrode and the base metal and between the electrode and the plasma torch nozzle. See Figure 16-6. Direct current electrode negative (DCEN) is used in both transfer methods so that most of the heat developed is on the base metal.

In the *nontransferred arc process,* the electrical circuitry is different from that of the transferred arc circuit. The arc is struck between the electrode and the torch nozzle. The workpiece is not part of the electrical circuit. A schematic of this system is shown in Figure 16-7.

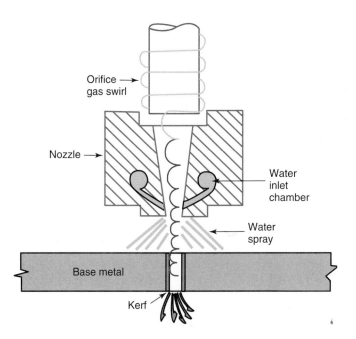

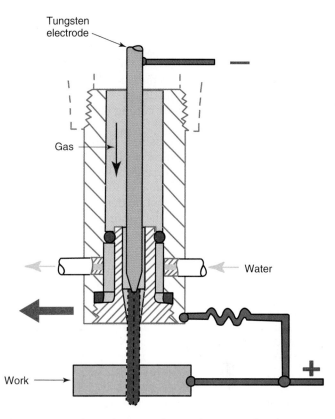

Figure 16-4. *A water injection plasma arc cutting torch nozzle. Water is injected into the plasma stream to constrict (reduce) its diameter.*

Figure 16-6. *This schematic shows the principle of operation of the transferred arc plasma cutting process.*

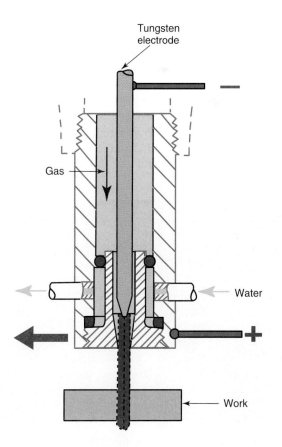

Figure 16-5. *A plasma arc cutting torch being used to cut steel plate under water. Cutting under water makes this process much quieter. (ESAB Welding and Cutting Products)*

Figure 16-7. *The principle of operation of the non-transferred arc plasma cutting process is shown in this schematic.*

Of the two methods, the transferred arc creates the greatest amount of heat and is used when cutting. The plasma arc cutting method may be used to cut almost all types of metals with great success and economy because of the extreme temperatures and the shielding qualities of the gases used. Cuts may be made dross (slag) free. Diver-welders have done plasma arc cutting underwater at great depths.

The plasma arc process has also been used to "machine" inconel and stainless steel metals, which rapidly wear down ordinary cutting tools. The metal is held and rotated in a lathe and the plasma arc torch is held in a special fixture on the movable tool holder. In this way, accurate cuts may be made over the entire length of the rotating metal workpiece. The plasma transfer arc process is more often used for surfacing operations. See Chapter 26.

16.1.1 Plasma Arc Cutting Equipment and Supplies

Plasma arc cutting may be done with automatic equipment, or manually as shown in Figure 16-8. The equipment, with the exception of the power source, is the same as that used for plasma arc welding. See Chapter 20. The manual-type torch for heavy-duty cutting is sometimes equipped with a heat shield to protect the operator's hands from the intense heat. For heavy-duty cutting of aluminum up to 5" (127mm) thick, it is necessary to have a power source capable of supplying dc power up to 700A (amperes) at 170V. Figure 16-9 shows an arc welding power source with power modules stacked up to provide enough amperage for plasma arc cutting. A power source of 400A may be required to cut 1 1/2" (38mm) aluminum or 1" (25mm) stainless steel. A high-frequency unit is required to start the pilot arc at the beginning of a cut. Cooling water pumps and a shielding gas or gases under pressure are also required. For heavy-duty cutting, cooling water pumps are required. They circulate enough water through the torch to prevent damage from the extreme temperatures of the plasma arc.

Figure 16-9. A portable arc welding machine used for plasma arc cutting. Power modules are stacked on top of each other to produce sufficient amperage. Note the plasma arc torch in the foreground. (Thermadyne Industries, Inc.)

Plasma arc torches may be either manual or machine-operated. Figure 16-10 shows several manual plasma torches and one machine-mounted plasma arc cutting torch. Machine-operated torches are usually designed with a rack and pinion gear adjustment similar to a oxyacetylene machine cutting torch.

Common shielding gases used for plasma arc cutting are nitrogen, CO_2, and varying mixtures of argon and helium. Air has also been used as the plasma gas with acceptable results.

Figure 16-8. A plasma arc cutting torch being used to cut an aluminum plate. (Thermadyne Industries, Inc.)

16.2 AIR CARBON ARC CUTTING (CAC-A) EQUIPMENT

The *air carbon arc cutting and gouging process* uses an arc between a carbon electrode and the base metal to heat the metal. A stream of pressurized air directed through the electrode holder blows the molten metal away to form a kerf (cut) or gouge. See Heading 4.9.

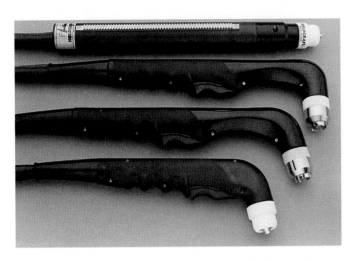

Figure 16-10. Plasma arc cutting torches. At the top is a torch for machine mounting. It has a rack-type gear on the body for height adjustment. The other three torches are for manual cutting. (Weldcraft Products, Inc.)

The equipment used for *air carbon arc cutting* (CAC-A) and gouging includes:

- An ac or dc arc welding machine. Constant current (cc) is used when cutting or gouging manually. Constant voltage (cv) or cc may be used when cutting or gouging mechanically.
- An air compressor or compressed air in cylinders (for small applications).
- An air carbon arc torch equipped with an air jet device.
- An electrode lead.
- A compressed air hose. This hose is usually combined with the electrode lead, as shown in Figure 16-11.
- A workpiece lead.
- Carbon or graphite electrodes designed for use with dc or ac.

The gloves, special clothing, booths, ventilation devices, benches, and other equipment as used in SMAW are required. See Chapter 11 and Headings 11.11 and 11.12. A #12 to #14 arc welding filter lens is required in the welding helmet. The polarity used with a dc power source should be direct current electrode positive, DCEP (DCRP). The machine must also be capable of delivering the required current at a 100% duty cycle. The normal arc voltage should be between 35V and 55V. An open circuit voltage of about 60V is sufficient. Special carbon electrodes are made for use with ac current. Refer to Heading 4.9.

The air carbon arc process may be used in conjunction with a motorized carriage. Figure 16-12 shows an air carbon arc holder mounted on a motorized carriage or tractor. The motorized carriage may be used in the horizontal, vertical, and overhead welding positions. In these positions, a special track is used, and the tractor is specially equipped and known as a "climber." The track is held to magnetic materials by magnets. On nonmagnetic materials, the track is held to the metal by suction cups or clamps. Specially equipped tractors may be operated by remote control. Figure 16-13 shows an air carbon arc machine in use.

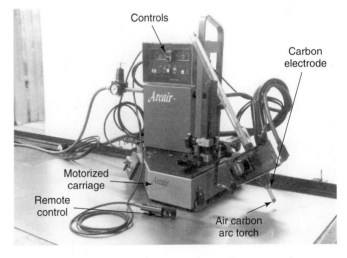

Figure 16-12. An air carbon arc electrode mounted on a motor-driven carriage. This carriage may be controlled remotely. (Arcair Div.,Thermadyne Industries, Inc.)

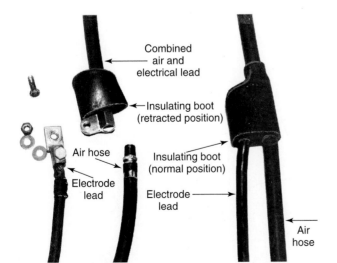

Figure 16-11. A combination electrode lead and air hose assembly used for air carbon arc cutting. (Arcair Div., Thermadyne Industries, Inc.

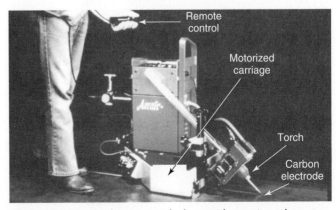

Figure 16-13. Gouging a steel plate with an air carbon arc torch mounted in a motorized carriage. The operator is using a remote control to steer the carriage. (Arcair Div., Thermadyne Industries, Inc.)

16.2.1 Air Carbon Arc Compressed Air Supply

Air is fed to the air carbon arc torch jet at 80 psig to 100 psig (552kPa to 690kPa). The air source should be capable of supplying sufficient air to produce a clean cut or gouge. Large air compressors are used to supply the air required. With light-duty carbon electrode holders, the pressure may be as low as 40 psig (276kPa) at a flow rate of 3 ft³/min (8.5L/min). The manufacturer's instructions should be referred to for the recommended air pressure. This air should contain a minimum of abrasives and moisture. A sufficient amount of air should flow through the jet to remove the molten metal. See Figure 16-14.

For light-duty electrode holders, the inside diameter of the air hose should be 1/4" (6mm). For medium-duty electrode holders, the inside diameter of the hose should be 3/8" (10mm), and for heavy-duty holders, the hose inside diameter should be 1/2" (13mm).

On the hand-held air carbon arc torch, an air control valve or button on the torch is used to control the on and off function of the air. On a motorized carriage, the air pressure is turned on and off by means of a switch on the carriage.

Compressed air in cylinders may be used when portability is desired. It may also be used when the length of the cut is relatively short. **Caution:** Oxygen should *never* be used in place of air for carbon arc cutting or gouging.

16.2.2 Air Carbon Arc Electrode Holder

The hand-held air carbon arc electrode holder or torch is similar to the standard electrode holder. The air carbon arc torch has an air passageway, an air on-off valve, and an orifice (hole) for the air jet. See Figure 16-14.

The electrode lead and air hose connections are usually made outside the handle of the torch. See Figure 16-11. The air valve is mounted in the handle and is either lever- or button-operated. Most of these valves have a "lock-open" feature. This allows the operator to keep the air flowing, and still have a comfortable hand position on the handle.

It is very important that the air jet be directed as close as possible to the arc crater. Any misalignment or abuse of the air orifice may produce below-average results.

16.2.3 Air Carbon Arc Electrodes

Different types of CAC-A electrodes may be obtained for use with ac or dc power sources. Carbon electrodes are available in the graphite form or as a carbon mixture. The three types of carbon electrodes are *dc plain, copper-coated,* and *ac copper-coated*. The copper coating helps to reduce the oxidation of the electrode body. It also helps to keep the electrode cool. The copper coating does not, however, add much to the conductivity of the electrode.

Carbon electrodes are usually round in cross section, but they may also be purchased in flat or half-round shapes. The flat and half-round shapes are used to gouge a rectangular groove.

Standard carbon electrodes are available in diameters from 1/8" to 3/4" (3mm to 19mm). The range of diameters available differs depending whether they are for use with dc or ac. Most carbon electrodes are available in 12" (305mm) lengths. Electrodes are also available with male and female ends which fit into each other to minimize loss. See Figure 16-15. These electrodes come in diameters of 5/16" to 1" (8mm to 25mm). Figure 16-16 shows a table of carbon electrode sizes and the approximate current settings for each. The table in Figure 16-17 shows cutting

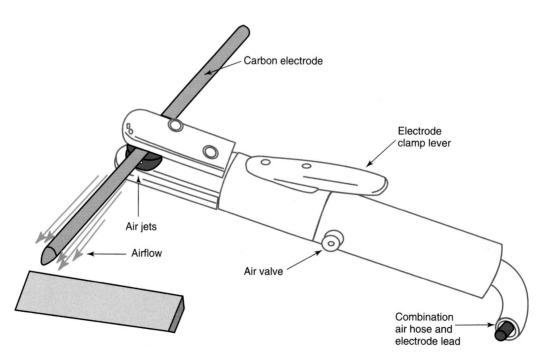

Figure 16-14. *An air carbon arc electrode holder with a carbon electrode installed. Note the air jet orifices under the electrode.*

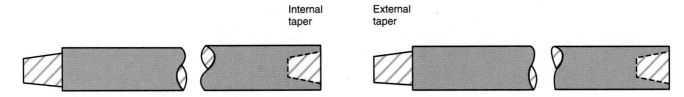

Figure 16-15. *Two carbon electrodes joined by matching tapers. New electrodes can be added to old ones to permit a continuous cutting or gouging operation. This system also reduces electrode waste.*

Electrode diameter		Amperage with DCEP (DCRP) electrode		Amperage with ac electrode	
in.	mm	Min.	Max.	Min.	Max.
5/32	4.0	90	150	--	--
3/16	4.8	150	200	150	200
1/4	6.4	200	400	200	300
5/16	7.9	250	450	250	300
3/8	9.5	350	600	300	500
1/2	12.7	600	1000	400	600
5/8	15.9	800	1200	--	--
3/4	19.1	1200	1600	--	--
1	25.4	1800	2200	--	--

Figure 16-16. *A table of suggested current settings for various diameters and types of air carbon arc electrodes.*

Electrode size for single groove diameter		Desired depth		Amperage	Travel speed	
in.	mm	in.	mm	DCEP (DCRP)	ipm	mm/min
5/16	7.9	1/4	6.4	425	36	914
5/16	7.9	5/16	7.9	450	33	838
3/8	9.5	1/4	6.4	500	46	1168
3/8	9.5	3/8	9.5	500	25	635
3/8	9.5	1/2	12.7	500	17	432
1/2	12.7	3/8	9.5	850	35	889
1/2	12.7	1/2	12.7	850	24	610
1/2	12.7	5/8	15.9	850	20	508
5/8	15.9	1/2	12.7	1250	28	711
5/8	15.9	5/8	15.9	1250	22	559
5/8	15.9	3/4	19.0	1250	17	432
3/4	19.0	3/4	19.0	1600	20	508
3/4	19.0	1	25.4	1600	15	381

Figure 16-17. *A table of operating conditions for N-Series, Arcair air carbon arc torches. (Arcair Div.,Thermadyne Industries, Inc.)*

speeds and electrode sizes recommended for various groove diameters and depths.

As shown in Figure 16-16, some current ratings are quite high and the arc welding machine must be of ample capacity. Two or more standard machines may be connected in parallel to give the current capacity needed as shown in Figure 16-18.

16.3 EXOTHERMIC CUTTING EQUIPMENT AND SUPPLIES

The *exothermic cutting process* is a fairly new process, introduced in 1978. The AWS has not given it a designation or abbreviation, but they consider it a derivative of lance oxygen cutting (LOC). This process is a cross between oxygen arc cutting and oxygen lance cutting. Its original use was for underwater cutting at offshore oil and gas platforms. It soon was used for cutting and gouging on land. Most metal cutting and gouging today is done using either an oxyfuel gas, arc, or exothermic cutting process.

The equipment and supplies required for exothermic cutting or gouging are shown in Figure 16-19. They include:

- A special hand-held torch.
- Special exothermic cutting rods.
- A copper striker plate.
- A 12V battery.
- An oxygen supply with a pressure regulator.
- An oxygen hose.
- Electrical cables.

An exothermic cutting torch has a collet to hold the electrode securely and an oxygen control lever to turn the oxygen on and off. See Figure 16-20.

The exothermic electrode is a hollow steel tube from 1/8" to 1/4" (3mm to 6mm) in diameter. One end of the tube is reduced in diameter to fit into the electrode holder. The hollow tube is filled with many alloy fuel wires to within 2" (51mm) of the electrode holder end. The outer

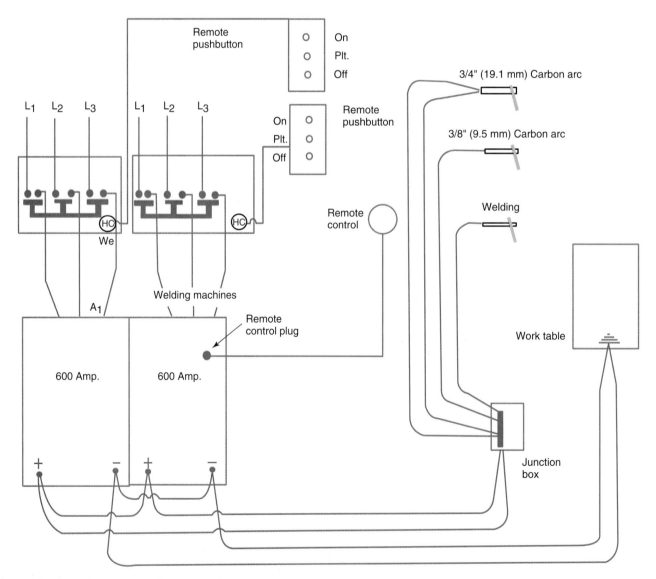

Figure 16-18. *Wiring diagram for connecting two dc arc welders in parallel.*

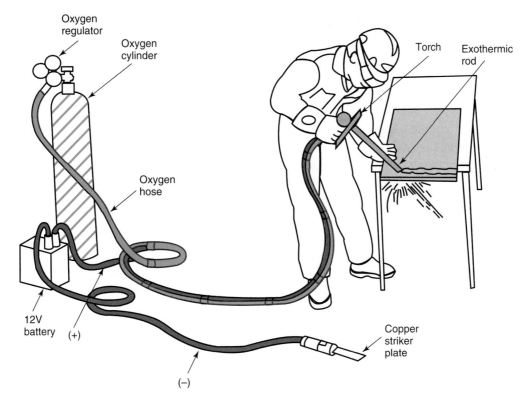

Figure 16-19. *An exothermic arc cutting outfit. (Broco, Inc.)*

Figure 16-20. *A basic exothermic cutting torch set. It consists of a torch, a leather shield for piercing, a copper striker plate and ground cable, and two sizes of collets. (Broco, Inc.)*

Figure 16-21. *Several exothermic electrodes. The crimping holds the exothermic wires in place in the copper-coated outer tube. Note: To get at hard-to-reach places, these electrodes may be bent. (Broco, Inc.)*

tube is crimped (creased) several times along its length to hold the fuel wires in place. See Figure 16-21.

The equipment required for exothermic cutting or gouging is light and very portable. No arc welding machine or fuel gas cylinder is required. If the cutting job is small, a small size oxygen cylinder can be used. All the equipment needed can easily be carried like a backpack.

This process is used by firefighters and rescue personnel to cut through doors and concrete walls. It has been used by the military to make cuts during field operations.

16.4 OXYGEN ARC CUTTING (AOC) EQUIPMENT

The *oxygen arc cutting process* is still recognized by the AWS, although it is not used to the extent that it once was in the United States, due to the introduction of the exothermic cutting process.

Oxygen arc cutting is done by striking an arc on the metal to be cut with a hollow (tubular) electrode. Oxygen under pressure is passed through the hollow electrode. The oxygen flow oxidizes the metal and blows the molten metal in the pool away, as in oxyfuel gas cutting. This action continues to form a kerf or cut in the base metal. Refer to Heading 4.10.

Oxygen arc cutting equipment is a combination of oxyfuel gas and arc welding stations. It uses these items of oxyfuel gas station equipment:
- Oxygen cylinder.
- Oxygen regulator.
- Oxygen hose.

It uses the following parts of an arc welding station:
- Arc welding machine.
- Workpiece lead (ground cable).
- Electrode lead (cable).
- Booth.
- Ventilation system.
- Bench.

The special equipment needed is:
- Special oxygen arc torch.
- Oxygen arc cutting electrodes.

16.4.1 Oxygen Arc Cutting Torch

The electrode holder or torch is designed to hold a tubular iron or steel electrode. See Figure 16-22. Most underwater electrode holders have a watertight collet assembly designed to fit over the end of the electrode. This

Figure 16-22. *An underwater cutting torch. (Broco, Inc.)*

seal keeps water from contacting the electrical connections in the torch. This special collet assembly also connects the electrode to the oxygen supply. Neoprene or silicone rubber gaskets are used under the collet nut and the control valve nut. See Figure 16-23. To control the oxygen flow, a cutting oxygen control lever is mounted on the torch.

16.4.2 Oxygen Arc Cutting Electrodes

The electrode consists of a steel tube covered with a flux coating. The tube carries the arc current, the inside of the tube is the oxygen passageway, and the covering serves as an insulator. With the heavy flux covering, the AOC electrode is dragged across the metal. The covering provides a means of maintaining the correct arc length. It also provides fluxing agents to improve the cutting action.

Electrodes for AOC are available in 3/16" and 5/16" (5mm and 8mm) outside tube diameters and 1/16" and 1/10" (2mm and 3mm) inside diameters respectively. These electrodes are normally produced in 18" (457mm) lengths.

16.5 SHIELDED METAL ARC CUTTING (SMAC) ELECTRODES

Shielded metal arc cutting is done using the same equipment as used with SMAW. A special cutting electrode is generally used, but a regular SMAW electrode may also be used. This method of cutting is no longer in common use. It is done when none of the other, more efficient arc methods are available. SMAC may be used in a home craft shop or in an emergency situation.

The SMAC special cutting electrode is designed to produce a high-velocity stream of gas and particles. As the electrode is consumed, a deep cavity is formed in the end of the electrode. This cavity causes a jet-like action of the gases and metal particles. That jet action produces a driving arc which melts and cuts the base metal. Figure 16-24 shows the approximate electrode sizes and current settings when using special shielded metal arc cutting electrodes.

When cutting mild steel, regular shielded metal arc welding electrodes may be used. However, a much higher welding current range will be required. Figure 16-25 lists the current values to use when doing various cutting operations with standard covered steel welding electrodes.

If regular mild steel SMAW electrodes are used for cutting, they will last a little longer if they are soaked in water for a maximum of 10 minutes before being used. The moisture in the coating slows the vaporizing of the coating material, thus producing a deeper "cup" (cavity) at the arc end of the electrode. This deeper cavity creates a stronger jet action which improves the cutting. Water-soaked electrodes should be used immediately. Water-soaked electrodes should *not* be used for welding because the moisture produces hydrogen embrittlement in the weld.

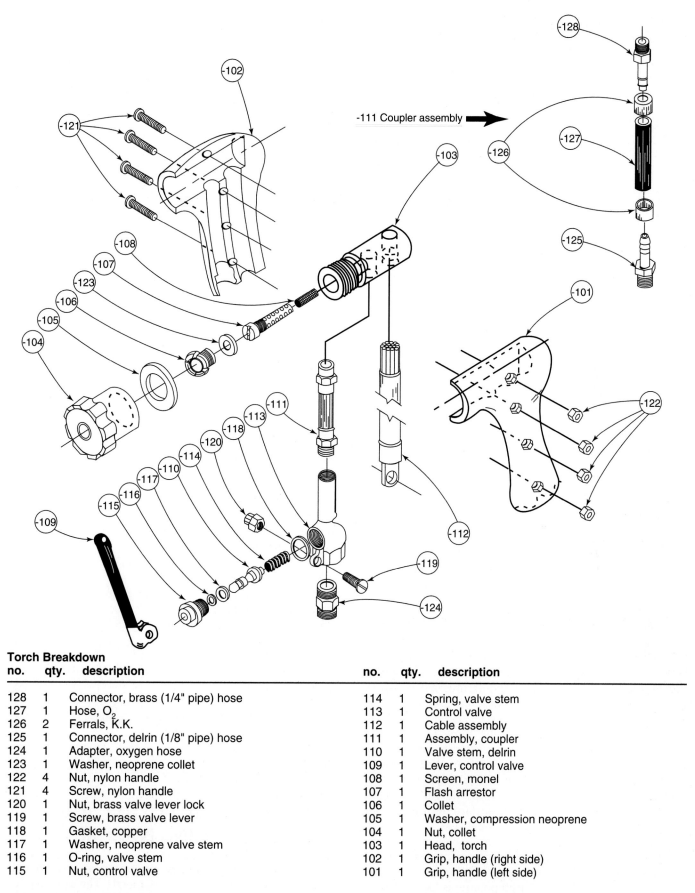

Torch Breakdown

no.	qty.	description	no.	qty.	description
128	1	Connector, brass (1/4" pipe) hose	114	1	Spring, valve stem
127	1	Hose, O_2	113	1	Control valve
126	2	Ferrals, K.K.	112	1	Cable assembly
125	1	Connector, delrin (1/8" pipe) hose	111	1	Assembly, coupler
124	1	Adapter, oxygen hose	110	1	Valve stem, delrin
123	1	Washer, neoprene collet	109	1	Lever, control valve
122	4	Nut, nylon handle	108	1	Screen, monel
121	4	Screw, nylon handle	107	1	Flash arrestor
120	1	Nut, brass valve lever lock	106	1	Collet
119	1	Screw, brass valve lever	105	1	Washer, compression neoprene
118	1	Gasket, copper	104	1	Nut, collet
117	1	Washer, neoprene valve stem	103	1	Head, torch
116	1	O-ring, valve stem	102	1	Grip, handle (right side)
115	1	Nut, control valve	101	1	Grip, handle (left side)

Figure 16-23. *An exploded view of an underwater oxygen-arc cutting torch assembly. The torch used for dry land exothermic cutting is constructed in a similar manner, but is not as watertight. (Broco, Inc.)*

Metal thickness		Electrode diameter		Ac current range amps	DCEN (DCSP) amps
in.	mm	in.	mm		
1/8	3.2	3/32	2.4	40-150	75-115
1/8 -1	3.2 - 25.4	1/8	3.2	125-300	150-175
3/4 - 2	19.1 - 50.8	5/32	4.0	250-375	170-500
1 - 3	25.8 - 76.2	3/16	4.8	300-450	
3 and over	76.2 and over	1/4	6.4	400-650	

Figure 16-24. A table of suggested electrode diameters and approximate current settings for cutting with shielded metal arc cutting electrodes.

Metal thickness		Electrode diameter		DCEN (DCSP) current range amps
in.	mm	in.	mm	
1/8	3.2	3/32	2.4	100 - 160
1/8 -1	3.2 - 25.4	1/8	3.2	160 - 350
3/4 - 2	19.1 - 50.8	5/32	4.0	250 - 400
1 - 3	25.4 - 76.2	3/16	4.8	350 - 500
3 and over	76.2 and over	1/4	6.4	400 - 650

Figure 16-25. Approximate current settings and electrode sizes for shielded metal arc cutting when using standard welding electrodes.

16.6 GAS METAL ARC (GMAC) AND GAS TUNGSTEN ARC CUTTING (GTAC) PRINCIPLES

The *gas metal arc cutting* and *gas tungsten arc cutting* processes are still recognized by the AWS. They are, however, not extensively used due to the introduction of better and less expensive forms of arc cutting. Aluminum, stainless steel, nickel, and most nonferrous metals up to 1/2" (13mm) may be cut using these processes. Refer to Heading 4.11.

The principle of cutting is the same for both the GMAC and GTAC processes. An arc is struck between the wire or tungsten electrode and the work. As the base metal melts, it is blown away by the flow of shielding gas. The shielding gas flow must extend through the full thickness of the metal being cut to keep the surface of the cut from oxidizing. Figure 16-26 shows the GTAC process in progress.

GTAC is best accomplished using a constant current (cc) dc welding machine with up to 600A and an open circuit voltage of 70V. The preferred polarity is DCEN to put most of the heat on the base metal. The shielding gases used with GTAC are argon/helium mixtures or nitrogen.

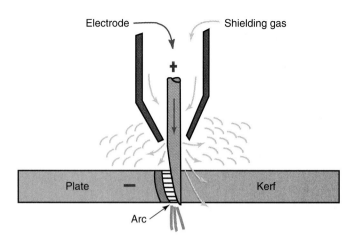

Figure 16-26. The basic operation of the GTAC process. Notice that DCEP (DCRP) is used and that the electrode extends through the thickness of the base metal.

The effectiveness of this cutting process is due to the action and velocity of the inert gas. The inert gas keeps the molten metal from forming oxides. These oxides are what make aluminum, stainless steel, nickel, and most nonferrous metals hard to cut. Sufficient gas must flow to prevent oxidation. The use of too much inert gas is wasteful and will not speed the cutting process.

GMAC is done with a constant voltage (cv) welding machine and with a DCEP polarity. Gas metal arc cutting is not often used, because it may require as much as 2000A of current. See Figure 16-27.

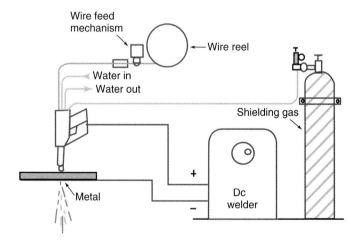

Figure 16-27. Schematic of a gas metal arc cutting outfit.

16.7 CARBON ARC CUTTING (CAC) AND GOUGING EQUIPMENT

The *carbon arc cutting* process was the original method used to cut or melt away metals. The same equipment can also be used for gouging. The base metal is melted by the heat of the arc struck between the carbon electrode and the base metal. Metal is removed from the kerf by the force of the arc and gravity. The quality of the cut is generally poor. This method is often used in small shops where more efficient and more expensive equipment is generally not available. The AWS abbreviation for carbon arc cutting is CAC.

The complete carbon arc cutting or gouging station consists of:

- An ac, dc, or an ac/dc arc welding machine.
- A booth with better-than-normal ventilation (1000 cfm or more).
- An electrode lead.
- A workpiece lead.
- A special carbon electrode holder.
- Carbon or graphite electrodes.
- Heavy-duty gloves.
- Heavy-duty clothes.
- A helmet with a lens of #12 to #14 shade.

The high amperage flow at the arc releases a considerable amount of heat. Heavy gloves and clothing must be worn to protect the welder from the intense heat of the carbon arc. The electrode holder often has a shield around the handle to protect the operator's hand from radiant heat. Some carbon electrode holders are water-cooled. When water-cooled equipment is used, less protection from heat is required. It is very important to have a good ventilation system.

See Chapter 11 for more information about the machines, cables, carbon electrodes, and electrode holders used in this process.

TEST YOUR KNOWLEDGE

Write your answers on a separate sheet of paper. Do not write in this book.

1. Name the arc cutting processes that have the following AWS letter abbreviations.
   - A. CAC
   - B. PAC
   - C. SMAC
   - D. CAC-A
   - E. GMAC
2. What is a plasma?
3. The temperature of a plasma arc may reach _____°F or _____°C.
4. In a water injection plasma arc cutting torch nozzle, what is the purpose of the water besides cooling?
5. What is a "dual-flow" plasma cutting process?
6. Using the PAC process, what cutting speed has been reached when cutting 1/4" (6mm) aluminum?
7. When plasma arc cutting, list all materials that may be used as a shielding medium.
8. What dc polarity is used when air carbon arc cutting?
9. When doing air carbon arc cutting, a carbon or _____ electrode can be used.
10. On nonmagnetic metals, how may the motorized carriage or climber be attached to the metal?
11. How much air pressure is supplied to a heavy-duty air carbon arc electrode holder?
12. What current is suggested for a 3/8" (10mm) air carbon arc electrode with DCEP (DCRP)?
13. Describe the exothermic cutting rod.
14. How does the electrode used in oxygen arc cutting differ from other cutting electrodes?
15. In shielded metal arc cutting, a high-velocity gas and particle stream is caused by a _____ _____ in the end of the electrode.
16. Regular arc welding electrodes will last longer when used for cutting if they are soaked in _____.
17. What amperage is suggested for use with a 1/4" (6mm) gas metal arc cutting electrode?
18. When doing GMAC and GTAC, what dc polarity is suggested?
19. Name four types of shielding gases that are used with GMAC and GTAC.
20. When carbon arc cutting, what shade filter lens should be used?

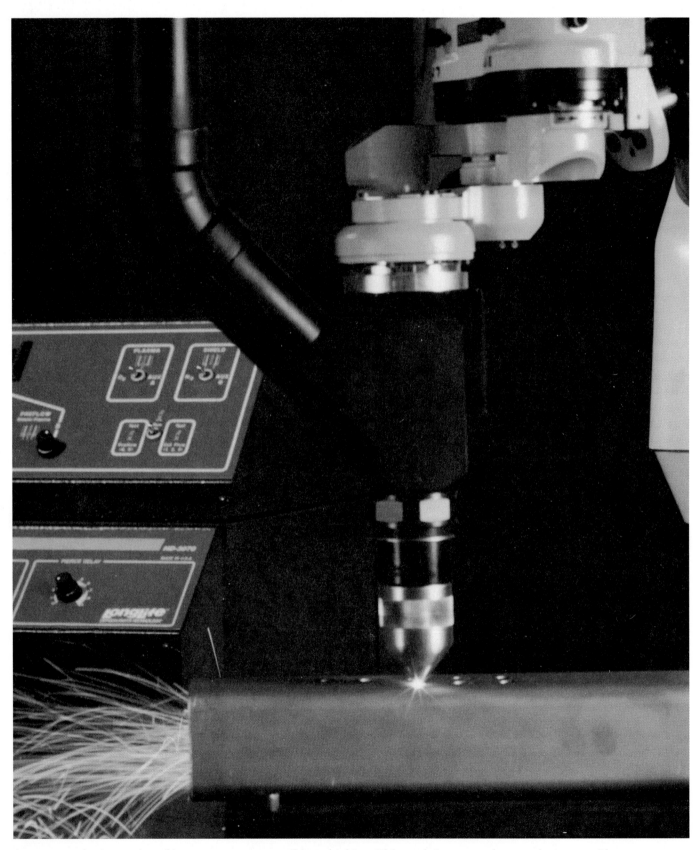

An oxygen plasma torch making a cut on square mild steel tubing. This torch is mounted on a robot arm, with movement computer-controlled for precision and repeatability. (Hypertherm, Inc.)

Chapter 17

ARC AND OXYGEN ARC CUTTING AND GOUGING

LEARNING OBJECTIVES

After studying this chapter, you will be able to:
* Select and set the recommended electrode size and current for cutting various plate thicknesses using the CAC, AOC, and/or PAC process.
* Perform an acceptable cut using the CAC, AOC, PAC, and/or SMA process.
* Perform an acceptable gouge on mild steel using the CAC, AOC, and/or SMA process.
* Recognize the safety and fire hazards that are involved in arc cutting and gouging.

The intense heat of the electric arc makes it possible to melt and make fluid very small areas of a metal workpiece. If the fluid metal can be made to flow away, either by gravity or by gas pressure, the metal will be removed, leaving a cavity or cut.

The development of technology in arc welding machine design, electrodes, and other devices involved in arc cutting have made arc cutting and gouging processes more efficient, accurate, and economical.

Plasma arc cutting (PAC), air carbon arc cutting (CAC-A), exothermic cutting (arc-started LOC), and oxygen arc cutting (AOC) are the most popular arc cutting and gouging processes in use today.

17.1 PLASMA ARC CUTTING AND GOUGING (PAC)

Plasma arc cutting or gouging can be done on plain carbon steel, stainless steel, aluminum, copper, and other metals. Refer to Headings 16.1 and 16.1.1 for information regarding PAC principles and the cutting or gouging equipment used.

The first step required in plasma arc cutting or gouging is to adjust the recommended power supply amperage and set the correct flow rate for the plasma cutting gas. If a secondary gas is used for cooling, that gas flow must also be set. See Figures 17-1, 17-2, and 17-3. Before the arc is started, the plasma gas or any secondary gas should be allowed to flow for 2 to 10 seconds to clear any moisture from the system.

If the torch is water-cooled, the cooling water must be turned on. A water flow safety device will prevent the arc from being started if the water is not flowing.

Thickness		Speed		Orifice diam.		Current	Power	Flow	Rate
in.	mm	in./min	mm/s	in.	mm	(DCSP), A	kW	Cu.ft/hr	L/min.
1/4	6.4	300	127	1/8	3.2	300	60	174	82
1/2	12.7	200	85	1/8	3.2	250	50	175	
1	25.4	90	38	5/32	4.0	400	80	200	83
2	50.8	20	8	5/32	4.0	400	80	200	94
3	76.2	15	6	3/16	4.8	450	90	200	94
4	101.6	12	5	3/16	4.8	450	90	200	94
6	152.4	8	3	1/4	6.4	750	170	250	118

Notes: 1. The gases used for the plasma gas are argon and nitrogen with additions of hydrogen of up to 35%.
2. Always consult the equipment manufacturer for their recommendation.

Figure 17-1. Suggested settings for the variables used when plasma arc cutting aluminum.

Thickness		Speed		Orifice diam.		Current		Flow	Rate
in.	mm	in/min	mm/s	in.	mm	(DCSP), A	Power kW	Cu.ft/hr	L/min
1/4	6.4	200	85	1/8	3.2	300	60	150	71
1/2	12.7	100	42	1/8	3.2	300	60	150	71
1	25.4	50	21	5/32	4.0	400	80	170	80
2	50.8	20	9	3/16	4.8	500	100	200	94
3	76.2	16	7	3/16	4.8	500	100	200	94
4	101.6	8	3	3/16	4.8	500	100	200	94

Notes: 1. Gases used are the same as used for aluminum. See Figure 17-1.
2. Refer to equipment manufacturer for their recommendations.

Figure 17-2. *Suggested values for the variables used when plasma arc cutting stainless steel.*

Thickness		Speed		Orifice diam.		Current		Flow	Rate
in.	mm	in/min	mm/s	in.	mm	(DCSP), A	Power kW	Cu.ft/hr	L/min
1/4	6.4	200	85	1/8	3.2	275	55	170	80
1/2	12.7	100	42	1/8	3.2	275	55	170	80
1	25.4	50	21	5/32	4.0	425	85	170	80
2	50.8	25	11	3/16	4.8	550	110	200	94

Notes: 1. Plasma arc gases are usually compressed air, nitrogen with up to 10% hydrogen added or nitrogen with oxygen added in a dual gas flow system.
2. Equipment manufacturers should be consulted for each application.

Figure 17-3. *Suggested values for the variables used when plasma arc cutting carbon steel.*

Earmuffs or earplugs should be worn, since this is a very noisy process. The noise is considerably less if the cutting is done on a water cutting table.

Eye protection recommended for use when plasma arc cutting or gouging is shown in Figure 17-4.

Cutting current amperes	Lens shade number
Up to 300	9
300-400	12
400-800	14

Figure 17-4. *Recommended protection lens shades for use when plasma arc cutting. (American Welding Society)*

To start the arc, the welder must press the start button or trigger. The arc welding machine will then create a high-frequency current to start the arc. Also, when the start button is pressed, the plasma gas begins to flow and the dc/cc arc welding machine starts the arc current. Once the plasma arc is established through the ionized starter arc, the starter arc is turned off. The cut may proceed as shown in Figure 17-5. Refer also to Heading 4.13.

When doing manual cutting, the torch is held at an angle of between 70° and 90° to the base metal. Automatic PAC torches are normally set at 90° to the base metal. The

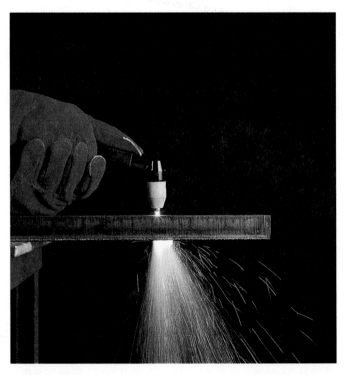

Figure 17-5. *A manual plasma arc cutting torch in use, cutting a 1" (25mm) steel plate. Note how dross-free (slag-free) the previous cut was. (ESAB Welding and Cutting Products)*

cut should be made at the speed recommended to produce the best cut. Cuts that are made using the PAC process are virtually free of *dross* (slag). Automatic machine cuts have a very smooth surface compared to those made with the OFC process.

A plasma jet tends to remove more metal at the top of the kerf than at the bottom. This results in an angled kerf. The kerf angle is about 4° to 6° when 1" (25mm) steel is cut. PAC will produce a kerf that is about 1 1/2 times wider than an oxyfuel gas cut; allowance must be made for the kerf size.

In the transferred arc process, the arc will go out when the cut is completed. Once the cut runs off the metal, there is no longer a complete circuit through the base metal. When the arc goes out, the gas will also stop flowing. If an automatic carriage is being used, it will also stop, due to the wiring of the circuits. In the nontransferred arc process, the operating switch must be turned off when the end of the cut is reached.

Plasma arc cutting may be done in any position, and it will cut virtually any metal. This makes it a very useful and versatile process. Plasma arc cutting may be done under several inches of water, as shown in Figure 17-6. This greatly reduces noise levels. Fumes and ultraviolet radiation are almost eliminated, as well. Even though the cutting is done under water, the cutting speed and quality are not affected.

17.1.1 Plasma Arc Gouging

The equipment used for plasma arc cutting may also be used for gouging. However, when used for plasma arc gouging, the arc welding machine must have a steep output curve and must provide enough voltage to sustain the long arc required. The amperage settings for plasma arc gouging are the same as those used for cutting. See Figures 17-1, 17-2, and 17-3.

The gas recommended for all gouging is argon mixed with 35% to 40% hydrogen. If a secondary cooling gas is used, it may be argon, nitrogen, or air. The gas chosen depends on cost-effectiveness and the quality of gouge desired.

When gouging, the torch angle is about 20° to 30° from the base metal. The depth of the gouge depends on the speed of travel. For a deep gouge, several passes are recommended. Figure 17-7 shows a PAC torch being used to gouge a steel plate.

Figure 17-7. A PAC torch being used to gouge a steel plate. Note the high quality of the gouged surface. (Hypertherm)

17.2 AIR CARBON ARC CUTTING AND GOUGING (CAC-A)

Air carbon arc cutting and gouging may be done manually or automatically. Study Heading 4.9. The manual torch is shown in Figure 17-8 removing a weld bead. Figure 17-9 shows an air carbon arc torch. Note the air holes below the carbon cutting electrode.

See Figures 16-17 and 16-18 for the suggested electrode diameter, amperage, and travel speeds for carbon arc cutting or gouging. Refer to Heading 16.2 for more information on the equipment needed for air carbon arc cutting.

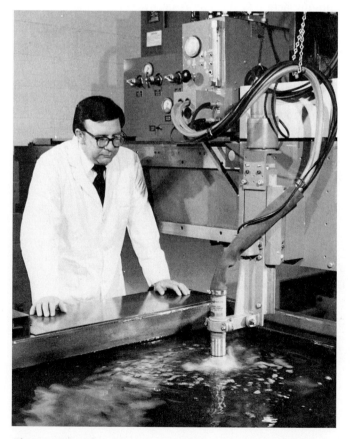

Figure 17-6. Plasma arc cutting being done under several inches of water. This procedure lowers the noise level. Fumes and ultraviolet radiation are almost eliminated. (ESAB Welding and Cutting Products)

Figure 17-8. *An air carbon arc electrode holder and electrode being used to remove a weld bead. Notice the flat, rectangular cross section of the electrode, the electrode angle, and the welder's clothing. (Tweco/Arcair, Division of Thermadyne Industries, Inc.)*

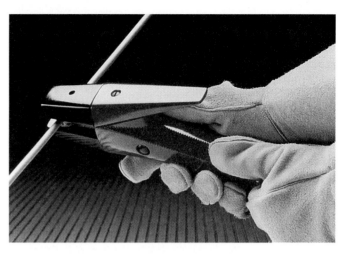

Figure 17-9. *An air carbon arc cutting torch. Notice the air discharge holes behind the carbon electrode. (Tweco/Arcair, Division of Thermadyne Industries, Inc.)*

When gouging, the airstream must be turned on before the arc is struck. Regardless of the position in which a gouge is made, the airstream must be from behind the electrode and between the base metal and the electrode. This permits the metal to be blown out of the arc pool, as shown in Figure 17-10. Figure 17-11 shows the air orifices in the jaws of the electrode holder. If the air carbon arc process is used to cut completely through a part, the electrode angle should be about 20° to 45° from vertical as shown in Figure 17-12. Figure 17-13 shows a V-groove gouged into a plate. Figure 17-14 illustrates the sequence of gouges used to remove a large fillet weld.

In the vertical cutting position, gouging should be done from the top down. This permits gravity to help remove the molten metal from the arc groove. Gouging in the horizontal position may be done from the right or left. When gouging overhead, the electrode should be placed into the electrode holder so that it is nearly parallel to the centerline of the holder. With this electrode position, the gouging action will occur as far away from the welder as possible. This may prevent hot metal from falling on the welder.

When manually gouging on steel, the electrode should extend from the electrode holder about 6" (150 mm). When gouging on aluminum or aluminum alloys, the electrode extension is about 4" (100mm). The welder must reposition the electrode in the holder regularly to maintain the ideal electrode extension. Once the arc is

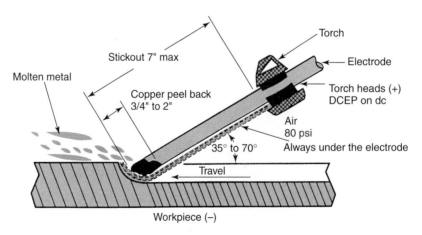

Figure 17-10. *A schematic of the air carbon arc cutting process in progress. Note how the compressed air blows under the carbon electrode to remove the molten metal. (American Welding Society)*

Figure 17-11. A view of the two air orifices in the jaws of the CAC-A electrode holder. Note the reflective gloves being worn. (Tweco/Arcair, Division of Thermadyne Industries, Inc.)

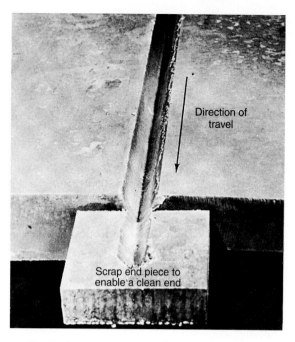

Direction of travel

Scrap end piece to enable a clean end

Figure 17-13. A V-groove gouged in 2" (51mm) carbon steel. This view illustrates the use of a run-off tab to get a complete gouge to the end of the plate.

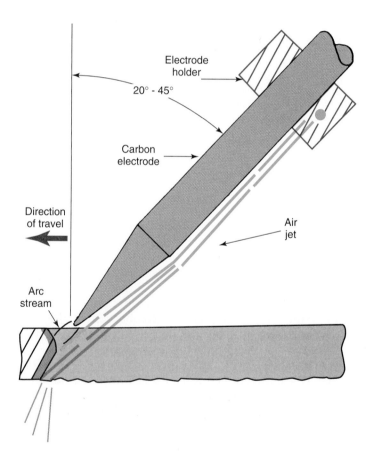

Electrode holder

20° - 45°

Carbon electrode

Direction of travel

Air jet

Arc stream

Figure 17-12. Air carbon arc cutting schematic. Note the air jet provided to blow away the molten metal.

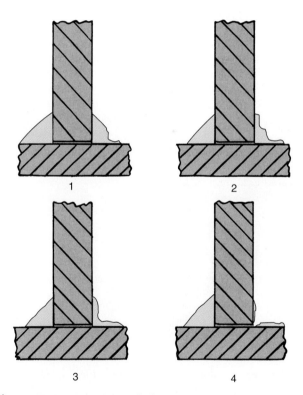

1

2

3

4

Figure 17-14. Recommended gouging sequence to be used when removing a fillet weld.

struck, the electrode is withdrawn a short distance to allow the arc to stabilize. After the arc is stabilized, the electrode should remain in close contact with the metal being gouged. The electrode angle is about 20° from the base metal. Forward speed and smoothness of travel will determine the quality of the finished gouge.

When automatic gouging machines are used, the amperage, air pressure, air volume, and travel speed are adjusted to approximately the same setting as used for manual gouging and cutting. See Figure 16-18. The electrode extension should be approximately the same as for manual gouging. In most automatic machines, the electrode is automatically fed out to maintain a uniform extension at all times. The operator must add new electrodes to permit continuous gouging. The electrode angle is about 45° or less when gouging automatically. The accuracy of the depth of the gouge may be held to ±0.03" (±0.76mm) with the automatic process.

Carbon deposits may indicate an improper air jet flow or excessive travel speed. A rough cut results from a slow travel speed. Poor cuts may also result from a poorly positioned air jet or a jet with low air pressure or volume.

17-3 EXOTHERMIC CUTTING, GOUGING, AND PIERCING

See Heading 16.3 for the equipment required for exothermic cutting or gouging. See also Figure 17-15.

In preparation to make an exothermic cut or gouge, the following steps must be taken:

1. The electric lead which runs to the exothermic rod is connected to the positive (+) terminal of the 12V battery and the negative (-) battery lead is connected to the copper striker plate.

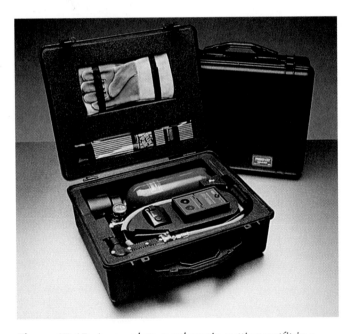

Figure 17-15. *A complete exothermic cutting outfit in a small carrying case. This unit is highly portable and contains gloves, goggles, a small oxygen cylinder and regulator, cutting torch, a small 12V battery, leads, and a copper striker plate. (Broco, Inc.)*

2. The exothermic rod is placed into the torch collet assembly and the oxygen hose is connected to the torch body fitting.
3. When the exothermic rod is to be ignited, it is scraped across the copper plate. Electricity between the exothermic rod and the copper plate causes the exothermic rod to spark and the fuel wires to begin to burn. See Figure 17-16. Once the exothermic fuel rods are ignited, the oxygen control lever on the torch is depressed. The fuel rods or wires and outer tube will continue to burn at the tip.

Figure 17-16. *Igniting the exothermic rod by striking it on the copper plate. The sparking means it is lighted. (Broco, Inc.)*

Exothermic rods burn with a temperature of 10,000°F (5540°C). As a comparison, the oxyacetylene cutting flame has a temperature of about 5600°F (3093°C).

To support combustion, oxygen is supplied from an oxygen cylinder and regulator to a fitting on the exothermic cutting torch. Oxygen flows through the hollow exothermic rod to support the combustion of the fuel rods/wires and to blow the molten metal away. Figure 17-17 shows an exothermic cut in progress.

To make a cut with this process, the exothermic rod is ignited as shown in Figure 17-16. The torch is held at an angle between 45° and 90° to produce a good cut, and the oxygen lever depressed. It is recommended that the tip of the exothermic rod be kept in the molten pool while making a cut. This is done to ensure that the molten metal is blown away through the far side of the base metal and to prevent molten metal from being blown upward from the pool.

Figure 17-17. *An exothermic cut in progress. (Broco, Inc.)*

Figure 17-18. *A diver/welder making a cut with an exothermic torch and rod. Note that the welder is holding the rod to steady it. (Broco, Inc.)*

To make a good cut, the required oxygen pressure must be set. The suggested cutting pressures for various metals and thicknesses are shown below:

Metal	Metal Thickness	Oxygen Pressure
Iron or steel	Up to 1/2" (13mm)	30 psig - 40 psig (207kPa - 276kPa)
Iron or steel	1/2"- 1" (13mm - 25mm)	40 psig - 50 psig (276kPa - 345kPa)
Nonferrous	Up to 1" (25mm)	40 psig - 50 psig (276kPa - 345kPa)
All metals	Up to 2" (51mm)	50 psig - 60 psig (345kPa - 414kPa)
All metals	Over 2" (51mm)	60 psig - 80 psig (414kPa - 552kPa)

Most arc cutting and gouging processes can also be done under water. See Figure 17-18. When doing exothermic cutting underwater, the copper-coated rod is wrapped with a waterproof tape. The underwater torch is better insulated than a similar torch used on the surface. The torch and exothermic rod are shown in Figure 17-19.

When piercing, a leather shield is placed on the cutting torch and held in place under the collet nut. This shield is used to protect the welder's hands from the hot metal that is deflected upward until the hole is cut through the base metal. See Figure 17-20.

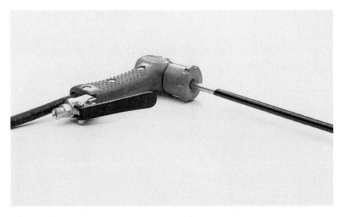

Figure 17-19. *An exothermic underwater cutting torch and insulated rod. (Broco, Inc.)*

To pierce a hole through metal, hold the electrode at 90° to the base metal surface and press down on the oxygen control lever. Move the electrode in and out to *deepen* the hole, and around to *enlarge* the diameter of the hole.

Caution: When piercing, remember that the molten metal will be spraying upward until the hole goes through the bottom of the base metal, so keep your face away from the hole.

The recommended oxygen pressure for gouging is 30 psig to 40 psig (207kPa to 276kPa) for shallow gouging and 40 psig to 50 psig (276kPa to 345kPa) for deep gouging. These pressures are set on the oxygen tank regulator.

To gouge, the exothermic rod is ignited by scraping it across the copper striking plate. It is held at a 45° angle to the base metal, and the oxygen lever is depressed. See Figure 17-21. Once a molten pool forms on the metal surface, the electrode is dropped down to an angle of 5° to 10° to the surface. The electrode is then pushed along the

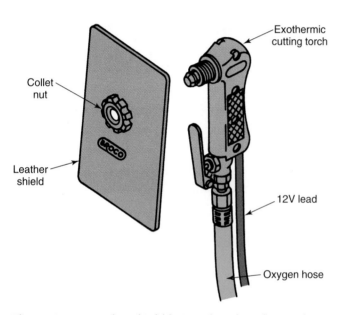

Figure 17-20. Leather shield being placed on the exothermic torch. (Broco, Inc.)

Figure 17-21. A gouge being made on carbon steel. Note the low angle of the rod. Most slag is easily wiped away. (Broco, Inc.)

gouging line to form the desired depth of gouge. The electrode may be moved from side to side to make a wider gouge.

17.4 OXYGEN ARC CUTTING AND GOUGING (AOC)

The oxygen arc cutting and gouging process is a combination of an electric arc and a jet of oxygen. Either a dc or an ac arc welding machine may be used. In addition, a special oxygen arc cutting torch, a special hollow electrode, and an oxygen cylinder are required. Refer to Heading 16.4. Oxygen arc cutting is used to cut alloy steels, aluminum, and cast iron. These metals are difficult to cut using the oxyfuel gas cutting process. This process is also used on carbon steel.

The oxygen arc cutting process usually makes use of a tubular metal electrode. An electric arc is struck between the hollow electrode and the work. Oxygen fed through the hollow electrode oxidizes the metal and blows it away. This action forms the cut or kerf. Study Heading 4.10. A covering is often used on the hollow cutting electrode, especially if fluxing ingredients are needed for cutting the base metal. When gouging, this covering allows the welder to hold the electrode against the base metal to maintain a uniform arc distance.

Figure 17-22 lists suggested amperages and oxygen pressure for use in cutting various metals and metal thicknesses. An oxygen cylinder and a high-volume regulator are required. The oxygen arc torch is connected to the regulator by means of a heavy duty oxygen hose. The pressures used for cutting may be as high as 75 psig (517kPa). The electrode lead from the welding machine is connected to the electrode, as shown schematically in Figure 17-23.

To make a cut or gouge, the welder first strikes the arc on the base metal, then depresses the oxygen lever on the cutting torch. The cut or gouge begins as shown in Figure 17-24. A continuous arc is maintained throughout the cutting operation. While cutting or gouging, the electrode may be dragged across the metal. For cutting, the electrode is held at an angle between 70° and 90° to the surface of the base metal. The electrode is angled slightly to cause the molten metal to blow forward from the far side of the kerf.

While gouging, the electrode is held at an angle of about 15° to 30° from the base metal surface. The forward speed and the angle of the electrode may be varied, depending on the desired depth of the gouge. The gouge may be widened by using a weaving motion with the electrode. The hollow electrode is consumed during the oxygen arc cutting or gouging operation.

Holes may be pierced easily with this process. The arc is struck and the oxygen lever depressed to begin the piercing cut. The electrode should be held at 90° to the surface of the base metal. **Caution: When piercing a hole with any cutting process, the welder should be aware that the molten metal will be directed upward until the hole reaches the far side of the base metal.**

Suggested Amperages and Oxygen Pressures for Oxygen Arc Cutting				
Cast Iron				
Thickness		**Amps**	**Oxygen pressure**	
in.	mm		psig	kPa
1/4	6.4	180	10	68.9
1/2	12.7	185	10-15	68.9-103.4
3/4	19.1	190	15-20	103.4-137.9
1	25.4	200	20-25	137.9-172.4
1 1/4	31.2	210	25-30	172.4-206.8
1 1/2	38.1	215	30-35	206.8-241.3
1 3/4	44.5	220	35-40	241.3-275.8
2	50.8	225	40-45	275.8-310.3
2 1/4	57.2	225	45-50	310.3-344.7
2 1/2	63.5	230	50-55	344.7-379.2
2 3/4	69.9	230	55-60	379.2-413.7
3	76.2	235	60	413.7
Aluminum				
Thickness		**Amps**	**Oxygen pressure**	
in.	mm		psig	kPa
1/4	6.4	200	30	206.8
1/2	12.7	200	30	206.8
3/4	19.1	200	30	206.8
1	25.4	200	30	206.8
1 1/4	31.2	200	35	241.3
1 1/2	38.1	200	35	241.3
1 3/4	44.5	200	40	275.8
2	50.8	200	40	275.8
2 1/4	57.2	175	45	310.3
2 1/2	63.5	175	45	310.3
2 3/4	69.9	175	45	310.3
3	76.2	175	45	310.3
Carbon and Low-Alloy High-Tensile Steel				
Thickness		**Amps**	**Oxygen pressure**	
in.	mm		psig	kPa
1/4	6.4	175	75	517.1
1/2	12.7	175	75	517.1
3/4	19.1	175	75	517.1
1	25.4	175	75	517.1
1 1/4	31.2	200	75	517.1
1 1/2	38.1	200	75	517.1
1 3/4	44.5	200	75	517.1
2	50.8	200	75	517.1
2 1/4	57.2	225	75	517.1
2 1/2	63.5	225	75	517.1
2 3/4	69.9	225	75	517.1
3	76.2	225	75	517.1

Figure 17-22. Amperages and oxygen pressure suggested for various metals and metal thicknesses when oxygen arc cutting.

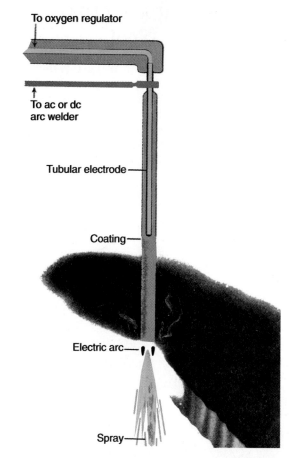

Figure 17-23. A schematic illustration of the main parts of a tubular electrode and holder as used in oxygen arc cutting. This process may be used on many different metals such as stainless steel, alloy steel, aluminum, and cast iron. (Cut-Mark, Inc.)

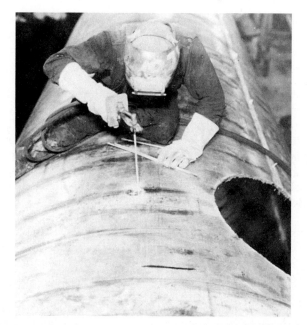

Figure 17-24. The oxygen arc process (AOC) being used to cut holes in a large pipe. A hollow, covered electrode is being used. Note the welder's clothing and the electrode position. (Cut-Mark, Inc.)

17.5 SHIELDED METAL ARC CUTTING AND GOUGING (SMAC)

Metals may be cut and gouged using the shielded metal arc cutting process. However, this process for cutting and gouging is used very little commercially, because other cutting and gouging methods have been introduced that are cheaper and more efficient. When cutting or gouging using this process, a thickly covered electrode with a slow-burning covering should be used. If the covering

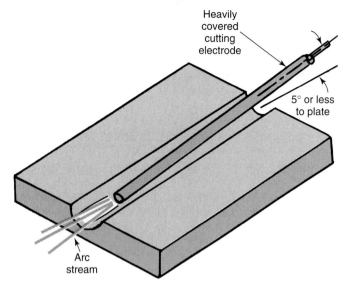

Figure 17-26. A typical gouge cutting operation, using a shielded metal arc cutting (SMAC) electrode.

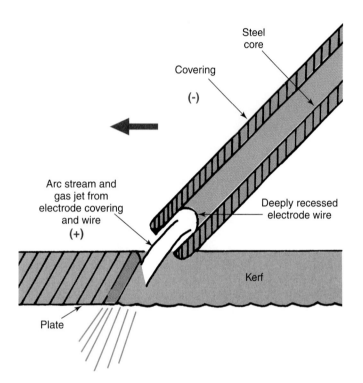

Figure 17-25. A shielded metal arc cutting electrode being used to cut a metal plate. Note how deeply the electrode wire is recessed within the special slow-burning covering.

burns slowly, a deep cup is formed in the end of the electrode. This deep cup causes the electrode metal to melt off the end with a jet action, creating a greater arc force. When cutting or gouging, the base metal is melted and blown away by the force of the arc, Figure 17-25. The electrode angle should be near 70° from the metal surface and tipped backward at the top to ensure that the molten metal is blown forward from the kerf.

When gouging, the electrode angle should be only about 5° to 15° from the metal surface. However, this angle should be varied as necessary to control the depth of the gouge. The thickly covered electrode may touch the base metal surface and be pushed along the gouge line without causing the electrode to short out, Figure 17-26.

Refer to Figure 16-25 for suggested amperages to use with various diameters of electrodes, both ac and DCEN (DCSP). Shielded metal arc cutting electrodes come in the following diameters:

- 3/32" or 2mm
- 1/8" or 3mm
- 5/32" or 4mm
- 3/16" or 5mm
- 1/4" or 6mm

This system is easy to setup. A constant current arc welding machine is used. The welder needs to use a shielded metal arc cutting electrode for good gouging or cutting results. Special SMAC electrodes have been developed and may be available from some suppliers.

17.6 GAS METAL ARC CUTTING (GMAC) AND GAS TUNGSTEN ARC CUTTING (GTAC)

Gas metal arc cutting (GMAC) and gas tungsten arc cutting (GTAC) have not been used commercially since the introduction of other more efficient and economical arc cutting processes.

One of the drawbacks to the use of GMAC is that it may require currents as high as 2000 A, which would require the stacking or interconnecting of arc welding machines to provide that much power.

GTAC may be used to cut stainless steel and many nonferrous metals. GTAC may be done manually or mechanically. A higher amperage is needed for cutting than for welding on the same thickness and type of metal. A 65% argon/35% hydrogen mixture at 60 ft³/hr. (28L/min.) is generally used for cutting. Best results are obtained with DCSP. The arc melts the metal and the force of the arc and the shielding gas pressure blows the molten metal out of the kerf. The electrode is held at an angle of near 70° from the base metal. Gouging is not normally done with the GMAC or GTAC processes.

17.7 CARBON ARC CUTTING AND GOUGING (CAC)

Cast iron is difficult to cut or gouge using the oxyfuel gas process, so an arc cutting process is used. Cast iron and other metals may be successfully cut or gouged using the carbon arc process. This process, however, is not used as extensively as it once was. Faster, more economical, and cleaner cuts are being made with other processes, such as, PAC, CAC-A, exothermic, and AOC. The manual carbon arc cut or gouge is generally very ragged and of poor appearance. The amperages used for carbon arc cutting are generally higher than for welding on the same metal thickness. Manufacturers or suppliers of carbon electrodes will normally supply amperage ranges for use with various electrode diameters. See Figure 17-27 for suggested electrode diameters and current settings.

Prior to use, the carbon electrode should be ground to a very sharp point. The length of the taper should be 6-8 times the electrode diameter. The electrode should stick out from the electrode holder a distance equal to 10 times the electrode diameter. This is necessary to reduce the electrical resistance and to reduce the heating effect on the electrode. If the carbon wears away too fast, shorten the electrode extension from the electrode holder to as little as 3" (76mm).

During the actual cutting, the carbon electrode should be manipulated in a vertical elliptical movement to undercut the metal. This motion facilitates the removal of the molten metal. In addition to the vertical motion, a side-to-side crescent motion is recommended along the line of the cut. The electrode angle is about 20° from vertical. Figure 17-28 illustrates the relative position of the electrode and the work when cutting cast iron.

The carbon arc method of cutting may be used quite successfully on cast iron because the temperature of the arc is sufficient to melt the iron oxides formed. It is important to undercut the cast iron kerf if you desire an even cut. Figure 17-29 shows a cast iron gear hub being gouged using a carbon electrode. The molten metal must flow away from the gouge or cutting area by gravity. Therefore,

Thickness of Plate		Current Setting and Carbon Electrode Diameter			
		300 Amps 1/2" Dia. (12.7mm)	500 Amps 5/8" Dia. (15.9mm)	700 Amps 3/4" Dia. (19.1mm)	1000 Amps 1" Dia. (25.4mm)
in.	mm	Speed of Cutting in Minute per Foot			
1/2	12.7	3.5	2.0	1.5	1.0
3/4	19.1	4.7	3.0	2.0	1.4
1	25.4	6.8	4.1	2.9	2.0
1 1/4	31.8	9.8	5.6	4.0	2.9
1 1/2	38.1	—	8.0	5.8	4.0
1 3/4	44.5	—	—	8.0	5.3
2	50.8	—	—	—	7.0

Figure 17-27. A table of recommended electrode sizes, current settings, and speeds for carbon arc cutting various thicknesses of steel.

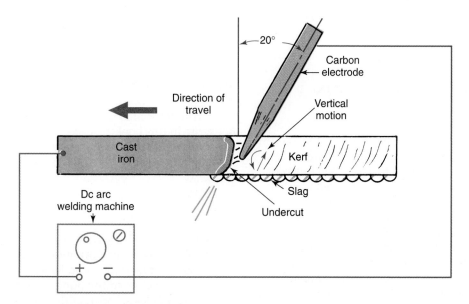

Figure 17-28. Carbon arc cutting. Note that the carbon electrode is connected to the negative terminal of the welding machine, and the workpiece to the positive terminal.

Figure 17-29. *Carbon arc gouging a cast iron gear hub. Note the type of electrode holder, and the position of the gear, which permits the molten cast iron to flow down and away from the gouge. (Tweco/Arcair, Division of Thermadyne Industries, Inc.)*

gouging is done best with the metal in a vertical position. The gouge should be made from top to bottom, or *down-hill* ("vertical-down"). This will keep the molten metal flowing out of the gouge area. The graphite form of carbon electrode is preferred because high amperages are required.

A direct current/constant current (dc/cc) arc welding machine and other items of arc welding station equipment may be used for carbon arc cutting and gouging. Refer to Heading 16.7. When using a dc arc welding machine with dc electrodes, DCEN (DCSP) is used. When using an ac arc welding machine, special ac cutting electrodes are used.

When carbon arc cutting or gouging, the light from the arc is more intense than when shielded metal arc welding. A lens shade #12 to #14 is recommended for carbon arc welding, cutting, or gouging.

Follow all safety precautions listed for arc welding procedures whenever cutting. Refer to Heading 17.8 before attempting carbon arc cutting.

17.8 SAFETY REVIEW: ARC CUTTING

Electric arc cutting does not present any great or new hazards. Some cutting operations may continue for a considerable length of time and at high current rates. Precautions must therefore be taken to avoid skin burns or eye damage. The following precautions are recommended:
- Be sure that all surfaces of the skin are well-shielded from the arc rays. Wear approved gloves, helmet, and approved clothing.
- Use a helmet filter lens a shade or two darker than would be used for welding with the same size electrode. A #12-14 lens is recommended for carbon arc and air carbon arc cutting.
- Use completely insulated electrode holders.
- Stand or work only in dry surroundings.

- A considerable amount of fumes and gases are liberated during all of the arc cutting operations. The ventilation of the working space *must* be sufficient to ensure that the welder is working in clean, fresh air at all times.
- Most of the arc cutting processes are accompanied by a great amount of sparking and showers of molten globules of metal. These hot particles may be thrown some distance from the arc. **Wear garments that are fire-resistant. Pockets, cuffs, and other clothing crevices that could trap sparks must be covered.**
- Be sure that the working area is fireproof. All flammable objects (such as wood benches, floors, or cabinets) should be removed from the vicinity or covered with flame resistant materials. Be sure there are no openings in the floor that might allow the sparks to travel to the floor below.
- To reduce the possibility of fire, some electric arc cutting stations use a "wet table." The wet table provides for a thin water film under the flame cutting area. The wet film quenches the sparks.
- **Check local fire codes before cutting a tank that has previously held a flammable material.**
- The oxygen arc cutting process uses a higher electrical current than is used for welding the same thickness of metal. Be certain that the welding machine has ample capacity for the current needed. Never exceed the arc welding machine's duty cycle rating.
- The plasma arc cutting process and most oxygen arc cutting processes are very noisy. Workstations where these processes are used should be located where the noise will not be objectionable. **The operator and others in the vicinity must wear industrial earmuffs to prevent hearing loss.**
- When piercing a hole, take precautions to avoid being burned. The molten metal will be projected upward until the hole goes through the metal.

TEST YOUR KNOWLEDGE

Write your answers on a separate sheet of paper. Do not write in this book.
1. Name the arc cutting processes that are identified by the following AWS abbreviations:
   CAC
   PAC
   SMAC
   CAC-A
   AOC
2. When plasma arc cutting or oxygen arc cutting, how must your hearing be protected?
3. How is the arc in the nontransferred and transferred arc processes stopped at the end of the cut?

4. How may the noise, fume, and radiation levels be reduced when doing plasma arc cutting?
5. When cutting with the air carbon arc process, in what position should air jets be in relation to the electrode?
6. When CAC-A, at which angle to the work should the electrode be?
7. When air carbon arc cutting with the metal in the vertical position, should the cutting be done from the top down or bottom up?
8. The depth of a gouge done with automatic CAC-A can be controlled to within _____" or _____mm.
9. What oxygen pressure is suggested for cutting 1/4" (6mm) thick steel with the exothermic process?
10. What amperage should be used with 1/8" (3mm) cutting electrodes when shielded metal arc cutting with dc and cutting electrodes?
11. What dc polarity is recommended by SMAC?
12. When piercing a hole with oxyfuel gas or an arc cutting process, what danger is there from molten metal?
13. What polarity should be used when gas metal arc or gas tungsten arc cutting?
14. How does the electrode used in oxygen arc cutting differ from the electrodes used in the other arc cutting processes?
15. What current range and oxygen pressure are suggested for cutting 3/4" (19mm) aluminum with the AOC process?
16. What number filter lens is recommended when CAC cutting?
17. When is the carbon arc process generally used?
18. The manual carbon arc cut is generally very _____ and of _____ appearance.
19. The length of the taper on a carbon electrode used for cutting should be _____ to _____ times the electrode diameter.
20. Prior to cutting into a tank that may have contained a flammable liquid, what must be done?

This gantry-type shape-cutting machine has four plasma arc torches that can make both straight and beveled cuts under computer numerical control (CNC). In addition to the plasma arc torches, up to eight oxyfuel gas cutting torches can be mounted on the gantry. The machine can handle workpieces up to 30 feet in length. (ESAB L-TEC Cutting Systems)

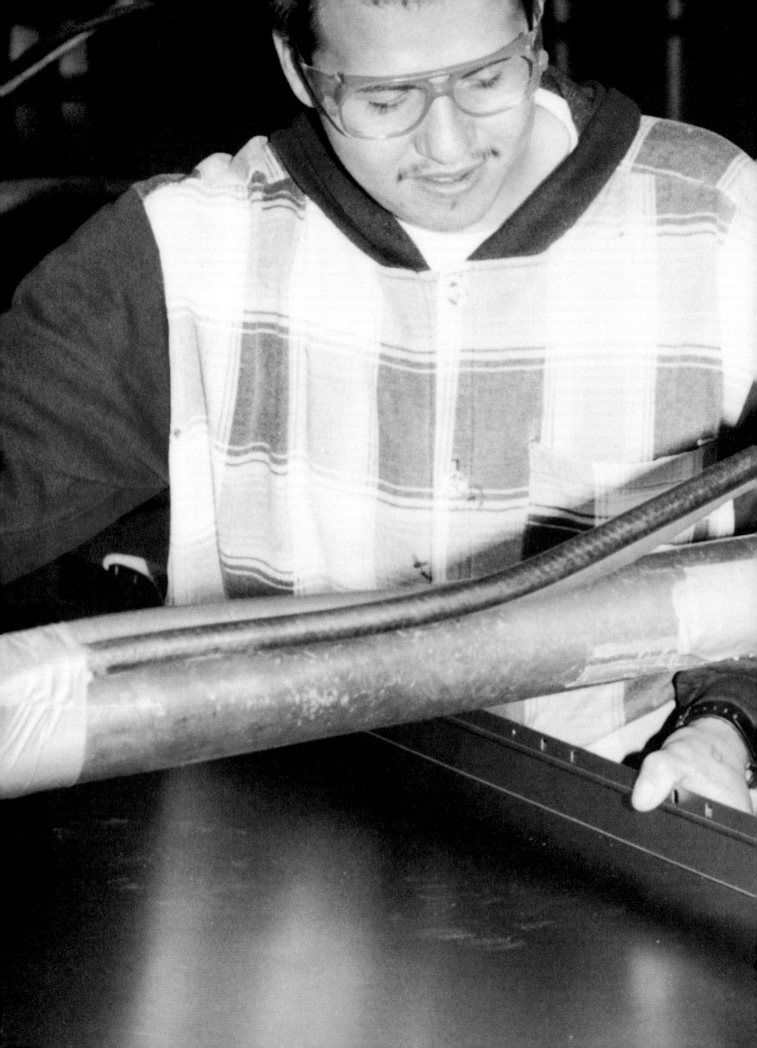

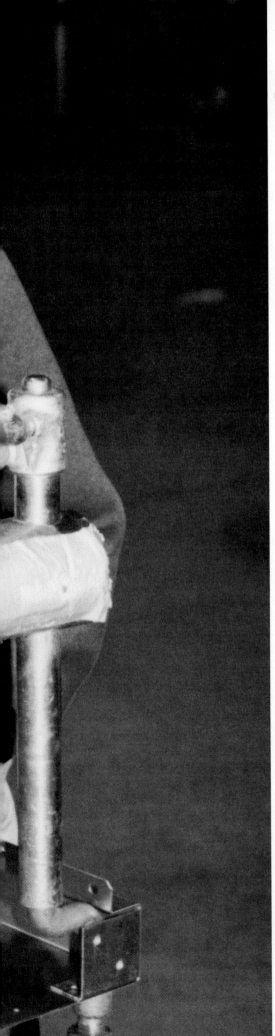

Part 6

RESISTANCE WELDING

Resistance spot welding is widely used to fabricate sheet metal products such as office furniture and appliance cabinets.

429

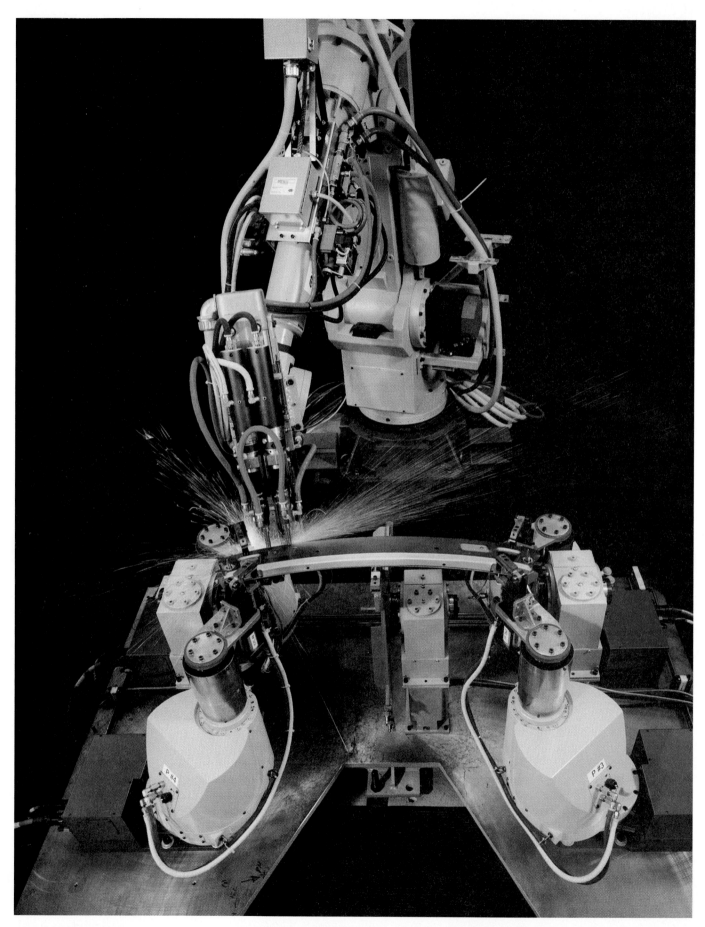

Robotic spot welding is widely used in industry for fabrication of sheet metal parts. (FANUC Robotics North America, Inc.)

Chapter 18

RESISTANCE WELDING EQUIPMENT AND SUPPLIES

LEARNING OBJECTIVES

After studying this chapter, you will be able to:
* Describe how a welding current is induced in the transformer of a resistance welding machine.
* Identify various types of resistance spot, seam, projection, upset, and flash-type welding machines.
* Describe and identify the major parts of a resistance spot, seam, or projection welding machine.
* Name and describe the variables involved in resistance welding.
* Name and describe the various weld time variables involved in a resistance welding cycle.
* Select, inspect, and change a spot welding electrode.
* Identify the safety hazards involved when resistance welding and describe methods to reduce or eliminate these hazards.

Designing and constructing resistance welding machines involves a combination of electrical design, machine structures, mechanisms, and controls. The machines vary from simple mechanisms to exceptionally complicated units. Refer to Headings 4.14, 4.15, 4.16, and 4.17.

Accurate control of the resistance welding variables is essential if good resistance welds are to result. The Resistance Welder Manufacturers' Association (RWMA) and the National Electric Manufacturers' Association (NEMA) have developed standards for resistance welding machines. These standards have enabled the manufacturers to produce machines of known rated capacity and rec-

ognized durability. The parts that make up a resistance welding station include a resistance welding machine, an electronic controller, and resistance welding electrodes.

18.1 ELECTRIC RESISTANCE WELDING MACHINES

In design, most electric resistance welding machines are quite similar. However, the manner in which the metal to be welded is held, and the appearance and position of the welding electrodes, varies from machine to machine. The methods used to control the various times, the welding current, and the pressure also vary between machines.

Automatic controls are used on all industrial resistance welding machines. An electronic controller is used to precisely control the welding process. Silicon-controlled rectifiers (SCRs) are usually used to start and stop the electrical flow to the electrodes. The controller also sends electrical signals to solenoids that apply or release the pneumatic or hydraulic forces on the electrodes.

Two types of manual resistance welding machines are used. On one type, the welder controls the welding times and the force on the electrode. This type of resistance welding machine may be found in home workshops and small repair shops. With the second type, the welder controls the welding times, but the electrode force is applied using hydraulic or pneumatic force. Since inexpensive controls are now available, very few manual resistance welding machines are in use.

The parts of a resistance welding machine are:
* Frame.
* Transformer.
* Welding arms and electrodes.
* Force mechanisms.
* Controller (not used in a manual-control welder).

The transformer is of special construction and usually has several tap settings or adjustments. It is usually water-cooled, but it may be air-cooled. The secondary winding usually consists of one loop or of several parallel loops. The ends of these loops are soldered, bolted, or brazed to the electrode arms of the machine.

The welding arms and operating mechanisms are different for each type of electric resistance welder. Each one usually uses a foot-operated switch in connection with pneumatic or hydraulic cylinders to press the parts together. Manual resistance welding machines use the welder's foot pressure and mechanical leverage to create the force on the electrodes that squeeze the pieces together for welding.

All automatically controlled welders have a controller. The controller regulates the timing of the weld cycle. It activates the pressure cylinder, turns the weld current on and off, and releases the force on the electrodes at the proper time. A resistance welding controller can be seen in Figure 18-1.

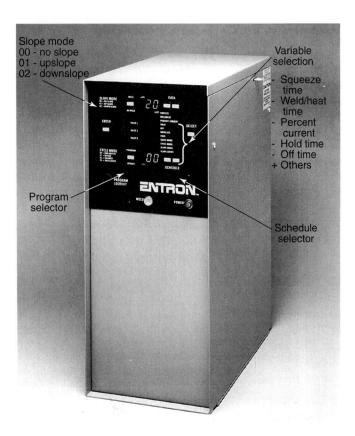

Figure 18-1. *Controls cabinet for a resistance welder. This unit has digital displays. Many different weld schedules can be stored and called up as required. (Entron Controls, Inc.)*

The part of the resistance welding machine that contacts the metal to be welded is called the *electrode*. Two electrodes are used and electric current is passed between them. The electrodes are usually copper. Electrodes are made in many different forms; each has its own purpose. The electrodes must be able to carry the large currents used in resistance welding. They must also be able to withstand the high force used.

18.2 TRANSFORMERS

A resistance welding transformer is designed and built to use the voltage and current provided by the electric utility companies. However, neither the current nor voltage supplied to the machine is suitable for resistance welding. Resistance welding requires high-amperage/low-voltage electrical energy. The transformer must lower the supplied voltage and increase the current. This type of transformer is called a *step-down transformer*, since it reduces the supplied voltage.

For example, an ideal step-down transformer will transform 50 amperes (A) at 230 volts (V) to 100A at 115V, or to 1000A at 11.5V, or to 10,000A at 1.15V. In each example, the wattage (volts × amps) equals 11,500 watts (W). The resistance welding transformer usually delivers secondary welding current to the electrodes in the range of 5000A to 100,000A. The secondary circuit voltage is usually about 10V open circuit and decreases to less than 1V when current flows during welding.

Most transformers are intermittent in operation. They deliver current only for short intervals of time. The amount of time the transformer delivers current, in ratio to the time the current is off, is called the *duty cycle*. Thus, if a transformer delivers current for 3 seconds out of each 60 (one minute) the duty cycle is 5%:

$$3 \div 60 = 0.05 \text{ or } 5\%$$

A transformer has two electrical coils that are wrapped or wound around an iron core. The two coils are not electrically connected. One coil is called the primary coil, the other is called the secondary coil. The **primary coil** is part of the primary circuit; the **secondary coil** is part of the secondary circuit. The secondary circuit is usually referred to as "the secondary."

Alternating current (ac) is made to flow in the primary coil. As the current flows, a magnetic field builds up. The current then stops as the ac changes direction. The magnetic field collapses when the current stops. This occurs 120 times a second. The magnetic field that is created and destroyed passes through the secondary windings. This induces (creates) a current in the secondary. The current in the secondary is also an alternating current.

The resistance welding transformer must be designed to carry the large currents required during welding. The secondary wiring is usually made of cast copper or rolled copper bars. These bars are usually water-cooled. Copper tubes are welded or brazed to the bars to carry the cooling water, as shown in Figure 18-2. Smaller transformers may be air-cooled.

The core of the transformer is made from laminated iron. A **laminated core** is made up of many thin metal sheets stacked together to the required thickness. It has been determined that a laminated core is more efficient than a solid core. The core is usually made of 4% silicon steel laminate.

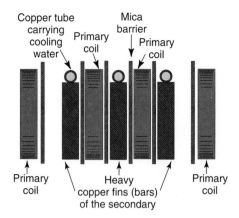

Figure 18-2. *Schematic cross section of a resistance welding transformer. Note that the secondary circuit bars have copper tubes brazed to them.*

The primary windings are insulated so that no accidental grounds can occur. Figure 18-3 illustrates a typical primary winding. Great care is taken to insulate the ribbon-type primary windings from each other and from the frame. A secondary winding that consists of three parallel loops is shown in Figure 18-4. This design is still considered *one* winding. The three parallel paths only make one winding or loop.

Figure 18-5 shows a schematic wiring diagram for a tap-type resistance welding transformer. This machine provides for eight current ranges. The amount of current is dependent on the ratio of the number of turns in the primary circuit to the number of turns in the secondary. Each current range on the tap-type transformer puts more and more primary windings into use and thus creates more transformer output. Each plug or *tap* is wired into a number of primary windings. The number of primary windings is changed by moving the tap selector switch to a different tap position. The highest tap setting provides the greatest number of primary windings and, therefore, the greatest secondary transformer output. An amperage control for a resistance welder with a high and low tap and an eight-step primary winding is shown in Figure 18-6.

Transformers vary in capacity and will suit varied welding needs. The capacity is usually listed as a ***KVA (kilovolt-ampere) rating***. Resistance welding transformers are rated on a basis different from most transformers. The KVA is the *input* or primary circuit rating. For example, with a 240V input and 130A flowing, the machine would be rated at 31KVA.

$$\frac{(240V \times 130A)}{1000VA} = \frac{31,200VA \times KVA}{1000VA} = 31.2KVA = 31KVA$$

Note: 1KVA=1000VA

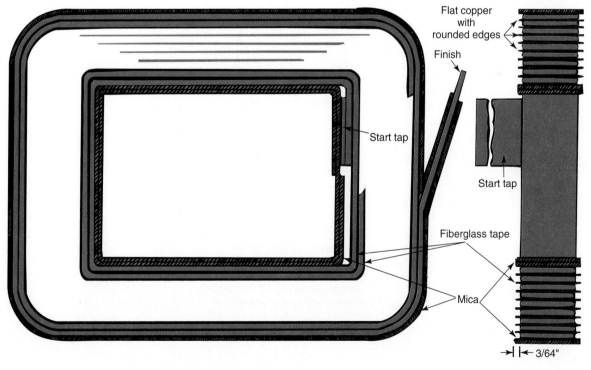

Figure 18-3. *The construction of the primary winding of a typical resistance welding transformer. The conductor is flat or ribbon-shaped and the turns are insulated from each other by fiberglass. (Taylor-Winfield Corp.)*

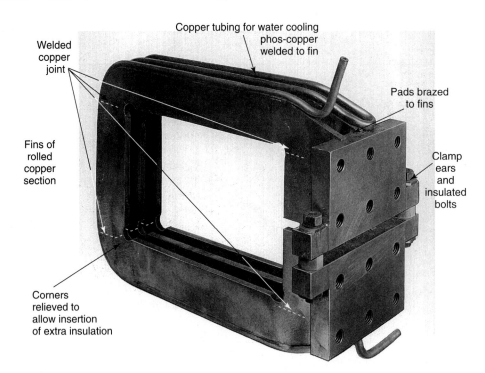

Figure 18-4. A typical secondary winding design for a resistance welding transformer. In this case, three bars are connected in parallel to increase current-carrying capacity and still retain the one-turn electrical function.

Eight-step winding.
Secondary has one turn.
Design for 10 percent secondary voltage steps.

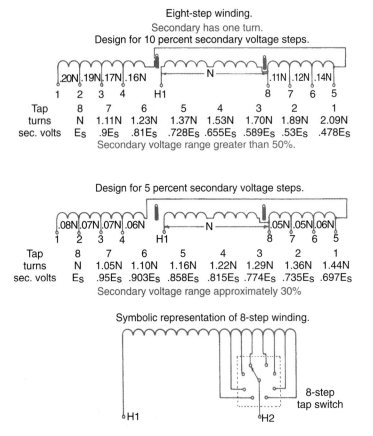

	.20N	.19N	.17N	.16N	N	.11N	.12N	.14N
	1	2	3	4	H1	8	7	6	5
Tap	8	7	6	5	4	3	2	1
turns	N	1.11N	1.23N	1.37N	1.53N	1.70N	1.89N	2.09N
sec. volts	E_S	$.9E_S$	$.81E_S$	$.728E_S$	$.655E_S$	$.589E_S$	$.53E_S$	$.478E_S$

Secondary voltage range greater than 50%.

Design for 5 percent secondary voltage steps.

	.08N	.07N	.07N	.06N	N	.05N	.05N	.06N
	1	2	3	4	H1	8	7	6	5
Tap	8	7	6	5	4	3	2	1
turns	N	1.05N	1.10N	1.16N	1.22N	1.29N	1.36N	1.44N
sec. volts	E_S	$.95E_S$	$.903E_S$	$.858E_S$	$.815E_S$	$.774E_S$	$.735E_S$	$.697E_S$

Secondary voltage range approximately 30%

Symbolic representation of 8-step winding.

8-step
tap switch

H1   H2

Figure 18-5. Schematic wiring diagram that shows adjustable primary windings of a resistance welding transformer. This unit has eight taps (adjustments) in the primary circuit.

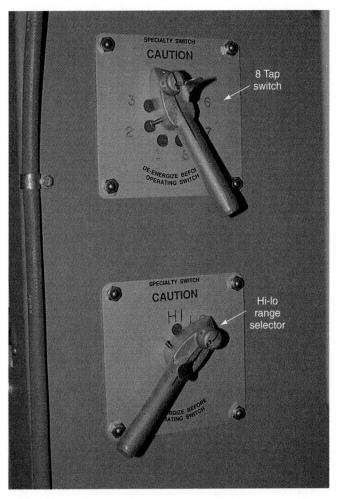

Figure 18-6. These switches provide a high/low amperage tap selection and a choice of eight transformer tap settings. (Taylor-Winfield Corp.).

A resistance welding transformer is usually rated at 50% **duty cycle**. This means that it can be used 30 seconds out of each 60 seconds. The duty cycle is based on a one-minute period. The transformer cannot be used for three minutes constantly and then cooled for three minutes. The transformer will overheat.

Duty cycles vary all the way from 1% to 100%, with a normal rating of 50%. If the duty cycle is high, the KVA setting must be smaller to ensure safe transformer temperatures. The manufacturer usually specifies the KVA rating based on the 50% duty cycle. Seam welding can be done continuously, so the transformer on a seam welding machine should have a 100% duty cycle.

The welding transformer is rated on a specified input, and not on the output like other transformers. The output of the transformer will vary with the conductivity of the metal being welded, the electrode circuit, and other factors. Therefore, the welding manufacturer rates only the primary input capacity of the transformer.

When a machine is used at its rated KVA, the temperatures of the transformer windings will stay within safe limits for the insulation used. When the KVA rating or duty cycle is exceeded, that transformer efficiency decreas-

es rapidly as the temperature rises. There is also the possibility of an insulation breakdown occurring.

Small spot welders may be rated in VA (volt-amperes). This is done when the KVA rating would be a decimal, as in the case where 300VA = 0.3KVA.

The primary current flow influences the electrical power service. For example, a 100KVA machine may draw as much as 400A. If this load is applied to the power service all at once, a large voltage drop may occur in the line voltage. This drop in line voltage will affect other machines on the same power line. For example, a 10% drop in line voltage will cause a 20% to 30% reduction in the welding heat available at the electrodes. Thus, automatic machines will produce substandard welds if the line voltage varies.

18.3 FORCE SYSTEMS

There are four methods of applying force to the electrodes or to the base metals when performing resistance-type welds. The four methods of applying force are magnetism, pneumatics (air), hydraulics (fluids), and mechanical leverage. Pneumatic cylinders are most often used. Pneumatic cylinders are less expensive than hydraulic cylinders, are fast-acting, and can exert large forces on the electrodes. The type of force mechanism employed depends on the machine and process used.

One common spot and projection welding machine is the rocker-arm type. It uses a mechanical leverage to multiply the force applied. Force is usually applied to the rocker arm by pneumatic or hydraulic cylinders.

A mechanical leverage system is simply a series of levers. When the welding operator presses on the foot pedal, leverage causes more force to be applied to the electrodes. A schematic of a mechanical leverage system is shown in Figure 18-7.

The most common force system is the air-operated or pneumatic system. The air-operated system is simple to use and requires little maintenance. Compressed air is used to supply the force. The desired air pressure is set on a valve. The valve setting is proportional to the force applied at the electrodes. Figure 18-8 shows an air-operated rocker arm welding machine that uses an air cylinder to supply the force. The major parts of a rocker arm spot welding machine are labeled.

Hydraulic and pneumatic cylinders can be used to supply the required force to the electrodes for resistance welding. Hydraulic or pneumatic systems are used on press-type spot or projection welding machines. A press-type resistance welding machine applies the force directly on the electrode holder and electrode with no leverage involved. A press-type resistance welding machine is shown in Figure 18-9.

When welding aluminum, a force system also must supply a forging force to the movable electrode. See Figure 19-19. The forging force is applied to the electrode after the welding current is turned off and while the weld area is still molten or plastic. It forces the electrodes down and

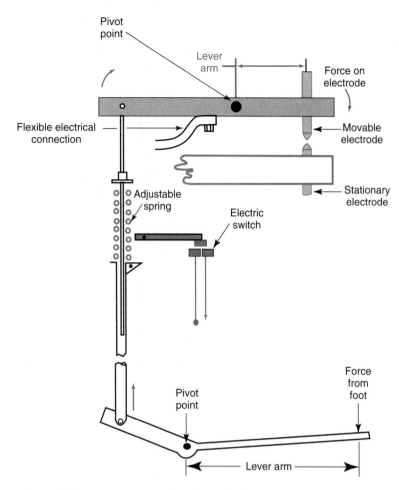

Figure 18-7. A mechanical leverage system. The electrode force is controlled by adjusting the adjustable spring and by the force applied by the welding operator.

holds them at a set position while the metal cools and solidifies.

Electric motors may be used to rotate the cam that brings the base metals together during the flash butt welding or upset welding processes. See Figure 4-21. Strong magnetic coils, similar to solenoids, may be used to bring the parts together quickly and with a hammerlike force when making a flash butt resistance weld. Refer to Heading 18.12.

The force on the resistance welding electrode may vary from a few pounds or newtons (N) to a few thousand pounds or newtons. Common electrode forces used in industry vary from 300 lb. (1335 N) to 1400 lb. (6228 N). Very thin gauge sheet metal can be welded with 100 lb. (445 N) of force on the electrode.

18.4 CONTROLLERS

Resistance welding is a fast operation. A weld is often completed in less than one second. The control of the welding operation must be very accurate. A controller is needed to control the welding operation. There are five variables in resistance welding. These variables are:

- Current.
- Time.
- Force applied to the metal.
- Electrode contact area.
- Welding machine.

The welding controller is used to control the time and the current required to make a weld. There are four different times required to complete one resistance welding schedule. A *resistance welding schedule* is the sequence of events that must occur to complete one weld.

These times are:
- Squeeze time.
- Weld time.
- Hold time.
- Off time.

Each of the times is set on the controller in cycles. One cycle is 1/60 of a second.

The *squeeze time* is the time required for the electrodes to close on the workpiece and apply the proper force. This force clamps the pieces and provides for a good electrical contact.

The *weld time* is the time the current flows through the workpiece. As the current flows, it heats the two pieces of metal where they are in contact with one another. The current is stopped by the controller.

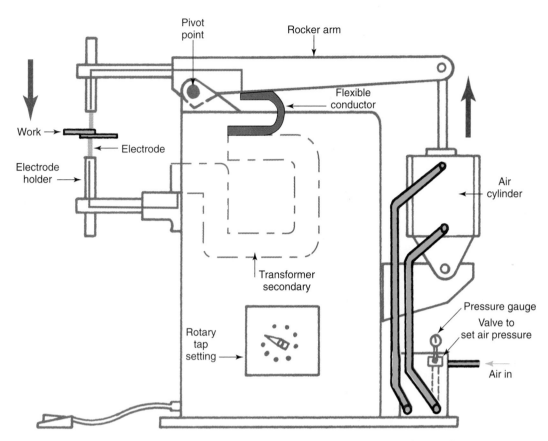

Figure 18-8. *A rocker arm welding machine that uses air pressure to supply the welding force. The air pressure is adjustable. The main parts of a rocker arm machine are labeled.*

Figure 18-9. *This spot welder uses a pneumatic cylinder to apply force to the movable electrode. Notice the flexible secondary connection to the upper arm.*

The **hold time** is required so the molten metal has time to cool. The two pieces of metal are now welded together.

A typical welding controller is shown in Figure 18-10. The welding operator can set the squeeze time, weld time, and hold time in cycles. An example of a complete welding schedule is: squeeze for 20 cycles, weld for 10 cycles, hold for 20 cycles. This weld cycle is shown in Figure 18-11.

There is also a setting for *off time*. Off time is used when the welding machine is making spot welds repeatedly. The off time allows the welding operator to reposition the parts being welded, or to place some new material between the electrodes. The weld schedule is then repeated automatically.

The welding controller is also used to adjust the current. Large changes in current are made by changing the tap setting on the transformer. This is similar to changing the range selector on an arc welding machine. Smaller changes within a current range are made by adjusting the percent heat control on the welding control panel. This control is marked from 0% to 100%. See Heading 12.5 for an explanation of setting this type of control.

The welding current does not have to remain a constant. When welding large thicknesses, *upslope* (an increasing current) is often required. The current is gradually increased from 10% heat, up to the desired welding percent heat. *Downslope* (a decreasing current) can also be used. It may also be required to use two or more pulses of current. A short cooling time separates the different pulses. See Figure 18-12 for an example of how these functions are used in a weld schedule.

An example of a complex weld schedule is as follows: squeeze for 20 cycles, slope the current up (upslope) from 10% heat to 50% heat in 15 cycles, weld at 50% heat for 30 cycles, cool for 5 cycles, weld a second pulse at 50% heat for 30 cycles, and hold for 20 cycles. This weld schedule is illustrated in Figure 18-12. To make use of these special upslope, downslope, and multiple-pulse functions, a more sophisticated controller must be used. An example of a welding controller designed to direct upslope, downslope, and multiple-pulse functions is shown in Figure 18-13.

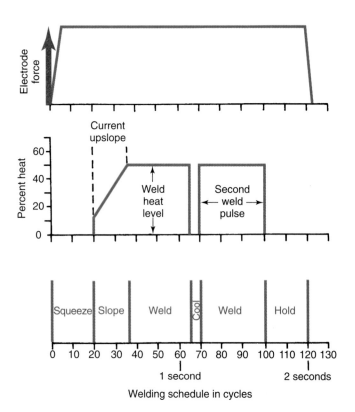

Figure 18-12. A complex welding schedule which includes upslope, two current pulses, and a cooling time.

18.5 CONTACTORS

The primary and secondary circuits of a resistance welding machine must carry very large amounts of electrical current. This high current must be switched on and off at specified times during a resistance welding cycle. The voltage in the primary circuit is so high that it causes arcing across a mechanical-type switch. Therefore, electronic *silicon-controlled rectifiers (SCRs)* are utilized as *contactors* (switches) in the primary circuits of welding controllers. An SCR is a heavy-duty diode. See Figure 18-14. It allows current to flow in one direction only. Two SCRs are used in each circuit that requires switching. One conducts current during the positive side of the ac cycle and the other conducts during the negative side of the ac cycle.

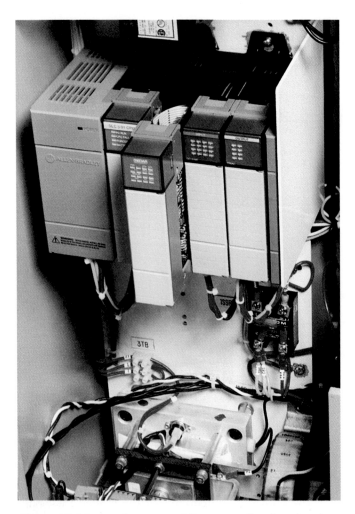

Figure 18-10. A resistance welding controller showing the electronic memory circuits and controls it contains. (MEDAR, Inc.)

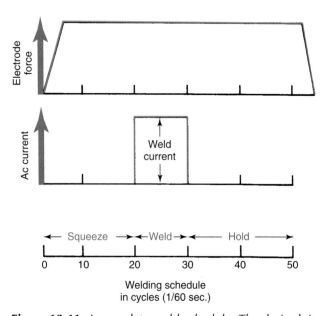

Figure 18-11. A complete weld schedule. The desired times for each portion of the weld schedule are set in cycles. Notice the time for the pressure system to react.

Figure 18-13. *This resistance welding timer will meet the needs of most pneumatic and hydraulic spot or projection welding machines. (British Federal)*

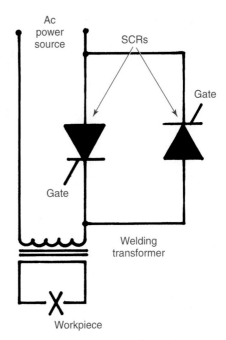

Figure 18-15. *Two silicon-controlled rectifiers (SCRs) connected in parallel. This allows current to flow through the transformer during the positive and negative sides of the alternating current (ac) cycle.*

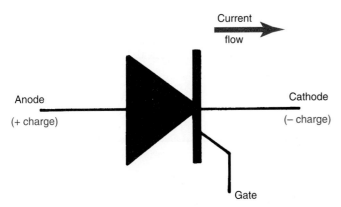

Figure 18-14. *The symbol for a silicon-controlled rectifier (SCR) as used on electronic circuit drawings. The SCR allows current flow in only one direction.*

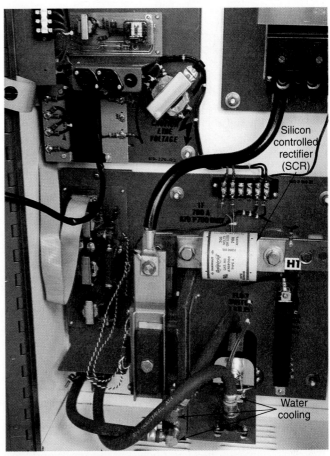

Figure 18-16. *Silicon-controlled rectifiers (SCRs) are used to control the electric current. Two SCRs are used to control the current in both half cycles. One SCR is hidden from view. (Robotron Div., Midland-Ross Corp.).*

See Figure 18-15. When the controller sends a signal to the gate side of the SCR, it allows current to flow. When the polarity changes in the ac cycle, the SCR stops conducting. Using two SCRs allows the current to flow continually. A welding controller using SCRs is shown in Figure 18-16.

Older welding controllers use devices called *ignitron tubes* as contactors. The ignitron tube is designed to allow current flow in one direction. Like the SCR, two ignitron tubes must be used in each circuit to allow the current to flow during both halves of the ac cycle. Liquid mercury within the ignitron is vaporized by an ignitor. Current then flows between the anode (+) and cathode (–) terminals. See Figure 18-17. Figure 18-18 shows a solid-state controller that uses ignitron tubes for contactors. Many ignitron controllers are still in use.

Since all newer welding controllers use SCRs, the term "SCR" will be used to refer to all contactors. SCRs can be made to begin conducting at various points in the ac cycle. The different times correspond to the percent heat required and set on the control panel. If a high percent heat is required, the SCRs will begin conducting very early in the ac cycle. When lower percent heat settings are used, the SCRs will begin conducting later in the ac cycle. See Figure 18-19.

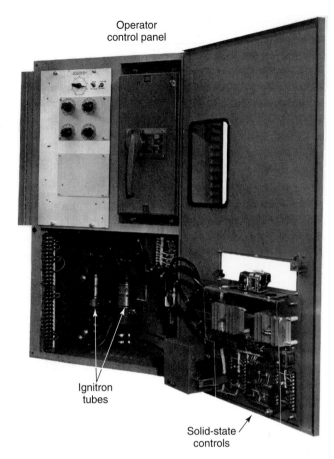

Figure 18-18. A resistance welding controller. The panel door is open to show the solid-state controls and the ignitron tubes.

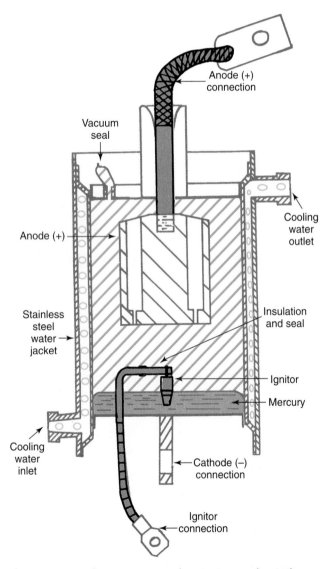

Figure 18-17. The main parts of an ignitron tube. When the ignitor fires, the mercury vaporizes. The vaporized mercury completes the circuit between the anode and the cathode. (General Electric Co.).

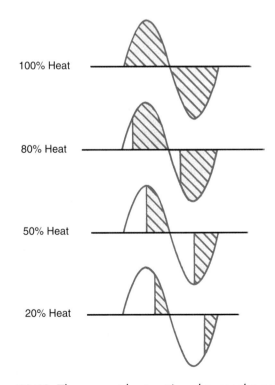

Figure 18-19. The percent heat setting changes the point at which the current begins to flow in each cycle. The higher the percent heat setting, the longer the current will flow in each cycle.

When a lower percent of heat is selected, the trigger voltage from the controller to the SCR gate is delayed until later in the ac cycle. The amperage to the electrodes as a result is lower. Figre 18-20 shows an SCR-type circuit.

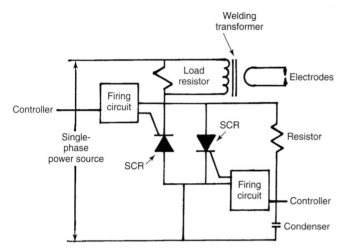

Figure 18-20. *The firing of the two SCRs is controlled by the controller which sends a signal to the firing circuit to allow the SCR to conduct current.*

18.6 SPOT WELDING ELECTRODES

Resistance welding electrodes conduct the current to the surfaces of the metals to be welded. There are certain requirements that these electrodes must possess. An electrode must:

- Be a good conductor of electricity.
- Be a good conductor of heat.
- Have good mechanical strength and hardness.
- Have a minimum tendency to alloy (combine) with the metals being welded.

Pure copper possesses good electrical and thermal properties; however, it is rather soft and does not wear well. Also it tends to soften with heat. Most electrodes are a copper alloy.

The electrode must be a good conductor of electricity so that the current can flow to the workpiece without overheating the electrode. It must be a good conductor of heat so that the heat generated at the point of contact with the weldment can be conducted away without causing the electrode to overheat. Typical spot welding electrodes are shown in Figure 18-21. Every electrode has a face and a shank. The *electrode face* is the part of the electrode that contacts the work. The face may have a number of different designs, as shown in Figure 18-22. The *electrode shank* must be large enough to carry the force and the welding current. In order to meet spot welding requirements, some electrode shanks are bent. Examples of bent electrodes can be found in Figure 18-26.

Some electrodes have two pieces. The two pieces are called the electrode cap and the adaptor. The electrode caps can be male or female, as shown in Figure 18-23.

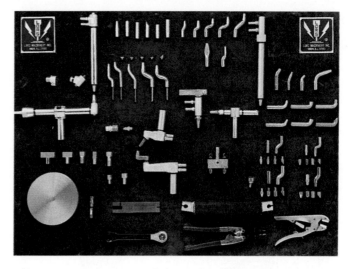

Figure 18-21. *A variety of electrodes, electrode holders, caps, adaptors, and electrode removing and dressing tools. Note the large round electrode used for seam welding. (LORS Machinery, Inc.)*

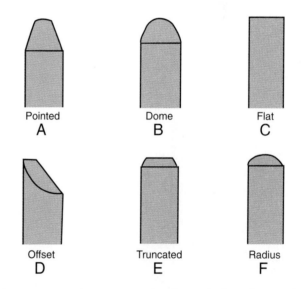

Figure 18-22. *The six common electrode face designs for spot welding. The letter designations A-F are those used by the Resistance Welder Manufacturers' Association.*

Most electrodes and electrode caps are water-cooled. There is a hole down the center of the electrode. A flexible water tube is placed inside the electrode. The cooling water flows through the tube and cools the electrode very near its face. The water returns around the tube and keeps the entire electrode cool. The path of the cooling water through a bent electrode is shown in Figure 18-24.

18.6.1 Electrode Materials

The Resistance Welder Manufacturers' Association (RWMA) and the Resistance Welding Alloy Association (RWAA) recognize two groups of materials used for resis-

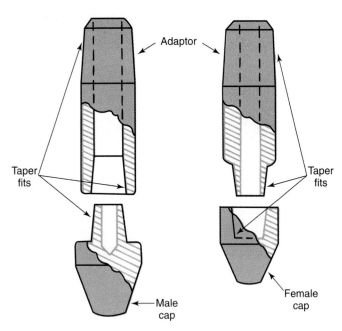

Figure 18-23. Two-piece electrode cap and adaptors. The electrode caps may be male or female and are usually held on the adaptor by a taper fit.

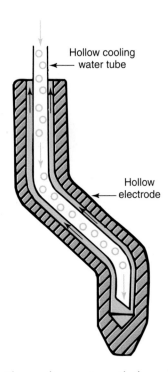

Figure 18-24. The cooling water path through a bent electrode.

tance welding electrodes. They are designated Group A and Group B.

Group A materials are copper-based alloys having good electrical and thermal (heat) conductivity, with improved hardness and wear qualities.

Group B materials are refractory metal alloys. The electrical and thermal properties of materials in this group

are not as good as those of the alloys in Group A. However, Group B materials have extremely high melting temperatures, and high compressive strength and wear resistance. They are usually mixtures of tungsten and copper.

When ordering or purchasing resistance welder electrodes, the following data may be referred to as a guide.

Group A, Copper-base Alloys

The most common materials used are Group A, Class 1 and 2. The Class 2 electrodes are the most widely used. These two types have the best conductivity with good hardness. They are also the least expensive and most available.

Characteristics and uses of various classes of Group A electrodes are:

Class 1. Highest conductivity (80%); recommended for spot welding aluminum alloys, magnesium alloys, galvanized iron, brass, or bronze. The electrode material is a copper-cadmium alloy.

Class 2. High conductivity (75%) and good hardness. For high production spot and seam welding, clean mild steel, low-alloy steels, stainless steels, nickel alloys, and monel metal. The electrode material is a copper-chromium alloy.

Class 3. Higher strength and hardness than Class 2 electrodes, with 45% conductivity. Recommended for projection welding electrodes, or for flash and butt welding electrodes. Recommended for stainless steel. The electrode material is a copper-zirconium alloy.

Class 4. This is a hard, high-strength alloy. This alloy is recommended for electrodes used in special applications where the forces are extremely high and the wear is severe. The conductivity is only 20%.

Class 5. This alloy is used chiefly as a casting. It has high mechanical strength. The electrical conductivity is only 15%.

Group B, Refractory Metal Compositions

Characteristics and uses of the refractory alloys in Classes 10-14 are:

Class 10. This electrode material has the best conductivity (45%) of the Group B electrodes. It is recommended for facings for projection welding electrodes and flash and butt welding electrodes.

Classes 11 and 12. This material is harder than the material in Class 10. It is used when exceptional wear resistance is required.

Classes 13 and 14. These electrodes are made of commercially pure tungsten and molybdenum, respectively. They are used to weld nonferrous metals that have very high electrical conductivity.

18.7 ELECTRODE HOLDER

The resistance welding electrodes and adaptors are held by electrode holders. The holders are clamped into the ends of the movable spot welding arm and the stationary arm. Refer to Figures 18-8 and 18-9. Most of the electrode holders are water-cooled.

The electrode holders hold the electrode in proper position, carry the welding current, and provide the electrode with water cooling. The water tube runs down the center of the electrode holder and continues into the electrode. Figure 18-25 shows typical electrode holders and tips.

It should be noted that, on most spot welding machines, electrode holders are adjustable for length and position. In general, the electrode holder should be adjusted to the shortest length at which the weld metal may be easily inserted. Electrode holders are made of a copper alloy that provides good current-carrying qualities and rigidity. Some examples of electrodes and electrode holder combinations are shown in Figure 18-26.

The electrode or adaptor can be attached to the electrode holder in three differentways:
- Taper fit.
- Threaded.
- Straight shank.

A tapered fit is the most common, especially for high-production spot welding. Replacing electrodes and electrode caps is quick and easy when a tapered fit is used. An ejector tube or knock-out bar is used to loosen tapered adaptors and electrodes. The RWMA-preferred taper is the *Jarno taper*. It is preferred for the following reasons:
- The taper numbers progress in order from #3 to #7.
- The Jarno taper is 0.600"/ft. for all sizes.
- The nominal major diameter is easily determined by multiplying the Jarno taper number by 1/8". Example: The major diameter of a #6 Jarno taper is: $6 \times 1/8" = 6/8"$ or 0.750".

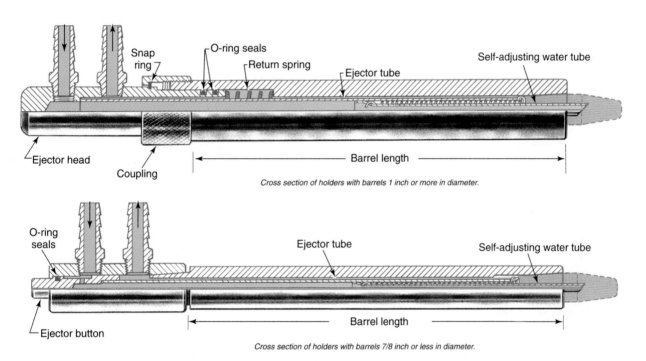

Cross section of holders with barrels 1 inch or more in diameter.

Cross section of holders with barrels 7/8 inch or less in diameter.

Figure 18-25. *Cross section of two electrode holders. (Tuffaloy Products, Inc.)*

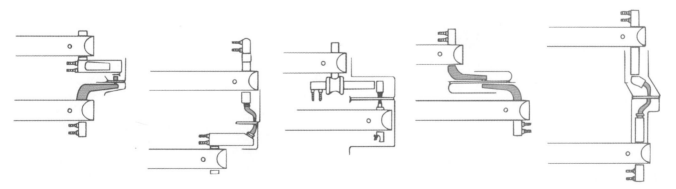

Figure 18-26. *Different types of electrodes and electrode holders can be combined to meet a job requirement. (Tuffaloy Products, Inc.)*

A threaded attachment is used when high welding forces are required. The straiht shank electrode or adaptor is mechanically connected to the holder by a separate coupling or collar.

18.8 SPOT WELDING MACHINES

Spot welding machines are the most common of the resistance welding machines. Spot welding machines are made in a great variety of sizes from small bench units to extremely large welding machines. A small bench model may be used to spot weld such items as costume jewelry and electronic components. Two bench-top welding machines are shown in Figures 18-27 and 18-28. Large, specially designed resistance welding machines are capa-

ble of welding hundreds of spots on large sheet metal products, such as automobile bodies, refrigerator cabinets, or computer panels, as shown in Figures 18-29 and 18-30. Some resistance welding machines are portable. Portable machines are covered in Heading 18.10. Most machines have only one set of electrodes.

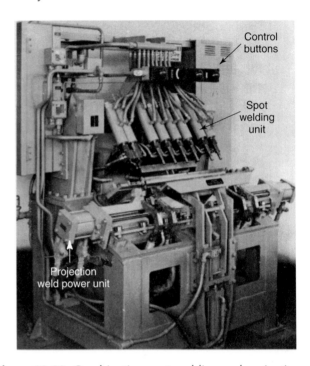

Figure 18-29. Combination spot welding and projection welding machine with four 55 KVA transformers. The input power is 440V, 60 cycle. The spot welding guns lower, weld, and retract in 1.5 seconds. There is one projection welding unit at each end. These units also advance, weld, and retract in 1.5 seconds. There are dual controls. The operator manually loads and unloads the parts. Note the pneumatic, hydraulic, and electric devices. The necessary parts are water-cooled using valves and electric timers to control the water flow. (Resistance Welder Corp.)

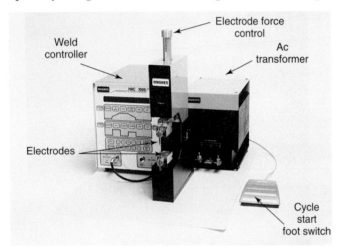

Figure 18-27. A miniature resistance spot welder. This machine has a welding control, electrode force control, ac transformer, and cycle start foot switch, just as larger machines do. (Hughes Aircraft Company, Technology Products Div.)

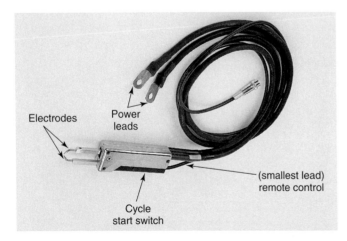

Figure 18-28. A miniature resistance spot welder with a cycle start switch, remote control connection, and tweezer-type electrodes. (Hughes Aircraft Company, Technology Products Div.)

Figure 18-30. This resistance spot welding machine uses sixteen individual electrodes to weld stiffeners onto a mainframe computer panel. (LORS Machinery, Inc.)

The basic purpose of the spot welder is to make a spot weld (fused nugget) between two or more lapped pieces of metal. The size of the spot weld is controlled by changing the resistance welding variables. The variables in resistance welding are:
- Current.
- Time.
- Force.
- Electrode contact area.
- Welding machine.

The selection of a resistance welding machine determines the following:
- The dimensions of the operating area.
- The KVA rating.
- The type of pressure system.

The operating area has two dimensions. These are called the throat depth and the horn spacing. The *throat depth* is the distance from the center of the electrodes to the frame of the welding machine. The throat depth determines the size (width) of a part which can be welded.

The *horn spacing* is the distance between the electrode arms when the electrodes are closed. The height of a part to be welded is controlled by this dimension. The horn spacing is usually adjustable. The throat depth and horn spacing are shown in Figures 18-31 and 18-32.

The throat depth should be kept to a minimum. As the throat depth increases, the KVA required to create a weld increases.

Resistance welding machines obtain their electrical energy in one of three ways:
- Single-phase machine.
- Three-phase machine.
- Stored energy machine.

All resistance welding machines demand a large current for a very short period of time. The single-phase machine is connected to one phase of the electricity supply. The large demand for electrical energy by the single-phase machine can cause a drop in the line voltage of the one phase. If several single-phase resistance welding machines are all connected to the same phase and are all fired at once, a large drop of voltage can affect other machines (welding and nonwelding) on the same phase. This decrease in voltage on the one phase causes an imbalance on the power system. To prevent an imbalance, a stored energy, or a three-phase machine can be used.

Stored energy welding machines are spot welding machines that have special electrical characteristics. A stored energy welding machine obtains the energy needed for welding from the service lines at a relatively slow rate. This stored energy, once it has reached the desired level, can be released at a high rate for welding. By taking the energy at a slower rate from the service line and storing it, the stored energy machine does not cause a voltage drop. There are two types of stored energy resistance welding machines:
- Electrostatic (capacitor-type).
- Electrochemical (battery-type).

The *electrostatic resistance welding machine* is the one most often used. Its principle of operation depends upon storing the electrical energy in a capacitor. When the

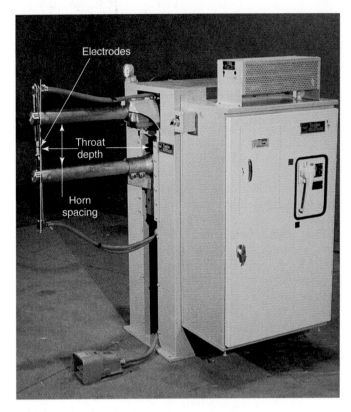

Figure 18-31. Throat depth and horn spacing measurements on a pneumatically operated rocker-arm type resistance spot welding machine. (Taylor-Winfield Corp.)

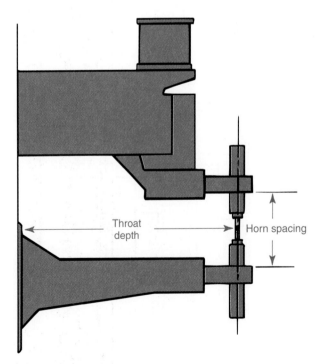

Figure 18-32. Throat depth and horn spacing on a press-type machine. The horn spacing is measured with the electrodes touching one another (power off).

energy is needed to weld, it is released by the capacitors. The current passes through the workpiece to form a weld.

The use of capacitors to store electrical energy enables the welding machine to use a smaller KVA transformer. The off-time part of the cycle of the machine is used to electrically load the capacitors or condensers.

The application of stored energy machines, both electrostatic and electrochemical, is in small precision welding. They are usually used on metal thicknesses of less than 0.03"(0.8mm) and are small in size. See Figures 18-27 and 18-49.

Large stored energy machines are being replaced by three-phase machines. The *three-phase resistance welding machine* draws energy equally from each of the three phases of the electricity supply. The three-phase machine does not cause an imbalance in the power supply.

Three-phase machines have three times as much electrical circuitry in the primary as a single-phase machine. A single-phase machine has two SCRs. A three-phase machine has two SCRs for each phase, for a total of six SCRs. A three-phase machine also has three primary transformer windings. Figure 18-33 shows a schematic of a three-phase resistance welding machine. Because of the additional electrical requirements and increased size, a three-phase resistance welding machine is more expensive than a single-phase machine.

18.9 PROJECTION WELDING EQUIPMENT

Projection welding equipment is similar to the equipment used in spot welding. In projection welding, one of the pieces to be welded has one or more small projections

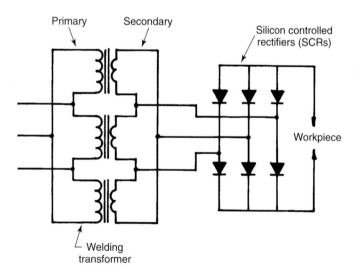

Figure 18-33. The electrical circuit drawing for a three-phase resistance welding machine. This three-phase machine has six SCRs and three transformers.

pressed into it. The current passes through the projection and forms a weld between the two metal pieces. Projection welding is used to accurately locate the welds on a part which is in high production.

A press-type or rocker-type welding machine can be used for projection welding. If there is only one projection weld, flat electrodes can be used in either type machine. When more than one projection weld is made at a time, special large *platens* (flat electrodes) must be used.

A platen is a large, generally flat surface through which welding current flows. The upper platen is movable. The lower platen is adjustable, but once adjusted, it remains stationary. An air or hydraulic cylinder forces the upper platen against the lower platen. See Figure 18-34. The two parts being joined must be flat. If they are not flat, a specially made die can be attached to the platen. Higher forces and currents are required to make multiple projection welds, because current is flowing through more than one weld area. See Figure 19-23.

18.10 PORTABLE WELDING MACHINES

The *portable resistance welding machine* has four main parts:

- A portable welding gun with electrode holders, electrodes, and force mechanism.
- A welding transformer.
- An electronic contactor and controller.
- A cable and hose to bring power and cooling water to the electrodes.

A portable resistance welding machine is shown in Figure 18-35. Such portable resistance welding machines are used to make spot welds on weldments, often on a moving assembly line. They are also called "resistance welding guns" in the shop. Some portable resistance welding machines have transformers mounted directly on the gun. This type of gun is called a *transgun*. Several transguns are shown in Figure 18-36. Since welding guns are heavy, they are hung by cables from an overhead support. They are often counterbalanced to lighten their weight. This arrangement makes them very easy to move to the weldment. It also allows the gun to move from weld to weld and into a variety of positions. Their range of movement is limited only by the length of the power cable and water cooling hose. Long power cables have a disadvantage. The longer the power cable, the higher the secondary output voltage must be.

Resistance welding guns are designed to do a specific job. They are used to assemble sheet metal parts in a variety of industries.

Portable resistance welding machines require a higher secondary voltage because of the increased length of the secondary power cable. The secondary power is increased by having two or three turns in the secondary circuit. Cooling water is circulated through the transformer and welding gun to remove excess heat.

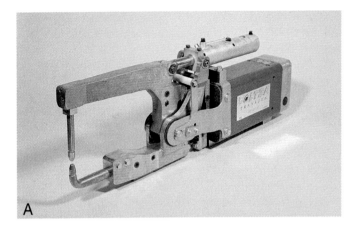

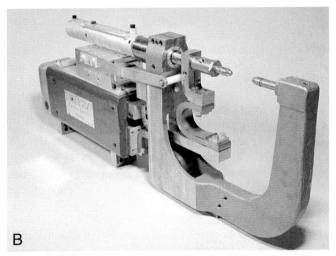

Figure 18-34. *A hydraulically actuated resistance spot welding machine. Note that when the electrode holders and electrodes are retracted, there is a set of platens in place for projection welding.*

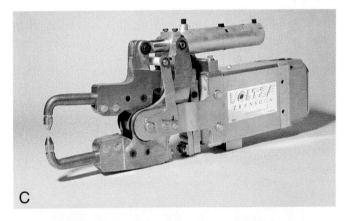

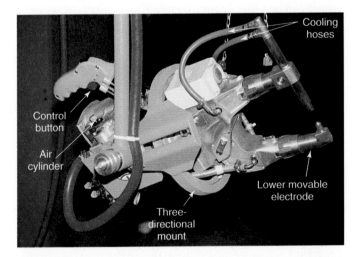

Figure 18-35. *A resistance welding machine. This unit is supported from overhead to relieve weight from the operator. It has a built-in transformer and electrodes that can be rotated as required. The lower electrode is pneumatically operated.*

Figure 18-36. *Three different sizes of portable resistance welding machines. These welding guns are called "transguns." Each has its own welding transformer and pneumatic actuating cylinder. Note the flexible secondary circuit connector in "B." (Marc-L-Tec, Inc.)*

18.11 SEAM WELDING MACHINES

A *seam welding machine* is a special form of a resistance welding machine. Seam welding uses two copper alloy wheels as electrodes. These wheels press the two pieces of metal together and roll slowly along the seam.

Either a continuous current is passed between electrodes and work or current is passed through the electrodes and work at timed intervals.

A seam welding machine has the following parts:

- A main frame which contains the transformer and tap switch.
- Secondary connections to the electrodes.
- Wheel electrodes.
- Stationary lower electrode arm.
- Movable upper electrode arm.
- Air cylinder and ram to apply required force.
- Electrode drive mechanism.

The main frame, air cylinder, and ram are similar to those used on a press-type welding machine. The electrode drive mechanism can be one of three types:

- Gear drive.
- Knurl drive.
- Friction drive.

Usually only one electrode wheel is driven; the second rotates freely. A gear drive turns the center of the wheel at a constant rpm. As the wheel wears down, the welding speed decreases.

A knurl or friction drive uses a small roller that contacts the outer edge of the wheel electrode. The small roller rotates at a constant rpm. The small roller drives the wheel electrode at a constant rpm. With a knurl or friction drive, the welding speed will remain constant as the electrodes wear.

A seam welding machine can be one of three types. The wheels can be parallel to the front of the machine (*transverse seam*); perpendicular to the front of the machine (*longitudinal seam*); or a combination of the two (*universal*). See Figure 18-37. The universal machine can be used to make both longitudinal and transverse seam welds. There are two lower arms that can be changed to perform longitudinal or lateral seams. One upper arm and wheel can be rotated as required. See Figure 18-38.

A seam welding machine is controlled by a controller. Current can be timed to make overlapping spot welds, spots at a given spacing, or current can run constantly to make a continuous seam.

Figure 18-38. *A universal seam welder. Note that these electrode wheels are set up for transverse welding. The upper wheel may be rotated 90° and the lower wheel may be changed to weld longitudinally. (Sciaky, Inc., Subsidiary of Phillips Service Industries)*

Pneumatic (air) pressure is usually used to control the force applied to the electrodes. The pressures required are not extremely high. The details of the upper electrode operating mechanism of a universal seam welding machine are shown in Figure 18-39.

Occasionally, seam welds are made on parts that do not allow a continuous seam to be made. A portion of a wheel can be used to get up close to the area which could not be welded by a complete wheel electrode. Figure 18-40 shows an example of a partial wheel and how it can be used.

The metal part being seam welded and the wheel electrodes get very hot. The electrodes and the work must be water-cooled. Flexible copper tubes carry cooling water to the joint. Both electrodes must be cooled by the water spray. A flexible water tube can be seen, in use with a shoe-type electrode, in Figure 18-40.

18.12 UPSET WELDING MACHINES

Upset welding is a resistance welding process that joins two pieces of metal over the entire area of surface

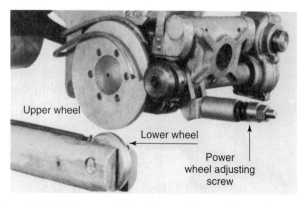

Figure 18-37. *Seam welding electrodes used to make a longitudinal seam. These electrode wheels are perpendicular to the front of the machine.*

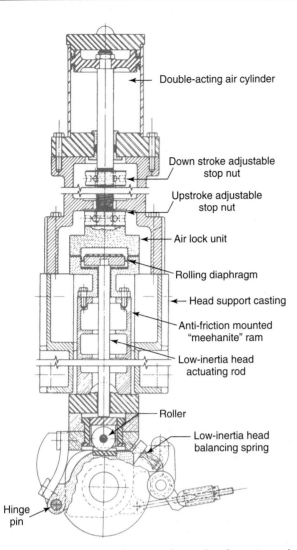

Figure 18-39. Air-operated upper electrode of a universal seam welder. It is used to raise and lower the upper roller and to develop the correct welding pressure. (Taylor-Winfield Corp.)

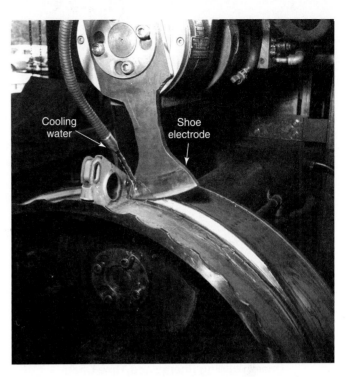

Figure 18-40. Shoe electrode used to weld a seam close to a part in the path of a continuous seam. Note the stream of water, which is cooling the electrode and the part.

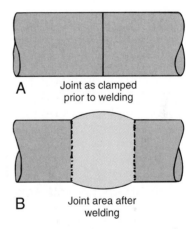

Figure 18-41. An upset welding joint. A—The two parts prior to welding. B—The metal after welding. Notice the enlarged or upset metal area at the weld joint.

contact. Upset welding is used to join the butt ends of rods and bars. The heat is obtained from resistance to the flow of electric current through the entire area of surface contact. Force is applied before heating is started and is maintained during the heating period. At the end of the heating period, a large force is applied to upset the metal at the joint. The metal cools and forms a weld. The finished weld produces an enlarged metal area at the weld. See Figure 18-41. This enlargement or upset metal is usually removed so the weld area is the same size as the original metal.

The machine uses the same type of transformer as a spot welding machine. The electrodes in this case are clamping dies or vises. One die is movable and one is fixed. The surfaces to be welded must be clean. Figure 18-42 shows an upset welding machine with the machine opened for loading or unloading. Figure 18-43 shows the power required and the time needed for making an upset weld on 1/2″ (13mm) square steel bars.

Flash butt welding machines are used to weld the ends of metal rods, rails, beams, and other shapes together. The process is similar to upset welding. However, in upset welding, the ends of the pieces are touching when the welding begins. To perform a flash butt weld, the parts are clamped in special fixtures that carry the electricity to the two parts. One fixture is fixed and one is movable. An arc is struck between the ends of the two parts to be welded. When the ends of the parts become molten, the parts are forced together by pneumatic or hydraulic cylinders and held until they cool. The parts are then removed from the clamping fixtures. See Figure 18-44.

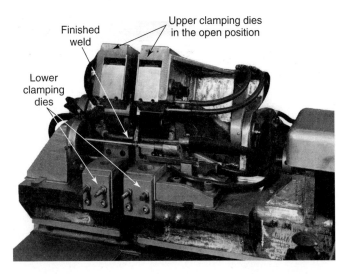

Figure 18-42. An upset welding machine. The completed weld is shown mounted in the open clamping dies.

Upset Welding Energy and Time Chart			
kW	Time in seconds	Distance between gripping dies	
		in.	mm
19	3	0.79	20.1
14	6	1.6	40.6
12	9	2.4	61.0
10	11	3.2	81.3
8	14	3.9	99.1
7.5	17	4.5	114.3
7	20	5.5	139.7

Figure 18-43. Recommended upset welding energy and timetable for steel bars 1/2" (12.7mm) square.

18.13 SPECIAL RESISTANCE WELDING MACHINES

There are numerous specially designed resistance welding machines. These machines consist of some special application of one or more of the basic-type machines. Two of the more popular models are the cross-wire welding machine, shown in Figure 18-45, and the forge welding machine.

Cross-wire resistance welding machines use special single electrodes or multiple electrodes that hold the wire or rod in correct alignment. The current passes through the diameters of the wires where they cross. The resistance at the contact spot creates enough heat to cause fusing of the metal. A strong welded joint is produced.

Forge welding is similar to the cross-wire and upset welding processes. The current heats the metals to a plastic and/or flow temperature. The metal is upset by a movement of the electrodes (press action) and the parts are welded together.

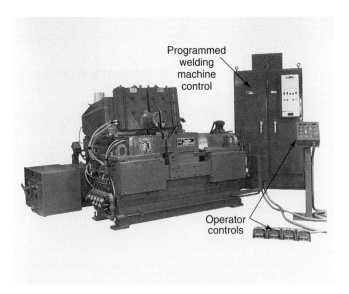

Figure 18-44. A flash butt welding machine with operator and computer controls. (Taylor-Winfield Corp.)

Figure 18-45. Welding cross wires with a resistance welding machine.

18.14 CARE OF RESISTANCE WELDING EQUIPMENT

There are several main areas of resistance welder maintenance:
- Mechanical.
- Electrical.
- Hydraulic.
- Pneumatic.
- Water-cooling.

Mechanical maintenance consists of:
- Lubrication.
- Checking moving parts for wear and alignment.
- Checking the force applied by the electrodes against the base metal being welded.
- Checking mechanical safety devices, such as guards and machine mounting.

Maintenance manuals should be consulted to determine the exact specifications to be used on each particular machine.

Electrical maintenance consists of:

- Checking all electrical connections.
- Checking primary voltage and current.
- Checking secondary voltage and current.
- Checking, cleaning, and installing fuses, switches, and relays.
- Loose shunts are a frequent source of trouble. All electrical connections should be checked daily. Either a voltmeter or an ohmmeter may be used. Oscilloscopes and oscillographs are needed to make a complete analysis of electrical equipment.

Caution: Use extreme caution when working with resistance welding machines. The high currents and voltages used in resistance welding can cause death.

Whenever working on the secondary of a welding machine, make sure the machine is off. The circuit should also be turned off. If work must be done on the primary, disconnect the machine from the electrical power. Disconnect the circuit breaker and remove the fuses from the machine. If the machine does not have its own circuit breaker, the electrical power must be disconnected at the substation.

Whenever a circuit is disconnected, proper lock-out or tag-out procedures must be followed for worker safety. Such procedures consist of locking the breaker in an "off" position (or locking the access cover to the box), if possible. If "locking out" cannot be done, a highly visible tag or sign must be placed on the circuit breaker. The tag or sign should state why the breaker is off and should include the date and the time. **This precaution will prevent someone from turning the power back on and possibly killing the person repairing the welding machine. All maintenance and repair work should be performed by a person knowledgeable in electrical circuits and power.**

Hydraulic maintenance consists of:

- Checking pressure.
- Checking quantity and condition of hydraulic fluid.
- Checking hydraulic lines and connections.
- Checking hydraulic valves for leaks.
- Checking hydraulic pumps and cylinders for leaks. Do not attempt to repair hydraulic leaks. Turn the equipment off and inform a supervisor or repair person. Hydraulic fluid is under very high pressure and can cause severe harm.

Pneumatic maintenance procedures are similar to those used for hydraulic maintenance. The water-cooling circuit is an important component of a resistance welding machine. Thermometers, pressure gauges, and volume measurements are necessary for checkup purposes.

An adequate supply of cool water should be provided to each welding machine, at a minimum of 30 psi (207kPa) line pressure. Cooling water should be at a temperature less than 85ºF (29ºC).

Be certain that the water hose connections are tight. Typically, the welding transformer, ignitron tubes, SCR mounts, electrode holder, and electrodes are all water-cooled. On small machines, however, some of these may not be water-cooled.

Use electrode holders that will pass at least two gallons of water per minute (7.6 L/min). Make sure the water inlet hose is connected to the electrode holder inlet. The water tube inside the electrode holder must extend to the electrode to keep the electrode cool.

Seam welding wheels and the work should be sprayed with cooling water. The water should be directed at the point where the wheel contacts the work.

The water hoses should be free of any deposits that might reduce water flow. Make sure none of the hoses or connections leak, since water in the machine could cause an electrical short.

18.15 PARALLEL GAP RESISTANCE SPOT WELDING

Parallel gap resistance spot welding is a variation of spot welding that allows spot welds to be made from one side only. Its application to date has been on metal up to .015"(0.38mm) thick.

Electronic circuits are normally built up on ceramic- or fiberglass-base boards. Breaks occasionally occur in the copper trace circuits on expensive electronic circuit boards. They cannot be repaired by the normal spot welding process, because fiberglass and ceramics will not conduct electricity. Spot welding can be used to weld a bridge over the gap in the trace (electrical circuit). Parallel gap resistance spot welding permits the spot weld to be made from one side.

Two parallel electrodes are mounted in a miniaturized resistance welding machine as shown in Figure 18-46. The upper arms that hold each electrode holder and electrode are pneumatically operated. The programmable power supply can be an ac transformer or a capacitor-discharge type. The output of capacitor-discharge power supplies is very low and measured in watt seconds. See Figure 18-47.

Figure 18-48 illustrates how the parallel gap resistance spot welder would make a repair weld on an electronic circuit board. The same equipment is used to weld a tab between the opposite terminals of two or more small batteries to make a battery power pack with higher voltage outputs. Figure 18-49 shows tabs being welded on two batteries. The welding current will not go through the battery. The current contacts only the very end of the battery terminal where the weld takes place, as shown in Figure 18-49.

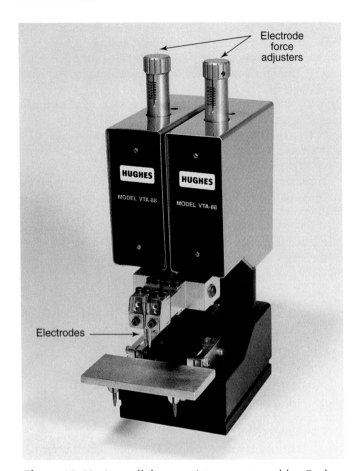

Figure 18-46. A parallel gap resistance spot welder. Each upper electrode holder has a pressure regulating scale on the top. (Hughes Aircraft Company, Technology Products Division)

Figure 18-47. Three capacitor-discharge power supplies with different output capacities. (Hughes Aircraft Company, Technology Products Division).

TEST YOUR KNOWLEDGE

Write your answers on a separate sheet of paper. Do not write in this book.

1. Resistance welding requires _____ amperage/_____ voltage electrical energy.
2. What happens to the voltage and current when it goes through a step-down transformer?

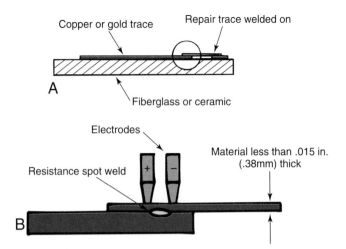

Figure 18-48. A parallel gap resistance spot welding repair. A—The broken original trace is bridged by another conductor and spot welded from one side. B—Enlarged view.

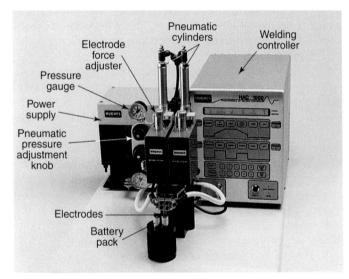

Figure 18-49. A complete parallel gap spot welding outfit. A tab is being welded across two batteries to make a battery pack. (Hughes Aircraft Company, Technology Products Division).

3. The current in the secondary of a resistance welder is _____ (ac or dc).
4. What is a typical duty cycle for a resistance welding machine?
5. On what period of time is the duty cycle based?
6. Name three types of electrode force systems used on a resistance welding machine.
7. Which force system can supply the highest force?
8. What are the five variables in resistance welding?
9. List the times required to make a complete spot weld schedule.
10. Name two ways to increase the current on a resistance welding machine.

11. What is required to make an ignitron tube conduct current?
12. What parts of a resistance welding machine are water-cooled?
13. When two-piece electrodes are used, what is the name of each piece?
14. The size of the pieces that can be welded by a resistance spot welding machine is limited by two dimensions that are called _____ _____ and _____ _____ .
15. What is the advantage of using a stored energy machine?
16. What type of machine is used for projection welding?
15. What is the advantage of using a stored energy machine?
16. What type of machine is used for projection welding?
17. What are the main parts of a resistance seam welding machine?
18. When seam welding, what is the advantage of using a knurl or friction drive, rather than a gear drive?
19. List two processes that use the following principle: Two metals are brought together and heated by passing electrical current through them. When they have been heated, a large force is applied to join the pieces.
20. If a leak in the hydraulic or pneumatic system is discovered, what should be done?

A rocker-arm type spot welding machine. Force is applied to the movable top electrode by a pneumatic cylinder. The lower electrode is stationary. (Berkeley-Davis, Inc.)

The gun-type resistance welding machine being used here has a long throat depth to reach to the back of the stainless steel hospital cabinet being spot-welded. (LORS Machinery, Inc.)

Chapter 19

RESISTANCE WELDING

LEARNING OBJECTIVES

After studying this chapter, you will be able to:

* Describe the variables involved in producing an acceptable resistance weld.
* Describe and be able to set or measure such items as tap setting, percent of heat, horn spacing, throat size, and contact size.
* Calculate the approximate contact tip diameter, current, weld size, electrode force, and weld time.
* Visually and destructively test a spot weld to determine its quality.
* Define the term "feedback control."

Resistance welding is based upon the fundamental principle that when an electrical current is passed through a metal, friction caused by *resistance* to this electrical flow heats the metal. The majority of the heat is developed where the two pieces of metal to be welded are in contact. By applying sufficient current, very high temperatures can result. When these temperatures reach the *fusion* (melting) temperature, welding occurs.

The term "resistance welding" includes a variety of welding applications. The resistance welding processes include spot welding, projection welding, seam welding, upset welding, and others. See Headings 4.14 to 4.17.

Resistance welding has a number of advantages. It is fast, there is very little warpage of the metal, and the process can be accurately controlled. This type of welding is particularly well-suited to all forms of automatic production activities.

19.1 PRINCIPLES OF RESISTANCE WELDING

When two pieces of metal are in contact, the area where they touch has a high resistance to the flow of electrical current. Due to surface roughness, the two or more pieces of metal to be welded cannot have their surfaces in perfect or complete contact. The roughness of the metals that form the contacting surfaces offer the highest resistance to current flow. When an electrical current is passed across the surfaces in contact, heat is generated. If enough current is used, for a long enough time, the metal surfaces heat until they become molten. If the two pieces are then pressed together while their surfaces are molten and allowed to cool, the pieces will fuse into one piece.

The welding current used in resistance welding machines can be either ac or dc. Direct current machines are best used when thick metal pieces are joined. Since there is no waveform as occurs with an ac current, dc has a steadier heat input to the weldment. Resistance welding is usually done on thinner sections of metal, so most resistance welding machines develop an ac welding current.

An alternating current (ac) resistance welding machine is basically an electric transformer. It is powered from an alternating current circuit. Resistance welding requires very high current (high amperage) at a relatively low voltage. This requires the resistance welding machine to have a step-down transformer. A *step-down transformer* decreases the voltage supplied and increases the current supplied. This is accomplished by having many turns or windings in the transformer primary. The secondary winding will ordinarily have only one turn. Sometimes two or three turns are used if the secondary leads are long, such as in portable resistance welding machines. Figure 19-1 illustrates an elementary electric circuit for a resistance welding machine.

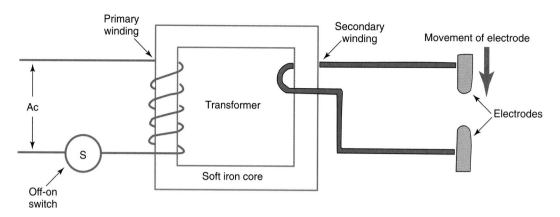

Figure 19-1. *A basic resistance spot welding machine electrical circuit. Note that the primary winding has many more turns than the secondary winding. This is a step-down transformer.*

19.1.1 Types of Electric Resistance Welding

There are several types of welding processes and equipment based on the resistance welding principle. Some of the more common types are:

- Spot welding (RSW).
- Gun welding (portable resistance welding).
- Projection welding (PW).
- Seam welding (RSEW).
- Upset welding (UW).
- Flash welding (FW).
- Butt joint seam welding.
- Metal fiber welding.
- High-frequency resistance welding.

All of these operations are fundamentally the same, but the preparation of the metal and the construction of the machine may be different.

19.2 RESISTANCE SPOT WELDING (RSW)

One way to join sheet metal parts is to drill holes and fasten the parts together with mechanical fasteners, such as rivets, machine screws, or sheet metal screws. Another way to join sheet metal parts is to spot weld them together. Although more than two pieces of metal can be spot welded together, two pieces are most commonly joined.

Spot welding is the most common resistance welding process. Refer to Heading 4.14. The process consists of lapping two pieces of metal and clamping them between two electrodes. A current is passed between the two electrodes. A small molten spot is formed. After the current stops, the electrodes continue to hold the metal, allowing the spot of molten metal to solidify. The two pieces of metal are now fused together by a *spot weld* or *weld nugget*. Figure 19-2 shows a diagram of spot welds. A cross section of a spot weld nugget is shown in Figure 19-3.

In resistance spot welding, there are five variables that must be controlled. These variables are:

- Time.
- Current.

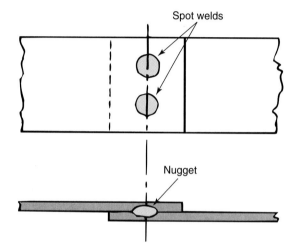

Figure 19-2. *Two pieces of metal, lapped and spot-welded in two places. Note the nugget where the fusion takes place between the two pieces.*

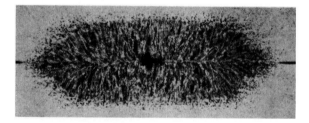

Figure 19-3. *A magnified cross section of a spot weld nugget. Notice the apparent void at the center of the nugget. (Taylor-Winfield Corp.)*

- Force.
- Electrode contact area.
- Machine selection.

The selection of the type of machine is covered in Heading 19.2.1. Refer also to Chapter 18. The proper settings for the first four variables are discussed in Heading 19.2.3.

The time required to make a spot weld is divided into three separate timing periods. These are:
- Squeeze time.
- Weld time.
- Hold time.

Squeeze time is the time required for the electrodes to close on the metal and apply the proper pressure. *Weld time* is the time that the current flows and heats the metal. *Hold time* is the period, after the current stops, when the pressure is still applied to allow the molten metal spot to solidify. These three times make up one *weld sequence* or a *welding schedule.*

One additional time, the off time, is sometimes used. *Off time* is the time between the end of one weld sequence and the beginning of the next.

Each of these times is set on the controller or control panel. The times are set in numbers of cycles or "hertz." One cycle or one *hertz (Hz)* is 1/60th of a second. Examples of resistance welding controllers are shown in Figures 18-10 and 18-13.

The current is an important variable. The amount of current used can be changed in one of two ways:
- Change the tap setting on the welding machine.
- Change the percent heat setting on the welding machine controller.

The *tap setting,* as shown in Figure 18-6, is used to make large changes in the current settings. A given tap setting has a range of currents. To vary the current within a given range, the *percent heat setting* is changed. The higher the percent heat setting, the larger the current (as measured in amperes). For example, a given tap setting may be able to supply between 4000A and 10,000A. If the percent heat setting is set at 75% heat, the current will be approximately 4000A + (0.75 × the current range). The current range in this case is: 10,000A – 4000A = 6000A. Therefore,

4000A + (0.75 × 6000A)

= 4000A + 4500A

= 8500A.

To operate a spot-welding machine, the operator must carefully and safely clean the electrodes, turn on the cooling water, and set the tap switch on the transformer to the correct current setting. The welding operator must also set the times and percent heat on the controller, and set the desired pressure or *force* on the electrodes.

To perform a weld or a series of welds, the welding operator places the pieces to be welded on the lower (stationary) electrode. The operator steps on the foot pedal or pushes the starting switch, the electrodes come together, and the weld is made.

If a series of welds is to be made, the operator moves the part to each new weld location, presses the foot pedal or pushes the start button, and makes the new weld. This process continues until the correct number of welds is made. When making a series of welds, the "off-time" may be set on the controller.

On electronically programmed machines, the weld schedule is completed automatically. At the end of the weld schedule, the two pieces are welded together and the electrodes separate. The part is removed and new pieces are loaded to begin a new weld schedule.

When using a completely manual spot welding machine, the welding operator controls the times of the weld schedule. The operator first pushes the foot pedal part way. The electrodes close on the metal to be welded and apply the proper force. This is the squeeze time. The operator then pushes the pedal a little bit further. This closes an electrical contact switch. The current flows and forms a molten nugget. This is the weld time. The operator then pushes the foot pedal all the way down. This releases the electrical contact, but pressure is still being applied. This is the hold time, when the molten metal becomes solid. Finally, the operator releases the pressure on the foot pedal. The electrodes separate and the welded metal is removed. A schematic of a manual resistance welding machine is shown in Figure 18-7.

Automatically controlled welding machines are shown in Figures 19-4 and 19-5. These machines have a pneumatic (air) force system. There is a valve to adjust the air pressure and a valve to turn the cooling water on. The machines illustrated in Figures 19-4 and 19-5 both have a foot pedal to control the weld operation.

Some resistance welding machines use two palm switches to start the welding cycle. Palm switches are

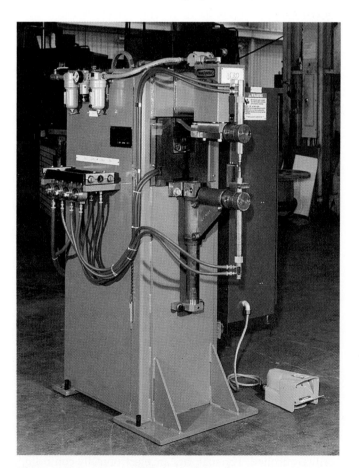

Figure 19-4. A spot welding machine that uses a pneumatic electrode force system. The controller is at the right of the welding machine. Note the foot-actuated start switch at lower right. At the left side of the welding machine, just above the hose connections, note the colored balls. The movement of these balls indicates that cooling water is flowing. (ACRO Automation Systems, Inc.)

Figure 19-5. *The operator of this spot welder is using a foot switch to begin the welding program. Note the amber-colored filter screen used to protect other workers from sparks. (LORS Machinery Inc.)*

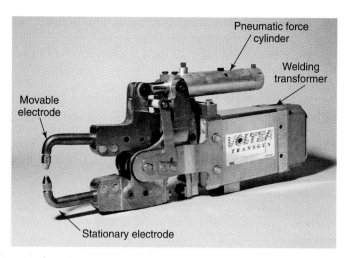

Figure 19-6. *A portable resistance welding machine. The transformer, pneumatic force cylinder, and movable and stationary arms are built as one unit. Such a unit is also known as a transgun. These units are often hung from an overhead support. A foot-operated switch is used to start the weld cycle. (Marc-L-Tec, Inc.)*

pushed by hand. Both palm switches must be pushed to start a weld. With this set-up, there is no danger of the welding operator getting a hand caught between the electrodes. The dual palm switches are used to satisfy Occupational Safety and Health Administration (OSHA) requirements.

19.2.1 Spot Welding Machines

There are many components and systems that make up a resistance spot welding machine. The three basic designs used are:
- Rocker arm machines (See Figure 18-8).
- Press-type machines (See Figure 18-9).
- Portable-type machines. The parts of a resistance welding machine are labeled in Figure 19-6.

Spot welding machines use different types of force systems. The different types of force systems in use are:
- Mechanical leverage.
- Pneumatic (air) pressure.
- Hydraulic (fluid) pressure.

Pneumatic types are the most common. These force systems are described in Heading 18.3.

Spot welding machines can be electronically or manually controlled. The majority are electronically controlled. Spot welding machines are also classified according to the type of electrical power the machine uses. There are three types:
- The single-phase machine.
- The three-phase machine.
- The stored energy machine.

Refer to Chapter 18 for additional information.

Two other welding machine variables are the KVA rating and the duty cycle rating. KVA means *kilovolt amperes* or thousands of watts. This is a measure of the power or heating potential of the welding machine. A 150 KVA machine would develop 150,000W of power or heat. Duty cycle is the percentage of time, in a one-minute period, that a welding machine can actually be used to perform welds without overheating. A 50% duty cycle would indicate that the welding machine could safely be used to make welds for 30 seconds in every one-minute period.

Each of these variables is important when identifying or selecting a spot welding machine. There is no machine that is best for all applications. Each machine has advantages and disadvantages. Select the best machine for a specific job. The most common spot welding machine is an automatically controlled, single-phase welding machine that uses a pneumatic force system. Press-type and portable-type machines are both very common. The KVA ratings vary from 10 KVA to 150 KVA. The most common duty cycle is 50%.

19.2.2 Portable Resistance Welding Machines

A *resistance welding machine* is portable and can be moved and positioned on the parts to be welded. Resistance welding machines are known in the shop as *gun welding machines* or *gun welders*. The resistance welding machine is moved to the parts to be welded instead of moving the parts to the welding machine. Resistance welding machines are popular in welding sheet metal fabrications, such as automobile bodies, washing machines, refrigerators, and microwave ovens. A pneumatic cylinder is normally used to apply force on the movable electrode. The timing and current are set on the weld controller. The pressure is set by adjusting an air pressure valve.

The welding operator selects the place to weld and places the nonmoving electrode against the metal. The operator then pushes the start switch (trigger) and the controller performs the weld schedule. When the weld is completed and the electrodes separate, the machine is moved to the next weld location.

19.2.3 Spot Welding Set-up

As noted earlier, there are five variables in resistance welding: time, current, electrode force, electrode contact area, and machine selection.

Machine selection involves choosing the best machine for the given application. The welding operator is often required to set up the weld schedule. The operator must know what type of material is being welded and the thickness of the metal. Refer to Heading 18.8.

When spot welding low-carbon steel, the equations in Figure 19-7 can be used to get an approximate value of the different variables. For example, if two sheets of low-carbon steel 0.060" (1.5mm) thick are to be welded, the approximate values for the various variables would be as follows:

- weld time: $120 \times (TT) =$ time in cycles
  $120 \times (0.120")$
  $= 14.4$ cycles (Hz) or
  $4.73\text{mm} \times 3.04\text{mm}$
  $= 14.38$ Hz
- required current (amperage): $125,000 \times (TT) =$ amperes (A)
  $125,000 \times 0.120" = 15,000\text{A}$ or
  $4935\text{mm} \times 3.04\text{mm}$
  $= 15,002\text{A}$

- electrode force: $8500 \times (TT) =$ force
  $8500 \times 0.120" = 1020$ lb. or
  $1500 \times 3.04\text{mm}$
  $= 4560$ newtons (N)
- contact tip diameter: $0.18" +$ (total metal thickness ["]) [TT]
  $0.18" + 0.120" = 0.300"$ or
  $4.54\text{mm} + 3.04\text{mm} = 7.58\text{mm}$

Weld size should be about 2/3 the diameter of the electrode tip: $0.300" \times 0.666 = 0.200"$ or $7.58\text{mm} \times 0.666 = 5.05\text{mm}$.

Refer to Figure 19-8 for a table of suggested values for spot welding variables. Several terms used in Figure 19-8 require an explanation. The letters used for the *electrode shape* refer to the shape of the tip. There are six electrode shapes identified by RWMA letter designations: **A** (pointed), **B** (dome), **C** (flat), **D** (offset), **E** (truncated), and **F** (radius). These electrode shapes may be referred to by name, or by letter designation. *Button diameter* is the diameter of the completed spot weld. *Minimum weld spacing* is the distance from the center of one spot weld to the center of the next. Weld spacing is an important factor to consider when making spot welds. If this spacing is too small, shunting may occur. *Shunting* takes place when some of the electricity intended for a spot weld travels through an adjacent spot weld that is too close. Spacing is different when two pieces (two-stack) or three pieces (three-stack) of metal are spot welded.

Additional information can be obtained from the Resistance Welder Manufacturers' Association (RWMA) *Resistance Welding Manual,* or from the American Welding Society (AWS).

Once the desired values are known, the welding operator must use the values to set the welding machine. Only

**Calculating approximate values
for resistance welding variables
for low-carbon steel up to 1/8" (3.2mm)**

Variable	Equation (U.S. conventional)	Units of measure	Equation (SI metric)	Units of measure
Contact tip diameter	0.18 + (total sheet thickness in inches)	inches	4.54 + (total sheet thickness in mm)	mm
Weld time	120 × (total sheet thickness in inches)	cycles	4.73 × (total sheet thickness in mm)	cycles
Current	125,000 × (total sheet thickness in inches)	amperes	4935 × (total sheet thickness in mm)	amperes
Electrode force	8500 × (total sheet thickness in inches)	lb.	1500 × (total sheet thickness in mm)	newtons
Weld size	The weld diameter should be approximately 2/3 the contact tip diameter.			

***Figure 19-7.** A table of equations used to calculate the approximate values of the variables used in resistance spot welding on low-carbon mild steel. The total metal thickness is the thickness of both pieces being welded.*

Single thickness in.	Electrode		Force, lb.	Weld time, (60Hz) cy	Welding current, (approx.) A	Minimum weld spacing		Minimum shear strength, lb.	Button dia., in.
	Face dia., in.	Shape				2 stack, in.	3 stack, in.		
0.020	0.188	E,A,B*	400	7	8,500	0.38	0.62	320	0.10
0.025	0.188	E,A,B	450	8	9,500	0.62	0.88	450	0.12
0.030	0.250	E,A,B	500	9	10,500	0.62	0.88	575	0.14
0.035	0.250	E,A,B	600	9	11,500	0.75	1.06	750	0.16
0.040	0.250	E,A,B	700	10	12,500	0.75	1.06	925	0.18
0.045	0.250	E,A,B	750	11	13,000	0.94	1.18	1150	0.19
0.050	0.312	E,A,B	800	12	13,500	0.94	1.18	1350	0.20
0.055	0.312	E,A,B	900	13	14,000	1.06	1.31	1680	0.21
0.060	0.312	E,A,B	1000	14	15,000	1.06	1.31	1850	0.23
0.070	0.312	E,A,B	1200	16	16,000	1.18	1.50	2300	0.25
0.080	0.312	E,A,B	1400	18	17,000	1.38	1.60	2700	0.26
0.090	0.375	E,A,B	1600	20	18,000	1.56	1.88	3450	0.27
0.105	0.375	E,A,B	1800	23	19,500	1.68	2.00	4150	0.28
0.120	0.375	E,A,B	2100	26	21,000	1.81	2.50	5000	0.30

* Shape Definitions: E = Truncated
A = Pointed
B = Dome (with 3" radius at tip)

Figure 19-8. *Suggested values for the variables used when resistance spot welding. The letters E, A, and B refer to the shape of the electrode recommended. See Figure 18-22 for electrode shapes.*

certain tip diameters are manufactured. Select a tip diameter that is close to the calculated value.

The welding time is set on the controller in cycles. The operator may select the time by adjusting a rotating knob or by pushing buttons on the controller. The buttons on a controller are used the same way that push buttons are used on a hand-held calculator. A push-button controller is shown in Figure 19-9. Three times must be set: the squeeze time, the weld time, and the hold time. The off time is optional.

Figure 19-9. *A controller that uses push buttons to set the time of the weld schedule and the percent heat. (Pertron Controls Corp.)*

To set the current requires prior knowledge about the particular welding machine, trial and error, or special equipment. A welding operator who is familiar with a welding machine will know what transformer tap setting and percent heat settings are required to weld two pieces of metal.

The ***trial and error method*** involves welding two pieces together, tearing them apart, and making an adjustment in the current setting. First, select a transformer tap setting. Use a higher setting for thicker material and a lower setting for thinner material. Set the percent heat control at 70%. This is a starting point to make a weld.

Examine the two pieces after they are torn apart. If the weld size (nugget) is too small, more current is required. Increase the transformer tap setting or the percent heat and make two more weld samples. Continue to make test samples and tear them apart until the correct size of weld is obtained.

When molten metal squirts out from a spot weld, it is called an ***expulsion weld.*** If an expulsion weld occurs when making test welds, the current is too high. In this case, decrease the tap setting or the percent heat. Weld two more pieces together and pull them apart. Again, make any necessary change in the current to increase or decrease the weld size. When the weld size equals the tip diameter, the proper current is being used.

A third way to set and adjust the current is to use a ***current analyzer***. A coil is placed around the electrode or electrode holder. The coil is connected to the current analyzer. When a weld is made, the current analyzer will display the current in ***kiloamperes*** (1 kiloampere = 1000A). The tap setting or percent heat is changed to increase or decrease the welding current. Two current analyzers and their pickup coils are shown in Figures 19-10 and 19-11

Some companies have experimented with a sound feedback system to determine when a spot weld has been properly made. A well-made spot weld will give off a particular sound while it is being made. A microphone is placed near the electrodes and the sound of each spot weld is analyzed electronically. If the wrong sound is picked up while a weld is being made, that weld is rejected.

The final variable to be set is the electrode force. The force is changed by adjusting a pressure valve. The valve adjusts the hydraulic or pneumatic (air) pressure that is applied to the electrodes. Some electronic controllers permit the pressure or force to be set by pushing buttons on the controller panel. To set the force, a *force gauge* is placed between the electrodes. The electrodes are forced to close over the force gauge. The gauge will read the force applied. Any change required in the force is made and another force reading is taken. This is repeated until the proper force is obtained. The use of a force gauge is shown in Figure 19-12. After setting all of the variables, the spot welding machine is properly set and should make good quality spot welds.

The machine may, however, make poor welds. The metal may be distorted, as shown in Figure 19-13, or it may have large indentations. This may be caused by improper electrode alignment.

Blow-holes and *cavities* can occur from improperly set controls, hold time that is too short, or improper positioning of parts with reference to the tips. See Figure 19-14.

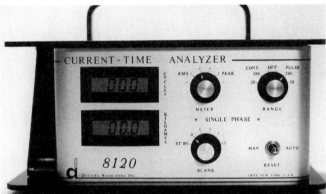

Figure 19-10. A pickup coil and a current analyzer are used to monitor the welding current and weld time. (Duffers Scientific)

Figure 19-11. A current analyzer or welding monitor. This unit displays the welding current and welding time on a digital monitor. Three current ranges—2K, 20K, 200K—may be selected. (British Federal)

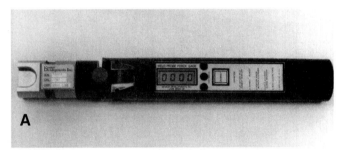

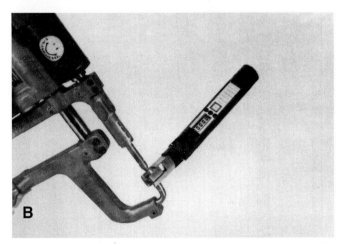

Figure 19-12. Force gauge. A—Gauge used to measure the force on the electrode tips. The measured force is read on digital readout. B—The gauge being used to measure the electrode tip force on a resistance welding machine. (Sensor Development, Inc.)

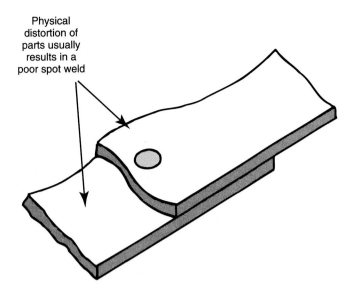

Figure 19-13. *Sheet metal warping or distortion will cause improperly made spot welds.*

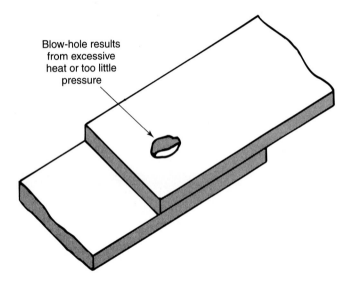

Figure 19-14. *A blow-hole in sheet metal, resulting from a faulty spot weld.*

Other physical conditions, such as dirt or burrs on the surface of the metal, can result in a quick, excessive concentration of welding current. If the heat developed is not absorbed or conducted to the surrounding metal, a blow-hole may form.

Tips that have parallel faces, but whose centers are not aligned, have a reduced effective tip area. See Figure 19-15A. When the welding tip surfaces are not parallel, pressure and current are confined to a fraction of the proper area, as shown in Figure 19-15B. When the tips are misaligned, the pressure (force/area) on the metal between the tips is excessive. The metal surface may be deeply indented by this excessive pressure. The electrodes must be realigned. With the electrodes properly aligned, the welding machine may be used to make consistent, high-quality spot welds.

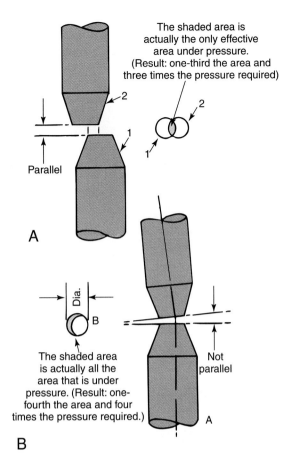

Figure 19-15. *Effects of using electrodes that are misaligned. A—These electrodes are off-center. B—These electrodes are aligned, but are not parallel. In either case, the pressure on the contact area is excessive.*

19.2.4 Proper Resistance Welding Practices

The welding machine and the controller must be properly set before any welding is done. The set-up of the resistance welding machine will be described in Heading 19.6.

Material to be welded is placed between the electrodes. Each welding process is initiated by pressing a foot pedal or a palm switch. Squeeze time, weld time, and hold time are preset and provided by the welding controller. After the pieces are welded together, the electrodes separate.

After welding, the metal is very hot and should be handled with care. Gloves should be worn to protect the operator from burns. Safety glasses must be worn because an occasional expulsion weld throws molten metal in every direction.

Some maintenance is required during spot welding, due to the heat developed. This heat is necessary to form the weld; however, the heat also damages the electrodes. Electrodes typically can be used to make 2000 or more spot welds on mild steel. After this, the electrode begins to flatten out, increasing the electrode contact area. To keep the contact area constant, the electrode must be reshaped. Reshaping is also called ***dressing*** the electrode. Figure 19-16 shows hand and power tools used to dress electrodes. These tools have cutters that clean and shape the electrode tip.

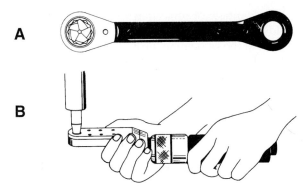

Figure 19-16. *Electrode cleaning and dressing tools. A— Manual dressing tool. B—Power dressing tool. (CMW, Inc.)*

If the electrodes or electrode tips cannot be dressed, they must be replaced. A tool for removing tapered electrode tips is shown in Figure 19-17. Some electrode holders have built-in electrode ejectors. See Figure 18-25. A rubber or lead hammer is used to place a new electrode tip in the tapered shank of the electrode holder. A steel hammer should not be used, because it could damage the electrode tip or the taper.

Some powered electrode dressers are used, as shown in Figure 19-18. When resistance welding is done using a robot, a tip dressing station is often added to the equipment in the robot cell. The robot is programmed so that the tips are automatically dressed after a preset number of welds have been made.

19.2.5 Spot Welding Aluminum and Stainless

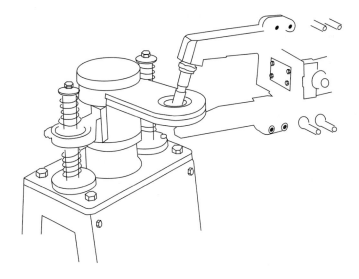

Figure 19-18. *A powered tip dressing tool. This tip dresser is moved to the spot welding machine and adjusted to properly dress the electrodes. (Hercules Welding Products)*

Steel

The spot welding of aluminum and stainless steel differs from the welding of mild steel. Spot welding aluminum requires more current for shorter times than mild steel. Aluminum and its alloys have a tendency to crack while they are cooling. To prevent cracking, additional pressure is applied. This increase in pressure is called a *forge force*. Only certain pressure systems are capable of applying this forge force. A weld schedule used to weld aluminum can be seen in Figure 19-19.

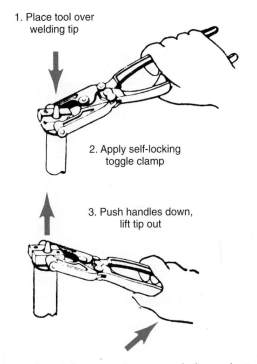

1. Place tool over welding tip

2. Apply self-locking toggle clamp

3. Push handles down, lift tip out

Figure 19-17. *A tool for removing tapered electrode tips.*

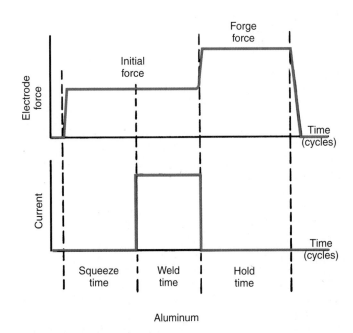

Figure 19-19. *Weld schedule used for aluminum. The forge force is applied during the hold time.*

Stainless steel is also welded faster than mild steel. Stainless steel cools very rapidly and forms martensite. Martensite is very hard and brittle. To improve the properties of the martensite, the finished weld is tempered. Tempering involves heating the metal to a prescribed temperature and then allowing it to cool slowly. (See Heading 29.6 for more information on martensite and tempering.) In spot welding, tempering is performed by passing additional current through the spot weld after it has cooled. This tempering heat lasts for 3-6 cycles. A complete weld schedule used to weld stainless steel is shown in Figure 19-20.

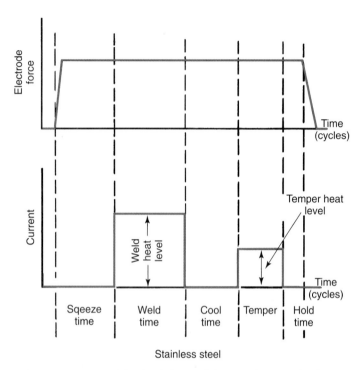

Figure 19-20. *Weld schedule used for stainless steel. The stainless steel weld is tempered by applying additional heat after the cool time.*

19.3 PROJECTION WELDING (PW)

Projection welding requires the forming of projections or bumps on one of the pieces of metal to be joined. The projections are accurately formed in precise locations by a special set of dies. After the projections are formed, the raised portions on one piece of metal are pressed into contact with another piece. Figure 19-21 shows a projection weld set-up. Current is passed through the two pieces. The current is concentrated at the projections. These points heat and form a spot weld. Refer to Heading 18.9.

An advantage of projection welding is that it locates the welds at certain desired points. Also, several spots may be welded at the same time. Projection welding electrodes are often in the shape of a flat plate called a *platen.* See Figure 19-22. Specialty electrodes or platens must be designed to adequately contact all the projections on the

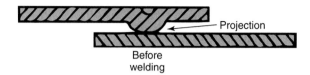

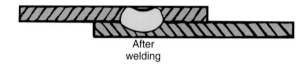

Figure 19-21. *A schematic drawing of metal before and after projection welding. Some of the depression formed when the projection is stamped remains on the top surface of the finished weld.*

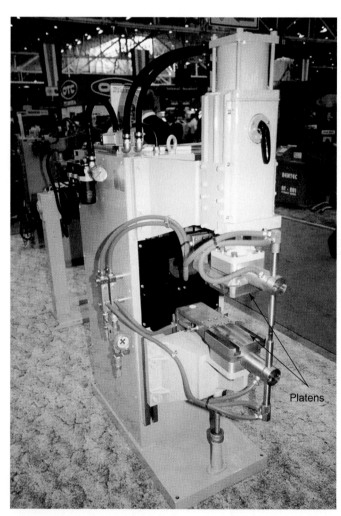

Figure 19-22. *Projection welding platens are installed on the stationary and movable arms of this welding machine. To use the platens, the spot welding electrodes are removed.*

punched parts, such as the nuts and bolts shown in Figure 19-23. These special electrodes may be inserted into custom-designed holders and attached to the platens on the resistance welding machine. The cost of the tooling, required to fabricate the special electrodes and their holders, makes this welding method practical only when high production is planned.

Macrographs of a projection weld in various stages of completion are shown in Figure 19-24. Each successive photograph illustrates the effect of the number of cycles of the electrical current flowing through the weld.

Figure 19-23. A variety of nuts, bolts, and pins with projections stamped into them so they can be projection-welded. Note the circular projection on the bolt in the upper left corner. (The Ohio Nut and Bolt Co.)

19.4 RESISTANCE SEAM WELDING (RSEW)

Seam welding produces a continuous or intermittent seam. Refer to Heading 4.17. The weld is usually near the edge of two overlapped metals. Two wheel electrodes travel over the metal, and current passes between them. Refer to Heading 18.11. The current heats the two pieces of metal to the fusion point. See Figure 19-25.

The current can be continuous or intermittent. *Continuous current* produces a continuous seam. *Intermittent current* produces spots that can be timed to overlap. Continuous or overlapping spots can produce a leakproof lap joint for use with gases and liquids. Examples of continuous and overlapping seam welds are shown in Figure 19-26.

Another form of seam welding is called *butt seam welding*. It is used to weld a longitudinal seam in pipe. Electric current heats the edges of the metal to a molten condition just before it passes between a set of rollers. The two rollers press the pipe edges together, and the metal is welded. The rollers also hold the welded joint in position while it cools to a safe temperature. This resistance welding method uses high-frequency current to heat the metal. Figure 19-27 shows a longitudinal seam weld being made on a pipe.

A seam welding machine must have a higher duty cycle than a spot welding machine. A machine that produces a continuous or overlapping spot weld seam must be rated at a 100% duty cycle.

19.5 UPSET WELDING (UW)

Upset welding involves clamping two pieces of metal that are to be welded in two separate clamps. One clamp is stationary; the other is movable. Metal to be upset-welded must be very clean and must be properly aligned. The clamps and the electrical wiring of an upset welding machine are shown in Figure 19-28. Refer to Heading 18.12.

In upset welding, the two pieces are forced together, and a large current is passed from one clamp to the other. The metal between the two clamps is heated. A majority of the heat is concentrated at the place where the two metal pieces touch. The ends of the metal are heated to the fusion (melting) temperature. After reaching the fusion temperature, the two pieces are pressed together (upset). This upsetting occurs while the current is flowing and after the current is turned off. The metals join together and, upon cooling, become one piece. After a short cooling period, the clamps are released and the part is removed from the welder. Figure 19-29 shows the final upset weld.

19.6 FLASH WELDING (FW)

Flash welding is a resistance welding process generally used to weld the butt ends of two pieces. Refer to Heading 4.16. The end surfaces of the two pieces are not cleaned or machined as in upset welding. The parts are held in two clamps. One clamp is stationary; the other is movable. See Figure 19-30.

When the weld process is started, the butt ends of the two pieces are brought into a light contact. Heat for welding is created when a high current is passed from one piece to the other. The high resistance of the uneven contacting surfaces causes these areas to heat. Because the surfaces are rough and uneven, small arcs occur across the contact surfaces. This action is called *flashing*. These small arcs also heat the contacting surfaces. During the arcing period, the parts are slowly and continuously brought together under a light pressure. When the surfaces are molten, the current is stopped. At this time, a very high pressure is applied, usually by an air or hydraulic cylinder. A motor-driven cam is sometimes used to apply the pressure. The parts are squeezed together, forcing some of the molten, unclean metal on the contacting surfaces outward. Figure 19-31 shows two pieces of metal before and after they have been flash-welded. The parts fuse together, and are held in position for a short time to allow them to cool. After the weld is completed, the clamps are released and the parts removed. Figure 19-32 shows a flash welder in operation.

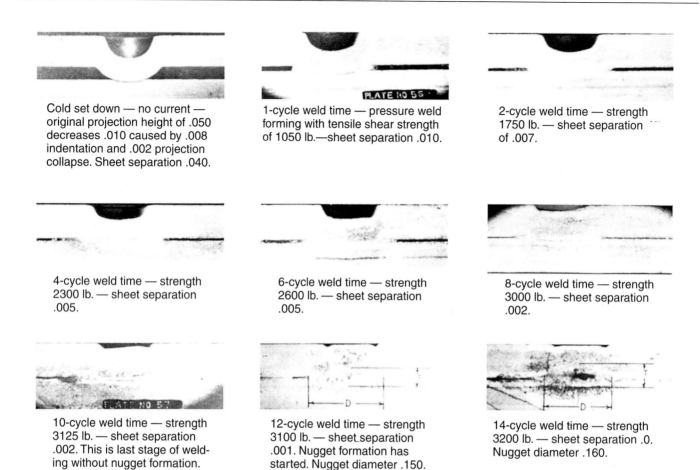

Cold set down — no current — original projection height of .050 decreases .010 caused by .008 indentation and .002 projection collapse. Sheet separation .040.

1-cycle weld time — pressure weld forming with tensile shear strength of 1050 lb.—sheet separation .010.

2-cycle weld time — strength 1750 lb. — sheet separation of .007.

4-cycle weld time — strength 2300 lb. — sheet separation .005.

6-cycle weld time — strength 2600 lb. — sheet separation .005.

8-cycle weld time — strength 3000 lb. — sheet separation .002.

10-cycle weld time — strength 3125 lb. — sheet separation .002. This is last stage of welding without nugget formation.

12-cycle weld time — strength 3100 lb. — sheet separation .001. Nugget formation has started. Nugget diameter .150.

14-cycle weld time — strength 3200 lb. — sheet separation .0. Nugget diameter .160.

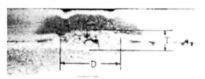

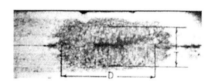

16-cycle weld time — strength 3700 lbs. — sheet separation .0. Nugget diameter .160. This is recommended weld time.

20-cycle weld time — strength 4000 lbs. — Nugget diameter .300. This last stage indicates increase in nugget size with increased weld time.

Figure 19-24. *Samples of projection welds as the weld progresses from 0 cycles to 20 cycles weld time. (1 cycle = 1/60 second) (Taylor-Winfield Corp.)*

Flash welding has two advantages over upset welding: flash welding is faster, and the metal surfaces do not have to be cleaned before welding.

19.7 PERCUSSION WELDING (PEW)

Percussion resistance welding uses an arc to heat the metals to be welded over their entire surface area. The arc is a short pulse of electrical energy.

The power supplies for percussion welding are of two types:
• Capacitor discharge.
• Magnetic force, which uses a transformer.

The parts to be welded are held apart by a small projection, or one part is moved toward the other. The energy in the capacitors is discharged and an arc heats the parts. Pressure is applied percussively (very rapidly) during or after the current flow, thus the name percussion welding.

This process can join parts of equal or unequal cross section. Examples include joining wires end to end, and welding wires to flat surfaces. This process is used in the electronics industry.

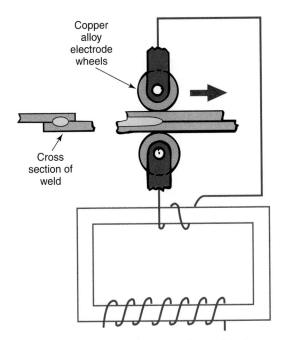

Figure 19-25. *A schematic drawing of metal being seam-welded.*

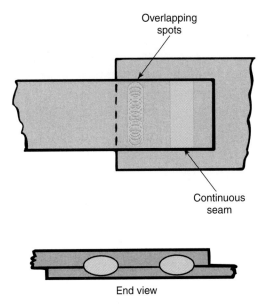

Figure 19-26. *This sketch shows metal that has been seam-welded. A seam weld is made with wheel-type electrodes. If the current to the electrode is turned off and on, a series of overlapping spots form the seam. An uninterrupted flow of current to the electrodes will form a continuous seam.*

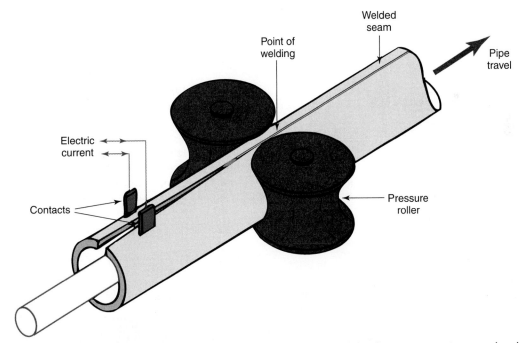

Figure 19-27. *High-frequency resistance welding used to butt-weld the longitudinal seam on a pipe or tube during production.*

19.8 BUTT JOINT SEAM WELDING

Butt joint seam welding is a method of joining metal, as shown in Figure 19-33. Sheet steel may be butt-welded with this process. The sheets to be welded are placed between wheel electrodes, and the joint is covered both top and bottom with a thin metal foil approximately 0.010″ (0.254mm) thick. The foil is the same material as the metal being joined.

The foil tends to concentrate the welding current in the immediate area of the weld joint. The weld has a slightly raised bead as compared to a slight indentation from the usual resistance welding operation. Figure 19-34 shows the weld in various stages of its formation.

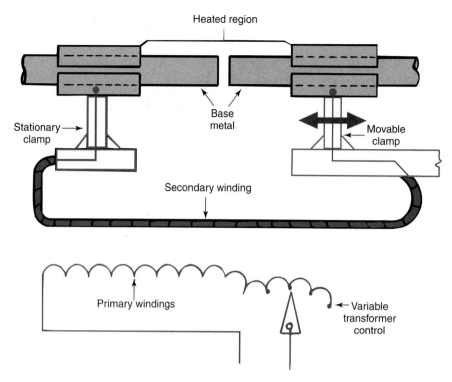

Figure 19-28. A diagram of an upset welding machine, showing the transformer primary and secondary and the clamping mechanism. Only the region between the clamps is heated.

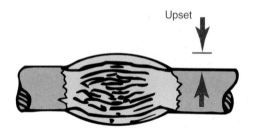

Figure 19-29. A completed upset weld. Note that the diameter of the rod is increased (upset) in the area of the joint.

19.9 METAL FIBER WELDING

A ***metal fiber weld*** is usually a lap joint resistance weld. Metal fiber sheets are formed by ***felting*** very tiny filaments of the metal to be used. The sheets are like a thin sheet of felt cloth. A strip of this felted fiber is then placed between the two pieces of metal to be joined. Figure 19-35 shows a cross section of the metal fiber and welding electrodes prepared to make a metal fiber resistance weld.

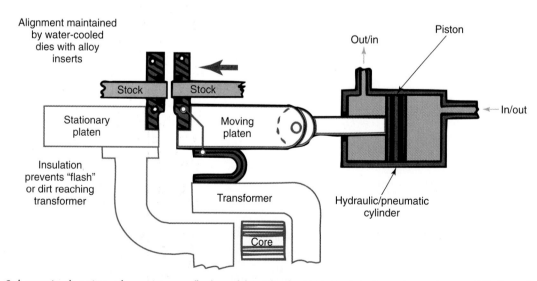

Figure 19-30. Schematic drawing of a resistance flash welder. The heated stock is forced together rapidly by a hydraulic or pneumatic (air) cylinder.

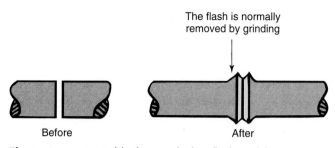

The flash is normally removed by grinding

Before

After

Figure 19-31. *Metal before and after flash-welding. The enlarged area is known as "flash."*

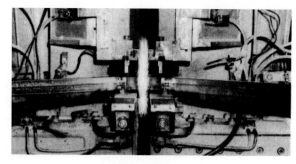

Figure 19-32. *A flash welder in operation. (Airmatic-Allied, Inc.)*

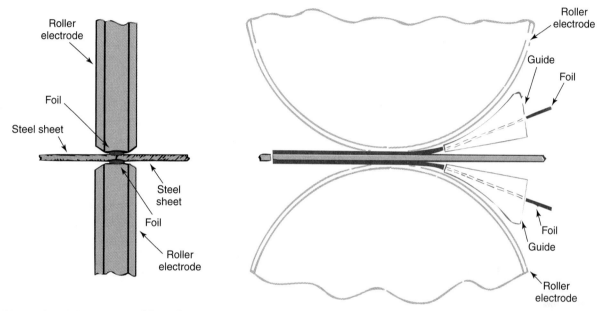

Roller electrode

Foil

Steel sheet

Steel sheet

Foil

Roller electrode

Roller electrode

Guide

Foil

Foil

Guide

Roller electrode

Figure 19-33. *Butt joint seam welding sheet metal, using metal foil. The foil serves to concentrate the weld current and to add reinforcement.*

Metal fiber welding is applicable to a variety of resistance welding jobs. The metal fiber used may be impregnated with various metals such as copper, brazing metals, silver, or an alloy. Such fibers may be used to join two different metals. Copper may be joined to stainless steel. Other similar joints also may be made using metal fiber welding.

When resistance welding with metal fiber, less electrode pressure is needed. The fiber offers a greater resistance to the flow of current than the solid metal, so it heats more quickly and to a higher temperature than the solid metal. The high temperatures at the surfaces of the metal to be joined assist in forming a good resistance weld. Also, since a lower electrode pressure is required, less indentation of the metal will be made.

19.10 HIGH-FREQUENCY RESISTANCE WELDING

When extremely high-frequency current flows through a conductor, it flows at or near the surface of the metal and not through the entire thickness. This phenomenon of high-frequency electricity has been used to weld sections as thin as 0.004" (0.102mm).

Low-amperage current with frequencies as high as 450,000 Hz (cycles per second) are used. This high frequency heats the surface of the metal. When pressure is applied to force the metal parts together, an excellent fusion weld occurs at the surfaces of the parts. This process is used to weld longitudinal seams in pipe, as shown in Figure 19-27.

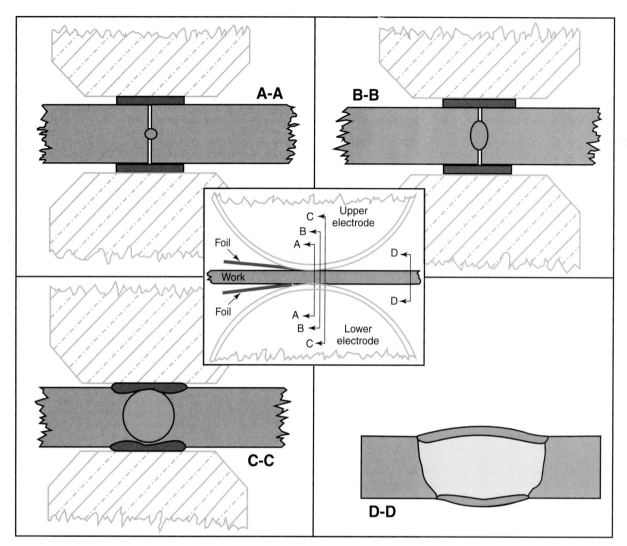

Figure 19-34. *Fusion action during the butt joint seam welding process. A–A—Start of weld. B–B—Nugget is approximately one-half developed. C–C—Weld approximately complete. D–D—This is a sample cross section of a completed weld.*

Copper has been welded to steel, and alloy steel to mild steel, by this method. Exotic metals can also be welded in an inert gas atmosphere.

The electronic equipment needed for this high-frequency resistance method of welding is complicated. Special training is needed to become a competent technician for maintaining, adjusting, and servicing these machines.

19.11 FEEDBACK CONTROL

To improve the quality of spot welds, feedback control and monitoring systems are used. A monitoring system measures and displays the welding variables. The welding operator must interpret the results and manually make changes to obtain the desired weld quality. A feedback controller also monitors the welding variables. The feedback controller changes the welding variables electronically to keep a constant weld quality.

The variables that can be monitored are the voltage, current, and acoustic emission. Using the voltage and current, the resistance of the weld can be calculated as $R = V/I$. (R = resistance, V = volts, and I = current).

Another method of examining the resistance welding process is to monitor the acoustic (sound) emission during the weld. *Acoustic emission* is sound that is emitted from the metal during welding. The sound is changed into an electrical signal. Figure 19-36 shows some of the typical sounds given off during a resistance spot weld.

The information from the acoustic emission and other variables is used to maintain a desired spot weld quality. The weld can be stopped if expulsion occurs. A feedback controller that monitors acoustic emissions is shown in Figure 19-37.

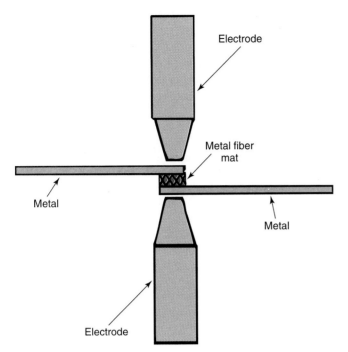

Figure 19-35. *The set-up for a metal fiber resistance weld. The metal fiber mat is placed between the base metals before they are welded.*

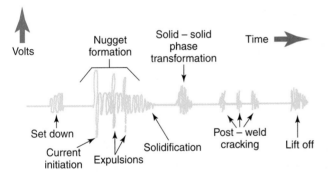

Figure 19-36. *Typical acoustic emissions (sounds) during a resistance spot weld. Not all of these signals are present in every weld. (Physical Acoustics Corp.)*

19.12 REVIEW OF SAFETY IN RESISTANCE WELDING

Resistance welding and resistance welding equipment, if properly handled, are very safe. The greatest dangers come from the following:

• Electrical shock and other electrical hazards.
• Cuts from sharp metal edges.
• Burns from hot metal and from flying sparks or molten metal.
• Crushing injuries caused by the high forces exerted by the electrodes.

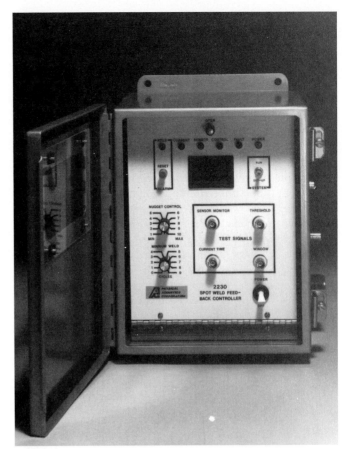

Figure 19-37. *A spot weld feedback controller used to maintain good quality spot welds by monitoring acoustic emissions. (Physical Acoustics Corp.)*

It is recommended that all operators of resistance welding equipment wear face shields or flash goggles. There may be some flying sparks or "flash" thrown from the joint being welded. Protective clothing is necessary.

The welding voltage across the electrodes is very low, but the secondary current is very high. The primary circuit wiring should be handled only by qualified electricians. All resistance welding equipment should be grounded.

Naturally, any surface being welded will be very hot. You should always wear protective gloves to handle the materials being welded. Gloves will also protect your hands against sharp or ragged sheet metal edges.

Most modern resistance welding machines have pneumatically operated electrodes. The electrode forces may be quite high. There is always danger of injury if the operator's hand or fingers should accidentally be caught between the electrodes as they come together on the work.

Safety devices should always be used to ensure that no part of your body can be in a danger area at the time a machine starts to operate.

TEST YOUR KNOWLEDGE

Write your answers on a separate sheet of paper. Do not write in this book. For all calculation questions, show your work for credit.

1. What is the basic principle of all resistance welding?
2. What does a step-down transformer do?
3. Name five variables used in resistance spot welding.
4. A welding schedule is composed of three different times. Name them in order.
5. Name two ways to change the welding current.
6. What is the difference between hold time and squeeze time?
7. Give another word for *pneumatic*.
8. Resistance welding machines are classified by the type of electrical power they use. What are the three types?
9. Calculate the approximate current required to resistance spot weld two pieces of 0.060" (1.52mm) mild steel.
10. Calculate the approximate weld time and electrode force required to resistance spot weld a 0.040" (1.02mm) piece to a 0.090" (2.29mm) piece.
11. When molten metal squirts out from a spot weld, it is called an _____ _____ .
12. How is the electrode force adjusted?
13. What does "dressing an electrode" mean?
14. What is a forge force? Tempering?
15. Name two advantages of projection welding.
16. Why does flashing occur during flash welding?
17. What are two advantages of flash welding over upset welding?
18. Name the two types of seams that can be welded by resistance seam welding.
19. What variables can be monitored by a feedback control device?
20. What are the dangers associated with resistance welding? How can injury be prevented?
21. Calculate the approximate electrode contact tip diameter required to spot weld a piece of 1/16" (1.58mm) steel to a piece of Steel Sheet Manufacturers' Standard #12-gage steel. Show your work for credit.
22. Calculate the approximate electrode contact tip diameter needed to weld the two pieces mentioned in Question 9. Show your work for credit.
23. Calculate the weld time for the pieces in Question 21.
24. Calculate the weld time for the pieces in Question 9.
25. Calculate the electrode force for the weld in Question 21.

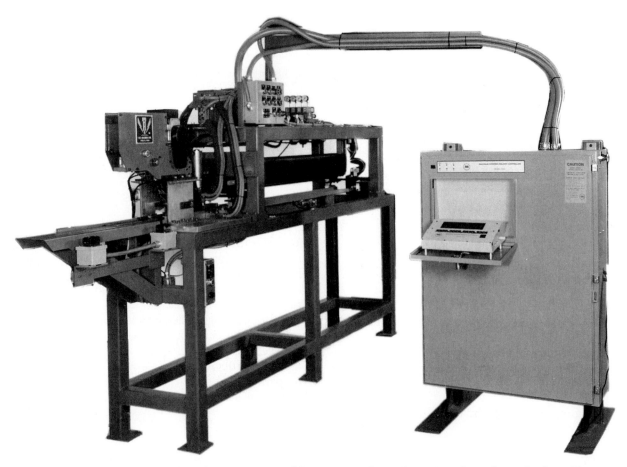

An automatic resistance welder used to do cross-wire welding on wire forms for point-of-purchase displays. The computerized control unit, at right, controls motion, welding cycles, and electrode force. (LORS Machinery, Inc.)

This electronic force gauge will measure the amount of "squeeze" being exerted by resistance welding electrodes. When activated, the hand-held display unit shows force as an easily visible bar against a numeric scale. (Medar, Inc.)

Part 7

SPECIAL PROCESSES

Stud welding is being used to attach bolts to the bottom of a large base to mount swivel-type wheels. (TRW Nelson Stud Welding Division)

This huge friction welder exerts up to 45 tons pressure on the workpieces as they are being rotated against each other. Note the video screen used for monitoring the process. (Manufacturing Technology, Inc.)

Chapter 20

SPECIAL WELDING PROCESSES

LEARNING OBJECTIVES

After studying this chapter, you will be able to:
* Describe the various special welding processes.
* Identify and specify these special welding processes, using the AWS abbreviations.
* Identify equipment that is used with many of the special welding processes.

The general classifications of welding includes various unusual or special methods that have been developed for fusing metals together. Some of these methods are highly specialized. Some are patented, and may only be used with the permission of the patent owner. Other welding methods have been developed as special adaptations of the standard oxyfuel gas, arc, or resistance processes.

20.1 SPECIAL WELDING PROCESSES

Three of the main American Welding Society classifications of welding processes are *oxyfuel gas welding, arc welding,* and *resistance welding*. Each of these three main classifications of welding has, within it, several related processes. See Figure 4-39. However, any welding method that can produce the intermingling of molecules (cohesion) is considered a welding process. Many welding processes have been developed. Some of the processes are closely related and are listed under the three main classifications.

Two additional classifications contain the rest of the welding processes. These two categories are *solid-state welding (SSW)* and *other welding*. Many of the welding processes that are considered special in this chapter are used regularly in industry. Some of the welding processes in the *arc, solid-state, other,* and *oxyfuel gas* classifications that may be considered "special" are listed below and explained within this chapter.

ARC RELATED (AW) PROCESSES
* Submerged Arc Welding (SAW)
* Electrogas Welding (EGW)
* Narrow Groove Welding (not a process separately recognized by AWS)
* Arc Stud Welding (SW)
* Arc Spot Welding (not a process separately recognized by AWS)
* Plasma Arc Welding (PAW)
* Underwater Shielded Metal Arc Welding (a special application of SMAW)

SOLID-STATE WELDING (SSW) PROCESSES
* Cold Welding (CW)
* Explosion Welding (EXW)
* Forge Welding (FOW)
* Friction Welding (FRW)
* Hot Pressure Welding (HPW)
* Ultrasonic Welding (USW)

OTHER WELDING PROCESSES
* Electroslag Welding (ESW)
* Thermite Welding (TW)
* Electron Beam Welding (EBW)
* Laser Beam Welding (LBW)

OXYFUEL GAS WELDING (OFW)
* Self-Generating Oxyhydrogen Welding (an application of OHW)
* Solid Pellet Oxyfuel Gas Welding (an application of OFW)

20.2 SUBMERGED ARC WELDING (SAW)

Submerged arc welding (SAW) is a welding process that has grown rapidly in popularity. It has some outstanding advantages. It is very fast. There is no visible arc, no splatter, and the welds are of high quality. Refer to Heading 4.18 and to Figure 4-23.

The SAW method involves striking an arc between a consumable electrode and the joint while the arc is buried in a granular flux (such as titanium oxide silicate). See Figure 20-1.

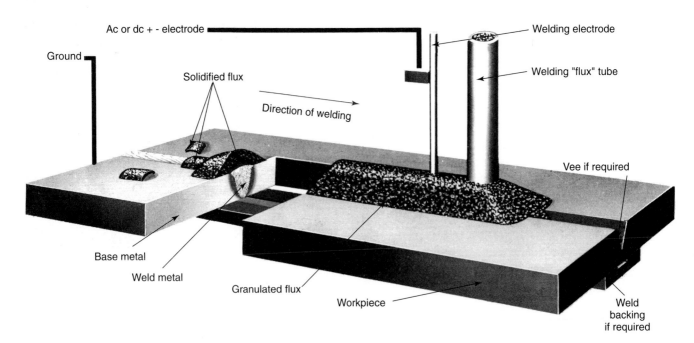

Figure 20-1. *Diagrammatic view of a submerged arc weld in progress. (ESAB Welding and Cutting Products)*

Some submerged arc machines are able to produce single-pass welds on butt joints up to 3" (75mm) in thickness, plug welds up to 1 1/2" (40mm) in thickness, and fillet welds with up to 3/8" (10mm) legs.

Common electrode sizes used with submerged arc welding range from 5/64" to 7/32" (2mm to 5.5mm). Typical currents are about 1000A for automatic submerged arc welding with a single electrode and a dc power supply. Higher currents are required when more than one electrode is used. Submerged arc welding may be done using either an ac or dc power source. Usually, a constant voltage (cv) power source is used. When using large-diameter electrodes, a constant current (cc) power source is preferred. The typical submerged arc welding machine, as shown in Figure 20-2, has a power-operated carriage. A universal variable speed motor controls the travel speed from 7" to 210" per minute (3mm/sec to 89mm/sec). A hopper feeds the granular flux to the joint just ahead of the arc. The heat generated by the arc melts the adjacent flux granules. As the flux solidifies, it covers the weld with an airtight slag that protects the weld until it cools. This slag may be readily removed. The unmelted flux granules can be used again. Special equipment in some applications picks up the unused flux and feeds it back into the hopper. See Figure 20-3.

Since the arc is submerged and therefore not visible to the operator, the correct setting is indicated by an ammeter reading and a voltmeter reading.

Submerged arc welding is excellent for production jobs like ship and bridge decks that require long, thick weld seams. The carriage may be used on a standard track or a template track, or the metal being welded may be mounted on a carriage and moved under the welding head.

Consumable electrodes are furnished in large coils. Variable speed electric motors control the electrode feed

Figure 20-2. *A submerged arc welding outfit. Automated flux recovery equipment is in use. A butt weld is being made on a cylindrical shape that is mounted below the welding fixture. (Invincible Airflow Systems)*

speed. The controls for the process are mounted on the welding outfit. See Figures 20-2 and 20-3.

Submerged arc welding may be done using a semiautomatic, automatic, or mechanized welding equipment. In each case, the welding operator must set the wire feed speed, travel speed along the joint, the voltage or amperage (depending on the type of power source used), and the

Figure 20-3. *A submerged arc weld in progress. Note the hopper and the tubes for flux distribution and pick-up. (Invincible Airflow Systems)*

Figure 20-4. *Semi-automatic submerged arc welding. The process is similar to GMAW, but the arc is beneath a layer of flux. (Hobart Brothers Co.)*

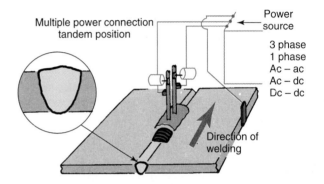

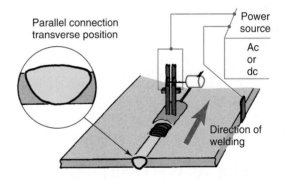

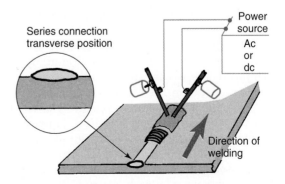

Figure 20-5. *Three different ways to use multiple electrodes in submerged arc welding.*

electrode extension. These are the same variables used when gas metal arc welding. When doing automatic or mechanized submerged arc welding, the operator must initially align the metal to be welded with the electrode and welding head.

When this process is done with mechanized equipment, the welding gun is normally driven along the seam by an electric motor. The operator must guide the welding head and keep it in line with the weld seam. Semi-automatic welding is being done in Figure 20-4. Note the small hopper for granular flux built into the electrode holder. A combination electrode and wire feed cable is attached to the electrode holder. The operator guides the electrode holder and is able to make fast welds on either straight or irregular joints.

Submerged arc welding may be done with more than one electrode. The use of more than one electrode may be desirable when the finished weld must be wide, or when higher welding speeds are desired. See Figure 20-5 for schematic drawings of three different methods of multiple-electrode welding.

The most common use is the single electrode. Two and three electrodes are usually used in a tandem position, as shown in Figure 20-5. Large welds can be completed quickly using this method. Figure 20-6 shows a comparison of the metal deposited by single and by multiple electrodes.

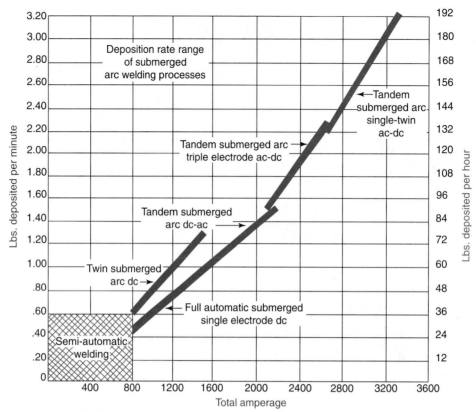

Figure 20-6. *A comparison of weld deposition rates for single and multiple electrode automatic submerged arc welding. Note that the automatic welding process deposits more filler metal than the semiautomatic processes.*

Submerged arc welding is usually done in the flat welding position. This is because there is a large amount of molten metal and molten slag. SAW can also be used to make horizontal fillet welds. Submerged arc welding is not normally used to weld other types of joints horizontally, vertically, or overhead. The flux and molten metal would not stay in place.

In submerged arc welding, the flux remaining over the completed weld acts as a heat insulator. As a result, the weld remains very hot for some time after it is completed. **To avoid burns or fire hazards, be sure to follow all safety precautions.**

20.3 ELECTROGAS WELDING (EGW)

The *electrogas welding (EGW)* process is an arc welding process. It was developed theoretically to weld sections of metal of virtually any thickness. Currently, welds are made on thicknesses ranging from 3/8" (10mm) up to about 4" (102mm). These welds can be made in one pass without an edge preparation. See Figure 20-7.

The EGW process is used almost exclusively to weld butt joints in the vertical welding position. However, T-joints, V-groove, and double V-groove joints are also welded. The equipment required to produce an electrogas weld is similar to that used for GMAW or FCAW. It includes:

- A dc power source for each electrode used.
- One or more solid or flux cored electrodes and electrode guide tubes.
- One or more wire feeders.
- Water-cooled retaining shoes.
- Shielding gas (if a solid electrode is used).

Solid electrodes may be used with a shielding gas fed into the joint area. Self-shielding flux cored electrodes also may be used. If self-shielding electrodes are used, no additional shielding gas is required.

Water-cooled retaining shoes are mounted on each side of the weld joint and move upward as the weld proceeds from the bottom to the top of the joint. Depending on the thickness of the joint, one or more flux cored electrodes will be fed into it. Standard wire drive units are used to move the electrode through a wire guide down to the weld surface.

The electrogas welding process is started by initiating the arc on a starter block that is tacked to the bottom of the joint. The cooling water and shielding gas (when used) begin to flow as the arc is struck. If a preflow time is set into the machine, the water and shielding gas (when used) will begin to flow *before* the arc is struck. To ensure an equal distribution of heat from the arc(s), the electrodes may be **oscillated** (moved from side to side) within the joint by oscillating equipment.

The shielding gas used with solid electrodes may be carbon dioxide (CO_2) or an argon-carbon dioxide (Ar-CO_2) mixture. Electrode diameters of 1/16", 5/64", 3/32", or 1/8" (1.6mm, 2.0mm, 2.4mm, or 3.2mm) may be used. A consumable or nonconsumable electrode guide may be used. The consumable guide will add 5% to 10% of the filler metal to the weld.

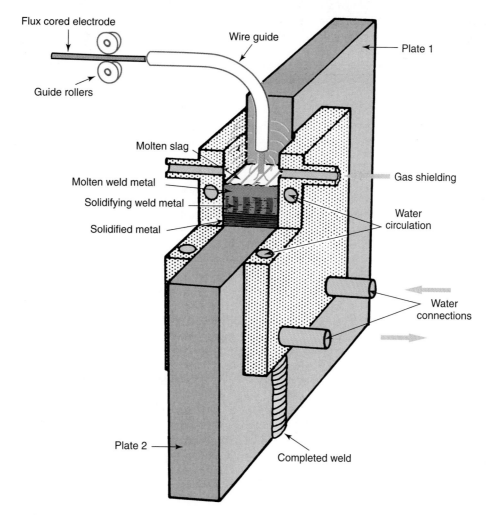

Flux cored electrode

Guide rollers

Wire guide

Plate 1

Molten slag

Molten weld metal

Solidifying weld metal

Solidified metal

Gas shielding

Water circulation

Water connections

Plate 2

Completed weld

Figure 20-7. *Schematic drawing of an electrogas weld in progress. The shoes are water-cooled and are moved up as the weld proceeds. A shielding gas or flux cored electrode is used to protect the molten metal from oxidation.*

20.4 NARROW GROOVE WELDING

Narrow groove welding (also called *narrow gap welding*) was developed to weld thick sections using a groove that is much narrower than normal. See Figure 20-8. This welding procedure saves materials and time. It requires some special equipment, however.

Normal-width V-groove, U-groove, and bevel-groove joints provide enough space for a welder to move an elec-

trode (such as a SMAW electrode) from one side of the groove to the other. This assures good penetration into the side walls of the groove. In order to obtain penetration into the walls of a *narrow* groove joint, the arc and electrode must be made to oscillate, or move from one side of the joint to the other. One method of causing the electrode to do this is to bend the electrode into a wave form as it is fed into the joint. See Figure 20-9. Another method is to use two electrodes that are twisted around each other before

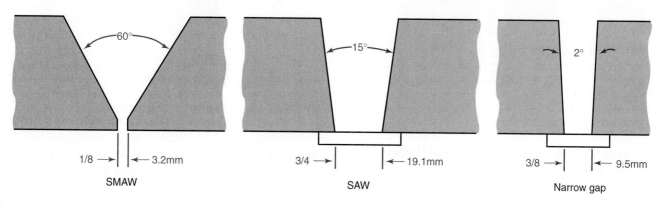

60°

1/8 →   ← 3.2mm

SMAW

15°

3/4   ← 19.1mm

SAW

2°

3/8 →   ← 9.5mm

Narrow gap

Figure 20-8. *A comparison of the weld joint preparation for shielded metal arc welding (SMAW), submerged arc welding (SAW), and narrow groove welding. Note that the narrow groove joint requires less weld metal to fill the groove.*

being fed into the joint. With this method, the arc occurs in a circular manner as the electrode melts. See Figure 20-10. An unbent electrode can also be moved from side to side by oscillating the electrode guide, as shown in Figure 20-11.

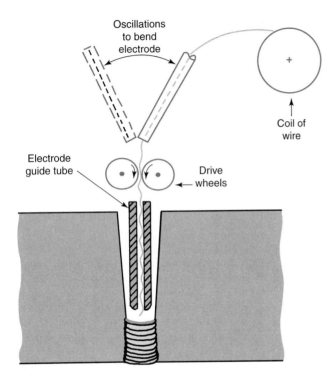

Figure 20-9. *A mechanism used to bend the electrode wire into a wave form and guide it into the weld joint. The wire will oscillate across the weld pool as the weld progresses.*

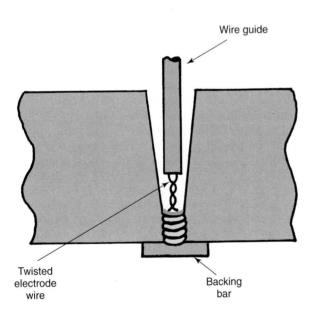

Figure 20-10. *Two wire electrodes are twisted before being fed into the weld pool to create an arc that circulates in the weld groove.*

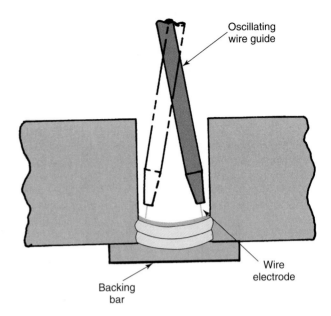

Figure 20-11. *To insure fusion at the sides of the groove, the wire guide is oscillated automatically.*

20.5 ARC STUD WELDING (SW)

Arc stud welding is a special welding development that quickly and efficiently welds studs and other fastening devices to plates and other surfaces. The process permits attaching an assembly to a structure, and fastening different parts to this structure without piercing the metal; Figure 20-12. Refer to Heading 4.20 and to Figure 4-25.

Figure 20-12. *A stud welded to a steel plate. Note the depth of fusion in this application, as shown in the stud that has been sectioned.*

Arc stud welding eliminates drilling or punching holes in the main structure. It saves the work of mechanically fastening an object to the main structure using bolts, rivets, or screws. Arc stud welding may be done using a dc power source similar to those used with the SMAW process. Melting of the end of the stud and weld pool occur as an arc is struck when the welding gun touches and raises the stud from the base metal.

The power source may also be a capacitor-discharge type. The capacitor-discharge type welding machine produces a rapid discharge of current that has been stored in an electrical device called a *capacitor*. With the capacitor-discharge system, the arc is struck by using one of two methods. The arc may be struck by lifting the stud away from the base metal, as explained above. The arc may also be struck while the stud is contacting the base metal. The current heats a small tip on the stud; an arc is struck when the tip melts.

A complete set-up for arc stud welding using a capacitor-discharge power source is shown in Figure 20-13. The equipment consists of:

- A capacitor-discharge power source with controls.
- An arc stud welding gun.
- Cables.
- Studs.
- An inert gas cylinder, regulator, and gauges, if required.

The control unit serves as a timer that controls the arc stud welding process. The control unit stops the arcing process at the set time. Times may vary from 0.5 seconds to 2 seconds, depending on the stud size.

An arc stud welding gun has the following main parts:

- Trigger.
- Device to hold the stud.
- Solenoid.
- Springs.

A cross section of an arc stud welding gun is shown in Figure 20-14.

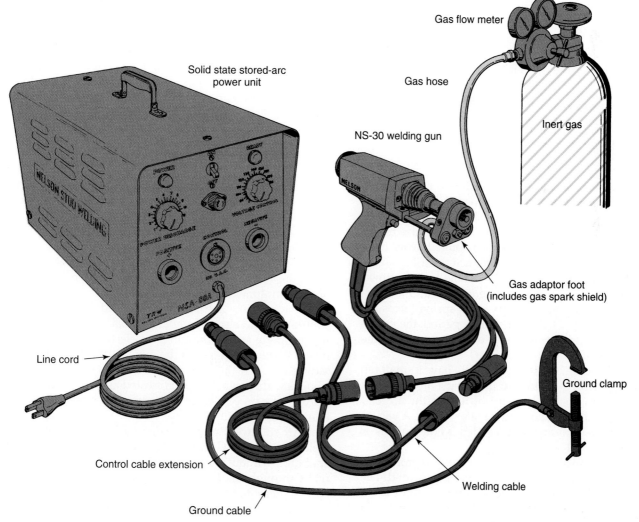

Figure 20-13. Equipment required for arc stud welding. The power cable to the gun and the ground cable are attached to a capacitor-type stored energy power source. Note the inert gas cylinder, regulator, and gauges that are required when nonferrous metals such as aluminum are welded. (TRW Nelson Stud Welding Division)

Arc stud welding is a rapid process. It can be done in any position. An arc stud weld is being made in Figure 20-15. The arc stud welding process is illustrated in Figure 20-16. The stud is placed in a chuck on the gun. A

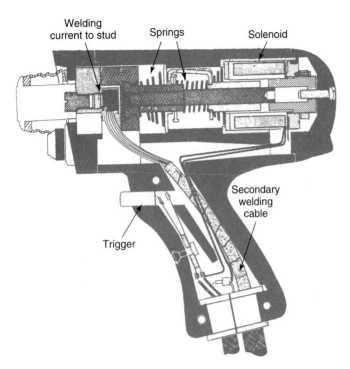

Figure 20-14. *Cross section of an arc stud welding gun. (TRW Nelson Stud Welding Division)*

Figure 20-15. *A welder attaching long bolts to an angle iron part using the arc stud welding process. Note the long extension on the gun. (American Welding Society)*

ceramic ferrule is placed on the end of the stud. The ferrule acts to contain the molten metal during arcing and serves as a mold while the metal solidifies. To begin the welding process, the stud is placed in contact with the base metal. When the trigger is pulled, the stud is pulled away from the part by a solenoid in the gun. This action strikes an arc between the stud and the base metal. At the end of a timed interval, the timer stops the welding current and the current to the solenoid. The arc stops and a spring forces the stud into the weld pool. After a cooling period, the gun is removed and the ceramic ferrule is broken away from the stud.

Prior to welding, the studs generally have a small nib or bump on them. This is to help initiate the arc. The stud also may contain a small amount of flux at the arc end under the small nib. Flux helps to stabilize the arc. A variety of parts that may be arc stud welded to assemblies is shown in Figure 20-17.

Aluminum can also be arc stud welded. To protect the aluminum from oxidizing, a shielding gas is provided. The process is similar to welding steel. Refer to Figures 20-13 and 20-18. Figure 20-19 shows an arc stud welding gun used to weld aluminum.

Some flash and sparking occurs in arc stud welding operations. The welder's eyes, face, and hands must be properly protected from this sparking.

20.6 ARC SPOT WELDING

The gas tungsten arc welding (GTAW) and gas metal arc welding (GMAW) processes may be used to produce spot welds between two sheets of metal or between a sheet of metal and a plate or structural metal member.

The spot welding process consists of striking an arc and holding that arc in one place. The top sheet of metal melts through and fuses with a molten portion of the sheet or structural member underneath.

The arc time during the spot weld is automatically controlled. The electrode holder usually has an insulator that properly spaces the tungsten electrode from the base metal. When the trigger in the gun is pulled, the electrode moves to touch the base metal. When the tungsten electrode is withdrawn, the arc is created and the timer is then started. The process is generally used on metal that is less than 1/8" (3mm) thick.

Arc spot welding can be done using the gas metal arc welding process (GMAW). An arc is struck at a chosen spot and, without moving the GMAW gun, the electrode melts through the top sheet and into the sheet below it. After a preset time, the welding current is stopped to complete the spot weld. See also Heading 15.13 and Figure 15-60.

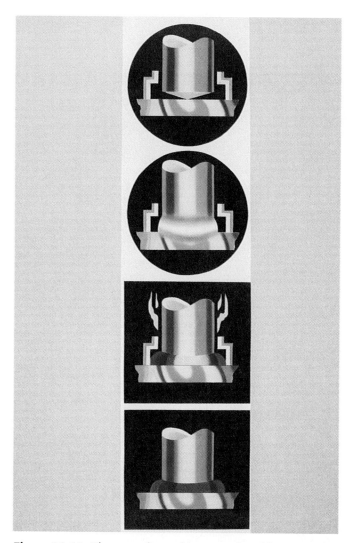

Figure 20-16. *The steps for making a stud weld. At top, the tip of the stud is placed into contact with the base metal. Then, the tip of the stud melts and an arc is formed. Next, the stud is forced into the weld pool. Bottom, the operation is complete. (TRW Nelson Stud Welding Division)*

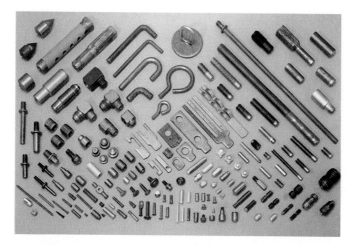

Figure 20-17. *All of the items shown may be arc stud welded by using special adapters in the arc stud welding gun. (TRW Nelson Stud Welding Division)*

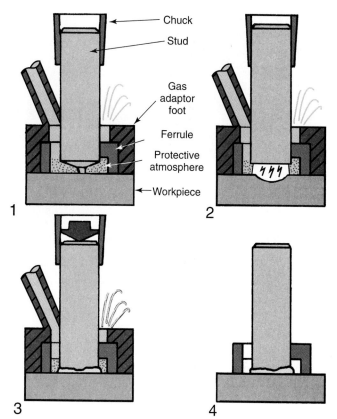

Figure 20-18. *Operations required when welding a stud to an aluminum plate. Notice the gas adaptor foot that confines the shielding gas.*

20.7 PLASMA ARC WELDING (PAW)

A cross section of a plasma arc torch nozzle is shown in Figure 20-20. This figure identifies the areas and critical dimensions of the torch nozzle. The orifice diameter may be cylindrical, convergent, or divergent, as shown in Figure 20-21.

Pressurized gases are used for two purposes for plasma arc welding. An orifice gas, which becomes a plasma, is fed into the plenum between the constricting nozzle and the electrode. A weld shielding gas is fed into the area between the constricting nozzle and the shielding gas nozzle.

The orifice gas becomes a *plasma* as it is heated while passing through the arc. This plasma is speeded up and propelled toward the work as it expands and passes through the exit orifice of the torch. The orifice gas is normally a mixture of argon and hydrogen (Ar–H), but argon and helium (Ar–He) may also be used. The orifice gas flow rate is usually between 0.5 ft³/hr. to 10 ft³/hr. (0.24L/min. to 4.8L/min.). Orifice gas cannot contain oxygen (O_2), because it will contaminate the tungsten electrode and may contaminate the weld. Carbon dioxide (CO_2) or argon and carbon dioxide (Ar–CO_2) are often used as the shielding gas. Shielding gas flow rates range between 20 ft³/hr. (10 L/min.) and 60 ft³/hr. (30L/min.), depending on the current used. A pure tungsten or a thoriated tungsten electrode is generally used in the PAW torch. The current used

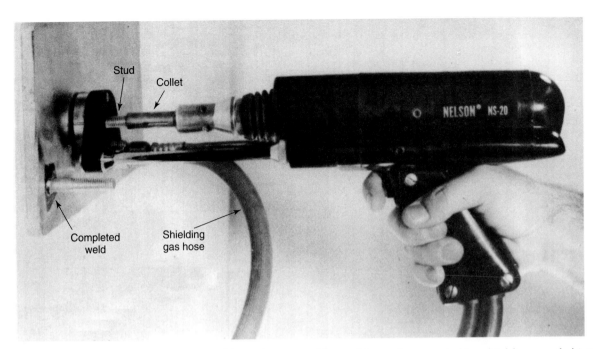

Figure 20-19. Arc stud welding gun used for welding aluminum. The weld area is protected by shielding gas fed into a chamber around the stud. (TRW Nelson Stud Welding Division)

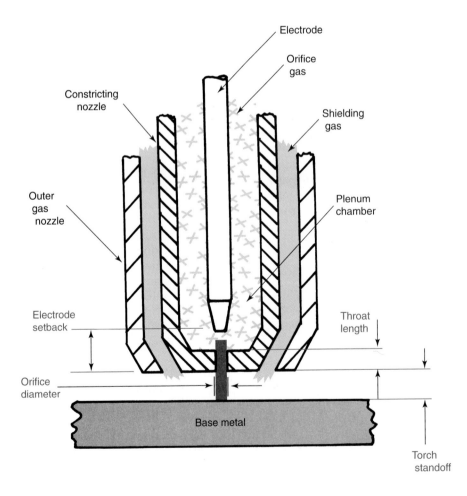

Figure 20-20. Cross section of a typical plasma arc torch. The various important dimensions of the torch nozzle are shown in red.

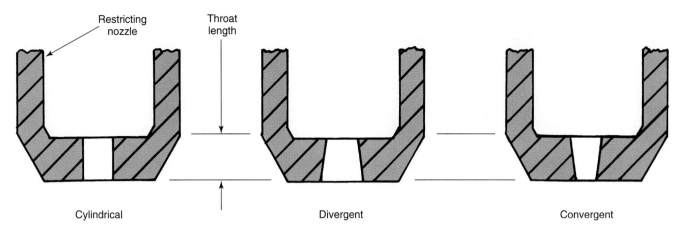

Figure 20-21. The plasma torch exit orifice may have a cylindrical, divergent, or convergent shape.

is generally DCEN. The advantages of PAW include its concentrated current flow, arc stability with large changes in arc length, and its low heat input and high welding speed.

The arc used in PAW may be a transferred or a nontransferred arc. A *transferred arc* has a negative electrode and the workpiece is positive (DCEN or DCSP). A *nontransferred arc* is an arc between the negative electrode and a positive constricting nozzle. In the nontransferred arc, the workpiece is not a part of the electrical circuit. Even though the arc is nontransferred, the plasma has sufficient energy to melt the workpiece surface. The transferred arc has better penetrating ability. The transferred and nontransferred arcs are illustrated in Figure 20-22. The transferred arc is most often used for welding with amper-

ages from 0.1A to 500A. The nontransferred arc can be used to weld or cut materials that do not conduct electricity.

The butt joint is most often used when plasma arc welding. Flanged-type butt joints may be welded without a filler wire. The flange is melted down and becomes the filler metal. Filler wire is required in most joint designs. When mechanized equipment is used, the filler metal is fed into the arc pool by a wire feed mechanism, as shown in Figure 20-23.

Welding of 1/4" (6.4mm) stainless steel or 1/2" (12.7mm) titanium can be completed in one pass. Arc voltage varies from 21V for thin material to 38V for thicker material. Currents range from 120A to 275A. Travel speeds can be as high as 30" per minute (13mm/second). Figure

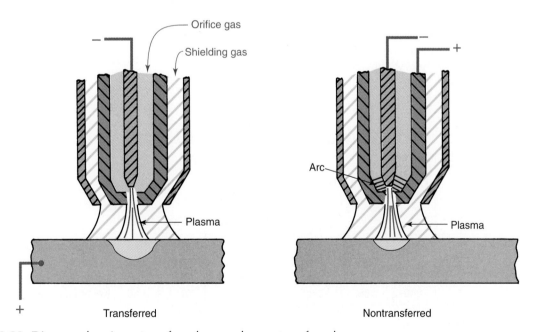

Figure 20-22. Diagram showing a transferred arc and a nontransferred arc.

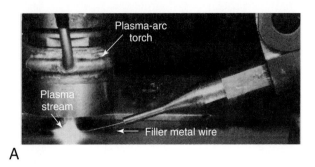

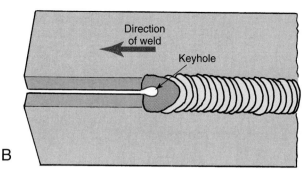

Figure 20-23. *Plasma arc welding. A—An automatic plasma arc torch with filler wire being fed into the weld pool automatically. (ESAB Welding and Cutting Products) B—A partially completed PAW weld showing the "keyhole" at the leading edge of the weld pool. The size of the keyhole indicates the amount of penetration.*

20-24 lists some variables used when welding stainless steel.

The keyhole welding method may be used when PAW to ensure complete penetration. Current and gas flow rates must be carefully regulated to prevent too large a keyhole. Because of the high welding speeds possible, PAW is most efficient when done with mechanized equipment. **Plasma arc welding and cutting produce intense sound. It is usually necessary to wear ear protection during the PAW process.**

20.8 UNDERWATER SHIELDED METAL ARC WELDING

Underwater shielded metal arc welding is done underwater, using a well-insulated electrode holder and special covered electrodes, as shown in Figure 20-25.

The electrode covering is protected from water damage by special waterproof outer coatings. About 10% higher current settings are used for welding underwater than when welding the same object in the atmosphere. Because of the problem of rapid heat loss in water, stringer beads should be used rather than wide weaving beads. Also, a short arc length and DCEN (DCSP) are recommended.

Due to the poor visibility underwater, a #4 to #8 welding lens is recommended for use on the mask or helmet. **When wearing the metal helmet for deep dives, the welder must be careful not to ground the helmet to any part of the welding circuit.** Figure 20-26 shows an underwater weld in progress.

It is recommended that a communication system be used between the diver and the equipment operator, who is on a diving tender (boat) on the surface. The arc welding current should be turned on by a helper on the diving tender after the welder signals that he or she is ready to begin welding. The welding current should be turned off as soon as welding is completed.

Extensive diving and welding training should be completed before attempting to weld underwater. **Follow all safety precautions for both welding and diving.**

Thickness		Travel speed		Current (DCSP) (DCEN)	Arc voltage	Nozzle type	Gas flow				Remarks
							Orifice		Shield		
in.	mm	in/min.	mm/sec.	A	V		ft³/hr.	L/min.	ft³/hr.	L/min.	
0.092	2.3	24	10	115	30	111M	6	2.8	35	16.5	Keyhole, square-groove weld
0.125	3.2	30	13	145	32	111M	10	4.7	35	16.5	Keyhole, square-groove weld
0.187	4.8	16	7	165	36	136M	13	6.1	45	21.2	Keyhole, square-groove weld
0.250	6.4	14	6	240	38	136M	18	8.5	50	23.6	Keyhole, square-groove weld

Notes:
  a. Nozzle type: number designates orifice in thousandths of an inch; "M" designates design.
  b. Gas underbead shielding is required for all welds.
  c. Gas used: 95% Ar-5% H_2.
  d. Torch standoff: 3/16" (4.8mm)

Figure 20-24. *Variables to be used when plasma arc welding stainless steel.*

Figure 20-25. Underwater welding electrode holder. Note that the holder is well-insulated, with no bare metal areas exposed. (Broco, Inc.)

Figure 20-26. A diver/welder making a weld on an underwater support for an ocean oil rig. Note the electrodes in the weighted holder and the wire brush attached to the rope. (AWS Welding Journal)

20.9 SOLID-STATE WELDING (SSW)

The American Welding Society defines *solid-state welding* as follows: "A group of welding processes that produce coalescence by the application of pressure at a welding temperature below the melting temperature of the base metal and the filler metal."

A number of welding processes are classified as solid-state welding processes. These processes do not use an arc, do not use a beam of energy, do not use a flame, and do not use resistance to heat the metal. They are done when the metal is cold, warm, or hot. However, the temperature does not exceed the melting points of the metals. The metal to be joined is heated (if necessary) to a point where the pieces can be forced together. No filler metal is needed.

The American Welding Society recognizes the following as solid-state welding processes:
- Coextrusion welding (CEW).
- Cold welding (CW).
- Diffusion welding (DFW).
- Explosion welding (EXW).
- Forge welding (FOW).
- Friction welding (FRW).
- Hot pressure welding (HPW).
- Roll welding (ROW).
- Ultrasonic welding (USW).

The most common of these processes are covered in the following paragraphs.

20.10 COLD WELDING (CW)

The theory of the *cold welding (CW)* process is that the pressure at the surface causes fusion only a few molecules deep. This is sufficient to hold the material together and provide good strength. This room-temperature pressure welding process works best on soft, ductile metals like aluminum and its alloys, copper, alloys of cadmium, nickel, lead, zinc, etc. No heat is needed. The cleaned metal pieces are forced together under considerable pressure, and the ductility of the metals produces a thin but true fusion condition. Refer to Heading 4.21 and Figure 4-26. Where the metals are pressed together, enough pressure must be applied to reduce their thickness to about one-fourth of the original thickness. Aluminum, when welded by this process, has a tensile strength of up to 22,000 psi (152MPa). Figure 20-27 shows three typical cold welds.

The metals to be joined by cold welding must be very carefully prepared. Oxides and other contaminants must be completely removed. An abrasive wire wheel, turning at high speed, has been found to be satisfactory. This method is considered to be better than chemical methods, because residual solvents often remain on the metal surface after chemical cleaning.

The design of the tool for imposing pressure on the metals is very important. Dies must be designed to compensate for varying hardness of the metals. For example, when cold welding aluminum to copper, the tool must

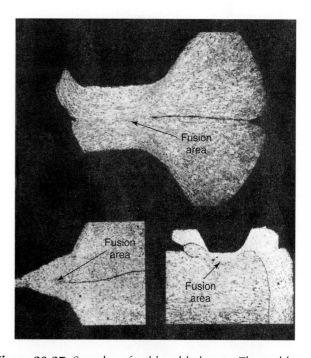

Figure 20-27. Samples of cold welded parts. The welds are aluminum to aluminum (1003). Notice the large indentation of the surface. (Kelsey-Hayes Co.)

have twice as much contact against the softer aluminum. Figure 20-28 shows a hand-operated cold welding tool. It can weld aluminum up to a combined thickness of 0.080" (2mm), or copper to 0.060" (1.5mm).

The tools can be either hand-, pneumatic-, or hydraulic-powered. Figure 20-29 shows a hydraulic-powered unit that can butt weld up to 3/4" (19mm) round or strip copper. It can also perform lap welds.

20.11 EXPLOSION WELDING (EXW)

Metals have been successfully welded with the energy from an explosion, using a process called *explosion*

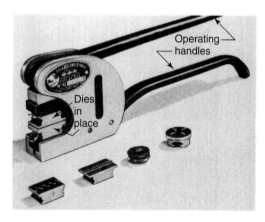

Figure 20-28. A hand-operated cold weld machine with extra dies.

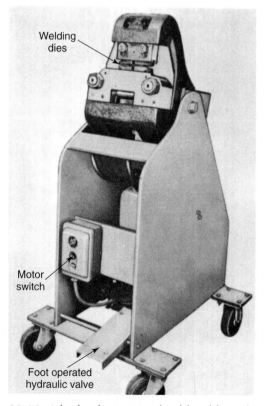

Figure 20-29. A hydraulic-powered cold weld machine.

welding (EXW). **Refer to Heading 4.22 and to Figure 4-27. This process is very dangerous. It should be performed only by explosive experts, and in specially designed or water-filled chambers.** Special permits must be obtained from local, state, and federal authorities before this type of work can be done.

To prepare an explosion weld, the metals are placed at an angle to each other. A protective material is placed over the metal and then the explosive is placed on the metal surface. An explosion welding set-up is shown in Figure 20-30. The explosion forces the plates together at high velocity, causing surface ripples in the metal. These ripples lock or weld the two metals together. This process has been successfully used to weld steel to steel, aluminum to aluminum, copper to steel, stainless steel to molybdenum, and many other metal combinations.

20.12 FORGE WELDING (FOW)

The oldest form of fusing two pieces of metal together is *forge welding (FOW).* Refer to Heading 4.23 and to Figure 4-28. This type of welding requires considerable skill on the part of the smith. It is usually limited to the joining together of pieces of solid metal stock. The two pieces of metal to be welded together are heated in a blacksmith's forge, like the one shown in Figure 20-31. This forge uses charcoal or coal and an air blast from below the coals to produce the heat necessary for forge welding. Metal to be welded is placed in the bed of hot coals. The metal pieces are heated to a *white heat*, just short of the rapid oxidizing or burning point. They are then placed together so that the surfaces may be forced against one another by pounding with a hammer. A blacksmith usually *swages* (enlarges) the ends to be joined. The pressure of the hammer blows, and the extra heat produced from the hammering, fuses the two pieces together. See Figure 20-32. This same type of welding may be done using power hammers, fixtures, and presses. It is often called *forge welding* or *hammer welding*.

A weld of this nature, when done correctly, has every quality of the original metal. However, because of the skill necessary to produce a successful joint, and the relative ease with which other processes accomplish the same task, forge welding has virtually been replaced by more modern welding processes. Forging and forge welding are still used today, however, to make some experimental and prototype parts. The method is also used to produce copies of antique hardware and utensils.

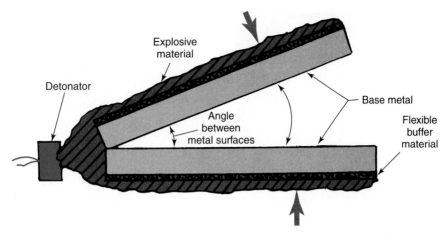

Figure 20-30. *Schematic drawing of a set-up for explosion welding. It shows how the explosive material and protective coating are placed on the base metal. The base metal pieces come together with great force and create a welded bond.*

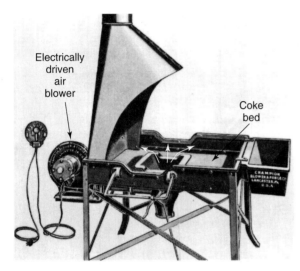

Figure 20-31. *A blacksmith's forge used to heat metal for forge welding.*

20.13 FRICTION WELDING (FRW)

Friction generates heat. If two surfaces are rubbed together, enough heat can be generated that the parts subjected to the friction will be fused together. This is the principle involved in *friction welding (FRW)*. Refer to Heading 4.24 and Figure 4-29.

A friction weld is prepared by mounting the pieces to be welded in a chuck or fixture. One fixture is stationary; the other revolves. The pieces to be welded are brought together under pressure. Friction between the parts raises the temperature to a welding heat. See Figure 20-33. The parts are then forced together. After being allowed to cool, the friction weld is complete. The weld is produced in about 15 seconds. The progression of a friction weld on solid steel rod is shown in Figure 20-34.

In friction welding, the spinning part of the chuck is driven by a motor. It is made to rotate while the metal pieces are in contact. A small pressure is applied, causing the metal to heat. Once the heating is completed, great

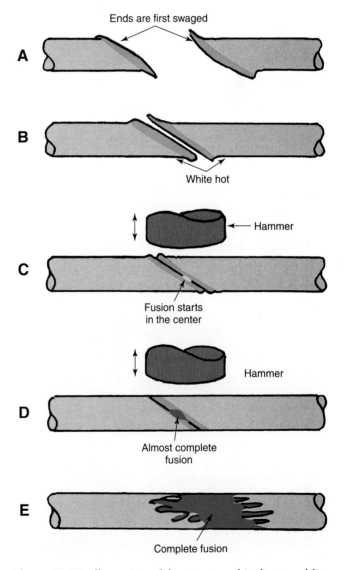

Figure 20-32. *Illustration of the steps used in forge welding.*

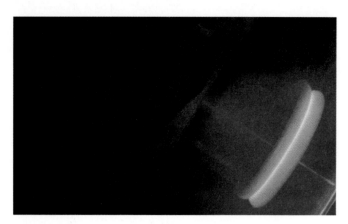

Figure 20-33. A just-completed friction weld. The base metal is heated to welding temperature, but never actually melts. (Manufacturing Technology, Inc.)

Figure 20-35. A friction welding machine with its controller. The power supply is in the tall cabinet at far left. (Manufacturing Technology, Inc.)

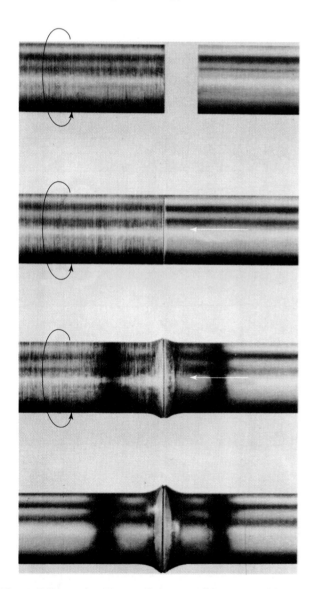

Figure 20-34. The steps in friction welding two rods of 1" (25mm) mild steel. The rod at the left is rotated at a high speed while the rod at the right is held stationary. The rods are then forced together. The friction creates enough heat to make a sound butt weld.

force is applied along the axis to join the surfaces. Figure 20-35 shows a large friction welding machine. The process is similar to upset or flash welding, but friction is used to generate the heat.

Most friction welding is used to weld ferrous parts. The process also is well-suited for joining dissimilar (unlike) materials to mild steel in production. The electrical, chemical, nuclear, and marine industries use friction welding to join dissimilar metals. No protective atmosphere is required, which saves a significant amount of time and materials.

20.14 ULTRASONIC WELDING (USW)

In *ultrasonic welding (USW)*, metals are clamped together under pressure and high-frequency vibrations are introduced into those metals through a welding tip or *sonotrode*. The vibrations, at frequencies above those human beings can hear, break up the surface films and cause the solid metals to bond tightly together. Refer to Heading 4.25 and also to Figure 4-30. This happens without the use of heat and without melting the metal. Careful cleaning of the metals is not necessary. The low clamping pressure reduces deformation.

Ultrasonic welding equipment contains one or more transducers that convert high-frequency electrical power into mechanical vibration at the same frequency. The welding frequency is usually between 10,000 Hz (cycles per second) and 75,000 Hz. The metals to be welded are clamped between the transducer and a small anvil. In addition to the transducers, there is a coupling system that transmits the mechanical vibration to the welding tip and thus into the metals being joined.

As a production tool, ultrasonic welding is used to weld semiconductors, foils, microcircuits, and fine wires. Ultrasonics has several advantages. There is no heat distortion in the parts. No fluxes or filler metals are required. Thin sheets can be welded to thick sections. The pressures used are less, and the welding times shorter, than those used in resistance welding. Many types of metal can be joined to themselves or to other metals. The equipment can be operated by semiskilled personnel with minimal training.

For spot welding, the metals are clamped together and transverse or lateral vibration is imposed on the assembly, as shown in Figures 20-36 and 20-37. An ultrasonic spot welding machine is shown in Figure 20-38.

Ultrasonic welding can also be used to produce seam welds. A seam welder transducer and wheel are shown in Figure 20-39. A part to be welded is drawn between two counter-rotating rollers. Vibrations are introduced through the upper roller. A continuous leaktight seam weld is produced between the two metals.

Ultrasonic welding can produce welds in similar or dissimilar metals. The maximum thickness that can be welded is 0.10" (2.5mm) for aluminum and 0.040" (1mm) for harder materials. This process produces welds very quickly.

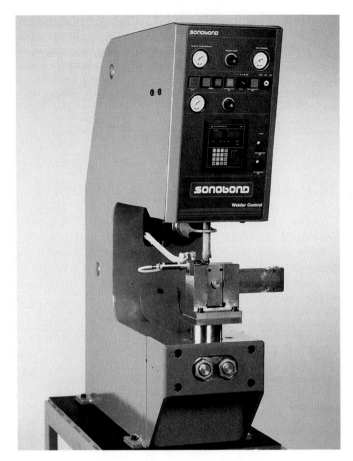

Figure 20-38. *An ultrasonic spot welding machine. This machine uses sound waves to rapidly vibrate the upper contact point. This creates heat and a weld. (Sonobond Ultrasonics, Inc.)*

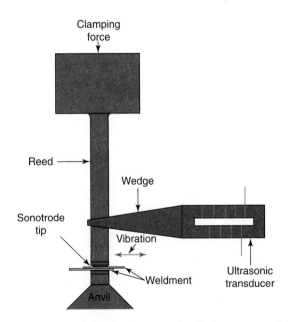

Figure 20-36. *Wedge-reed principle of ultrasonic welding. (Sonobond Ultrasonics, Inc.)*

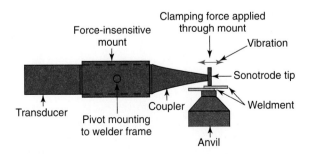

Figure 20-37. *Lateral-drive principle of ultrasonic welding. (Sonobond Ultrasonics, Inc.)*

Figure 20-39. *An ultrasonic seam welder. The wheel-shaped contact is vibrated to create a seam weld as the wheel rolls along the weld joint. (Sonobond Ultrasonics, Inc.)*

20.15 OTHER WELDING PROCESSES

Other welding processes are those that do not meet the definition for oxyfuel gas welding, arc welding, resistance welding, or solid-state welding. All these "other welding processes" are listed in Figure 4-39, *AWS Welding Processes*. The most common of these processes will be explained on the following pages.

20.16 ELECTROSLAG WELDING (ESW)

The *electroslag welding (ESW)* method has been developed to weld very thick sections or joints. The process eliminates the need for multiple passes and for bevel-, V-, U-, or J- grooves. Refer to Heading 4.19 and to Figure 4-24. Metal thicknesses up to 30" (76cm) have been welded with this process.

The process is used to weld joints in a vertical welding position. The equipment used for electroslag welding includes:

- A dc power source.
- One or more electrodes and electrode guide tubes.
- One or more wire feeders and oscillators.
- Retaining shoes (molds).
- Flux. See Figure 20-40.

Prior to beginning the weld, a starter block is tacked to the bottom of the joint. U-shaped braces are tacked across the joint to hold it together while welding is being performed. The electroslag process is started by producing an arc between the electrode(s) and the joint bottom. Flux is added and forms a layer of molten slag. When a large layer of slag has formed, the arc is no longer needed. Since the arc is used only for starting the process, ESW is *not* considered to be an arc welding process. The resistance to electrical current flow through the flux creates the heat necessary to melt the electrode(s) and base metal and to

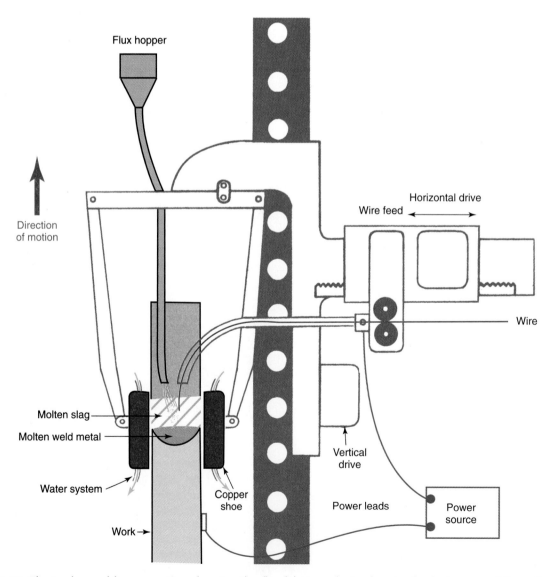

Figure 20-40. *Electroslag welding operation showing the flux hopper, electrode wire feed, shoes, and the rail on which everything moves as the weld progresses.*

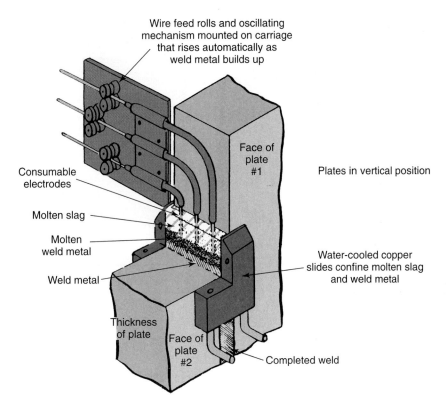

Wire feed rolls and oscillating
mechanism mounted on carriage
that rises automatically as
weld metal builds up

Consumable
electrodes

Molten slag

Molten
weld metal

Weld metal

Thickness
of plate

Face of
plate
#2

Face of
plate
#1

Plates in vertical position

Water-cooled copper
slides confine molten slag
and weld metal

Completed weld

Figure 20-41. *A schematic drawing of the electroslag welding process. In this application, three electrodes are used. The molten slag floats above the weld metal and prevents oxidation.*

keep the weld pool molten. The electrodes used are either solid wire or flux cored. The process is fast and requires no edge preparation of the metal. More than one electrode may be used. This permits a thick joint to be welded faster, as shown in Figure 20-41.

Water-cooled copper shoes are used to contain the molten metal and slag. As the weld is made, the molds move up the joint. The weld is completed in one pass. Butt joints, T-joints, corner joints, and other types of joints may be made with this process. See Figure 20-42. Butt welds are the most common. Specially shaped shoes are required for T-joints and corner joints. Figures 20-41 and 20-43 show three electrodes being used to fill the joint. The extreme heat produced by the molten slag and metal in the weld causes the base metal to melt away from the original joint gap, as shown in Figure 20-44.

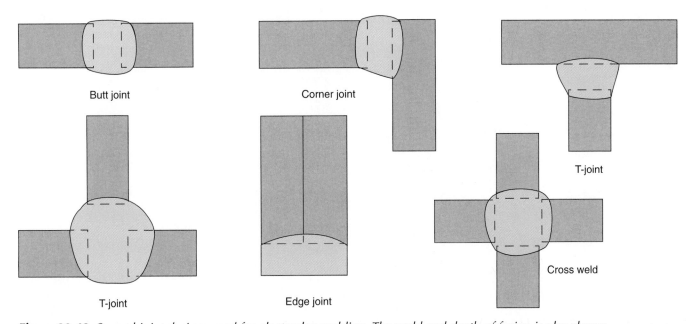

Butt joint

Corner joint

T-joint

T-joint

Edge joint

Cross weld

Figure 20-42. *Several joint designs used for electroslag welding. The weld and depth of fusion is also shown.*

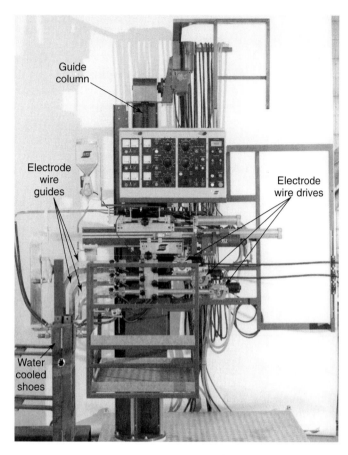

Figure 20-43. *An electroslag welding station for three electrodes. Each electrode has a separate drive and guide. (ESAB Welding & Cutting Products)*

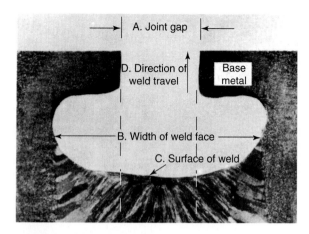

Figure 20-44. *A macrograph (1.5 x) of a manganese molybdenum steel joint welded by electroslag process. Note how the base metal melts away from the edges of the original joint gap.*

In a special application of electroslag welding, pipe sections are butt welded. Figure 20-45 shows a schematic of a circumferential weld as it begins and before it ends. The welding is done at the 3 o'clock position. Figure 20-46 shows ESW equipment after the weld is completed.

20.17 THERMITE WELDING (TW)

Thermite welding (TW), is based on the fundamental chemical principle that aluminum is a more chemically

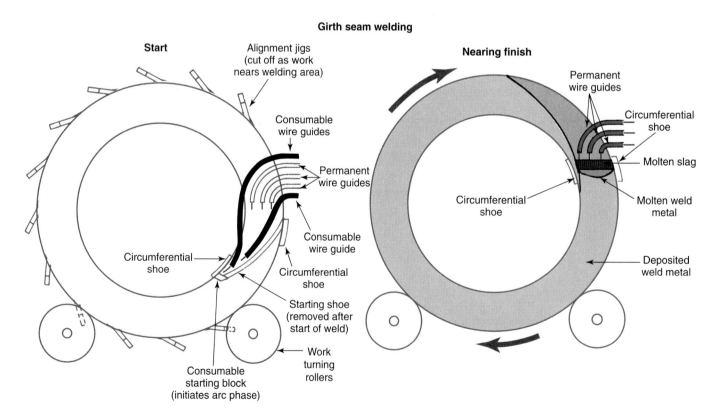

Figure 20-45. *The use of the electroslag process for welding butt joints in cylindrical objects. In this case, two consumable wire guides are used at the beginning of the weld. Three wires are used to fill the joint.*

Figure 20-46. A completed electroslag circumferential weld on a thick-walled pipe. Notice the copper shoes. (ESAB Welding and Cutting Products)

active metal than iron. The process consists of mixing iron oxide and aluminum, both in powdered form. They are placed in a hopper above a mold that surrounds the area to be welded. This mixture is ignited. The aluminum combines chemically with the oxygen molecules in the iron oxide, producing a temperature of approximately 5000°F (2760°C). An aluminum slag (aluminum oxide) and liquid iron are produced. The high-temperature molten iron is heavier than the aluminum oxide, so it settles to the bottom of the mold. There, it comes into contact with the steel joint, melting the surface of the steel and fusing with it. This method may be used to weld both large and small steel parts. It is very popular where large sections are to be welded together. Preheating is often used if the pieces to be welded are large.

Thermite welding may be roughly compared to a foundry casting operation. One difference is that metal being poured is of a considerably higher temperature than metal melted in a furnace.

Welding railroad rails together, welding new teeth on large gears, and welding large fractured crankshafts are common applications of thermite welding. Thermite welding is used to join sections of castings when size prevents casting them in one piece. It is also used to repair large steel structures that are made on special order and would be costly to replace. Thermite welding has been applied successfully in almost every industry.

Thermite welding is also used for welding pipe. However, in this application, the molten iron does not mix with pipe metal, but merely furnishes the heat to melt the pipe ends. The pipe ends are forced together while molten.

Since thermite welding is a specialized form of casting, molds are necessary to control the flow and to shape the liquid iron. The metal parts to be welded together are firmly and accurately set up for welding. A joint is usually machined to provide a V-groove all around it so that the molten metal can gain access to all parts. Another method

is to use a wax pattern placed in the joint. It has the same shape as the final weld. A sand mold is built around the wax pattern and the work to be welded. Vent holes are necessary on larger jobs. During the preheating of the metal to be welded, the wax used as a pattern melts away, leaving the correctly shaped cavity to receive the molten

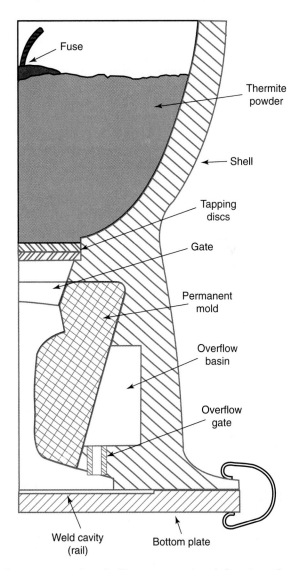

Figure 20-47. One-half a cross-sectional drawing of a setup for using thermite welding to join railroad rails together. This setup uses a permanent mold and a shell to contain the reaction.

metal. A third process uses a permanent mold. Figure 20-47 shows a section of a mold assembled around a rail joint to be welded.

In most applications, the mold includes a container at the top to hold the aluminum and iron oxide powders. This container holds enough aluminum and iron oxide to perform the weld. Before the mold is placed around the joint to be welded, it is necessary to clean the joint surfaces. They must be bright and clean. It is especially important to remove any oil, grease, or water from the metal being welded. These materials could vaporize and build up a dangerously high pressure during the weld, causing the mold to burst. Some molds have a preheating gate, which provides for heating the metal just prior to the pouring of the iron. This preheat also ensures that all moisture is removed—any moisture present when molten metal is poured could cause a sudden creation of steam and an explosion. An ignition powder, which ignites at a low temperature but burns at a high temperature, must be used in order to start the thermite reaction. The aluminum and iron oxide mixture will ignite only after it has been brought to a temperature of approximately 2000°F (1100°C). Once the mixture starts to burn, it is self-propagating. The chemical reaction is allowed to go to completion before the pouring is started. The process is considered safe because of the very high temperature necessary to begin the thermite reaction. This eliminates the chance of accidental ignition of the mixture.

After the weld is completed and the metal has been allowed to cool, the mold is removed. After the removal of the mold, the weld may need trimming because of the excess metal clinging to it. The pouring gate, all risers, and metal in the vent holes must be cut off and ground down.

20.18 ELECTRON BEAM WELDING (EBW)

Electron beam welding (EBW) uses a concentrated high-energy beam of high-velocity electrons to melt the metal to be welded. Refer to Heading 4.27 and to Figure 4-32. The parts that are to be welded are placed very close together, almost touching. The edges to be welded must be very straight. An electron beam can make a complete penetration weld in a 10" (254mm) thick steel plate. See Figure

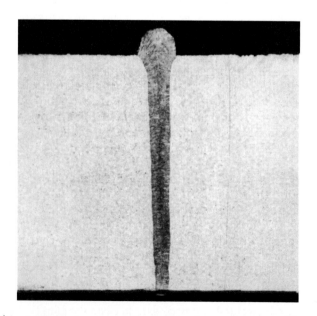

Figure 20-48. *A cross section of an electron beam weld. Note how narrow the weld is in relation to the thickness of the base metal. (PTR-Precision Technologies, Inc.)*

20-48. Electron beam welding is also very fast. Steel that is 0.25" (6mm) thick can be welded at a rate of over 200"/min. (5m/min.).

A schematic of the principal parts of an electron beam gun is shown in Figure 20-49. Current is passed through a tungsten filament that emits electrons. The electrons pass through the bias electrodes, which shape the beam. Electrons are then accelerated toward the anode, which has a positive charge. A voltage difference of up to 175,000V exists between the emitter and the anode. This is

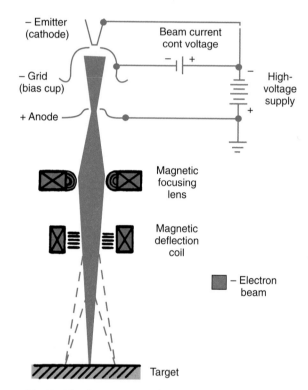

Figure 20-49. *The basic parts of an electron beam welder. (PTR-Precision Technologies, Inc.)*

known as the *accelerating voltage*. An electromagnetic lens is used to focus the electron beam. A cutaway view of an electron beam welder is shown in Figure 20-50.

Electron beam welding is usually done in a vacuum. In a high vacuum, there are no particles in the air to interfere with the electron beam. In a partial vacuum, there are some particles, which causes the electron beam to spread out. In a partial vacuum, the parts to be welded must be closer to the source of the electrons to maintain a high energy density.

Electron beam welding can also be done without a vacuum. The parts must be very close to the beam source (1/4" [6mm] or less). Figure 20-51 shows how the beam spreads out as the pressure increases. Figure 20-52 illustrates three types of electron beam systems. The best results are obtained in a high vacuum.

A part being welded must be placed on a carriage that moves it under the electron beam. An optical system is used to align the part. The optical system principle is shown in Figure 20-52A. Figure 20-53 shows an operator

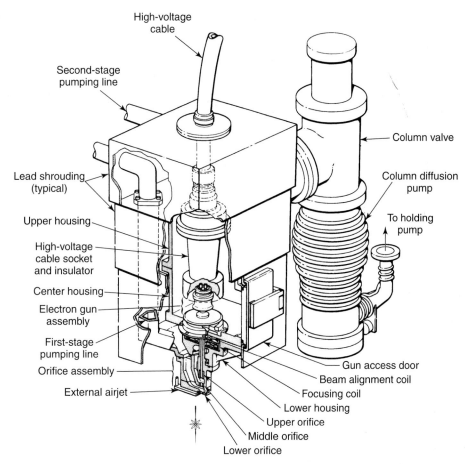

High-voltage
cable

Second-stage
pumping line

Column valve

Lead shrouding
(typical)

Column diffusion
pump

Upper housing

To holding
pump

High-voltage
cable socket
and insulator

Center housing

Electron gun
assembly

First-stage
pumping line

Orifice assembly

External airjet

Gun access door

Beam alignment coil

Focusing coil

Lower housing

Upper orifice

Middle orifice

Lower orifice

Figure 20-50. *A cutaway of a nonvacuum electron beam welder. (PTR-Precision Technologies, Inc.)*

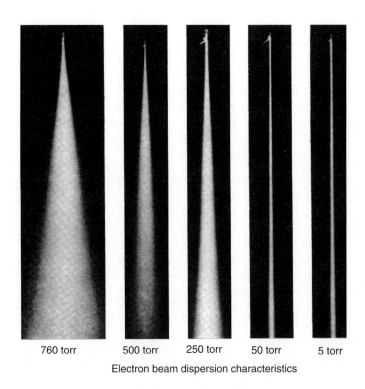

| 760 torr | 500 torr | 250 torr | 50 torr | 5 torr |

Electron beam dispersion characteristics

Figure 20-51. *The electron beam spreads out as the pressure increases. Atmospheric pressure is 760 torrs; 5 torrs is a high vacuum. (PTR-Precision Technologies, Inc.)*

aligning a piece inside a high-vacuum electron beam machine. To ensure coverage of the beam on the weld joint, the beam may be programmed to *oscillate* (move back and forth). Several electron beam patterns are shown in Figure 20-54. Figure 20-55 shows a high-production use of a partial-vacuum electron beam welder. The parts are moved to the electron beam welder because the electron beam welder itself is too large to move.

20.19 LASER BEAM WELDING (LBW)

The term *laser* stands for **l**ight **a**mplification by **s**timulated **e**mission of **r**adiation. A laser beam is merely high-energy-density light. The laser light beam has one wavelength and is in-phase. *In-phase* means all the particles or waves move together and are concentrated to have a higher energy level.

The laser beam may be produced by one of three laser-producing materials. These materials are:

- CO_2 (carbon dioxide) gas.
- Nd-YAG (neodymium-doped, yttrium aluminum garnet) crystal rods.
- Nd-glass or ruby rods.

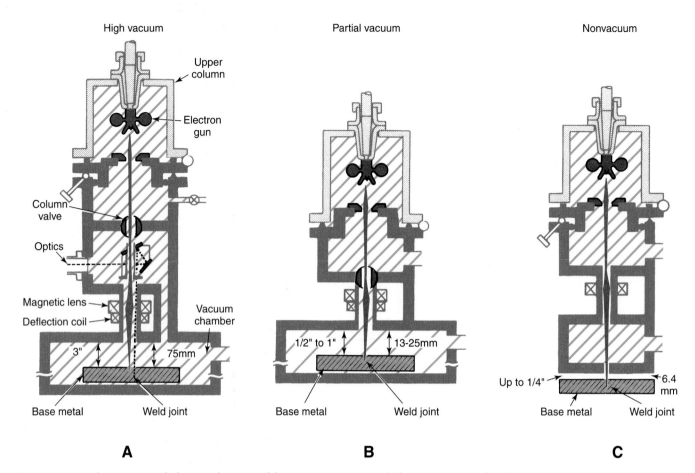

Figure 20-52. *Three types of electron beam welding systems. A—Weld being made under high vacuum. B—Weld being made under partial vacuum. C—Weld being made under atmospheric pressure. (PTR-Precision Technologies, Inc.)*

Figure 20-53. *Electron beam operator aligning a part for welding. (PTR-Precision Technologies, Inc.)*

The laser is created by exciting the atoms or molecules of these laser materials within the laser welding machine. The excited atoms emit a high-energy light of a single wavelength. The excited atoms are reflected back and forth between specially coated mirrors until a desired light energy level is reached. When that energy level is reached, the light energy beam (laser beam) is released to strike the metal being welded. The light energy striking the base metal releases heat energy and melts the metal. The laser material is excited again and the process repeats itself. Laser pulse durations may last for as little as 50 μs (millionths of a second) to as long as 50 ms (thousandths of a second).

Most welding is done with the carbon dioxide or Nd-YAG laser. Since the Nd-glass or ruby laser cannot be used continuously, it is not well-suited for most welding applications. A schematic of an Nd-ruby laser is shown in Figure 4-31. Refer also to Heading 4.26

The CO_2 and Nd-YAG lasers can recycle rapidly and can furnish a pulse or continuous laser beam output. A CO_2 laser operates in a similar manner to the Nd-glass or ruby laser. Helium and nitrogen gases are used with carbon dioxide in the CO_2 laser. A long tube is filled with

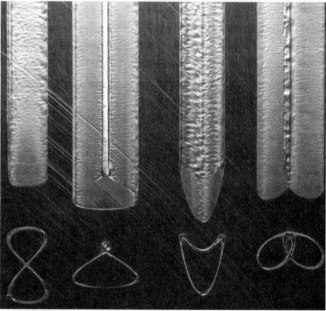

Figure 20-54. Using oscillatory beam deflection capability with the electron beam. At top, the beam is being moved in a simple "bow tie" motion; at bottom, the electron beam is being moved in a variety of other repetitive motion patterns. The view from above shows the effect that these beam patterns have on the bead shape produced. (PTR-Precision Technologies, Inc.)

CO_2, He, and N_2 gases. The atoms in the gases are excited electrically; they, in turn, excite other CO_2 atoms. High-density light energy is created as these atoms are excited and rapidly return to their normal state. The light energy

Figure 20-55. A partial vacuum electron beam welder with a four-station rotating table for continuous loading, welding, and unloading. Notice the control panels to the left of the welder. (PTR-Precision Technologies, Inc.)

is reflected back and forth within the laser machine until the desired energy level is reached. They are released, striking the base metal and causing it to melt. The main parts of a CO_2 laser machine are shown schematically in Figure 20-56.

A CO_2 gas laser can produce a full penetration weld in 3/4" (19mm) steel. Most solid-state lasers can only penetrate 1/4" (6mm) of steel. A micrograph of a laser weld is shown in Figure 20-57.

The laser beam is focused and aimed using lenses and mirrors. Laser welding is done at atmospheric pressure and does not require a vacuum. Shielding gas is used to protect the weld from oxidation. The parts to be welded must have straight edges and be very close together. See

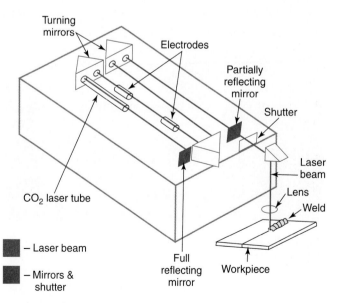

Figure 20-56. A CO_2 laser. Mirrors are used to reflect and aim the laser beam. A lens is used to focus the beam.

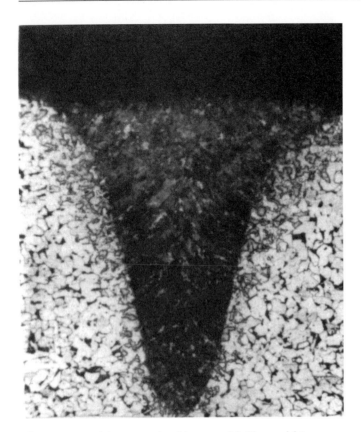

Figure 20-57. *Macrograph of laser weld. The weld is narrower than an arc weld.*
(Charles Albright, Ohio State University)

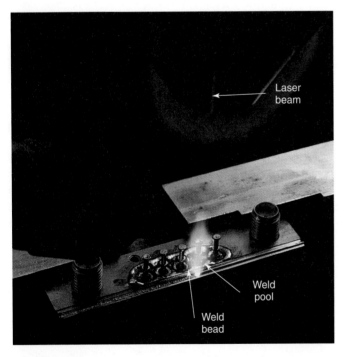

Figure 20-58. *An Nd-YAG laser welding an aluminum microelectronics package. Note the nozzle of the laser gun at the top. The laser beam is visible at the leading edge of the weld bead. (Raytheon Company)*

Figure 20-58. The travel speed depends on the thickness of the material being welded, the depth of penetration desired, and the kilowatt power of the laser machine. With a 1 kW laser, welding on 1/8" (3mm) thick steel, complete penetration can be obtained at a travel speed of about 50"/min. (1.3 m/min.).

Lasers may be used to drill extremely small holes in metal parts. Laser-drilled holes may be very precisely located and may be drilled to varying depths and at various angles. See Figure 20-59.

The laser is also used for cutting and for heat treating. In heat treating, the laser beam is not focused to a spot. The beam is made slightly divergent, so it will cover and heat-treat a small area.

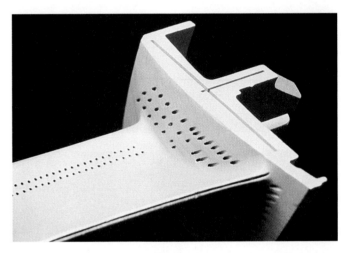

Figure 20-59. *Laser-drilled holes (0.020") in a turbine stator blade. Some holes are drilled at a 20° angle. Modern turbine engines may have more than 100,000 laser-drilled cooling holes. (Convergent Energy)*

20.20 SELF-GENERATING OXYHYDROGEN GAS WELDING (OFW)

Self-generating oxyhydrogen gas welding (OFW) is an unusual form of oxyfuel gas welding in which the process creates its own oxygen and fuel gas supply, and then uses them to produce the heat for welding. The production of oxygen and hydrogen gases by the electrolysis of water has been done for many years. The burning of hydrogen in air or in pure oxygen has been used as a welding flame or gas welding process for several decades.

Only recently, however, have these two processes been combined. Distilled water is broken down into its two component gases by electrolysis. These gases are then fed to a torch and burned. Refer to Figure 5-2 for a schematic of how hydrogen and oxygen are separated from water by electrolysis.

A self-generating oxyhydrogen gas welding is shown in Figure 20-60. The unit operates on 115V ac. An ac to dc converter provides power for the electrolytic reaction. The

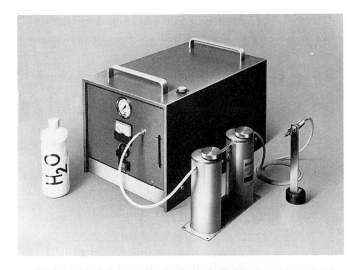

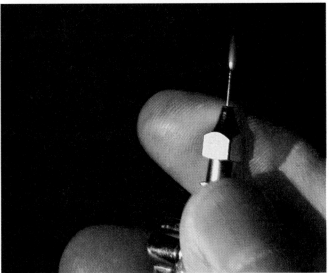

Figure 20-60. *Self-generating oxyhydrogen gas welding. A—A self-generating oxyhydrogen gas welding outfit. The methyl alcohol booster is being used and the tiny torch is in a stand at the right. B—The size of the torch used with this process is compared to a finger. (E. Spirig-Switzerland, supplied by Solder Absorbing Technologies, Inc.)*

mixed gases travel along a single hose to the torch. The unit also has a booster that contains methyl alcohol. As the gas mixture bubbles through the booster, alcohol vapor is mixed with the gas and it also burns in the torch flame. The alcohol vapor provides a slightly reducing flame for soldering, brazing, or welding. Figure 20-61 shows a self-generating gas welding outfit.

All the safety precautions recommended in the previous chapters on oxyfuel gas welding, soldering, and brazing, should be followed.

Figure 20-61. *A self-generating oxyhydrogen welding outfit being used to solder a small copper box. (E. Spirig-Switzerland, supplied by Solder Absorbing Technologies, Inc.)*

20.21 REVIEW OF SAFETY

- In submerged arc welding, the flux remaining over the completed weld acts as a heat insulator. The weld, as a result, remains very hot for some time after it is completed.
- **The usual eye protection and protective clothing should be worn when doing electroslag or electrogas welding.**
- There is some flash and sparking from arc stud welding operations. **Eyes, face, and hands must be protected from this sparking. Similar protection is necessary when performing arc spot welding.**
- Laser welding uses light energy. The beam of light from the laser welding equipment must never strike any object except the parts to be welded. **Always wear laser beam eye protection.**
- Plasma arc welding or cutting produces intense sound. **In addition to eye, face, and body protection, it is usually necessary to wear hearing protection during such welding or cutting operations.**
- There is usually considerable sparking during the hammering together of parts to be joined in forge welding. **Face, eyes, and hands must be adequately protected.**
- Friction welding generates considerable sparking. **Protective clothing, goggles, and gloves should be worn.**
- Handling explosives is always dangerous. They must be carefully stored, applied, and detonated. Follow manufacturer's instructions.
- A considerable quantity of molten metal is formed in the crucible during the thermite

welding process. When this molten metal enters the mold, it must be carefully managed so that the metal does not come in contact with moist or wet surfaces. **Protect eyes, face, hands, and body.**

- Oxygen and hydrogen gas produced in a self-generating unit may form a very explosive mixture. Hydrogen is a highly explosive gas.
- The underwater welder should be in touch with the surface operator by means of a communication system. **Precautions must be taken if a metal helmet is worn so that no part of it comes in contact with the welding circuit.**

TEST YOUR KNOWLEDGE

Write your answers on a separate sheet of paper. Do not write in this book.

1. How are the arc and molten metal protected from oxidation in submerged arc welding?
2. Can more than one electrode wire be used when submerged arc welding? Why?
3. What is the purpose of the copper shoes used during electroslag or electrogas welding?
4. What is the difference between electroslag welding and electrogas welding?
5. What is the advantage of narrow gap welding?
6. What is the problem with narrow gap welding?
7. In arc stud welding, what is placed over the stud prior to welding?
8. The tungsten electrode in plasma arc welding is surrounded by a(n) _____ gas. This gas may be an inert gas like _____ or _____, but it cannot contain _____ .
9. What is the difference between a transferred arc and a nontransferred arc in plasma arc welding?
10. List three things a welder can do to prevent rapid heat loss while underwater welding.
11. How are the solid-state welding processes different from all other welding processes?
12. What is used in cold welding, explosion welding, and forge welding to join two pieces of metal?
13. How is heat generated in friction welding?
14. What are some advantages (at least four) of ultrasonic welding?
15. List two advantages of electron beam welding. List the main disadvantages.
16. How high an accelerating voltage may be used in electron beam welding?
17. Why is a CO_2 laser better than a ruby laser for welding?
18. How is a laser beam focused on the part being welded?
19. How is the oxygen produced in self-generating oxyhydrogen gas welding?
20. What must welders do while PAW to protect their hearing?

Chapter 21

SPECIAL FERROUS WELDING APPLICATIONS

LEARNING OBJECTIVES

After studying this chapter, you will be able to:
* Define low-, medium-, and high-carbon steel, as well as plain carbon and alloy steel.
* Describe preheat treatment, interheating, and postweld heat treatment, and why each is done.
* Describe the proper procedure for welding stainless steel, tool steel, and cast iron.

Practically all metals can be successfully welded. This chapter will explain some of the factors that must be considered and applied when welding ferrous metals other than plain carbon steels. *Ferrous metals* are metals and their alloys that contain large amounts of iron (Fe) and include the following:

- Medium- and high-carbon steels.
- High-strength low-alloy (HSLA) steels.
- Chromium-molybdenum steels (Cr-Mo).
- Nickel-based alloys.
- Precoated steels.
- Stainless steels.
- Cast iron.

Many of the special ferrous metals listed above require preheating, interpass heating, and/or postweld heating. Such heating is done to ensure that the steel in the weld and weld *heat-affected zone (HAZ)* are left as strong as possible when the weld is completed and cooled. Chapter 29 describes the heat treatment processes in greater detail.

When preheating, interpass heating, or postweld heating, some means of monitoring the temperature is required. Temperature-indicating crayons are often used for this purpose. See Heading 29.15.

21.1 FERROUS METALS AND ALLOYS

The two large classifications of metals are:
- Ferrous metals.
- Nonferrous metals.

Chapter 22 covers the special welding applications for nonferrous metals and alloys.

Ferrous metals are iron and steel. Wrought iron is easily hammered and formed. It is almost pure iron and contains only about 0.2% carbon. Cast iron is **brittle** (easily fractured) and contains between 2.0% and 4.5% carbon. Steel is pure iron to which small quantities of carbon are added to provide its superior strength. To obtain special properties in steel, alloys are added in varying amounts. An alloy is a chemical element such as chromium (Cr), sulfur (S), phosphorous (P), nickel (Ni), titanium (Ti), manganese (Mg), molybdenum (Mo), cobalt (Co), or columbium (Cb). See Figure 21-1. Various alloys added to plain carbon steel can make it stronger, tougher, more corrosion-resistant, and easier to machine. As the carbon content and percentage of certain alloys increases, the metal becomes harder to weld.

To make steels more resistant to corrosion, they may be coated with aluminum (*aluminized*) or zinc (*galvanized*). Such coatings, however, make these steels harder to weld.

21.2 WELDING MEDIUM- AND HIGH-CARBON STEELS

To weld medium- or high-carbon steels, or steel alloys, the metals must be accurately identified so that the correct welding procedure can be used. The following information must be known in order to make an acceptable weld:
- The correct preheat temperature (if applicable).
- The proper interpass heating temperature (if applicable).

Alloying Element	Chemical Abbreviation	Characteristic given to steel
Aluminum	Al	Helps with deoxidization.
Boron	B	Improves hardenability.
Carbon	C	Increases hardness, strength, and improves wear quality.
Chromium	Cr	Increases hardenability and corrosion resistance.
Cobalt	Co	Increases hardness and wear qualities.
Columbium	Cb	Helps to eliminate carbide precipitation.
Copper	Cu	Improves strength and corrosion resistance.
Lead	Pb	Improves machinability.
Manganese	Mg	Increases strength and hardenability. Makes the steel more responsive to heat treatment.
Molybdenum	Mo	Improves hardenability and high-temperature strength.
Nickel	Ni	Increases toughness and strength.
Phosphorus	P	Improves strength.
Silicon	Si	Helps to deoxidize the steel and improves the hardenability.
Sulfur	S	Improves machinability.
Titanium	Ti	Helps with the elimination of carbide precipitation.
Tungsten	W	Improves high-temperature strength and improves wear qualities.
Vanadium	V	Increases toughness and helps to create a fine grain structure.

Figure 21-1. *A list of commonly used elements that may be added to alloy steel to produce a variety of special characteristics. The name and chemical abbreviation is given for each element.*

- The proper postweld heat treatment temperature (if applicable).
- The proper welding process and procedure.
- The correct electrode or filler metal to use.

A medium-carbon steel contains between 0.30% and 0.50% carbon. Medium-carbon steels are often used to manufacture hand tools, cutting tools, and machines. They are generally hard and tough, and yet will not break or wear easily. See Figures 21-2 and 21-3.

Preheating may be required to control the rate of cooling, which in turn reduces the amount of brittle, high-carbon martensite formations. The need for preheating increases as the percentage of carbon increases. The welding process also affects the need for preheating. A fast welding process that has a low heat input to the metal will have too-fast a cooling rate and will require preheating. Medium-carbon steel with 0.45% to 0.60% carbon should be preheated to between 200°F and 400°F (93°C and 204°C).

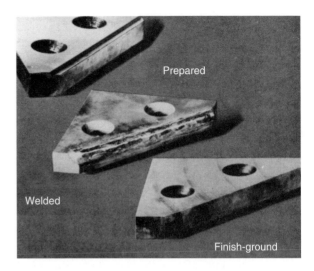

Figure 21-2. *A broken metal shear blade prepared for welding, welded, then finished by grinding. (Eureka Welding Alloys, Inc.)*

Figure 21-3. *A boring mill cutter. A—Before welding, showing damaged cutting edges. B—After welding. When the buildup is completed, the cutter is reground to the proper shape. (Eureka Welding Alloys, Inc.)*

If the joint is thick, interpass heating may be required at the same temperatures. Postweld heat treating may also be required. This is done to relieve stresses. Stress-relieving is especially necessary on thick metal sections or when the weldment is restrained in a fixture during welding.

High-carbon steels contain between 0.50% and 1.0% carbon. They are used in the manufacture of cutting tools and dies. See Figure 21-4. High-carbon steels are preheated to 400°F and higher to reduce the formation of martensite. Postweld heating is required to relieve stresses. Interpass heating is required on thick sections of metal.

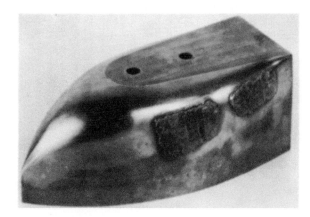

Figure 21-4. A die repaired by rebuilding the worn areas with overlapping beads. The surface will be ground to match the original finish. (Eureka Welding Alloys, Inc.)

One of the greatest problems to making a successful weld on all high-carbon and alloy steels is the introduction of hydrogen (H_2) into the weld and heat-affected zone. Hydrogen can come from the atmosphere around the weld and some electrode coatings. Hydrogen in the weld area can cause cracks to occur in the weld or heat-affected zone.

When using the SMAW process to weld a high-carbon steel, low-hydrogen electrodes should be selected to eliminate or reduce the hydrogen pickup in a weld. Electrodes like the EXX15, EXX16, EXX18, EXX28, and EXX48 are low-hydrogen electrodes. Low-hydrogen electrodes must be kept in an electrode oven until just before they are used. This practice reduces the chance of the electrode picking up hydrogen from the atmosphere. A short arc length is recommended to eliminate oxidation.

Gas metal arc welding (GMAW) can be used to weld medium- or high-carbon steel. The recommended shielding gases are carbon dioxide (CO_2), argon and oxygen, or argon and carbon dioxide. Gas tungsten arc welding (GTAW) using argon (Ar) can be used to join medium- or high-carbon steels. The FCAW process may also be used successfully on medium- and high-carbon steels with fluxes that are low in hydrogen. Oxyfuel gas processes may be used, but are slower than the arc processes.

21.3 WELDING ALLOY STEELS

Alloy steels have been produced for hundreds of years. New alloys are constantly being developed to meet special needs in industry. Each commercially produced steel alloy has special properties and contains varied alloying elements. Figure 21-5 shows what types of steel alloys

are used in the construction of containers that hold low- and very low-temperature gases. These steel alloys, which maintain their strength at very low temperatures, are known as *cryogenic steels*. A chart of electrodes used for low-alloy steel welding is shown in Figure 21-6.

High-strength low-alloy (HSLA) steels are finding greater use in sheet metal fabrication, automobiles, and steel building construction. Low-alloy steels usually contain less than 3% alloying elements. HSLA steels are 10% to 30% stronger than straight carbon steel. Because of their increased strength, they permit the design of thinner and lighter parts that are as strong as the parts previously used.

Welding of HSLA steels can be done using virtually any welding process, but preheating is normally required to reduce martensite formation. See Figure 21-7. When welding different thicknesses of HSLA steels, use the higher preheat temperature. The edges of the joints should be preheated to a distance equal to the metal thickness or 3" (76mm), whichever is greater. To eliminate the absorption of hydrogen, all low-hydrogen procedures should be followed. Low-hydrogen electrodes should be used with SMAW. FCAW fluxes should have a zero hydrogen content. Interpass heating of thick metal sections is recommended. Postweld heat treatment is seldom required.

Stamping and forging presses and machine cutting tools are used in production. The tools and dies used in manufacturing are capable of producing thousands of duplicate parts. Higher-quality tools and dies are generally made from high-carbon alloy steels. If they break or wear, they are difficult to replace. They are, however, weldable. Most welding processes can be used to repair breaks and fill in worn surfaces. See Figures 21-2, 21-3, and 21-4.

21.4 WELDING CHROME-MOLYBDENUM STEELS

Chrome-molybdenum steels (often called *chrome-moly steels*) are relatively corrosion-resistant. They contain from 0.5% to 9.0% chromium and from 0.5% to 1.0% molybdenum. They may also contain up to 0.35% carbon. Chrome-moly (Cr-Mo) steels may also contain small amounts of vanadium, titanium, or columbium for special applications. They are used when strength at high temperatures is required. These steels are also resistant to oxidation and sulfur corrosion.

Preheat temperatures of between 300°F and 500°F (150°C and 260°C) are used when the carbon content is above 0.15%. Postweld heating is done by raising the temperature of the finished weld 100 degrees and holding at that temperature for a time to diffuse the hydrogen in the weld.

Most forms of arc welding can be used successfully on Cr-Mo steels. A short arc length is best to eliminate oxidation and the loss of chromium in the weld. If high temperatures are a consideration, braze welding can be used.

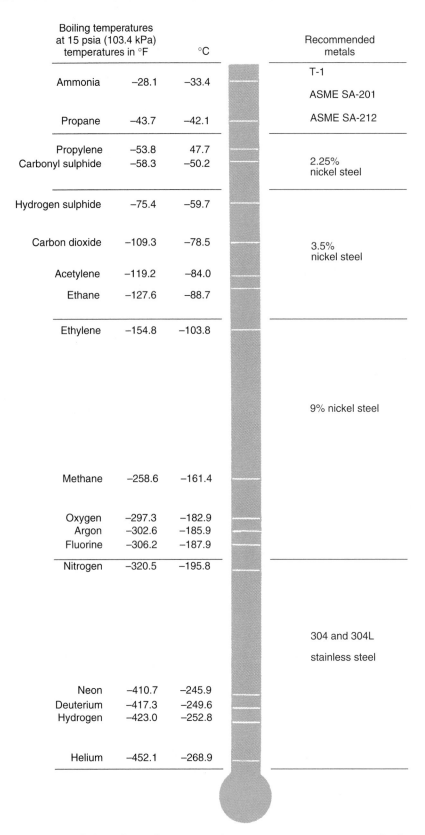

Figure 21-5. *A table of recommended steels used to contain low-temperature (cryogenic) fluids. (Welding Design and Fabrication)*

AWS classification[a]	Type of covering	Capable of producing satisfactory welds in positions shown[b]	Type of current[c]
E70 series – Minimum tensile strength of deposited metal, 70,000 psi (480 MPa)			
E7010-X	High cellulose sodium	F, V, OH, H	DCEP
E7011-X	High cellulose potassium	F, V, OH, H	ac or DCEP
E7015-X	Low hydrogen sodium	F, V, OH, H	DCEP
E7016-X	Low hydrogen potassium	F, V, OH, H	ac or DCEP
E7018-X	Iron powder, low hydrogen	F, V, OH, H	ac or DCEP
		H-fillets	ac or DCEN
E7020-X	High iron oxide	F	ac or dc, either polarity
		H-fillets	ac or DCEN
E7027-X	Iron powder, iron oxide	F	ac or dc, either polarity
E80 series – Minimum tensile strength of deposited metal, 80,000 psi (550 MPa)			
E8010-X	High cellulose sodium	F, V, OH, H	DCEP
E8011-X	High cellulose potassium	F, V, OH, H	ac or DCEP
E8013-X	High titania potassium	F, V, OH, H	ac or dc, either polarity
E8015-X	Low hydrogen sodium	F, V, OH, H	DCEP
E8016-X	Low hydrogen potassium	F, V, OH, H	ac or DCEP
E8018-X	Iron powder, low hydrogen	F, V, OH, H	ac or DCEP
E90 series – Minimum tensile strength of deposited metal, 90,000 psi (620 MPa)			
E9010-X	High cellulose sodium	F, V, OH, H	DCEP
E9011-X	High cellulose potassium	F, V, OH, H	ac or DCEP
E9013-X	High titania potassium	F, V, OH, H	ac or dc, either polarity
E9015-X	Low hydrogen sodium	F, V, OH, H	DCEP
E9016-X	Low hydrogen potassium	F, V, OH, H	ac or DCEP
E9018-X	Iron powder, low hydrogen	F, V, OH, H	ac or DCEP
E100 series – Minimum tensile strength of deposited metal, 100,000 psi (690 MPa)			
E10010-X	High cellulose sodium	F, V, OH, H	DCEP
E10011-X	High cellulose potassium	F, V, OH, H	ac or DCEP
E10013-X	High titania potassium	F, V, OH, H	ac or dc, either polarity
E10015-X	Low hydrogen sodium	F, V, OH, H	DCEP
E10016-X	Low hydrogen potassium	F, V, OH, H	ac or DCEP
E10018-X	Iron powder, low hydrogen	F, V, OH, H	ac or DCEP
E110 series – Minimum tensile strength of deposited metal, 110,000 psi (760 MPa)			
E11015-X	Low hydrogen sodium	F, V, OH, H	DCEP
E11016-X	Low hydrogen potassium	F, V, OH, H	ac or DCEP
E11018-X	Iron powder, low hydrogen	F, V, OH, H	ac or DCEP
E120 series – Minimum tensile strength of deposited metal, 120,000 psi (830 MPa)			
E12015-X	Low hydrogen sodium	F, V, OH, H	DCEP
E12016-X	Low hydrogen potassium	F, V, OH, H	ac or DCEP
E12018-X	Iron powder, low hydrogen	F, V, OH, H	ac or DCEP

a. The letter suffix "X" as used in this table stands for the suffixes A1, B1, B2, etc. (see Figure 11-47) and designates the chemical composition of the deposited weld metal.
b. The abbreviations F, V, OH, H, and H-fillets indicate welding positions as follows: F = Flat; H = Horizontal; H-fillets = Horizontal fillets.
    V = Vertical   }   For electrodes 3/16" (4.8mm) and under, except 5/32" (4mm) and under for classifications
    OH = Overhead }   EXX15-X, EXX16-X, and EXX18-X.
c. DCEP means electrode positive (reverse polarity). DCEN means electrode negative (straight polarity).

Figure 21-6. Alloy steel covered electrodes for shielded metal arc welding (SMAW). (From AWS A5.5)

Steel classification	Metal thickness		Temperature	
	in.	mm	°F	°C
A242 A441	up to 0.75	up to 19.1	32	0
A572 (Gr 42, 50) A572 (Gr 60, 65 under 2.50")	0.81 to 1.50	5.3 to 38.1	50	10
A588	1.51 to 2.50	38.4 to 63.5	150	66
A633 (Gr A, B, C, D) A633 (Gr E under 2.50")	2.51 and above	63.8 and above	225	121
A572 (Gr 60, 65) A633 (Gr E)	2.51 and above	63.8 and above	300	149

Figure 21-7. *Minimum preheat and interpass temperatures for high-strength low-alloy steels suggested by the American Society of Tool Manufacturers.*

21.5 NICKEL-BASED ALLOYS

The nickel-based alloys contain between 32% and 82% nickel (Ni). Other alloying elements include carbon (C), chromium (Cr), molybdenum (Mo), iron (Fe), cobalt (Co), and copper (Cu).

These alloys are used for corrosion resistance and high-temperature applications. Two alloys, Inconel 601 and Hastelloy X, are used to resist oxidation up to 2200°F (1204°C).

Most nickel-based alloys may be welded by using the SMAW, GMAW, GTAW, FCAW, or SAW processes. However, there are many alloys that are very difficult to weld.

Hastelloy D, an alloy containing 10% silicon and 3% copper, is considered unweldable by means of arc welding processes. It can, however, be welded using the oxyacetylene process.

The presence of small amounts of sulfur may cause hot cracking in the weld and heat-affected zone. Aluminum, silicon, and phosphorus also may create hot cracking problems. Lead (Pb) causes hot shortness, a weakness that causes the molten weld metal to fall away from the base metal.

Some nickel-based alloys require preheat and postweld heat treatments to relieve stresses, reduce cracking, and renew their original characteristics.

The filler metal or electrode should be selected to match the base metal chemical composition. Examples of SMAW electrodes used to weld nickel-based alloys are ENiCu-2, ENiCrFe-3, and ENiMo-2. Two of the filler metals used with GMAW, GTAW, and SAW are ERNi-3 and ERNiCrFe-6. In order to better select a suitable filler metal or electrode, the chemical composition of the metal to be welded should be known. If unknown, it may be obtained from the manufacturer.

21.6 WELDING PRECOATED STEELS

Carbon steels and alloy steels that need to be protected from *oxidation* (rust) are often coated with a different metal. *Aluminizing* is a process used to coat a surface with aluminum. *Galvanizing* is a process that involves coating steel with zinc. Both processes make welding these coated metals more difficult.

When welding coated metals, the surface coating melts and may contaminate the base metal and filler metal in the molten weld pool. Zinc in the weld will cause unacceptable cracks called "zinc penetration cracking," particularly in the area of the weld. It also reduces the coating's effectiveness.

Most welding processes that would normally be used on uncoated steels also can be used on coated steels. Filler metals that are low in silicon must be used, because welds that contain less than 0.2% silicon are generally free from zinc penetration cracking. When SMAW is used, E6012 or E6013 rutile-covered electrodes are recommended. An ER70S-3 (low-silicon) electrode wire is recommended when welding galvanized steel with the GMAW process.

Following are several other recommendations for welding aluminized or galvanized steels:
- Use single- or double-bevel-groove joints.
- Maintain a root opening of about 1/16" (1.6mm).
- Remove the coating material from the area around the bead with an oxyfuel gas torch or by shot blasting.
- Run procedure qualification tests with the electrodes proposed for use on the weld.

With the GMAW process, if the coating is not removed before the weld is started, use a slower travel speed to melt and vaporize the coating. With SMAW, use a forward-and-back motion of the electrode to melt the coating ahead of the weld pool.

21.7 WELDING MARAGING STEELS

Maraging steels contain less than 0.03% carbon. They contain less than 0.10% manganese and silicon and less than 0.010% phosphorus and sulfur. They contain about 18% nickel, about 10% cobalt, up to 5% molybdenum, and small amounts of titanium and aluminum. One alloy contains 5% chromium, which replaces the cobalt.

Maraging steels have high strength and toughness and have good weldability. The word *maraging* is a combination of two words, *martensite* and *aging*. Martensite forms when this steel cools from the higher-temperature austensite phase. The steel is then held at about 900°F (482°C) for up to 12 hours to age the martensite. This aging increases the strength, toughness, and hardness of the steel.

These steels can be welded with most arc welding processes. The filler metal must be of the same composition as the base metal. To reduce oxidation, a short arc length is recommended.

Preheating is not required, but postweld heat treating is needed to age the martensite again across the weld and heat-affected zone (HAZ).

21.8 STAINLESS STEELS

A large number of steel alloys are known as ***stainless steels***. They all contain carbon (0.03% to 0.45%), chromium (11% to 32%), nickel (0.60% to 37%), molybdenum (0.35% to 4.0%), and smaller amounts of manganese, phosphorus, silicon, sulfur, and copper.

Stainless steels are generally corrosion-resistant, which accounts for the "stainless" description. However, corrosion of some stainless steels can occur under adverse conditions. They have good physical characteristics due to the alloying elements used and generally have a very clean, bright appearance. Completed weld beads often need polishing to renew their bright appearance.

Stainless steels are generally divided into the following four groups:
- Chromium martensitic.
- Chromium ferritic.
- Austenitic.
- Precipitation-hardening.

The American Iron and Steel Institute (AISI) has classified the alloy types and assigned a three-digit number to them. They are the 200, 300, and 400 series. Examples of actual alloys are *201, 304, 316L, 410, 420,* and *430*. The "L" in 316L means that this steel has a low carbon content.

The designation numbering system for precipitation-hardening stainless steels is a combination of three or more numbers and letters. Refer to the AWS *Welding Handbook,* Volume 4, 1982 edition.

21.8.1 Welding Chromium Martensitic Steels

Chromium martensitic steels are very hard and not very ductile. They contain about 11.5% to 18% chromium (Cr) and carbon contents of from 0.15% to 1.20%. The chromium content in these steels permit them to form martensite when they are cooled rapidly (quenched). Martensitic steels are in the 400 or 4XX series of stainless steels. See Figure 21-8.

These steels may be welded using any of the arc processes normally used to weld steel. Preheating is required at about 500°F (260°C) to prevent cracking. Refer to Figure 21-9. Postweld heat treatment may be required to temper or anneal the weld and heat-affected zone. This is done to decrease hardness and increase strength and toughness. The procedure is to heat the weld area to about 1300°F (704°C), then slowly cool the weldment in the furnace to 1000°F (593°C). Finally, the weldment is slowly air-cooled. Figure 21-10 shows the annealing temperatures for martensitic steel. Mertensitic steels are normally welded using austenitic filler metal. Austenitic filler metal must be used when no postweld heat treatment is done.

Alloy	C	Mn	Si	Cr	Ni	P	S	Other Elements
403	0.15	1.00	0.50	11.5-13.0		0.04	0.03	
410	0.15	1.00	1.00	11.5-13.0		0.04	0.03	
414	0.15	1.00	1.00	11.5-13.5	1.25-2.50	0.04	0.03	
416	0.15	1.25	1.00	12.0-14.0		0.04	0.03	
420	0.15	1.00	1.00	12.0-14.0		0.04	0.03	
422	0.20-0.25	1.00	0.75	11.0-13.0	0.5-1.0	0.025	0.025	0.15-0.30V; 0.75-1.25 Mo; 0.75-1.25W
431	0.20	1.00	1.00	15.0-17.0	1.25-2.50	0.04	0.03	
440A	0.60-0.75	1.00	1.00	16.0-18.0		0.04	0.03	0.75 Mo
440B	0.75-0.90	1.00	1.00	16.0-18.0		0.04	0.03	0.75 Mo
440C	0.95-1.20	1.00	1.00	16.0-18.0		0.04	0.03	0.75 Mo
*CA-6NM	0.06	1.00	1.00	11.5-14.0	3.5-4.5	0.04	0.04	0.40-1.0 Mo
CA-15	0.15	1.00	1.50	11.5-14.0	1.0	0.04	0.04	0.5 Mo
CA-40	0.20-0.40	1.00	1.50	11.5-14.0	1.0	0.04	0.04	0.5 Mo
*CA-casting alloy								

Figure 21-8. *Maximum composition percentages for martensitic stainless steels.*

Percent carbon	Preheat temperature		Welding heat input	Postweld recommendation
	°F	°C		
Under 0.10	Min. 400	Min. 204	Normal	Optional
0.10 to 0.20	400 to 500	204 to 260	Normal	Cool slowly. Heat treatment optional
0.20 to 0.50	500 to 600	260 to 316	Normal	Required
Over 0.50	500 to 600	260 to 316	Higher than normal	Required

Figure 21-9. Maximum recommended preheat, welding, and postweld temperatures for martensitic stainless steels.

Alloy	Subcritical annealing temperature		Full annealing temperature	
	°F	°C	°F	°C
403	1200 to 1400	649 to 760	1525 to 1625	829 to 885
410	1200 to 1400	649 to 760	1525 to 1625	829 to 885
414	1200 to 1350	649 to 732	Not recommended	
420	1250 to 1400	677 to 760	1550 to 1650	843 to 899
431	1150 to 1300	621 to 704	Not recommended	
440A,B,C	1250 to 1400	677 to 760	1550 to 1650	843 to 899
*CA-6NM	1100 to 1150	593 to 621	1450 to 1500	788 to 816
CA-15	1150 to 1200	621 to 649	1550 to 1650	843 to 899
CA-40	1150 to 1200	621 to 649	1550 to 1650	843 to 899
*CA - casting alloy				

Figure 21-10. Recommended annealing treatments for martensitic stainless steels.

A Type 410 filler metal is used to weld 403, 410, 414, and 420 martensitic stainless steels. To match the carbon content in Type 420 stainless steel, a ER420 filler metal is used. Type CA-6NM castings and other similar alloys are welded using a 410NiMo filler metal.

No unusual welding techniques are required with any welding process used. A short arc length is recommended when welding any of the stainless steels to reduce oxidation and eliminate the loss of chromium. Martensitic steels which contain more than 0.20% carbon are more difficult to weld, as are all high-carbon steels, because cracking is more of a problem.

21.8.2 Welding Chromium Ferritic Steels

Chromium ferritic steels contain from 10.5% to 30% chromium. Wrought forms contain between 0.08% and 0.20% carbon. Cast forms may contain up to 0.50% carbon. Aluminum (Al), molybdenum (Mo), titanium (Ti), and columbium (Cb) are added as ferrite stabilizers. These steels are also found in the 400 or 4XX series of stainless steels. See Figure 21-11.

Alloy	C	Mn	Si	Cr	Ni	P	S	Other elements
405	0.08	1.00	1.00	11.5-14.5		0.04	0.03	0.10 - 0.30 Al
409	0.08	1.00	1.00	10.5-11.75		0.045	0.045	Ti - 6 x C% minimum
429	0.12	1.00	1.00	14.0-16.0		0.04	0.03	
430	0.12	1.00	1.00	16.0-18.0		0.04	0.03	
434	0.12	1.00	1.00	16.0-18.0		0.04	0.03	0.75 - 1.25 Mo
436	0.12	1.00	1.00	16.0-18.0		0.04	0.03	0.75 - 1.25 Mo; (Cb + Ta) = 5 x C%
442	0.20	1.00	1.00	18.0-23.0		0.04	0.03	
444	0.025	1.00	1.00	17.5-19.5	1.00	0.04	0.03	1.75 - 2.5 Mo; 0.035 N maximum; (Cb + Ta) minimum = 0.2 + 4(C% + N%)
446	0.20	1.50	1.00	23.0-27.0		0.04	0.03	0.25 N
26-1	0.06	0.75	0.75	25.0-27.0	0.50	0.04	0.02	0.75 - 1.50 Mo; 0.2 - 1.0 Ti; 0.04N; 0.20 Cu
29-4	0.010	0.30	0.20	28.0-30.0	0.15	0.025	0.02	3.5 - 4.2 Mo; 0.020 N; 0.15 Cu
29-4-2	0.010	0.30	0.20	28.0-30.0	2.0-2.5	0.025	0.020	3.5 - 4.2 Mo; 0.020 N; 0.15 Cu
*CB-30	0.30	1.00	1.50	18.0-21.0	2.0	0.04	0.04	
*CC-50	0.50	1.00	1.50	26.0-30.0	4.0	0.04	0.04	

*CB + CC - casting alloys

N - nitrogen (helps to form austenite)

Figure 21-11. Maximum composition percentages for ferritic stainless steels.

These steels may be arc welded using all the processes normally used to weld steel. Type 409 and 430 ferritic stainless steels are welded with Type 409 or 430 filler wire. Type 409 filler wire is available only as flux cored electrodes. Type 430 filler metal is available as covered electrodes, solid or flux cored wire, and welding rods.

Preheating is normally not necessary except on Types 430, 434, 442, and 446. These steels contain high percentages of carbon and chromium and are subject to weld cracking.

The heat-affected zone in ferritic welds is subject to grain growth. This grain growth in the steel results in a loss of toughness. The amount of the grain growth depends on the highest temperature reached and the length of time it is held at that high temperature. To reduce this effect, welding should be done with a small weld pool and at the highest possible speed. This will reduce the *time-at-temperature* factor, which affects grain growth.

Preheating at temperatures from 300°F (149°C) to 450°F (232°C) may be required to eliminate cracking and stresses. Interpass heating should be done at, or only slightly above, the preheat temperature to reduce the possibility of grain growth. Postweld heat treatment is done at temperatures between 1300°F (704°C) and 1550°F (843°C). The steel should then be cooled rapidly to between 1000°F (537°C) and 700°F (371°C).

Many electrodes are made for welding chromium ferritic steels. When a chromium stainless steel is welded with an austenitic electrode, the electrode should have a higher chromium content than the base metal. This will compensate for any dilution of the chromium in the weld. An E309 electrode is commonly used.

When GTAW, compatible filler metals must be used. The shielding gases used are He, Ar, or combinations of the two. When GMAW, various metal transfer methods may be used. The recommended shielding gas for spray transfer is argon (Ar) with 1% oxygen (O_2). A combination of helium, argon, and 2.5% carbon dioxide (CO_2) is used with the short-circuiting transfer method.

21.8.3 Welding Austenitic Stainless Steel

Austenitic stainless steels contain 16% to 30% chromium, 3% to 37% nickel, and 2% manganese. Small amounts of silicon (Si), phosphorous (P), sulfur (S), and other elements are also included in these steels. The carbon content of wrought austenitic stainless steels range from 0.03% to 0.25%. See Figure 21-12. Austenitic steels have very good strength and corrosion resistance at high and low temperatures. Typical 300 series austenitic stainless steels contain 18% chromium (Cr) and 8% nickel (Ni).

Austenitic stainless steels can be welded using all the normally used arc welding processes. Figure 21-13 lists the recommended filler metals for welding wrought or cast austenitic stainless steels (3XX). If the correct electrode, heat treatment, and welding procedure are not used, the metallurgical microstructure of the weld and heat-affected zone will be negatively affected. Intergranular corrosion and changes in grain microstructures may occur.

Procedure qualification tests should be made and closely adhered to when welding austentic stainless steels.

21.8.4 Welding Precipitation-Hardening Stainless Steels

Precipitation-hardening stainless steels (PH) develop their strength through heat treatment processes. Chemically, the term "precipitation" means to separate a solid or solid phase of a material from a liquid substance. Precipitation hardening occurs when martensite is separated out from the liquid steel. There are three types of PH steels. These are martensitic, semiaustenitic, and austenitic. Examples of martensitic types are 17-4 PH and 15-5 PH; semiaustenitics are 17-7 PH, PH 15-7 Mo, and AM 350, and austenitics are A-286 and 17-10 P.

These steels can be welded using any arc welding process normally used for steel. The weldability of these steels depends on the joint design and the conditions under which they are welded. Filler metal with the same composition as the base metal must be used, except when welding austenitic PH stainless steels. A nickel-based filler metal is used when welding austenitic PH stainless steels. When welding one PH stainless steel to another, or when high strength is not required, an austenitic stainless steel electrode Type 308, 309, or 310 can be used. Refer to the AWS *Welding Handbook*.

A postweld heat treatment consisting of a solution heat treatment followed by an aging process will obtain the best mechanical properties and corrosion resistance. Quite often, only aging is required.

Argon is used for manual GTAW. Helium (He) or a helium and argon mixture are used when automatic GMA welding. This is done to take advantage of the greater heat input and penetration possible with helium shielding gas. The root of the weld should be protected from contamination with an inert gas during welding.

When using GMAW, argon with 1% to 2% oxygen is used and argon and helium mixtures can be used to produce higher heat input to the weld. In the flat position, the spray transfer method is preferred. Short-circuiting or pulsed-spray transfer is used for welds in other positions.

21.9 WELDING DISSIMILAR FERROUS METALS

There are many iron and steel fabrications that require welding metals together even though their compositions are different. Welding stainless steels to mild steels, low-alloy steels to mild steels, and nickel-based alloys to stainless steel are typical examples.

It is very important that the composition and the properties of each metal be known. While welding, the composition of the electrode and filler metal will change due to the intermixing and dilution of the various alloying elements in each base metal. The filler metal used must be chosen carefully. The characteristics of the resulting alloy

Alloy	C	Mn	Si	Cr	Ni	P	S	Other elements
201	0.15	5.5-7.5	1.00	16.0-18.0	3.5-5.5	0.06	0.03	0.25N
202	0.15	7.5-10.0	1.00	17.0-19.0	4.0-6.0	0.06	0.03	0.25N
301	0.15	2.00	1.00	16.0-18.0	6.0-8.0	0.045	0.03	
302	0.15	2.00	1.00	17.0-19.0	8.0-10.0	0.045	0.03	
302B	0.15	2.00	2.0-3.0	17.0-19.0	8.0-10.0	0.045	0.03	
303	0.15	2.00	1.00	17.0-19.0	8.0-10.0	0.20	0.15 minimum	0-0.06 Mo
303Se	0.15	2.00	1.00	17.0-19.0	8.0-10.0	0.20	0.06	0.15 Se minimum
304	0.08	2.00	1.00	18.0-20.0	8.0-10.5	0.045	0.03	
304L	0.03	2.00	1.00	18.0-20.0	8.0-12.0	0.045	0.03	
305	0.12	2.00	1.00	17.0-19.0	10.5-13.0	0.045	0.03	
308	0.08	2.00	1.00	19.0-21.0	10.0-12.0	0.045	0.03	
309	0.20	2.00	1.00	22.0-24.0	12.0-15.0	0.045	0.03	
309S	0.08	2.00	1.00	22.0-24.0	12.0-15.0	0.045	0.03	
310	0.25	2.00	1.50	24.1-26.0	19.0-22.0	0.045	0.03	
310S	0.08	2.00	1.50	24.0-26.0	19.0-22.0	0.045	0.03	
314	0.25	2.00	1.50-3.0	23.0-26.0	19.0-22.0	0.045	0.03	
316	0.08	2.00	1.00	16.0-18.0	10.0-14.0	0.045	0.03	2.0-3.0 Mo
316L	0.03	2.00	1.00	16.0-18.0	10.0-14.0	0.045	0.03	2.0-3.0 Mo
317	0.08	2.00	1.00	18.0-20.0	11.0-15.0	0.045	0.03	3.0-4.0 Mo
317L	0.03	2.00	1.00	18.0-20.0	11.0-15.0	0.045	0.03	3.0-4.0 Mo
321	0.08	2.00	1.00	17.0-19.0	9.0-12.0	0.045	0.03	Ti - 5xC% minimum
329	0.10	2.00	1.00	25.0-30.0	3.0-6.0	0.045	0.03	1.0-2.0 Mo
330	0.08	2.00	0.75-1.5	17.0-20.0	34.0-37.0	0.04	0.03	
347	0.08	2.00	1.00	17.0-19.0	9.0-13.0	0.045	0.03	
348	0.08	2.00	1.00	17.0-19.0	9.0-13.0	0.045	0.03	0.2 Cu; (Cb+ta) = 10xC% (Ta - .10% max.)
384	0.08	2.00	1.00	15.0-17.0	17.0-19.0	0.045	0.03	

Figure 21-12. *Chemical compositions for austenitic stainless steels.*

in the weld must be considered, so that a weld at least as strong as the weakest of the base metals will result.

The welding parameters must be controlled to insure the solidified weld metal will meet physical and chemical requirements. The filler metal or electrode must be selected to properly mix with the base metals being welded and produce a high-quality weld. Controlling the dilution of the welds and weld zone is an important consideration. Quite often, a small weld pool must be maintained to minimize dilution. Multiple stringer beads are often used, instead of a wide weaving bead, to fill a weld joint. This is done to minimize dilution.

If the melting temperatures of the dissimilar metals are within about 200°F (93°C) of each other, normal welding procedures may be used and no difficulties normally occur. However, if the melting temperature span is greater then this, welding becomes more difficult. It may be nec-

essary to weld a layer of a metal with an intermediate melting temperature onto the face of the metal with the higher melting temperature. This process is called *buttering*. Buttering will allow a weld to be made between two metals of closer melting temperatures.

If one of the metals requires preheating, it must be preheated independently and in isolation from the second piece. Postweld heat treatment of dissimilar metals also can be a problem, if one piece requires a treatment different from the other.

Another solution to the problem of joining dissimilar metals is to braze them. Brazing can be done at temperatures that will not melt either part of the joint. When there is no melting of the two base metal, there can be no problem of dilution or intermixing of the two metals. Joint strength may be sacrificed, however, if brazing is used instead of welding.

Type of stainless steel		Recommended filler metals		
Wrought	Cast[a]	SMAW[b]	GMAW, GTAW, PAW, SAW	FCAW[4]
201 202	—	E209 E219 E308	ER209 ER219 ER308	E308T-X
301 302 304 305	CF-20 CF-8	E308	ER308	E308T-X
304L	CF-3	E308L E347	ER308L ER347	E308LT-X E347T-X
309	CH-20	E309	ER309	E309T-X
309S	—	E309L E309Cb	ER309L	E309LT-X E309CbLT-X
310 314	CK-20	E310	ER310	E310T-X
310S	—	E310 E310Cb	ER310	E310T-X
316	CF-8M	E316	ER316	E316T-X
316L	CF-3M	E316L	ER316L	E316LT-X
316H	CF-12M	E16-8-2 E316H	ER16-8-2 ER316H	E316T-X
317	—	E317	ER317	E317LT-X
317L	—	E317L	ER317L	E317LT-X
321	—	E308L E347	ER321	E308LT-X E347T-X
330	HT	E330	ER330	—
347	CF-8C	E308L E347	ER347	E308LT-X E347T-X
348	—	E347	ER347	E347T-X

a. Castings higher in carbon but otherwise of generally corresponding compositions are available in heat resisting grades. These castings carry the "H" designation (HF, HH, and HK, for instance). Electrodes best suited for welding these high-carbon versions are the standard electrodes recommended for the corresponding lower carbon corrosion-resistant castings shown above.
b. Covered electrodes for shielded metal arc welding (SMAW).
c. Bare welding rods and electrodes for gas metal arc (GMAW), gas tungsten arc (GTAW), plasma arc (PAW), and submerged arc welding (SAW).
d. Tubular electrodes for flux cored arc welding (FCAW). The suffix -x can be -1, -2, or -3.

Figure 21-13. Recommended filler metals for welding wrought and cast austenitic stainless steels.
(American Welding Society)

21.10 WELDING CAST IRON

Cast iron contains between 1.8% and 4% carbon which makes the metal brittle. There are four types of cast iron:

- Gray cast iron.
- White cast iron.
- Malleable cast iron or malleable iron.
- Nodular or ductile cast iron.

All cast irons are difficult to weld, because they contain a great deal of carbon. Welding temperatures and an uncontrolled cooling rate can produce undesirable microstructures in the metal. Carbides tend to form at the weld boundaries and high-carbon martensite also tends to form. Both of these are brittle and have very low strength. Welding also produces high stresses in this brittle metal. Ductile and malleable iron are more *ductile* (not brittle) and are easier to weld.

Gray cast iron is cooled slowly and forms graphite in a flake form as it cools. It is brittle and lacks ductility. A broken section has a gray appearance.

Gray cast iron can be arc welded using a nickel or nickel alloy electrode such as ENi-CI or ENiFe-CI. The nickel content in the filler metal permits the weld to move or "creep" after welding. This relieves stresses and prevents cracks. Nickel-bearing filler metals also make the completed weld easier to machine. Pure iron filler metal which contains very low-carbon contents can be used. Carbon from the base metal will be diluted in the weld area by the pure iron filler metal and create a less brittle weld. The American Welding Society specification AWS A5.15 describes cast iron electrodes and their recommended uses. Gray cast iron can also be braze welded successfully.

White cast iron forms when cast iron cools very rapidly. The high carbon content remains throughout the microstructure. As a result, it is brittle and has virtually no ductility. It is considered unweldable. The fractured end of white cast iron has a white appearance.

Preheating of cast iron produces slower cooling rates. Postweld heating will slow the cooling rate and reduce the amount of hardness in the weld and heat-affected zone . See Figure 21-14 for recommended preheat and interpass temperatures for use when welding cast iron. Figure 21-15 shows an example of how preheating, interpass heating, and postweld heat treating can relieve stresses and eliminate the tendencies for a weld or the base metal to crack.

Malleable cast iron is made by heat treating white iron. See Heading 29.12. The heat treatment causes nodules of graphite to form. This nodular form is called *temper carbon.* The resulting malleable iron is more ductile because of the temper carbon.

Malleable iron may be welded using nickel alloy filler metals or electrodes. Electrodes such as ENi-CI or ENi-CI-A may be used. See Figure 21-14 for suggested preheat and interpass heating temperatures.

Ductile cast iron is similar to gray cast iron except in the form of the graphite that develops when they are produced. The graphite in ductile cast iron is in the form of spheres. As a result, this type of cast iron is more ductile.

Type of cast iron	Matrix microstructure	Welding process			
		Arc		Oxyacetylene	
		°F	°C	°F	°C
Gray	—	70-600	21-315	800-1200	427-649
Malleable	Ferritic	70-300	21-149	800-1200	427-649
Malleable	Pearlitic	70-600	21-315	800-1200	427-649
Ductile	Ferritic	70-300	21-149	400-1200	204-649
Ductile	Pearlitic	70-600	21-315	400-1200	204-649

Figure 21-14. *Recommended preheat and interpass temperatures for welding cast iron.*

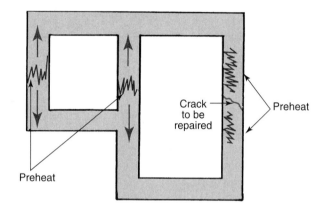

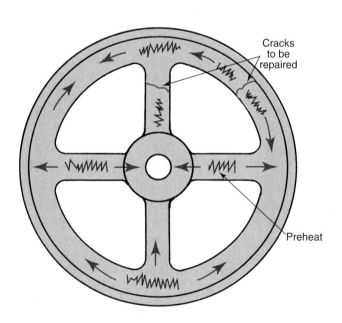

Figure 21-15. *Localized preheating of cast iron parts is used to relieve stresses, slow the cooling rate, and eliminate cracking.*

Spheroidized graphite is helped to form by adding magnesium (Mg).

Welding can be done using nickel alloy electrodes such as ENi-1. Ductile cast iron may also be welded using a carbon steel electrode like E70S-2. The welding process used must minimize the heat input to the metal or the magnesium will be vaporized and lost. If this happens, the graphite form will become flakes in the weld area. The resulting weld metal will become gray cast iron which is more brittle and has no ductility. See Figure 21-14 for preheat and interpass temperatures.

Cast iron is prepared for welding much like steel. The surfaces on each side of the joint must be cleaned for some distance away from the joint. The edges must be ground or machined if a bevel weld is used. If braze welding is to be done, the graphite particles (dust) need to be removed to insure a better braze weld.

Cast iron joints may be strengthened using the methods shown in Figure 21-16. Backing strips may be used to support the weld area during welding. See Figure 21-17.

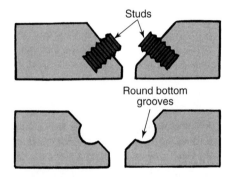

Figure 21-16. *Methods of strengthening a cast iron joint prior to welding.*

21.10.1 Oxyfuel Gas Welding Cast Iron

The tip size used for welding cast iron should be similar to the size used for welding steel of the same thickness. A neutral flame should be used along with a cast iron welding rod of the proper size. A flux is also required.

When welding cast iron, the backhand technique is usually used. Backhand welding also tends to slow the cooling rate of the weld and reduce stresses. The torch should be held at a 60° angle to the plate. The inner cone should not touch the metal.

Flux is added by coating the welding rod with it. The flux enters the weld as the welding rod is consumed. The flux must have the correct constituents. It must be fresh, clean, and free from moisture.

In cast iron welding, the molten pool is not very fluid. It is important that gas pockets and oxides be worked to the surface of the weld. This may be done by stirring the molten pool with the filler rod. The oxides or gases will then be removed by the flux. The weld must have thorough penetration, and a slight crown is preferred. An oscillating torch and filler rod motion are usually used. Figure 21-18 shows cast iron being welded by the oxyfuel gas process. Notice how the flux is added, and that the backhand technique is being used.

The oxyfuel gas process is also used to braze and braze weld cast iron. Brazing and braze welding cast iron are described in Heading 10.6.6.

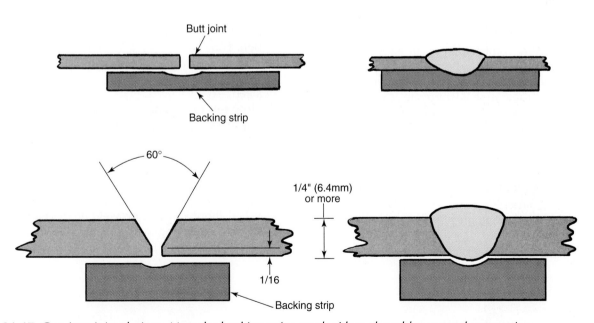

Figure 21-17. *Cast iron joint designs. Note the backing strips used with each weld to control penetration.*

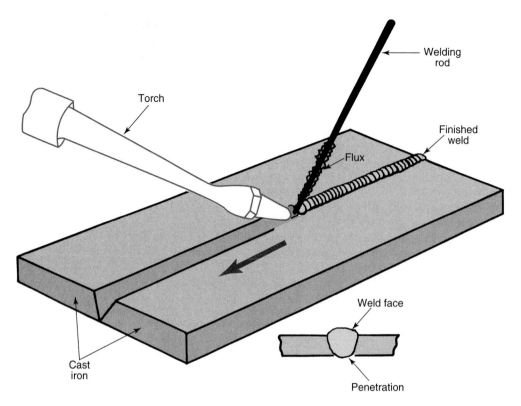

Figure 21-18. Welding cast iron with the oxyfuel gas process. Note that the backhand method of welding is used on thicker sections.

TEST YOUR KNOWLEDGE

Write your answers on a separate sheet of paper. Do not write in this book.
1. What is a ferrous metal or a ferrous-based alloy?
2. List three popular alloy steels.
3. Low-alloy steels contain less than _____ % alloying elements. They are _____ % to _____ % stronger than a straight carbon steel.
4. How is the alloy composition of an alloy electrode designated?
5. What letter is used to designate a molybdenum alloy electrode? A chrome-molybdenum electrode?
6. List the three general classifications of stainless steel and give an example of each.
7. List four electrodes used to weld stainless steel. One electrode classification should be a low-carbon, and one should be a GMAW electrode.
8. Two common problems with the welding of stainless steel are _____ and _____ .
9. What should be done to prevent the problems listed in question 8?
10. A typical stainless steel used in industry is sometimes called an 18-8 stainless steel. This stainless steel, which contains 18% chromium and 8% nickel, is a(n) _____ type stainless steel. See Figures 21-8, 21-11, and 21-12.
11. What are nickel-based alloys used for?
12. What type of heat treatment is given to a tool steel after welding?
13. Maraging steels contain _____ _____ amounts of carbon. They have _____ _____ strengths.
14. Why is a maraging steel aged?
15. What is the problem with arc welding galvanized steel?
16. How can the strength of a cast iron groove weld be improved?
17. Give two examples of electrodes used to weld cast iron.
18. Why is cast iron preheated prior to welding?
19. What preheat temperature is used when SMAW cast iron?
20. When should a cast iron weld be placed in a furnace for stress relieving?

Chapter 22

SPECIAL NONFERROUS WELDING APPLICATIONS

LEARNING OBJECTIVES

After studying this chapter, you will be able to:
* Define a nonferrous metal and alloy.
* Prepare and arc weld both wrought and cast aluminum.
* Describe how to prepare and weld die cast metal with the oxyfuel gas process.
* Describe how to prepare and GTA weld brass, bronze, titanium, and other nonferrous metals.
* List and identify the equipment that makes up a plastics welding station.
* Describe how to make an acceptable weld on plastic.

As stated in Chapter 21, practically all metals can be welded. Chapter 21 covered the welding of the ferrous alloys. This chapter is devoted to welding nonferrous metals and alloys. Nonferrous metals and alloys do not contain large amounts of iron. They may contain small amounts of iron, however, as alloying elements.

Some of the special welding applications used on nonferrous metals and alloys require preheat treatment and postweld heat treatment. See Chapter 29 for descriptions of heat treatment processes.

Many types of plastics, some of which are weldable, are being used in industry today. This chapter also shows how plastics can be welded.

22.1 NONFERROUS METALS AND ALLOYS

There are two large classifications of metals. These are:
* Ferrous metals.
* Nonferrous metals.

Ferrous metals are those bearing a substantial iron content. The *nonferrous metals* consist of all metals that do not contain much iron. Many metals are classified as nonferrous. Some of the more popular of these metals are:
* Aluminum and it alloys.
* Magnesium and its alloys.
* Copper and its alloys.
* Titanium and its alloys.
* Zirconium and its alloys.
* Lead and its alloys.
* Zinc and its alloys.
* Beryllium and its alloys.

22.2 ALUMINUM

Aluminum is one of the more common and popular metals. It is available in all standard shapes and forms, and can be shaped and formed by all standard methods. Parts made of aluminum are joined by all conventional methods.

Aluminum is light in weight and has good strength. It is well-suited to low-temperature applications, and has good resistance to corrosion. (Additional information on aluminum may be obtained in Heading 28.7.4.)

Aluminum is available in a commercially pure form. It is also alloyed with many other metals. Aluminum can be obtained in many forms: rolled, stamped, drawn, extruded, forged, or cast. All forms of aluminum, other than *cast aluminum*, are called *wrought aluminum*. The most common forms are plates, sheets, rolled forms, extruded forms, and aluminum pipe.

The Aluminum Association has classified aluminum alloys. Some are three-digit classifications; others are four-digit. Materials with a three-digit code number (XXX) are castings, while those with four-digit codes (XXXX) are wrought aluminum alloys. The three-digit series has seven main classifications (0XX through 7XX). The four-digit code number has eight main classifications, ranging from 1XXX through 8XXX. Wrought aluminum designations are listed in Figure 22-1.

The Aluminum Association alloy group designation	Major alloying element	Example	Heat-treatable
1XXX	99% aluminum (minimum)	1100	no
2XXX	copper	2024	yes
3XXX	manganese	3003	no
4XXX	silicon	4043	yes
5XXX	magnesium	5052	no
6XXX	magnesium and silicon	6061	yes
7XXX	zinc	7075	yes
8XXX	other	—	—

Figure 22-1. *Aluminum alloy designations, alloying elements, and heat treatability.*

22.3 PREPARING ALUMINUM FOR WELDING

As noted previously, aluminum is used commercially in two principal forms: wrought and cast. Welding procedures are similar for the two forms, but some differences do exist. Certain characteristics of aluminum make it rather difficult to weld:

- The ease with which the aluminum oxidizes at high temperatures.
- The aluminum melts before it changes color.
- The oxide melts at a much higher temperature than the metal.
- The oxide is heavier (more dense) than the metal.

Despite these difficulties, welds that are just as strong and ductile as the original metal can be made on aluminum. The filler metal, if used, should be of proper aluminum composition. Figure 22-2 lists the characteristics of various filler metals used to weld different base metal combinations. Various filler metals are also listed. Different physical characteristics are listed in columns. Each combination of base metals, filler metals, and physical characteristics is rated. From this chart, the best filler metal can be selected, depending on what physical characteristic is most important in the finished weld. The welder must know the composition of the alloys being joined before selecting the best welding filler metal to use. See Figure 22-2.

The metal must be cleaned before welding. It can be mechanically cleaned using a clean stainless steel wire brush or clean stainless steel wool. The metal can be chemically cleaned by dipping it in a cleaning solution and then rinsing it in water. Follow all safety precautions when using cleaning solutions.

One problem that occurs when welding aluminum that is not present when welding ferrous metals or alloys is a characteristic called *hot shortness*. As aluminum approaches its melting temperature, it loses strength. The weld pool and the area around the weld pool may fall through, leaving a large hole in the metal. Heading 22.3.1 discusses hot shortness.

When welding aluminum, the welder must obtain the same results as when welding steel:

- Good fusion.
- Good penetration.
- A straight weld.
- Buildup over the joint.
- No defects.

When oxyfuel gas welding aluminum, or when shielded metal arc welding, the base metal may require preheating. If thick sections are to be welded, preheat is required because the aluminum will conduct the heat away from the weld area too fast. The recommended preheat temperature is between 300°F to 400°F (149°C to 204°C).

Preheating is not necessary when GTAW or GMAW aluminum. It is important to tack weld aluminum parts prior to welding. Aluminum expands and contracts more than other metals as it is welded and cooled back to room temperature. This causes distortion in and around the weld.

Sometimes, tack welds are not strong enough to prevent the expansion and contraction from affecting the final joint configuration. It is often necessary to tack weld the parts in such a way that after welding, the finished weldment is in the correct orientation. Figure 22-3 shows how parts can be tack welded to allow for contraction from welding. This technique can be used for any metal being welded.

22.3.1 Hot Shortness

Aluminum may need to be supported before, during, and immediately after the weld is made. This is especially necessary when welding a wide bead in thin sheet metal. Aluminum, like other nonferrous metals, has a characteristic called *hot shortness*. Hot shortness is a condition in which a metal loses its strength prior to melting. Hot shortness will allow the weld pool and the area around the pool to fall away and leave a large hole in the base metal. The problem of the weld area falling through only occurs when making a full-penetration weld. Thus, backing strips or backing rings may be necessary to support the metal while welding. Stainless steel is usually used as the backing material when welding aluminum. If a backing is not used, then a small weld pool is necessary. Welds that are not full-penetration do not need a backing strip or backing ring.

22.3.2 Use of a Backing Strip

Metal that extends beyond the back side (side opposite the bead) of a weld is called *penetration*. When it is necessary to control the shape and size of this penetration, a backing strip is often used. Backing strips also are used to control the molten metal in the weld pool and to reduce the possibility of hot shortness failure.

A groove is often machined into the metal backing strip. This groove ensures that the penetration has the desired shape and size when it cools. Preformed backing

Base alloys		1060 1350	1100	2014 2036	2219	3003 ALCLAD 3003	3004	ALCLAD 3004	5005 5050	5052 5652	5083 5456	5086 5356	514.0, A514.0 B514.0, F514.0 5154, 5254	5454	6005, 6063 6101, 6151 6201, 6351 6951	6061 6070	7005, 7021 7039, 7046 7146, A712.0 C712.0	413.0, 443.0 444.0, 356.0 A356.0, A357.0 359.0	319.0, 333.0 354.0, 355.0 C355.0, 380.0	
Characteristics / Filler Alloys		WSDCTM	WSDCTM	WSDCTM	WSDCTM	WSDCTM	WSDCTM	WSDCTM	WSDCTM	WSDCTM	WSDCTM	WSDCTM	WSDCTM	WSDCTM	WSDCTM	WSDCTM	WSDCTM	WSDCTM	WSDCTM	
319.0 333.0 354.0 355.0 C355.0 380.0	2319			BAAAAA	BAAAAA														BAAAAA	2319
	4043	BAAAAA	BAAAAA	CCBCAA	CCBCAA	BBAAAA	BBAAAA	BBAAAA	BBAAAA	AAAAAA	AAAA A	AAAA A	AAAA A	AAAAAA	BBAAAA	BBAAAA	BBAAAA	BBAAAA		4043
	4145	AABAAA	AABAAA	ABCBAA	ABCBAA	AABAAA	AABAAA	AABAAA	AABAAA						AABAAA	AABAAA	AABAAA	AABAAA	ABBBAA	4145
413.0 443.0 444.0 356.0 A356.0 A357.0 359.0	4043	AAAAAA	AAAAAA	BBAAAA	BBAAAA	AAAAAA	AAAAAA	AAAAAA	AAAAAA	ABAAAA	ABBA A	ABBA A	ABBA A	ABBAAA	ABAAAA	ABAAAA	ABBAAA	ABAAAA	4043	
	4145	AABBA	AABBA	AABAA	AABAA	AABBA									AABBA	AABBA	AABBA	AABBA	4145	
	5356									BABB A	AAAA A	AAAA A	AAAB A	AAAB A			AAAA B		5356	
7005, 7021 7039, 7046 7146, A712.0 C712.0	4043	AACAA	AACAA	BBAAA	BBAAA	ABCAA	ADCBA	ADCBA	ABCBA	BDCBA					ADCBA	ADCBA	BDCBA	4043		
	4145			ABAA	AABAA													4145		
	5183	BABA A	BABA A			BABA A	BABA A	BABA A	BABA A	AABA A	AABA A	AABA A	AABA A	AABA A	AABA A	AABA A	AABA A	5183		
	5356	BAAA A	BAAA A			BAAA A	BBAA A	BBAA A	BAAA A	ABAA A	ABAA A	ABAA A	ABAA A	ABAA A	ABAA A	ABAA A	ABAAAA	5356		
	5554						CCAAAA	CCAAAA	CAAAAA	BCAAAA			BCAA A	BCAAAA	BCAAAA	BCAAAA	BCAA A	5554		
	5556	BABA A	BABA A			BABA A	BABA A	BABA A	BABA A	AABA A	AABA A	AABA A	AABA A	AABA A	AABA A	AABA A	AABA A	5556		
	5654						CCAA B	CCAA B	CAAA A	BCAA A			BCAA A	BCAA A	BCAA A	BCAA A	BCAA A	5654		
6061 6070	4043	AACAA	AACAA	BBAAA	BBAAA	ABCAA	ADCAA	ADCAA	ABCAA	ADCAA	ADCA	ADCA	ADCA	ADCBA	ACBAA	ACBAA	4043			
	4145	AADBA	AADBA	AABAA	AABAA	AADBA	BCDBA	BCDBA	ABDBA								4145			
	5183	BAB A	BAB A			BAB A	BAB A	BAB A	BAB A	BABC B	AABA A	AABA A	BABC B	BABC A	BAAC A	BAAC B	5183			
	5356	BAA A	BAA A			BAA A	BBA A	BBA A	BAA A	BBAC A	ABAA A	ABAA A	BBAC A	BBAC A	BAAC A	BBAC A	5356			
	5554									CCABAB	BCAA A	BCAA A	CCAB B	CCAAAA	CBABBA	CBABBB	5554			
	5556	BAB A	BAB A			BAB A	BAB A	BAB A	BAB A	BABC B	AABA A	AABA A	BABC B	BABC A	BAAC A	BAAC B	5556			
	5654									CCAB A	BCAA B	BCAA B	CCAB A	CCAB B	CBAB B	CBAB B	5654			
6005 6063 6101 6151 6201 6351 6951	4043	AACAA	AACAA	BBAAA	BBAAA	ABCAA	ADCAA	ADCAA	ABCAA	ADCAA	ABCA	ABCA	ABCA	ABCBA	ACBAA	4043				
	4145	AADBA	AADBA	AABAA	AABAA	AADBA	BCDBA	BCDBA	ABDBA							4145				
	5183	BAB A	BAB A			BAB A	BAB A	BAB A	BAB A	BABC B	AABA A	AABA A	BABC A	BABC A	BAAC A	5183				
	5356	BAA A	BAA A			BAA A	BBA A	BBA A	BAA A	BBAC A	AAAA A	AAAA A	BAAC A	BAAC A	BAAC A	5356				
	5554									CCABAB	BAAA A	BAAA A	CAAB A	CAAAAA	CBABBA	5554				
	5556	BAB A	BAB A			BAB A	BAB A	BAB A	BAB A	BABC B	AABA A	AABA A	BABC A	BABC A	BAAC A	5556				
	5654									CCAB A	BAAA B	BAAA B	CAAB B	CAAB B	CBAB B	5654				
5454	4043	ABCCA	ABCCA		AAAAA	ABCCA	ADCCA	ADCCA	ABCCA	ADCCA					4043					
	5183	BABB A	BABB A			BABB A	BABB A	BABB A	BABB A	AAAB A	AABB A	AABB A	AABB A	AABB A	5183					
	5356	BAAB A	BAAB A			BAAB A	BBAB A	BBAB A	BAAB A	ABAB A	ABAB A	ABAB A	ABAB A	ABAB A	5356					
	5554	CAAAAA	CAAAAA			CAAAAA	CAAAAA	CAAAAA	CCAAAA	CAAAAA	BCAA A	BCAA A	BCAA A	BCAAAA	5554					
	5556	BABB A	BABB A			BABB A	BABB A	BABB A	BABB A	AABB A	AABB A	AABB A	AABB A	AABB A	5556					
	5654									BCAB B			BCAA B	BCAB B	5654					
514.0 A514.0 B514.0 F514.0 5154 5254	4043	ABCC	ABCC		AAAA	ABCC	ADCC	ADCC	ABCC	ADCC				4043						
	5183	BABB A	BABB A			BABB A	BABB A	BABB A	BABB A	AABB B	AABA A	AABA A	AABB B	5183						
	5356	BAAB A	BAAB A			BAAB A	BBAB A	BBAB A	BAAB A	ABAB A	ABAA A	ABAA A	ABAB A	5356						
	5554	CAAA A	CAAA A			CAAA A	CCAA A	CCAA A	CAAA A	CCAA B	BCAA A	BCAA A	BCAA B	5554						
	5556	BABB A	BABB A			BABB A	BABB A	BABB A	BABB A	AABB B	AABA A	AABA A	AABB B	5556						
	5654	CAAA B	CAAA B			CAAA B	CCAA B	CCAA B	CAAA B	BCAA A	BCAA B	BCAA B	BCAA A	5654						
5086 5356	4043	ABCB	ABCB		AAAA	ABCB	ACCB	ACCB	ABCB				4043							
	5183	AABA A	AABA A			AABA A	AABA A	AABA A	AABA A	AABA A	AABA A	AABA A	5183							
	5356	AAAA A	AAAA A			AAAA A	ABAA A	ABAA A	AAAA A	ABAA A	ABAA A	ABAA A	5356							
	5554									CCAA A			5554							
	5556	AABA A	AABA A			AABA A	AABA A	AABA A	AABA A	AABA A	AABA A	AABA A	5556							
	5654									BCAA B			5654							
5083 5456	4043	ABCB	ABCB		AAAA	ABCB	ACCB	ACCB	ABCB			4043								
	5183	AABA A	AABA A			AABA A	AABA A	AABA A	AABA A	AABA A	AABA A	5183								
	5356	AAAA A	AAAA A			AAAA A	ABAA A	ABAA A	AAAA A	ABAA A	A AA A	5356								
	5554									CCAA A		5554								
	5556	AABA A	AABA A			AABA A	AABA A	AABA A	AABA A	AABA A	AABA A	5556								
	5654									BCAA B		5654								
5052 5652	4043	ABCAA	ABCAA	AAAAA	AAAAA	ABCAA	ABCAA	ACCAA	ABCAA	ADCBA	4043									
	5183	BAB A	BAB A			BAB A	BAB A	BAB A	BAB A	AABC B	5183									
	5356	BAA A	BAA A			BAA A	BAA A	BBA A	BAA A	ABAC A	5356									
	5554									CCAAAB	5554									
	5556	BAB A	BAB A			BAB A	BAB A	BAB A	BAB A	AABC B	5556									
	5654									BCAB A	5654									
5005 5050	1100	CBAAAA	CBAAAA			CCAAAA			B AAAA	1100										
	4043	AACAA	AACAA	BBAAA	BBAAA	ABCAA	ABCAA	ABCAA	ABDAA	4043										
	4145	BADBA	BADBA	AABAA	AABAA	BBDBA				4145										
	5183	CAB B	CAB B			CABC B	BAB A	BABB A	BAC B	5183										
	5356	CAB B	CAB B			CABC B	BAA A	BAAB A	BAB B	5356										
	5556	CAB B	CAB B			CABC B	BAB A	BABB A	BAC B	5556										
ALCLAD 3004	1100	DBAAAA	DBAAAA			CCAAAA			1100											
	4043	AACAA	AACAA	BBAAA	BBAAA	ABCAA	ADDAA	ADDAA	4043											
	4145	BADBA	BADBA	AADAA	AABAA	BBDBA			4145											
	5183	CABC B	CABC B			CABC A	BACC A	BACC A	5183											
	5356	CABC B	CABC B			CABC A	BBBC A	BBBC A	5356											
	5554						CCABAA	CCABAA	5554											
	5556	CABC B	CABC B			CABC A	BACC A	BACC A	5556											
3004	1100	DBAAAA	DBAAAA			CCAAAA		1100												
	4043	AACAA	AACAA	BBAAA	BBAAA	ABCAA	ABDAA	4043												
	4145	BADBA	BADBA	AABAA	AABAA	BBDBA		4145												
	5183	CAB B	CAB B			C BC A	BACC A	5183												
	5356	CAB B	CAB B			CABC A	BBBC A	5356												
	5554						CCABAA	5554												
	5556	CAB B	CAB B			C BC A	BACC A	5556												
3003 ALCLAD 3003	1100	BBAAAA	BBAAAA			BBAAAA	1100													
	4043	AABAA	AABAA	BAAAA	BAAAA	AABAA	4043													
	4145	AACBA	AACBA	AABAA	AABAA	AACBA	4145													
2219	2319			BAAAAA	AAAAAA	2319														
	4043	BAAAA	BAAAA	BCBCA	BCBCA	4043														
	4145	AABAA	AABAA	ABCBA	ABCBA	4145														
2014 2036	2319			CAAAAA	2319															
	4043	BAAAA	BAAAA	BCBCA	4043															
	4145	AABAA	AABAA	ABCBA	4145															
1100	1100	BBAAAA	BBAAA	1100																
	4043	AABAA	AABAA	4043																
	5356			5356																
1060 1350	1100	BBAAAB	1100																	
	1060	CCAAAA	1060																	
	4043	AABAA	4043																	

Aluminum Filler Alloy Chart

How To Use the Filler Alloy Chart

First, select the base alloys to be joined, one from the gray column on the exterme left and the other from the gray row running along the top of the chart. Then, from the base alloy in the left column, move to the right until you reach the block directly under the base alloy from the top row. Or, from the base alloy in the top row, move down until you reach the block directly across from the base alloy in the left column.

This intersecting block contains horizontal lines of letters (A thru D) which represent the filler alloys directly across from them in the blue box at the ends of each row. The letters in each line give the A to D rating of the characteristics tested at the top of each column—W, S, D, C, T, and M.

You will find that by choosing the different filler alloys in each block, you can vary the characteristics of the weld, that is, "trade-off" one characteristic for another until you find the filler that best meets your needs.

For example, when joining base alloys 3003 and 1100, find the intersecting block. Now, note that filler alloy 1100 provides excellent ductility (D), corrosion resistance (C), performance at elevated temperatures (T) and color match after anodizing (M), with good ease of welding (W) and strength (S). However, if ease of welding and shear strength are *utmost* in importance, and ductility and color match can be sacrificed slightly, filler alloy 4043 can be used advantageously.

Legend

Filler alloys are rated on the following characteristics:

Symbol	Characteristic
W	Ease of welding (relative freedom from weld cracking.)
S	Strength of welded joint ("as-welded" condition). (Rating applies particularly to fillet welds. All rods and electrodes rated will develop presently specified minimum strengths for butt welds.)
D	Ductility. (Rating is based upon free bend elongation of the weld.)
C	Corrosion resistance in continuous or alternate immersion in fresh or salt water.
T	Recommended for service at sustained temperatures above 150° F (65.5°C).
M	Color match after anodizing.

- A, B, C, and D are relative ratings in decreasing order of merit. The ratings have relative meaning only within a given block.
- Combinations having no rating are not usually recommended.
- Ratings do not cover these alloys when heat-treated after welding.

Figure 22-2. *Chart used to select the best filler metal when welding two aluminum alloys.*

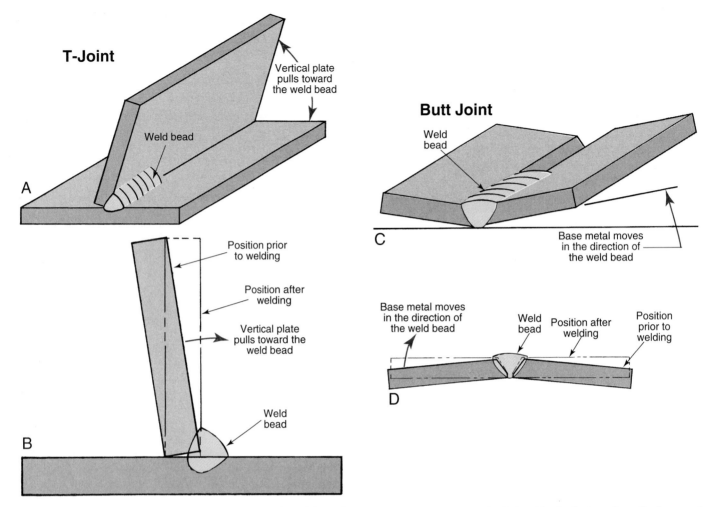

Figure 22-3. *Allowing for contraction when tack welding aluminum. A—Unrestrained T-joint. Vertical piece is pulled toward the weld bead. B—T-joint that is tack welded a few degrees from vertical. Offset compensates for movement of the vertical piece. C—Unrestrained butt joint. Outer edges bend toward the weld bead. D—Butt joint in which the pieces are tack welded with a reverse angle. The amount of offset is determined by experimentation.*

strips can be purchased in roll form. These preformed strips are taped onto the metal.

Backing strips may be removed after welding is completed. However, they are occasionally welded in place and become part of the joint. If the backing strips is to remain a part of the joint, it is made of a similar metal or alloy to that being welded. If the backing strip is to be removed, it is usually made of a metal dissimilar from the metal being welded. This prevents the backing from being welded into the joint. Backing used for aluminum is usually 300-series stainless steel. Backing material for stainless steel is usually copper.

When round parts such as pipes are welded, a backing ring may be used. See Heading 23.11 concerning the use of a backing ring.

22.4 ARC WELDING WROUGHT ALUMINUM

The most common methods for joining aluminum and its alloys are the gas tungsten arc welding (GTAW) and the gas metal arc welding (GMAW) processes. These gas-shielded processes are desirable because the inert gas protects the aluminum from oxidation.

Gas tungsten arc welding can be done using ac, DCEP, or DCEN. When using ac or DCEP, the oxides on the aluminum are broken up. Aluminum is usually welded using alternating current. The finished weld has a clean, shiny appearance. When using DCEN, there is no cleaning action. The aluminum oxide on the surface must be removed before welding. The finished weld appears dull, but stainless steel brushing will brighten its appearance.

When ac welding, the electrode should be pure tungsten or contain zirconium. Tungsten electrodes containing ceria or lanthana may also be used. Thoriated tungsten electrodes are not recommended for ac welding. Thoriated electrodes are recommended when using direct current. Remember however, that DCEN is not recommended, because there is no cleaning action.

The electrode tip should be formed into a ball shape when ac welding. Argon or an argon and helium mixture is recommended as the shielding gas. GTAW produces high-quality welds on aluminum.

Gas metal arc welding (GMAW) is also used to weld aluminum. DCEP is used and provides the necessary cleaning action to remove the aluminum oxide from the base metal. The shielding gas is either pure argon or an argon-helium mixture.

Gas metal arc welding should be done with spray transfer or pulsed spray transfer. Spray transfer can only be used in the flat welding position or on horizontal welds. Pulsed spray transfer can be used to weld out-of-position joints. High welding currents (wire feed speed) and high welding speeds are used to produce very high-quality welds in aluminum with small heat-affected zones. Short circuiting transfer and globular transfer are not recommended, because the penetration is poor. The ability to melt the base metal and to obtain good fusion is also poor with these methods.

Shielded metal arc welding (SMAW) can also be used on aluminum. A flux coated aluminum electrode is used. The current is DCEP. The quality of a shielded metal arc weld is not as high as welds produced with GTAW or GMAW. The SMAW process should be used to weld aluminum only when GTAW or GMAW are not available. Weld quality obtained with GTAW or GMAW is better than the weld quality obtained with SMAW.

Oxyfuel gas welding of aluminum can be done successfully. However, GTAW and GMAW are preferred. Welding aluminum using an oxyfuel gas process is not recommended for several reasons. A few of these reasons are:

- Oxyfuel gas welding will create a large weld pool, which may fall through because of aluminum's hot shortness.

- There is no cleaning action and fluxes must be used. If not properly cleaned, these fluxes will continue to attack the aluminum.
- Oxyfuel gas welding puts more heat into the base metal than any other process. This causes a large heat-affected zone and more distortion than the arc welding processes.

Flux is always used when oxyfuel gas welding aluminum. Both the metals being joined and the welding rod, if used, must be flux coated. The welder usually mixes the flux with water to form a paste. When welding is finished, this flux must be washed from the weld area with water or with a mild sulfuric acid-water solution as soon as possible. **Safety goggles and rubber gloves need to be worn when using acid solutions.** If flux is left on the weld, it will have a corrosive effect.

Due to the high heat conductivity of aluminum, a larger torch tip is needed for aluminum than for a corresponding thickness of steel. Acetylene is preferred as the fuel gas because it produces the most heat when burned. A neutral or reducing (slightly carburizing) flame is used. Hydrogen fuel gas can be used on thin sheet.

Aluminum can be spot welded. Because aluminum conducts heat very well, the current used to resistance weld is about three times the current used to weld a similar thickness of steel. Figure 22-4 list the parameters required to spot weld aluminum. Also refer to Chapters 18 and 19 for additional information on resistance spot welding. A resistance spot welding machine with forge force ability is recommended. The forge force reduces cracking in aluminum as it cools. Figure 22-5 shows an aluminum spot welding machine.

22.4.1 Welding Cast Aluminum

Welding of cast aluminum and aluminum alloys is not significantly different from welding wrought aluminum. Both sand castings and permanent mold castings are weldable. Aluminum die castings are not usually weldable, because of certain ingredients in the aluminum alloys.

Castings that have been heat-treated lose the heat treatment properties when welded. If a casting requires heat treatment, it should be welded before being heat-treated, if possible.

| Thickness | | Electrode face dia. | | Force | | Weld time | Weld current | Weld spacing (min.) | | Weld diameter | |
inch	mm	inch	mm	pounds	Newton	cycles	amps	inch	mm	inch	mm
0.025	0.63	0.625	15.88	390	1735	6	21800	0.38	9.5	0.14	3.6
0.032	0.81	0.625	15.88	500	2224	6	26000	0.38	9.5	0.16	4.1
0.040	1.02	0.625	15.88	600	2669	8	30700	0.44	11.3	0.18	4.6
0.050	1.27	0.625	15.88	660	2936	9	33000	0.50	12.7	0.21	5.3
0.062	1.57	0.625	15.88	750	3336	10	35900	0.50	12.7	0.25	6.3
0.093	2.36	0.875	22.23	950	4226	12	46000	0.75	19.0	0.33	8.4
0.100	2.54	0.875	22.23	1050	4670	15	56000	0.75	19.0	0.36	9.1
0.125	3.17	0.875	22.23	1300	5782	15	76000	1.00	25.4	0.42	10.7

Figure 22-4. *Variables recommended for resistance spot welding aluminum.*

Figure 22-5. *A spot welding machine designed to weld aluminum.*

The arc welding processes are preferred to oxyfuel gas welding for cast aluminum, but both can be used. When oxyfuel gas welding thin sections of metal, a stainless steel rod can be used to control and shape the weld pool. Figure 22-6 shows two rods. These rods are usually made from 1/4″ (6mm) stainless steel flattened at one end into a flat spoon-like shape.

When welding large and/or thick-section aluminum castings, the metal should be preheated to 400°F to 500°F (204°C to 260°C). Thicker sections should be beveled as is done with steel.

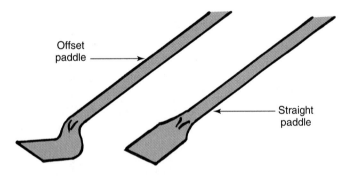

Figure 22-6. *Rods used for weld pool control when oxyfuel gas welding cast aluminum.*

22.5 WELDING MAGNESIUM

Magnesium and its alloys are often confused with aluminum. Welders frequently try unsuccessfully to use aluminum welding techniques on magnesium. The metal is shaped into various wrought forms and is also cast in sand or permanent molds.

There are several magnesium alloys on the market. These metals can usually be arc welded. Gas tungsten arc welding (GTAW) and gas metal arc welding (GMAW) are the most widely used processes. Oxyfuel gas welding is recommended only for emergency repairs. Brazing and soldering methods can be used to join magnesium. Figure 22-7 lists some of the weldable magnesium alloys and their weld strengths.

Magnesium oxidizes very rapidly when heated to its melting point. In fact, when small shavings are heated to the melting point, the magnesium will burn spontaneously and leave a white ash. This burning and white ash is one way to identify the metal. **Be careful to use only a very small amount of magnesium scrapings or shavings to try this test.**

The most popular magnesium alloys are those with aluminum or zinc added. These are the AZ series (A for aluminum; Z for zinc). The high zinc alloys are weldable by the resistance spot and resistance seam welding processes. These alloys are the ZE, ZH, and ZK series. The usual arc processes cannot be used on these alloys.

22.6 WELDING DIE CASTINGS

Die castings are metal alloy castings (sometimes called *white metal castings*) that are made in a steel mold under pressure. The mold, which can also be called a *die*, is used many times (often from ten thousand to over a hundred thousand times). Die casting accuracy can be better than 0.005″ (0.13mm). Die casting alloys are usually alloys of high-zinc, high-aluminum, or high-magnesium content. Other alloys include tin and lead. These castings are brittle and break easily. Many articles are made using this method, however, because of the ease of manufacture. Failure due to brittle nature of the material produces considerable demand for repair welding of die castings.

ASTM alloy	Filler rod alloy	Base metal strength		Welded strength	
		1000 psi	MPa	1000psi	MPa
AZ31B-H24	AZ61A	42	290	37	255
AZ10A-F	AZ92A	35	241	33	228
AZ61A-F	AZ61A	45	310	40	276
AZ92A-T6	AZ92A	40	276	35	241
HK31A-H24	EZ33A	38	262	31	214
HM21A-T8	EZ33A	35	241	28	193
ZE10A-H24	AZ61A	38	262	33	228
ZK21A-F	AZ61A	42	290	32	221

Figure 22-7. *Table of weldable magnesium alloys, recommended filler metal, and weld strengths.*

To successfully weld a die casting, you should know the constituent metals in the alloy. Zinc castings are heaviest and magnesium castings lightest, while aluminum castings are in-between in weight. Zinc die castings are the most common. The zinc die castings melt at about 800°F (427°C), while magnesium and aluminum die castings melt at about 1100°F to 1200°F (593°C to 649°C). Zinc alloy die castings are very difficult to weld because of their low melting temperature and high rate of oxidation.

For casting repair, welding rod should be of a composition similar to the original metal, if possible. Welding rods designed for the general repair of die castings may be purchased.

A die casting is prepared for welding just as other metal would be. It must be beveled if there is a thick section. It must be thoroughly cleaned. Any plating must be ground away in the area to be welded. The parts must be firmly supported before, during, and after welding.

Die casting alloys are very fluid when melted. To control the weld metal, carbon paste or blocks are used. Figure 22-8 shows a die casting with a carbon paste mold formed around the fracture. Figure 22-9 shows carbon paste and carbon plates used to form molds for controlling weld metal.

The oxyacetylene or oxyfuel gas method is usually used to repair die castings. A heavy carburizing flame should be used and the usual welding procedure followed. A very small tip is recommended. You must be careful, because the metal melts before it changes color. Use the welding rod to break the surface oxides as the welding metal is being added. The welder may use a stainless steel or brass paddle, as shown in Figure 22-6, to smooth the surface of the weld and to remove oxides.

Another successful method of repairing die castings is to use a soldering iron or soldering copper to melt the metal. An oxyfuel gas torch is used to keep the body of the soldering copper at a red heat while the point of the copper heats the die casting and fuses the two pieces together. The torch flame is not put on the die casting at all, unless the casting requires preheating. This method is especially suitable for use when repairing small sections.

22.7 COPPER AND COPPER ALLOYS

Copper and most of its alloys can be welded. There are two groups of copper:
- Oxygen-bearing (99.9% copper minimum, 0.04% oxygen).
- Deoxidized (99.5% copper minimum, phosphorous 0.015% to 0.040%).

The oxygen-bearing copper is difficult to fusion weld. Copper has a high *specific heat*, so it heats at about half the rate of aluminum. Both copper and aluminum have much higher heat conductivity than steel. (Additional information on copper may be found under Heading 28.7.1.)

Copper is alloyed with many different elements. The two most common are zinc and tin. Copper-zinc alloys are

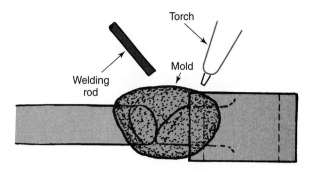

Figure 22-8. *Preparing a die casting for welding by forming a mold of carbon paste around the fracture.*

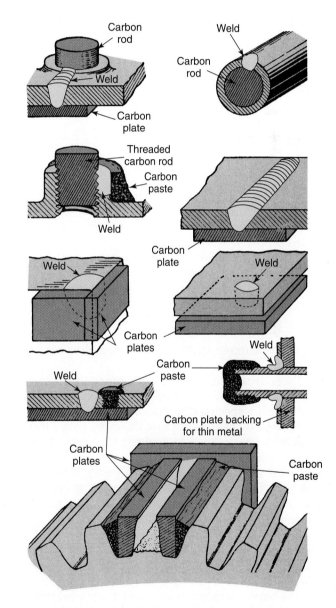

Figure 22-9. *Uses of carbon block and paste backing in typical welding applications.*

called *brass*. (Refer to Heading 28.7.2.) Copper-tin alloys are called *bronze*. (Refer to Heading 28.7.3.) Additional elements can be added to form different brasses and bronzes.

Copper and zinc alloys (brass)

- Gilding – 94% to 96% copper (zinc remainder)
- Commercial bronze – 89% to 91% copper (zinc remainder)
- Red brass – 84% to 86% copper (zinc remainder)
- Low brass – 80% copper (zinc remainder)
- Cartridge brass – 70% copper (zinc remainder)
- Yellow brass – 65% copper (zinc remainder)
- Admiralty brass – 71.5% copper, 1.1% tin (zinc remainder)
- Naval brass – 61% copper, 0.75% tin (zinc remainder)
- Manganese bronze – 58.5% copper, 1% tin, 1.4% iron, 0.5% manganese (maximum) (zinc remainder)
- Aluminum brass – 77.5% copper, 2.2% aluminum (zinc remainder)

Copper and tin alloys (bronze)

- Grade A – 0.19% phosphorus, 94% copper, 3.6% tin
- Grade C – 0.15% phosphorus, 90.5% copper, 8% tin
- Grade D – 0.15% phosphorus, 88.5% copper, 10% tin
- Grade E – 0.25% phosphorus, 95.5% copper, 1.25% tin

Alloys of copper and tin (bronze) may crack if not welded properly.

Alloys of copper are used a great deal, mainly because of their ductility, which enables them to be worked easily and shaped into many complicated patterns. These alloys are used in both cast and wrought forms. They are resistant to certain kinds of corrosion and are excellent conductors of heat and electricity. Copper alloys may be recognized by their characteristic red or yellow color.

Pure deoxidized copper is comparatively easy to weld, while some alloys of copper are difficult to weld. In order to test a specimen to determine if it may be easily welded, quickly heat a sample of the metal with a torch to the molten state. If the weld pool remains quiet, clear, and shiny, it indicates the metal is comparatively pure copper and that it will be easy to weld. However, if the molten weld pool boils vigorously and gives off a quantity of gaseous fumes, this indicates the presence of ingredients in the copper that will make it difficult to weld.

Annealed deoxidized copper has a tensile strength of 30 ksi to 35 ksi (207MPa to 241MPa) and may be welded with arc or oxyfuel gas processes to produce full-strength welds.

The tensile strength of annealed oxygen-bearing copper is also 30 ksi to 35 ksi (207MPa to 241MPa). When welding oxygen-bearing copper, the cuprous oxide redistributes itself in the heat-affected zone and weakens it. It is difficult to obtain a weld with a strength greater than 70% to 85% of the annealed base metal strength. The heat-affected zone also suffers a loss of ductility and corrosion resistance.

If a copper sample is brittle and breaks easily, this is an indication of alloying elements or impurities. One alloy that makes copper susceptible to cracking, and thus difficult to weld, is lead. Phosphorus in small quantities makes welding copper easier. Most copper, copper alloys, brass, and bronze can be joined by brazing and soldering.

22.7.1 Welding Copper

Gas tungsten arc welding (GTAW) and gas metal arc welding (GMAW) are the preferred processes when welding copper. Shielded metal arc welding is used only for repair welding or to weld thin sections when GTAW or GMAW is not available. Oxyacetylene welding also can be used to join copper.

GTAW and GMAW Copper

Gas tungsten arc welding (GTAW) is used with direct current electrode negative. A 2% thoriated tungsten electrode is recommended. Forehand welding is preferred, but both forehand and backhand methods will work. Stringer beads are recommended. Wide beads may allow the edges of the weld pool to become oxidized. Filler metal of a composition similar to the base metal should be used. Filler metals used for copper begin with the designation ERCu.

Argon, helium, or a mixture of argon and helium are used as shielding gases. Argon is used when welding thin copper up to 0.062" (1.6 mm) in thickness. Helium, which is used on thicker sections, provides a higher arc voltage that gives better penetration. A helium-argon mixture of 2:1 or 3:1 is often used.

Gas metal arc welding (GMAW) is used to weld copper. ERCu-type electrode wire is used. To obtain sufficient heat to melt the copper and obtain good penetration, spray transfer or pulsed spray transfer is recommended. Forehand welding is used when welding in the flat or horizontal welding positions. When welding in the vertical welding position, the direction of travel is uphill. Overhead welding using GMAW is not recommended. As with GTAW, stringer beads should be used, rather than wide weaving beads,.

Argon shielding is used for metal thicknesses up to 0.25" (6.4 mm). A mixture of 75% helium and 25% argon in used for thicknesses above 0.25" (6.4 mm). Helium allows for more heat into the base metal, better penetration, and lower preheating. It deposits the electrode wire faster.

Preheating of the copper base metal is not required for thin sheets of copper. GTAW can weld sheet up to 0.125" (3.2 mm) thick without preheat. GMAW can weld up to 0.25 (6.4 mm) without preheat. Above these thicknesses, preheat should be used. Thicker sections of copper pull the heat away from the welding area and can cause problems when welding. Preheat reduces stresses in the welded metal and also allows for faster welding. More of the heat from the arc can be used for melting the base metal and not heating the surrounding area.

Preheating temperature increases as the metal thickness increases. Temperatures can range from 200°F to 1000°F (95°C to 540°C). Preheating requirements for GTAW are higher than those required for GMAW. Using helium as the shielding gas also reduces the amount of preheat required. As an example, if welding 3/8" (9.5mm) deoxidized copper with argon shielded GTAW, the preheat needs to be about 700°F (370°C). To weld the same thickness using argon-shielded GMAW requires a preheat temperature of about 550°F to 600°F (290°C to 315°C). Welding the same thickness with either process and helium shielding reduces the required preheat to 450°F to 500°F (230°C to 260°C).

SMAW Copper

Shielded metal arc welding uses ECu-type covered electrodes. The electrode diameter needs to be as large as possible. This allows a higher current to be used, which puts more heat into the base metal. Welding should be done in the flat welding position. When welding copper thicker than 0.125" (3.2 mm), preheat temperatures of 500°F (260°C) or higher should be used.

OFW Copper

Copper and its alloys may be welded with oxyacetylene or other oxyfuel gas processes. However, the high heat of the oxyacetylene flame is preferred. The general procedure is similar to that followed in welding steel.

Because of the rapid heat conduction of copper, a larger tip must be used. A tip one or two sizes larger than would be used on the same thickness of steel is sufficient. When welding thicknesses over 0.125" (3.2mm), preheat is required. Preheat temperatures are higher that those required for GTAW. Concurrent heating may be necessary to keep the base metal properly preheated while welding thick copper.

When welding copper and its alloys, a flux is required to prevent oxidation. The flux is applied to the weld joint and to the welding rod. The backhand method is used. This prevents molten metal from running ahead of the torch, becoming oxidized, and getting trapped in the weld.

22.7.2 Welding Brass

Arc welding brass is similar to welding pure copper. One major difference is caused by the presence of zinc. When the zinc is heated by the arc, it becomes a gas. This zinc gas forms fumes around the weld and can cause porosity in the weld. Shielding gas must be used to remove any zinc fumes.

The electrode or filler metal used does not contain zinc. A copper-tin or copper-silicon alloy electrode is used. The most common electrodes are classified as ECuSn-C (copper-tin) and ECuSi-A (copper-silicon). The base metal is preheated to 200°F to 600°F (95°C to 315°C) for GTAW and GMAW. When SMAW, the preheat temperature is 400°F to 700°F (200°C to 370°C).

When gas tungsten arc welding, the zinc fumes can be reduced by directing the arc onto the filler rod. The filler rod is kept in contact with the base metal. The arc heats and melts the filler rod as it is used to fill the weld joint. Using this method, the arc is not directed onto the base metal and the zinc fumes are greatly reduced. This process is illustrated in Figure 22-10. To minimize zinc fumes when

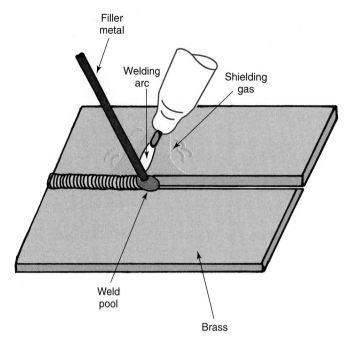

Figure 22-10. *Gas tungsten arc welding brass. The welding arc is directed at the filler rod to reduce zinc fumes.*

GMAW or SMAW, direct the arc at the leading edge of the weld pool.

The method used to oxyfuel gas weld brass are similar to those used when welding deoxidized copper. However, filler metals with matching compositions are not available. A copper silicon filler rod (RCuSi-A) is used. Brass can also be joined by braze welding. Brass brazing rods RCuZn-A and RCuZn-C are used. A braze weld will not be as corrosion-resistant as a fusion weld. A good flux is required when welding or braze welding brass.

22.7.3 Welding Bronze

Elements other than tin are added to copper to create a number of different bronze alloys. Each group of bronze welds differently. It is important to know what type of bronze is being welded, so a proper electrode or filler metal can be selected. Also, preheat and postweld heat treatments are different for the various types of bronze.

Copper-Tin Alloys (Phosphor Bronze)

The most common arc processes for welding phosphor bronze are SMAW and GMAW. Gas tungsten arc welding is used for repairs of castings and for joining thin sheet metal. Covered electrodes are available as ECuSn-A and ECuSn-C. They are used with direct current electrode positive (DCEP). Gas metal arc welding electrodes are available as ECuSn-C and used with DCEP. These filler metals contain some phosphorus to help prevent cracking.

Stringer beads are recommended for both processes. Argon gas or an argon-helium mixture is usually used for shielding with GMAW and GTAW. The bronze base metal should be preheated to 400°F (204°C). Interpass heating should be used to keep the phosphor bronze about 400°F

(204°C). After welding each pass, peen the weld and heat-affected zone. *Peening* is hitting the metal with a hammer or with hard round balls (like ball bearings) propelled by air pressure. After welding is complete, a postweld heat treatment should be used to obtain the best ductility and corrosion resistance. Heat the weldment to 900°F (482°C), then quench (rapidly cool) the weldment to room temperature.

Oxyfuel gas processes are not recommended for welding bronze. The high heat input can cause cracking and porosity. Arc welding is preferred. Adequate quality welds can be produced with practice. Filler rods are available as RCuSn-A and RCuSn-C. If a good color match is not necessary, the RCuZn-C filler metal can be used. Flux should be used when oxyfuel gas welding bronze.

Copper-Aluminum Alloys (Aluminum Bronze)

Aluminum bronzes with less than 7% aluminum are difficult to weld. They tend to crack in the heat-affected zone. Aluminum bronzes with 7% aluminum or more (or that have more than one alloying element) can be welded, but techniques to minimize cracking must be used.

Preheat is used when welding metal over 3/16" (4.8mm). The preheat and interpass heating temperatures vary, depending on the alloy being welded. The temperature for alloys with less than 10% aluminum should be 200°F to 300°F (95°C to 150°C). After welding, allow the weldment to air cool to room temperature. Alloys with 10% or more aluminum require a preheat temperature of 500°F (260°C). After welding, these bronze alloys should be rapidly cooled to room temperature.

Filler metal for GTAW and GMAW aluminum bronzes are designated ERCuAl-A2 or ERCuAl-A3. Electrodes for SMAW are designated ECuAl-A2 or ECuAl-B. Argon gas or an argon-helium mixture can be used. Helium increases the heat into the bronze and allows for faster welding speeds. Oxyacetylene is not recommended as a welding process for these alloys.

Copper-Silicon Alloys (Silicon Bronze)

The silicon bronzes can all be welded. The silicon acts as a good deoxidizer and can form a thin layer over the weld to protect the weld metal from oxidation. Silicon bronzes have a lower thermal conductivity, so preheating is not necessary. Preheating can be used when welding thicker metals, but only to about 200°F (95°C).

GTAW, GMAW, and SMAW can all be used to weld silicon bronze alloys. Electrode filler metal is designated ERCuSi-A. Aluminum bronze filler metal (ERCuAl-A2) can be used sometimes. The SMAW electrode designation for this type alloy is ECuSi. The electrode ECuAl-A2 can also be used. Oxyacetylene welding can be used to join these alloys, but the arc welding processes do a better job. A flux and an oxidizing flame are used when oxyacetylene welding.

Copper-Nickel Alloys

The copper-nickel alloys are very corrosion-resistant and have good hot strength. These alloys have many uses.

A primary use for copper-nickel alloys is to contain sea water or in other saltwater applications.

The alloys vary from 5% to 30% nickel, with the 10% and 30% nickel alloys being most common. These alloys are weldable by the arc welding processes and by oxyacetylene welding. Oxyacetylene should be used, however, only if one of the arc welding processes is not available. Preheating is not necessary, and interpass heating should be limited to 150°F (65°C). Filler metals for copper-nickel alloys are designated ERCuNi; electrodes for SMAW are designated ECuNi.

The copper-nickel weld pool is sluggish. A slight weaving motion is required to help mix the weld pool. As when welding any other base metal, argon shielding gas gives better control of the arc; however, helium produces more heat and more penetration. Argon or an argon-helium mixture are used for both GTAW and GMAW.

22.8 TITANIUM

Titanium and titanium alloys have very good strength and ductility. Titanium has excellent resistance to corrosion. (Additional information may be obtained in Heading 28.10.1.)

Titanium is heavier than aluminum, but lighter than steel. It is available in many forms, such as forgings, bar, sheet, or plate. When titanium is annealed, it is relatively soft and ductile. It obtains its maximum strength when it is solution heat treated and then aged. Strengths can exceed 200 ksi (1380MPa).

Titanium is available in four different commercially pure grades, and is alloyed with many different elements. The four commercially pure grades are classified as ASTM Grade 1, Grade 2, Grade 3, and Grade 4. When titanium is alloyed with other elements, the major alloying elements are included as part of the classification. Examples are: Ti-5Al-2.5Sn, Ti-2Al-11Sn-5Zr-1Mo, Ti-6Al-4V, Ti-8Mn, and Ti-8Mo-8V-3Al-2Fe. The numbers indicate the percentage of each alloying element. Titanium filler metal and welding electrodes are designated in the same way.

22.8.1 Welding Titanium

The biggest problem with titanium and titanium alloys is that, at elevated temperatures, titanium will dissolve oxygen, nitrogen, and hydrogen. Small amounts of oxygen and nitrogen increase the strength and hardness of the metal. Large amounts of these elements or of such impurities as grease, oil, or moisture cause titanium to become brittle.

To prevent titanium from being contaminated, an inert gas shielding must be used. The inert gas must be very pure with no added moisture. The weld requires shielding gas on both sides. The molten metal and heat-affected zone must be protected until they have cooled to 800°F (427°C) or lower.

One method used to shield the weld metal as it cools is to cover the weld and surrounding area with shielding gas. Figure 22-11 illustrates two side shields around the weld area. Shielding gas completely covers the weld and heat-affected zone. In addition to shielding the top surface, the bottom surface must also be protected with a gas called a *backing gas*. Figure 22-12 shows a channel used to protect the root side of the weld. When an automatic machine is used, a trailing shield protects the titanium as it cools. The trailing shield is attached to the welding torch and moves with the torch. A trailing shield is illustrated in Figure 22-13. The welding torch always provides the primary gas shielding of the molten metal. The shielding gas used to

protect the weld metal and heat-affected zone as they cool is called *secondary gas shielding*.

Another method used to protect the metal from contamination is to weld in an inert gas atmosphere. The welding is performed in a chamber completely filled with an inert gas. Two types of chambers are used:

*   Small units where the welder is outside the chamber with arm and glove manipulators to handle the electrode holder. See Figure 22-14.

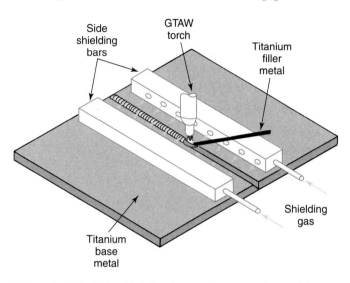

Figure 22-11. Side shielding is used to cover the weld metal and heat-affected zone with shielding gas as they cool.

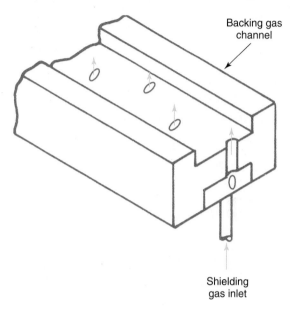

Figure 22-12. A backing gas channel is used to protect the root side of a weld.

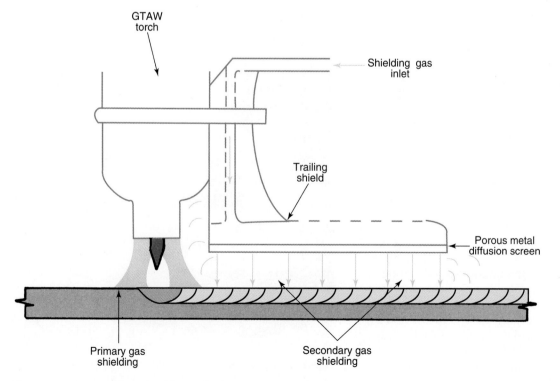

Figure 22-13. A trailing shield attached to an automatic welding torch. The porous screen allows the shielding gas to cover the entire weld area.

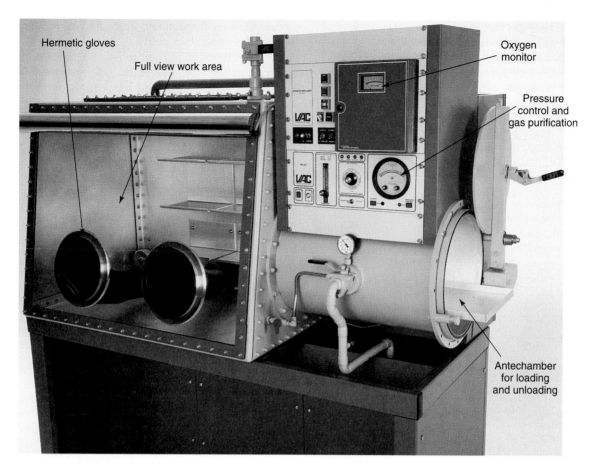

Hermetic gloves

Full view work area

Oxygen
monitor

Pressure
control and
gas purification

Antechamber
for loading
and unloading

Figure 22-14. A controlled-atmosphere shielding gas cabinet. Difficult-to-weld metals may be welded inside this cabinet. Oxygen and moisture levels are held to less than 1 ppm (part per million) (Vacuum Atmospheres Co.)

- Large units with gas locks. The chamber is full of shielding gas and the welder enters the chamber to do the welding. The welder must wear breathing apparatus.

Whenever you enter a gas-filled area, you must wear a supplied air respirator or self-contained breathing apparatus. In cases where workers without breathing equipment enter a chamber filled with a shielding gas, such as argon, unconsciousness can occur in less than ten seconds, with death following rapidly.

An inflatable, flexible bag type of shielding gas chamber is shown in Figure 22-15. The bag is inflated with an inert gas. This bag is designed to allow the welder to access the parts being welded through different sets of access holes.

The filler metal or electrode used when welding titanium should match the base metal composition. Classification of filler metals and electrodes is the same as for the base metal. The major alloying elements and their percentages are listed as part of the designation. Examples are: ERTi-1, ERTi-2, (commercially pure grades 1 and 2), ERTi-3Al-2.5V, ERTi-8Al-1 Mo-1V, and ERTi-6Al-4V. In the examples, E stands for electrode, R for solid rod; Ti for titanium; Al for aluminum, Mo for molybdenum; and V for vanadium.

The most common methods of welding titanium are GMAW and GTAW. Other processes include plasma arc

Figure 22-15. A bag bell jar atmosphere system. The many air-tight arm access ports allow the welder to reach all areas of the weldment while welding in an inert atmosphere. (Vacuum Atmospheres Co.)

welding, electron beam welding, laser welding, resistance welding, friction welding, and brazing.

Welding titanium requires that the base metal be extremely clean. All oxides, grease, oil, and even fingerprints must be removed prior to welding. Before welding, parts must be cleaned, then handled with gloves. Any contamination on the parts can cause oxygen, hydrogen, nitrogen, or carbon to combine with the titanium and weaken the metal.

When GTAW, a 2% thoriated electrode (EWTh-2) is used with direct current electrode negative (DCEN). The electrode should not stick out too far from the gas nozzle. This will help prevent tungsten inclusions. The arc should *not* be touch-started on the titanium. A starting tab can be used to strike the arc. The starting tab is not part of the weld. It is an extra piece of metal placed next to the weld on which the weld is started. The tab is removed once the weld is complete.

When GMAW, it is important to protect the titanium base metal from oxygen, nitrogen and hydrogen. Extremely pure shielding gas must be used. Additional trailing shield attachments or side shielding, similar to those described earlier, must be used. Backing gas is also required. Even when welding a T-joint or corner joint, a backing gas is required to keep the back side of the joint from contamination. The weld and heat-affected zone must be protected with shielding gas until the metal temperature is below 800°F (427°C).

It is also highly important that the electrode wire remain very clean. The electrode wire cannot have any grease, oil, or dirt on it. Cover the spool of wire to prevent dust and dirt from getting on the wire. When not using a spool of titanium wire, remove it from the welding machine and store it in a bag in a clean area or a clean container.

22.9 WELDING ZIRCONIUM

Zirconium is a rare metal used for some commercial applications. Its unique properties make it useful for astronautic (travel beyond the earth's atmosphere) applications and for chemical and nuclear energy uses. (Additional information may be obtained in Heading 28.10.3.)

Zirconium is weldable using the gas tungsten and gas metal arc welding processes. Overall, the cleaning and welding requirements for zirconium are very similar to those of titanium. The metal must be cleaned as thoroughly as titanium. Welding is done in a 100% inert gas atmosphere. Zirconium can also be joined by electron beam welding, laser beam welding, plasma arc welding, resistance welding, or brazing.

22.10 WELDING BERYLLIUM

Beryllium is a commercial metal that is lighter than aluminum and has a fairly high melting temperature. Its tensile strength is approximately 55 ksi (380MPa).

(Additional information may be found in Heading 28.10.4.)

Beryllium can be gas tungsten arc welded, electron beam welded, solid-state welded, resistance spot welded, or brazed. Beryllium is very susceptible to cracking. High heat inputs and high cooling rates must be avoided.

Beryllium is very toxic. Do not allow any beryllium shavings or dust to be taken internally (by breathing or eating).

22.11 WELDING DISSIMILAR METALS

It is possible to weld together dissimilar nonferrous metals, such as aluminum to copper, aluminum to lead, copper to lead, and many other combinations. Steel and some of the stainless steels can be welded or brazed to such metals as aluminum, copper, and others.

In most cases, arc welding and oxyfuel gas welding have not proven as successful for welding dissimilar metals as resistance welding, friction welding, and/or the solid-state welding processes

22.12 PLASTICS

Plastics are used for a large variety of component parts in manufactured articles. These materials vary greatly in chemical composition and physical characteristics. However, regardless of their properties or form, plastics all fall into one of two groups. The two groups are the thermoplastic and the thermosetting types.

The *thermoplastics* become soft when heated, and are capable of being formed. These plastics harden when cooled. Thermoplastics may be heated and reformed repeatedly.

The following plastics fall into the thermoplastic classification: acrylic, cellulosic, acetal resin, nylon, vinyl, polyethylene, polystyrene, polyvinylidene, polypropylene, polycarbonate, and polyfluorocarbon.

The *thermosetting plastics*, once formed and cooled, cannot be reheated and reformed. Thermosetting plastics include: aminoplastic, phenolic, polyester, silicone, alkyd, epoxy, casein, and allylic.

22.12.1 Plastic Welding Principles

Thermoplastics may be welded in the same manner as metals. All joint designs may be used, and successful welds are possible in various positions. Figures 22-16 and 22-17 show a completed butt weld and a fillet weld on plastics.

The welding rod must be of the same composition as the plastic being welded. Plastic welding rods may be obtained in round, oval, triangular, or flat strip forms. The welder's choice of welding rod shape is affected by the shape of the joint, the thickness of the plastics to be welded, and the welding equipment to be used. The joints and

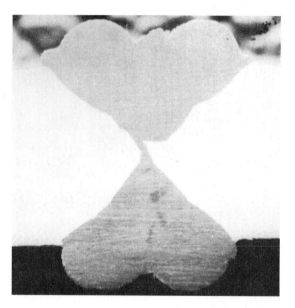

Figure 22-16. *Cross section of plastic butt joint welded on both sides.*

Figure 22-17. *Cross section of a plastic T-joint welded on both sides. (Laramy Products Co., Inc.)*

the welding rod must be clean to produce a high quality weld.

22.12.2 Welding Plastics

Heat for plastic welding is supplied by a heated gas, either compressed air or a shielding gas. The gas is heated as it passes through the welding torch, then is directed onto the surface of the joint. No flame touches the joint, only the heated gas.

Electric heating coils are usually used to heat the gas as it flows through the torch. Figure 22-18 illustrates a typical torch used for plastic welding. This torch uses an electric coil element to heat the welding gas. Figure 22-19 shows a complete plastic welding outfit.

When welding, the torch is held in one hand and the plastic welding rod is fed to the weld area with the other hand. This is similar to oxyfuel gas welding. Figure 22-20 illustrates hand welding of a butt joint on plastic material.

Figure 22-21 shows a tip which is called a speed welding tip. The plastic filler rod is inserted in the guide tube. There are two openings for the heated gas. One opening is used to preheat the plastic, the second opening heats and melts the plastic and the filler rod. An electric welding torch with a speed welding tip installed is shown in Figure 22-22. A light, uniform pressure on the plastic rod is

Figure 22-18. *A torch designed for welding plastics. It uses an electric heating coil to heat the welding gas. (Laramy Products Co., Inc.)*

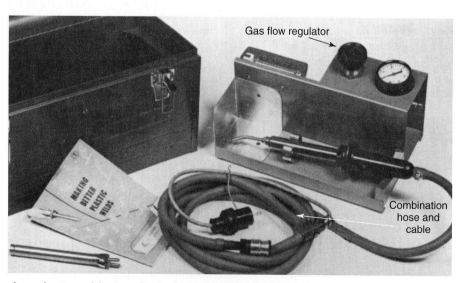

Figure 22-19. *A complete plastic welding outfit. (Laramy Products Co., Inc.)*

required. A light downward pressure is also exerted on the torch to press the rod into the heated weld area.

If the weld seam warrants it, the welding rod may be in a strip form. The strip is applied in a manner similar to the rod shape using a special *speed welding tip*. See Figure 22-23.

The welding gas may be a shielding gas or compressed air. To regulate the welding gas temperature, adjust the gas flow. If the gas flow over the heating element is slow, the gas picks up more heat. If the gas flows quickly over the heating element, it will pick up less heat. Therefore, the welder decreases the welding gas flow to *increase* the temperature, or increases the welding gas flow to *decrease* its temperature.

To change the heating capacity of the torch, the tip size is changed. The torch heats the welding gas to 450°F to 800°F (232°C to 427°C). This heated gas is directed to the weld area where it heats the joint and the welding rod to a soft (plastic) state. While both the joint and the rod are soft, they are forced together with light pressure. The surfaces of the joint and welding rod bond. As the surfaces cool, they harden, and a good weld results.

Four things are necessary to accomplish a good weld on plastics:

- Correct welding temperature.
- Correct pressure on the welding rod.
- Correct welding rod angle.
- Correct welding speed.

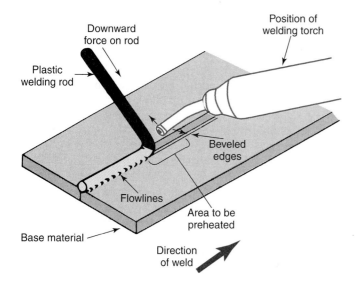

Figure 22-20. *Correct procedure for hand-welding a butt joint on plastic material. Note the position of the plastic filler rod and the torch tip. (Kamweld Products Co., Inc.)*

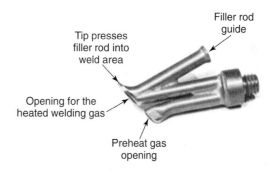

Figure 22-21. *Speed welding tip for welding plastics. (Laramy Products Co., Inc.)*

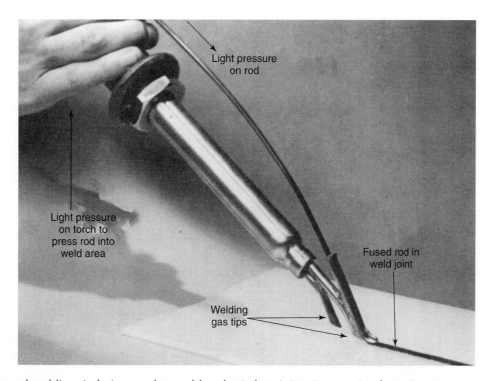

Figure 22-22. *A speed welding tip being used to weld a plastic butt joint. (Laramy Products Co., Inc.).*

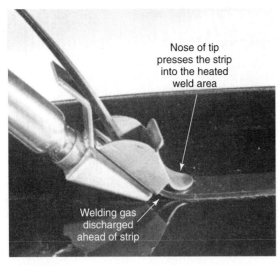

Nose of tip
presses the strip
into the heated
weld area

Welding gas
discharged
ahead of strip

Figure 22-23. Plastic butt joint being welded using a plastic strip as the joining material. Note the special shape of the speed welding tip. (Laramy Products Co., Inc.)

TEST YOUR KNOWLEDGE

Write your answers on a separate sheet of paper. Do not write in this book.

1. Nonferrous metals and alloys can contain iron as an _____ element.
2. All forms of aluminum, other than cast aluminum, are called _____ aluminum.
3. What organization has a standardized system for classifying aluminum alloys?
4. What is the major alloying element in a 4XXX series alloy? A 5XXX series alloy?
5. What filler metal should be used to obtain the best strength and ductility when welding a piece of 5052 aluminum to a piece of 3003 aluminum? See Figure 22-2.
6. What preheat temperature is used when oxyfuel gas welding aluminum? When arc welding aluminum?
7. Aluminum is usually welded using the _____ or _____ process.
8. What types of welding currents produce a cleaning action on the aluminum?
9. What type of flame is recommended when oxyfuel gas welding aluminum?
10. Flux left on an aluminum weld will have a _____ effect.
11. What alloys are added to magnesium to form the most popular magnesium alloy?
12. Why are carbon blocks and carbon paste often used when welding die castings?
13. Which type of copper is comparatively easy to weld?
14. Why does a weld in oxygen-bearing copper have less strength than the base metal?
15. Does the heat in a copper weld leave the weld faster or slower than the heat in a steel weld?
16. What is the biggest problem when welding titanium? Why?
17. List two ways to help eliminate porosity and contamination when welding titanium.
18. When GTAW titanium, what type of electrode and current is used?
19. What classification of plastics can be welded?
20. How is the plastic heated when welding?

Chapter 23

PIPE AND TUBE WELDING

LEARNING OBJECTIVES

After studying this chapter, you will be able to:
* Define the differences between a pipe and a tube.
* Name and recognize a variety of pipe and tube fixturing tools.
* Describe the welding procedures used when welding pipe or tubing with oxyfuel gas, SMA, GMA, and/or GTA welding process.
* Cite several nondestructive evaluation methods for pipe weld quality.

Pipes and tubes are used to carry fluids, both gases and liquids, from one point to another. Pipes of various sizes may carry gasoline, oil, and natural gas thousands of miles from the source to the user. Tubing carries refrigerants in refrigerators and air conditioners; gasoline, oil, and brake fluids in automobiles, and water and natural gas within our homes.

A *pipe* is normally a hollow cylinder with thick walls. It may be made of metal, plastic, or other materials. The wall of a pipe is generally thick enough to permit a thread to be cut into the outer surface. Pipes may be assembled mechanically with threaded joints or with compression fittings, or they may be welded or brazed. The pipe wall is thick enough and strong enough to permit it to carry high-pressure fluids or to support heavy loads when used in construction.

A *tube* is a hollow conduit with a relatively thin wall. In cross section, tubing may be round, square, hexagonal, oval, or almost any other shape required. The walls are too thin to be threaded. Tubing is assembled with mechanical connectors or may be welded, soldered, or brazed.

23.1 TYPES OF PIPE

Generally, the larger the pipe diameter, the greater the wall thickness. The following materials are often used to make pipe:
* Cast iron.
* Low-carbon steel.
* Low-alloy steel and high-strength/low-alloy (HSLA) steel.
* Stainless steel.
* Copper or brass.
* Plastic.

Cast iron pipe is seamless pipe formed in a mold. The properties of cast iron are such that the sections of the pipes are difficult to thread. Cast iron pipe is usually joined mechanically. This limits the use of cast iron pipe to low-pressure installations, such as drainage work. Figure 23-1 shows two types of mechanical pipe butt joints.

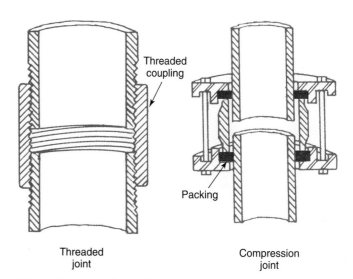

Figure 23-1. Mechanical pipe butt joints.

535

Low-carbon, low-alloy, and HSLA steel pipe may be manufactured in either seamed or seamless form. Seamed pipe is fabricated by arc (often SAW) welding or by high-frequency resistance welding. Refer to Figure 19-27. The seamless form of pipe is fabricated from solid bar stock drawn through special dies. Seamless pipe has considerably more strength and can be used for work involving much higher pressure than seamed pipe.

Whenever corrosive chemicals are carried in a piping system, pipe made from stainless steel and other noncorrosive steel alloys is used. Plastic pipe and tubes are being used where highly corrosive chemicals are being transported at low pressure. Plastic pipes are also used in low-pressure systems that are laid in water or in the ground, since the plastic will not corrode.

Copper pipe and brass pipe are used for water and refrigerant lines. This pipe is usually connected by soldering. Figure 23-2 shows three types of welded, brazed, and soldered pipe butt joints.

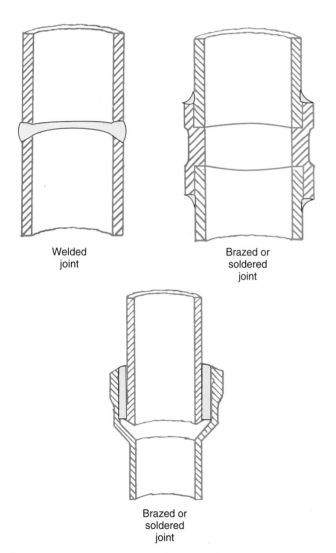

Figure 23-2. *Cross sections of pipe and tube butt joints joined by welding, brazing, and soldering.*

Aluminum alloy pipe and tubing is also finding wide use. Aluminum pipe and tubing may be soldered, brazed, or welded relatively easily. Aluminum pipe may be welded using almost any welding process normally used with aluminum. Welding may be done using manual, semiautomatic, or automatic methods.

23.1.1 American Standards Association (ASA) Pipe Schedules

The size of a pipe is specified by its approximate inside diameter (ID). However, a 1/4" or 1" pipe does not actually measure 1/4" or 1" on its inside diameter. The *American Standards Association (ASA) Pipe Schedule* is used to refer to and order pipe. A pipe schedule gives the *nominal* (approximate) ID size of a pipe, its *schedule number*, the inside and outside diameters, and the wall thickness. See Figure 23-3. A given size pipe, such as 8" pipe, may be manufactured with as many as 11 different inside diameters and wall thicknesses. A 3/8" pipe has only two schedule numbers, or two different wall thicknesses. To specify a pipe size, the approximate inside diameter and schedule number must be given.

23.1.2 Methods of Joining Pipe

Pipes may be joined together or to fittings by the following methods:
- Pipe threads.
- Flange fittings.
- Welding.
- Adhesives (usually used only on plastic pipe).

To assemble pipe using threads, the ends of pipe and a variety of fittings are threaded using the *American Standard Taper Pipe Thread*. As the pipe and fitting are screwed together, they get tighter due to the taper in the threads. Properly assembled, the threaded joint is air- and water-tight.

A *flange fitting* has a collar around the end of the fitting. Flange fittings are assembled to a pipe by using threads or by welding. The flange on the fitting has several holes in it. To connect two pipes with flanges, a sealing gasket is placed between the mating flanges. Bolts are placed into the holes and the flanges are drawn tight to compress the gasket. Flanged joints are easily disassembled by removing the flange bolts.

When pipes are to be permanently connected, welding is the most popular method of joining. The advantage of welding pipe are neatness, compactness, rapidity, and lower cost. A variety of pipe joints prepared or manufactured for assembly by welding are shown in Figures 23-4 and 23-5. Pipe welding may be performed using the shielded metal arc, gas metal arc, gas tungsten arc, or submerged arc welding processes. They are also joined by resistance welding using the flash or friction welding processes. Refer to Headings 19.6 and 20.13. Oxyfuel gas welding of pipe is not done commercially. Any process used may be done manually, semiautomatically, or automatically.

Nominal pipe size, in	Schedule no.	Wall thickness, in	Inside diameter, in
1/8 (0.405)	10S	0.049	0.307
	40ST, 40S	.068	.269
	80XS, 80S	.095	.215
1/4 (0.540)	10S	.065	.410
	40ST, 40S	.088	.364
	80XS, 80S	.119	.302
3/8 (0.675)	10S	.065	.545
	40ST, 40S	.091	.493
	80XS, 80S	.126	.423
1/2 (0.840)	5S	.065	.710
	10S	.083	.674
	40ST, 40S	.109	.622
	80XS, 80S	.147	.546
	160	.188	.464
	XX	.294	.252
3/4 (1.050)	5S	.065	.920
	10S	.083	.884
	40ST, 40S	.113	.824
	80XS, 80S	.154	.742
	160	.219	.612
	XX	.308	.434
1 (1.315)	5S	.065	1.185
	10S	.109	1.097
	40ST, 40S	.133	1.049
	80XS, 80S	.179	.957
	160	.250	.815
	XX	.358	.599
1 1/4 (1.660)	5S	.065	1.530
	10S	.109	1.442
	40ST, 40S	.140	1.380
	80XS, 80S	.191	1.278
	160	.250	1.160
	XX	.382	0.896
1 1/2 (1.900)	5S	.065	1.770
	10S	.109	1.682
	40ST, 40S	.145	1.610
	80XS, 80S	.200	1.500
	160	.281	1.338
	XX	.400	1.100

Nominal pipe size, in	Schedule no.	Wall thickness, in	Inside diameter, in
2 (2.375)	5S	.065	2.245
	10S	.109	2.157
	40ST, 40S	.154	2.067
	80XS, 80S	.218	1.939
	160	.344	1.687
	XX	.436	1.503
2 1/2 (2.875)	5S	.083	2.709
	10S	.120	2.635
	40ST, 40S	.203	2.469
	80XS, 80S	.276	2.323
	160	.375	2.125
	XX	.552	1.771
3 (3.500)	5S	.083	3.334
	10S	.120	3.260
	40ST, 40S	.216	3.068
	80XS, 80S	.300	2.900
	160	.438	2.624
	XX	.600	2.300
3 1/2 (4.0)	5S	.083	3.834
	10S	.120	3.760
	40ST, 40S	.226	3.548
	80XS, 80S	.318	3.364
4 (4.5)	5S	.083	4.334
	10S	.120	4.260
	40ST, 40S	.237	4.026
	80XS, 80S	.337	3.826
	120	.438	3.624
	160	.531	3.438
	XX	.674	3.152
5 (5.563)	5S	.109	5.345
	10S	.134	5.295
	40ST, 40S	.258	5.047
	80XS, 80S	.375	4.813
	120	.500	4.563
	160	.625	4.313
	XX	.750	4.063

Nominal pipe size, in	Schedule no.	Wall thickness, in	Inside diameter, in
6 (6.625)	5S	0.109	6.407
	10S	.134	6.357
	40ST, 40S	.280	6.065
	80XS, 80S	.432	5.761
	120	.562	5.501
	160	.719	5.187
	XX	.864	4.897
8 (8.625)	5S	.109	8.407
	10S	.148	8.329
	20	.250	8.125
	30	.277	8.071
	40ST, 40S	.322	7.981
	60	.406	7.813
	80XS, 80S	.500	7.625
	100	.594	7.437
	120	.719	7.187
	140	.812	7.001
	XX	.875	6.875
	160	.906	6.813
10 (10.75)	5S	.134	10.482
	10S	.165	10.420
	20	.250	10.250
	30	.307	10.136
	40ST, 40S	.365	10.020
	60XS, 80S	.500	9.750
	80	.594	9.562
	100	.719	9.312
	120	.844	9.062
	140, XX	1.000	8.750
	160	1.125	8.500
12 (12.75)	5S	0.156	12.438
	10S	0.180	12.390
	20	0.250	12.250
	30	0.330	12.090
	ST, 40S	0.375	12.000
	40	0.406	11.938
	XS, 80S	0.500	11.750
	60	0.562	11.626
	80	0.688	11.374
	100	0.844	11.062
	120, XX	1.000	10.750
	140	1.125	10.500
	160	1.312	10.126
14 (14)	5S	0.156	13.688
	10S	0.188	13.624
	10	0.250	13.500
	20	0.312	13.376
	30, ST	0.375	13.250
	40	0.438	13.124
	XS	0.500	13.000
	60	0.594	12.812
	80	0.750	12.500
	100	0.938	12.124
	120	1.094	11.812
	140	1.250	11.500
	160	1.406	11.188

Nominal pipe size, in	Schedule no.	Wall thickness, in	Inside diameter, in
16 (16)	5S	0.169	15.670
	10S	0.188	15.624
	10	0.250	15.500
	20	0.312	15.376
	30, ST	0.375	15.250
	40, XS	0.500	15.000
	60	0.656	14.688
	80	0.844	14.312
	100	1.031	13.938
	120	1.219	13.562
	140	1.438	13.124
	160	1.594	12.812
18 (18)	5S	0.165	17.670
	10S	0.188	17.624
	10	0.250	17.500
	20	0.312	17.376
	ST	0.375	17.250
	30	0.438	17.124
	XS	0.500	17.000
	40	0.562	16.876
	60	0.750	16.500
	80	0.938	16.124
	100	1.156	15.688
	120	1.375	15.250
	140	1.562	14.876
	160	1.781	14.438
20 (20)	5S	0.188	19.624
	10S	.218	19.564
	10	.250	19.500
	20, ST	.375	19.250
	30, XS	.500	19.000
	40	.594	18.812
	60	.812	18.376
	80	1.031	17.938
	100	1.281	17.438
	120	1.500	17.000
	140	1.750	16.500
	160	1.969	16.062
24 (24)	5S	0.218	23.564
	10, 10S	0.250	23.500
	20, ST	0.375	23.250
	XS	0.500	23.000
	30	0.562	22.876
	40	0.688	22.624
	60	0.969	22.062
	80	1.219	21.562
	100	1.531	20.938
	120	1.812	20.376
	140	2.062	19.876
	160	2.344	19.312
30 (30)	5S	0.250	29.500
	10, 10S	0.312	29.376
	ST	0.375	29.250
	20, XS	0.500	29.000
	30	0.625	28.750

Figure 23-3. *A condensed Standard Pipe Schedule which shows the schedule number, wall thickness, and inside diameter for each nominal pipe size. (©1992, TRI-Tool, Inc.)*

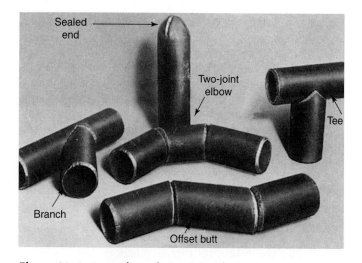

Figure 23-4. *A number of pipe joint designs.*

Adhesives are often used to join plastic tubing and pipe. The resulting joints are water-tight and, when the adhesive is properly applied, have adequate strength and reliability.

23.2 TYPES OF TUBING

Tubing is made from steel, copper, aluminum, brass, or plastic. In the manufacture of automobiles, aircraft, refrigerators, and air conditioners, tubing is used for moving fluids. Tubing is also an excellent choice for transmitting forces in hydraulic systems.

Tubing is available in flexible (soft) or rigid (hard) forms. The size of tubing is measured and specified by its outside diameter (OD). Whenever a change in direction is necessary, *flexible tubing* is easily bent to redirect it. *Rigid tubing*, like pipe, requires the use of appropriate fittings to

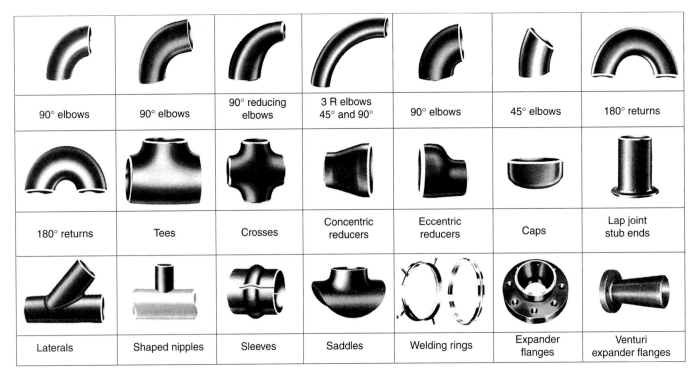

90° elbows	90° elbows	90° reducing elbows	3 R elbows 45° and 90°	90° elbows	45° elbows	180° returns
180° returns	Tees	Crosses	Concentric reducers	Eccentric reducers	Caps	Lap joint stub ends
Laterals	Shaped nipples	Sleeves	Saddles	Welding rings	Expander flanges	Venturi expander flanges

Figure 23-5. *Pipe joint fittings. Such fittings are commercially made in various sizes and thicknesses of steel and steel alloys. The edges are generally beveled and ready for welding.*

change its direction. Tubing is obtained in different strengths and is manufactured in either seamed or seamless forms.

Seamless tubing is made by piercing a piece of metal and drawing that metal through a die and over a *mandrel*. A similar method is to force a pierced piece of metal over a mandrel by means of rollers, as shown in Figure 23-6.

Seamed tubing is produced by rolling flat stock into a round cylindrical form or any other form required, then welding the seam. The joint formed is normally a butt joint. Resistance butt seam welding is often used to weld a seamed tube in production. See Figure 19-27

Tubing standards have been established by the American Society of Mechanical Engineers (ASME), the American Society for Testing and Materials (ASTM), the Society of Automotive Engineers (SAE), and the U.S. Government (MIL specifications).

Tubing is available in various cross-sectional shapes:
- Round.
- Square.
- Rectangular.
- Hexagonal.
- Octagonal.
- Streamline.
- Oval.
- Irregular (various shapes).

23.2.1 Methods of Joining Tubing

Tubing may be joined by using:
- Mechanical compression or flared fittings.
- Mechanical quick disconnect fittings or couplers.
- Soldered fittings.
- Brazed fittings.
- Welding.
- Adhesives.

Compression fittings and *flared fittings* are used on small copper or steel tubing. They are easily assembled and disassembled. Compression fittings may be used on relatively low-pressure systems like fuel and water lines. Flared or double-flared fittings are used with higher-pressure systems. They are found on automotive brake fittings, hydraulic systems, and refrigeration lines.

Quick disconnect couplers are spring-loaded and gasketed fittings that enable quick assembly and dismantling of tubing systems without loss of the fluid. These quick disconnect fittings must be joined to tubing by threads, soldering, brazing, or welding. Figure 23-7 shows a quick disconnect coupling.

Soldered connections and *brazed connections* use special fittings similar to those shown in Figure 23-5. The inside diameter of a solder or braze fitting is made slightly larger than the outside diameter of the tubing. This allows space for the solder or braze metal that joins the assembly. The solder or braze metal is applied to the heated joint and drawn into the fitting by capillary action. The solder or braze metal forms a thin film, joining the tube and fitting by adhesion. Soldered or brazed connections may be used on seamed and seamless tubing and on smaller diameter copper, brass, or bronze pipe.

Welding is also done to join tubing, using the SMAW, GMAW, and GTAW processes. The welding of tubing is discussed in Headings 23.6 and 23.7.

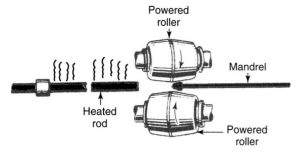

Start of operation

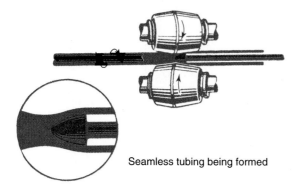

Seamless tubing being formed

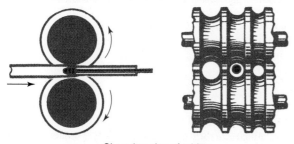

Various steps in operation

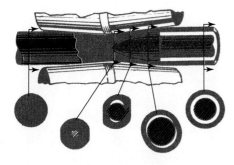

Changing size of tubing

Figure 23-6. *Forming seamless tubing from a solid rod. (Michigan Seamless Tube Co.)*

23.3 PREPARING PIPE JOINTS FOR WELDING

Pipe must be carefully prepared for welding to ensure properly aligned joints and strong, neat welds. Ready-to-use welded pipe fittings are available, many with the joint

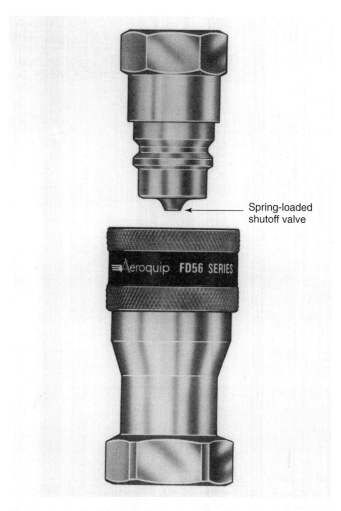

Figure 23-7. *Quick disconnect coupling used for hose lines that need easy and fast connecting and disconnecting, such as hydraulic, pneumatic, or water lines. The gas or liquid flow is shut off by a spring-loaded valve when the coupling is disconnected. (Aeroquip Corp.)*

ends already beveled and ready for welding. Refer to Figure 23-5. These fittings are available in various pipe diameters and schedule sizes. When these fittings are ordered, the pipe diameter and schedule size must be given to match the pipe being welded.

Pipes are seldom welded using a square-groove butt joint. Because the walls are relatively thick, the edges are always prepared for some type of groove weld. The groove preparation may be done by oxyfuel gas cutting, grinding, or machining. When preparing high-pressure pipe for welding, the shape of the groove and the angles are often very complex, as shown in Figure 23-8.

Special bevel-cutting machines are used to accurately cut these complex angles on the pipe. These machines are powered by electric or hydraulic motors and come in a variety of sizes for use on different sizes of pipe. See Figure 23-9. The machines can be used both to cut and to bevel the pipe. Figure 23-10 shows a large, hydraulically powered pipe beveling machine. Special cutters are used in these machines to cut any desired shape as they move around

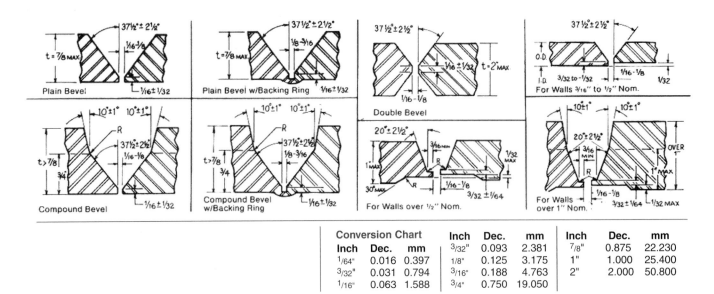

Conversion Chart			Inch	Dec.	mm	Inch	Dec.	mm
Inch	Dec.	mm	3/32"	0.093	2.381	7/8"	0.875	22.230
1/64"	0.016	0.397	1/8"	0.125	3.175	1"	1.000	25.400
3/32"	0.031	0.794	3/16"	0.188	4.763	2"	2.000	50.800
1/16"	0.063	1.588	3/4"	0.750	19.050			

Figure 23-8. *Several typical weld joint designs for pipe. (E.H. Wachs Company)*

Figure 23-10. *A technician demonstrates the use of a large hydraulically powered pipe machining tool. The tool cuts the proper angles to prepare a pipe for welding. (TRI-Tool, Inc.)*

Figure 23-9. *Several pipe machining tools for use on various sizes of pipe. (TRI-Tool, Inc.)*

the pipe. The finished cut has perfectly formed angles and is clean and ready for welding.

Single-angle bevels may be cut using an oxyfuel gas cutting torch mounted on a semiautomatically or automatically operated track around the pipe. The pipe beveling machine shown in Figure 23-11 has the torch set at the desired angle. The torch is rotated around the pipe manually. The pipe shown in Figure 23-12 is rotated past a stationary cutting torch.

After the ends of the pipe are prepared, they must be accurately aligned prior to welding. A large variety of tools have been designed to align pipes and hold them in position to prevent distortion. Figure 23-13 shows several fixtures used for aligning pipe. Figure 23-14 is a clamp used to align a butt joint while it is welded with a rotating GTAW torch. Figure 23-15 shows a tool that is used to align a flange with the center of the pipe to which it will be welded. A pipe elbow is aligned using a clamp like that shown in Figure 23-16.

Prior to welding, the joint should be tack welded every 6" to 8" (15cm to 20cm) around the circumference of the pipe to hold its alignment.

Figure 23-11. This oxyacetylene cutting torch is moved around the pipe by a hand crank and gears. (Mathey/Leland International Ltd.)

Figure 23-12. A computer-controlled pipe cutter. On this machine, the torch remains stationary as the pipe is rotated under it. (American Welding Society)

23.3.1 Layout of Pipe and Tube Joints

Many times, a pipe welder is required to fabricate fittings or to prepare specially shaped joints when connecting pipes. Several shop practices have been developed to solve the problems involved when fitting pipes together at different angles. Problems also occur when assembling pipes of different diameters.

Paper or sheet metal patterns may be developed on the drafting board and used in the shop. Examples of developed patterns are shown in Figures 23-17 and 23-18.

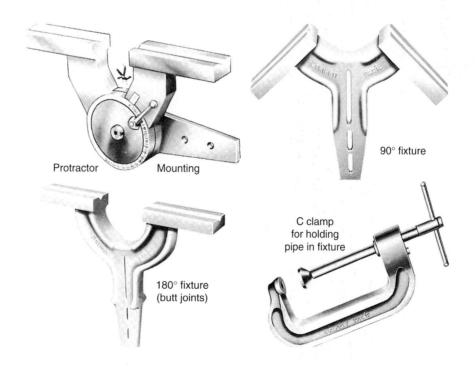

Protractor   Mounting

90° fixture

180° fixture
(butt joints)

C clamp
for holding
pipe in fixture

Figure 23-13. Fixtures for holding pipes while they are being welded. (Strippit, Unit of IDEX Corp.)

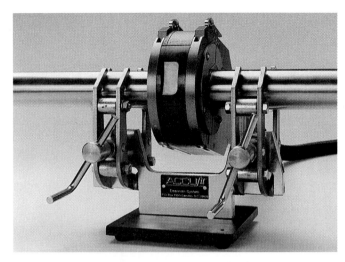

Figure 23-14. *This double clamp holds the pipe in alignment while it is being welded by a rotating GTAW torch. (Dearman, a division of Cogsdill Tool Products, Inc.)*

Figure 23-15. *This tool is used to align the inside diameters of the pipe and flange prior to welding. (Dearman, a division of Cogsdill Tool Products, Inc.)*

Another easy method for developing the shape of an angled pipe connection is shown in Figure 23-19.

Many special devices have been developed to aid in producing accurate pipe layouts quickly and easily. Figure 23-19 is a mechanical pipe layout device used with soapstone to mark a pipe. Many automated pipe cutting machines have computerized programs that can be called up to cut every conceivable shape on various pipe diameters and fittings.

Figure 23-16. *A large elbow is aligned to the pipe with a chain-type alignment tool. (Dearman, a division of Cogsdill Tool Products, Inc.)*

23.4 ARC WELDING PIPE JOINTS

The most common joint used in pipe welding is the single-V-groove butt joint. The position of the pipe and joint to be welded varies greatly. The most common position or orientation for a pipe weld is when the pipe is horizontal and the joint is vertical. The American Welding Society refers to this position as the *5G position*. The various AWS pipe welding positions are shown in Figure 23-20. A butt joint in the 5G position, when the pipe is not rotated, involves welding in the vertical, overhead, and flat welding positions. See Figure 23-21. Common arc welding processes used for pipe welding include SMAW, GTAW, and GMAW. The shielded metal arc welding process is the most often used for pipe welding.

Before welding begins, the pipe must be cut to size, the edges prepared, the joint aligned (and often held in place with fixtures and clamps), and the joint tack welded.

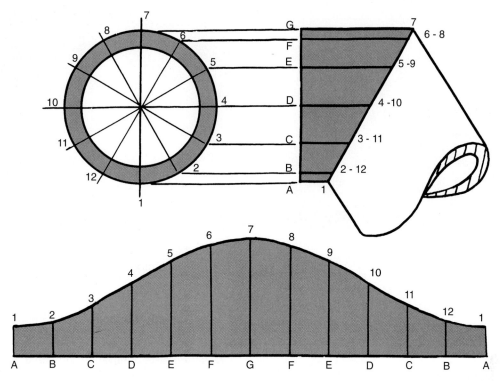

Figure 23-17. *Preparing a paper or metal template (layout) for a 60° angle pipe joint.*

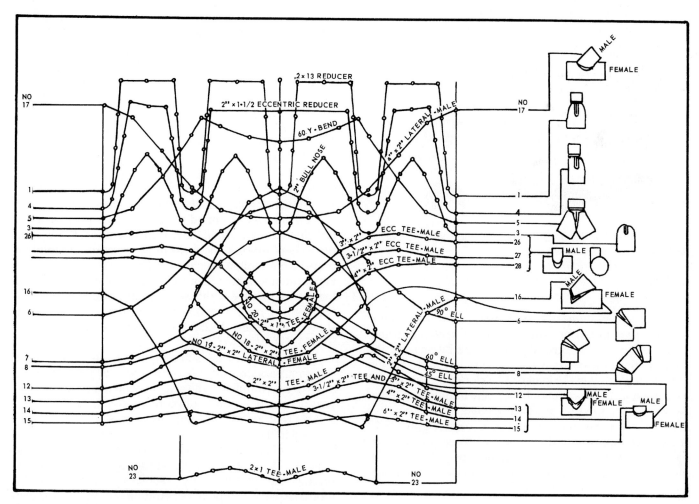

Figure 23-18. *Geometric curves for joints in 2" (50.8mm) diameter pipe.*

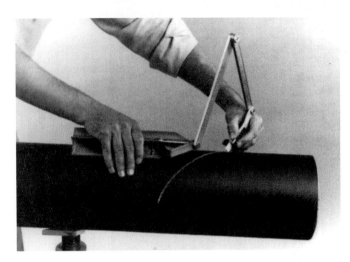

Figure 23-19. *A mechanical pipe layout tool that uses soapstone to mark a line. This line will be cut with an oxyfuel gas torch in preparation for welding. (Dearman, a division of Cogsdill Tool Products, Inc.)*

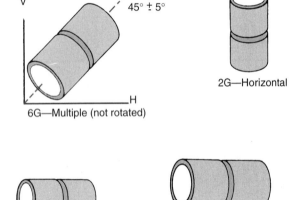

Groove welds in pipe

Figure 23-20. *Pipe joint configurations and their AWS position designations.*

A pipe joint in the 5G position may be welded in one of two ways: uphill or downhill. *Uphill welding* (sometimes referred to as vertical up) is done by starting the weld at the bottom or 6 o'clock position and moving upward to the top of the joint or 12 o'clock position. Uphill welding normally produces better penetration, and fewer passes are required to complete a weld. This is possible because it can be done at a slower pace. Figure 23-22 shows a comparison of the number of passes that may be required to complete a weld using these two welding techniques. The uphill method is used on thicker-walled pipe and on high-pressure pipe welds.

Figure 23-21. *This welder is arc welding in the overhead position. The pipe is in the 5G position.*

Typical Number of Weld Passes Required to Complete a Weld			
Wall thickness of pipe		Downhill	Uphill
in.	mm		
1/4	6.4	3	2
5/16	7.9	4	2
3/8	9.5	5	3
1/2	12.7	7	3
5/8	15.9	–	4
3/4	19.1	–	6
1	25.4	–	7

Figure 23-22. *This table shows that fewer weld passes are required to complete a weld joint when welding uphill. Downhill welding is not used on wall thicknesses greater than 1/2" (12.7mm).*

When *downhill welding* (sometimes called vertical down), the weld is started at the 12 o'clock position and it is welded downward to the 6 o'clock position. When using the SMAW process, the weld must move more rapidly to prevent the molten slag from rolling into the weld pool. Penetration is better when welding uphill. Downhill welding is used on pipe with a wall thickness thinner than 1/2" (12.7mm). Uphill and downhill welding techniques are shown in Figure 23-23. Both methods are used with SMAW, GTAW, and GMAW.

The different passes in a pipe weld are given names. The first pass on a pipe is called the *root pass*. After the root pass is completed, the slag must be removed. The root pass is usually ground out partially to remove any crown in the root pass. This will prevent slag from getting trapped in the weld. The root pass is usually made with an E6010 or E7010 electrode. These electrodes have good penetration. The root pass is sometimes done with GTAW to produce the highest quality possible. It is essential that the root pass have complete penetration. To obtain complete

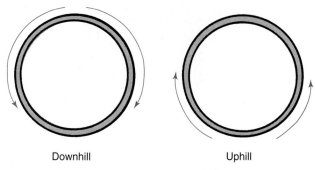

Figure 23-23. *Downhill welding begins at the top of the pipe and progresses towards the bottom. In uphill welding, the weld starts at the bottom and progresses up the pipe to the top.*

penetration, the keyhole method is used. Heading 12.7.5 explains the keyhole method.

After the root pass is completed, cleaned, and ground, the second pass is welded. This second pass is called the *hot pass*. The hot pass uses more current than the root pass. This hot pass must fuse well with the root pass and the pipe walls. The hot pass must also melt any slag left from the root pass. The hot pass should be welded as soon as possible after the root pass is completed. The hot pass is usually welded within five minutes after the root pass is completed.

Filler passes are used to fill the weld joint. Several filler passes are usually required. Filler pass welds may be stringer beads (straight beads) or a slight weave may be used. Each filler pass must fuse (melt) into the previous weld passes and into the pipe walls.To prevent slag inclusions, each pass must be cleaned prior to welding the next pass.

The final pass is the *cover pass*. This pass is used to cover the weld joint. A weaving motion is used to produce a wide bead. Some specifications do not allow large cover passes, so smaller passes are used. The final pass or passes should have 1/32" to 1/16" (1mm to 1.5mm) buildup above the pipe. There must be good fusion into all previous weld passes and into the pipe walls and pipe surface. These weld passes are shown and labeled in Figure 23-24.

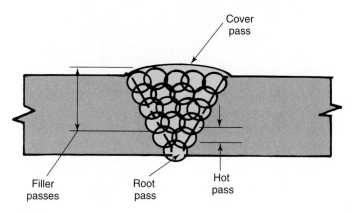

Figure 23-24. *The specific names used for the various passes in a weld joint.*

The hot pass, filler passes, and cover pass are made with E6010 or E7010 electrodes or an E7018 electrode. E6010 or E7010 electrodes are used when downhill welding; E7018 electrodes are usually used when uphill welding. When an alloy pipe is welded, an alloy electrode must also be used. When a backup ring is used, an E7018 electrode can be used for the root pass.

Before attempting to vertical pipe weld (5G position), you must be able to weld satisfactorily in the flat, vertical, and overhead welding positions. Horizontal pipe welding (2G) is similar to horizontal welding on plate.

When large-diameter pipe is used, two welders may work on the same pipe. See Figure 23-25. Each welder works on one side of the pipe. When two (or more) welders work on the same pipe joint, the weld joint is completed faster.

When welding alloy steel pipe, stainless steel pipe, or aluminum pipe, the technique used is the same as described above. The proper procedures for welding these base metals (electrode selection, preheat, postweld heat treatment, etc.) were described earlier in Chapters 21 and 22 (alloy steel in Heading 21.3, stainless steel in Heading 21.8, and aluminum in Headings 22.2 through 22.4).

Figure 23-25. *Two welders welding on a large-diameter pipe that will carry high-pressure natural gas.*

23.5 GTAW AND GMAW PIPE JOINTS

Gas tungsten arc welding is used to weld pipe. The GTAW process produces a very high quality weld, but more slowly than other methods. This process is used to obtain an excellent quality root pass. Shielding gas must be used to protect the weld joint. The root side of each weld should be protected by a shielding gas. It is possible, but not cost-effective, to fill the entire pipe section with shielding gas. For this reason, the volume to be shielded is reduced by sealing the pipe on each side of the joint with water-soluble "purge dams." One of the dams is punctured and shielding gas is injected into the root area through this hole. The other dam is also punctured to allow the air and contaminants to escape. The shielding gas purges the root side of the joint of any possible contaminating atmosphere. Since the dams are placed on each side of the joint, they cannot be removed easily after the weld is completed. However, when water is sent through the pipe, the water-soluble dams are washed away. See Figure 23-26. Welding can be done uphill or downhill, as when using SMAW. To ensure complete penetration, the root pass is the most critical. The keyhole method is used to obtain complete penetration. Refer to Chapters 14 and 15 for information on gas tungsten and gas metal arc welding. Figure 14-35 shows a technique used to help control the arc length when welding the root pass.

The welds are inspected by eye from the outside. They may also be visually inspected on the inside by using a small video camera on a motorized carriage. See Figure 30-3. Such a camera may be mounted on a long flexible tube when inspecting tubing or small-diameter pipe. Welds may also be inspected by examining negatives developed from X-ray or radiation isotope exposures.

Gas metal arc welding is also used to weld pipe joints. The electrode filler metal should match the base metal composition. GMAW can fill a weld joint faster than GTAW.

The procedure for welding alloy pipe or aluminum pipe is the same as described for welding alloy steel or aluminum base metal. GTAW is commonly used to weld aluminum. Shielding gas and purge dams are also used when welding aluminum or alloy pipe.

23.6 WELDING TUBE JOINTS USING SMAW

Tubing may be obtained in many shapes other than round. The welding methods described in this heading will deal with round tubing. These same techniques, however, may be used when welding tubes of other shapes.

Tube welding is similar to thin sheet metal welding, except the weld joint is a three-dimensional curve, as in pipe welding. Further, since the root of the weld is not accessible, and because the weld root may be in contact with flowing fluids, the amount of penetration is critical. *Too little* penetration will create a weak joint. *Too much* penetration will reduce the inside diameter of the pipe and restrict the flow. These two faults are shown in Figure 23-27. If the faults are beyond acceptable limits and are considered defects, they must be removed and rewelded.

Pipes and tubing may be welded using similar techniques. Figure 23-28 shows the uphill welding technique used with an arc welding process. Figure 23-29 illustrates the downhill welding technique. If the pipe or tube can be continually rotated, the weld may be made as a continuous, nearly flat position weld at or near the 12 o'clock position, as shown in Figure 23-30. The technique for welding a T-joint on pipe or tubing when one tube is vertical and

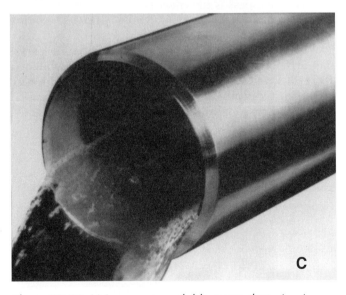

Figure 23-26. *Using a water-soluble purge dam. A—A purge dam in place inside a pipe. B—The pipe joint being welded using the GTA process. C—Water being used to dissolve and remove the purge dam. (CMS/Gilbreth Packaging Systems)*

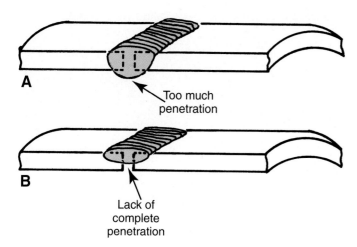

Figure 23-27. *Tube joint weld showing two common penetration defects. A—Too much penetration. B—Not enough penetration.*

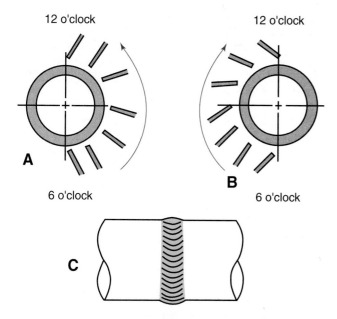

Figure 23-28. *The uphill welding procedure used to weld in the 5G position with the SMAW, GTAW, and GMAW processes. A and B—Start at the 6 o'clock position and weld up to the 12 o'clock position as the pipe or tube is rotated.*

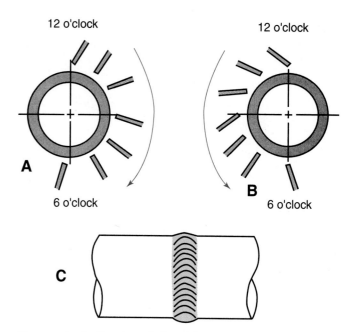

Figure 23-29. *The downhill welding procedure used to weld in the 5G position with the SMAW, GTAW, and GMAW processes. A and B—Start at the 12 o'clock position and weld down to the 6 o'clock position as the pipe or tube is rotated.*

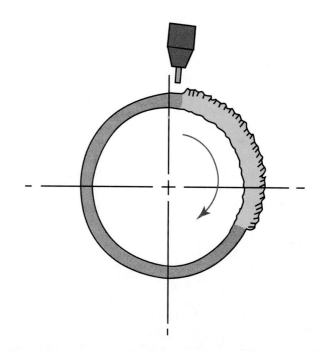

Figure 23-30. *When the pipe or tube can be rotated, the weld is made in a nearly flat position. Welding is done near the 12 o'clock position as the pipe or tube is turned.*

the other is horizontal is shown in Figure 23-31. Since tubing walls are thin, considerable skill is required to SMAW tube joints. Low amperages must be used and the weld pool kept small. Expert manipulation of the electrode is necessary to produce high-quality welds. The same tacking procedures, aligning procedures, and welding sequences are used for tube joints as for pipe joints. GTA and GMA welding processes are generally used to weld tubing joints.

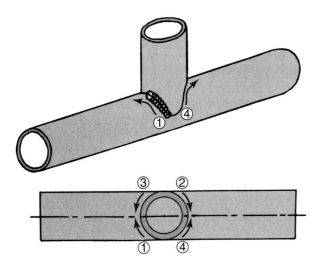

Figure 23-31. *A method of welding a T-joint pipe assembly with one pipe in the vertical position. The recommended sequence is used to prevent contraction from pulling the pipes out of alignment as they cool.*

23.7 WELDING TUBE JOINTS USING GMAW AND GTAW

The shielding gases used for tube welding are the same as those used to GMA or GTA weld regular joints made from the same metal as the tubing. To conserve the shielding gas, purge dams may be used. The pressure of the shielding gas should be kept relatively low so that the molten metal will not be forced out of the weld pool.

Gas tungsten arc welding may be done on pipes and tubing using a special very short GTAW torch, mounted on a powered carriage that moves around the tube or pipe on a track. Figure 23-32 shows such a fully automatic GTAW machine mounted on a 4″ (102mm) tube. A similar GTAW machine, as shown in Figure 23-33, may be used to make perfect welds on steam condensers and other pipe assemblies where the clearances are extremely small.

Figure 23-32. *A rotating pipe welding machine. Note the oscillator device that moves the welding gun from side to side to form a weaving bead. Note also the excellent weld bead shape. (Magnatech Limited Partnership)*

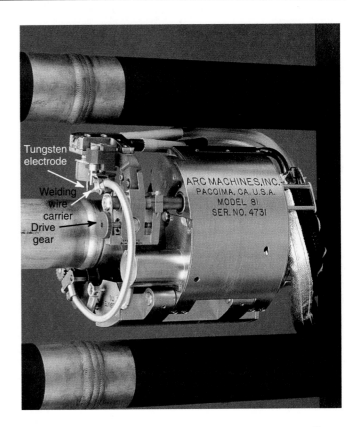

Figure 23-33. *This mechanized GTAW machine is small enough to rotate around pipes that have very little clearance. Note the drive gear, electrode, and welding wire carrier. (Arc Machines, Inc.)*

23.7.1 Making Tubing Repairs

Tubing is not always used to carry fluids. It is often fabricated to form the frame or support for some structural assembly, such as a race car safety cage or an ultralight aircraft.

When structural tubes or clusters of tubes break or are weakened by corrosion, they may be repaired and strengthened by welding on a well-designed repair patch. When a break occurs in a straight run of tubing, it can be repaired using a *fishtail* repair, as shown in Figure 23-34. The larger repair tube may be used in one piece or cut lengthwise to make two pieces. Plug welds and lap welds are used to make this repair.

A repair for a broken cluster tube assembly is shown in Figure 23-35. Tube clusters may also be repaired as shown in Figure 23-36.

23.8 HEAT TREATING PIPE AND TUBE WELDED JOINTS

Heat treating a pipe or tube joint presents special problems. The assembly is usually too awkward to put into a furnace. Gas flame heating has been used, but maintaining temperature accuracy along the weld is difficult.

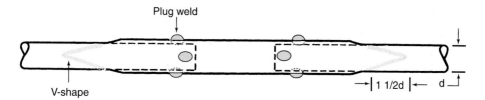

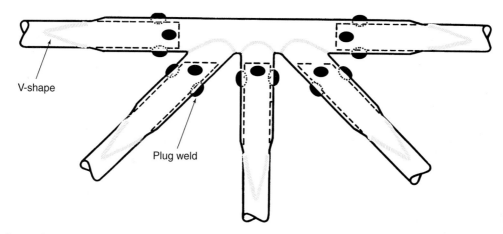

Figure 23-34. Fishtail repair on straight tubing. Note the V-shape at each end of the larger-diameter tube.

Figure 23-35. Tubing cluster repair. Note the V-shape at the end of each larger diameter tube.

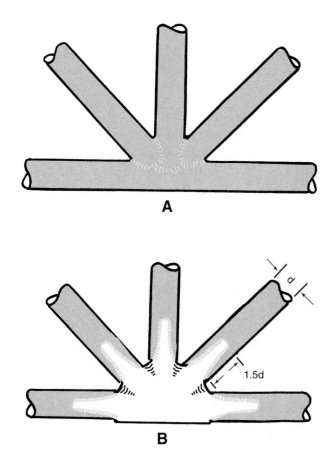

Figure 23-36. A cluster of tubes may be welded and strengthened as shown. A—First, the tubes are welded to each other. B—A metal reinforcement piece is cut and welded into place.

Two methods that have been successfully used are electric heating methods:

- Resistance heating.
- Induction heating.

Resistance heating can be used for preheating, for heating during the welding operation, or for postheating. The process involves wrapping resistance wire around the pipes or tubes at the proper places. Current is then passed through the wire, heating the wire and the pipe. The temperature can be controlled with a thermostat. Special units have been developed that are easily installed (quick clamp and toggle method). Chapter 29 describes additional methods used to heat metals for heat treatment.

Preheating is often required. Temperatures up to 600ºF (316ºC) may be used. Postweld heat treating can be as high as 1500ºF (816ºC).

The time that the weld is held at the specified temperature, and the rate of cooling, are important factors when postweld heat treating. Some alloys require a two-step postweld heat treatment. You should always check with the manufacturer or the supplier of the metal to determine the exact preheat and postweld heat treatments. Refer to Chapter 29 for additional information.

23.9 INSPECTING PIPE AND TUBE WELDS

It is important to know whether a completed pipe or tube weld has sufficient strength, ductility, and toughness. The *welding procedure specification (WPS)* must be carefully established and checked by inspection and testing facilities. The welder must be qualified to make sure he or she can weld to the accepted procedure. In determining the ability of an individual welder, standard procedures have been developed for testing sample welds produced by that welder. Methods have been devised to determine the quality of the weld joint without destroying it. One example of this type of testing is called visual inspection. This is just one of many tests classified as *nondestructive evaluation.* See Chapter 30.

Some of the common faults that must be avoided in pipe welding (and all welding) are:
- Too much penetration.
- Lack of penetration.
- Slag inclusions.
- Porosity.
- Lack of fusion.
- Undercutting.

Practically all high-pressure pipe welds are tested under pressure, using gas (air, nitrogen, or carbon dioxide), water, or a halogen. Figure 23-37 shows a section of pipe being prepared for a test with water. The water (*hydrostatic*) pressure imposed is usually one-and-one-half times the pressure the weld is expected to be under in actual use. The water pressure method checks the integrity of the complete assembly. This test is known as a *proof test.*

Figure 23-37. *This long section of gas pipe has a dome fitting welded at each end. Water is pumped into the pipe to fill it. The pipe is then pressurized to several thousand pounds per square inch to test all the welds along its length for leaks or failure. After the test, the domes are cut off and the pipe is welded to the next section.*

The gas pressure method of testing a pipe or tube is used mainly to detect leaks. A soap solution is placed around the joint; a leak is indicated by the appearance of bubbles.

The *halogen gas leak test* will locate smaller leaks than can be found with any other known method. A halogen gas (refrigerant) is put into the pipe or tube under pressure. Any leak is detected by a very sensitive electronic ionizing sniffer. It is indicated by a meter reading or by

a buzzer sound. See Figure 23-38. This method is an adaptation of leak-detecting methods used in the heating, refrigeration, and air conditioning (HVAC) field. It will signal a leak of an extremely small size.

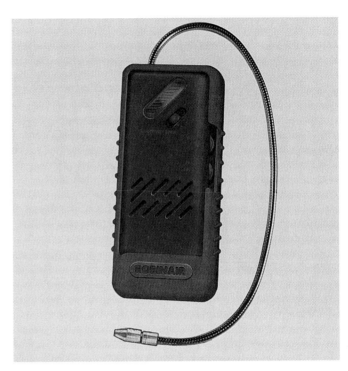

Figure 23-38. *A battery-powered electronic leak detector that can be used with the halogen gas testing method. (CPS Products, Inc.)*

23.9.1 Code Requirements

Almost all pipe and tube welding is covered by a *welding code* or codes. All the requirements of the code or codes must be met. A commonly used code is the one developed by the ASME (American Society of Mechanical Engineers). The codes require that the welding procedure and the welders be qualified. Refer to Chapter 31 on codes and procedures for more information.

23.10 HOT SHORTNESS

When heated to the melting point for welding, aluminum, cast iron, and some other metals become extremely weak in the weld pool area. During this time, the metal has a tendency to suddenly fall away from the weld pool, leaving a hole. This tendency is called *hot shortness.*

Metals that exhibit hot shortness must be supported during welding. This is usually done with backing strips or backing rings.

23.11 WELD BACKING RINGS

Metal that extends beyond the root of the weld is called *root reinforcement.* When it is necessary to control the shape and size of this root reinforcement, some form of *weld backing* is typically used. In a pipe joint, the root of the weld is usually not accessible. Because of this, only one of the four weld backing methods is practical. The backing method used on a pipe joint is the *backing ring*.

The root reinforcement contour is often machined into the surface of the backing ring. Backing rings are placed across the root opening and tack welded to hold them in place. The rings may have pins or studs on them to properly space the root opening. These rings also serve to align the inside diameters of the pipes or tubes being joined. See Figures 23-39 and 23-40. These rings should be used only when they will not interfere with the designed flow volume in the pipe. Backing rings must be made of a metal that is compatible with the base metal and electrode used. Preformed backing rings are purchased to exactly fit the inside diameter of the pipe or tube.

Figure 23-39. *Backing ring for pipe butt joints. This device aligns the pipe and also helps control the penetration. The ring shown is for 6" (15.2cm) ID pipe.*

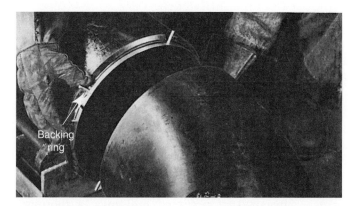

Figure 23-40. *A pipe and an elbow being assembled prior to arc welding. Note backing ring used here for alignment and to control penetration.*

The pins or studs on the backing ring are usually removed after the pipe is aligned and tack welded. Backing rings used in a pipe joint are very difficult to remove. They are often welded in place by the root pass and become part of the joint. Backing rings can be removed if they are accessible and have not been welded in place.

23.12 WELDING PLASTIC PIPE

Plastic pipe is often used to carry natural gas at relatively low pressures from a well site to the distribution point. Such plastic pipe may be joined by heating the joining surfaces until they are almost molten, then forcing the ends together under pressure. Figure 23-41 shows a portable machine which will prepare the edges of the pipe, heat the ends, and force the ends together to create the weld. The machine has two clamps that align and hold the pipe. It also has a hydraulically powered rotating cutter, which cuts both ends of the pipe at the same time to a flat and smooth surface. This cutter is swung in and out of the way as needed. See Figure 23-42. Once the butt ends are cut smoothly, the heating element is swung into position between the pipe ends. The pipe is brought into contact with the heater and heated for a specific amount of time. See Figure 23-43. Finally, the heater is quickly swung away and the pipe ends forced together under pressure. The mating surfaces of the butt joint are upset as the weld is formed. Figure 23-44 shows what the completed weld looks like. Figure 23-45 shows a section of pipe that has several completed butt joints and a pipe reducer fitting welded as a T-joint.

Figure 23-41. *This portable plastic pipe preparation and welding machine will trim the ends of the pipe, heat the two ends, then force them together to weld them.*

Figure 23-42. *The sharp knives on this rotating wheel trim the ends of the pipe to make them straight and clean.*

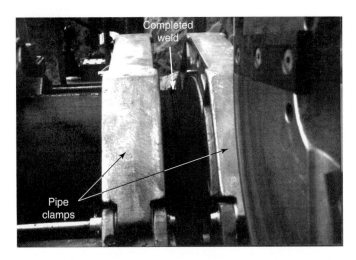

Figure 23-44. *A completed butt weld. Note that the weld seam is upset when the weld is made.*

Figure 23-43. *This resistance heater will heat the ends of both pipes at the same time to the correct temperature to create a sound weld when the ends are pressed together under hydraulic pressure.*

Figure 23-45. *A section of welded plastic pipe with a T-joint and reducer welded into place.*

23.13 REVIEW OF SAFETY IN PIPE AND TUBE WELDING

All of the safety rules listed in previous chapters also apply to pipe and tube welding. Of particular importance to the pipe and tube welding operator are the following points:

- The pipe and tubing must be absolutely free of flammable liquids or gases to prevent an explosion. If the pipes or tubes have been used to carry flammable or explosive substances, the lines must be steam-cleaned several times to ensure that no hazardous materials remain. A shielding gas should flow through the piping system while the weld is being made

- The welding of pipes and tubing often requires the welder to work in an awkward position or climb into almost-inaccessible places. A firm

footing, excellent ventilation, and a quick escape route are essential. Fume removal units should be used in confined spaces. Whenever carbon dioxide or a shielding gas is used in a confined space, the welder must wear a supplied air welding helmet and hood. Shielding gases force breathable air from confined spaces — a welder could become unconscious in seconds and die within minutes from lack of oxygen.

- Because pipes and tubing are frequently installed in occupied buildings, a thorough investigation must be made to locate and remove any and all flammables from near the place of work. Firefighting equipment must be handy and a fire protection person should be standing by during the welding and/or cutting.

- Whenever using a gas under high pressure, always install a pressure relief valve in the test line. It should be adjusted to 5 psi to 10 psi (34.5 kPa to 68.9 kPa) above the testing pressure.

• Pipe welds should be vigorously inspected and tested. Sufficient penetration is absolutely essential to a successful weld. It requires considerable skill to weld pipe correctly. A pipe welder's ability should be periodically assessed by checking and testing samples to destruction. Pipe welding, when properly done, is stronger than the threaded assembly method.

• Most pipe and tube welding procedures are subject to code regulations. Always inquire of the building and safety departments at the local, county, and state levels of government for the applicable codes before accepting and undertaking a structural, pipe, or tube welding project.

TEST YOUR KNOWLEDGE

1. List three ways to produce pipe.
2. For typical use, does seamed pipe or seamless pipe have the greater strength?
3. What is the wall thickness of a 6" diameter, schedule 40 pipe?
4. The size of a tube is measured by its _____ diameter.
5. What are two ways to prevent or reduce distortion in pipe welding?
6. What are the two ways that a butt joint in a horizontal pipe can be welded?
7. Downhill welding is used only when the pipe wall thickness is less than _____.
8. What are the names given to the first, second, and final passes in a pipe weld?
9. What type of electrodes are commonly used when welding the root pass? Why?
10. How soon after the root pass is welded should the hot pass be welded?
11. What are two common faults in pipe and tube welding that must be avoided?
12. When GTAW or GMAW tube joints, how is the root of the weld protected from oxidation?
13. What is the inside diameter of an 8" diameter, schedule 80 pipe?
14. What is the danger of welding in a confined space using an inert gas?
15. List at least two ways that a pipe differs from a tube.
16. What provides the force to upset the plastic pipe described in Heading 23.12?
17. List two purposes for a backing ring.
18. List two electric heating methods used to heat treat pipe and tube welds.
19. What is a proof test used for?
20. What safety device is used to limit the amount of pressure in a pipe that is being tested?

This diver is using a specialized cutting machine to precisely cut a large pipe prior to making a welded repair.

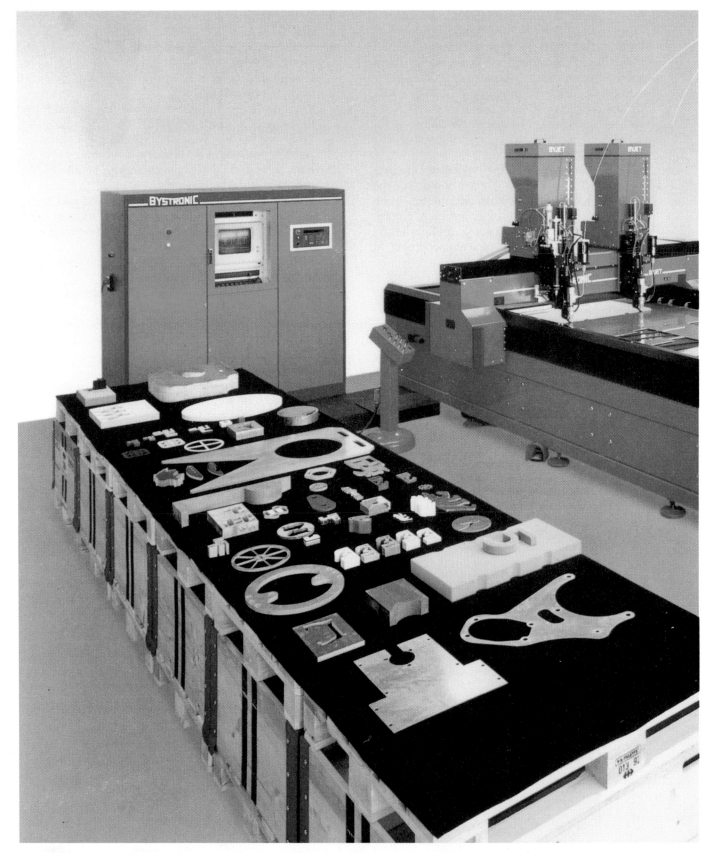

Water jet cutting is widely used in industry to cut complex shapes from various types of flat stock. Since it is a cold cutting process, the abrasive water jet method makes clean, oxide-free cuts with no heat-affected zone. (Bystronic, Inc.)

Chapter 24

SPECIAL CUTTING PROCESSES

LEARNING OBJECTIVES

After studying this chapter, you will be able to:

* Name and explain six metal cutting processes other than oxyfuel gas cutting.
* Identify various cutting processes by means of their AWS abbreviations.
* Explain when a particular cutting process would be chosen over another.
* List the safety hazards involved when using the cutting processes explained in this chapter.

This chapter will discuss several cutting processes that are considered special or unusual. These processes include the following:

- Oxygen lance cutting (LOC).
- Oxyfuel gas underwater cutting (OFC).
- Oxygen arc cutting underwater (AOC).
- Metal powder cutting (POC).
- Chemical flux cutting (FOC).
- Laser beam cutting (LBC).
- Water jet cutting.

24.1 OXYGEN LANCE CUTTING (LOC)

The principle of oxyfuel gas cutting is that once the metal is heated and the oxygen for cutting is turned on, the metal becomes a fuel and burns (oxidizes). The addition of oxygen supports the *oxidation* (burning) of the metal. With the ordinary oxyfuel gas cutting torch and tip, the preheating flame and the oxygen blast for cutting is combined in one torch. The maximum thickness of material that the regular hand-held cutting torch is able to cut is about 24" (610mm).

Steel up to 8' (approximately 2.5m) thick may be cut using an oxygen lance and a heating torch. The *oxygen*

lance is a long length of ordinary steel tubing or pipe. This tubing is connected to an oxygen hose and an on-off control valve. The metal to be cut is heated to a white-hot temperature by means of an oxyfuel gas torch. Normally, an oxyacetylene cutting torch is used. However, other fuel gases may be used. Once the metal is heated up to its *kindling temperature* (temperature at which it will burn), the air valve on the oxygen lance is turned on and the cutting begins. See Figure 24-1.

To do oxygen lance cutting, two people are generally required. One person keeps the metal at the kindling temperature with the cutting torch. The second person controls the oxygen lance.

Figure 24-1. *A large casting being cut using the oxygen lance. The preheating is supplied by a heavy-duty cutting torch.*

The oxygen lance is used for such operations as cutting risers from large castings. It may be used to cut through thick pieces of steel in scrapping operations or to open the pouring gates in large furnaces. Refer to Heading 4.29 and to Figure 4-34 for basic information on the oxygen lance cutting (LOC) process.

24.1.1 Oxygen Lance Cutting Equipment

The equipment required for the oxygen lance cutting operation is listed below:

- A complete oxyfuel gas welding or cutting outfit to provide the preheat for lancing. For extremely thick sections, a special long torch is required to provide a constant preheat temperature. These special torches do not have a cutting head and a tip set at 90° to the torch body. Instead, the torch body, head, and cutting tip are all in a straight line. This type torch may be several feet long. An example of such a torch is shown in Figure 24-10.
- One or more oxygen cylinders. For large-volume cutting, several cylinders may be manifolded to the regulator.
- A heavy duty high-volume, two-stage oxygen regulator.
- Oxygen hose of an appropriate length, complete with fittings for the oxygen lance.
- Lengths of 1/8" or 1/4" (3.2mm or 6.4mm) ID pipe to serve as the lance. This pipe must be long enough to penetrate the desired thickness. It also must be sufficiently long to permit consumption of some of the pipe as the cutting progresses.
- **Safety equipment, including goggles, helmet, leather jacket, leggings, apron, and gloves. Considerable sparking accompanies the oxygen lance cutting operation. The workers as well as the equipment and surroundings must be protected against hot particles.**

24.1.2 Oxygen Lance Cutting Procedures

To start a cut with the oxygen lance, first preheat the edge of the piece with a heavy-duty welding or cutting torch. Once the metal reaches a cherry red or white-hot temperature, turn on the oxygen for the lance. On extremely thick pieces, the preheat temperature is maintained with a special long torch that can reach well down into the kerf. Generally, the cut is started at an angle and gradually brought to a vertical position by manipulating the lance. The cut is much easier to start and maintain using this principle. Figure 24-2 is a schematic drawing showing the oxygen lance principle.

Lancing may also be used to pierce holes through pieces up to 8' (2.5m) thick. This makes it possible to pierce holes in parts that would be expensive if not impossible to machine.

As the pipe lance extends down into the cut, it is subjected to excessive temperature in the presence of oxygen. The end of the pipe will be burned away continually. As pipe is consumed, lengths of new pipe are added. To join

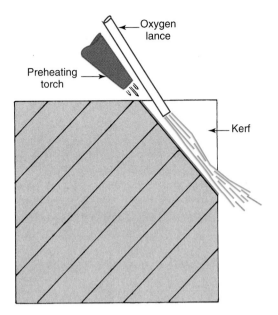

Figure 24-2. Drawing of an oxygen lance and cutting torch being used to cut a piece of thick metal.

the new pipe to the old, standard threaded pipe couplings are used. The cutting lance has been used for such tasks as piercing the center of a freight car axle shaft throughout its length without burning through the side of the shaft. Considerable expense is eliminated by using this method to cut large steel or cast iron sections.

Figure 24-3 shows welders using a standard cutting torch and an oxygen lance to cut a large casting.

24.2 OXYFUEL GAS CUTTING (OFC) UNDERWATER

The oxyfuel gas method of cutting ferrous metals underwater is much the same as the method used on the surface, with one exception. To keep the flame lighted

Figure 24-3. The oxygen lance being used with a regular cutting torch to cut a large riser from a steel casting. Note the long length of iron pipe used for the cutting lance. (ESAB Welding and Cutting Products)

underwater, the cutting tip and combustion area must be surrounded by compressed air. This keeps the water away from the combustion area. Refer to Heading 4.30 and to Figure 4-35.

The use of acetylene below 15' (4.6m) of water is not recommended, since acetylene is not safe to use at pressure over 15 psig (103.4 kPa). Hydrogen should be used as the fuel gas below a depth of 15' (4.6m) when cutting underwater. The hydrogen flame has been used to a depth of 200' (61m) underwater. A proprietary fuel mixture called

MAPP gas may also be used for underwater cutting. MAPP gas is stable under high pressures.

The underwater torch has a special high-pressure air system. The purpose of the high-pressure air is to keep the water away from the flame by maintaining an air pocket at the point of cutting. The construction of an underwater oxyfuel gas torch is shown in Figure 24-4. Notice how the pressurized air, shown in green, surrounds the cutting tip. Special training in underwater welding and cutting techniques is required. The operator should also be a qualified diver.

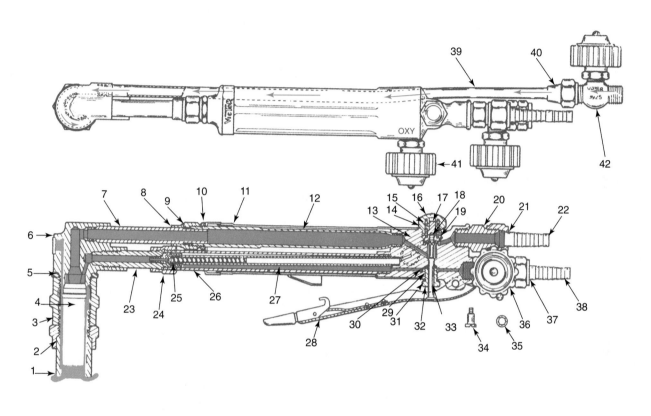

1–Air jacket	14–H. P. valve plug	30–"O" ring (friction)
2–Lock nut	15–Valve spring	31–"O" ring (sealing)
3–Tip nut	16–Seat holder assembly	32–"O" ring retainer
4–Oxygen-hydrogen tip	17–Seat holder	33–Valve stem
Sizes 2 to 7	18–Seat screw	34–Bolt for lever
Oxygen acetylene	19–Seat	35–Lock washer
Sizes 1 to 4	20–Oxygen connection	36–Acetylene control valve assembly
5–Gasket	21–Nut	Valve stem assembly
6–Torch head	22–Tailpiece	Control valve body
7–H. P. oxygen tube	23–Mixing chamber tube	37–Nut
8–H. P. oxygen tube coupling nut	24–Mixing chamber nut	38–Tailpiece
9–Ferrule	25–Spiral mixer	39–Compressed air tube
10–Lock nut	26–Acetylene tube	40–Compressed air tube coupling nut
11–Barrel	27–Inner oxygen tube	41–Oxygen valve stem assembly
12–Rear H. P. oxygen tube	28–Lever	42–Compressed air valve assembly
13–Body	29–H. P. valve "O" ring retainer assembly	Valve body
		Valve stem assembly

Figure 24-4, *A cross section of an underwater oxyfuel gas cutting torch. The air valve, part #42, controls the air flow to the air jacket, part #1. The air flow and air pocket around the cutting tip are shown in green. (Victor Equipment Co.)*

Oxyfuel gas torches are usually constructed as compactly as possible. The compressed air sheathing around the tip is adjustable to enable the operator to hold the torch head against the metal being cut. Too-high an air pressure will force the air jacket and torch away from the metal being cut. Slots in the air jacket, #1 in Figure 24-4, permit the escape of the products of combustion and air. The cutting tip must have exceptional preheating capacity to overcome the excessive heat loss encountered through conductivity and convection when cutting underwater. The cutting tip orifice must be large enough to overcome the cooling effect of the water and the corrosion on the metal. Cutting tips must be large to save the operator time and to reduce the need to remain at great depths for long intervals. The air pressure, and therefore the amount of air, is controlled by a separate air valve designed to be a part of the cutting torch. See Figure 24-4, part #42. Figure 24-5 illustrates a typical underwater oxyhydrogen cutting torch.

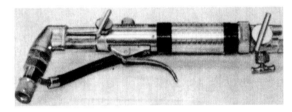

Figure 24-5. *A typical oxyhydrogen underwater cutting torch. (Craftsweld Equipment Corp.)*

24.2.1 Oxyfuel Gas Underwater Cutting Equipment

The equipment needed for underwater cutting is the same as for surface cutting, with some exceptions. A special torch is needed. It has a special slotted air jacket surrounding the cutting tip. The torch also has an additional connection and passages for compressed air, as shown in Figure 24-4. A compressor and long lengths of air hose are required.

The fuel gas and oxygen cylinders and the air compressor are usually carried on a boat or barge. A compressor for the welder's air supply, if used, is carried aboard the diver's support vessel.

The oxyfuel gas underwater cutting torch has four control valves:
- The fuel gas valve.
- The preheat oxygen valve.
- The cutting oxygen valve (lever-operated).
- The compressed air valve.

Vision is generally very poor underwater. Therefore, the operator must know the positions of the valves well enough to use them skillfully without being able to see them.

24.2.2 Oxyfuel Gas Underwater Cutting Procedures

A diver who is thoroughly familiar with surface cutting will not find underwater cutting too difficult. The oxygen and fuel gas working pressures are set at the cutting site. Because of the long hose lengths, the gauge pressures will be much higher than the working pressures at the torch. The diver/welder should be in contact with the surface by radio or rope signal. This is necessary for safety and for making gas and oxygen pressure changes when needed. The compressed air pressure is controlled on the surface tender vessel. The volume is controlled by the welder on the torch. A special electric lighter is built into the air jacket. This device will light the fuel gas and oxygen mixture underwater.

To start a cut, the operator holds the torch against the metal and watches the color of the oxyfuel gas flame. When a bright yellow flame is seen, the preheat temperature has been reached. The cut is then started by turning on the cutting oxygen and proceeding in the same manner as on the surface. The welder is unable to see the cutting area, however, because it is covered by the air jacket. Because of the pressure created by the compressed air within the air jacket on the torch, an extra effort must be made by the operator to keep the torch against the work. The air tends to push the torch away from the metal being cut. A distinct rumbling sound is heard when the torch is operating. When the sound stops, the diver knows the flame has gone out.

The torch angle used is 90°. Once the fuel gas and oxygen pressures are set, the welder must move the torch at a uniform speed and keep the torch against the metal. Since the welder cannot see the cut, the forward speed is all that can be controlled.

24.3 OXYGEN ARC UNDERWATER CUTTING (AOC)

The oxygen arc process was originally designed to be used for cutting underwater. The process was then adapted for use out of water. The equipment required for use in and out of water are essentially the same. Refer to Headings 16.4, 16.4.1, and 16.4.2. The major difference is that the underwater electrode is wrapped or coated to make it waterproof. In the underwater oxygen arc cutting process, an arc is struck on the metal to be cut, using a hollow steel electrode that has high-pressure oxygen flowing through it, as shown in Figure 24-6. The arc heats the metal and the oxygen jet quickly oxidizes the metal to perform the cut.

Oxygen arc underwater cutting has proven to be very successful and is widely used in industry. The electrode is usually 5/16" (7.9mm) tubular steel, flux-covered and waterproofed. A fully insulated special cutting torch (elec-

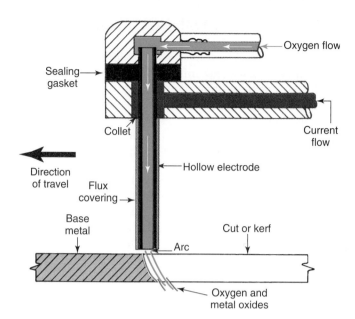

Figure 24-6. *A schematic of an oxygen arc cutting torch. Note the collet that holds the hollow electrode and makes the electrical contact.*

trode holder) is shown in Figure 24-7. The oxygen arc underwater torch shown has a cutting oxygen lever and a special watertight electrode collet (clamp). Water is kept from contacting all electrical connections and the cutting oxygen passages in the torch.

This system works well because the moment the arc is struck, a very high temperature spot is produced on the

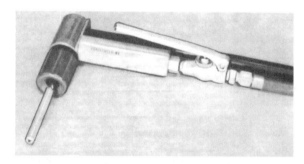

Figure 24-7. *A fully insulated arc-oxygen underwater cutting torch. (Craftsweld Equipment Corp.)*

metal. With the oxygen turned on, the cutting starts instantly.

The operator then draws the cutting electrode along the line of cut while holding an arc. This method of cutting is usable at all depths. Oxygen arc cutting electrodes last only a few minutes and may be changed underwater. The welding current is usually shut off when electrodes are changed. All metals may be cut with this system, including steels, alloy steels, cast iron, and all the nonferrous metals.

Direct current electrode negative (DCEN) is the polarity used. Figure 24-8 illustrates the internal construction of

an oxygen arc underwater cutting torch. Hollow iron or steel electrodes are used. Iron and steel electrodes have replaced hollow carbon and ceramic electrodes, because those electrodes were too brittle.

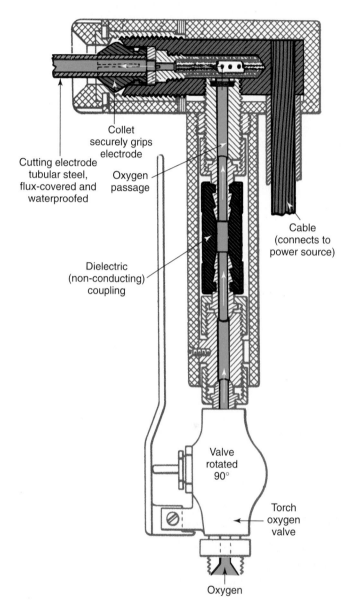

Figure 24-8. *A cross-sectional drawing of an arc-oxygen underwater cutting torch. (Craftsweld Equipment Corp.)*

24.3.1 Oxygen Arc Underwater Cutting Equipment

The equipment used in an oxygen arc underwater cutting outfit includes the following:
- A standard constant current ac or dc power source.
- An oxygen cylinder with a high-volume, two-stage oxygen regulator.

- An extra-heavy-duty oxygen hose with a larger than normal inside diameter. This is necessary because of the long length of hose used and the high volume of oxygen.
- A welding lens a few shades lighter than required on the surface may be used when the visibility is restricted.
- A waterproof and electrically insulated oxygen arc cutting torch. See Figure 24-9.
- An electrical off-on switch for the welding leads on the tender.
- A waterproofed workpiece lead.
- A waterproofed electrode lead.
- Waterproofed hollow cutting electrodes.
- Rubber gloves.

Electrodes used are tubular iron or steel with a water-proofed flux covering. They are 5/16" (7.9mm) in diameter and 14" (356mm) long. The passage in the electrode through which the oxygen travels is 1/8" (3.2mm) in diameter.

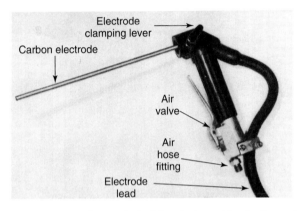

Figure 24-9. *An electrode holder designed for hollow electrodes used when oxygen arc cutting. Note the air hose fitting and the air control lever.*

24.3.2 Oxygen Arc Underwater Cutting Procedures

With the ac or dc power supply turned on, the welder strikes the arc by touching the base metal and withdrawing the electrode about 1/8" (3.2mm). The welder then depresses the oxygen lever to start the cut. The technique used for underwater arc cutting is to drag the electrode along the line to be cut. A constant arc length is maintained by this method due to the thickness of the covering on the electrode.

Skill is required to move the electrode at a constant speed. The welder must move the electrode slowly enough to ensure cutting through the full thickness of the metal. If the movement is too slow, however, the cut may be stopped due to the lack of proper preheating of the metal. A slow forward movement will also cause excessive electrode and oxygen consumption during the cutting operation.

24.4 METAL POWDER CUTTING (POC)

Metal powder cutting is used to cut alloy steels, cast iron, bronze, nickel, and aluminum. Most materials—even reinforced concrete—may be cut with the metal powder cutting technique.

In metal powder cutting, iron powder is introduced into a standard oxyfuel gas cutting torch flame. The heat of combustion of the iron powder increases the total heat of the flame.

When cutting stainless steel with an ordinary oxyfuel gas torch, a very difficult-to-melt refractory material, chromium oxide, is formed on the surface of the metal in the cutting zone. This oxide insulates the base metal and prevents further oxidation by the cutting oxygen. By introducing iron powder into the oxyfuel gas flame, the insulating chromium oxides are kept from forming. The welder is able to continue cutting with no hindrance. The iron particles oxidize or burn in the oxygen stream at the torch tip, producing a very high temperature and concentrated heat. The powder also produces a reducing action on the metal, and a high-velocity cutting action. Figure 24-10 shows a large casting being cut using the metal powder cutting process.

Cutting speeds comparable to the conventional oxyfuel gas cutting of low-carbon steels may be obtained when using the metal powder cutting torch on stainless steels. Ferrous metals up to 5' (approximately 1.5m) thick have been cut using metal powder cutting torches mounted on automatic cutting machines.

When cutting a nonferrous metal, it is helpful to add a more active ingredient than iron powder to the oxyfuel gas flame. From 10% to 30% powdered aluminum is used with iron powder when cutting nonferrous metals.

Reinforced concrete up to 18" (46cm) thick may be cut at speeds of 1" to 2 1/2" (25mm to 64mm) per minute. Walls as thick as 12' (3.7m) may be cut with a metal powder cutting lance, as shown in Figure 24-11.

Several methods have been used to feed the iron powder to the flame:
- Introduction into the oxygen orifice of the tip.
- Introduction into a separate orifice in or near the tip. See Figure 24-12.
- Introduction into the preheating orifices of the tip.

If the powder is introduced into the oxygen orifice, the oxygen pressure must be lower than the air or nitrogen pressure feeding the metal powder to the tip. If a separate orifice is used, there is a delay in ignition of the iron powder and the tip must therefore be held 3" or 4" (75mm or 100mm) from the work.

The extremely high temperature that results from the ignition of the powder practically eliminates the need for preheating the metal prior to the cutting action. Eliminating the requirement for preheating saves a great deal of time. The metal powder cutting process has a considerable penetrating and carry over action. *Carry over action* is the ability of the cutting action to maintain itself in slag pockets and over spaces between plates.

***Figure 24-10**. A metal powder cutting torch being used to cut a riser from a large casting. (ESAB Welding and Cutting Products)*

***Figure 24-11**. Cutting concrete using the metal powder cutting process. (ESAB Welding and Cutting Products)*

24.4.1 Metal Powder Cutting Equipment

The equipment required for metal powder cutting includes a cutting torch or lance, oxygen and a fuel gas supply, and a method of introducing the metal powder into the flame.

Some special metal powder cutting torches have an additional tube to carry the powder to the torch head. The metal powder is added to the preheating flames at the torch head. Such a torch is shown in Figure 24-12. The oxygen and metal powder tubes on the torch are both opened when the oxygen control lever is depressed. Normally, there is a slight delay in the opening of the oxygen valve. This allows the iron oxide to coat the surface before the oxygen reaches it. The delay helps to start the cut faster, particularly on nonferrous metals.

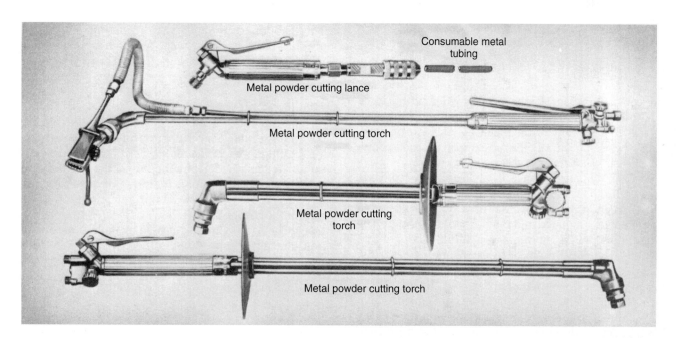

***Figure 24-12**. Several types of metal powder cutting torches. Consumable metal tubing is inserted into the end of the metal powder cutting lance. (ESAB Welding and Cutting Products)*

Powder cutting adapters are made that may be attached to the conventional machine-type cutting torch, as shown in Figure 24-13. Powder is carried to the adapter through tubing, and escapes through multiple openings around the conventional cutting tip. It is melted by the preheat flames and carried to the molten metal by the cutting oxygen stream. Since the powder is directed into the oxygen stream from all sides, the multiple opening attachment is suitable for form cutting and straight cutting. When powder-cutting nonferrous metals, or when cutting extra-thick ferrous metals, an attachment with a single powder tube is used. The powder is carried to a point near the conventional tip, where it is discharged into the flame. More powder may be discharged from the single tube opening than can be released from the small openings in the multiple opening attachment. The single tube discharge is usually considered to be better suited for thick sections than the multiple opening attachment. The single tube attachment works best on straight cuts. It is not as well suited for shape cutting as the multiple opening attachment.

Two types of powder dispensers are presently in use. In one, an enclosed hopper is used as shown in Figure 24-14. The hopper is filled and then pressurized. Powder is fed

Figure 24-14. A pressurized powder dispenser for use with metal powder cutting equipment. Notice the pressure regulator used to control the pressure for the powder feed. Note also that only air or nitrogen are to be used to pressurize the powder.

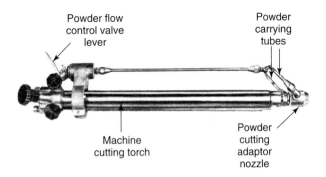

Figure 24-13. A machine cutting torch with a metal powder cutting nozzle adaptor, a powder valve, and manifold tubing attached. (ESAB Welding and Cutting Products)

under constant pressure to the metal powder cutting torch or adaptor. The metal powder cutting control valve on the torch controls the flow of the metal powder. Air or nitrogen may be used to supply the pressure on the metal powder dispenser. **Caution: Never use oxygen to pressurize the powder dispenser, since it can form an explosive mixture with the metal powder.**

The second type of dispenser uses vibrating action to keep the powder agitated and moving along a trough. The powder falls from the vibrating trough into a hopper and is fed to the torch by air pressure. Vibrating-type dispensers are used where a more constant and accurately adjusted powder flow is required.

When the thickness of metal to be cut is greater than 18" (46cm), it is advisable to use the powder cutting lance. It has a specially designed mixing handle to which long lengths of pipe are attached as shown in Figure 24-15. Oxygen and powder are mixed near the end of the handle

and carried to the cutting area through the iron pipe. The pipe is slowly consumed during the operation. Reinforced concrete walls up to 12' (3.7m) thick may be cut using the metal powder lance.

24.4.2 Metal Powder Cutting Procedures

The techniques and torch angles used for metal powder cutting are approximately the same as those used for oxyfuel gas cutting. An oscillating motion is sometimes required when cutting cast iron with an oxyfuel gas torch. This motion overcomes the insulating effects of the graphite particles in the metal. No torch motion is required when metal powder cutting cast iron. To allow the metal powder to reach its ignition temperature before it hits the work, the torch should be held about 1 1/2" (38mm) away from the work.

A technique called the *flying start* is often used for beginning a cut faster on cast iron and thick metal sections. The cutting torch is used to apply iron oxide powder, aluminum oxide powder, or a mixture of the two to the metal in the area where the cut is to start. This is done before the oxygen is turned on or the flame lighted. The torch is lighted and the cutting oxygen lever depressed to start the cut. Due to the additional heat created by the combustion of the iron and aluminum powder, the cut starts immediately. Before the introduction of this method, 15% to 25% of cutting time was spent bringing the base metal up to a temperature where it would oxidize and could be cut.

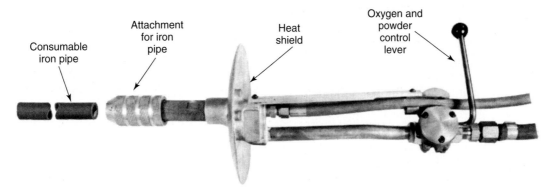

Figure 24-15. *A metal powder cutting lance handle. Consumable black iron pipe is attached to the handle. Clean cuts in reinforced concrete over 12' (3.6m) thick can be made with 70% iron and 30% aluminum powder. (ESAB Welding and Cutting Products)*

24.5 CHEMICAL FLUX CUTTING (FOC)

The *chemical flux cutting* process is very similar to the metal powder cutting process. Both use very similar equipment. The difference is that in the FOC process, a *chemical* flux powder is used instead of a metal powder. During the ordinary oxyfuel gas cutting process, difficult-to-cut oxides are formed on the surface of the base metal and in the kerf. Some of these oxides melt at a much higher temperature than the base metal. These oxides make cutting the base metal very difficult.

When chemical flux cutting, a chemical compound is added to the flux, rather than iron or aluminum powder as in metal powder cutting. These chemical fluxes have a reducing reaction on the oxides which tend to form on the base metal. In other words, the chemical fluxes cut down the amount of these undesirable oxide formations.

Figure 24-16 is a schematic of a chemical flux cutting outfit. Note that a separate supply line feeds the flux into the work. In this cutting process, the chemical flux is fed into the oxygen stream. This causes a reducing action on the alloy metal oxides, permitting their easy removal from the kerf. The carry-over action of the chemical flux method enables the cutting of multiple layers of metal plates without tight clamping.

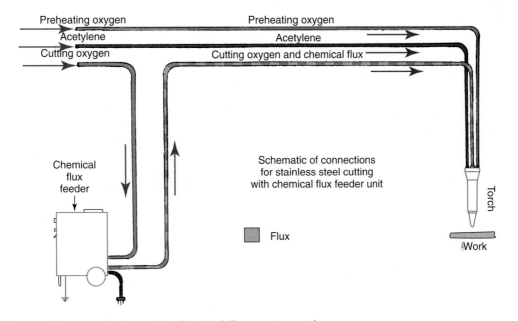

Figure 24-16. *A drawing showing the parts of a chemical flux cutting outfit.*

24.6 LASER BEAM CUTTING (LBC)

The fundamental principles of the laser beam as used for welding are shown in Heading 4.26 and Figure 4-31. A more detailed description of the various types of lasers which may be used for cutting may be referred to in Heading 20.19.

The *laser beam cutting* method uses a constant beam. Some of the advantages of laser beam cutting are:

- The laser beam generator does not have to be located near the cutting operation.
- A laser beam can be deflected by mirrors and can be directed into hard-to-reach areas.
- The metal or other material being cut is not part of the electrical circuit.
- A laser creates an extremely narrow kerf and a small heat-affected zone.
- Extremely deep, small-diameter holes may be cut.
- Compared to other cutting processes, LBC has lower noise and fume levels.
- The equipment may be automated, so highly reproducible cuts may be made.

Oxygen or an inert gas may be used as an assist gas to remove molten metal from the cut or kerf before it can solidify. The assist gas surrounds the laser beam, as shown in Figure 24-17. Cuts have been made on 1" (25mm) carbon steel using the carbon dioxide (CO_2) laser. Since the beam spreads out quickly and the laser energy is harder to focus, cuts in material less than 3/8" (9.5mm) thick are most common. Refer to Figure 20-51. As the metal thickness increases, the required laser power increases and the cutting speed decreases.

Laser cutting and piercing is done with automatic-type equipment. Because of the precision usually required, the cut is often computer-programmed.

Lasers can also be used to drill holes in materials. This process is called *laser beam drilling (LBD)*. This process provides a method of drilling extremely small diameter holes to exact depths. The drilling laser is normally a pulsed beam. Holes with diameters from 0.0001" to 0.060" (0.0025mm to 1.5mm) can be drilled in most materials very accurately. See Figure 20-59. Since the depth of the hole created with one laser beam pulse is usually only six times the hole diameter, many pulses must be used to drill completely through a material.

Laser cutting and drilling processes have been adapted for use in medical surgery. The laser beam is also used in engraving metals and other materials.

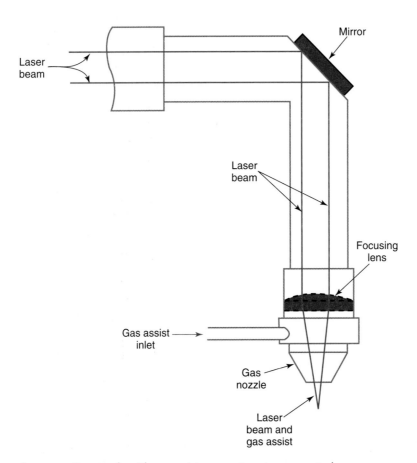

Figure 24-17. A laser beam cutting torch with an assist gas system incorporated.

24.7 WATER JET CUTTING

Water jet cutting is a process that uses a high-pressure jet of water, with or without an abrasive added, to cut a wide variety of materials. A sampling of some of the materials that have been cut with the water jet process is shown below,

- 7.5" (191mm) carbon steel
- 3" (76mm) 7075 T-6 aluminum
- 10" (254mm) titanium
- 2.5" (64mm) thick epoxy/graphite with 470 plies
- 0.08" (2.03mm) ABS plastic
- 0.103" (2.62mm) circuit board
- 0.050" (1.8mm) rubber
- 0.125" (3.2mm) wood

The water jet cutting process competes favorably with the bandsaw, oxyfuel gas, plasma, and laser cutting processes. Cardboard and rubber-backed carpeting can also be cut with this process.

During the water jet cutting process, water is filtered and its pressure raised to between 30,000 psi and 50,000 psi (207MPa and 414MPa). The water flow rate is about 3.5 gpm (13.3 L/min.). The high-pressure water is forced through a sapphire orifice, forming a high-velocity jet as it exits. The velocity of the water jet depends on the water pressure. When hard materials are to be cut, abrasives are often added to the jet of water. The abrasives are fed from a hopper into a mixing chamber. The pressurized water mixes with the abrasive in the mixing chamber prior to exiting the nozzle orifice. Figure 24-18 shows metal being cut with a water jet and an abrasive. Dimensional tolerances of ± 0.004" (0.102mm) are possible. The process is easily adapted to robotics where location tolerances may be ± 0.002" (0.051mm). A complete water jet cutting station is shown in Figure 24-19.

Water under this high pressure can do extreme damage to a person, so stay well away from the water jet cutting operation and the piping that is carrying this high pressure.

Figure 24-19. A complete water jet cutting station. (Ingersoll-Rand Waterjet Cutting Systems)

24.8 REVIEW OF SAFETY IN SPECIAL CUTTING PROCESSES

Using the cutting methods discussed in this chapter, you may be dealing with very high temperatures, high pressures, high currents, and high voltages. The welder must be particularly alert for the types of hazards which these conditions may produce. Burns, flying sparks, and electrical shock are ever-present hazards.

Reread the review of safety for Chapters 8 and 17. In addition, be sure to observe the following precautions;:

- Always wear protective clothing, face, eye, and hand protection.
- Due to the fairly high oxygen pressures carried in cutting hoses, be sure all hoses used are in good condition and that all fittings are tight.
- Never weld or cut on a container which may have a fuel gas mixture in it, since it may explode!
- Check with the local fire marshal for recommended procedures before cutting on tanks, cylinders, or other containers which may at

Figure 24-18. A water jet cutting apparatus seen as it cuts an intricate pattern. Note the high quality of the completed cut on the segment to the right. (Ingersoll-Rand Waterjet Cutting Systems)

some time have held flammable materials. Some of the procedures recommended for preparing a container that may have held a combustible liquid or gas may include:

♦ Using live steam under pressure to continuously steam out the container for at least a half hour (until the entire container is heated to the boiling temperature) before beginning the cutting operation.

♦ If it is an edge or dome that is to be cut, fill the container with water up to the point where the cut is to be made.

♦ In some cases after washing out with live steam, argon or nitrogen is flowed through the container. The gas is continually flowed for some time before the cut is made in order to replace all of the air in the container with inert gas. Argon is preferred to helium for this purpose, since it is heavier and will not flow away as readily as helium.

• Be sure that the area in which all cutting operations are performed is well-ventilated. There will be both small metal particles and metal oxides produced in the cutting operations.

• When performing underwater cutting operations, be sure approved diving equipment is provided. If possible, provide a two-way telephone communication system between the underwater welder and the surface staff assisting the welder.

• With most cutting operations, there is considerable sparking and throwing of molten slag. Therefore, the area in which the cutting is performed must be of fireproof construction. Combustibles that cannot be moved must be covered with fireproof materials. A firewatch should also be posted. Always have an approved fire extinguisher at hand when performing cutting operations.

• If performing heavy oxyfuel gas cutting operations for a considerable length of time, be sure to manifold two or more acetylene cylinders together. In cold weather particularly, the acetylene cylinders cannot provide a heavy flow of acetylene over a long time because of low vapor pressures.

• Due to the great amount of splatter when flame- and arc-cutting, be sure your eyes are well-protected with goggles or a helmet, even though rays from the flame or arc may not seem to require such protection.

• Use an inert gas to pressurize and carry the metal powder to the cutting oxygen discharge area when metal powder cutting. Fine powder when confined with a high concentration of oxygen can be explosive.

• Oxygen should not be used for pressurizing any operating or test procedure. Use carbon dioxide, nitrogen, helium, or argon. Use a pressure relief valve on pressurized powder dispensers. Adjust the relief valve to the maximum pressure to be used.

• Wear special protective lenses where working around laser equipment.

TEST YOUR KNOWLEDGE

Write your answers on a separate sheet of paper. Do not write in this book.

1. Name the procedures that are listed below as AWS abbreviations:
   a. OFC.
   b. LOC.
   c. AOC.
   d. POC.
   e. FOC.
   f. LBC.
2. What method can be used to cut steel 8' (2.4m) thick?
3. What material is used for the lance in oxygen lance cutting?
4. For large-volume oxygen lance cutting, several oxygen cylinders may be _____ to the pressure regulator.
5. When cutting underwater using oxyfuel gas, what fuel gas is generally used below 15' (4.6m)?
6. What keeps the water away from the cutting flame when OFC underwater?
7. Why are large cutting tips used for OFC underwater?
8. Name the four control valves used on an underwater oxyfuel gas cutting torch.
9. How is the flame ignited underwater?
10. Since cuts using LBC are limited in depth to about six times the hole diameter for each pulse of the laser, how are deeper cuts made?
11. Which polarity is used when oxygen arc cutting?
12. When cutting underwater with an arc process, what number filter lens should be used?
13. What type gloves are suggested for an underwater arc welder?
14. How is a constant arc length maintained when doing oxygen arc underwater cutting?
15. Why is metal powder used with the oxyfuel gas flame in the POC process?
16. What flame or arc cutting process can be used to cut concrete walls?
17. Never use oxygen to pressurize the powder dispenser when metal powder cutting, or an _____ mixture may result.
18. Describe the "flying start" cutting process.
19. How does chemical flux cutting differ from metal powder cutting?
20. How is it possible to cut and pierce holes in hard-to-reach spots with laser beam cutting?

Chapter 25

AUTOMATIC AND ROBOTIC WELDING

LEARNING OBJECTIVES

After studying this chapter, you will be able to:

* State the advantages of automatic or robotic welding.
* Explain the functions of flow switches, solenoid valves, and solenoid switches.
* List and identify the major parts of a robotic system.
* List and identify the parts of a robotic welding workstation.
* Contrast the uses of electrically and hydraulically operated robots.
* Name and identify the axes of 4-, 5-, and 6-axis robots.
* Define the terms working volume, positioner, operator controls, and teach pendant.
* List and discuss the safety hazards encountered in and around a robotic workstation.

Virtually every industry and business is moving towards using more automation and robotic equipment. The welding industry is converting many manual welding operations to either semiautomatic or fully automatic processes.

The American Welding Society defines *automatic welding* as follows: "Welding with equipment that requires only occasional or no observation of the welding, and no manual adjustment of the equipment controls." This definition indicates that a welder or operator is not needed when welding automatically. However, a very knowledgeable technician is needed to set up the process initially.

A welding technician must select the parameters or guidelines to be used for the automatic welding operation. The parameters include voltage, amperage, wire feed speed, travel speed, proper electrode diameter and alloy, and correct shielding gas.

In addition, various controls, electrical monitors, timers, gauges, and instruments are needed to monitor an automated process. Automatic welding has many forms, from motorized carriages to versatile robots, and can be used with most welding processes. Automation can be used to perform tasks or operations such as cutting, welding, and surfacing.

The robot is modernizing the welding industry. The Robotic Industries Association defines a robot as "A reprogrammable, multifunctional manipulator, designed to move material, parts, tools, or specialized devices, through variable programmed motions to accomplish a variety of tasks."

Stated another way, a robot is a programmable machine that can move material or tools (like a welding torch) through a set of motions to accomplish a desired task. In robotic welding, the robot simply holds the welding torch and moves it along or around the weld joint. An automatic welding process is used to make the weld.

25.1 ADVANTAGES OF AUTOMATIC WELDING

The most common robotic welding processes are resistance welding, gas metal arc welding (GMAW), and gas tungsten arc welding (GTAW). *Resistance welding* is done automatically, because the process involves too many variables and happens too fast for the operator to control. Resistance welding is covered in Chapters 18 and 19.

Some advantages of automatic GMAW and GTAW over the manual SMAW process are:

* No electrode stub loss due to the continuous feed from a reel of welding wire.
* Relief from the labor of concentrating on the arc length, speed, and other variables that must be controlled for a good quality weld.
* Much higher currents may be maintained with any given electrode size, compared to manual welding.

- Weld height, width, fusion, and penetration will be uniform once the automatic controls are adjusted correctly.
- Weld rates are higher in automatic welding than in manual welding.

The gas metal arc welding (GMAW) process is covered in Chapters 13 and 15. The gas tungsten arc welding (GTAW) process is covered in Chapters 13 and 14.

The advantages of automatic welding and robotic welding are:

- Increased productivity.
- Improved weld quality.
- Reduced cost.

25.2 FEEDBACK CONTROLS

As a welder prepares to weld, he or she uses certain feedback controls to make sure the equipment is in good working order. A welder turns on the power supply and can hear it make noise. The welder turns on the shielding gas and sees the floating ball in the flowmeter rise. When the trigger on a GMAW torch is pressed, the wire feeds out and the shielding gas flows. These types of feedback are seen or heard by the welder.

A machine or a robot cannot *see* if the shielding gas is flowing. It cannot *hear* if the power source is on. A machine or a robot must also know if the cooling water is on. In addition, a machine or robot must know if the air pressure system, hydraulic fluid, or electric motor is working properly to make it move.

An automatic welding machine or robot needs *feedback controls* to determine if everything is working properly. All feedback controls for machines are in the form of an electrical signal. As an example, a robot checks to determine if cooling water is flowing. If an electrical feedback signal is received, this means the cooling water is flowing. If an electrical feedback signal is not received, the cooling water is not flowing, so the robot will turn itself off to prevent damage.

Another type of control is used to turn something — like the wire feeder or the shielding gas — on or off. This type of control is a solenoid-operated valve or relay. These electrical controls are discussed in the following headings.

25.2.1 Flow Switches

Flow switches are used to check if gas or fluid is flowing through a pipe or tube. A flow switch can be used to light a warning light to tell the welding operator that the flow is insufficient or stopped. A flow switch can also be used to turn off equipment if fluid *stops* flowing.

An example of a fluid flow switch is shown in Figure 25-1. This switch uses a flexible reed or paddle that is placed in the path of fluid flow. When fluid is flowing, the reed will bend. The bending movement is electrically connected to a safety switch or to an indicator light. The same principle is used for an air- or gas-flow switch. Larger paddles are used to detect gas flow. An air or gas flow switch is shown in Figure 25-2.

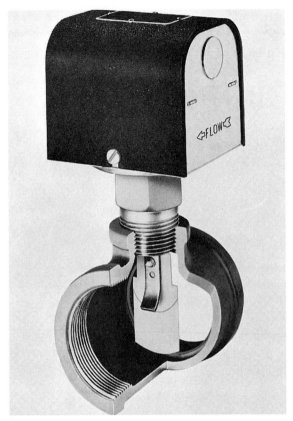

***Figure 25-1**. A fluid flow switch. This switch may be wired to turn on a warning light and/or turn off the electrical power to the welding machine if the water flow stops or becomes insufficient for cooling. (McDonnell & Miller, Inc.)*

Another type of switch depends on fluid pressure for activation. The fluid pressure keeps the electrical circuit closed. If the pressure falls, the switch will open and illuminate a warning light, sound an alarm, or turn off the equipment. An example of a fluid pressure switch is shown in Figure 25-3.

25.2.2 Solenoid-operated Valves and Relays

Solenoid-operated valves and relays are very similar. A *solenoid-operated valve* is used to start and stop the flow of a gas or liquid. A *solenoid-operated relay* is used to start and stop the flow of electrical current. These solenoid-operated valves and relays are essential to the operation of semiautomatic and automatic welding equipment.

A solenoid-operated valve is shown in Figure 25-4. Gas- or fluid-carrying hoses are connected to the valve at the inlet and outlet. The electrical coil is connected to an electrical circuit. A solenoid may be wired to operate on ac or dc current.

A solenoid-operated valve will not allow gas or fluid to pass through unless a certain voltage is supplied to the coil. The *plunger*, which is also called the valve, is forced against the valve seat by a spring. The solenoid plunger and seat are machined so that they mate and form a gas-tight seal. When the solenoid valve and seat are forced

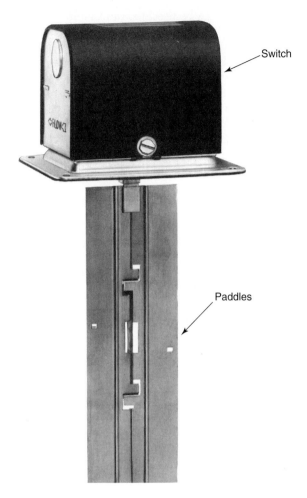

Figure 25-2. An air flow switch with paddles that fold back when air or gas is flowing. (McDonnell & Miller, Inc.)

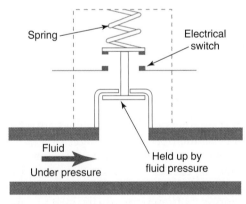

Figure 25-3. A safety switch that depends on fluid pressure. If the fluid pressure decreases, the electrical circuit will close. The welding machine may be turned off or an alarm activated when the switch closes.

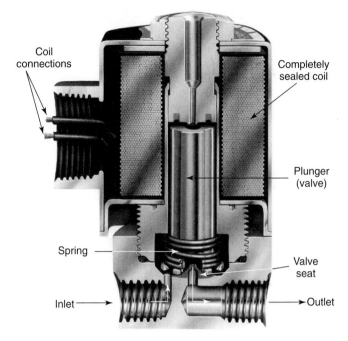

Figure 25-4. A solenoid valve used to control shielding gas flow. The plunger is forced against the valve seat by the spring. The valve shown is in the closed position. (Airmatic-Allied, Inc.)

proper voltage and current, a magnetic field is established. The magnetic field pulls the plunger upward, off the valve seat. Gas or fluid may now pass through the valve.

To stop the flow of gas or fluid, the power to the coil is shut off. The magnetic field that holds the plunger up will cease to exist. The spring will force the plunger against the valve seat, so no gas or fluid will be able to pass through the valve.

One common use of a solenoid-operated valve is to turn the shielding gas on and off during gas tungsten or gas metal arc welding. The shielding gas can also be made to flow by manually pushing the purge switch. Pressing the purge switch will send power to the solenoid gas valve, allowing shielding gas to flow. See Figure 25-5.

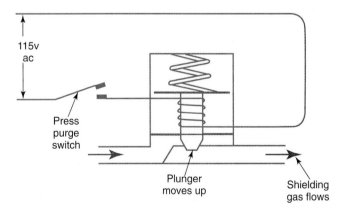

Figure 25-5. This circuit is used to purge the lines on a GMAW or GTAW torch. When the purge switch is pressed, the solenoid lifts the plunger and allows shielding gas to flow.

together no gas or fluid can flow through the solenoid valve.

To allow gas or fluid to flow, the plunger must be lifted off the valve seat. The plunger is lifted by supplying the required voltage to the coil. When the coil receives the

When GMAW or GTAW, the shielding gas must flow for a short period after the welding is stopped. This extended gas flow protects the electrode, weld, and base metal while they cool.

The extended gas flow is provided electronically. The solenoid-operated gas valve is connected to a second electrical circuit. This circuit will send electricity to the solenoid, after the electricity in the main circuit is stopped.

This additional gas flow is set by adjusting a knob on a GMAW or GTAW machine. The knob can be used to vary the resistance of the second electrical circuit. The postflow (extended gas flow) will vary as the resistance of the circuit is changed. Some welding machines control the postflow by counting the cycles in the ac wave. Alternating current has 60 cycles per second. When the postflow is set for 10 seconds, the machine electronically counts 60 cycles per second. After 600 cycles (60 cycles/second times 10 seconds) the current in the second circuit stops and so does the postflow. This electronic circuit timing method is very exact.

Another use for solenoid-operated valves is to turn cooling water on and off. Water is used to cool some GTAW and GMAW torches. Water is also used to cool resistance welding equipment.

A solenoid-operated relay is like a solenoid-operated valve. In a solenoid-operated relay, an electrical coil is wrapped around an iron core. One set of electrical contacts is on the base of the relay. The second set of contacts is on a movable arm. The contacts on the arm are kept away from the contacts on the base by a spring. To operate the relay, proper voltage is sent to the electrical coil. Current passing through the coil will set up a magnetic field. This magnetic field pulls the arm down to close the electrical contacts. When voltage to the coil is turned off, the magnetic field stops. The spring pulls the arm up and the electrical switch opens. A relay is shown in Figure 25-6. The big advantage of a relay is that a smaller voltage (6V, 12V, 24V, or 115V) can be used to control a larger voltage (like 24V, or 115V) can be used to control a larger voltage (like 115V or 230V).

One common example of the use of a relay is in a GMAW machine, where it is used to operate the wire feeder. When the trigger on the GMAW torch is pressed, the solenoid receives a small voltage and the relay switch closes. After the relay switch closes, the wire feeder receives 115V ac and begins to feed wire. See Figure 25-7. Thus a small voltage of 24V, can control a large voltage of 115V. Another way to operate the wire feeder is to push the inch switch. This will also supply 115V ac to the wire feeder as seen in Figure 25-7.

All automatic and semiautomatic welding operations depend on the use of solenoid-operated valves and relays. Figure 25-8 shows a gas metal arc welding circuit. The circuit contains a relay, a solenoid-operated gas valve, a transformer, and three switches. The transformer is used to reduce the voltage from 115V ac to 24V ac. Only 24V goes to the gun switch, since it is dangerous to have 115V in a welder's hand.

A welder can check to see if everything is operating properly by pressing the purge switch and the inch switch. When the welder presses the purge switch, electricity goes to the solenoid-operated gas valve. The plunger is raised allowing shielding gas to flow. See Figure 25-5. In the same way, when the inch switch is pressed, the wire feeder receives 115V ac and feeds electrode wire.

When the welder is ready to weld, the gun switch is pressed. The 24V circuit will cause the relay switch to close. When the relay closes, both the solenoid-operated gas valve and wire feeder will get 115V ac and begin to operate. Both shielding gas and electrode wire will be delivered to the welding torch. The welder can continue welding until the gun switch is released. When the gun switch is released, the solenoid-operated relay will no longer receive 24V. The relay spring will open the electrical switch. The gas valve will close and the wire feeder will stop. Note that the postflow gas circuitry is not shown in Figure 25-8.

25.2.3 Three-way Valves for Hydraulic or Pneumatic Systems

Three-way solenoid-operated valves are often used to open and close the operating passages of a hydraulically or pneumatically operated cylinder on an automatic welding machine. Figure 25-9 shows a three-way valve. This valve is normally in the closed position. The passage of hydraulic pressure from port A on the inlet to port C on the outlet side of the valve is blocked by the main valve.

Follow the action in Figure 25-9. When the solenoid is energized, the solenoid pilot valve is moved down. A passage is then completed from port D through port E to the top of the main valve. This pressure on top of the main valve forces it down, thus opening the passage from A to C and to the operating cylinder connected to passage C.

When the solenoid is deenergized, the solenoid pilot valve is moved up by spring pressure blocking the passage from port D to port E. Pressure is relieved from above the main valve by passing through ports E, F, G, and to release

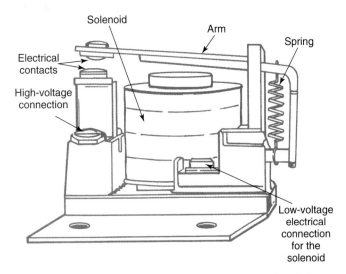

Figure 25-6. The parts of a solenoid-operated relay. When the solenoid receives power, it will pull the arm down and connect the contacts of the switch.

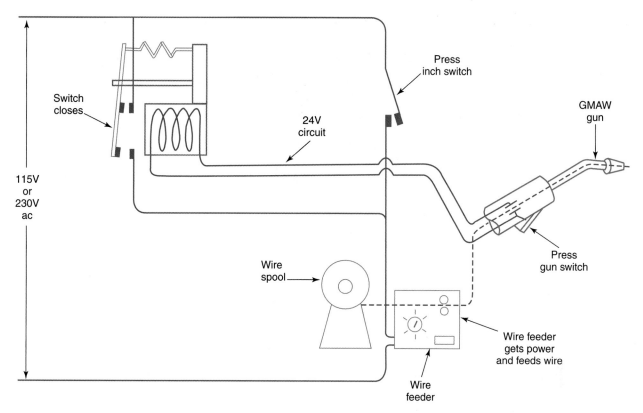

Figure 25-7. *To operate the wire feeder, a welder can press the inch switch or press the gun switch. The gun switch will cause the relay to close, which will allow electricity to flow to the wire feeder.*

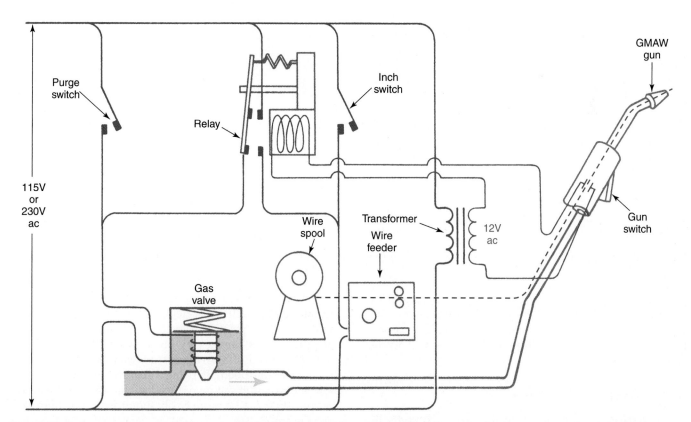

Figure 25-8. *A gas metal arc welding circuit that uses a transformer, a solenoid-operated relay, a solenoid-operated gas valve, and three switches. When the purge switch is pressed, the gas valve allows shielding gas to flow. When the inch switch is pressed, the wire feeder supplies electrode wire. When the gun switch is pressed, the relay closes; as a result, the gas valve supplies shielding gas and the wire feeder begins to feed the electrode wire.*

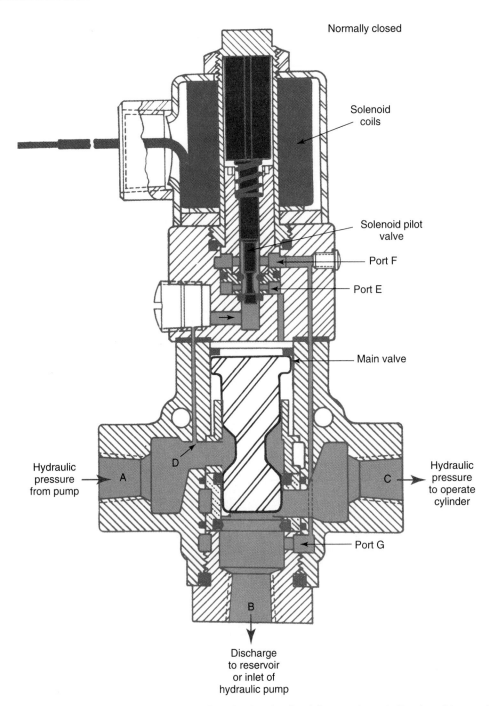

Normally closed

Solenoid
coils

Solenoid pilot
valve

Port F

Port E

Main valve

Hydraulic
pressure
from pump

A

D

C

Hydraulic
pressure
to operate
cylinder

Port G

B

Discharge
to reservoir
or inlet of
hydraulic pump

Figure 25-9. *A three-way solenoid valve for controlling hydraulic fluid flow. (Airmatic/Beckett-Harcum)*

port at B. When pressure is relieved from the top of the main valve, it is moved up by pressure from below. This closes the passage from A to C, and opens the passage from C to B, allowing the fluid in the C passages to discharge through opening B.

25.2.4 Electrical Actuators

Electrical actuators, which are electric motors, are most-often used to move and position the various parts of a robot about their rotational axes. The two types of motors used are the stepping motor and the permanent magnet dc motor.

A **stepping motor** has the ability to rotate a specific number of revolutions or degrees of rotation. This rotation will move the specific distance. The speed of rotation of a stepping motor can be very precisely controlled.

The speed of a **permanent magnet dc motor** can also be closely controlled. By using a gear reduction system, the speed of rotation can be reduced so that the rotation of the output shaft can be more easily monitored and controlled. See Figure 25-10.

Sensors are used to provide feedback information to the robot controller. These sensors can tell the controller the exact number of revolutions that the stepping motor or permanent magnet-type motor has turned. Other sensors

Figure 25-10. A rotary actuator, consisting of a dc servomotor and gear train. It provides precise closed-loop control. (Jordan Controls, Inc.)

can tell the controller the rotational speed of the motor. With such a precisely operating motor and with the feedback information, the controller can also cause the motors to move the welding torch or gun along a weld joint, regardless of whether it is a straight line or a curve.

Stepping motors and permanent magnet dc motors may be used to:

- Move the robot arm into position and rotate the arm as required.
- Move the positioner and work under the welding torch or gun.
- Feed the electrode to the gun, or welding wire to the joint at a controlled rate.
- Move a welding, cutting, gouging, or metal-spraying torch or gun along a prescribed path at the correct speed.

A variable speed motor is often used in a motorized carriage to move the carriage and torch or gun along the weld joint or cut line. A knob on the carriage is used to set the travel speed of the carriage. The knob is attached to a variable resistor. When the resistance of the variable resistor is increased, the amperage to the field windings in the variable speed motor is decreased. The amperage through the armature winding of the motor will increase and the motor speed will increase. As the motor speed increases, the carriage speed increases. The opposite is true when the resistance of the variable resistor is decreased. The speed of the carriage is therefore varied by changing the resistance of the variable resistor.

25.3 AUTOMATIC WELDING PROCESSES

As noted earlier, *automatic welding* is defined by the AWS as "welding with equipment that requires only occasional or no observation of the welding, and no manual adjustment of the equipment controls."

Semiautomatic welding is defined by the AWS as: "Manual welding with equipment that automatically controls one or more of the welding conditions."

Mechanized welding is defined by the AWS as: "Welding with equipment that requires manual adjustment of the equipment controls in response to visual observation of the welding, with the torch, gun, or electrode holder held by a mechanical device."

Gas metal arc welding (GMAW) and flux cored arc welding (FCAW) are semiautomatic welding processes. The welding power source used is a constant voltage machine. The wire feeder automatically delivers the electrode to the weld pool at a preset and constant speed. The welder must control the welding gun manually.

By allowing a machine to control the torch, instead of the welder controlling it, the process becomes mechanized. Only when feedback controls are used does a process become automatic. See Figure 25-11.

Figure 25-11. Submerged arc welding (SAW) being done automatically. Once all the variables are set by the operator, no changes need to be made throughout the weld operation. (Invincible Air Flow Systems)

Many semiautomatic, automatic, and mechanized welding and cutting process have been described in this book. These processes and the location where more information can be found are shown below:

- Arc stud welding   Heading 20.5
- Cutting   Heading 8.9
- Electron beam welding   Heading 20.18
- Electroslag welding   Heading 20.16
- Electrogas welding   Heading 20.3
- Friction welding   Heading 20.13
- Furnace brazing   Heading 10.6.7
- Gas tungsten arc welding   Heading 23.7
- Gouging   Chapter 17
- Laser welding   Heading 20.19
- Resistance welding   Chapters 18 and 19
- Submerged arc welding   Heading 20.2

Even though each of the processes listed above may be an *automatic* process, the processes involved are quite different. In submerged arc welding, the torch is moved mechanically along the weld joint and the flux and electrode wire are automatically fed to the joint. In laser and electron beam welding, all the variables may be set, then monitored as these welds proceed. Since the equipment is too large to be moved, the parts are mechanically moved under the beam. See Figure 20-49 through Figure 20-54. A resistance spot weld is made by placing the two pieces of metal between the electrodes. All of the welding variables are set and performed automatically. After the cycle start button is pushed, the entire welding sequence is completed automatically. For a friction weld, the parts to be joined are clamped in fixtures. After this clamping operation, the process is completely automatic.

From these brief examples, the variety of processes and applications can be seen. Almost every process can be automated. Figure 25-12 shows the submerged arc welding processes being done automatically. GMAW is widely used in industry. Most arc welding robots use the automatic GMAW process. Other robots may use automatic GTAW or resistance welding. All automatic welding or cutting machines rely on feedback controls, solenoids, relays, and variable speed motors to control and direct the automatic welding process.

25.4 INTRODUCTION TO ROBOTS

The major reasons to use an automatic welding process are to:

- Increase productivity.
- Improve quality.
- Reduce cost.

Industrial robots, which are usually just called *robots*, are used to accomplish all three of these objectives. A robot used with an automatic welding process can weld faster and more accurately, and produce welds of higher quality, than a human welder using manual methods. This increases productivity. A robot will perform the same sequence of moves over and over. The robot will maintain the same arc length, same voltage, and same travel speed every time.

Figure 25-12. A close-up view of an automatic SAW in progress. Note the flux handling equipment and wire feeder. (Invincible Air Flow Systems)

This will improve the quality of the welding operation and tend to produce fewer defects. Since the robot welds faster, makes better quality welds, and welds continuously, this reduces the cost of the welding operation.

Robots are not suited for every welding application. A robot is a complex, expensive piece of machinery. It requires special maintenance and care. When used properly, a robot can be an asset to large production companies and to the smaller welding job shop.

Before attempting to use robotic equipment, refer to Heading 25.5 on Safety Practices in Automatic and Robotic Welding.

25.4.1 Robotic System

The robot itself is only part of a complete *robotic system*. The robotic system includes all the equipment and software needed to accomplish the welding, cutting, or surfacing task. This equipment is normally assembled and located within what is called the *robot workcell*. Required equipment for a robot workcell includes:

- A robot.
- The electronic robot controller.

- Automatic welding, cutting, or surfacing equipment.
- A positioner.
- Fixtures to hold the part to be welded.
- Material handing equipment.
- Operator controls.
- Safety screens and/or electronic barriers.

Much of this equipment is shown in Figure 25-13.

Some other items which may be needed to properly set up and maintain a robotic system include:

- A master part that is used to "teach" (program) the robot.
- A program listing for the robot.
- A device to save the program after it has been completed.
- A torch or gun set-up jig to "zero" the robot.
- A torch cleaning station.
- An inspection jig to check the program and finished parts.
- A set-up and troubleshooting guide.

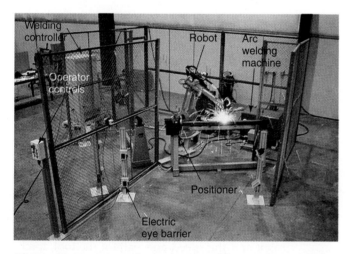

Figure 25-13*. A complete robotic welding system with the major parts of the system shown. (ABB Flexible Automation, Inc.)*

25.4.2 Robotic Movements

An industrial robot is required to move the welding torch through a set of motions. The robot needs some system to enable it to move. Most robots are connected to a 240V or 480V electrical service. This electricity operates electric motors used to move the robot about its various axes.

Instead of using electric motors to move, a robot may be moved or rotated on its axis by means of hydraulic cylinders or hydraulic motors. Electricity is still needed, however, to operate the hydraulic pumps in the system.

A robot may move in three planes and may also rotate about each of these planes. See Figure 25-14. There are two common systems of robot motion. One is the ***rectilinear robot***; the other is the ***articulated robot***. The rectilinear robot moves in straight lines in the "X", "Y", and "Z" axes, with rotation possible around two of these three axes. See

Figure 25-14. The articulated robot rotates around all three axes and swings in arcs. It can place a tool or torch at any position in space and it may approach that spot in space from almost any direction. See Figure 25-15.

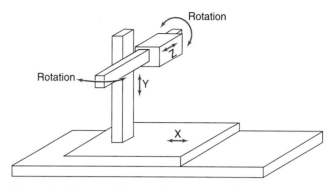

Figure 25-14*. A rectilinear robot showing the motions that can be made in the "X," "Y," and "Z" axes.*

Most welding robots are of the articulated type. Welding robots commonly have four, five or six axes. The six-axis robot is the most versatile, but it is more expensive than a four- or five-axis robot. There are many mechanical joints in a six-axis robot. Rotation is possible around each mechanical joint.

The axes of rotation on robots are generally referred to by *number*, as shown in Figure 25-15. Each axis of rotation on the robot may also be referred to by *name*. The axes are named after similar moving parts of the human body. A six-axis robot with the name and movement of each axis is shown in Figure 25-16. The numbers and names assigned to the various axes and their functions on a six-axis robot are:

1. Waist rotation (rotation around vertical axis).
2. Shoulder bend (rotation about the horizontal axis).
3. Elbow bend (a second rotation about the horizontal axis).
4. Arm roll (rotation).
5. Wrist pitch (up and down rotation).
6. Wrist roll (rotation of the tool, torch, or gun).

Each axis or joint must have its own mechanism to allow it to move. An electric robot must have an electric actuator (motor) for each axis. The motors for the waist, shoulder, and elbow are usually larger than the motors used to move the arm and the wrist pitch and roll axes. In the same way, a hydraulic robot must have an actuator for each axis. A hydraulic actuator is used to change hydraulic energy into mechanical energy. Two common hydraulic actuators are the hydraulic cylinder and a hydraulic motor.

A hydraulic motor operates like a water wheel. Hydraulic fluid causes a turbine wheel with blades on it to rotate. A gear is attached or machined at one end of the turbine shaft. The gear on the turbine shaft meshes with a second gear. When the first gear rotates, the second gear also rotates. This second gear causes some part of the robot to move. See Figure 25-17.

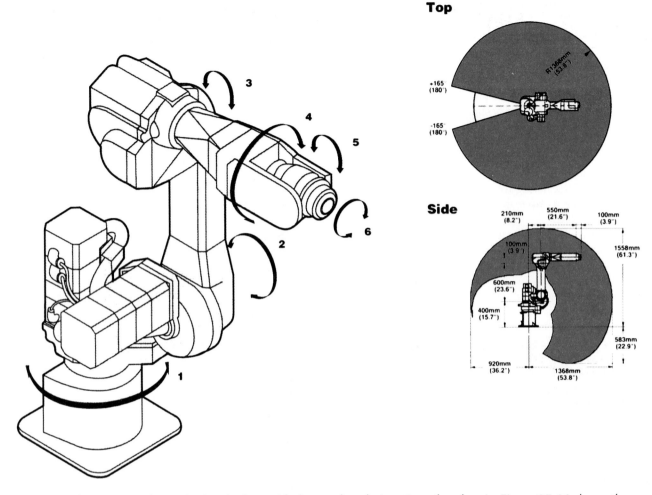

Figure 25-15. *The motions of an articulated robot, with the number designation of each axis. Figure 25-16 shows the names given to these axes. The robot's working volume is shown in color. (FANUC Robotics North America, Inc.)*

The robot has a limited area that it can reach. This area in which the robot can move and work is called the robot's *working volume*. Any parts that are to be welded must be placed within the working volume. Each type of robot has its own working volume. Some working volumes are much larger than others. A working volume for one type of robot is shown in Figure 25-18.

Never enter the robot workcell and especially, the robot's working volume, while it is operating. The robot moves so rapidly and with such great force that it can do great injury to a human being.

25.4.3 Robot Controller

A *robot controller* controls the entire robotic system. It is the "brains" of the operation. The robot controller performs the following functions in the robotic system:

- Contains the computerized program to be performed.
- Turns the shielding gas on and off and monitors its flow.
- Turns the cooling water on and off and monitors its flow.
- Turns the welding or machining power source on and off.

- Controls and monitors the movement of the robot and the speed and path of the tool, torch, or gun.
- Controls and monitors the variables required for the welding, cutting, or surfacing process to be used.
- Controls the movements of the material handling equipment, if used.
- Controls and monitors the safety interlocks in the robot workcell.

The main part of the robot controller is a computer. Every part of the robotic system is connected to this computer. In addition to controlling the operation of the system, the computer monitors all the parts of the system to make sure everything is running properly. The computer also stores the program or sequence of actions that the robotic system will follow when performing the welding operation.

The basic responsibilities of the robot controller can be classified as follows:

- Monitoring the robotic system.
- Storing the program.
- Executing the program.

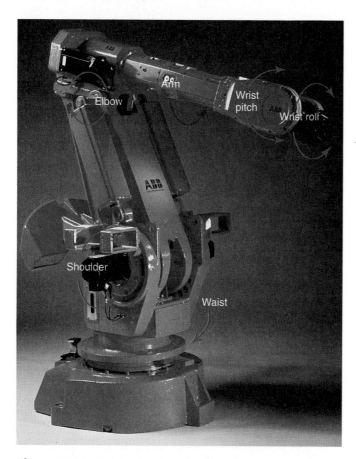

Figure 25-16. *A six-axis robot showing the names and numbers assigned to the axes and their motions. (ABB Robotics, Inc.)*

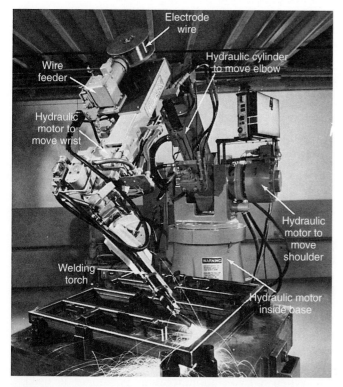

Figure 25-17. *A six-axis hydraulic robot. Note the hydraulic motor used to move the shoulder and wrist. Also note the hydraulic cylinder used to move the elbow.*

Figure 25-18. *The working volume of this robot is very large, as is the part it is welding. The robot moves on its "X" axis on an overhead I-beam. Its "Y" axis is also overhead. These movements are rectilinear. The other parts of the robot move with an articulated motion. Note the large positioner in use. (ABB Robotics, Inc.)*

The robot controller uses feedback control to monitor the robotic system. As discussed in Headings 25.2 and 25.2.1, feedback controls and flow switches are used to monitor the operation of the system. Checks are constantly being made to make sure the motors or hydraulic pumps do not overheat. Checks are made to make sure cooling water is flowing. If something is not operating properly, the controller will illuminate a warning light. If the problem is serious, the controller may shut down the robotic system to prevent damage.

A *program* is a series of operations. The most important function of the robot controller is to execute this program. Through the series of operations defined in the program, the robotic system performs the welding operation. During a complete welding program, the following functions may be accomplished by the controller:

1. The material handling equipment, or a worker, places the materials into fixtures on the positioner.
2. The positioner moves the fixtured pieces into the robot's working volume.
3. With the water flowing and the welding power source turned on, the shielding gas is allowed to flow.
4. The weld is started and the welding speed, wire feed, and all other variables are controlled and monitored as the robot moves the welding gun or torch along the weld joint.
5. With a flexible robotic welding system, feedback devices send signals to the controller so that corrections for joint fit-up or location may be made as necessary.
6. When the welding is completed, the shielding gas is shut off.

7. While welding on one set of parts, material handling equipment unloads the completed weldment and loads new parts. When welding on a set of parts is completed, the process is repeated.

During the weld, the controller may shut down the welding operation if the water or shielding gas stops flowing, or if someone enters the robot working volume. Programming the robot is explained in Heading 25.4.7.

A robot controller uses electrical signals to perform the sequence of operations called the program. A controller usually sends small electrical signals to solenoid-operated valves and relays. Through these valves and relays, the controller controls every part of the robotic system.

A hydraulically actuated robot uses solenoid valves. To move the robot, hydraulic fluid must be allowed to enter a hydraulic cylinder or flow through a hydraulic motor. Solenoid-operated valves control the flow of hydraulic fluid. Refer to Figure 25-9. In order to move one axis of the robot, the controller sends a signal to the solenoid-operated hydraulic valve. The solenoid lifts the plunger and allows hydraulic fluid to flow. When hydraulic fluid flows, the robot moves. When the controller determines that the robot has moved to the desired location, the electrical signal to the solenoid is stopped. No more hydraulic fluid will flow and the robot's movement stops.

Electric robots use solenoid switches to control the robot's movements. To move one axis of an electrically actuated robot, the controller sends an electric signal to a solenoid-operated relay. The solenoid will move the plunger and close the electrical switch. Current from the electrical service will operate an electric motor to move a part of the robot. When the controller turns off the signal to the solenoid-operated relay, the relay switch will open. No current will flow to the electric motor, so the robot will not move.

It must be understood that each axis of a robot has its own electric motor, hydraulic motor, or hydraulic cylinder. Each electric motor has a relay. Each relay is controlled by the robot controller. Likewise, each hydraulic motor and hydraulic cylinder has its own solenoid-operated hydraulic valve. The robot controller controls each valve. Through the use of solenoid-operated valves and relays, the robot controller controls the movements of the robot.

25.4.4 Automatic Welding Equipment

Automatic welding equipment used in robotic welding is connected to the robot controller. As stated in Heading 25.4.3, the robot controller operates the entire robotic system. This includes the welding equipment.

The controller can set the desired welding voltage and wire feed speed. This is done through electrical signals and electric circuits. When resistance spot welding, the controller can set the current level and electrode force. The resistance welding sequence of squeeze time, weld time, hold time, and off time is also set by the controller. When the welding start button is pressed, the robot moves the welding gun to the correct location. The electrodes come together with the correct pressure. The resistance welding

sequence is performed and the weld is made. The robot moves the resistance welding gun to a new location and another weld is made. This process continues until the program in the controller is completed.

When performing a weld with gas metal arc equipment, the controller sends signals to the welding equipment to start the shielding gas flow. Relays close and the wire feed and welding begins.

Automatic welding equipment mounted on a robot is exactly the same equipment used in manual and semiautomatic welding, as described in other headings in this book. However, the welding equipment is now mounted on a robot arm and operated by a controller which sets and monitors all the welding variables.

The controller must turn the shielding gas and cooling water on and off, start and stop the welding sequences according to the program, and execute the programmed welding conditions. The controller should also be able to control and monitor the wire feed and power supply, and the travel speed of the welding gun or torch. Some robots with feedback sensors can even change the welding conditions as required to meet changes in the welding joint alignment. Welding power sources used with robotics systems must be able to operate at a nearly 100% duty cycle.

25.4.5 Welding Positioners

A *welding positioner* is used to hold the assembly of parts to be welded and move that assembly into the best welding position possible within the working volume of the robot. Parts are normally clamped into fixtures on a positioner. When clamped into fixtures, the parts will be held in a predetermined fixed position.

Another use of a positioner is to allow the robot to reach all areas of a part. When a large assembly is placed on a positioner, part of the assembly may be outside the robot's working volume. Welds can be made on the weld joints that are within the robot's working volume. The positioner then moves the assembly so that various areas are inside the working volume and more welds are completed. The assembly is continually moved until all welds are completed. By using a positioner, the robot's working volume can effectively be increased.

If welds must be made on the top and bottom of an assembly, the positioner can rotate the assembly so that it can be welded in the flat welding position. Figure 25-19 shows an articulated robot being used as a positioner.

Positioners are available that can hold two or more parts to be welded. When one part is being welded, a second assembly can be placed in the fixture on the positioner. After the robot completes welding one part or assembly, the positioner moves the second part into the robot's working volume. The completed weldment is removed from the fixture and another assembly is installed into the fixture to await welding. Through the use of a two-station welding positioner, the robot can weld with less downtime, which increases productivity. See Figure 25-20.

When a two-station rotating positioner is used, a metal screen is normally mounted between the two stations. This screen will protect the operator, who is on the

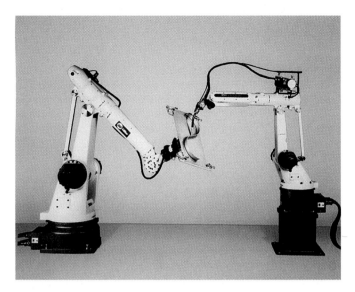

Figure 25-19. An articulated robot is being used in this welding application as a welding positioner. (Miller Electric Mfg. Co. – Automation Div)

Figure 25-20. A two-position, horizontally rotating positioner. The worker at the left is unloading parts from, and reloading parts into, the welding fixture. (ABB Robotics, Inc.)

other side of the screen from the welding operation, while loading and unloading parts from the fixtures. Refer to Figure 25-20.

Positioners usually have two axes about which they rotate. Some positioners have as many as five axes. A positioner is not a required part of a robotic system. Positioners are, however, very common in high-production robotic workcells.

25.4.6 Operator Controls

The person directing a robot controls the operation by using the operator's control panel to begin the robot pro-

gram and to turn off the robot in an emergency. An operator's control panel is usually used when an operator is required to load and unload parts. When a robot completes the welding operation on one part, the robot arm and attached welding equipment move back out of the way. It will not move from this "parked" position until the welding operator pushes the "cycle start" button.

While the robot arm and the attached welding equipment are out of the way, the part that was just welded is removed from the welding fixture. A new part is placed in the welding fixture. The operator then presses the cycle start button on the operator's control panel. When the cycle start button is pressed, the robot will perform the complete welding operation. When the robot finishes welding, the robot arm and welding torch or gun will again move out of the way and wait for the cycle start button to be pressed again.

If something goes wrong during the operation of the robotic system, the operator can stop the process. An "emergency stop" button is on the operator's control panel. To interrupt the program, this button is pressed.

Not all robotic systems use an operator's control panel. Some systems have the emergency stop and cycle start buttons on the robot controller. An operator's control panel is shown in Figure 25-21. Refer to Heading 25.5 for Safety Practices in Automatic and Robotic Welding.

25.4.7 Programming the Robotic System

As noted in preceding sections, the robotic welding system is controlled by a robot controller. The robot controller stores the sequence of steps and functions that the robot will perform. This sequence of steps and functions is stored in a computer as a "program." Before the controller can store a program, the controller must be *taught* each step and function in the program. This process is called *programming* the robotic system.

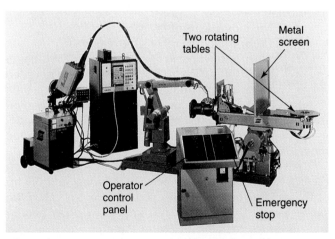

Figure 25-21. A positioner with two rotating tables separated by a metal screen. The screen protects the operator from viewing the arc and from metal spatter. The operator's control panel is used to begin the welding operation. An operator can stop the operation by pressing the emergency stop button. (ESAB Welding and Cutting Products)

When teaching or programming the robotic system, the welding operator uses a ***teach pendant***. A teach pendant is shown in Figure 25-22. The teach pendant has control over each axis of the robot and each axis of the positioner (if used). The welding operator can move the robot using the teach pendant. One axis at a time is moved by pressing the proper axis button.

Figure 25-22. *A teach pendant used to program a six-axis robot and its positioner. (FANUC Robotics – North America)*

There are many steps and functions within one program. Some of the more important steps and functions are listed below.

- The movements of the robot.
- The speed of the robot's movement.
- When and where to begin the welding sequence.
- When and where to stop the welding sequence.
- The welding variables.

See Heading 25.4.3 for other functions of the controller that the program should include.

One of the most important parts of a program is the movement of the robot.

All programs must have a point in space from which they can measure the movements of the robot and the tip of the welding electrode. This point in space is called the ***set-up point***. This point is also known by other names, such as the "zero point" or "home point."

A tool listed in Heading 25.4.1, called the ***torch or gun set-up jig***, is used in the robotic workcell to locate a predetermined point in space: the *set-up point*. A specific point on the robot arm is then aligned with the set-up point. After this alignment is completed, the robot and welding torch or gun can be moved through the required welding sequence and the program placed into memory. All robot

movement will be measured by the computer from the set-up point.

There are two ways to program the motion of the robot: point-to-point or continuous path. ***Point-to-point*** motion involves programming a series of points. The robot will move in a straight line, directly from one point to the next.

In point-to-point movement, the welding operator uses the teach pendant and moves the robot to a desired location. After the new location is reached, the "program" button is pressed. This point becomes one point in the program. The welding operator then moves the robot to the next point to be in the program. Again, the "program" button is pressed. This process is continued until the robot's movements for a complete cycle or sequence of actions are completed.

Sometimes, it may be necessary to delete a point in a program. This is done the same way as when programming a point. The robot is moved to the point to be removed, and the "delete" button is pressed. That point is no longer a part of the program.

An example of point-to-point movement is shown in Figure 25-23. A robot operator wants to move the welding torch on the end of the robot arm from point 1 to point 4. Four points may be programmed, so the robot's movements will miss the obstacle between points 1 and 4. The robot will move from point 1 to point 2, from point 2 to 3, and then from point 3 to point 4. Each movement will be a straight line.

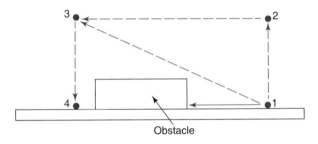

Figure 25-23. *Point-to-point movement. To move a robot from point 1 to point 4, two paths can be followed. Either the path 1-2-3-4, or the path 1-3-4 can be used. The robot cannot move directly from point 1 to point 4 or it will hit the obstacle.*

If the robot operator decides to delete point 2, the robot will move from point 1, to point 3, and then to point 4. This movement is faster than moving from 1 to 2 to 3 to 4. The obstacle is still missed. If point 3 were deleted, however, the robot would attempt to move directly from point 1 to point 4. During this movement, the welding torch on the robot arm would hit the obstacle. The welding torch might be damaged. The robot might also require maintenance after such a collision.

If a robot or the welding torch should hit something as it travels in the workcell, an emergency stop sensor will usually shut down the robotic system. After a collision, the robot must be realigned with the set-up point.

The other type of robot movement is the *continuous path*. In continuous path movement, the robot will follow the exact path that it was taught. If the robot moved along a curve while being programmed, it will repeat the curve. The weld locations are the programmed points. Every movement of the robot will be part of the program.

If the robot was taught to follow the path 1-2-3-4 in Figure 23-23, it would always perform the same move. Point 2 cannot be deleted. To change a continuous path program, the robot must be reprogrammed. This involves deleting or erasing the old program, and programming a new path.

Another important function that must be programmed is the speed at which the robot moves. Speeds can be relatively slow, such as welding speeds of 10"/min. to 30"/min. (4mm/sec to 13mm/sec). Maximum speeds depend on the robot being used. Speeds can exceed 2000"/min.(850mm/sec); speeds of 800"/min. (340mm/sec) are common.

Each point-to-point movement can be made at a different speed. When a robot is moving through open areas away from any obstacles, high travel speeds can be used. When the robot moves toward the workpiece, or near an obstacle, the travel speed is usually reduced. The travel speed can be reduced in steps. See Figure 25-24. If the movement from point 3 to point 5 was all done at 60"/min. (25.4mm/sec), the movement would be quite slow. Moving from point 3 to point 4 at 400"/min. (169mm/sec) requires less time. Reducing the travel speed in steps allows a faster speed to be used for all but the last few inches (millimeters) of travel.

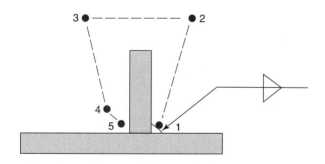

| From | To | Speed | |
		in/min	mm/sec
1	2	400	169
2	3	800	339
3	4	400	169
4	5	60	25.4

Figure 25-24. Robot speeds used to go from point 1 to point 5. Note the speed is faster from point 3 to point 4 than from point 4 to point 5.

For the robotic system to know when to begin welding, it must be programmed. In addition to knowing when to weld, the controller must be programmed to turn on the shielding gas and wire feeder (when used). The controller must set the current, weld sequence times, and electrode force in resistance welding.

Each of these operations is a step in the program. A typical set of functions required to begin GMAW is described below:

1. The first function is to set the voltage and wire feed speed. Welding parameters are set by entering the values while programming the robot.
2. The second function is to turn on the shielding gas.
3. Next, add a delay function for 0.5 seconds. This allows the shielding gas to cover the welding area before the weld begins.
4. The next function is to begin feeding electrode wire.
5. Then, add another delay function of 0.5 seconds. This is to allow the arc to stabilize and a proper-sized molten pool to develop.
6. The final function is for the robot to begin moving along the joint to be welded.

A similar set of functions is required to stop welding. Time may also be programmed to allow the positioner to be rotated and the part unloaded before the weld cycle begins again. Setting the welding parameters is usually done using the teach pendant. A teach pendant is shown in Figure 25-22.

The robotic system has two modes of operation, manual and automatic. All programming is done in the *manual mode*. Any change to the robot's movements or change in the welding parameters is done in the manual mode.

To execute the program, the robotic system is placed in the *automatic mode*. The robotic system is then completely automatic. It will perform the program exactly. The program can be repeated over and over.

25.4.8 Robotic System Operation

When the programming of a robotic system is completed, it can then be used to perform the desired welding operations. Before making any welds, the welding operator should inspect the robotic system. This inspection includes the following:

- Check to be sure the shielding gas cylinder contains gas.
- Check to see if the electrode wire spool has wire.
- Check all electrical connections.
- Check hydraulic lines for leaks.
- Check cooling water lines for leaks.

In addition to checking the equipment, the welding operator should check the welding parameters. For GMAW, these include:

- Voltage setting.
- Wire feed speed.
- Proper electrode wire composition.
- Proper electrode diameter.
- Proper distance from contact tube (or nozzle) to work.
- Proper travel speed for the robot when it is doing the welding.

Finally, the robot operator places the parts to be welded in their proper place. This usually involves placing the parts in a fixture. See Figure 25-25.

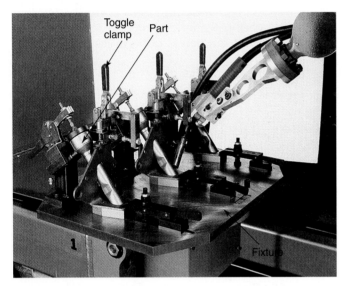

Figure 25-25. A four-position fixture. Three parts are loaded. Only the toggle clamp on the left is in the locked position. The welding process used is GMAW. (ABB Robotics, Inc.)

The robotic system must be in the automatic mode. To begin the program, the "cycle start" button is pressed. The robot controller will direct the robotic system through the program steps. Remember that all parts of the robotic system — the robot, the positioner, and the automatic welding equipment — are controlled by the controller.

When a robotic system has completed a program, the robot arm will usually move back out of the way. The system will not move until the "cycle start" button is pressed again or a sensor tells the controller that everything is ready to weld the next part. **Never let any part of your body be near a moving welding torch.**

Never enter the robot's working volume when the robot is operating. To prevent entry into the robot workcell, fencing or an electronic barrier is installed. Sensors may be mounted on the fencing. If anyone enters the workcell, the sensors or electronic barrier will cause the robotic system to shut down.

25.4.9 Comparing Robotic Systems

There are many manufacturers producing different robots made for different purposes. Industry uses many of these robots on assembly lines. The robots shown in Figure 25-26 are using the resistance spot welding process. Not all robots are made for welding, however.

Figure 25-26. A number of robots equipped with resistance spot welding guns welding on an assembly line. (ABB Flexible Automation, Inc.)

Before any robotic system is purchased, the manufacturer should be consulted. Each robotic system has advantages and disadvantages. The two types of robotic systems available for welding are the hydraulic robot and the electric robot.

A hydraulic robot is much larger than an electric robot, and often weighs over 5000 lb. (2300kg). A hydraulic robot is capable of lifting more weight than an electric robot. Hydraulic robots usually cost more than electric robots, however.

Electric robots can be large or small. Small electric robots often weigh less than 1000 lb. (450kg); some weigh under 300 lb. (140kg). Large electric robots can weigh as much as, or more than, a hydraulic robot.

One advantage of an electric robot is that it has better accuracy in placing the weld and in following a programmed weld line. Robot accuracy is also called *repeatability*. Robot accuracy is the ability of a robot to go from one point to another point exactly. There is always some chance of error involved. The robot may not go exactly to the programmed point, but will be very close. An electric robot usually has an accuracy (repeatability) of ± 0.004" (0.10mm) or less. This means that the robot will move to within 0.004" (0.10mm) of the programmed point during every repeated cycle. Hydraulic robots are generally not as accurate as electric robots.

25.5 REVIEW OF SAFETY PRACTICES IN AUTOMATIC AND ROBOTIC WELDING

All of the safety practices discussed in previous chapters relative to GMAW, GTAW, resistance welding, cutting processes, and related topics also apply to automatic and robotic welding and cutting processes.

Be very alert when around power-driven devices and machines. **Never let any part of your body come near a moving welding torch or gun. Never enter the robot workcell or working volume when the robot is operating.** Some type of fencing or an electronic light barrier should be placed around the robot workcell to prevent entry.

Always check the electrical leads and gas hoses so they do not get caught during the operation of automatic equipment or robots. Before operating a system at full speed, the program (sequence of moves) should be executed at slow speeds. During this test run, any interference of cables or hoses can be seen and corrected.

If automatic welding equipment or a robotic system is not functioning properly, press the "emergency stop" button. Do not attempt to fix the equipment while it is operating. An automatic machine or robot has a lot more power and strength than a person.

TEST YOUR KNOWLEDGE

Write your answers on a separate sheet of paper. Do not write in this book.

1. How does automatic welding differ from manual welding or semiautomatic welding? (Refer to the definition of automatic welding.)
2. What is a welding operator required to do in automatic GMAW?
3. List two advantages of automatic welding.
4. What are flow switches used for?
5. What type of electrical current is required to operate a solenoid?
6. How is fluid or gas prevented from flowing through a solenoid-operated gas valve?
7. How is a relay switch made to close?
8. What is the advantage of using a relay?
9. What is meant by a robot's set-up point?
10. What is the greatest advantage of an electrically actuated robot over an hydraulically actuated robot?
11. What do automatic welding and cutting machines rely on to perform and to control the automatic process? List at least three items.
12. Name the two types of electric motors normally used to move a robot around its axis.
13. List the names given to the six axes of a six-axis robot.
14. What is a hydraulic actuator used for?
15. What is the working volume of a robot?
16. What is the best duty cycle rating for an industrial robot?
17. List two advantages of using a positioner.
18. What does a welding operator use to teach or program a robotic system?
19. Name the two types of robot motion systems discussed in this chapter.
20. What should be done to prevent people from entering the working volume of a robot while it is in operation?

Hardfacing is a surfacing method used on equipment that is subject to abrasive wear, such as this huge excavator bucket. The hardfacing reduces wear, greatly extending service life of the equipment. (Lincoln Electric Co.)

Chapter 26

METAL SURFACING

Surfaces wear because of abrasion, fatigue, chemical or atmospheric corrosion, and metal transfer or adhesion. Bearing surfaces, metal rock crusher surfaces, and the surfaces of earthmoving equipment are examples of machine areas that wear rapidly. If they could not be repaired, these large and expensive parts would have to be replaced.

The American Welding Society defines *surfacing* as "The application by welding, brazing, or thermal spraying of a layer of material to a surface to obtain desired properties or dimensions, as opposed to making a joint."

Cladding is a variation of surfacing that deposits or applies surfacing materials to improve corrosion- or heat-resistance. *Hardfacing* is a special form of surfacing in which surfacing material is deposited to reduce wear.

Metal parts may be surfaced by adding pure metal, alloyed metal, or nonmetallic materials such as ceramics. These metal or nonmetal materials may be applied using a powdered or solid form of surfacing material. The processes used to apply the surfacing materials are welding, brazing, or thermal spraying.

Surfacing, cladding, and hardfacing may be done using a variety of welding processes. These processes include the following:

* Oxyfuel gas welding or brazing (OFW).
* Shielded metal arc welding (SMAW).
* Flux cored arc welding (FCAW).
* Submerged arc welding (SAW).

Surfacing materials are also applied using processes called thermal spraying. *Thermal spraying* is defined by the AWS as "A group of processes in which finely divided metallic or nonmetallic surfacing materials are deposited in a molten or semimolten condition on a substrate to form a thermal spray deposit. The surfacing material may be in the form of powder, rod, cord, or wire."

Surfacing may be done using one of the following thermal spraying processes:

* Flame spraying.
* Arc spraying.
* Detonation flame spraying.
* Plasma spraying.

Today, in the face of rising equipment costs, more and more industries are having worn parts repaired using hardfacing or thermal spraying. Manufacturers are making new parts that have wear-resistant materials applied by means of hardfacing or thermal spraying to increase their service life.

26.1 PRINCIPLES OF SURFACING

Surfacing material is normally applied to metal parts. However, concrete, stone, and many synthetics have had surfacing materials applied to them by one of the thermal spraying methods. The surfacing metal may be applied to improve appearance, to improve physical strength, or to protect the base material from chemical corrosion.

By applying a layer of the correct alloy to the surface of a piece of equipment, the part may be made more resistant to corrosion by chemicals. Parts that come into contact with abrasive materials may be made more resistant to wear by hardfacing them with a more wear-resistant material. Some of the advantages of surfacing a part with a desired metal alloy or other material include:

- Certain dimensions may be maintained under adverse conditions, abrasion, corrosion, or impact shocks.
- The service life of a part may be greatly increased.
- Parts may be returned to the correct size when worn.
- Production costs may be lowered since less-expensive low-alloy materials may be used to make the part. Only the areas of high wear on such a part are coated with a more-expensive alloy required to stand up under the service conditions. Low-alloy steel surfaces that contact chemically corrosive gases or liquids may be surfaced with stainless steel to resist the corrosive action. The resulting cost per square foot is appreciably lower than if stainless steel had been used to manufacture the entire item.
- Costs also would be lowered because fewer replacement parts would need to be carried in stock, due to the increased service life of each part.

Surfacing treatments may improve the surface's resistance to corrosion, impact breaks, or abrasive wear. However, no single surface treatment will give maximum resistance to all of these types of deterioration at the same time. When consideration is given to the surfacing of a particular part, four factors should be considered:
- The nature and cause of the wear problem.
- The surfacing material needed to reduce this wearing condition.
- The surfacing process which will most economically apply the selected surfacing material.
- The proper technique to be used for depositing the surfacing material.

26.1.1 The Nature of Wear Problems

As mentioned previously, one of the causes of *wear*, or the deterioration of metals, is chemical action.. The gradual consumption of the metal may be caused by external rusting or corrosion, which are chemical reactions. It may also be caused by the *chemical corrosion* of parts, pipes, or tanks that carry certain corrosive chemicals.

It is possible to combat any wear problem by putting a layer of material on the surface to resist the factors that are causing the wear, as shown in Figure 26-1. Such materials as nickel-based, copper-based, or cobalt-based alloys, as well as a variety of tungsten carbide mixtures, are available. Lead and zinc metals may also be applied. The surfacing material used will depend on the type of wear that is expected.

Parts may wear due to constant impacts with hard materials. Road-grader blades, tool and die surfaces, power shovel buckets, and engine valve faces are examples of parts that suffer *impact wear*. Impact wear tends to change the original dimensions of a part by pounding it out of shape. Impact forces may also cause a part to break.

To reduce impact wear or breakage, a base material that has greater toughness should be used. Harder

Figure 26-1. Arc spraying being used to rebuild and hardface a brake rotor. (Miller Thermal, Inc.)

surfacing material will permit a softer inner base metal to withstand surface wear. The softer inner metal will permit the hard surface to withstand impact loads.

Another cause of wear is contact with abrasive materials. Bulldozer blades, rock-crushing rollers and housings, and rock quarry conveyors are examples of parts that suffer *abrasive wear*. Hard materials such as tungsten carbide, or ceramics such as chromium and aluminum oxide compounds, may be used to reduce abrasive wear. Soft metals such as copper, brass, bronze, or aluminum have been sprayed on steel parts to prevent abrasive wear where two surfaces come into a bearing contact.

A material used to prevent a wear problem must be chosen for its ability to prevent wear from one or more of the following causes:
- Abrasive wear – a rubbing or scraping action.
- Surface fatigue – the loss of areas of the metal surface from pitting or flaking.
- Corrosive wear – usually chemical, heat, or atmospheric attacks on the surface.
- Galling or adhesion – the transfer of material from one surface to another.

26.1.2 Hardness Determination

Various characteristics of metals and the importance of each characteristic are explained in Chapter 28.

Hardness is the metal property which is given the greatest degree of attention when considering surfacing applications. *Hardness* is defined as "the property of a material to resist indentation or the action of cutting tools or cutting materials." It is important to know the hardness of the surfacing material to make sure that it meets the requirements of a particular job. The three most common standards for measuring hardness are the Brinell scale, the Rockwell scale, and the Shore Scleroscope scale. Test procedures for these measurements are given in Chapter 30.

If testing machines are not available, a reasonably accurate test for hardness may be conducted by using a mill file. Follow these guidelines:

- If metal is removed easily, the hardness is about 100 Brinell or 60 Rockwell B. An example of this hardness would be low-carbon steel.
- If metal is readily cut with moderate pressure exerted, the hardness is about 200 Brinell or 15 Rockwell C. Medium carbon steel is an example of this hardness.
- If metal is difficult (though possible) to cut, the hardness is approximately 300 Brinell or 30 Rockwell C. An example is high-alloy steel.
- If metal is cut with great pressure only, the hardness is probably about 400 Brinell or 40 Rockwell C. Tool steel has this hardness.
- If metal is nearly impossible to cut, the hardness is about 500 Brinell or 50 Rockwell C. An example of this hardness is tool steel.
- If metal cannot be cut, the hardness is about 600 Brinell or 60 Rockwell C. Hardened tool steel has this hardness.

26.2 SELECTION OF A SURFACING PROCESS

The oxyfuel gas process for depositing hardfacing material has many desirable features:

- The oxyfuel gas process may be used where a surface is to be applied which will require a minimum of final surface finishing.
- Carbon may be added to the surface by using a carburizing or reducing flame. Added carbon will improve the abrasion-resistance of the surface.
- Preheating and postheating can be carefully controlled using the oxyfuel gas flame to prevent cracking on the surfacing material.
- Nonferrous metals are more easily applied by this method.
- Oxyfuel gas works well in fusing hardfacing materials.

When a flawless surface deposit is required, thermal spraying processes are preferred. However, the metal being surfaced must be very clean. This requires additional equipment and labor costs. Because of the high quality of the deposit, thermal spraying processes are being used more and more in the aircraft and aerospace fields for surfacing new parts subjected to some form of extreme wear. One application for surfacing is in building up larger dies that become broken or require changes.

Semiautomatic continuously fed wire processes are often used during production to add a desired surface material to a new part. Large volumes of surfacing material can be deposited in a short period of time using the FCAW or SAW processes. These processes have many advantages:

- The equipment is portable and affordable.
- Surfacing materials may be applied easily and quickly with these processes. FCAW may be used to apply surfacing materials in any position. SAW is best used in the flat welding position.
- Great varieties of surfacing materials may be applied because of the large variety of electrodes available.
- Austenitic stainless steels, austenitic manganese steels, and nickel-chromium ferritic steels may be surfaced using these processes.

Automatic and *mechanized surfacing* processes normally give the highest quality of surfacing deposits. These processes are not as portable as the semiautomatic processes, and are best done in the flat position. The advantages of speed and accuracy of the deposit are best utilized in the hardfacing of long straight or curved surfaces in new production. The automatic and mechanized processes work best for repairs when parts can be dismantled and brought to the automatic machinery.

Thermal spraying may be used to deposit almost all virgin metals and many alloys to other metal surfaces. The deposit is usually thin, and the thickness may be held to close tolerances. The surfaces being treated must be cleaned and slightly roughened. Sandblasting may be used for this purpose. In some instances, multiple passes may be made to build up a surface with hardfacing material without danger of cracking the previous layers.

Virtually any material may be bonded to any other with the plasma arc process. Ceramics may be sprayed on the surfaces of metals to increase their resistance to corrosion from heat and/or chemicals. Stainless steel may be sprayed on low-alloy steel to increase resistance to corrosion and abrasive wear.

The electric arc spraying process is the most economical thermal spraying process. It is capable of depositing a large variety of surfacings with high efficiency.

26.2.1 Surfacing Using the Oxyfuel Gas Process

To obtain the best results, the surface should be cleaned and the metal preheated to eliminate warpage. Both thermal spraying and hardfacing (using a solid rod), may be done with the oxyfuel gas process. When surfacing, use a tip that is one to two sizes larger than used with the same-diameter rod when welding. A slightly reducing (carburizing) flame is preferred. Any carbon added to the surface will aid in the hardfacing process. The reducing flame will reduce the oxides on the surface of the work. Refer to Figure 4-2. The angles for the torch and rod are the same as for welding, as shown in Figure 26-2.

When surfacing steel, the metal should be brought up to approximately 2200°F (1204°C) before the rod is touched to the work. This step is shown in Figure 26-3. The preheat temperature is important. It is best to use a temperature-indicating crayon or some other means for determining the temperature. The process of surfacing is very similar to brazing as it applies to the welder's techniques and manipulation. Figure 26-4 shows steel being surfaced using the

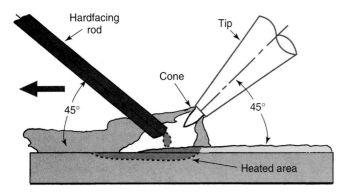

Figure 26-2. *Surfacing material being applied with the use of an oxyfuel gas torch. The outer flame area preheats the base metal in advance of the deposited surfacing material. The base metal melts the surfacing rod, as in brazing. A carburizing flame is used when hardfacing.*

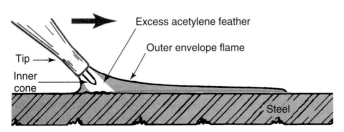

Figure 26-3. *Flame and tip position to use when preheating a metal prior to hardfacing. (Haynes Stellite Co.)*

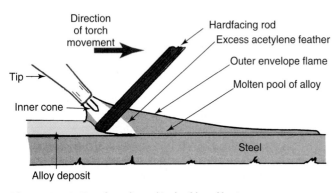

Figure 26-4. *Forehand method of hardfacing.*

forehand process. Figure 26-5 shows the backhand technique. The techniques used for holding the torch and the surfacing rod are shown in Figures 26-6 and 26-7.

When surfacing cast iron, the surfacing material will not flow easily. To produce a good job, you must break through the surface crust of the iron by rubbing the heated surface with the rod.

An oxyfuel gas torch may be fitted with a hopper assembly, as shown in Figure 26-8, to do flame spraying. A variety of surface hardening and cladding materials are available for use with this flame spraying torch combination. Surfacing or cladding powder is placed into the hopper on the torch. When the base metal is preheated to the proper temperature, the lever on the torch is pressed to

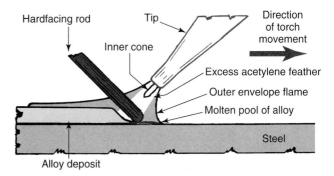

Figure 26-5. *Backhand method of hardfacing. Note that the torch points in the direction opposite its forward movement.*

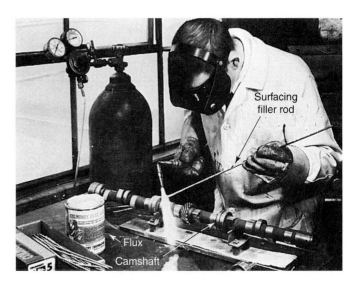

Figure 26-6. *Hardfacing the wear areas of an automobile engine camshaft. (Wall Colmonoy Corp.)*

Figure 26-7. *Helix-type conveyer being rebuilt using a hardfacing rod and oxyacetylene flame. (Wall Colmonoy Corp.)*

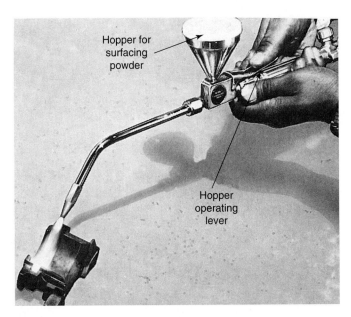

Figure 26-8. *An oxyfuel gas torch equipped with a hopper to hold surfacing material. A silicon-boron-nickel powder is being used to repair a mold for a glass container. (Wall Colmonoy Corp.)*

release the powder in the hopper. The surfacing powder is released into the torch tube and melted in the flame at the torch tip. The molten surfacing or cladding powder is transferred to the base metal, where it fuses with the surface.

26.2.2 Surfacing Using the Manual Shielded Metal Arc Process

Direct current is normally used with electrode negative (DCEN) polarity. A higher current setting is used than is normal for the electrode diameter. The high current setting will permit a long arc, which will preheat the metal ahead of the arc. This will keep the surfacing material molten long enough for the impurities to float to the surface. An angle of about 45° is correct for the electrode. A long arc is necessary to spread the heat over a large area. This will prevent deep penetration, localized heating, and warpage.

Preheating is not necessary for all metals before applying surfacing materials, but it is advisable with alloy steels. When in doubt about preheating, the following guide rules may be helpful:

- Except for heavy sections, low- and mild-carbon steels do not require preheating.
- Preheat high-carbon and medium-carbon high-alloy steels.
- Steel parts with 12% to 14% manganese should be preheated to 200°F (93°C) to relieve stresses. However, the part temperature should never exceed 500°F (260°C).
- Cast iron should be preheated to 500°F to 700°F (260°C to 371°C) at a slow and uniform rate.

- When in doubt, preheat to 500°F to 700°F (260°C to 371°C), but be certain the part being preheated is not a manganese steel.

If narrow beads are desired, an oscillating motion should be used in the direction of travel with no sideways motion. A good technique is to lay a bead about 1" (25mm) long and then start the next stroke of the same length 3/4" (20mm) back on the first stroke. Extend it 1/4" (6mm) past the first stroke. The third stroke should start 3/4" (20mm) back on the second stroke and extend 1/4" (6mm) beyond the second stroke. This procedure is continuously repeated until the desired length bead is laid.

Wider beads require a circular motion. A circular motion is started at the center and continued in spirals until the desired width of bead is obtained. The operator then continues to make circular motions and extend each

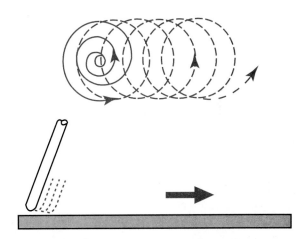

Figure 26-9. *Circular motion being used to obtain bead widths from 3/4" to 1 1/4" (20mm to 30mm).*

circle about 1/4" (6mm) beyond the last circle in the desired direction. Each time a circular motion is made, the weld moves forward 1/4" (6mm), as shown in Figure 26-9.

The recommended height of a deposit is 1/16" to 1/4" (1.5mm to 6mm), and the recommended bead width is 3/4" to 1 1/4" (20mm to 30mm). When laying a second bead, the second bead should overlap one-fourth to one-third the width of the first bead. This process will ensure a fairly uniform surface to the deposited metal, as shown in Figure 26-10.

If more than one layer of surfacing material is required, the beads of the first layer should be cleaned before depositing the second layer. This will minimize slag inclusion. Each bead should be made in the same manner as described above for laying beads from 3/4" to 1 1/4" (20mm to 30mm) wide.

An interesting technique for depositing hardfacing beads is to lay the beads in a basketweave pattern. This leaves low spots between the beads. If the part being hardfaced is used in abrasive material such as sand, the abrasive material will build up in the low spots between the beads and act as an added protection for the base metal, as shown in Figure 26-11.

It should be noted here that small beads made with small-diameter rods cool fastest and mix the least with the

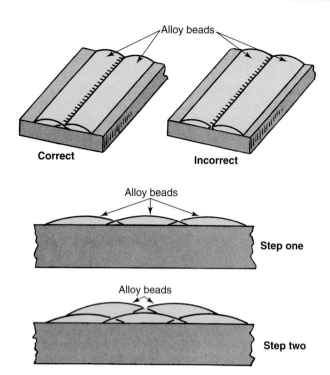

Figure 26-10. *The proper way to overlap adjacent surfacing beads.*

Figure 26-11. *Examples of hardfacing bead patterns used on earthmoving equipment. A—A basket weave pattern used on a tooth adapter. B—Two different patterns used on a ripper tooth. (Stoody Co.)*

base metal. Wide beads made with high current and large electrodes cool slowly and mix with the base metal to a greater extent.

Before the rod diameter, current setting, and surfacing procedures are decided, the welder must determine:
- How wide the bead should be.
- How fast the bead should cool.
- How harmful the mixing of the surfacing material and base metal will be to the finished job.

In many situations, better results may be obtained by using superimposed layers of different hardfacing alloys. Figure 26-12 shows a cutting tool that has had hardfacing material applied.

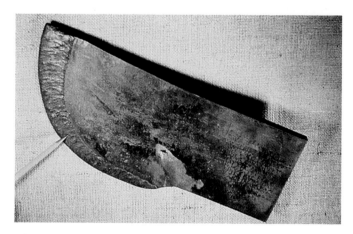

Figure 26-12. *The cutting edge of this bean cutter has been hardfaced to increase its useful life. (Stoody Co.)*

26.2.3 Surfacing Using the Flux Cored Arc Welding Process

Flux cored arc welding equipment may be used to hardface or clad metals. The hollow electrode wire used for these applications will have the desired powders in its core to add the required characteristics to the metal's surface.

The procedure used to apply surfacing material is the same as used for welding with this process. A weaving motion is required if the area to be surfaced is large. Each bead and layer should be overlapped, as shown in Figure 26-10.

26.2.4 Surfacing Using the Submerged Arc Welding Process

The equipment used when surfacing with the submerged arc welding process is the same as used when welding with this process. The electrode wire is chosen for the surface characteristics desired. As an example, plain carbon steel may be surfaced using stainless steel electrode wire. This will give a corrosion-resistant surface to the low-cost plain carbon steel part. A steel part may be surfaced using a brass or bronze electrode to provide a better bearing surface to the part. Figure 26-13 shows submerged arc welding being used to add a surfacing material to a part.

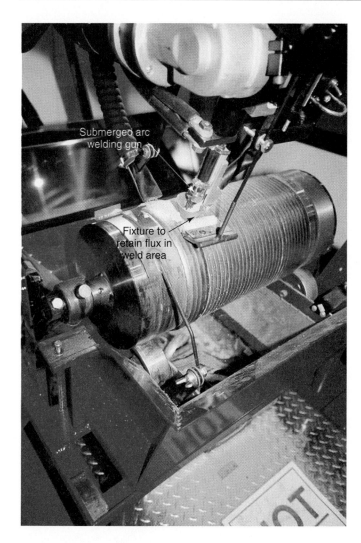

Figure 26-13. *SAW being used to surface a worn steel roller. Note the fixture used to hold the flux around the arc area while the drum is rotated.*

26.3 PRINCIPLES OF FLAME SPRAYING

The *flame spraying process* may be used to spray pure or alloyed metals onto a part or assembly that requires a surfacing buildup. Flame spraying is accomplished by melting surfacing or cladding materials in an oxyfuel gas flame. These molten materials are then transferred to the surface of the part being surfaced. The surfacing or coating material may be in the form of wire, rod, or powder. The surfacing materials may be ceramics, hardening metals and alloys, or soft metals like brass or bronze. The fuel gas used in the oxyfuel gas process may be acetylene, propane, natural gas, hydrogen, or methylacetylene-propadiene (MAPP). The surfacing material may be used "as sprayed." It may also be fused to the base metal by reheating with an oxyfuel gas torch, or by using electric induction heating coils.

Worn or inaccurately machined parts may be repaired by having their surfaces built up to acceptable dimensional limits by one of the flame spraying processes. This process works well with most inorganic materials (those not containing carbon) that melt without decomposition. Flame-sprayed surfacing materials, for example, can be used to build up scored surfaces on engine crankshaft journals, so they can be machined to a diameter that will receive a standard main bearing replacement.

Wire flame spraying may use any material that can be formed into a wire shape. The wire is usually a metal or a metal alloy. Ceramic material may also be formed into wire and flame sprayed. Powders may be placed inside hollow metal wire or thin ceramic tubes.

Attaching a wire feed flame spraying torch to a lathe and its crossfeed mechanism will ensure an even layer of surfacing material on a cylindrical part. See Figure 26-14. In original production, surfaces occasionally are flame sprayed with a harder alloy material to reduce wear on those surfaces.

Surfacing material may also be in a powdered form. Figure 26-15 shows a part being flame sprayed with a torch that is supplied with a high-pressure carrier gas to feed the powder to the torch. **Oxygen cannot be used as the carrier gas. Oxygen, in combination with a fine powder, can create an explosive mixture.** Another method used to supply the powdered metal for flame spraying is shown in Figure 26-16. The powder is fed to the flame by gravity from the hopper mounted on top of the torch. Powder spraying may be done with any material that melts below the flame temperature of about 4000°F to 5500°F (2205°C to 3038°C). Some close-fitting parts of jet engines, for instance, require airtight seals and operate at extremely high temperatures. The contacting surfaces of these parts are metal sprayed with a soft metal. As the rotating parts expand and contract, the soft metals contract and conform to the rotating parts to produce the required airtight seal.

In production brazing of certain alloys, the flow of brazing material between the parts to be joined (capillary flow) is difficult to obtain. To ensure that the brazing material will cover the entire surface to be joined, the surfaces may be first flame sprayed with brazing metal. They are assembled, then the assembly is heated in an oven to complete the brazing operation.

Flame spraying may also be used on a surface to improve its resistance to wear or corrosion. The surface produced by flame spraying normally requires very little finishing prior to use. Flame spraying is often used to refinish worn shaft bearing journals.

An interesting application of flame spraying is to use layers of blended metals. This blending allows a material that normally would not adhere to another material to be applied successfully. One material could be applied to the other in several graded layers. The beginning layer would have little of the final surfacing material in the powder mixture. As following layers are applied, the content of the final surfacing material is increased. The pure material is applied in the topmost layers.

26.3.1 Flame Spraying Torch

The flame spraying torch must feed oxygen, fuel gas, a high-pressure carrier gas (not oxygen), and the surfacing wire or powder to the torch nozzle. Each must be provided

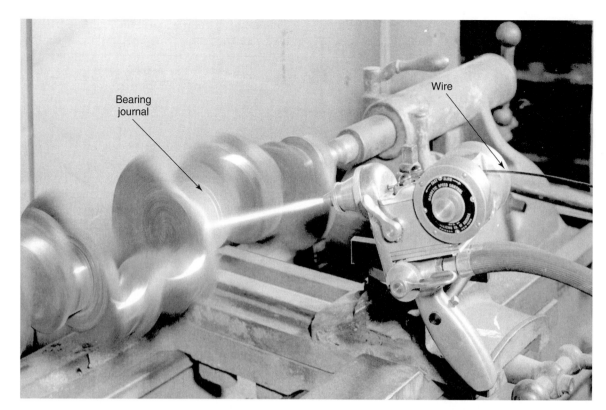

Figure 26-14. A flame spraying torch with a built-in wire feed mechanism being used to surface a crankshaft journal. The torch is mounted on the lathe crossfeed for better spray control. (Metallizing Co. of America, Inc.)

Figure 26-15. Metal surfacing a part as it revolves in a lathe. Metal for spraying is fed to the torch in a powder form. A carrier gas feeds the powder metal to the torch. This flame spraying process is followed up with an oxyacetylene-flame fusing operation. (Wall Colmonoy Corp.)

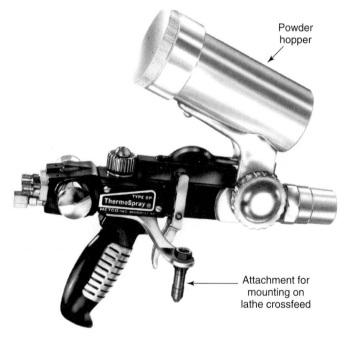

Figure 26-16. Powder is gravity fed from a hopper to this flame spraying torch. The torch is equipped with an electric vibrator to improve the powder flow. (Metco, Inc.)

in the correct amount, so that the flame spraying process will be successful.

Figure 26-17 shows a drawing of the wire-feed type flame spraying torch. The oxyfuel gas flame at the nozzle melts the surfacing material. The carrier gas pressure propels the molten powder or metal to the part being surfaced. The wire feed mechanism on the torch in Figure 26-18 is driven by an air turbine motor. Figure 26-19 shows the locations and names of the essential parts of a wire-feed flame spraying torch.

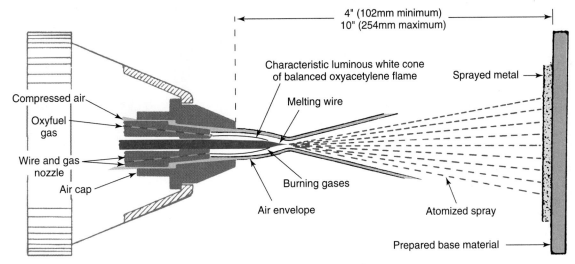

Figure 26-17. *A schematic cross section of a wire-feed flame spraying torch.*

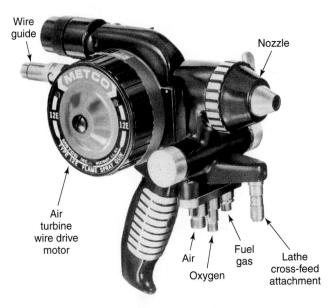

Figure 26-18. *A flame spraying torch. An air turbine wire drive motor is built into this torch. (Metco, Inc.)*

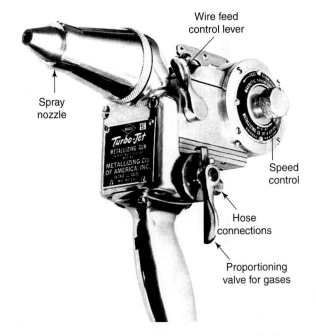

Figure 26-19. *A flame spraying torch with wire feed. (Metallizing Co. of America, Inc.)*

The high-temperature oxyfuel gas flame completely surrounds the wire as it is fed through the torch nozzle. The pressurized carrier gas atomizes the molten wire material and sprays it on the surface to be coated.

Figure 26-20 is a drawing illustrating the operation of a powder-feed flame spraying torch. The surfacing material may be fed to the torch nozzle by gravity, as shown in Figure 26-21. Figure 26-22 illustrates a flame spraying torch that uses a pressurized carrier gas to carry the surfacing powder to the torch. The carrier gas also propels the molten powder onto the material being surfaced.

26.3.2 Flame Spraying Process

Flame spraying is a process in which metals, alloys, or ceramic materials in wire or powdered form are fed through a special torch (spray gun). They are melted in an oxyfuel gas flame, then atomized by high-pressure gases. These gases carry the atomized particles to the cleaned and prepared surface. Figure 26-23 illustrates a flame spraying outfit used to spray metal in a wire form. The equipment needed for wire-feed flame spraying includes:

- Air compressor. The unit must be able to provide at least 30 ft³/min. (850 L/min.) of free air.
- Compressed-air drying unit. Any amount of moisture will interfere with the metallizing bond. If there is an excess amount of moisture in the air supply, it is generally advisable to use an air drying unit.
- Air receiver tank. This unit smoothes out the flow of compressed air to the compressed air

regulator by compensating for compressor pumping pulsations.

- Fuel gas and oxygen pressure regulators. Two-stage regulators are recommended.
- Fuel gas and oxygen cylinders.
- Gas flowmeters. Flowmeters keep a constant volume of gas flowing to the gun nozzle.
- Compressed air control. A single-stage regulator may be used to control the air pressure. Air filters are normally used to further purify the air. If the air supply is used to provide air to an operator's mask, an air filter is required.
- Wire feed control. This unit will straighten the wire as it comes from the coil and feed it at a uniform rate to the gun nozzle.
- Flame spraying gun. The flame spraying gun mixes the fuel gas, oxygen, and the desired powder or alloy wire. Some guns contain mechanisms to draw the wire from the wire coils.

Figure 26-24 illustrates a combination flow control for oxygen, fuel gas, and powder. The hopper on the control unit holds the powder.

The width of the spray cone varies with the wire diameter, the type of metal being sprayed, and the spraying speed (wire feed rate). The visible spray cone is considerably larger than the actual cone of spray metal, due to the sparks that surround the actual cone. About 95% of the metal being sprayed is in a smaller inner cone that is not visible when spraying, as indicated in Figure 26-25.

Different types of nozzles are used for various fuel gases; these should not be interchanged. If a propane nozzle with a cupped end is used with acetylene gas, backfiring may occur. A complete automated flame spraying installation is shown in operation in Figure 26-26. A solid ceramic rod is used with this torch to spray a ceramic surfacing material onto a part.

Another process used to flame spray surfaces is powdered surfacing material carried to the work by an inert gas, compressed air, or gravity. The powder in the pressure-fed units is carried to the torch by a third hose from a hopper where the powdered material is stored under pres-

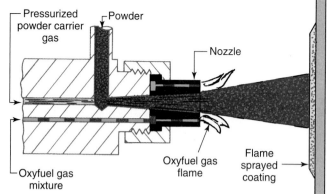

Figure 26-20. This schematic shows operation of a flame spraying torch that uses powdered surfacing material. A carrier gas (not oxygen) is used to move the powder to the torch. The same gas propels the molten powder to the surface being coated.

Figure 26-21. A gravity feed flame spraying torch in operation.

Figure 26-22. This flame spraying torch is designed for use on a lathe.

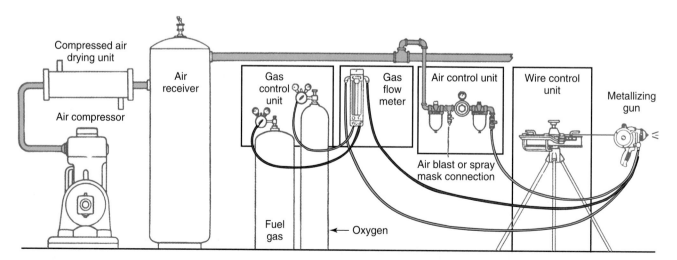

Figure 26-23. A schematic of the equipment used in a typical wire feed flame spraying outfit. (Metco, Inc.)

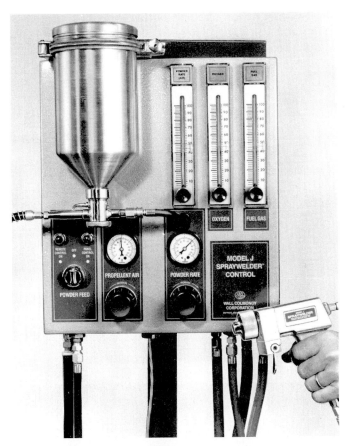

Figure 26-24. *A combination flow control for oxygen, fuel gas, and powder. The powder is stored in the hopper. (Wall Colmonoy Corp.)*

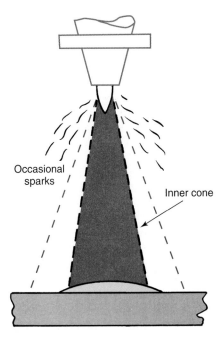

Figure 26-25. *As shown in this drawing, approximately 95% of the sprayed metal is in the inner cone, and is deposited on the surface being sprayed.*

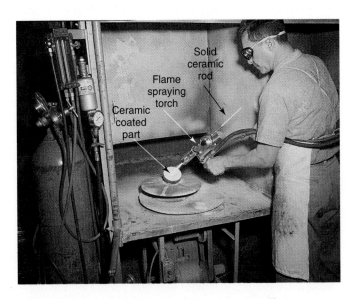

Figure 26-26. *A complete flame spraying installation. A solid ceramic rod is being used in this torch to spray ceramic surfacing on a small part. (Metallizing Co. of America, Inc.)*

sure. When this torch is used with powdered ceramics, an electric vibrator helps the ceramic powder to flow. One advantage of using powdered material for flame spraying is that flux may be mixed with the surfacing material. Flux will help the surfacing to adhere better.

With powders, the spraying rate is critical. If the spray rate is too slow, the small particles stay in the flame chamber too long and may vaporize. If the spray rate is too fast, the material may not melt completely. If this happens, it will not properly bond with the surface or other particles. The material being surfaced is sometimes preheated by using an oxyacetylene flame to improve the bonding of the surfacing material with the surface.

The spray must be applied in smooth even layers. The gun is held perpendicular to the surface being sprayed. Material layers about 0.001" to 0.002" (0.025mm to 0.050mm) are sprayed. Layer after layer may be sprayed until the total desired thickness of surfacing material is obtained.

The spray gun-to-surface distance should be about 4" to 10" (100mm to 250mm). This distance will depend on the process, surfacing material, and rate of deposit used. In the case of thin parts or small-diameter shafts, parts may have to be cooled by an air stream to prevent overheating.

26.3.3 Starting and Adjusting the Flame Spraying Station

The safety precautions recommended for use with all gas welding equipment apply also to the use of the flame spraying torch. Review safety precautions in Heading 6.14. Check all hoses, connections, and gauges for leaks before using the equipment. The procedures for adjusting and lighting the wire feed type torch are as follows:

1. Open the air valve and adjust the air pressure regulator to the correct pressure. Refer to the torch manufacturer's data sheets for recommended wire size, air cap size, air pressure, oxygen and fuel gas pressures, and deposit or spraying speeds.
2. Close the air valve.
3. With the drive roll knob loose, insert the wire into the rear wire guide and through the gun to the nozzle.
4. Tighten the drive roll knob until the wire begins to feed.
5. Adjust the wire feed speed.
    If equipped with an air turbine wire drive:
    a. Adjust the speed control ring that controls the speed of the turbine in the air motor. When the speed control ring is moved clockwise, the wire will speed up. When the control ring is turned counterclockwise, the wire will slow down.
    If equipped with an electric motor wire drive:
    b. Adjust the motor speed controls to feed the wire at the recommended rate.
6. Open the valve handle to the run position and adjust the fuel gas and oxygen pressures to the values specified by the torch manufacturer. Close the gas proportioning valve handle.

Before actually beginning to spray, set the wire feed rate accurately on a test piece. If the wire feeds too fast, its tip will extend beyond the hottest part of the flame. If this occurs, the surfacing material will leave the torch in large droplets. This will form a coarse, bumpy surface on the part being sprayed. If the wire is fed too slowly, the surfacing material will oxidize more easily. The sprayed surface will be highly oxidized and weak.

26.4 ELECTRIC ARC SPRAY SURFACING

Another process used to spray materials is the *electric arc spray method*. This method consists of maintaining an arc between two current-carrying wires. The wires are automatically fed to the arc position in the spray gun by electric drive mechanisms similar to those used in gas metal arc welding. The wire material melted by the arc is then sprayed onto the part to be surfaced. An air jet or an inert gas jet directed across the arc propels the molten wire to the material being surfaced, as shown in Figure 26-27. Figure 26-28 illustrates a commercial arc spraying torch. See also Figure 26-29. Figure 26-30 shows arc spraying being used to surface a part.

It is claimed that production rates can be tripled using the arc spraying process. Arc spray guns can spray between 3 lbs. and 200 lbs. (1.4kg to 90kg) of wire per hour. Different materials may be used in each of the wires fed to the arc. Using two different wires allows an alloying of materials at the torch. This alloy or material mixture is then sprayed onto the part to be surfaced.

Because of the low cost of electricity as an energy source, arc spraying is more economical than any other surfacing process. Figure 26-31 is a table that compares the cost of energy for various thermal spraying processes. **Industrial ear muffs, goggles, and protective clothing should be worn when using this process.**

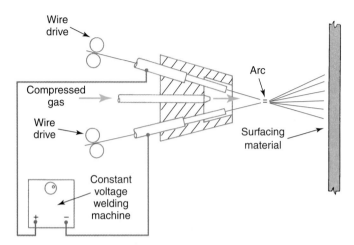

Figure 26-27. A schematic of the electric arc spray method of surfacing a part.

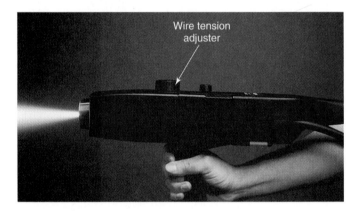

Figure 26-28. An arc spraying torch. Tension on the surfacing material wire is adjusted with the knob on top of the torch. See Figure 26-29 also. (TAFA, Inc.)

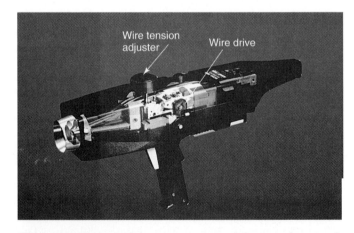

Figure 26-29. A partial cutaway view of the arc spray torch shown in Figure 26-28. (TAFA, Inc.)

Figure 26-30. A cylindrical part being surfaced using arc spraying. Note that the areas that are not to be surfaced are covered with a cream-colored protective material. *(Miller Thermal, Inc.)*

Powder flame spray:		
Oxygen	95 ft³/hr. x $0.074/ft³	= $7.03/hr.
Acetylene	59 ft³/hr. x $0.213/ft³	= $12.57/hr.
	Total = $19.60/hr.	
Wire flame spray, acetylene fuel:		
Oxygen	90 ft³/hr. x $0.074/ft³	= $6.66/hr.
Acetylene	32 ft³/hr. x $0.213/ft³	= $6.82/hr.
	Total = $13.48/hr.	
Wire flame spray, propane fuel:		
Oxygen	185 ft³/hr. x $0.074/ft³	= $13.69/hr.
Propane	30 ft³/hr. x $0.042/ft³	= $1.26/hr.
Air	2100 ft³/hr. x $0.00028/ft³	= $0.59/hr.
	Total = $14.95/hr.	
Rod flame spray:		
Oxygen	194 ft³/hr. x $0.074/ft³	= $14.36/hr.
Acetylene	57 ft³/hr. x $0.213/ft³	= $12.14/hr.
Air	2100 ft³/hr. x $0.00028/ft³	= $0.59/hr.
	Total = $27.09/hr.	
Electric arc spraying:		
Electricity	3kWh x $0.0824/kWh	= $0.25/hr.
Air	2100 ft³/hr. x $0.00028/ft³	= $0.59/hr.
	Total = $0.84/hr.	
Plasma arc spray:		
Argon	75 ft³/hr. x $0.20/ft³	= $15.00/hr.
Electricity	40 kWh x $0.0824/kWh	= $3.30/hr.
	Total = $18.30/hr.	

Figure 26-31. A comparison of the energy costs for various spray type surfacing processes, based on one hour of operation at typical spray rates. Note the cost of arc spraying compared to the other methods.

26.5 DETONATION SPRAYING

Detonation flame spraying is done with a gun that resembles a small cannon. See Figure 26-32. Air or oxygen, a fuel gas, and a charge of surfacing materials mix in the combustion chamber of the *detonation flame spraying gun*. An ignition system ignites the mixture, setting off the charge several times per second. The surfacing material in a molten condition is propelled from the end of the gun at 2500'/sec (762 m/sec). Nitrogen is sometimes used to purge (clean) the chamber after each explosion. Temperatures within the detonation spray gun reach 6000°F (3315°C).

Very high bond strength and surfacing material densities are obtained with this process. **Industrial-type ear muffs must be worn when using this process. Goggles and other protective clothing should be worn to protect the operator from sparks and ultraviolet burns.**

26.6 PLASMA ARC SPRAYING PROCESS

The plasma arc or jet is the result of forcing a gas, such as nitrogen, hydrogen, or argon into an enclosed electric arc. The gas is heated to such a high temperature by the electric arc that its molecules become ionized atoms or a plasma containing a great deal of energy. *Plasma* is considered by physicists to be a fourth state of matter (it is neither gas, nor liquid, nor solid). Chapter 20 explains the plasma arc process in detail.

Temperatures of up to 30,000°F (16,650°C) may be obtained with plasma arc equipment. This process can be very successfully used for surfacing by spraying. Typically, any inorganic material that will not decompose may be sprayed. Examples of materials that have been applied to a base metal are: ferrous metals, ceramics, tungsten, tungsten carbides, tantalum, zirconium diboride, platinum, columbium, hafnium, and vanadium carbides.

Densities of surfacing materials applied by the plasma arc process are up to 98% of the theoretical density. Pure tungsten and tungsten carbides may also be applied to almost any base material with up to 95% densities.

A schematic of a typical plasma arc torch is shown in Figure 26-33. The principle of the process is explained in the following paragraphs.

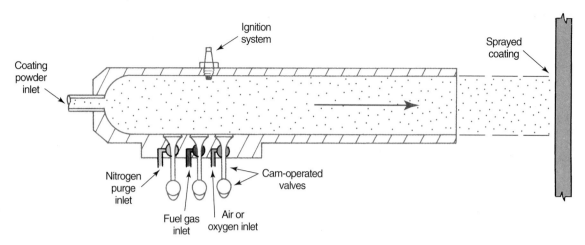

Figure 26-32. *A drawing of a detonation spray gun in use. The cam-operated valves allow nitrogen for purging and oxygen and fuel gas for combustion to enter the chamber several times per second. The powdered surfacing material enters at the back of the gun.*

Using direct current, an arc is formed between the internal electrode and the nozzle. The fuel gas is ionized within the nozzle and becomes plasma. This is known as a nontransferred plasma arc. The material to be sprayed is carried to the arc area in the nozzle. The surfacing material is melted and atomized by the heat of the plasma. The atomized surfacing material is then carried to the surface to be sprayed by the high velocity of the plasma jet. Speeds as high as 20,000′/sec. (6100 m/sec) may be reached with the plasma jet. Normally, much lower speeds are required. A plasma arc spraying torch is shown in Figure 26-34 applying material to a weldment.

Nitrogen with from 5% to 10% hydrogen gas is generally used for the plasma gas. Pure nitrogen is used as the

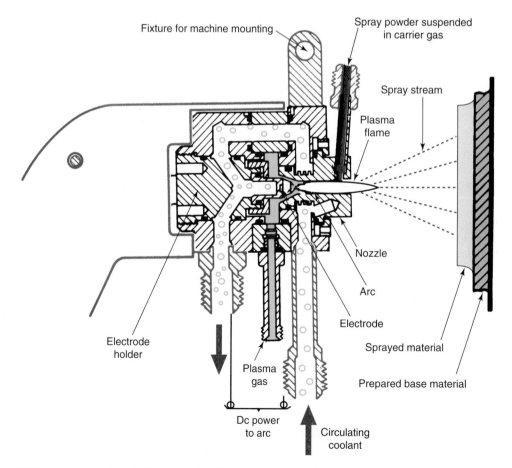

Figure 26-33. *A cross-sectional drawing of a plasma arc material spraying torch. Note the unusual electrode design and the powder and cooling water flow.*

carrier gas to propel the powdered surfacing material to the melting zone within the nozzle. The entire plasma jet nozzle is usually water-cooled. Distilled water is often used in the cooling circuit to prevent mineral deposits and thereby lengthen the life of the equipment.

A complete plasma arc station would include:
- A plasma jet spray gun.
- A control unit with an on-off switch, starter arc controls, gas flow controls, ammeter, and voltmeter.

Figure 26-34. *A plasma arc material spraying torch applying a surfacing material to a weldment. (Plasma-Giken)*

- A heat exchanger in which water is used to cool the distilled water circulating in the gun. Water flow controls are also included in this unit.
- A powder control unit to control the pressure and quantity of the powder fed to the gun.
- A power supply unit. The current supplied to the electrodes is controlled and stabilized by this unit.
- Gas and water hoses and electrical cables.
- Safety equipment. **This includes a face shield, heat-resistant and reflecting clothing, gloves, and ear plugs or muffs to protect the operator's ears from the intense noise caused by the plasma jet.**
- A water wash spray booth may be used to protect other workers from overspray of both flame and material.

An interesting application of the plasma arc material spraying process is used when making aerospace parts, such as rocket nozzles. The parts must stand up to extremes of temperature and erosion wear due to the high velocities and temperatures to which they are exposed. Rocket nozzles are made from zirconium, platinum, or other rare and expensive metals. These materials are too rare, in fact, to be *machined* to shape—the machining chips would have to be salvaged and remelted.

In the plasma arc spraying process, the desired metal is sprayed onto an aluminum or brass form (mandrel) to the desired thickness. The mandrel or form is then removed from the shell of rare metal by dissolving it with acid; only the rare metal part remains.

26.7 SURFACE PREPARATION

The bonding of the surfacing materials relies on mechanical interlocking of the cooled sprayed particles. Figure 26-35 shows how an arc- or flame-sprayed surface might appear under a powerful microscope.

Surface preparation is essential to any thermal spray surfacing application. Before the surfacing material is applied, the surface must be cleaned thoroughly. Nonporous surfaces may be cleaned by steam, vapor degreasing, hot detergent washing, or industrial solvents. Porous materials such as cast iron must be baked at 400ºF to 600ºF (200ºC to 320ºC) to vaporize any grease or oil that is in the pores.

Abrasive cleaning methods, such as grit blasting, may be used to remove mill scale. White metals such as aluminum and magnesium are cleaned using aluminum oxide or quartz, as shown in Figure 26-36. Figure 26-37

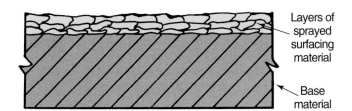

Layers of sprayed surfacing material

Base material

Figure 26-35. *A cross section of arc- or flame- sprayed surfacing. The sprayed surfacing material particles interlock and bond to the surface and to each other as they cool.*

shows a comparison of the bond strengths of materials to the cleaning methods used.

Machine cutting is used to remove contaminated (dirty) material. Material is also removed to provide for the thickness of the new surfacing material. On cylindrical parts, like shafts and pistons, the parts are reduced in diameter on a lathe. Material is removed to level out a worn surface and to provide for the thickness of the spray surfacing. Cylindrical parts are often rough threaded to a depth of 0.003" to 0.006" (0.08mm to 0.15mm) after they are straight-turned. Surfacing materials generally adhere or bond better to cylindrical parts that have been rough threaded to the shallow depths stated above. See Figure 26-38.

Surfaces to be coated are often preheated to drive off moisture. Any moisture on the surface will interfere with the sprayed-material bonding process. Preheating at 200ºF

Figure 26-36. Cleaning a gear by grit blasting. A special enclosed chamber is used.

Material	Deposition method	Bond-strength lb/in.^{2}	Surface* preparation
Molybdenum	Flame	5,500	Grit blast
9 Al-bronze	Electric arc	6,000	Clean only
9 Al-bronze	Electric arc	6,700	Grit blast
95 Ni-5 Al	Electric arc	9,100	Clean only
95 Ni-5 Al	Electric arc	9,700	Grit blast
95 Ni-5 Al	Plasma	9,500	Clean only
Aluminum	Flame	1,400	Grit blast
Aluminum	Electric arc	4,900	Grit blast

*Special bond coats are sometimes needed.

Figure 26-37. How bond strengths compare for different materials and deposition methods. Note the beneficial effect of grit blasting. (TAFA, Inc.)

to 250°F (93°C to 121°C) is helpful when flame spraying materials.

Some sprayed materials will not readily bond with some surfaces. To ensure good bonds for such materials, a bond coat is applied before the final surfacing layer is sprayed. The *bond coat* is an intermediate surfacing between the base material and the finished surfacing. Materials such as molybdenum, 9% aluminum-91% bronze, and 95% nickel-5% aluminum, are used as bond coats under other sprayed materials.

26.8 SELECTING THERMAL SPRAY SURFACINGS

When selecting the best material to use for surfacing, the following variables must be considered:
- The material being surfaced. Is the material metal, stone, ceramic, or plastic?
- Is the metal hard or soft?

10X magnification

Figure 26-38. Appearance of steel stock after machine cutting the surface prior to flame or arc spraying. The 10X magnification shows the roughness needed for good spraying results.

- Is the material being surfaced porous or nonporous?
- Is the finished surface to be hard or soft?
- Is the surfacing material to be in the wire or powder form?
- Should the surfacing material resist corrosion, wear, fatigue, or galling?

Figure 26-39 is a useful guide for the selection of thermal spray surfacing materials.

26.9 TESTING AND INSPECTING THERMAL SURFACINGS

Completed thermal sprayed surfaces are generally inspected using visual methods. The operator should inspect each layer of material before continuing. He or she should look for uniformly applied, fine- to medium-sized granules. There should be no blisters, cracks, chips, or loose particles. No oil or other contaminants should be visible on the surface. If defects are visible, they must be removed and the surfacing material applied again. Defects are generally removed by grit blasting.

The strength of the surfacing material applied may be tested by scratching two lines on the final layer of surfacing material. The distance between the two lines is ten times the thickness of the applied surfacing material. The amount of surfacing material that flakes off between the two lines is compared with other samples. The more surfacing material that comes off, the weaker the bond strength.

The 90° bend test may also be used to test the bond strength. Surfacing materials that flake off indicate a weak or poor bond. In this test, it is acceptable if the surfacing materials crack or craze, since they generally are hard and have little flexibility.

Dye penetrants and magnetic particle inspections may be used to reveal cracks and porosity. Eddy current instruments can be used in production to measure the thickness of the surfacing materials. This test is used to

Classification	Hardness	Hot hardness	Impact strength	Oxidation resistance	Corrosion resistance	Abrasion resistance	Metal-to-metal wear resistance	Machinability	Applications
Fe5-A,B,C	C55-60[1]	H<1000°F	H[2]	L	L	M	H[3]	H	Cutting tools, dies, guides
FeMn-A & B	450-550BHN[6]	H<500-600°F	H	L	L	M	H	L	Soft-rock-crushing equipment
FeCr-A	C51-62	H<800-900°F	L	H	L	L-H[7]	L-H[7]	L	Farm equipment
CoCr-A	C42	H>1200°F	M	H	H[4]	L	H	L	Engine exhaust valves
-B	C46	H>1200°F	M	H	H[4]	L	H	L	
-C	C55	H>1200°F	L	H	H[4]	M	H	L	
CuAl-A2	115-140BHN[5]	H<400°F	M	M	M[4]	L	H	H	Bearings and gear surfaces
CuAl-A3	140-180BHN[5]	H<400°F	L	M	M[4]	L	H	H	
-B	140-180BHN[5]	H<400°F	L	M	M[4]	L	H	H	
-C	180-220BHN[5]	H<400°F	L	M	M[4]	L	H	H	
-D	230-270BHN[5]	H<400°F	L	M	M[4]	L	H	H	
-E	280-320BHN[5]	H<400°F	L	M	M[4]	L	H	H	
CuSi	80-100BHN	H<800°F	M	M	M[4]	L	H	H	Bearing surfaces
CuSn-A	70-85BHN	H<800°F	L	L	M[4]	L	H	H	Bearing surfaces
-C	85-100BHN	H<800°F	L		M[4]	L	H	H	
NiCr-A	C28	H<800°F	M	H<1750°F	4	L-H[7]	H	M	Seal rings, cams, screw conveyors
-B	C38	H<800°F	M	H<1750°F	4	L-H[7]	H	M	
-C	C45	H<800°F	L	H<1750°F	4	L-H[7]	H	M	

Key:

C	-	Rockwell C hardness number
BHN	-	Brinell hardness number
L	-	Low
M	-	Medium
H	-	High

1. Anneal to Rockwell C30
2. After tempering
3. After annealing
4. Consult authority for each application
5. Hardness depends on welding process used
6. After work hardening
7. Depends on application

Figure 26-39. A guide to the selection of surfacing materials. Refer to AWS A5.13 or a corrosion authority for specific applications.

thickness of the surfacing materials. This test is used to ensure uniform thickness of the surfacing material layers.

26.10 REVIEW OF SURFACING SAFETY

All of the safety precautions specified for gas welding, arc welding, welding on containers, and similar applications, also apply to metal surfacing. Review all safety precautions before attempting any metal surfacing operations.

Excellent ventilation is essential at all times because of the fluxes used, and because of the toxic effects of some of the alloys in the surfacing materials. The best protection is provided by an air-supplied purifier with air supplied from a main source. Powered air purifiers would be a second choice; an air purifying respirator would be a less appropriate third choice. Thermal spraying stations should have ventilation systems capable of exhausting 200 ft³/min. to 300 ft³/min. (5600 L/min. to 8500 L/min.) of air from the spray area. This exhaust should be passed through a wet collector to capture all dust that contains metal and surfacing material.

Protective clothing should be kept clean and in good condition. Your eyes should be protected at all times. Ear plugs or muffs are necessary when using the arc spraying, detonation spraying, and plasma arc spraying torch, since they create an ear-damaging sound pitch and intensity.

You must be constantly alert to avoid spraying metal on flammables or on another person.

TEST YOUR KNOWLEDGE

Write your answers on a separate sheet of paper. Do not write in this book.

1. List at least four causes of surface wear.
2. List the methods used to perform hardfacing and thermal spraying.
3. List two advantages of surfacing a part.
4. What is meant by hardfacing?
5. Name two unusual materials which are not metals that have been coated by flame spraying.
6. The loss of areas of a metal surface by flaking or pitting is called _____ _____.
7. If a metal is difficult (though possible) to file or cut, the hardness is probably _____ Brinell or _____ Rockwell C.
8. When using powder in flame spraying, name two ways of delivering the powder to the flame.
9. When flame spraying with powdered surfacing materials, _____ should not be used as the carrier gas. It may cause an explosion if it is combined with fine powder.
10. When thermal spraying, how thick is each layer of the surfacing material?
11. When thermal spraying, if the surface of the applied surfacing is bumpy and coarse, what is the probable cause?
12. Which thermal spraying process is the most economical to use? What does it cost per hour?
13. Name six materials that may be applied to a base metal by plasma arc spraying.
14. When thermal spraying using the plasma arc, which type electrical circuit is recommended: the transferred or nontransferred arc?
15. Which thermal spraying process propels the surfacing material to the surface with the highest velocity? What is the velocity in ′/sec. (m/sec)?
16. What is the temperature of the plasma arc spraying process?
17. When a cylindrical part is rough threaded, how deep are these threads cut?
18. Name three materials suggested for use as a base or bonding coat for other surfacing layers.
19. Name four defects that may be detected by visually inspecting a thermal sprayed surface.
20. Industrial ear muffs should be worn when using which three thermal spraying processes?

Arc spraying being used to apply surfacing material to the compressor case of a commercial jet engine. (TAFA, Inc.)

Part 8

METAL TECHNOLOGY

Tapping a "heat" from an electric arc furnace. The molten metal is being poured from the furnace into a large ladle. (Inco Alloys, Inc.)

A worker uses an electric hoist to move a glowing hot copper casting to a cooling stand. The casting later will be machined to finished dimensions for use in a large electric motor. (The Electric Materials Company)

Chapter 27

PRODUCTION OF METALS

It is helpful for a welder to understand the origin, production, and refinement of metals in use today. Having a thorough understanding of metals is an asset to the welding technician. The processes used to change ore to commercial metals, and then to produce alloys, have been greatly improved over the past few decades.

Steel and steel alloys are the most common metals in industrial use. Great quantities are produced each year. Aluminum is also produced in large quantities. Magnesium and titanium have many applications, particularly in the aerospace industries. Other metals and metal alloys are produced to meet the needs of specific industries. This chapter will discuss several metals that have industrial applications.

27.1 MANUFACTURING IRON AND STEEL

Steel is produced by adding small amounts of carbon to pure iron. The first step to produce steel is to remove from the iron most, if not all, of the carbon and other impurities. The alloy (mixture) of iron and carbon is called *straight carbon steel* or *plain carbon steel*.

Alloy steels are produced by adding additional elements, such as nickel, chromium, or manganese to plain carbon steel. The alloying elements are added in controlled amounts to obtain desired properties in the alloy steel.

Follow the steps in steelmaking in Figure 27-1. The first step in the production of steel is to refine the iron ore. Iron ore contains many unwanted impurities. The refinement of iron ore is usually done in the reducing atmosphere of a blast furnace. When the iron ore is refined it is called *hot iron*. Most hot iron is taken directly to a basic oxygen furnace. Hot iron contains some impurities and a relatively large amount of carbon. Hot iron is sometimes poured into molds and allowed to cool. When solid, it is called *pig iron*. The high carbon content makes pig iron brittle. In order to produce steel, the carbon in the iron must be removed in the oxidizing atmosphere of a *steelmaking furnace*.

In steelmaking furnaces, the oxidizing atmosphere "burns off" and eliminates the carbon and the impurities in the iron. Once the impurities and carbon have been eliminated or reduced to a minimum, controlled amounts of carbon and other alloying elements are added to the iron to produce the desired type of steel.

Once the desired amounts of carbon and other elements have been added, the molten steel is either continuously cast into a slab or is cast into an ingot. The ingot requires additional processing to turn it into a bloom, billet, or slab. These shapes differ in cross-sectional shape or area, as seen in Figure 27-1. From these rough stock shapes, the steel may later be formed into more exact shapes and products.

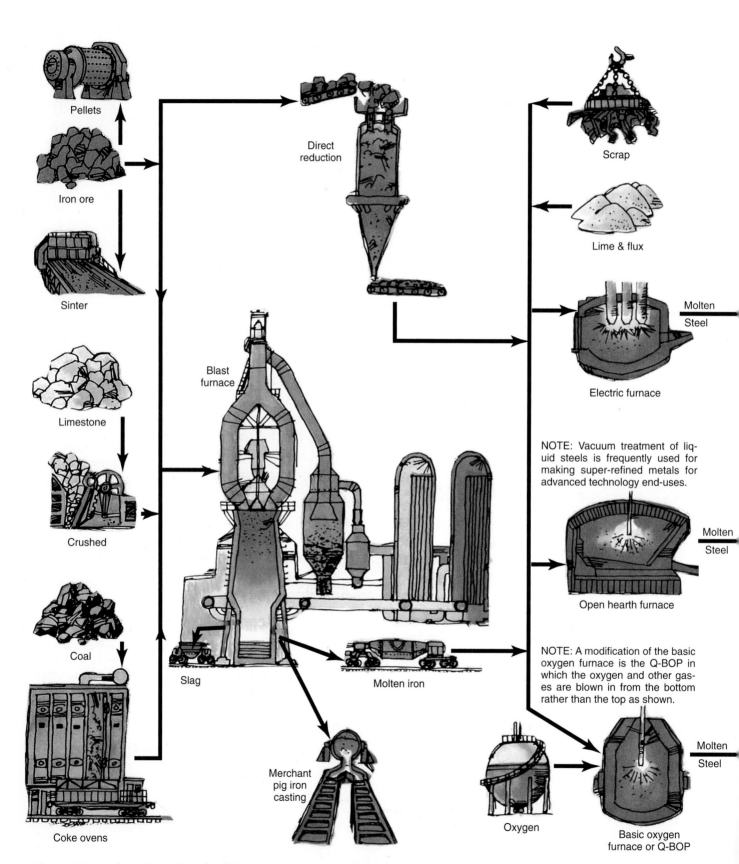

Pellets

Iron ore

Sinter

Limestone

Crushed

Coal

Coke ovens

Direct reduction

Blast furnace

Slag

Molten iron

Merchant pig iron casting

Scrap

Lime & flux

Electric furnace

Molten Steel

NOTE: Vacuum treatment of liquid steels is frequently used for making super-refined metals for advanced technology end-uses.

Open hearth furnace

Molten Steel

NOTE: A modification of the basic oxygen furnace is the Q-BOP in which the oxygen and other gases are blown in from the bottom rather than the top as shown.

Oxygen

Basic oxygen furnace or Q-BOP

Molten Steel

Figure 27-1. *A flowchart of steelmaking. (American Iron and Steel Institute)*

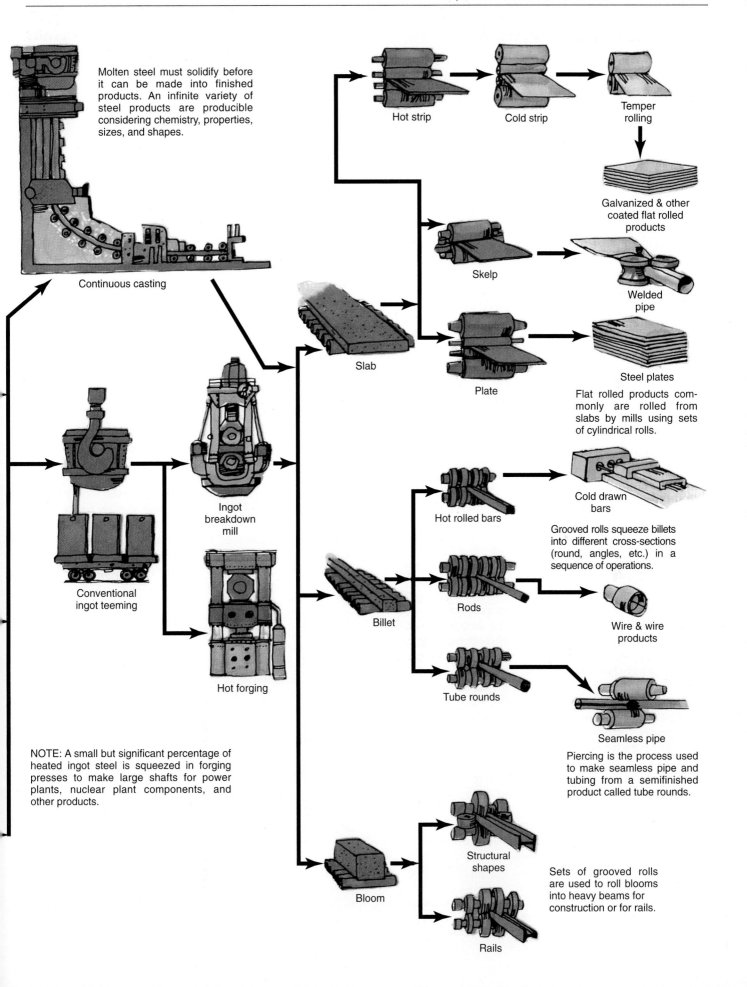

Molten steel must solidify before it can be made into finished products. An infinite variety of steel products are producible considering chemistry, properties, sizes, and shapes.

Continuous casting

Conventional ingot teeming

Ingot breakdown mill

Hot forging

NOTE: A small but significant percentage of heated ingot steel is squeezed in forging presses to make large shafts for power plants, nuclear plant components, and other products.

Slab

Hot strip

Cold strip

Temper rolling

Galvanized & other coated flat rolled products

Skelp

Welded pipe

Plate

Steel plates

Flat rolled products commonly are rolled from slabs by mills using sets of cylindrical rolls.

Billet

Hot rolled bars

Cold drawn bars

Grooved rolls squeeze billets into different cross-sections (round, angles, etc.) in a sequence of operations.

Rods

Wire & wire products

Tube rounds

Seamless pipe

Piercing is the process used to make seamless pipe and tubing from a semifinished product called tube rounds.

Bloom

Structural shapes

Rails

Sets of grooved rolls are used to roll blooms into heavy beams for construction or for rails.

27.1.1 Materials Used in Ironmaking

Iron seldom exists free in nature. It is mined from the earth in the form of *iron ore* or iron oxides mixed with impurities in the form of clay, sand, and rock. The most important types of iron ore are:

Hematite (red iron)	Fe_2O_3	70% iron
Magnetite (black)	Fe_3O_4	72.4% iron
Limonite (brown)	$Fe_2O_3H_2O$	63% iron
Siderite (iron carbonite)	$FeCO_3$	48.3% iron

Taconite (Fe_3O_3) has an iron content of only 20%-30%. It is made commercially useful by refining it up to 70% iron prior to shipping.

A good *flux* that will melt and combine with the impurities in the molten iron ore must be used in the blast furnace. Limestone is used as the flux in almost all blast furnaces.

One of the best fuels for the blast furnace is *coke*. The coke furnishes enough heat to reduce the iron ore. Coke is produced by heating soft (bituminous) coal in a closed container until the gases and impurities are driven off. What remains is coke, which is practically pure carbon and is low in such impurities as sulfur and phosphorus.

27.1.2 Blast Furnace Operation

The blast furnace has five major operations to perform:
- Deoxidize the iron ore.
- Melt the slag.
- Melt the iron.
- Carbonize the iron.
- Separate the iron from the slag.

The *blast furnace* is a huge tubular furnace made of steel and lined with firebrick. Figure 27-2 shows a typical blast furnace in cross section. The right proportions of iron ore, limestone, and coke are regularly dumped in at the top of the furnace through a bell-shaped opening (*hopper*). Around the bottom of the furnace are openings (*tuyeres*) through which hot air is blown. Openings near the top capture exhaust gases. The exhaust gases are used to preheat incoming air to the blast furnace. The exhaust gases are cleaned and filtered before being released to the atmosphere.

The operation of a blast furnace is continuous. Coke burns and produces enough heat to melt the iron. The excess carbon from the coke unites with the iron and lowers its melting temperature. Melted iron forms at the bottom of the furnace.

Limestone, acting as the flux, melts and combines with the impurities. The limestone floats the impurities on top of the molten iron as slag. Molten slag is drained from above the iron just before the iron is tapped or removed from the furnace.

As noted earlier, the iron coming out of the blast furnace is called hot iron. In modern practice, most molten hot iron is taken directly to a basic oxygen furnace to make steel. Some iron is poured into molds and cooled. The solid iron is called pig iron. Usually, cast iron foundries buy this pig iron. They further process the pig iron using cupola or induction furnaces. It is then cast and is called *cast iron*. For information on blast furnace chemistry, see Chapter 33.

27.2 STEELMAKING METHODS

Steel may be defined as iron combined with 0.1% to 1.86% of carbon. Other alloying elements are added to obtain different types of steels. See Chapter 28 for more technical specifications of steel.

Steel is produced by reducing hot iron and/or scrap steel in one of several types of furnaces. Types of steel production furnaces are:
- Basic oxygen.
- Electric.
- Open hearth.
- Vacuum.
- Crucible.
- Induction.

27.2.1 Basic Oxygen Process

Most of the steel in the United States is produced using the *basic oxygen furnace* process. The basic oxygen furnace and its operation are shown in Figure 27-3. To begin the process, the furnace is tipped and scrap and hot iron from a blast furnace are poured into the mouth of the furnace. See Figure 27-4. The furnace is then rotated to a vertical position under an exhaust hood. A water-cooled oxygen lance is lowered to a position about 6' (2m) above the molten metal. Oxygen is then blown into the furnace at supersonic speeds. Lime and other materials are added as flux to combine with the carbon and other impurities. Oxygen from above burns off the carbon and impurities in the molten metal. This produces steel. In 40 to 60 minutes, up to 300 tons (272 metric tons) of quality steel is produced.

At the end of the prescribed time, the oxygen is turned off and the oxygen lance is removed. The basic oxygen furnace is then tipped to pour out the steel produced into a *ladle*. Controlled amounts of carbon and alloying elements are often added to the ladle. Usually the molten steel in the ladle is continuously cast.

27.2.2 Electric Furnace

The second most popular method used to produce steel and steel alloys in North America is the electric furnace. The electric furnace was initially used to make special, high-quality steels, including stainless steels. Now the electric furnace is used to produce about 40% of all steel and steel alloys in the United States. In this type of furnace, the chemical composition of the metal may be closely controlled. Various alloying elements are added to create a steel with predetermined characteristics.

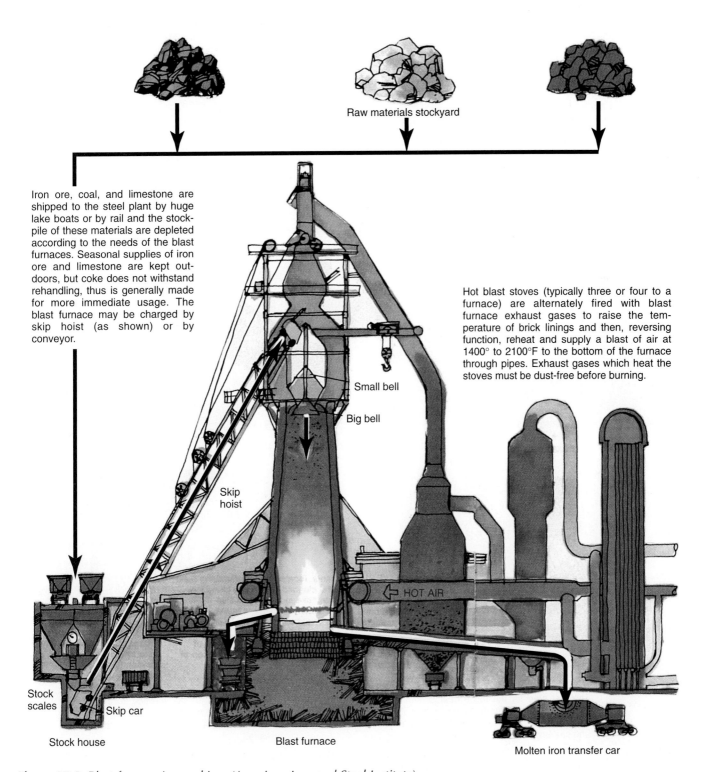

Raw materials stockyard

Iron ore, coal, and limestone are shipped to the steel plant by huge lake boats or by rail and the stock-pile of these materials are depleted according to the needs of the blast furnaces. Seasonal supplies of iron ore and limestone are kept out-doors, but coke does not withstand rehandling, thus is generally made for more immediate usage. The blast furnace may be charged by skip hoist (as shown) or by conveyor.

Hot blast stoves (typically three or four to a furnace) are alternately fired with blast furnace exhaust gases to raise the tem-perature of brick linings and then, reversing function, reheat and supply a blast of air at 1400° to 2100°F to the bottom of the furnace through pipes. Exhaust gases which heat the stoves must be dust-free before burning.

Small bell

Big bell

Skip hoist

HOT AIR

Stock scales

Skip car

Stock house

Blast furnace

Molten iron transfer car

Figure 27-2. *Blast furnace ironmaking. (American Iron and Steel Institute)*

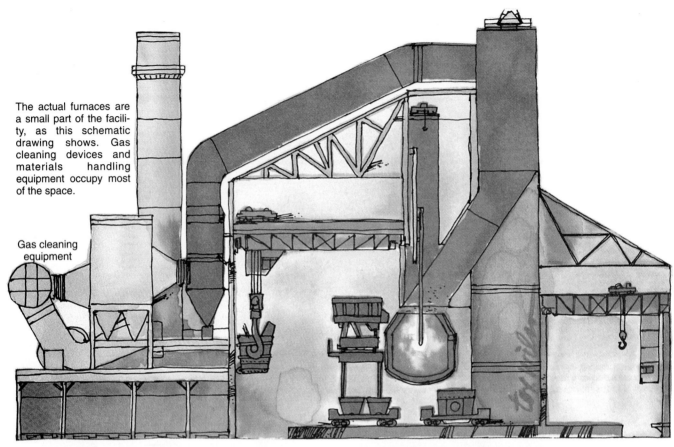

The actual furnaces are a small part of the facility, as this schematic drawing shows. Gas cleaning devices and materials handling equipment occupy most of the space.

Gas cleaning equipment

Diagram of basic oxygen furnace facility

The first step for making a heat of steel in a BOF is to tilt the furnace and charge it with scrap. The furnaces are mounted on trunnions and can be rotated through a full circle.

Hot metal from the blast furnace accounts for up to 80 percent of the metallic charge and is poured from a ladle into the top of the tilted furnace.

Figure 27-3. *The basic oxygen process of making steel. (American Iron and Steel Institute)*

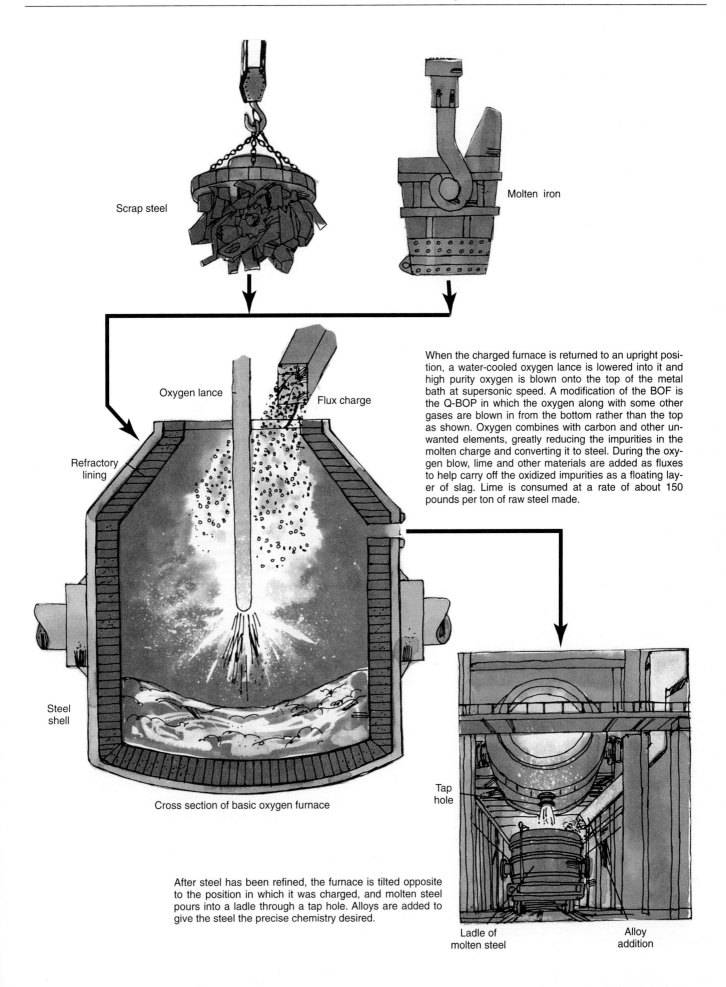

Scrap steel

Molten iron

Oxygen lance

Flux charge

Refractory lining

When the charged furnace is returned to an upright position, a water-cooled oxygen lance is lowered into it and high purity oxygen is blown onto the top of the metal bath at supersonic speed. A modification of the BOF is the Q-BOP in which the oxygen along with some other gases are blown in from the bottom rather than the top as shown. Oxygen combines with carbon and other unwanted elements, greatly reducing the impurities in the molten charge and converting it to steel. During the oxygen blow, lime and other materials are added as fluxes to help carry off the oxidized impurities as a floating layer of slag. Lime is consumed at a rate of about 150 pounds per ton of raw steel made.

Steel shell

Cross section of basic oxygen furnace

Tap hole

After steel has been refined, the furnace is tilted opposite to the position in which it was charged, and molten steel pours into a ladle through a tap hole. Alloys are added to give the steel the precise chemistry desired.

Ladle of molten steel

Alloy addition

Figure 27-4. *The three basic oxygen furnaces (BOF) shown here each produce 290 tons (263 metric tons) of steel in approximately 50 minutes. (Bethlehem Steel Corporation)*

Generally, selected and sorted scrap steel is used to charge the electric furnace. This is a big advantage of the electric furnace over the basic oxygen furnace, which requires mostly molten hot iron. After the furnace is charged with scrap steel, it is then closed to the atmosphere.

The heat required to melt the metal in this furnace is produced by an electric arc. Refer to Figure 27-5. Large-diameter movable electrodes are installed in the top of the furnace above the metal. See Figure 27-6. An arc is struck between the carbon electrodes and the metal. The arc, plus the heat created by the resistance of electricity flowing in the steel, furnishes the heat required to melt the metal. As the electrodes are consumed, they are moved down to keep a constant distance above the molten metal. The carbon electrodes may be changed as required from the top of the furnace.

Test samples of molten metal may be removed through small inspection ports. These samples are analyzed and alloying elements are added to the furnace as required.

Electric furnaces vary in capacity from 5 tons to 300 tons (4.5 metric tons to 272 metric tons). The time required to complete a *heat* (term for a furnace-full of molten metal) varies from three to six hours. The entire furnace is built to tilt, so that the molten steel can be poured out. As with the basic oxygen process, the molten steel is first poured into a ladle. Then, it is usually continuously cast to form solid steel slabs. The furnace is loaded or charged for the next heat with scrap steel and the process is repeated.

27.2.3 Open Hearth Process

The *open hearth furnace* method of making steel is an older process. It is no longer used in the United States. Some other countries still use this process to make steel, however.

In the open hearth furnace, metal is contained in a large shallow basin, holding up to 350 tons (317 metric tons) of metal. The furnace is charged with molten iron, solid pig iron, scrap, and fluxes. Up to 50% scrap can be used.

At each end of the basin or metal container, there is a preheating stove. The preheating stove is made of firebricks arranged in a checkerboard pattern. Air and a fuel gas enter the furnace at one end after passing through one of the preheating stoves. As the fuel gas and air burn above the metal in the furnace, they heat the metal. When the metal is heated, the unwanted elements and impurities oxidize or burn off.

Pure oxygen is used in some open hearth furnaces to speed the production of steel. Since more steel can be made in a given time, the use of oxygen is economically feasible. The higher temperatures obtained when using oxygen aid in burning out the carbon and impurities in the iron more quickly.

By taking periodical chemical analyses of the metal in the heat, the composition may be determined. By adding alloying ingredients, the heat may be held to close chemical tolerances.

27.2.4 Vacuum Furnaces

The melting of steel, steel alloys, titanium, and other pure metals in a *vacuum furnace* greatly reduces the amount of gas in the metal. There is only a small amount of gas present in the furnace. Since there is very little gas present, the absorption of gases by the molten metal is very small. Steels that are melted by more conventional methods absorb gases that cause porosity and inclusions in the metal when it solidifies. Gases formed in a vacuum furnace are pulled away from the molten metal by the vacuum pumps. The absence of these gases improves many of the qualities of the steel including ductility, magnetic properties, impact strength, and fatigue strength. *Vacuum stream degassing* and *ladle degassing* are shown in Figure 27-7. These processes work to remove gases from steel produced by basic oxygen or electric furnaces.

There are two main types of vacuum furnaces used today. These are the vacuum induction-type and vacuum arc furnaces.

The *vacuum induction furnace* is used when close control of the chemistry of the metal is of prime importance. The induction furnace is airtight and attached to vacuum pumps so that contaminating gases are constantly removed. Provisions are made to allow alloying materials to be added to the furnace without destroying the vacuum. The heating of the metal, the pouring of the metal into ingots, and the cooling of ingots are done under vacuum conditions to prevent contamination, as shown in Figure 27-7.

In the *vacuum arc furnace*, also called the *consumable electrode process*, the metal is first produced by another method. The metal is then formed into long round or square cylinders. These metal cylinders are then melted as huge consumable electrodes in the furnace. The metal elec-

Electrodes

At left, this cutaway drawing shows an electric furnace with its carbon electrodes attached to support arms and electrical cables and extending into the furnace. Molten steel and the rocker mounting on which the furnace may be tilted is also shown.

Electric furnace facility

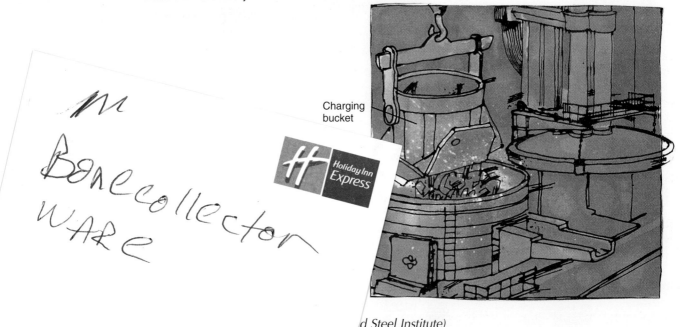

Charging
bucket

d Steel Institute)

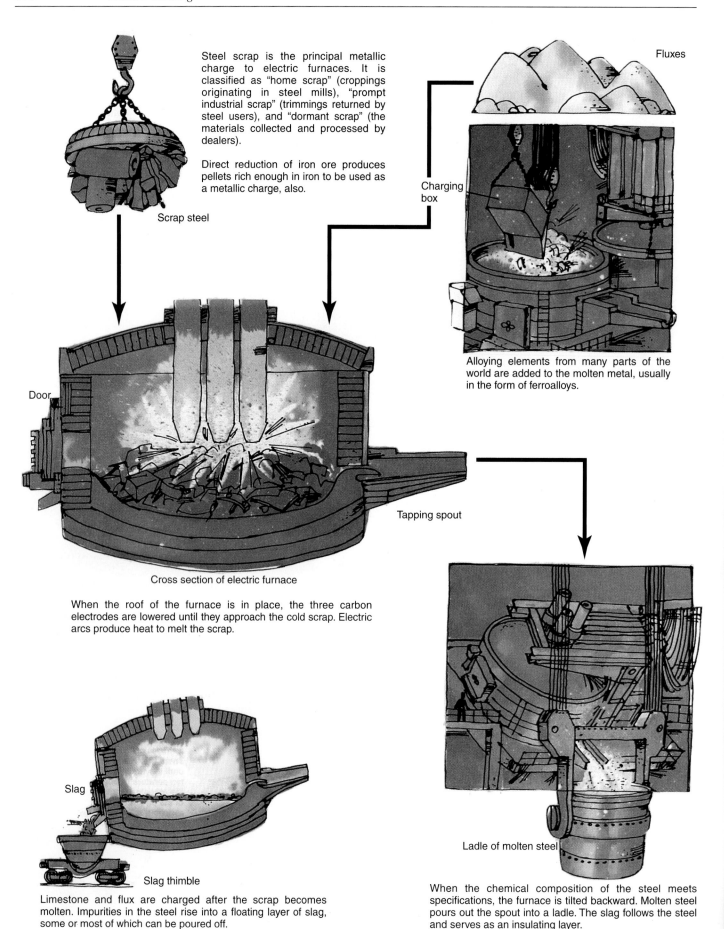

Steel scrap is the principal metallic charge to electric furnaces. It is classified as "home scrap" (croppings originating in steel mills), "prompt industrial scrap" (trimmings returned by steel users), and "dormant scrap" (the materials collected and processed by dealers).

Direct reduction of iron ore produces pellets rich enough in iron to be used as a metallic charge, also.

Scrap steel

Fluxes

Charging box

Alloying elements from many parts of the world are added to the molten metal, usually in the form of ferroalloys.

Door

Tapping spout

Cross section of electric furnace

When the roof of the furnace is in place, the three carbon electrodes are lowered until they approach the cold scrap. Electric arcs produce heat to melt the scrap.

Slag

Slag thimble

Limestone and flux are charged after the scrap becomes molten. Impurities in the steel rise into a floating layer of slag, some or most of which can be poured off.

Ladle of molten steel

When the chemical composition of the steel meets specifications, the furnace is tilted backward. Molten steel pours out the spout into a ladle. The slag follows the steel and serves as an insulating layer.

Figure 27-6. *Heat for this electric arc furnace is made available by three large retractible electrodes. The electrodes are lowered into the furnace near the charge of selected scrap iron and steel. (American Iron and Steel Institute)*

trodes are fed into the furnace at a controlled rate of speed to control the arc length.

As the metal drops from the end of the electrode, it falls into a water-cooled steel crucible and solidifies. The steel crucible is the grounded part of the electrical circuit. Air and contaminating gases are constantly pumped out of the furnace by vacuum pumps.

This process is frequently used when the uniformity and purity of the metal is very important. No provision is made in this furnace for adding alloying elements. Another process called the *electroslag remelting process* is very similar to the vacuum arc process. The difference is that the surface of the molten metal in the water-cooled mold is covered by molten flux. See Figure 27-8. Extremely low contamination is obtained, because impurities are caught and remain in the flux, not the resolidified metal.

27.2.5 Continuous Casting Process

The *continuous casting process* for the manufacture of steel is shown in Figure 27-9. Liquid steel from a furnace is moved in a ladle and is poured into a reservoir or *tundish*. From the tundish, the metal flows vertically into a water-cooled mold. The molten metal that is in contact with the sides of the mold cools quickly and shrinks away from the mold surface. This forms a shell around the molten metal in the center of the mold. This shell is supported by the withdrawing rolls as the column of steel is pulled from the mold. As the column of steel leaves the mold, jets of water are sprayed on the metal to cool and solidify the entire column. As it comes from the mold, metal may be in the form of a slab or a square bar, as seen in Figure 27-10. As the metal leaves the withdrawing rolls, it is cut to desired lengths for further processing.

The continuous casting process is more efficient and has a number of advantages over the older method of ingot processing, which is described in the following section. More than 80% of steel in the United States is produced using this continuous casting process. Continuous casting is also known as *strand casting*.

27.2.6 Ingot Processing

Before continuous casting was used, molten steel was poured into molds. The large mass of solidified steel removed from a mold is called an *ingot*. Today, only a small portion of steel produced in the United States is poured into ingots. Continuous casting eliminates the need for all processing of ingots and thus allows for a great savings in cost and time.

To produce an ingot, the molten steel from a furnace is poured into a ladle. The ladle is lined up over an ingot mold or a line of ingot molds. The steel is poured into the molds, where it solidifies. Figure 27-11 shows how ingots are produced and processed.

As soon as the steel solidifies, the mold is stripped from the ingot. Next, the ingot is placed into a heated *soaking pit*. The ingot may be placed into the soaking pit while still hot, or may first be allowed to cool completely. The purpose of a soaking pit is to allow the steel to reach a uniform temperature throughout before additional processing occurs.

After soaking, the ingot is removed from the soaking pit, as shown in Figure 27-12, and is brought to a rolling mill. The rolling mill forms the large ingot into shapes that can be used for final processing. These semifinished shapes are called *blooms*, *billets*, or *slabs*. Figures 27-13 and 27-14 show steel being rolled into different shapes.

The processes of ingot pouring, stripping, soaking, and rolling are required to produce semifinished shapes. By comparison, the same shapes are produced by the continuous casting process without all the intermediate steps.

27.2.7 Manufacturing Stainless Steel

Most stainless steel is produced using an electric furnace. Selected scrap stainless steel is a major part of the charge. The composition of the metal in the furnace is care-

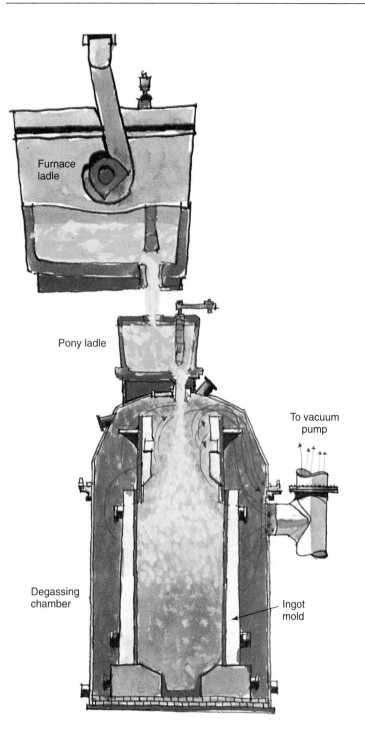

Furnace
ladle

Pony ladle

Degassing
chamber

To vacuum
pump

Ingot
mold

In vacuum stream degassing (left), a ladle of molten steel from a conventional furnace is taken to a vacuum chamber. An ingot mold is shown within the chamber. Larger chambers designed to contain ladles are also used. The conventionally melted steel goes into a pony ladle and from there into the chamber. The stream of steel is broken up into droplets when it is exposed to vacuum within the chamber. During the droplet phase, undesirable gases escape from the steel and are drawn off before the metal solidifies in the mold.

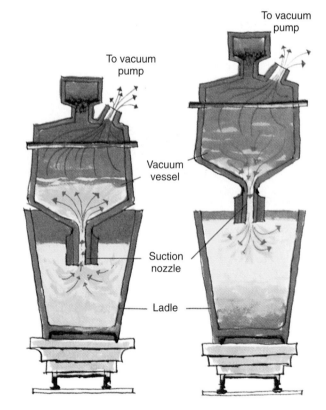

To vacuum
pump

To vacuum
pump

Vacuum
vessel

Suction
nozzle

Ladle

Ladle degassing facilities (right) of several kinds are in current use. In the left-hand facility, molten steel is forced by atmospheric pressure into the heated vacuum chamber. Gases are removed in this pressure chamber, which is then raised so that the molten steel returns by gravity into the ladle. Since not all of the steel enters the vacuum chamber at one time, this process is repeated until essentially all the steel in the ladle has been processed.

Figure 27-7. *The vacuum process of making steel. (American Iron and Steel Institute)*

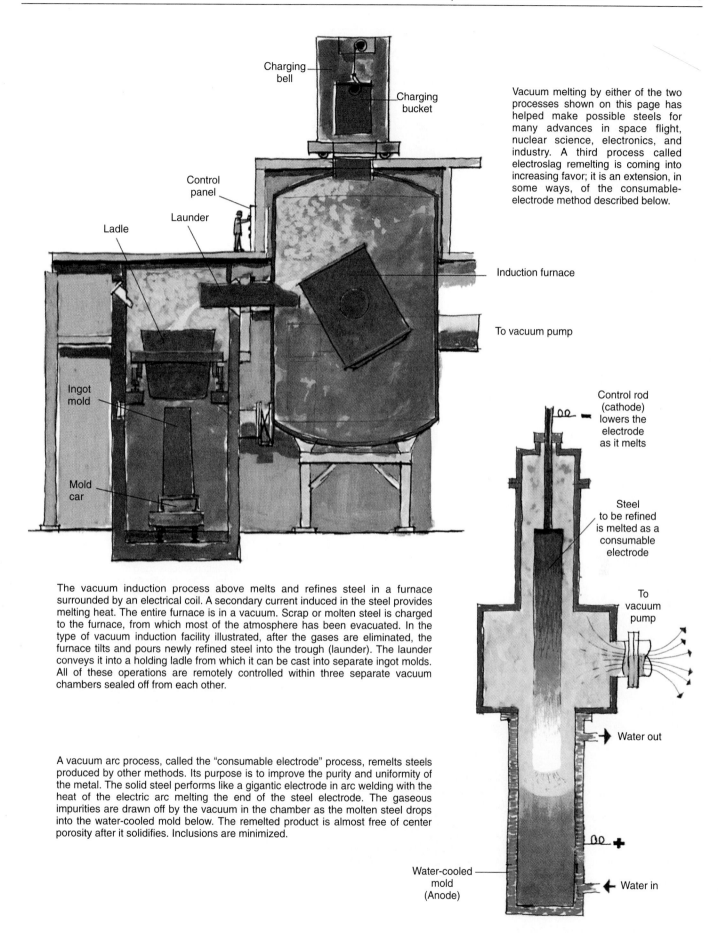

Charging
bell

Charging
bucket

Control
panel

Launder

Ladle

Induction furnace

To vacuum pump

Ingot
mold

Mold
car

Control rod
(cathode)
lowers the
electrode
as it melts

Steel
to be refined
is melted as a
consumable
electrode

To
vacuum
pump

Water out

Water-cooled
mold
(Anode)

Water in

Vacuum melting by either of the two processes shown on this page has helped make possible steels for many advances in space flight, nuclear science, electronics, and industry. A third process called electroslag remelting is coming into increasing favor; it is an extension, in some ways, of the consumable-electrode method described below.

The vacuum induction process above melts and refines steel in a furnace surrounded by an electrical coil. A secondary current induced in the steel provides melting heat. The entire furnace is in a vacuum. Scrap or molten steel is charged to the furnace, from which most of the atmosphere has been evacuated. In the type of vacuum induction facility illustrated, after the gases are eliminated, the furnace tilts and pours newly refined steel into the trough (launder). The launder conveys it into a holding ladle from which it can be cast into separate ingot molds. All of these operations are remotely controlled within three separate vacuum chambers sealed off from each other.

A vacuum arc process, called the "consumable electrode" process, remelts steels produced by other methods. Its purpose is to improve the purity and uniformity of the metal. The solid steel performs like a gigantic electrode in arc welding with the heat of the electric arc melting the end of the steel electrode. The gaseous impurities are drawn off by the vacuum in the chamber as the molten steel drops into the water-cooled mold below. The remelted product is almost free of center porosity after it solidifies. Inclusions are minimized.

Figure 27-8. *Two vacuum arc furnaces in operation. Note the electrode in the furnace on the left and the hydraulic mechanism above that feeds the electrode into the furnace. (INCO Alloys International)*

fully controlled. After the electric furnace processing is complete, another operation is performed to remove as much carbon as possible. Removing carbon is called *decarburization*. Figure 27-15 briefly discusses these decarburization processes and shows the process of producing stainless steels.

Some stainless steel is continuously cast; however, most stainless is poured into an ingot. The steps required for processing an ingot, as described in Heading 27.2.6, are followed with some modifications. The surface finish on stainless steel is important, so operations are required to provide a high quality finish.

27.2.8 Cast Iron

Gray cast iron is the most common form of cast iron. *Gray cast iron* is simply a casting that has been cooled slowly. This allows some of the carbon to separate, forming free graphite (carbon) flakes. This graphite causes the gray appearance. Gray cast iron can be machined.

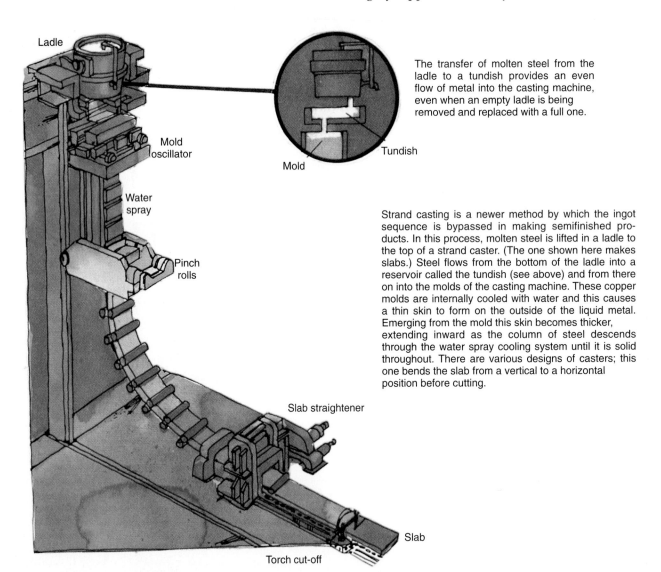

The transfer of molten steel from the ladle to a tundish provides an even flow of metal into the casting machine, even when an empty ladle is being removed and replaced with a full one.

Strand casting is a newer method by which the ingot sequence is bypassed in making semifinished products. In this process, molten steel is lifted in a ladle to the top of a strand caster. (The one shown here makes slabs.) Steel flows from the bottom of the ladle into a reservoir called the tundish (see above) and from there on into the molds of the casting machine. These copper molds are internally cooled with water and this causes a thin skin to form on the outside of the liquid metal. Emerging from the mold this skin becomes thicker, extending inward as the column of steel descends through the water spray cooling system until it is solid throughout. There are various designs of casters; this one bends the slab from a vertical to a horizontal position before cutting.

Figure 27-9. *The continuous (strand) casting process. (American Iron and Steel Institute)*

Figure 27-10. *A thick slab exiting from the rolls in the continuous casting or strand casting process. (American Iron and Steel Institute)*

Figure 27-11. *Steps involved in ingot processing. (American Iron and Steel Institute)*

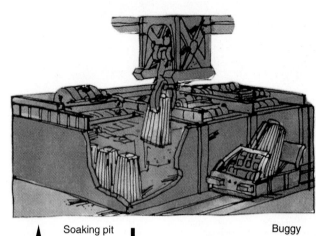

Soaking pit

Buggy

Stripped ingots are taken to furnaces called soaking pits. There they are "soaked" in heat until they reach uniform temperature throughout. Then the reheated ingots are lifted out of the pit and carried to the roughing mill on a buggy.

Ladle

Stripper crane

Ingot molds

The traditional method of handling molten steel is to position the ladle via an overhead crane above a line of ingot molds (left). Then the operator opens a stopper rod within the ladle, and a stream of steel flows through a hole in the bottom of the ladle to "teem" or fill the cast iron molds which rest on special ingot railroad cars.

Molten steel in an ingot mold cools and solidifies from the outside towards the center. When the steel is solid enough, a stripper crane lifts away the mold while a plunger holds the steel ingot down on the ingot car.

Roughing mill

Roughing mills are the first stage in shaping the hot steel ingot into semifinished steel – usually blooms, billets, or slabs. Some roughing mills are the first in a series of continuous mills, feeding sequences of finishing rolls.

Figure 27-12. *This ingot has been heated to an even temperature throughout and is being removed from the soaking pit.*

Figure 27-13. *A hot slab of steel exiting from a set of rolls. (American Iron and Steel Institute)*

Figure 27-14. *A hot ingot takes the shape of a bloom as it exits from a set of blooming rolls. (American Iron and Steel Institute)*

White cast iron is made by cooling the casting quickly. White cast iron is very hard and brittle. It is very difficult to machine white cast iron. *Cast iron* is usually made by melting and deoxidizing pig iron in a cupola furnace or an electric induction furnace. After processing in one of these furnaces, the iron is poured into a mold where it solidifies. Figure 27-16 shows an example of what can be produced with cast iron. Other types of cast iron are produced with alloy additions and heat treating. These are discussed in Chapter 29.

Cupola Furnace

The *cupola furnace* is used to produce cast iron. The cupola furnace resembles a small blast furnace. Coke is used as fuel to heat the furnace. Limestone is used as a flux. Pig iron is added, along with scrap cast iron and steel.

The cupola furnace eliminates the excess carbon and impurities as the metal and flux melt. When a large quantity of molten metal is formed, the furnace is ready to be tapped. *Tapping* is the term used to describe drawing off the molten metal from the furnace. The cupola furnace operates continuously as does a blast furnace. First the molten slag is drawn off. Then the furnace is tapped. The molten metal, as it comes from the cupola furnace, is ready to be cast into a mold. When it has solidified, it is called cast iron.

27.2.9 Induction Heating Process

Induction heating is defined as "raising the temperature of a material by means of electrical generation of heat within the material and not by any other heating method such as convection, conduction, or radiation." In an induction furnace, the metal to be heated is contained in a vessel called a *crucible;* electrical conductors are wound around the vessel to form a coil.

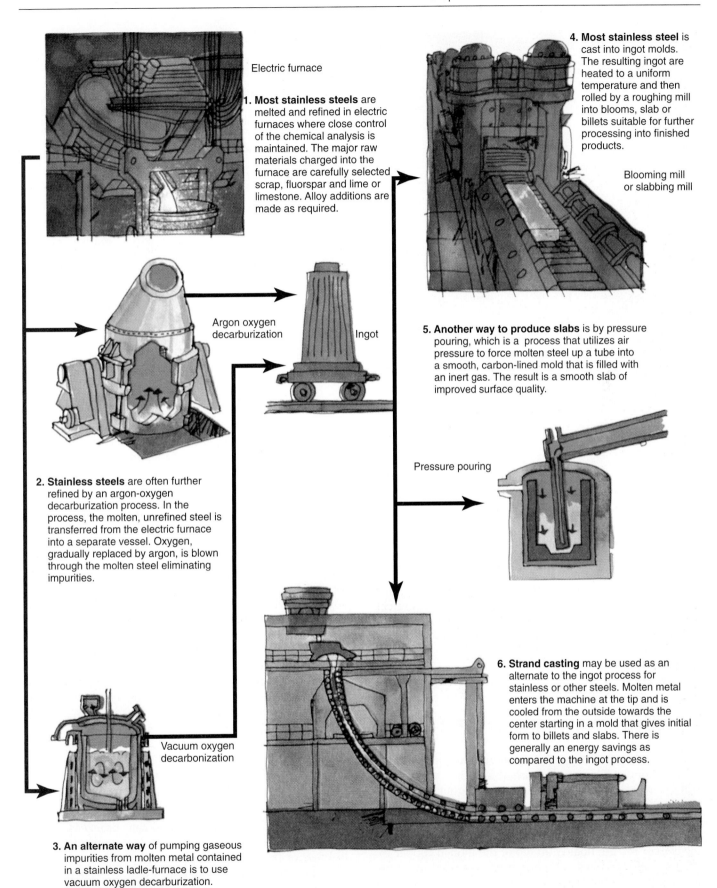

Electric furnace

1. **Most stainless steels** are melted and refined in electric furnaces where close control of the chemical analysis is maintained. The major raw materials charged into the furnace are carefully selected scrap, fluorspar and lime or limestone. Alloy additions are made as required.

Argon oxygen decarburization

Ingot

2. **Stainless steels** are often further refined by an argon-oxygen decarburization process. In the process, the molten, unrefined steel is transferred from the electric furnace into a separate vessel. Oxygen, gradually replaced by argon, is blown through the molten steel eliminating impurities.

Vacuum oxygen decarbonization

3. **An alternate way** of pumping gaseous impurities from molten metal contained in a stainless ladle-furnace is to use vacuum oxygen decarburization. The metal is heated and stirred by an induced electrical current. Oxygen is introduced through a water-cooled lance and solid additions are made through a hopper.

4. **Most stainless steel** is cast into ingot molds. The resulting ingot are heated to a uniform temperature and then rolled by a roughing mill into blooms, slab or billets suitable for further processing into finished products.

Blooming mill or slabbing mill

5. **Another way to produce slabs** is by pressure pouring, which is a process that utilizes air pressure to force molten steel up a tube into a smooth, carbon-lined mold that is filled with an inert gas. The result is a smooth slab of improved surface quality.

Pressure pouring

6. **Strand casting** may be used as an alternate to the ingot process for stainless or other steels. Molten metal enters the machine at the tip and is cooled from the outside towards the center starting in a mold that gives initial form to billets and slabs. There is generally an energy savings as compared to the ingot process.

Figure 27-15. The processing of stainless steel. (American Iron and Steel Institute)

Figure 27-16. *An eight-cylinder diesel engine block made from cast iron. (Central Foundry, Division of General Motors Corporation)*

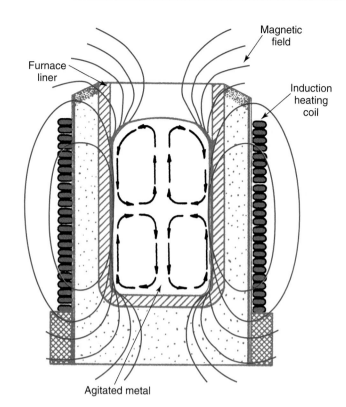

Figure 27-17. *Schematic cross section of a small induction furnace. Metal in the furnace is agitated and mixed by magnetic eddy currents to produce a more homogeneous metal.*

High-frequency alternating current is passed through the coil. This creates or induces a current in the metal in the crucible. The metal resists the flow of electrical current; this resistance causes heat to be generated. The metal melts quickly. Figure 27-17 shows a schematic of an induction furnace. The electrical currents set up a *stirring motion* in the molten metal to make it uniform. Electricity in the coil around the furnace induces or creates heat in the furnace.

By accurately controlling the frequency and amperage of the alternating current passing through the induction coil, it is possible to accurately control the temperature of the metal being heated. An induction furnace is shown in Figure 27-18.

27.3 MANUFACTURING COPPER

A great deal of copper ore contains a high percentage of pure copper. This ore is crushed and then washed in water to remove the lighter-weight earth particles from the heavier copper ore. The ore is then mixed with coke and limestone and placed into a small blast furnace. The liquid copper settles to the bottom of the furnace. Impurities in the copper ore are floated on top of the molten copper in the form of slag. This is the same process used to refine iron ore in a blast furnace. See Figure 27-19.

Once removed from the blast furnace, the copper is further refined by the electrolysis process. *Electrolysis* may be defined as a chemical change or decomposition. The decomposition is created by passing electricity through a solution of the material or through the substance while it is in a molten state.

The electrolytic cell used to refine copper is shown in Figure 27-20. Pure copper bars are used as the **cathodes** (negative poles). Impure copper to be refined forms the **anodes** (positive poles) in the cell. Copper sulfate with some sulfuric acid is used as the **electrolyte**, or fluid, in the cell.

When the electric current is turned on, pure copper leaves the anode and deposits on "plates" the cathode. The impurities fall to the bottom of the electrolytic cell. When the impure copper anode is consumed, it is replaced. The heavily plated pure copper cathodes are removed and replaced with thinner bars.

27.4 MANUFACTURING COPPER ALLOYS, BRASS AND BRONZE

Brass is an alloy of copper and zinc. Bronze is an alloy of copper and tin. When a third or fourth element is added to brass or bronze to improve its physical properties, an **alloy brass** or an **alloy bronze** is created. Some of the alloying elements added to brass are: tin, manganese, iron, silicon, nickel, lead, and aluminum. Alloying elements added to bronze are: nickel, lead, phosphorous, silicon, and aluminum.

Figure 27-18. *A complete induction melting furnace installation. (Ajax Magnethermic Corp.)*

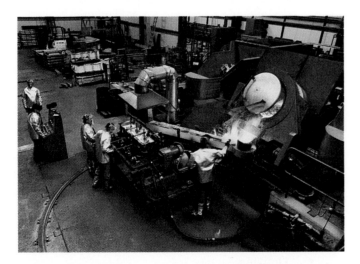

Figure 27-19. *Tapping a heat from a copper furnace and pouring ingots. This furnace produces a silver-copper alloy used to manufacture commutators for electric motors. (The Electric Materials Company)*

Copper alloys, brass, and bronze are made by melting solid materials in an induction furnace. Selected scrap copper, brass, and/or bronze as appropriate are placed in the furnace. Induction heating is used to melt the metals. Samples are taken from the furnace for analysis. Alloying elements are added as appropriate to obtain the desired alloy. When the *charge* (batch) of molten metal is completely processed, it is poured from the induction furnace into a holder. From the holder, the molten metal is usually continuously poured or cast into slabs. Typical slab sizes are from 6" to 8" (152mm to 203mm) thick, 24" to 48" (610mm to 1229mm) wide, and can be cut or cast to any length. A common length is about 25' (7.6m).

The term "bronze" is often used with copper alloys that contain no tin or only small quantities of tin in the alloy. *Hardware bronze*, as an example, contains approximately 90% copper, 8% zinc, and 2% lead. *Manganese bronze* contains approximately 58.5% copper, 1.0% tin, about 39% zinc, 0.28% manganese, and 1.4% iron.

27.5 MANUFACTURING ALUMINUM

Aluminum is normally produced by separating it from the oxide (Al_2O_3) found in bauxite ore. It may also be found in many other forms. After the aluminum oxide is removed from the ore, it is dissolved in a molten bath of sodium-aluminum fluoride (cryolite). An electric current is passed through the molten bath and pure aluminum is obtained by an electrolysis process, called the ***Hall process***.

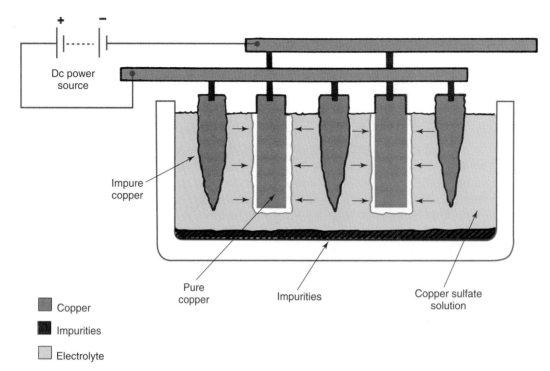

Figure 27-20. Schematic drawing of an electrolytic cell used to refine copper.

The electrolytic cell used in this process is a carbon-lined open-top furnace. The electrolytic cell has carbon electrodes suspended in a solution of aluminum oxide and cryolite. As the current passes through the solution, the aluminum oxide is reduced. Pure aluminum is deposited

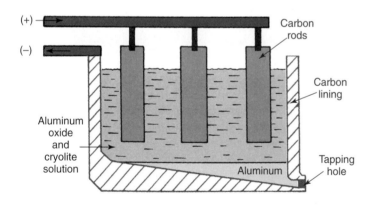

Figure 27-21. Schematic drawing of the Hall process for producing aluminum.

at the cathode, or negative terminal, which is the lining of the furnace. The pure aluminum collects at the bottom of the cell, as shown in Figure 27-21. Such a furnace is in continuous operation. Periodically, the molten aluminum is poured from the cell into ingot molds that are stored for further processing.

27.6 MANUFACTURING ZINC

Zinc is principally produced by a *distilling process*. The zinc ore is heated with coke in a clay crucible and the zinc vapor is then condensed in a clay condenser. It may also be refined by an electrolytic process. Zinc is used mainly as an alloying metal and for galvanizing.

27.7 PROCESSING METALS

Most metals, other than steel, alloy steel, and cast iron, are originally cast into ingots or molds for further processing. As needed, the castings are reheated to a specific temperature depending on the metal. The metal is then formed into a finished or semifinished shape by one of the following methods:

- Casting.
- Rolling (hot and cold).
- Forging.
- Extruding.
- Drawing.

Casting a metal in a sand or permanent mold is a popular method of producing objects with intricate shapes. A stationary or spinning (*centrifugal*) mold may also be used.

To improve the physical properties of a metal, and to form the metal into more usable rough stock shapes, ingots are often rolled. In a steelmaking plant, the ingots are rolled between large powerful rollers in a *rolling mill*. The ingots are reduced to blooms, billets, slabs, and even plate and sheet stock as required. These forms may be further processed in a rolling mill to produce rails, T-beams, I-beams, angles, bar stock, etc. Figure 27-22 illustrates some typical shapes that are formed by rolling. Numerous operations are required to form some of the shapes.

Forging, either drop or press, is used to obtain shapes that are stronger than castings and that cannot be easily rolled into shape. Forge hammers and/or forming dies are used to pound the metal into the shape desired.

Extrusion is a process in which a metal, normally in its plastic state, is pushed with great force through dies that are cut in the shape of the desired cross section. This process produces long lengths of metal with a uniform cross-sectional shape.

Drawing is a process of pulling metal through dies to form wires, tubing, and moldings. Drawing is a popular method of shaping metal to meet a certain requirement.

Many intricate shapes that would be difficult or impossible to make by any other method are now being

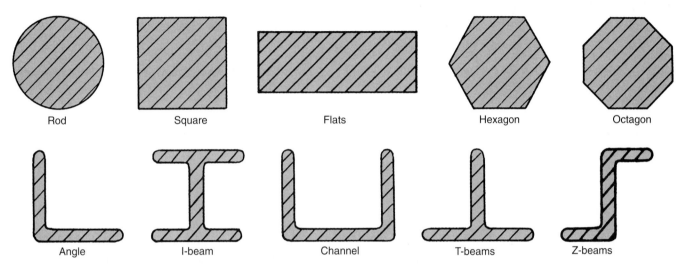

Rod   Square   Flats   Hexagon   Octagon

Angle   I-beam   Channel   T-beams   Z-beams

Figure 27-22. *Standard cross-sectional shapes of metals. The welding process may be used to combine the principal forms into complex structures.*

Figure 27-23. *Pump rotors made from powdered metals in a male-female die set.*

produced by the *powdered metal* process. The metal to be used is reduced to a fine powder, then forced under pressure into a heated steel mold. The properties of the finished part approximate those of the original solid metal. Figure 27-23 shows a die set being used to form gears from powdered metal. These parts are made to very close tolerances. Most can be used without any additional machining or grinding.

27.8 REVIEW OF SAFETY

All furnaces that contain molten metal should be approached with great caution. Remember that molten iron is over 2800°F (1540°C). **Careless handling can cause great bodily harm or death. Careless handling can also cause damage to equipment. Spilled molten metal will spread with great speed and will burn almost anything combustible. Special clothing, spats, gloves, and special shoes must be worn. Proper face guards should be worn. The eyes and face must be protected from flying particles and glare.**

OSHA regulations for workers in metal refining plants and foundries, and company safety policies, are there to keep people and the work environment safe. These procedures must be followed.

TEST YOUR KNOWLEDGE

Write your answers on a separate sheet of paper. Do not write in this book.
1. What is coke?
2. What is the purpose of limestone?
3. Is pig iron ductile or brittle?
4. What materials are used in a blast furnace to produce pig iron?
5. Why is a flux needed in a blast furnace?
6. What happens to most of the hot iron from blast furnaces?
7. Steel contains between _____ % and _____ % carbon.
8. Which type of steel furnace produces the greatest amount of steel?
9. Which type of furnace is charged with mostly scrap steel? (Hint: This type of furnace produces the most *stainless* steel.)
10. Describe how the electric arc furnace is used to produce steel.
11. What are the advantages of a vacuum furnace?
12. Most steel produced in the United States is made directly into slabs, not ingots, using the _____ _____ process.
13. Which process offers cost and time savings in the production of semifinished forms, continuous casting or ingot pouring?
14. The most common form of cast iron is _____ cast iron.
15. What is done with the molten metal produced in a cupola furnace?
16. How is a magnetic material in an induction furnace heated?
17. The production of copper requires two processes. First, copper ore is reduced in a _____ _____. It is further refined by the _____ _____.
18. How is aluminum produced?
19. What forming processes can be used to produce long pieces with a constant cross section?
20. What is done to a metal during a forging process?

Chapter 28

METAL PROPERTIES AND IDENTIFICATION

LEARNING OBJECTIVES

After studying this chapter, you will be able to:

* List and describe eight physical properties of metal.
* Find, from tables, the tensile strength and yield strength for various plain carbon steels.
* Define the terms eutectic point, solidus, liquidus, and critical temperature.
* Determine the structure of a steel at various points on the iron-carbon diagram.
* Know how to identify cast iron, plain carbon steel, and alloy steels by making a spark test on a grinder and using comparison charts.
* Describe plain carbon steels from their SAE/ANSI composition numbers.

It is necessary for a welder to have some accurate means of identifying metals. The welder must also have a good understanding of the constituents of metals in order to intelligently solve welding problems. Metals are divided into two major groups. These are:

* Ferrous metals.
* Nonferrous metals.

Iron and its alloys are classified as *ferrous metals.* "Ferrous" comes from *ferrum,* the Latin word for iron. These ferrous metals include all steels, tool steels, stainless steels, and cast irons. A great amount of welding is done on ferrous metals. The most common ferrous metals that are welded are the low-carbon steels. Other ferrous metals that are welded include medium- and high-carbon steels, alloy steels, stainless steels, and cast iron.

The *nonferrous metals* include metals and alloys that contain either no iron or insignificant amounts of iron. Some of the more popular nonferrous metals that a welder encounters are copper, brass, bronze, aluminum, magnesium, titanium, zinc, and lead. The various welding processes available now make it possible to satisfactorily weld practically all nonferrous metals.

28.1 IRON AND STEEL

Iron is produced by reducing iron oxide, commonly called iron ore, to hot iron or pig iron by means of the blast furnace. Many types of furnaces are used to change the hot iron or pig iron into the various steels. The two most common furnaces are the basic oxygen furnace and the electric furnace. See Chapter 27 for more information.

Carbon steel is an alloy of iron and controlled amounts of carbon. *Alloy steel* is a combination of carbon steel and controlled amounts of other desirable metal elements. The percentage of carbon content determines the type of carbon steel. For example, wrought iron has 0.003% carbon, meaning three-thousandths of 1%. Low-carbon steel contains less than 0.30% carbon. Medium-carbon steel varies between 0.30% and 0.55% carbon content. High-carbon steel contains approximately 0.55% to 0.80% carbon, and very high-carbon steel contains between 0.80% and 1.70% carbon. Cast iron contains from 1.8% to 4% carbon.

The carbon in steel and cast iron generally combines with the iron to form *cementite,* a very hard, brittle substance. Cementite is also known as *iron carbide.* As the carbon content of the steel increases, the hardness, the strength, and the brittleness also tend to increase.

Various heat treatments are used to enable steel to retain its strength at the higher carbon contents, and yet not have the extreme brittleness usually associated with high-carbon steels. Also, alloying metals such as nickel, chromium, manganese, vanadium, and other alloying metals, may be added to steel to improve certain physical properties.

A welder must also have an understanding of the impurities occasionally found in metals, and how those impurities affect the weldability of the metal. Two of the detrimental impurities sometimes found in steels are phosphorus and sulfur. Their presence in the steel may be due to their presence in the ore, or may be a result of the method of manufacture. Both of these impurities are detrimental to the welding qualities of steel. Therefore, during the manufacturing process, extreme care is always taken to keep the impurities at a minimum (0.05% or less). Sulfur improves the machining qualities of steel, but it is detrimental to its hot-forming properties.

During a welding operation, sulfur or phosphorus tends to form a gas in the molten metal. After the metal solidifies, the resulting gas pockets in the welds cause brittleness. Other impurities are dirt or *slag* (iron oxide). The dirt or slag may become embedded in the metal during rolling. Some of the dirt may come from the byproducts of the refining process used for the metal.

These impurities may also produce *blow holes* in the weld and reduce the physical properties of the metal in general. See Heading 28.4.2 for a procedure used to test a steel for impurities.

28.1.1 Physical Properties of Iron and Steel

A *physical property* is a characteristic of a metal that may be observed or measured. As mentioned previously, the physical properties of steel are affected by the following:

- Carbon content.
- Impurities.
- Addition of various alloying metals.
- Heat treatment.

Chapter 30 describes various machines used for determining a metal's physical properties. Testing machines are used in the welding shop to enable operators to identify metals and to check on the physical properties of weldments.

Some of the more important physical properties of steels are:

- Tensile strength.
- Compressive strength.
- Hardness.
- Elongation.
- Ductility.
- Brittleness.
- Toughness.
- Grain size.

Tensile strength is the ability of a metal to resist being pulled apart. This property may be measured on a tensile testing machine that puts a stretching load on the metal. Figure 28-1 illustrates the types of loads imposed on structures. Figure 28-2 shows how the tensile strength, elongation (explained below), and yield point are affected by the carbon content of steel. As the carbon content increases, the tensile strength and yield point first increase, then decrease. (The yield point is the point on the stress-strain curve when the metal begins to plastically deform.)

The *compressive strength* of a metal is a measure of how much squeezing force it can withstand before it fails. The metal is tested in a similar way to a tensile test, but instead of pulling on the metal, a squeezing (compression) force is applied.

Hardness is the quality which allows a metal to resist penetration. The Rockwell, Brinell, Shore Scleroscope or other tests may be used to identify hardness in a metal. See Heading 30.14 for information on hardness testing.

Elongation is a measure of how much a metal will stretch before it breaks. Elongation, or the percent of elon-

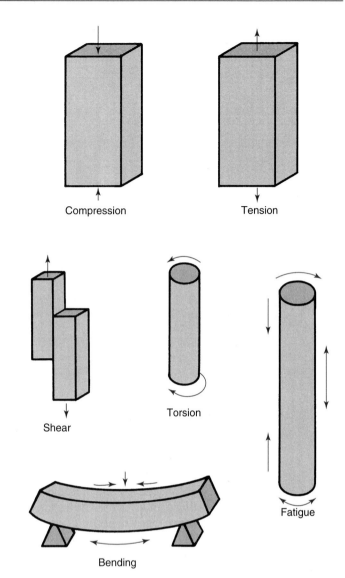

Figure 28-1. *Types of stresses or loads imposed on structures: compression, tension, shear, torsion, bending, and fatigue. In bending, the lower surface is in tension and the upper surface is in compression. In fatigue, a vibration or repeated reversal of the load is applied. The fatigue load may be either compression, tension, torsion, shear, or a combination of these loads.*

gation, is measured during a tensile test. Two marks are placed on the tensile sample an exact distance apart, usually 2" (50.8mm).The distance between the marks is again measured after the sample breaks. The original distance between the marks is compared to the increase in length to determine the percent of elongation. A metal that has over 5% elongation is said to be ductile. One with less than 5% elongation is considered brittle. The formula for determining percent elongation is:

$$\% \text{ Elongation} = \frac{L_f - L_i}{L_i} \times 100$$

L_f is the final length between marks. L_i is the initial length (2" or 50.8mm). Figure 28-2 shows how the elongation of a steel decreases as the carbon content increases.

SAE AISI no.	Carbon content in percentages	Tensile strength		Yield point		Elongation percent
		psi	MPa	psi	MPa	
1006	0.06	43,000	300	24,000	170	30
1010	0.10	47,000	320	26,000	180	28
1020	0.20	55,000	380	30,000	210	25
1030	0.30	68,000	470	37,500	260	20
1040	0.40	76,000	520	42,000	290	18
1050	0.50	90,000	620	49,500	340	15
1060	0.60	98,000	680	54,000	370	12
1070	0.70	102,000	700	56,000	390	12
1080	0.80	112,000	770	56,000	390	12
1090	0.90	122,000	840	67,000	460	10
1095	0.95	120,000	830	66,000	460	10

Figure 28-2. *Approximate physical property changes of carbon steel as the carbon content changes.*

Ductility is the ability of a metal to be stretched. Other terms that refer to this same property include formability, malleability, and workability. A very ductile metal, such as copper or aluminum, may be pulled through dies to form wire.

Brittleness is the opposite of ductility. A brittle metal will fracture if it is bent or struck a sharp blow. Most cast iron is very brittle.

Toughness is the ability to prevent a crack from propagating (growing). A metal is said to be tough if it can withstand an impact or shock loading.

The *grain size* and microstructure of a metal can be viewed under a microscope. The microstructure of a metal is the structure observed through a microscope. Microscopic views give a good indication of a metal's heat treatment, tensile strength, and ductility.

Before these properties are studied in detail, the welder should have an understanding of the effect of carbon on the properties of steel and a knowledge of alloys in general.

28.1.2 Alloy Metals

An alloy metal may be defined as an intimate mixture of two or more elements. Any ferrous or nonferrous metal may be alloyed to form an alloy metal with new and desirable characteristics.

Steel is a combination of iron and controlled amounts of carbon. Alloy steels are created by adding other elements to plain carbon steel. Some elements that are alloyed with carbon steel and the qualities imparted to steel by each are:

- Chromium – increases resistance to corrosion; improves hardness and toughness; improves the responsiveness to heat treatment.
- Manganese – increases strength and responsiveness to heat treatment.
- Molybdenum – increases toughness and improves the strength of steel at higher temperatures.
- Nickel – increases the qualities of strength, ductility, and toughness.

- Tungsten – produces dense, fine grains; helps steel to retain its hardness, helps steel to retain strength at high temperatures.
- Vanadium – retards grain growth and improves toughness.

The melting temperature of a metal is changed somewhat whenever an alloying metal is added. This may be illustrated by examining the cooling curve for a simple alloy. See Figure 28-3.

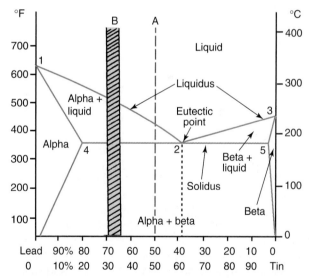

Figure 28-3. *Lead-tin diagram. Line A represents the popular 50-50 solder; Line B represents the range of body solders. An alloy solidifies over a range of temperatures. The alloy begins to solidify at the liquidus. It is completely solid at the solidus.*

A *simple alloy* consists of two metals in any proportion. An example of a simple alloy is the combination of tin and lead that is called solder. The melting temperature of the lead is 621°F (327°C). Tin has a melting temperature of 450°F (232°C). However, any mixture of the two results in a lower melting temperature than 621°F (327°C). At a certain proportion of the metals, the lowest melting temperature is reached. At this point, the metal alloy *solidifies* (changes from a liquid to a solid) at one temperature, not over a range of temperatures. This point is called the *eutectic point*, as shown in Figure 28-3.

In this diagram, at temperatures below line 1-2 generally, some of the lead becomes solid and most of the tin is liquid. As the alloy cools, all of the remaining lead and tin become solid at line 4-2. Likewise, under line 2-3, some tin solidifies and most of the lead remains a liquid. At line 2-5, all the remaining tin and lead become solid. Between points 4 and 5, the alloys have the same solidus (starting-to-melt) temperature. In the area to the left of 1-4, the alloy remains solid up to line 1-4, and is called *alpha*. In the area to the right of 3-5, the alloy remains solid up to the line 3-5, and is called *beta*. The other regions on the diagram are mixtures of alpha, beta, and liquid. All the regions are labeled on Figure 28-3.

28.2 COOLING CURVES

At atmospheric pressure, the temperature of a pure or "virgin" metal (not an alloy), remains constant during the change from the solid to the liquid phase. As a virgin metal is heated to its melting temperature with a constant heat input, a thermometer will show a constant rise in temperature per unit of time. The temperature increases until the metal reaches its melting temperature. As the metal melts, the temperature will remain constant for a length of time. During this time, the metal absorbs heat energy. This heat energy is required to give the atoms of the metal the energy required to break away from the solid and become liquid. After the metal is melted, the temperature will again rise as the metal is heated.

As a virgin metal cools from the liquid phase, the temperature may be observed to drop until the point where the metal solidifies. At this point, the temperature will remain constant while the atoms in the metal molecules change back to the solid state molecular structure. During this period of time while the atoms are changing position, the metal releases heat. After the metal has solidified, the temperature will again drop until it finally reaches room temperature.

The melting or solidifying of an *alloy* differs slightly from that of a pure metal. The alloy does not solidify at one specific temperature. It solidifies over a range of temperatures. This is shown in Figure 28-3. The alloy represented by line A begins to solidify when it crosses the liquidus line 1-2-3. It is not completely solid until it crosses the solidus line 4-5.

A change other than solid-to-liquid or liquid-to-solid is possible. One form of a solid may change to another form of a solid. This is true for iron and steel. There are three typical structures.

One of these is a *body-centered cubic structure*. There is one atom at each corner of a cube and one in the center. See Figure 28-4. This is the structure of steel at room temperature. A second form is the *face-centered cubic structure*. There is again one atom at each corner of the cube.

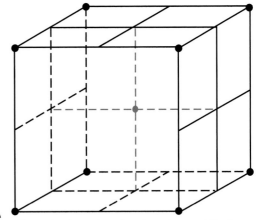

A

Examples: iron, chromium, tungsten, and titanium.

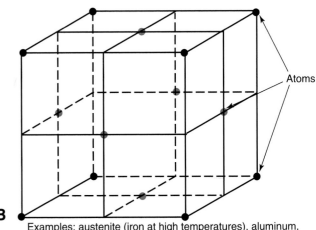

B

Atoms

Examples: austenite (iron at high temperatures), aluminum, copper, nickel, silver, gold, and lead.

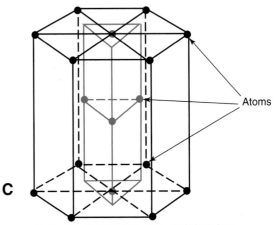

C

Atoms

Examples: magnesium, zinc, and zirconium.

***Figure 28-4.** Three types of crystal structure. A—Body-centered cubic. B—Face-centered cubic. C—Hexagonal close-packed.*

There is also one in the center of each face. This is the structure of austenite, a form of steel above 1341°F (727°C). Aluminum also has a face-centered cubic structure. A third type is a *hexagonal close-packed structure*. All three of these structures are illustrated in Figure 28-4.

When an alloy changes structure, there is a delay in the heating or cooling similar to a phase change. The temperatures at which phase changes or structural changes occur greatly depend on the alloy content. These temperatures are important when heat treating a metal and are often called **critical temperatures**. The critical temperatures for steel are shown on the iron-carbon diagram, Figure 28-5.

28.3 IRON-CARBON DIAGRAM

The **iron-carbon diagram**, Figure 28-5, shows the important temperatures for steel and cast iron. The iron-carbon diagram is very important in producing, forming, welding, and heat treating steel and cast iron. Before the diagram can be used, each region of the diagram must be fully understood.

Each of the regions on the diagram is labeled. There are also red capital letters on the diagram to help locate the various regions as they are discussed in the text. (These red letters are not normally a part of an iron-carbon diagram.)

Ferrite, the region marked A, contains very small amounts of carbon. Ferrite is also known as alpha ferrite or alpha iron. Ferrite can contain a maximum of 0.02% of carbon at 1341°F (727°C). As the temperature is increased to 1674°F (912°C), the amount of carbon in ferrite decreases to zero. Also, as the temperature of ferrite decreases to room temperature, the amount of carbon in ferrite decreases.

Ferrite is present in all steel and cast iron. It always contains the same amount of carbon, even though the carbon content of the steel or cast iron varies. Ferrite forms alone in regions A and E. Ferrite also forms along with cementite and is called pearlite in regions E, F, and G. Ferrite is formed when steel or cast iron is cooled below 1341°F (727°C). Ferrite has a body-centered cubic structure and is both ductile and tough.

Pure **cementite** has a molecular formula of Fe_3C. Cementite contains 6.69% carbon. Because it has such a high carbon content, pure cementite is not shown on Figure 26-5. As mentioned previously, as the carbon content in steel increases, the hardness and brittleness also increase. Cementite, with its high carbon content, is very hard and brittle. Cementite is also known as *iron carbide*.

Wherever the cementite is on the iron-carbon diagram, it always contains 6.69% carbon. At temperatures between 2106°F (1152°C) and 1341°F (727°C), region K, the cementite is mixed with austenite. Below 1341°F (727°C), the cementite can be found alone (regions F and G), or in combination with ferrite in the form of pearlite (regions E, F, and G). Cementite is thus formed in all steels and cast irons.

Pearlite is a combination of ferrite and cementite. The ferrite and cementite occur in alternating layers in the microstructure. The microstructure is shown in Figure 28-6. Pure pearlite, line B, is formed at 1341°F (727°C) and

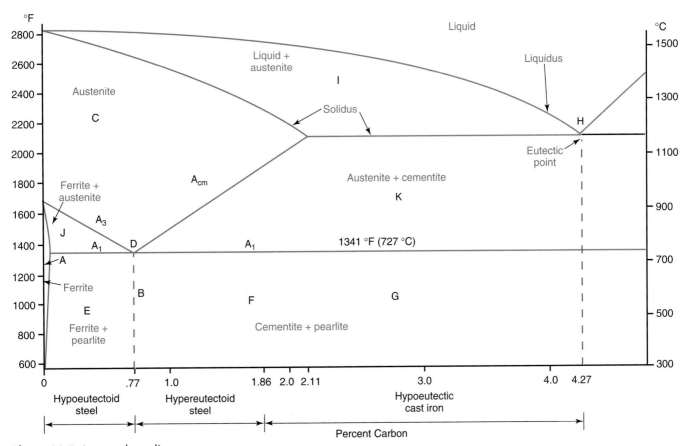

Figure 28-5. *Iron-carbon diagram.*

Figure 28-6. *Microstructure of pearlite. The dark regions are cementite, the light regions are ferrite.*

0.77% carbon. Pearlite always contains 0.77% carbon. If a steel contains less than 0.77% carbon, the pearlite is found along with ferrite, region E. If a steel or cast iron contains more than 0.77% carbon, cementite and pearlite will form (regions F and G).

Austenite, region C, is the final important region on the iron-carbon diagram. Austenite is stable above 1341°F (727°C) and below 2800°F (1538°C). Austenite can contain 1.86% carbon. Even though austenite does not occur at room temperature, it is an important region when heat-treating steels. Austenite has a face-centered cubic structure, which is different from the body-centered cubic structure of ferrite. See Figure 28-4 for the different structures. Austenite is also called *gamma iron*.

The *eutectoid point*, D, is an important point on the iron-carbon diagram. It occurs at 0.77% carbon and at 1341°F (727°C). When a steel is cooled through this point, pearlite is formed.

Very few steels have exactly the eutectoid composition. Most steels contain less than the eutectoid composition and are called *hypoeutectoid steels*, region E. The microstructure of these steels has a combination of ferrite and pearlite. Those steels which have a higher carbon content than the eutectoid point are called *hypereutectoid steels*, region F. The microstructure of a hypereutectoid steel has a combination of cementite and pearlite.

The *eutectic point*, H, is another important point on the diagram. It occurs at 4.27% carbon and 2095°F (1146°C). The reaction at the eutectic point is similar to the reaction at the eutectoid point. At the eutectic point, a liquid transforms into two solids, austenite and cementite. At the eutectoid point, a solid (austenite) transforms into two new solids, ferrite and cementite, called pearlite.

Cast iron is formed when the amount of carbon is between 1.8% and 4.0%. These cast irons have less than the eutectic composition and are called *hypoeutectic cast irons*, region G.

There are three remaining regions on the iron-carbon diagram. These are the *liquid plus austenite* region, the *ferrite plus austenite* region and the *austenite plus cementite* region. As a steel or cast iron is cooled through the liquid plus austenite region, I, the liquid transforms into austenite. As a steel is cooled through the ferrite plus austenite region, J, the austenite transforms into ferrite. Also, as a steel or cast iron is cooled through the austenite plus cementite region, region K, the austenite transforms into cementite.

The important *lines* on the iron-carbon diagram are also labeled. The *liquidus line* begins at 2800°F (1538°C) and goes down to 2095°F (1146°C) at 4.27% carbon, then increases again. The *solidus line* also begins at 2800°F (1538°C) and slopes down to 2095°F (1146°C) at 2.11% carbon. The solidus remains at 2095°F (1146°C) as the carbon content increases above 2.11%. The horizontal line at 1341°F (727°C), the eutectoid temperature, is the lowest critical temperature for steels. This temperature is known as the A_1. The upper critical temperature for hypoeutectoid steel is called the A_3. It begins at 1674°F (912°C) and slopes down to the eutectoid point. The upper critical temperature for a hypereutectoid steel is called the A_{cm}. This line extends from the eutectoid point up to the solidus line at 2.11% carbon. The lines A_1, A_3, and A_{cm} are found on most iron-carbon diagrams.

Most steels contain less than 1% carbon. A small change in carbon can greatly change the characteristics of a steel. The carbon content is specified in hundredths of 1%. Each 1/100 is called one point of carbon. Thus a steel containing 0.10% carbon is called a 10 point steel. One that has 0.85% is called an 85 point steel.

As an example of how a steel solidifies, follow the changes when a 20 point carbon steel is cooled through the complete temperature range. Begin with the steel as a liquid above 2800°F (1538°C). The molten metal cools until it reaches the liquidus line. At this point, solid austenite begins to form. The steel is now in the liquid plus austenite region, region I. It continues to cool until the solidus, where the steel becomes completely solid. The solid steel cools through the austenite region, region C. The steel crosses the A_3 line and begins to transform to ferrite. Ferrite continues to form until the eutectoid temperature is reached, the A_1. Now the remaining austenite forms pearlite. No further changes occur upon cooling to room temperature. The final structure has grains of ferrite and colonies of pearlite. The microstructure of a 20 point carbon steel is seen in Figure 28-7.

The iron-carbon diagram is also very important when heat-treating steel. The strength and ductility of a steel can vary over a wide range, depending on the heat treatment. See Chapter 29.

28.4 IDENTIFICATION OF IRON AND STEEL

A welder must be able to determine quite accurately the composition of the steel being handled. The most desirable method to determine the composition of a particular steel is to obtain the manufacturers' specifications

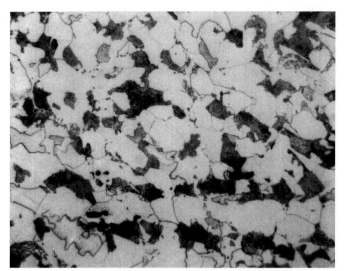

Figure 28-7. Microstructure of a 1020 steel. The white areas are ferrite grains. The other areas are pearlite.

and keep them on file. The metal should be marked to correspond with the information on file. In almost all manufacturing businesses, raw material is labeled to accurately identify it.

Sometimes, manufacturers' specifications are not available. This can often occur when doing repair welding. When specifications are not available, other methods may be used to determine the composition of the metal. Many tests have been developed to identify iron and steel. The following are the most common for shop use:

- Spark test, using a power grinder.
- Oxyacetylene torch test.
- Fracture test.
- Color test.
- Density or specific gravity test.
- Ring or sound of the metal upon impacting with some other metal.
- Magnetic test.
- Chip test.

Of these tests, the color, density, ring, and magnetic checks should take place almost subconsciously, over a period of time, as the welder works on various metals. The spark test and the gas torch test must be done under carefully prepared conditions. These tests indicate, to a remarkably accurate degree, the properties and constituents of the metal.

28.4.1 Spark Test

A spark test method of identifying metals is extensively used by welders to identify irons and steels. A power grinder is used as the test equipment. **When grinding, you must always wear goggles.** The grinder must be inspected to see that it is in good condition before you proceed with the test. A sample is tested by lightly touching the rim of the revolving wheel with the metal. The friction of the wheel surface, as it contacts the metal, heats the particles removed to the incandescent (glowing) and

burning temperature. The sparks resulting from the contact are found to differ in character for different steels. The lighter the contact, the better. You should use a black background, if possible, to better identify the spark.

The theory of the spark test is that — when a metal is heated — the different parts of the metal oxidize at different rates and the oxidation colors are different. Relatively pure iron, when heated by the grinding wheel, does not oxidize quickly. The sparks are long and fade out on cooling. The varying compounds of carbon and iron have different ignition temperatures when touched to the grinder. Therefore, the characteristics of the sparks differ as the carbon content of steel or cast iron increases. Four characteristics of the spark generally denote the nature and condition of the steel. These characteristics are:

- Color of spark.
- Length of spark.
- Number of explosions (spurts) along the length of the individual sparks.
- Shape of the explosions (forking or repeating).

For example, a mild steel containing 20 point carbon (0.20%) will show a long white spark that will jump approximately 70" (1.8m) from the power grinder (using a 12" [30cm] wheel). Some of these sparks will suddenly explode, shooting off smaller sparks at approximately 45° angles to the direction of travel of the original spark. A 30 point carbon steel will have sparks almost identical to the 20 point carbon steel, with the exception that more spark lines will explode. The total length of these spark lines will decrease slightly. This serves to show that as the carbon content of steel increases, the explosions of the sparks become more frequent. Also, as the carbon content increases, the length of the spark is decreased. In Figure 28-8, for example, a high-carbon tool steel containing 80 points of carbon has a very short spark with explosions occurring very rapidly. Sparks dissipate themselves very quickly. See Figure 28-9 for a table of spark test characteristics.

When higher-carbon-content metals are tested, you may become confused between extremely high-carbon steel and cast iron, because the spark is somewhat the same. Heat treatments have some effect on the nature of the spark. The cast iron spark, when leaving the point of contact, is a dull red and jumps only 20" to 25" (51cm to 64cm) from the wheel. The spark from a high-carbon steel of approximately 130 point carbon is white when it leaves the grinding wheel. The length of the spark is usually a little longer than the cast iron spark. Both of these metals show a small spark explosion at the end of the spark flight. The number of spark explosions in these two cases is considerably less than one would expect from such a high carbon content.

28.4.2 Oxyacetylene Torch Test

Even if you know the physical composition and the chemical composition of a metal, you must also know whether the metal has good welding properties. For example, some cold-rolled sheet steels may show very good physical and chemical properties. However, during some part of the manufacturing process, impurities have been

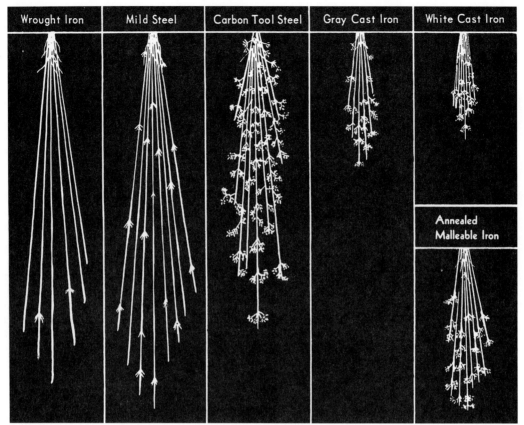

Figure 28-8. *Spark test for common cast irons and steels. (Norton Co.)*

Metal	Volume of stream	Relative length of stream*	Color of stream close to wheel	Color of streaks near end of stream	Quantity of spurts	Nature of spurts
Wrought iron	Large	65" (1.65 m)	Straw	White	Very few	Forked
Machine steel (AISI 1020)	Large	70" (1.78 m)	White	White	Few	Forked
Carbon tool steel	Moderately large	55" (1.40 m)	White	White	Very many	Fine, repeating
Gray cast iron	Small	25" (0.64 m)	Red	Straw	Many	Fine, repeating
White cast iron	Very small	20" (0.51 m)	Red	Straw	Few	Fine, repeating
Annealed mall. iron	Moderate	30" (0.76 m)	Red	Straw	Many	Fine, repeating

*Figures obtained with 12" (30 cm) wheel on bench stand are relative only. Actual length in each instance will vary with grinding wheel, pressure, etc.

Figure 28-9. *Table of spark test characteristics of common cast irons and steels. (Norton Co.)*

added to the steel or certain work was done to the metal, affecting its properties. The metal will not melt and fuse readily. The final weld may be unsatisfactory. The usual cause of this condition is impurities embedded in the metal. The impurities are usually slag and roller dirt or excessive sulfur and phosphorus. For these reasons, a welder should subject steel to the *oxyacetylene torch test.*

The actual test consists of melting a pool in the steel. If the metal is thin, the pool penetrates through the thickness of the steel until a hole is formed. This melting should be done with a neutral flame, held at the proper distance from the metal. The molten pool should not spark excessively or boil. It should be fluid and should possess good surface tension. The appearance on the edge of the pool or

hole indicates the weldability of the steel. If the metal that was melted has an even, shiny appearance upon solidification, the metal is generally considered as having good welding properties. However, if the molten metal is dull or has a colored surface, the steel is unsatisfactory for welding. The steel is also considered unsatisfactory for welding if the surface is rough (perhaps even broken up into small pits or porous spots).

This test is accurate enough for most welding. It is very easily applied with the equipment on the job. The test determines the one thing that is fundamentally necessary in any welding job: the weldability of the metal. Figure 28-10 shows how this test is conducted. While performing the weldability test of the metal, it is important to note the

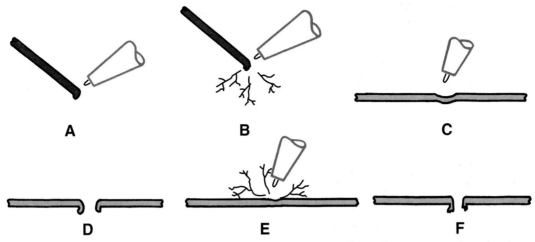

Figure 28-10. *Torch flame test: A—Good-quality filler rod. B—Poor-quality filler rod. C, D—Good-quality base metal. E, F—Poor-quality base metal.*

amount of sparking emitted from the molten metal. A metal that emits few sparks has good welding qualities.

28.4.3 Miscellaneous Identification Tests

In addition to the spark and torch tests, six other tests are occasionally used in the welding shop. The *fracture test* is used extensively and consists of breaking a portion of the metal in two. If it is a repair job, the fractured surface may be inspected. The appearance of the surface where the metal is cracked shows the grain structure of the metal. If the grains are large, the metal is ductile but weak. If the grains are small, the metal is usually stronger and has better toughness. Small grains are usually preferred. The fracture shows the color of the metal, which is a good means of distinguishing one metal from another. The test also indicates the type of metal by the ease with which it may be fractured. Figure 28-11 shows the appearance of the fractured surface for several different metals.

The *color test* separates two main divisions of metals. The irons and steels are indicated by their typical gray-white color. Nonferrous metals come in two general color classifications, yellow and white. Copper may be rather easily identified by the welder; the same applies to brass and bronze. Aluminum is a white metal. Aluminum alloys, zinc, and the like, are all somewhat silver-gray in color, although they may vary in shade. The metals may also be differentiated by means of the weight or density test of the specimen. A good example of identification by a *density* or *specific gravity test* is distinguishing between aluminum and lead. Roughly speaking, their colors are somewhat similar, but anyone may readily distinguish between the two metals because of their respective weights. Lead weighs about three times as much as aluminum.

The *ring test*, or sound of the metal test, is an easy means of identifying certain metals after some experience. It is used extensively for separating heat-treated steels

Figure 28-11. *Fractured surfaces of several different metals. (ESAB Welding and Cutting Products)*

from annealed steels. It is also used to distinguish alloys from the virgin metal. An example is the difference between aluminum and duralumin, an alloy of aluminum and copper. The pure aluminum sheet has a duller sound, or ring, than the duralumin, which is somewhat harder and has a more distinct ringing sound.

The *magnetic test* is an elementary test used to separate iron and steel metals from the nonferrous metals. Generally speaking, all steels are affected by magnetism while the nonferrous metals are not. However, some *stainless steels* are not magnetic.

A test that must be accompanied by considerable experience is the *chip test* of a metal. In this test, the cutting action of the chisel indicates the structure and heat treatment of the metal. Cast iron, for example, when being chip-tested, breaks off in small particles, whereas a mild steel chip tends to curl and cling to the original piece. Higher-carbon, heat-treated steels cannot be tested this way because of hardness.

A rough test to distinguish between mild carbon steel and chrome-moly steel is to compare the relative hardness of the metals while being cut with a hacksaw.

28.5 IDENTIFICATION OF IRON AND STEEL ALLOYS

Extensive research has been carried on with iron and carbon steels to which other elements have been added to improve or bring out certain properties. The two main fields in which this development has taken place are in the stainless and low-carbon alloy steels. These metals are sometimes difficult to identify from clean ordinary steels. The type of metal being welded must be known since its composition greatly affects its welding properties.

Alloys are added to plain iron-carbon steel to develop or improve the various physical properties of the metal. A common example includes the stainless steels. Alloying elements are used in these steels to make them corrosion-proof, and to develop other desired physical properties, such as toughness and strength. These metals consist of various combinations of chromium and nickel, along with molybdenum, titanium, zirconium, aluminum, or tungsten.

These other elements are added to the steel to increase its strength, hardness, and/or toughness. A good example of an alloying element is the use of tungsten. The addition of a small amount of tungsten to steel produces an extremely hard metal without sacrificing its other properties to any appreciable extent. Low-carbon steels are improved by adding various alloying elements. These alloys serve the purpose of producing stronger steels at a minimum cost. They are called high-strength, low-alloy (HSLA) steels. Heat treatment is often of great importance to obtaining the best physical properties of alloy steels.

28.5.1 Spark Test for Alloy Steels

The appearance of the power grinder spark for alloy steels will vary considerably, depending on the alloying metals included. The basic iron-carbon spark test is identical to that used with the straight iron-carbon steels. The number of spark explosions increases as the length of the spark flight decreases. The color of the spark becomes brighter and brighter as the carbon content increases. The major variations the alloy elements give to a particular steel are shown in the multitude of angled, branching sparks. These branching sparks are the result of the different alloys in the metal. These alloys also change the color of the spark.

As an example, a high-speed steel has only a slight indication of the spark explosion. Manganese in the steel causes the sparks to shoot out at 45° angles to the flight of the original spark. These sparks also tend to explode (repeating sparks) producing the appearance of a leafless tree branch, as shown in Figure 28-12.

Tungsten-chromium steels, which are used for high-speed work, show the typical high-speed-steel spark, with the exception that the spark turns to a chromium yellow (straw color) at the end of its flight. The chromium and the tungsten have a deteriorating effect on the carbon spark, causing it to be very fine or thin with repeating spurts. Figure 28-13 lists the spark characteristics for alloy steels. The chromium and tungsten spark is also an interrupted one. That is, it disappears for a portion of its flight and then appears again.

28.5.2 Torch Test for Alloy Steels

It is difficult to give rigid specifications of the torch test for all alloy steels. However, it is generally known that the higher the alloy content of a metal, the more difficult it is to weld that metal. Under the influence of the torch flame, metals have a tendency to boil because of the action of the alloys in them. Special fluxes are often used when welding alloy steels.

Stainless steels are difficult to oxyfuel gas weld; a flux is needed to produce good welds. Without the flux, the metal boils and a porous weld pool is produced.

Manganese steel is often used for surfaces that are exposed to abrasive wear, such as power shovel buckets. The usual application of welding in this case is to build up the worn surfaces. The steel melts readily under the torch flame and presents few difficulties in welding.

Nickel steels can be identified under the torch flame by the metal boiling action. Nickel-chrome steels behave in somewhat the same manner.

28.5.3 Miscellaneous Tests for Alloy Steels

The color test, ring test, magnetic test, fracture test, and chip test are all applicable to alloy steels, but perhaps the easiest one is the color test. The different alloy constituents tend to change the color of the metal, which is very noticeable and characteristic in certain compositions.

Stainless steels, for example, have a distinct silvery color that sets them apart from the other steel alloys. Some

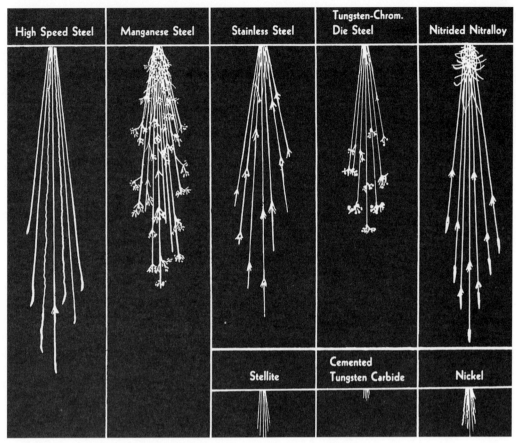

Figure 28-12. Spark test for some alloy steels and alloying metals. (Norton Co.)

Metal	Volume of stream	Relative length of stream*	Color of stream close to wheel	Color of streaks near end of stream	Quantity of spurts	Nature of spurts
High-speed steel (18-4-1)	Small	60" (1.52 m)	Red	Straw	Extremely few	Forked
Austenitic manganese steel	Moderately large	45" (1.14 m)	White	White	Many	Fine, repeating
Stainless steel (Type 410)	Moderate	50" (1.27 m)	Straw	White	Moderate	Forked
Tungsten-chromium die steel	Small	35" (0.89 m)	Red	Straw**	Many	Fine, repeating
Nitrided nitralloy	Large (curved)	55" (1.40 m)	White	White	Moderate	Forked
Stellite	Very small	10" (0.25 m)	Orange	Orange	None	
Cemented tungsten carbide	Extremely small	2" (0.05 m)	Light orange	Light orange	None	
Nickel	Very small***	10" (0.25 m)	Orange	Orange	None	
Copper, brass, aluminum	None					

*Figures obtained with 12" (30 cm) wheel on bench stand and are relative only. Actual length in each instance will vary with grinding wheel, pressure, etc.
Blue-white spurts.   * Some wavy streaks.

Figure 28-13. Table of spark test characteristics of alloy steels. (Norton Co.)

of these metals are magnetic; others are not. Identification can be accomplished to some extent by using the magnetic test. The chip test is not used extensively for identification because of the similarity in the appearance of practically all the stainless steel alloys. However, some of these alloys are considerably harder than others, and the chip test will bring this out quite clearly.

The ring or metallic sound of the metal is another test that can be applied to steel alloys with considerable accuracy. However, it is recommended that this test be used in conjunction with other tests.

28.6 NUMBERING SYSTEMS FOR STEELS

The Society of Automotive Engineers (SAE) has long been a leader in standardizing in industry. The SAE has developed a numbering system for identifying practically all steels. The code is based on a number with four digits, for example, 2315.

The first digit classifies the steel, with the number **2** representing nickel steels. The second digit represents the percent of the alloy metal in the steel, that is, **3** represents 3.25% to 3.75% nickel.

The last two digits represent the carbon content of the metal in points (hundredths of one percent), in this case, **15** or 0.15% carbon. Figure 28-14 is a table of sample SAE steels. The SAE first-number designations for steel and steel alloys are as follows:

1XXX	Carbon steels.
11XX	Special sulfur-carbon steels that have free cutting properties.
12XX	Phosphorus-carbon steels.
13XX	Manganese steels.
2XXX	Nickel steels.
3XXX	Nickel-chromium steels.
4XXX	Molybdenum steels.
5XXX	Chromium steels.

6XXX	Chromium-vanadium steels.
7XXX	Tungsten steels.
9XXX	Silicon-manganese steels.

Note: The Xs represent the space for the second, third, and fourth digits of the SAE designation.

Some special divisions of corrosion-resistant and heat-resistant steels are as follows:

30XXX	Nickel-chromium steels.
51XXX	Chromium steels.

The American Iron and Steel Institute (AISI) has also prepared standards for steels and steel alloys. In general, their code system is identical with that of the SAE code system. The American Society for Testing Materials (ASTM) also has developed a standardized code system for identifying and labeling steels.

28.7 NONFERROUS METALS

Nonferrous metals and their alloys are metals that do not contain iron as a major ingredient or component. They may contain small quantities of iron as an alloying element, but they are not considered ferrous metals. This group includes copper, brass, bronze, aluminum, solder,

SAE-AISI Carbon Steel Composition Numbers

SAE	AISI	Carbon range	Manganese range	Phosphorus (max.)	Sulfur (max.)
1010	C1010	0.08-0.18	0.30-0.60	0.040	0.050
1015	C1015	0.13-0.18	0.30-0.60	0.040	0.050
1020	C1020	0.18-0.23	0.30-0.60	0.040	0.050
1025	C1025	0.22-0.28	0.30-0.60	0.040	0.050
1030	C1030	0.28-0.34	0.60-0.90	0.040	0.050
1035	C1035	0.32-0.38	0.60-0.90	0.040	0.050
1040	C1040	0.37-0.44	0.60-0.90	0.040	0.050
1045	C1045	0.43-0.50	0.60-0.90	0.040	0.050
1050	C1050	0.48-0.55	0.60-0.90	0.040	0.050
1055	C1055	0.50-0.60	0.60-0.90	0.040	0.050
1060	C1060	0.55-0.65	0.60-0.90	0.040	0.050
1065	C1065	0.60-0.70	0.60-0.90	0.040	0.050
1070	C1070	0.65-0.75	0.60-0.90	0.040	0.050
1075	C1075	0.70-0.80	0.40-0.70	0.040	0.050
1080	C1080	0.75-0.88	0.60-0.90	0.040	0.050
1085	C1085	0.80-0.93	0.70-1.00	0.040	0.050
1090	C1090	0.85-0.98	0.60-0.90	0.040	0.050
1095	C1095	0.90-1.03	0.30-0.50	0.040	0.050

Resulfurized Carbon Steels

SAE	AISI	Carbon range	Manganese range	Phosphorus (max.)	Sulfur (max.)
1115	C1115	0.13-0.18	0.60-0.90	0.040 max.	0.08-0.13
1120	C1120	0.18-0.23	0.70-1.00	0.040	0.08-0.13
1125	C1125	0.22-0.28	0.60-0.90	0.040	0.08-0.13
1140	C1140	0.37-0.44	0.70-1.00	0.040	0.08-0.13

Figure 28-14. Typical SAE-AISI carbon steel composition numbers.

Rephosphorized and Resulfurized Carbon Steels

SAE	AISI	Carbon range	Manganese range	Phosphorus (max.)	Sulfur (max.)
1211	C1211	0.13 max.	0.60-0.90	0.07-0.12	0.08-0.15
1213	C1213	0.13 max.	0.70-1.00	0.07-0.12	0.24-0.33

Manganese Steels

SAE	AISI	Carbon range	Manganese range	Phosphorus (max.)	Sulfur (max.)	Silicon (max.)
1330	1330	0.28-0.33	1.60-1.90	0.040	0.040	0.20-0.35
1335	1335	0.33-0.38	1.60-1.90	0.040	0.040	0.20-0.35
1340	1340	0.38-0.43	1.60-1.90	0.040	0.040	0.20-0.35
1345	1345	0.43-0.48	1.60-1.90	0.040	0.040	0.20-0.35

Nickel Steels

SAE	AISI	Carbon range	Manganese range	Phosphorus (max.)	Sulfur (max.)	Nickel range
2315	- - - -	0.10-0.20	0.30-0.60	0.040	0.050	3.25-3.75
2330	- - - -	0.25-0.35	0.50-0.80	0.040	0.050	3.25-3.75
2340	- - - -	0.35-0.45	0.60-0.90	0.040	0.050	3.25-3.75
2345	- - - -	0.40-0.50	0.60-0.90	0.040	0.050	3.25-3.75
- - - -	2515	0.10-0.20	0.30-0.60	0.040	0.050	4.75-5.25

Nickel-Chromium Steels

SAE	AISI	Carbon range	Manganese range	Phosphorus (max.)	Sulfur (max.)	Nickel range	Chromium range	Silicon
3140	3140	0.38-0.43	0.70-0.90	0.040	0.040	1.10-1.40	0.55-0.75	0.20-0.35
3310	E3310	0.08-0.13	0.45-0.60	0.025	0.025	3.25-3.75	1.40-1.75	0.20-0.35

Molybdenum Steels

SAE	AISI	Carbon range	Manganese range	Phosphorus (max.)	Sulfur (max.)	Chromium range	Nickel range	Molybdenum range	Silicon
4130	4130	0.28-0.33	0.40-0.60	0.040	0.040	0.80-1.10	- - - -	0.15-0.25	0.20-0.35
4140	4140	0.38-0.43	0.75-1.00	0.040	0.040	0.80-1.10	- - - -	0.15-0.25	0.20-0.35
4150	4150	0.48-0.53	0.75-1.00	0.040	0.040	0.80-1.10	- - - -	0.15-0.25	0.20-0.35
4320	4320	0.17-0.22	0.45-0.65	0.040	0.040	0.40-0.60	1.65-2.00	0.20-0.30	0.20-0.35
4340	4340	0.38-0.43	0.60-0.80	0.040	0.040	0.70-0.90	1.65-2.00	0.20-0.30	0.20-0.35
4615	4615	0.13-0.18	0.45-0.65	0.040	0.040	- - - -	1.65-2.00	0.20-0.30	0.20-0.35
4620	4620	0.17-0.22	0.45-0.65	0.040	0.040	- - - -	1.65-2.00	0.20-0.30	0.20-0.35
4815	4815	0.13-0.18	0.40-0.60	0.040	0.040	- - - -	3.25-3.75	0.20-0.30	0.20-0.35
4820	4820	0.18-0.23	0.50-0.70	0.040	0.040	- - - -	3.25-3.75	0.20-0.30	0.20-0.35

Chromium Steels

SAE	AISI	Carbon range	Manganese range	Phosphorus (max.)	Sulfur (max.)	Chromium range	Silicon
5120	5120	0.17-0.22	0.70-0.90	0.040	0.040	0.70-0.90	0.20-0.35
5140	5140	0.38-0.43	0.60-0.90	0.040	0.040	0.70-0.90	0.20-0.35
5150	5150	0.48-0.53	0.60-0.90	0.040	0.040	0.70-0.90	0.20-0.35
52100	E52100	0.95-1.10	0.25-0.45	0.025	0.025	1.30-1.60	0.20-0.35

Figure 28-14. (Continued)

stellite, lead, zinc, nickel, etc. These metals can be identified by various means. They have distinctive colors. They are nonmagnetic and are usually relatively soft metals.

Nonferrous metals will not spark when touched to the grinding wheel. Nonferrous metal will generally clog the grinding wheel and prevent it from grinding properly.

Chromium-Vanadium Steels

SAE	AISI	Carbon range	Manganese range	Phosphorus (max.)	Sulfur (max.)	Chromium range	Vanadium	Silicon
6118	6118	0.16-0.21	0.50-0.70	0.040	0.040	0.50-0.70	0.10-0.15	0.20-0.35
6120	6120	0.17-0.22	0.70-0.90	0.040	0.040	0.70-0.90	0.10 min.	0.20-0.35
6150	6150	0.48-0.53	0.70-0.90	0.040	0.040	0.80-1.10	0.15 min.	0.20-0.35

Nickel-Chromium-Molybdenum Steels

SAE	AISI	Carbon range	Manganese range	Phosphorus (max.)	Sulfur (max.)	Chromium range	Molybdenum range	Nickel range	Silicon
8115	8115	0.13-0.18	0.70-0.90	0.040	0.040	0.30-0.50	0.08-0.15	0.20-0.40	0.20-0.35
8615	8615	0.13-0.18	0.70-0.90	0.040	0.040	0.40-0.60	0.15-0.25	0.40-0.70	0.20-0.35
8720	8720	0.18-0.23	0.70-0.90	0.040	0.040	0.40-0.60	0.20-0.30	0.40-0.70	0.20-0.35
8822	8822	0.20-0.25	0.75-1.00	0.040	0.040	0.40-0.60	0.30-0.40	0.40-0.70	0.20-0.35

Silicon-Manganese Steels

SAE	AISI	Carbon range	Manganese range	Phosphorus (max.)	Sulfur (max.)	Silicon range	Nickel range	Chromium range	Molybdenum range
9260	9260	0.55-0.65	0.70-1.00	0.040	0.040	1.80-2.20	- - - -	- - - -	- - - -
9840	9840	0.38-0.43	0.70-0.90	0.040	0.040	0.20-0.35	0.85-1.15	0.70-0.90	0.20-0.30

Chromium-Nickel Steels

SAE	AISI	Carbon	Manganese (max.)	Silicon (max.)	Phosphorus	Sulfur	Chromium range	Nickel range	Molybdenum
30304	304	0.08 max.	2.00	1.00	0.040	0.040	18.00 min.	8.00 min.	- - - -
30317	317	0.10 max.	2.00	1.00	0.040 max.	0.040 max.	16.00-18.00	10.00 min.	2.00 max.
51410	410	0.15 max.	1.00	1.00	0.040 max.	0.040 max.	11.50-13.50	0.60 max.	0.60 max.

Figure 28-14. *(Continued)*

Caution: Grinding of nonferrous metals is not recommended except for the extremely short intervals of time required for spark testing. The oxides of some nonferrous metals are toxic; therefore, the operator must wear an air-filtering breathing apparatus and protective clothing. The grinder must be equipped with an adequate exhaust system.

28.7.1 Copper

Copper is an element of the metal family that has many uses because of its electrical and thermal conductivity and its ability to resist corrosion. Most copper has a reddish-brown color. Copper melts at a temperature of 1981°F (1083°C). This is higher than the melting temperature of silver and considerably lower than the melting temperature of iron, as shown in the chart in Figure 28-15.

Metal	Melting temperatures	
	°F	°C
Aluminum	1217	659
Bronze 90 Cu 10 Sn	1562-1832	850-1000
Brass 90 Cu 10 Zn	1868-1886	1020-1030
Brass 70 Cu 30 Zn	1652-1724	900-940
Copper	1981	1083
Iron	2786	1530
Lead	621	327
Mild steel	2462-2786	1350-1530
Nickel	2646	1452
Silver	1761	960
Tin	450	232
Zinc	786	419

Figure 28-15. *Melting temperatures of some common metals.*

Manufacturing processes are such that commercial copper usually contains sulfur, phosphorus, and silicon as its impurities. Each of these impurities has a tendency to make the copper more brittle and to reduce its weldability. However, a very small amount of phosphorus in copper is an aid to the welding of the metal. It is helpful because the dissolving property that phosphorus has for copper oxides permits it to act as a flux.

The only copper that is recommended for fusion welding purposes is *deoxidized copper*. This copper has had a very small amount of silicon added to it. Silicon has the property of dissolving whatever copper oxides are present in the metal. Enough silicon is added to the copper during its manufacture that an excess will be left in the copper after deoxidizing action takes place. If this amount is too much, as mentioned previously, the copper tends to become brittle.

Another feature of copper, which is typical of practically all nonferrous metals, is its behavior called *hot shortness*. As copper is heated to its melting temperature, the metal becomes very weak at a certain temperature, even though it is still a solid. The slightest shock or weight will tend to distort the metal unless it is very firmly supported and firmly clamped. Proper support prevents distortion while the metal is passing through this "hot shortness" temperature.

28.7.2 Brass

Brass is an alloy of copper and zinc. Small amounts of other metals are frequently added. The amount of the zinc in the alloy may vary from 10% to 40%. A common alloy is 70% copper and 30% zinc. This metal is used principally because of its acid-resistant qualities, its appearance, and its properties as a good brazing alloy. There are two common types of brass: one type is called machine brass and contains 32% to 40% zinc; the other is red brass, with from 15% to 25% zinc. Some additional metals that are added to improve the physical properties of brass are tin, manganese, iron, and lead. These make brass a triple alloy. Brass may be identified by its color which is an opaque yellow.

28.7.3 Bronze

Bronze is an alloy of copper and tin. A common ratio is 90% copper and 10% tin. Bronze is more coppery in color than brass. It behaves much like brass when being welded. Generally speaking, the welder can use the same filler rod and the same flux for both bronze and brass. Like brass, bronze is highly resistant to corrosion. Because of its attractive appearance, it is often used for decorative parts and objects.

28.7.4 Aluminum

Aluminum is an element of the metal family known for its electrical conductivity, heat conductivity, resistance to corrosion, and light weight. It is obtainable either in rolled (wrought) or cast form. Aluminum may be combined with many other metals to form alloys. The different alloys that are added to aluminum are listed in Figure 28-16, along with the Aluminum Association designation for those alloys.

The Aluminum Association alloy group designation	Major alloying element	Examples
1XXX	99% Aluminum	1100
2XXX	Copper	2014, 2017, 2024
3XXX	Manganese	3003
4XXX	Silicon	4043
5XXX	Magnesium	5052, 5056
6XXX	Magnesium and silicon	6061
7XXX	Zinc	7075
8XXX	Other elements	

Figure 28-16. *Alloying elements added to aluminum and the designation for each. The 1XXX group contains at least 99% pure aluminum.*

In its pure form, aluminum metal has a white color. It is very ductile in the rolled aluminum sheet form. Cast aluminum is very brittle. The strength of the pure metal is considerably less than that of steel. Its melting temperature is approximately 1220°F (660°C), as shown in Figure 28-15. This metal also has a critical point called hot shortness. For that reason, it must be carefully supported when being welded. An element that may be added to aluminum to decrease its hot shortness is silicon.

Aluminum sheet metal may be obtained in several qualities and grades. The purest commercial aluminum contains 99.5% aluminum, while the more popular commercial grades contain 99% aluminum. Two metals that are added to aluminum to increase its desirable physical qualities are manganese and magnesium. Amounts of these are relatively small, varying from 1% to 5%.

Due to the activity of the metal, aluminum oxidizes very readily upon being heated. Aluminum oxide melts at about 5000°F (2760°C). Therefore, wire brushing or chemicals must be used to remove the oxide. The oxide is more dense than the molten metal and settles into the weld pool, causing a porous weld. Therefore, special fluxes must be used at all times during the welding process, unless an inert gas process (GTAW or GMAW) is used.

Another feature of aluminum that adds to the difficulty of welding is that it does not change in color before it reaches the melting temperature. In other words, the metal upon being heated maintains the same color, but when reaching the melting point, it suddenly becomes liquid. When welding aluminum, the welder can determine the melting temperature of the metal by using the filler rod to scratch the surface to reveal any softening.

Aluminum may be identified by its silvery white color when fractured, and by a comparison of weights with other metals. Aluminum is about one-third the weight of iron for a given volume.

Another test for aluminum is to burn some chips of the metal. Aluminum will burn to a black ash. Magnesium may be mistaken for aluminum. However, magnesium chips when heated will actually ignite and will form a white ash. **Caution: Magnesium chips and powder burn violently. When igniting magnesium, use only a very small quantity of the chips.**

28.8 HARD SURFACING METALS

Certain metal alloys have the property of extreme hardness. There are several different types:
- Ferrous metal with up to 20% alloying elements, such as chromium, tungsten, and manganese.
- Ferrous metal with over 20% alloying elements, such as chromium, tungsten, and manganese. (Cobalt and nickel are sometimes added.)
- Nonferrous metal with alloying elements, such as cobalt, chromium, and tungsten.
- Tungsten carbide (fused), with some other alloying elements.
- Granules of tungsten carbide. See Chapter 26 for information on metal surfacing.

These metals are difficult to identify. It is best to keep them in their identifying packages. Figure 28-12 shows the characteristic sparks for stellite and for cemented tungsten carbide, two hard surfacing materials.

These metals range from extremely hard to very tough. Their best application, therefore, is as a thin coating on metal of a more ductile nature. This combination produces a long-wearing surface and also one of great strength.

28.9 SOFT SURFACING METALS

The soft surfacing metals are usually nonferrous metals. These may be identified by the methods described earlier in this chapter.

The properties of the deposited surface metal depend upon the surfacing metal and the method that will be used to apply the metal to the surface of the parent metal. Surfacing metals can be applied by:
- Dipping.
- Electroplating.
- Spraying.
- Brazing or soldering.

28.10 NEWER METALS

Many of the metals formerly considered exotic are used as alloying elements to improve the properties of more common metals. A few of these newer metals are: columbium, titanium, lithium, barium, zirconium, tantalum, beryllium, nobelium, and plutonium.

The manufacture and use of these metals is increasing. Titanium, zirconium, and columbium are widely used as alloys in steel. Welding engineers are constantly developing and improving methods to weld and braze these metals. See Chapter 22 for information on the welding of some of these metals.

28.10.1 Titanium

Titanium is an important structural metal. It is the fourth most abundant metal, exceeded only by aluminum, iron, and magnesium. Commercially pure titanium melts at about 3035°F (1668°C). Many of the alloys are weldable under certain conditions. This metal has good weight-to-strength ratio and high-temperature properties, and it is very corrosion-resistant. The aircraft, chemical, and transportation industries use titanium in various applications. It is about 67% heavier than aluminum and about 40% lighter than stainless steels. It retains its strength very well up to 1000°F (540°C).

Titanium-carbon alloys find wide acceptance. The carbon content varies from 0.015% to 1.1%. The titanium-carbon alloys become more brittle as the carbon content increases, but they also become more corrosion-resistant. The maximum tensile strength is reached at about 0.4% carbon and is approximately 112,000 psi (772MPa). However, 0.04% carbon alloy has a strength of 92,000 psi (634MPa).

Titanium alloys using tungsten and carbon have a tensile strength of about 130,000 psi (896MPa). In combination with aluminum, it has a tensile strength of about 114,000 psi (786MPa). Some of the other alloys added to titanium and the resulting tensile strengths are given below:
- Chromium-nickel – 195,000 psi (1345MPa)
- Chromium-molybdenum – 175,000 psi (1207MPa)
- Chromium-tungsten – 150,000 psi (1034MPa)
- Manganese-aluminum – 160,000 psi (1103MPa)

Titanium-manganese alloys, such as 8% manganese, are popular in aircraft construction. Some other alloys consist of 3% aluminum and 0.5% manganese, 6% aluminum and 4% vanadium, and 5% aluminum and 2.5% tin.

28.10.2 Lithium

The welding industry is interested in lithium. When this metal is alloyed with some other brazing metal, the brazing can usually be performed without flux. For example, titanium can be successfully brazed in an inert atmosphere using an alloy of about 98% silver and 2% lithium. Lithium has a melting temperature of 367°F (186°C).

28.10.3 Zirconium

Zirconium is a rare metal found in nature combined with silicon as zirconium silicate ($ZrSiO_4$), commonly known as the semiprecious jewel *zircon*. Zirconium oxide melts at 4892°F (2700°C) and is used as a lining for high-temperature furnaces. This metal oxidizes easily. Its strength is affected by oxygen, nitrogen, and hydrogen, with which it may combine. Zirconium is used as an alloying element in alloy steels.

Zircalloys are a family of zirconium alloys containing about 1.5% tin, nickel, or chromium and up to 0.5% iron. These alloys have improved corrosion resistance and higher strength than unalloyed zirconium.

Because zirconium readily unites with oxygen and nitrogen when heated, it is welded best in an inert gas-filled chamber. The inert shielding gases used are argon and helium.

28.10.4 Beryllium

Beryllium is a light metal with a density (1.845 g/cc) that is slightly higher than the density of magnesium. It is often used as an alloying element with other metals. It has been used as an alloying element in copper and nickel to increase elasticity and strength characteristics.

Beryllium has a high thermal and electrical conductivity and a high heat absorption rate. Because of these qualities, it is difficult to weld. Beryllium has been welded by the inert-gas, resistance, ultrasonic, electron-beam, and diffusion welding processes. The brazing and soldering methods of joining have also been used.

Some beryllium copper alloys are very strong and are used to replace forged steel tools in places when an explosive atmosphere may be present, since these alloys are nonsparking.

Beryllium is used with magnesium to reduce the tendency of magnesium to burn during melting and casting. **This metal and its compounds have dangerously poisonous (toxic) properties, and special precautions must be taken when working with the metal and with alloys that use this metal.**

28.11 REVIEW OF SAFETY

Spark-testing metals for identification requires the tester to wear goggles and/or a face shield. The grinder must be in good condition, including balanced grinding wheels; wheel rest clearance must not to be greater than 1/16" (1.6mm).

When a torch flame or an arc is used to identify a metal by its melting and sparking behavior, all the safety precautions for gas welding and/or arc welding must be followed. See Chapters 1, 5, 6, and 11-15.

TEST YOUR KNOWLEDGE

Write your answers on a separate sheet of paper. Do not write in this book.

1. What is meant by an alloy steel?
2. How much carbon does a medium-carbon steel contain?
3. Carbon usually combines with iron to form _____, which is also called _____ _____.
4. What is elongation?
5. What characteristics does nickel give to steel?
6. Does a noneutetic alloy solidify at one temperature, or over a range of temperatures?
7. What is meant by a critical temperature of a metal?
8. How do the properties of steel change as carbon content increases?
9. Pearlite is a combination of _____ and _____.
10. The carbon content of ferrite is _____; the content of cementite is _____; the carbon content of pearlite is _____.
11. Using the iron-carbon diagram, describe the different regions a 1050 steel passes through as it cools from a liquid to a solid at room temperature.
12. What is present in the microstructure of a 1050 steel at room temperature?
13. Name five methods of identifying metals.
14. What can be learned from the spark test?
15. What can be learned from the oxyacetylene torch test?
16. What is the percent of carbon or points of carbon in an SAE 1045 steel?
17. What do the initials ASTM, SAE, and AISI represent?
18. Describe what is meant by the term "hot shortness."
19. What element may be added to aluminum to reduce hot shortness?
20. What alloying element is added to the aluminum series 2XXX? 5XXX?

Robotic arc welding of an automotive component. After fabrication, welded components often must be stress-relieved or given other heat treatment. (FANUC Robotics)

Chapter 29

HEAT TREATMENT OF METALS

LEARNING OBJECTIVES

After studying this chapter, you will be able to:

* List at least seven reasons for heat treatments.
* Name the methods used to heat metals for heat treatment.
* Describe the crystalline structure of steel when it is cooled quickly, slowly, or in different stages.
* Describe the weld heat-affected zone (HAZ) and what may happen in that zone as the rate of cooling of the weld is changed.
* Define the following terms: annealing, normalizing, tempering, spheroidizing, hardening, and casehardening.
* List several methods used to measure high temperatures.

The welder should be particularly interested in the heat treatment of metals. It is important to know what effect welding will have on the heat-treated material. A good understanding of the effects of welding on the physical properties of the metal is helpful. It is necessary to know if a metal must be preheated for welding, and if the metal should be heated during the welding operation. Finally, the welder should learn what heat treatment procedure to use after welding (postweld heat treatment) to bring back, as nearly as possible, the original properties of the metal.

Many common metals are welded without any heat treatment. Other metals require heat treatment. Heat can be applied at three different times. Heat applied before welding is called *preheating*. Adding heat during welding is called concurrent or *interpass heating*. Heating after welding is called *postweld heat treating* of a metal. In large shops and in manufacturing plants, these heat treatments are a part of the procedure determined by the engineering staff.

The most common application of postweld heat treating in a welding shop is stress relieving. This process relieves the metal of internal stresses and strains caused by the expansion and contraction during welding. It also improves the properties of the metal in the weld and in the heat-affected zone. Most structural welding involves only the knowledge of how a metal may be annealed or stress relieved. In repair welding, an extensive and accurate knowledge of all phases of heat treating is required by the welder.

29.1 THE PURPOSES OF HEAT TREATMENT

All metals can be heat treated. Some types of metals are affected very little by heat treating, but others, particularly steels, are greatly affected. Heat treating may serve the following purposes:

* Develop ductility.
* Improve machining qualities.
* Relieve stresses.
* Change grain size.
* Increase hardness or tensile strength.
* Change chemical composition of the metal surface, as in casehardening.
* Alter magnetic properties.
* Modify electrical conduction properties.
* Induce toughness.
* Recrystallize metal that has been cold-worked.

When heat treating metals, there are four factors of great importance:

* The *temperature* to which the metal is heated.
* The *length of time* that the metal is held at that temperature.
* The *rate* (speed) at which the metal is cooled.
* The *material surrounding* the metal when it is heated, as in casehardening.

When a weld is made, the metal in and around the weld joint is heated to various temperatures, depending upon its distance from the weld joint. This drop in the metal temperature from the weld joint outward is called a *temperature gradient*. In steels, some of the metal in the area of the weld is heated through only one critical point, others through two. Some spots may be heated to only 500°F (260°C), as shown in Figure 29-1.

Because of such uneven heating, strength, ductility, grain size, and other metal properties may vary greatly in the weld and in the *heat-affected zone (HAZ)*.

The welding operator may use preheating and/or concurrent heating (heat added while welding) to reduce temperature gradients in the weld area. Heating the metal before welding will help to prevent internal stresses and strains that may cause weld failure. The welder may use continuous heating for the same purpose. Postweld heating is used to relieve stresses and bring all the metal in the area of the weld to the same heat treated condition.

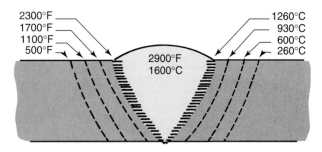

Figure 29-1. *Temperature zones in the weld area.*

29.1.1 Methods of Heating

Heating metals is a complex operation. All metals expand when heated and contract on cooling. A rather drastic change in volume takes place as metals are heated above their critical temperatures, or are cooled below their critical temperatures. Recall that a *critical temperature* is a point where a phase change occurs or a crystalline structure change takes place. Refer to Heading 28.2.

Unequal heating or cooling causes unequal expansion and contraction. Such expansion and contraction often will cause a warping of the structure. The ability to warp metals by heating them is not always a *disadvantage*. Warped articles can be straightened by the proper local heating of the metal.

A metal's surface may be heated by using heat from an:
- Air-fuel gas torch.
- Oxyfuel gas torch.

Or, the metal may be heated by:
- Electrical resistance heating.
- Induction heating.
- Furnace heating.

Electrical resistance heating may be used to heat a part by placing electrical resistance units against the material to be heated, as shown in Figure 29-2.

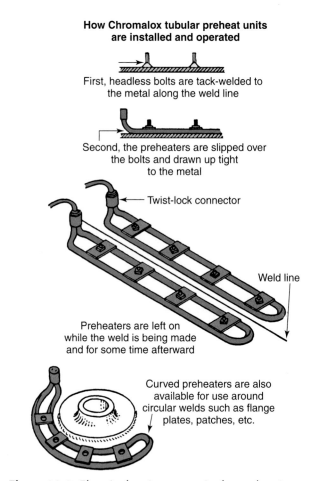

How Chromalox tubular preheat units are installed and operated

First, headless bolts are tack-welded to the metal along the weld line

Second, the preheaters are slipped over the bolts and drawn up tight to the metal

Twist-lock connector

Weld line

Preheaters are left on while the weld is being made and for some time afterward

Curved preheaters are also available for use around circular welds such as flange plates, patches, etc.

Figure 29-2. *Electrical resistance units for preheating.*

Induction heating is a means of heating that uses high frequency alternating current to create eddy currents in the metal. This method heats a metal throughout its thickness. See Figure 29-3.

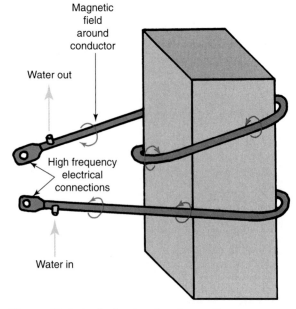

Magnetic field around conductor

Water out

High frequency electrical connections

Water in

Figure 29-3. *An induction heating coil.*

Heating an entire part or assembly is usually done in a *furnace.* A uniform temperature can be obtained throughout. A furnace may be heated by air-fuel gas, by electrical induction, or by electrical resistance units.

29.1.2 Methods of Cooling

The speed and evenness of cooling a metal object determines to a great extent the physical properties of the metal. Cooling may be accomplished in various ways. The heat loss in cooling is usually a combination of:
- Convection.
- Conduction.
- Radiation.

Cooling in a gas occurs mainly by convection and radiation. Remember that air is a gas. Cooling in a liquid occurs mainly by convection and conduction. Cooling can only be done to approximately the temperature of the cooling medium. The colder the cooling medium, the faster the cooling will take place. Cooling in a gas (by convection) is the slowest. Cooling in a cold liquid is fastest. The liquid can be any desired temperature. For example: water at room temperature (70°F, 21°C), ice water (32°F, 0°C), or liquid nitrogen (–230°F, –146°C).

Air is a common gas cooling medium. Two methods of cooling in air are:
- Thermal convection (cooling in still air).
- Forced convection (cooling with fans).

Figure 29-4 shows parts being cooled in air.

Figure 29-4. *Threaded rods being removed from a vertical-type heat treating furnace. They will be air-cooled. (J. W. Rex Company)*

Water is a common liquid cooling medium. Two methods of cooling in water are:
- Submerging the metal in a water tank.
- Spraying water on the metal to be cooled. Oil, molten low-temperature metals, liquid air, or liquid nitrogen are other cooling liquids.

Another means of cooling and controlling the cooling of metals is to clamp the part in either a water-cooled or refrigerant-cooled fixture. One problem in cooling a metal is that the surface of the metal cools faster than the inside of the piece or part.

29.2 CARBON CONTENT OF STEEL

Steels may be obtained with various carbon contents and alloying metals. Figure 29-5 shows the carbon content of some common items made of steel. The family of steels starts with low-carbon steel (almost wrought iron). As the carbon content increases, the steel becomes harder, stronger, and more brittle until a content of approximately 1.86% carbon is reached. This is the maximum percent of carbon that will combine with iron to form steel. Iron with carbon content that is higher than 1.86% forms cast iron.

The physical properties of iron and steel are described in Heading 28.1.1.

Article	Carbon Content
Axles	.40
Boiler plate	.12
Boiler tubes	.10
Castings, low-carbon steel	Less than .20
Casehardening steel	.12
Cold chisels	.75
Files	1.25
Forgings	.30
Gears	.35
Hammers	.65
Lathe tools	1.10
Machinery steel	.35
Metal tools	.95
Nails	.10
Pipe, steel	.10
Piano wire	.90
Rails	.60
Rivets	.05
Set screws	.65
Saws for wood	.80
Saws for steel	1.55
Shaft	.50
Springs	1.00
Steel for stamping	.90
Tubing	.08
Wire, soft	.10
Wood cutting tools	1.10
Wood screws	.10

Figure 29-5. *Steels and how the carbon content varies, depending on the use of the steel.*

29.3 CRYSTALLINE STRUCTURE OF STEEL

Practically all types of heat treatments deal with the crystalline structure of steel or the grain size. The heat treatments are also concerned with the distribution of cementite and ferrite in the metal. Cementite is commonly called *iron carbide*. In 10-point carbon steel, which contains 1/10 of 1% carbon, most of the carbon is in chemical combination with iron, forming *cementite*, Fe_3C. Since there is not much carbon in a 10-point steel, there is very little cementite. There is, however, a considerable amount of free iron which is called *ferrite*. Ferrite is very ductile. Steel containing a large amount of ferrite cannot be hardened. The only heat treatment possible is to alter the grain structure and size. Heating this steel to the **upper transformation temperature** (the A_3 critical temperature), does not greatly affect the steel's hardness. This is because it is the amount of cementite that determines the hardness. The A_3 critical temperature may be seen on the iron-carbon diagram, Figure 28-5.

Steels with a carbon content above 35 points can be hardened. The final microstructure and hardness of the steel greatly depends on the cooling rate from the austenite region. The steel microstructures that result from different cooling rates are shown in a **time-temperature-transformation (T-T-T) diagram**. See Figure 29-6.

To help you understand the T-T-T diagram, three different cooling rates for steel will be examined. In Figure 29-6, the steel is cooled from the austenite region. Steel can be cooled at different rates to produce different microstructures. The first line crossed as the metal cools is where the austenite microstructure begins to transform. The second line crossed indicates the point at which the transformation is complete.

If a steel is cooled slowly (line A on Figure 29-6) from the austenite region, *pearlite* forms. The microstructure is shown in Figure 29-7. If the steel is quenched in ice water (line C) *martensite* is formed. See Figure 29-8. If the steel is initially quenched to a moderate temperature of 675°F (357°C), then slow-cooled, *bainite* is formed. Each of these microstructures has different characteristics. The microstructures are listed in Figure 29-9. Martensite, which is cooled the fastest, is the hardest microstructure. Pearlite, which is cooled slowly, has the best ductility.

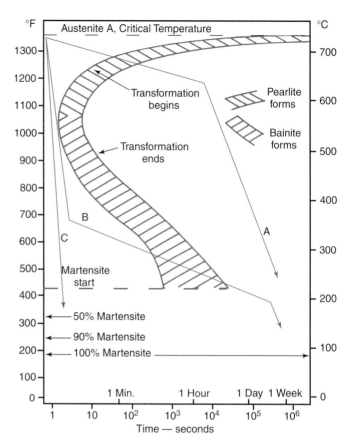

Figure 29-6. *Time-temperature-transformation (T-T-T) diagram for a eutectoid steel. If the steel is cooled slowly (line A), pearlite forms. If the steel is first quenched and then cooled slowly, bainite forms (line B). If the steel is fully quenched (line C), martensite forms. Note the time is in powers of ten. 10^4 means $10 \times 10 \times 10 \times 10 = 10,000$.*

The steels with higher carbon contents and higher alloy contents form martensite more easily than low-carbon, low-alloy steels. This also holds true for weldments as the weld metal cools.

Weld metal may contain pearlite or martensite, or both. The grains may be large or small. The carbides may be large, or they may be small and dispersed. The carbon and alloy content, and cooling rate of the weld metal and heat-affected zone, determine the microstructure.

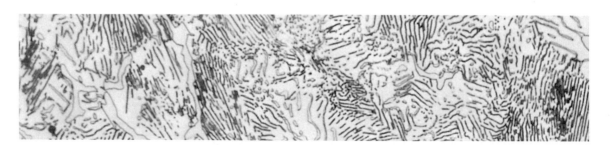

Figure 29-7. *Microstructure of a high-carbon steel with pearlite grain structure. This is magnified 1000 diameters.*

Figure 29-8. *Microstructure of martensite magnified 500 diameters.*

Microstructure	Characteristics
Martensite	Extremely hard, strong, and brittle.
Bainite	Good strength, not as hard as martensite, ductile and tough.
Pearlite	Good ductility, not as hard as bainite, fairly strong and tough.

Figure 29-9. *A table of the characteristics for the various types of steel microstructures.*

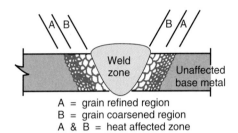

A = grain refined region
B = grain coarsened region
A & B = heat affected zone

Figure 29-10. *The heat-affected zone (HAZ) has two regions: the grain-refined region (A) and the grain-coarsened region (B). While welding, the grain-coarsened region is heated to a higher temperature than the grain-refined region.*

The heat-affected zone has both large and small grains. The small-grain region is next to the unaffected base metal. This area is heated to just above the A_3 critical temperature. New refined grains are formed. After it cools, the region is called the *grain-refined region.* The next region, next to the weld zone, is the *grain-coarsened region.* See Figure 29-10.

The temperature in the grain-coarsened region was well above the A_3 critical temperature. The grains had time to grow to a large size. Grain growth is a function of time

and temperature. The higher the temperature and the longer the time, the larger the grains will be. The grains are the smallest as they cross the A_3 critical temperature upon heating. This is illustrated in Figure 29-11. Examples 2 and 3 show a steel that is heated to a temperature well above the A_3 critical temperature. The grain size is large. In

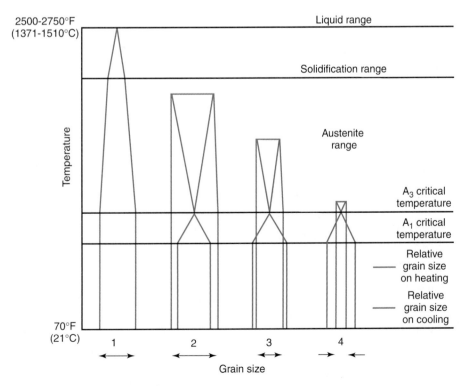

Figure 29-11. *Influence of heating on the grain size of steel. A steel that is heated to a high temperature has large grains. A steel that is heated to just above the A_3 critical temperature and then cooled has small grains.*

Example 4, the steel is heated to a temperature just above the A_3 critical temperature. This produces a very small grain size. Steel, upon cooling below the A_3 critical temperature, will retain the largest grain size achieved.

The various regions in the heat-affected zone and the weld metal have different properties. Postweld heat treating is used to make the different regions more uniform. There are a number of different heat treatments used in industry today. These include annealing, normalizing, quenching and tempering, thermal stress relieving, and spheroidizing.

29.4 ANNEALING STEEL

The term *annealing* is common in heat treating and includes several types of heat treating operations. Generally speaking, annealing is considered to be that type of heat treatment which leaves the metal in the softest condition.

An annealing operation may be done for several reasons:
- To soften metal so it may be cold-worked.
- To make steel machinable.
- To relieve the internal stresses and strains produced while metal is being shaped or welded.

The general procedure for fully annealing most steels is as follows:
1. Heat the steel 50°F to 100°F (28°C to 56°C) above the A_3 critical temperature. See Figure 29-12.
2. Hold for a period of time. A general rule is to allow one hour at the desired temperature for every inch of thickness.
3. Furnace-cool the steel at a controlled rate. Cool to 50°F (28°C) below the A_1 critical temperature.
4. Air-cool to room temperature. This process is shown graphically in Figure 29-13A.

As mentioned previously, the final grain size depends on time and temperature. The grain size affects the physical properties of the steel. As seen in Figure 29-14, the *elastic limit* (yield strength) and *ductility* increase as the grain size decreases. Note that, as the number of grains per inch increases, the grain size decreases. The *elastic limit* is the maximum load that will still produce elastic (linear) deformation in the steel. The *yield strength* is the load (in ksi or MPa) where plastic deformation begins.

Another type of annealing is the *process anneal*. It is usually performed on low-carbon steels. The steel is heated but remains below the A_1 critical temperature. It is held at this temperature long enough for softening to occur, then is air-cooled. Because of the lower temperatures involved and quicker cooling rates, it is faster and more economical than a full anneal.

29.5 NORMALIZING STEEL

If the extremely soft, full-annealed structure is not required, the normalizing heat treatment can be used. It saves time and money compared to annealing.

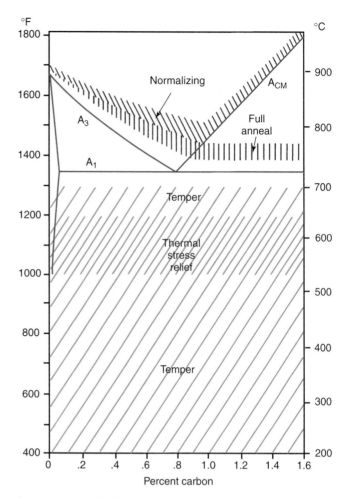

Figure 29-12. *The heat treatment temperatures for various carbon steels. To use the chart, find the percent carbon of the steel to be heat treated on the lower horizontal axis. Go directly up to the desired heat treatment. The proper heat treatment temperature is found on the vertical axis.*

Normalizing is used to make the internal structure of a steel more uniform. A normalized structure will have more uniform mechanical properties and better ductility than a part with internal stresses. It is not as soft and free from stresses as a fully annealed part.

The procedure for normalizing is as follows:
1. Heat the steel to 100°F (56°C) above the upper critical temperature. See Figure 29-12.
2. Hold the steel at the desired temperature until the temperature is uniform throughout.
3. Cool in still air to room temperature. The normalizing process is shown in Figure 29-13B.

29.6 QUENCHING AND TEMPERING STEEL

A steel that is quenched very fast from the austenite region forms martensite. Martensite is very hard and strong, but it is brittle. Figure 29-15 shows a quenching tank used to cool metal parts.

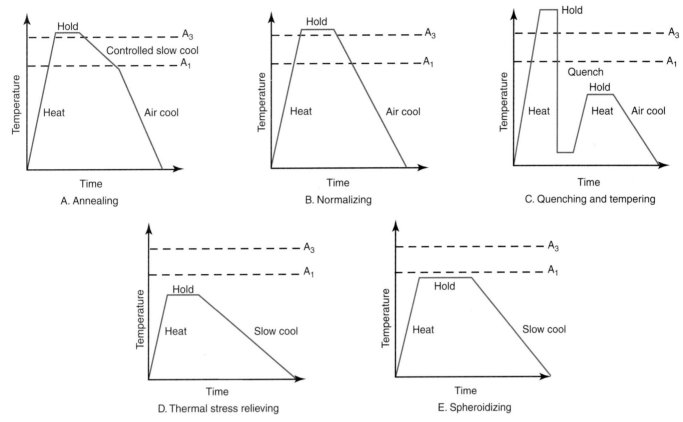

Figure 29-13. *Diagrams showing the five heat treatment processes. The hold time is long enough for the steel to obtain a uniform temperature. The hold time for spheroidizing is much longer to allow the cementite to form spheres.*

NO. of Grains per Linear inch (25.4mm)	Elastic Limit		Reduction of area percent
	psi	MPa	
210	44,000	303	20
222	44,500	307	22
322	47,000	324	35

Figure 29-14. *Effects of grain size on elastic limit and ductility. The reduction in area is an indicator of a metal's ductility or its ability to stretch before breaking. The higher the percent of area reduction, the higher the ductility of the metal.*

Tempering is used to improve the ductility and toughness of martensite. Some strength and hardness is lost. The final structure has very good strength and hardness, along with good toughness and ductility.

A *quench and tempered structure* is produced as follows:
1. Heat the steel into the austenite region. (The steel may already be in this region due to welding.)
2. Quench very rapidly to form martensite.
3. Heat to 400°F to 1300°F (204°C to 704°C) and hold to temper the martensite.
4. Cool in air to room temperature.

This process is shown in Figure 29-13C. Notice that the tempering temperatures are below the A_1 critical temperature. This is shown also on Figure 29-12.

Figure 29-15. *A number of cylinders being lowered into a quench tank. (J. W. Rex Company)*

29.7 THERMAL STRESS RELIEVING

Thermal stress relieving is used to reduce the residual stresses in a steel. These stresses may be the result of cold working or welding.

Thermal stress relieving is similar to tempering. The steel is heated to 1000°F to 1200°F (537°C to 649°C). This is below the A_1 critical temperature. The results of this heat treatment are that the residual stresses are reduced and ductility is improved.

The usual procedure for thermal stress relieving is as follows:

1. Heat the steel to the desired temperature.
2. Hold at this temperature.
3. Slowly cool back to room temperature. This process is shown in Figure 29-13D.

Figure 29-16 shows resistance coils in place on a large tank. These will be used to preheat the metal before welding and to stress relieve it after welding.

29.8 SPHEROIDIZING

Spheroidizing is a process to improve the ductility of a high-carbon steel. The cementite forms into small, separate spheres. The majority of the structure is ferrite. The microstructure is shown in Figure 29-17. The result of spheroidizing a steel is that a high-carbon steel has good ductility and can be formed or machined.

The procedure for spheroidizing is as follows:

1. Heat the steel to a temperature just below the A_1 critical temperature.
2. Hold at this temperature to allow the cementite to form spheres.
3. Slowly cool back to room temperature. This process is shown in Figure 29-13E.

29.9 HARDENING STEEL

Steels of less than 0.35% (35-point carbon content) cannot be successfully hardened by heat treatment. The maximum amount of carbon that a steel contains is 1.86%. Between these extremes, it is possible to obtain almost any degree of hardness by using the proper carbon content and the proper heat treatment. It is also common practice to

Figure 29-17. Microphotograph of the grain microstructure of spheroidized high-carbon steel. The spheres are cementite; the background is ferrite.

Figure 29-16. Resistance heating coils used on a large storage tank. The heating coils are used to preheat the joint and to thermally stress-relieve the welded structure after welding. (Duraline Div. of J. B. Nottingham & Co., Inc.)

obtain a refinement of the grain structure along with the hardness.

To obtain maximum hardness, the steel is heated just above its A$_3$ critical temperature, then quenched rapidly. The heat treatment temperatures for carbon steels with various carbon contents are shown in Figure 29-18. As mentioned in Heading 29.3, the rate of cooling determines to a great extent the hardness and brittleness of the metal. Cold water, air, and molten metal baths are used as mediums for cooling metal. Water-cooling results in extreme brittleness along with maximum hardness. Air-cooling and molten metal bath cooling tend to relieve much of the brittleness, but at a sacrifice of some hardness.

Hardness of steel depends on the distribution and structure of the cementite throughout the metal. When steel is cooled rapidly, there is no time for the cementite to clump. As the rate of cooling is slowed, the cementite has time to clump, leaving ductile ferrite which is softer than cementite.

Percent Carbon	Critical Temperature for Hardening and Full Annealing		Process Annealing	
	°F	°C	°F	°C
.10	1675-1760	913-960		
.20	1625-1700	885-927		
.30	1560-1650	849-899		
.40	1500-1600	816-871		
.50	1450-1560	788-849	1020	549
.60	1440-1520	782-827	to	to
.70	1400-1490	760-810	1200	649
.80	1370-1450	743-788		
.90	1350-1440	732-782		
1.00	1350-1440	732-782		
1.10	1350-1440	732-782		
1.30	1350-1440	732-782		
1.50	1350-1440	732-782		
1.70	1350-1440	732-782		
1.90	1350-1440	732-782		
2.00	1350-1440	732-782		
3.00	1350-1440	732-782		
4.00	1350-1440	732-782		

Figure 29-18. *Heating temperatures for hardening and annealing various carbon steels.*

29.9.1 Tempering a Cold Chisel

A common tempering operation in a laboratory is to heat treat a cold chisel. Most cold chisels are made of approximately 75-point carbon steel and must have the following properties:

- The cutting edge must be extremely hard.
- The metal just back of the cutting edge must be hard and tough, but not brittle.
- The body of the chisel and the hammering end must also be tough.

To obtain these various properties from the same carbon content steel, heat treat as follows:

1. Heat the cold chisel slowly to just above its critical temperature (1350°F or 732°C, evidenced by a cherry red color).
2. Allow enough time for the body of the chisel to assume this temperature throughout its thickness; then quench about one inch of the cutting edge end of the chisel in cold water.
3. Wait until the end in the water turns dark; then withdraw the chisel quickly from the water while the other end is still cherry red.
4. Heat from the cherry red end will travel to the chilled cutting edge, heating it slowly. With a small pad faced with emery cloth, polish the flat surfaces near the cutting end of the chisel. Be careful to avoid burning yourself.
5. Watch the polished surface carefully. As the heat from the shank of the chisel travels into this part, the polished surface will gradually change color. It will first become straw, then bronze. As the temperature rises, the surface will turn purple because of the oxidation of the metal. The purple color indicates the steel has been heated to approximately 600°F (316°C).
6. Now, slowly place the cutting edge in the water. If the shank of the chisel has cooled below a cherry red color (as it should have done), the complete chisel may be immersed in a bucket of cold water.

If the timing is correct on the heat treatment, and if the original temperature is not too high, the chisel will have a hard cutting edge, with a fairly hard and tough body. If the original temperature was too high, the chisel will be brittle and will crack easily because of the large crystalline structure. The cutting edge will be too hard and will chip off when used. If the cutting edge was requenched before it should have been, the cutting edge will be hard and the edge will break. If the shank of the chisel is quenched too soon (before its color has become a dark red), the shank will be too hard and brittle and will crack when being used. A chisel with brittle qualities is an extremely dangerous tool, because if it cracks or chips, the flying particles of metal may inflict injuries.

29.9.2 Surface Hardening

Surface hardening is widely used today in manufacturing operations. In many applications, such as gears, shafts, and connecting rods, it is advisable to produce as hard-wearing a surface as possible. The internal portion of the structure must remain ductile and tough. The hard surface provides long wear and maintains an exact contour, while the tough interior ensures that the part will withstand shocks without breaking.

One method that is commonly used to surface-harden a metal is *flame hardening*. A part that is to be flame hardened is first heat treated to produce a tough, ductile structure throughout. The surfaces to be hardened are then put in a flame hardening machine. A multiple-tipped oxyfuel flame is passed over the surface to heat it quickly to a high temperature. Following the flame, a

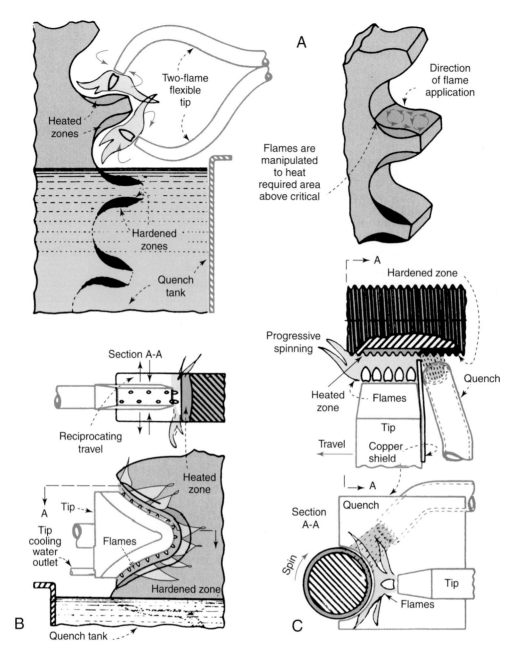

Figure 29-19. *Some popular flame hardening applications. A—Manual flame hardening allows concentration of heat at areas requiring maximum depth of hardness. B—Root and opposing faces of adjacent teeth are heated with contour tip. C—Progressive spinning is employed for fine threads, using water-cooled nonquenching tip.*

heavy flow of water quickly cools the entire surface, Figure 29-19. The depth of the hardness can be controlled by controlling the temperature of the surface and the cooling rate. Flame hardening is a rapid and economical means of hardening a metal surface.

Two additional methods of surface hardening a metal are to use a laser or an electron beam gun. The process is similar to flame hardening, except that a laser or an electron beam is used instead of a flame. The laser or electron beam is directed at the surface to be hardened. Both systems can be accurately focused so that small or difficult areas can be surface hardened. The depth of hardening can be controlled better with the laser or electron beam process

than with oxyfuel gas torch heating. These processes also are much faster than oxyfuel gas torch heating, but the equipment is very expensive.

29.9.3 Casehardening

Another solution to the problem of producing a metal article with a tough interior and a hard, wear-resistant surface is known as *casehardening.*

The article is made out of low-carbon steel, so any machining may be done easily. The surfaces to be casehardened are exposed to carbon while they are at a high temperature. This is achieved by placing the part that is to be casehardened in a furnace. The furnace contains a carbon atmosphere, usually *carbon monoxide* (CO). In the high temperature of the furnace, the low-carbon steel absorbs some of the carbon into its surface. The carbon penetrates into the part at about 1/64" (0.4mm) per hour. The surface of the steel part now contains more carbon than the body of the part.

In some cases, it is not desirable to caseharden a particular portion of the part that is placed in the furnace. To prevent an area from absorbing carbon, that area is plated with copper.

The casehardened part is heat treated using the quench and tempering process. The steel is heated to its critical temperature, then quenched. This produces a very hard surface. The interior of the part is still low-carbon steel. It has refined (small) grains. As stated in Heading 29.4 and Figure 29-14, as the grain size decreases, the strength and ductility of a steel increases. The interior of a casehardened part is thus strong and tough.

The part is finally tempered to improve the toughness of the hard outer surface. Figure 29-20 is a photograph of a casehardened gear tooth.

Figure 29-20. *A cross section of a casehardened steel gear tooth, showing casehardened surface.* (General Motors Corp.)

29.10 HEAT TREATING TOOL STEELS

Tool steels are all high-carbon steels. Some examples of the use of tool steels are chisels, hammers, saws, springs, lathe and mill tools, and punches and dies for punching and forming metals. The carbon content of these steels is shown in Figure 29-5.

When these tool steels are formed, the metal must be ductile. To obtain good ductility, the tool steel is either annealed or the spheroidizing process is used. Spheroidized steel is more ductile than an annealed tool steel.

These tool steels must all be very hard when used. To produce the hardness, the steel is heat treated using the quench and tempering process. Different tempering temperatures are used to obtain the desired combination of hardness and toughness.

To properly heat treat a tool steel, the welder must know the carbon content of the steel and the heat treatment process to be used. Figure 29-12 can be used to determine the correct heat treating temperature for the steel.

To determine the correct heat treatment temperature in Figure 29-12, find the carbon content of the steel on the lower (horizontal) axis. Follow a vertical line up to the desired process. Then, follow across horizontally to the temperature scale on the left or right. For example, determine the proper temperature to fully anneal a 0.90% carbon metal tool. Find .9 on the horizontal scale. Go directly up to the full anneal region, then directly across to the temperature scale. The correct temperature to fully anneal a 0.90% carbon tool steel is slightly above 1400°F (760°C).

29.11 HEAT TREATING ALLOY STEELS

As alloying elements are added to steel, the critical temperature of the steel is changed. Alloying elements usually lower the critical temperatures. When heat treating an alloy steel, the welder cannot just consider the amount of carbon in the steel. The alloy content of the steel must also be known.

It is almost impossible for a welder to know what alloys are in a steel. The only sure way to know is to obtain the percentages from the steel manufacturer. Even with the alloying elements known, it is very difficult to select the correct temperature for a heat treatment. The best way to heat treat an alloy steel is to request and follow the manufacturer's recommendation.

29.12 HEAT TREATING CAST IRONS

There are four main types of cast irons. These are gray cast iron, white cast iron, ductile cast iron, and malleable cast iron. These four types are similar in reference to the carbon content of each, but their properties are quite different.

Gray cast iron is formed when a casting is cooled slowly. The carbon forms graphite (carbon) flakes, as seen in Figure 29-21. Gray cast iron can also be created by heating a casting to a temperature of 1850°F to 2050°F (1010°C to 1121°C) and cooling the casting slowly. This heat treatment is used after a gray cast iron is welded. Welding often causes the weld metal and the heat-affected zone to be much harder and more brittle than the rest of the casting. Heat treating makes the microstructure and the properties of the casting more uniform.

White cast iron is formed when a casting is cooled quickly. White cast iron is harder than gray cast iron and is very brittle. White cast iron can also be formed by heat treating. The cast iron is heated to a temperature of 1650°F to 2050°F (899°C to 1121°C) and quenched (rapidly cooled).

The most common heat treatment for white cast iron is heat treating it to create *malleable cast iron*. Malleable cast iron is ductile and also machinable, whereas white cast iron is not. To obtain malleable cast iron, a white cast iron is heated to 1450°F to 1650°F (788°C to 899°C). It is held at this temperature for 24 hours for each inch of

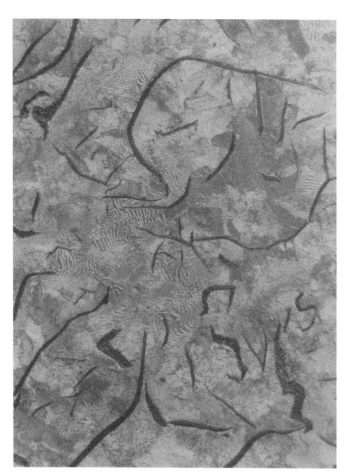

Figure 29-21. *Microstructure of gray cast iron. (Central Foundry, Division of General Motors Corporation)*

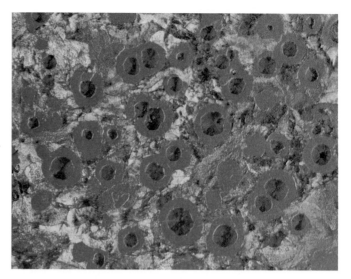

Figure 29-22. *Microstructure of ductile cast iron. (Central Foundry, Division of General Motors Corporation)*

casting thickness. While the casting is at this temperature, the carbon separates into irregular spheres of carbon. The casting is slowly cooled to room temperature. The microstructure of malleable cast iron contains small spheres of carbon; the remainder of the casting is low-carbon ferrite. This microstructure is ductile or malleable.

A fourth type of cast iron is **ductile cast iron,** which is sometimes called *nodular cast iron.* This type of cast iron is not obtained by heat treating, but by adding magnesium or cerium as an alloying element. When the iron is poured or cast, the magnesium or cerium attracts the carbon and forms graphite spheres. See Figure 29-22. Since the carbon is gathered into spheres, the remaining iron is low in carbon and therefore very ductile. Ductile cast iron has good strength, hardness, and ductility.

29.13 HEAT TREATING COPPER

Copper and copper alloys become hard and brittle when they are mechanically worked. However, the metal can be made ductile again by annealing. To anneal copper, it should be heated to a temperature between 1400°F and 1800°F (760°C and 982°C). It is then water-quenched. This heat treatment brings back the copper's original ductile qualities.

It is important to be careful when heating copper to its annealing temperature, because copper undergoes a physical phenomenon or change, called *hot shortness.* At this temperature, the copper suddenly loses its tensile strength. If the copper is not properly supported or if it is subjected to strain, it will fracture easily.

Some copper alloys are known for their high tensile strength and high hardness. To obtain the high strengths, the copper alloy is first annealed. It is then cold-worked, usually rolled. The alloy is then age-hardened. This age hardening involves heating the alloy to 600°F to 950°F (316°C to 510°C) for a number of hours. One copper-beryllium alloy has a tensile strength of 173,000 psi (1193 MPa).

29.14 HEAT TREATING ALUMINUM

Aluminum is similar to copper in a number of respects. Aluminum becomes harder and more brittle when it is cold-worked. Aluminum also has the characteristic of *hot shortness,* so the welder must support the part as it is being heated.

There are a number of heat treatments that are used for aluminum and its alloys. A letter-number code is used to signify the various heat treatments. The common heat treatments and their letter-number codes are shown in Figure 29-23. The letter-number code follows the alloy designation. See Figure 28-16 for the aluminum alloy designations. Some examples are as follows: 1100-0, 2017-T4; 6061-T6; 7075-W.

Strain hardening is a cold-working process. Some examples of strain hardening are rolling, stretching, and forming. Strain hardening increases the hardness of an aluminum alloy.

Annealing of an aluminum alloy is done by heating the alloy to 650°F (343°C) or 775°F (413°C), depending on the alloy. The alloy is heated for two to three hours and

H	strain hardening
O	annealed
W	solution heat treated only
T	thermally heat treated
T1	naturally aged
T2	annealed (for cast products only)
T3	solution heat treated and cold worked
T4	solution heat treated and naturally aged
T5	artificially aged (from a cast condition)
T6	solution heat treated and artificially aged
T7	solution heat treated and stabilized
T8	solution heat treated, cold worked and artificially aged
T9	solution heat treated, artificially aged and then cold worked
T10	artificially aged and then cold worked

Figure 29-23. *Letter-number code for the heat treatments used on aluminum alloys. Each "T" number is a subcategory of the broad classification "T."*

then cooled. The cooling rate is not important as it will not affect the final structure.

A solution heat treatment is similar to annealing. The procedure to solution heat treat an aluminum alloy is to heat the alloy to 940°F to 970°F (504°C to 521°C). Hold at this temperature and then quench in water. This produces a uniform distribution of the alloys in the aluminum. The aluminum alloy is then cold-worked or aged to achieve the desired strength and hardness.

Aging can be done naturally or artificially. Natural aging occurs at room temperature. Artificial aging is done by heating the aluminum alloy to an elevated temperature. An aluminum alloy that is aged has better strength and hardness than when annealed or solution heat treated. The temperature for T6 artificial aging is between 320°F and 360°F (160°C and 182°C) for 6 to 12 hours.

29.15 TEMPERATURE MEASUREMENTS

Correct heat treatment can be obtained only if the temperature of the metal can be accurately measured.

One of the more difficult measurements is to determine if the surface temperature is also the inner temperature of the structure.

Several ways used to measure high temperatures are:
- Optical pyrometers.
- Gas thermometers.
- Thermoelectric pyrometers, Figure 29-24.
- Resistance thermometers.
- Radiation pyrometers.
- Color change of the metal.
- Temperature-indicating crayons, cones, pellets, and liquids. Temperature-indicating pellets are shown in Figure 29-25.

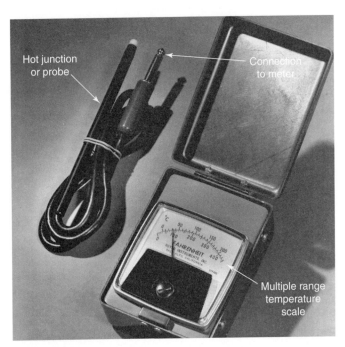

Figure 29-24. *Thermoelectric temperature measuring instrument. The electric probe draws small quantities of heat energy from the surface to be measured. (Royco Instruments, Inc.)*

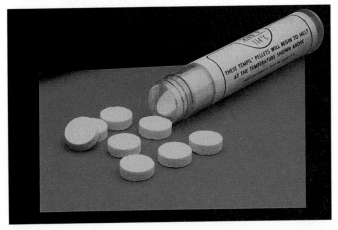

Figure 29-25. *Temperature-indicating pellets. These pellets, as marked, will begin to melt at 238°F (114°C). (Tempil° Division, Air Liquide America Corporation)*

One of the most important factors in welding is the proper preheating and postheating of the metal. The temperature to which the metal is heated is of vital importance, since a few degrees too low or too high can give undesirable results. Automatic furnaces with heat source controls and thermocouple temperature indicators are desirable. However, their expense makes them impractical for many small shops. An excellent solution to the problem for small shops is the use of temperature indicators, such as crayons, pellets, or liquids.

A liquid temperature indicator is shown in Figure 29-26. Figure 29-27 shows a temperature-indicating crayon and Figure 29-28 shows a crayon in use. These materials indicate a great variety of temperatures. They indicate temperatures in 12.5°F (7°C) steps in the 113°F to 400°F (45°C to 204°C) range. They are in 50°F (28°C) steps in the 400°F to 2000°F (204°C to 1093°C) range. Pellets exist with 100°F (56°C) steps from 2000°F to 2500°F (1093°C to 1371°C). Figure 29-29 shows a test kit that contains a number of temperature-indicating crayons. These crayons are used to mark the metal surface. When the correct temperature crayon is applied, the crayon will melt and change color. Figure 29-30 shows one way a welder can use such a crayon to measure the preheat temperature of a pipe prior to welding.

Figure 29-28. *A temperature-indicating crayon in use. The 250°F (121°C) mark will melt when the surface temperature reaches 250°F (121°C). (Tempil° Division, Air Liquide America Corporation)*

Figure 29-26. *This liquid temperature indicator is painted onto the metal. After it dries, it will melt at the temperature indicated on the bottle. (Tempil° Division, Air Liquide America Corporation)*

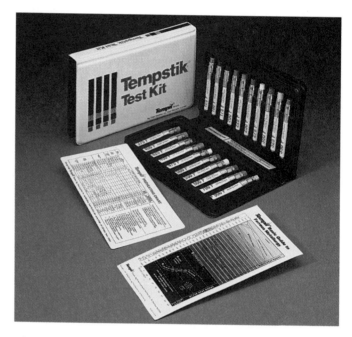

Figure 29-29. *A temperature test kit that contains temperature crayons from 125°F to 800°F (51.7°C to 426.7°C). (Tempil° Division, Air Liquide America Corporation)*

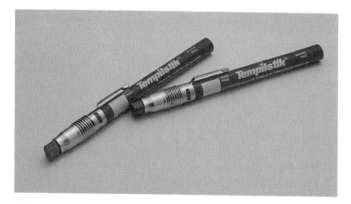

Figure 29-27. *Two temperature-indicating crayons are often used to hold the metal temperature between two temperature limits. (Tempil° Division, Air Liquide America Corporation)*

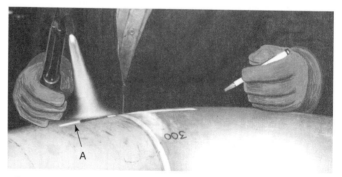

Figure 29-30. *A welder preparing a pipe joint to 300°F (149°C) prior to welding. Note the crayon mark at A, and how it has melted in the area where the torch is pointing. (Tempil° Division, Air Liquide America Corporation)*

29.16 REVIEW OF SAFETY

Heat treatment of metals requires the welder to use safe working habits with the hot metal. **Gloves are required and goggles are very strongly recommended, in most cases.** Hot metals should be safeguarded and marked to prevent other persons from injuring themselves.

Quenching metals requires a safety shield to protect the face from high-temperature flying fluids. Since the heat source used for heat treating varies, the welder should investigate the required safety steps for each heat source.

TEST YOUR KNOWLEDGE

Write your answers on a separate sheet of paper. Do not write in this book.

1. The most common postweld heat treatment is _____.
2. Heating can occur at three different times while welding: before, during, and after. What name is given to each period of heating?
3. What four factors are important when heat treating a metal?
4. What microstructure in steel determines the hardness of the steel? What microstructure in steel determines the ductility of a steel?
5. How is martensite formed? How is pearlite formed?
6. Describe the regions in the heat-affected zone and how they were formed.
7. What is the purpose of annealing?
8. How does normalizing differ from annealing?
9. Describe the quench and tempering process and the properties of a steel that is quench and tempered.
10. Thermal stress relieving reduces the _____ _____ and improves the _____ of a steel.
11. What happens when a high-carbon steel receives a spheroidizing heat treatment?
12. What does the hardness of a steel depend on?
13. Describe one way to produce a gear with a hard surface and a tough interior.
14. Describe the heat treatment that is most-often used on cast irons.
15. What heat treatment is used to soften copper?
16. What process is used to give copper high strength and hardness?
17. How is an aluminum alloy hardened?
18. What heat treatment did an aluminum alloy receive that has a designation T6?
19. What does a solution heat treatment produce in an aluminum alloy?
20. How can temperature-indicating crayons, pellets, and liquids be used to properly identify the temperature of a metal?

PROFESSIONAL WELDING

X-ray inspection is an important method of nondestructive evaluation for critical welds such as those used on the hull of a ship. (Seifert X-ray Corp.).

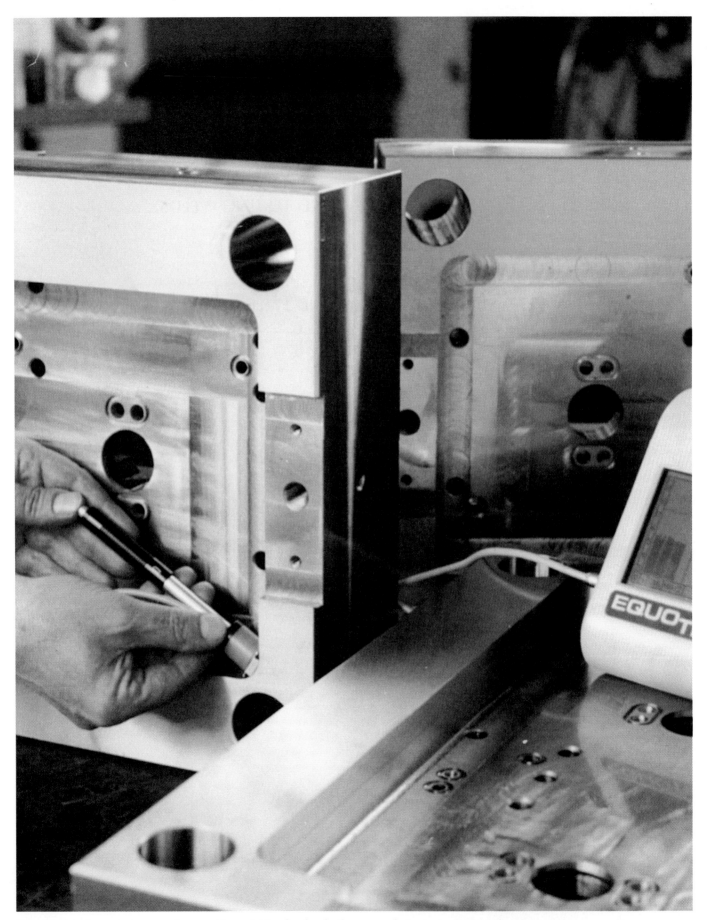

Portable hardness testers can be used to quickly check whether a workpiece meets specifications. Measurements are displayed graphically on a screen. (Foerster Instruments)

Chapter 30

INSPECTING AND TESTING WELDS

LEARNING OBJECTIVES

After studying this chapter, you will be able to:
* Differentiate between destructive testing and nondestructive examination methods.
* Describe various nondestructive examination methods and what flaws or defects they are capable of identifying.
* Identify and specify various nondestructive examination methods using the correct AWS abbreviation.
* Make a visual inspection on a weld.
* Perform a properly conducted die penetrant test on a weld.
* Make a safe and properly done bend test on a groove-type weld.
* Describe and identify various laboratory and shop-type equipment used to make destructive tests on welds.
* Visually inspect and perform a peel test on a spot weld.

All welds have flaws. Another name for a flaw is a discontinuity. Discontinuities are interruptions in the normal crystalline lattice structure of the metal. However, a discontinuity is not always a defect. They are only defects when the weld becomes unsuitable for the service for which it was intended.

The role of weld inspection is to locate and determine the size of any discontinuities. This nondestructive examination (NDE) must be done without damaging the weld. Nondestructive testing (NDT) is also used to check a weld's quality without destroying the weld.

The role of destructive testing is to determine the effects of various discontinuities. Destructive testing is used to determine the physical properties of a weld and to predict the service life of a weld. To perform this type of test, it is necessary to destroy the weld.

30.1 NONDESTRUCTIVE EXAMINATION (NDE)

Nondestructive examination (NDE) is used on a weld or on an assembly without damaging the part. The part will continue to be usable after the testing has been done.

Manufacturers are interested in the soundness of welds or their fitness for service. This has made NDE an important phase of welding. Basically, *fitness for service* is the concept that all welds contain flaws. Hopefully the flaws, also known as discontinuities, are few and are very small. Small discontinuities will not affect the performance of the welded assembly. If the flaws or discontinuities are too large, or there are too many of them, they will affect the performance of the weldment. The weld may fail because of the flaws. Flaws or discontinuities that are too large are called *defects*.

An important responsibility for engineers is to determine the point at which a discontinuity becomes a defect. When a weld joint is designed, it is welded and then evaluated nondestructively. All discontinuities are located, then the weld is tested to destruction. This is done on several samples. The purpose of these tests is to determine how large a discontinuity must be to become a defect. When this defect size or *critical size* is determined, all discontinuities of this size and above will be considered defects. On all similar welds, nondestructive examinations may be made to locate all discontinuities above the critical defect size.

The methods used for nondestructive examination (NDE) include:
* Visual inspection (VT).
* Magnetic particle inspection (MT).
* Liquid penetrant inspection (PT).
* Ultrasonic testing (UT).
* X-ray inspection (RT).
* Eddy current inspection (ET).
* Mass spectrometer detection (LT).
* Air pressure leak inspection (LT).
* Halogen gas leak inspection (LT).

The most popular NDE methods are: visual, magnetic particle, liquid penetrant, ultrasonic, and x-ray.

30.2 DESTRUCTIVE TESTS

Destructive tests are used to determine the physical properties of a weld. For every type of weld and material used, a Welding Procedure Specification (WPS) is normally written. To prove that a Welding Procedure Specification will produce a good weld, a Welding Procedure Qualification Record (WPQR) is used to document the weld and the tests performed on it. Heading 31.3 discusses the Welding Procedure Specification and the Welding Procedure Qualification Record. Different codes and specifications will identify the type and number of tests required to check the properties of a weld.

Destructive tests render the weld unfit for further service. Destructive-type tests include:

- Bend test.
- Tensile test.
- Impact test.
- Hardness test.
- Microscopic test.
- Macroscopic test.
- Chemical analysis.
- Fatigue.
- Hydrostatic test to destruction.
- Peel test (used on spot and projection welds).
- Tensile shear test (used on spot and projection welds).

30.3 VISUAL INSPECTION (VT)

The most common type of inspection is the visual inspection. *Visual inspection* checks for size, contour, and location of the welds. This type inspection is made to reveal defects like cracks, inclusions, undercut, and lack of penetration. Visual inspection looks for spatter, for excessive buildup on a weld, and to see if all slag has been removed from the weld.

A visual inspection can involve use of templates and gauges, as shown in Figures 30-1 and 30-2. Some companies have established welding workmanship standards that an inspector uses to decide if the weld is acceptable.

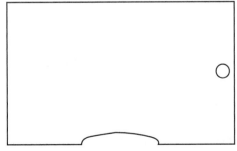

Figure 30-1. *A template for testing the bead contour of welds.*

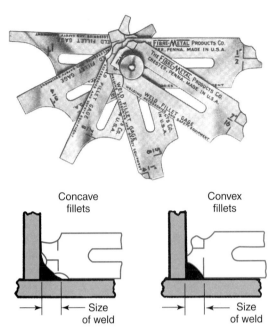

Figure 30-2. *Top—Gauges used to measure and inspect both convex and concave fillet welds from 1/4" to 1" (6.4mm to 25.4mm). Bottom—Blades of gauge used to check concave and convex weld contours. (Fibre Metal Products Co.)*

Visual inspection is limited to the external surfaces of the weld. This is a very good test when appearance and size are important features of the weld. Welds used in critical applications require additional inspections to find any internal defects.

Other tools used to visually inspect welds include: a magnifying glass, miniature TV cameras, calipers, and scales. See Figures 30-3 and 30-4. Tape measures, calipers, and scales are used to measure the overall structure to determine if the welded structure is within the specified dimensions. See Figure 30-5.

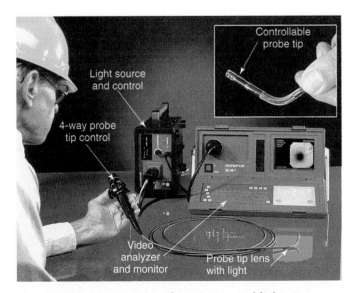

Figure 30-3. *A miniature video camera and light source used to visually inspect and record weld quality in areas that cannot be reached without this type of equipment. (Olympus America, Inc., Industrial Fiber Optics Div.)*

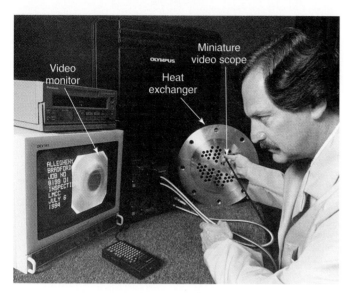

Figure 30-4. A technician visually inspecting the head end of a heat exchanger for defects. (Olympus America, Inc., Industrial Fiber Optics Div.)

Figure 30-5. Calipers, tape measures, and other manual tools are used to assist in the visual inspection of welds.

Each weld should be visually inspected by the welder. A good welder can identify most problems and make corrections in future welds. If a defect is created while welding, it should be identified and a supervisor or inspector notified. This is important, because the defect might otherwise go undetected.

30.4 MAGNETIC PARTICLE INSPECTION (MT)

Magnetic particle inspection (MT) is most effective in checking a weld for surface or near-surface flaws. It is used only on materials that can be magnetized. Prior to applying the magnetic particle material, the surface must be cleaned with a special cleaner furnished with the magnetic particle kit. A material containing fine magnetic particles is then applied to the weld being inspected. Fluorescent magnetic particles may also be used. These magnetic particles are applied as a liquid or in a dry form.

The metal is then subjected to a strong magnetic field. Because of the magnetic field, the sides of any small cracks and pits become the north and south poles of miniature magnets. The magnetic particles in the applied powder or liquid are attracted to the cracks and pits by their magnetism. These particles are colored red, black, or gray, and may be suspended in a liquid. The choice of red, black, or gray is dependent on which color gives the best contrast against the part being tested. When the magnetic field is removed, the inspector will find a concentration of magnetic particles in the area of every flaw. If fluorescent magnetic particles were used, a "black light" (ultraviolet light) is used to show the location of flaws.

If defects are found, the defect and the metal around it are removed. The part is rewelded and retested. Figure 30-6 shows the magnetic field that is created around the weld area being tested.

Two methods are used to create the magnetic field in the part being inspected:
- Passing current through the object being inspected by means of test prods.
- Placing a powerful electromagnetic yoke or permanent magnet against the object to allow the magnetic field to pass through the material being inspected.

The flux lines in the magnetic field should be as perpendicular to the crack as possible. Because the location of the crack is not known, the article should be magnetized

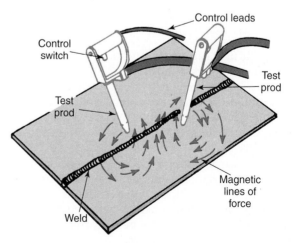

Figure 30-6. A magnetic field is created around a weld as current is passed through the weld between the two test prods. (Magnaflux Corp.)

twice. One inspection should be made with the test prods or yoke in one direction. A second inspection is done with the test prods or yoke turned 90° from the first inspection. Figure 30-7 shows magnetic particle inspection being used to inspect an assembly. The material to be tested should be as clean and bright as possible prior to the test.

An electromagnetic yoke, the cleaner, and magnetic powders are shown in Figure 30-8. Figure 30-9 shows magnetic particle inspection being used to inspect for flaws.

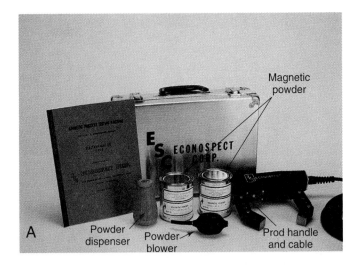

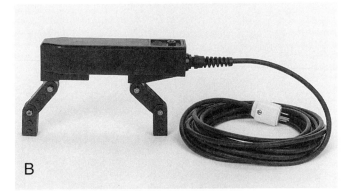

Figure 30-8. *Magnetic particle inspection. A—A kit for making a magnetic particle inspection. B—The electromagnetic prod and handle used for making tests. (Econospect Corp.)*

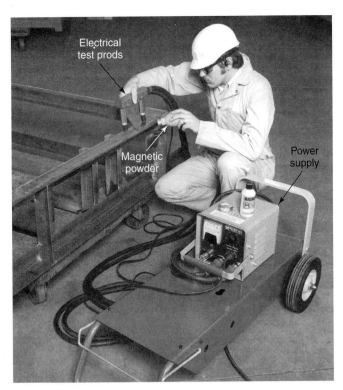

Figure 30-7. *Magnetic powder and electrical test prods are used to locate surface defects in a weld. A special dual prod assembly frees one of the inspector's hands and allows the application of powder while current is flowing. (Magnaflux Corp.)*

A magnetic field can also be produced by using a permanent magnet, as shown in Figure 30-10. The electromagnet and permanent magnet systems are light and convenient. Permanent magnets are also used where an electrical spark might be dangerous. Another type of electromagnet, shown in Figure 30-11, is used when inspecting pipes and shafts.

Magnetic particle testing is very popular in the manufacture of transportation equipment (aircraft, automobiles, trucks, buses, and ships), where it is used to check for cracks. Locating cracks in axles, shafts, gears, crankshafts, and landing gear parts before performance failure occurs is of considerable importance.

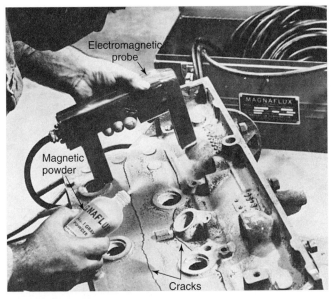

Figure 30-9. *Magnetic particle inspection being used to check an engine block for cracks. Note the dark lines indicating cracks. (Magnaflux Corp.)*

Figure 30-10. A magnetic particle test kit with a permanent magnet. The squeeze bottles hold red, black, and gray powders that contain magnetic particles. (Magnaflux Corp.)

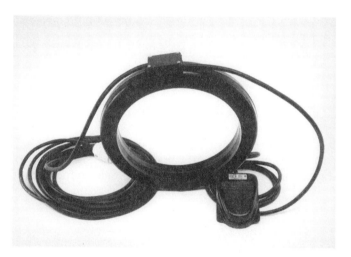

Figure 30-11. A ring or torus-shaped electromagnet used to make magnetic particle inspections on pipes or shafts. A foot-operated on/off switch is at the bottom right. (Econospect Corp.)

30.5 LIQUID PENETRANT INSPECTION (PT)

The *liquid penetrant inspection* method uses colored liquid dyes and fluorescent liquid penetrants to check for surface flaws. This system can be used to show surface flaws in metals, plastics, ceramics, or glass. This method will not detect subsurface discontinuities or defects.

The steps required to perform a satisfactory penetrant inspection using a colored liquid dye or a fluorescent liquid are:

1. Clean the part with the special cleaner provided in the kit.
2. Apply the liquid penetrant and allow time for it to penetrate into any cracks or pits.

3. Remove excess penetrant with water, solvent, or an emulsifier. The type of penetrant used determines what is used to remove the excess.
4. Apply the developer liquid.
5. Inspect for discontinuities (flaws) and defects.
6. Clean the surface after inspection is complete.

Prior to applying the developer, the excess penetrant must be removed. Some excess penetrants are removed by washing with water; others by using a solvent. A third type of penetrant removal uses an emulsifier. The emulsifier makes the penetrant water-soluble. Water is used to wash away the excess penetrant and emulsifier.

The developer is then applied. Some of the penetrant that is in a flaw will be drawn out. The dye can be seen and the discontinuity located. Surface discontinuities (flaws) and defects will show up as shown in Figure 30-12.

Fluorescent liquids are used in a manner similar to dye penetrants. A fluorescent liquid is applied to the surface being inspected. After a short time, the excess fluorescent liquid is removed with a cleaner. The developer is applied and an ultraviolet (black light) source is then brought to the surface. Any area where the fluorescent liquid has penetrated, and has been drawn out by the developer, will show up clearly under the ultraviolet light, as shown in Figure 30-11.

The dye, cleaner, and developer are available in aerosol spray cans for convenience. Some solvents used in the cleaners and developers contain high percentages of chlorine, to make the liquids nonflammable. Chlorine is a known health hazard. Solvents and developers containing chlorine should be used with great care. Newer cleaners and developers are available that do not contain chlorine.

The penetration ability of the dye varies according to the materials being tested. The penetrant activity varies with the temperature. It is important to allow sufficient time to permit accurate inspection. This time may vary

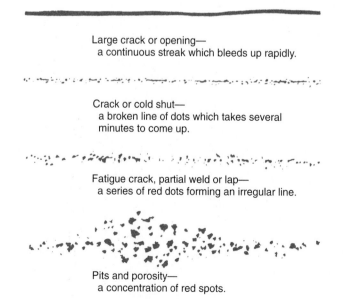

Figure 30-12. Indications of surface flaws found by the liquid dye or fluorescent penetrant method of inspection.

from 3 to 60 minutes. At room temperatures, the time usually recommended is from 3 to 10 minutes.

30.6 ULTRASONIC TESTING (UT)

Ultrasonic testing (UT) is a nondestructive testing (NDT) method of locating and determining the size of discontinuities (flaws) by means of sound waves. It can be used on virtually any type of material. The sound waves are passed through the material. Ultrasonic testing uses high-frequency sound waves at more than one million hertz (MHz). One *hertz* is one cycle per second.

An electronic device called a *transducer* is placed on the surface of the material being tested. Ultrasonic waves are sent into the material through the transducer. Ultrasonic waves will not travel through the air. Therefore, an excellent contact must be made between the transducer and the surface of the material. To ensure good contact, a material called a *couplant* is used. This couplant can be water, oil, or glycerol. The only specification for a couplant is that it is not harmful to personnel or the part being tested.

Ultrasonic waves are sent into the material for very short periods of time. This time period is only one millionth to three millionths of a second (1 to 3 microseconds). Once the ultrasonic wave is sent through the material, it is reflected from the surface boundaries or flaws where density changes occur. The reflected wave is then received by a transducer. Often the receiving transducer is the same transducer that started the wave. A second transducer with the same frequency rating may be used. After a wave is sent and received back, another wave is sent for 1 to 3 microseconds. This process is repeated approximately

500,000 times per second. During the test, the transducer is moved along the material surface.

Each reflected wave signal is amplified and shown on a display. See Figure 30-13. The display is calibrated to show the distance between the transducer located on the top surface of the material being checked and the bottom surface. Multiple echoes from the top and bottom surfaces may be displayed. Once the display is calibrated, any additional reflections indicate a flaw. The location of the flaw is determined by its distance from the top or bottom surface. The equipment must be calibrated for each type of material and thickness tested. Operators must be skilled to obtain consistent results. See Figure 30-14.

The equipment used for ultrasonic testing is light and portable. Ultrasonic testing is done easily on the job site. The advantages of UT are:
- It is fast.
- It gives immediate results.
- It is usable on most materials.
- Access from two sides is not required.

The disadvantages of UT are:
- A couplant is required.
- Defects parallel to the sound beam are difficult to detect.
- The display requires training to be interpreted accurately.
- Equipment must be calibrated regularly.

30.7 EDDY CURRENT INSPECTION (ET)

Eddy current inspection (ET) uses an ac coil to induce eddy currents into the part being inspected. An ac coil produces a magnetic field. The magnetic field induces an ac

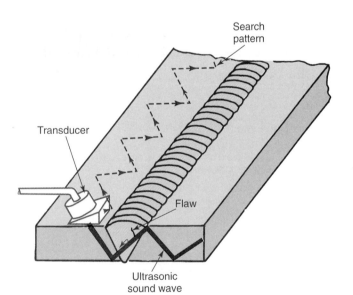

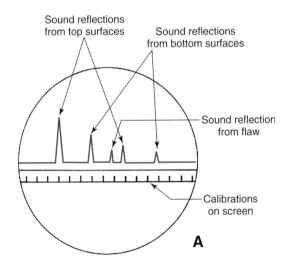

Figure 30-13. A schematic drawing showing the path followed by the ultrasonic search unit and the path of the sound waves as they move through the metal being tested. Sound reflections from the top and bottom surfaces occur at calibrated points on the oscilloscope (CRT) as shown in detail A. Any reflection that is not from the top or bottom surfaces indicates a flaw. The location of the flaw is determined by the distance from the top surface echo shown on the CRT.

Figure 30-14. An angle transducer with its control and display being used to test a weld on a pipe. (Panametrics, Inc.)

current into the part. This induced current flows in a closed circular path. These induced currents are known as *eddy currents.*

To perform an eddy current test, an ac coil is placed on the part to be tested. The coil is calibrated or adjusted to obtain a set value of impedance. *Impedance* is the resistance to the flow of alternating current. The set value of impedance is viewed on an oscilloscope.

The coil is moved over the surface of the part to be inspected. If a discontinuity is present in the material, it will interrupt the flow of the eddy currents. The change in the eddy currents affects the impedance of the coil. This change can be observed on the oscilloscope.

Eddy currents only detect discontinuities near the surface of the part. The depth of inspection depends on the alternating current frequency. With common frequencies, the depth of inspection for common metals will not exceed 1/8" (3.2mm). Eddy currents can be used to inspect both flat parts and round parts like pipes. Eddy currents are used to check for porosity and cracks. They can also be used to inspect for proper heat treatment.

30.8 X-RAY INSPECTION (RT)

Welds may be checked for internal discontinuities (flaws) by means of *x-ray inspection*. An x-ray is a wave of energy that will pass through most materials and reproduce their image on film. Figures 30-15 and 30-16 show the principles of x-ray inspection and the results of an x-ray inspection.

X-ray energy is produced electronically in an x-ray machine. Figure 30-17 shows a portable x-ray machine. X-ray energy occurs naturally in radioactive isotopes. Some radioactive isotopes can be used to inspect materials, but their use is very hazardous.

A radiograph (x-ray picture), is a permanent record of welds made on critical construction sites or critical components. Examples where permanent records are often required include nuclear power plants, pipelines, ships, submarines, and aircraft. Figure 30-18 shows equipment being set up to take an x-ray.

The energies required to x-ray different materials and different thicknesses vary considerably. Thicker materials require greater energies and longer times. Complex assemblies and access or distance to the desired area also affect the energies required.

When looking at a radiograph, the location of a flaw can be identified. See Figure 30-16. However, the depth at which the flaw is located in the metal cannot be determined by using an x-ray made from only one direction. When a flaw is identified, a second x-ray taken at an angle helps to determine its depth.

Figure 30-16. An x-ray radiograph of a weld in 1 1/2" (38.1mm) steel plate. Note how the undersurface crack at the left shows up in the x-ray radiograph.

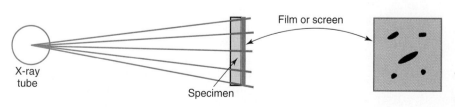

Film or screen

Equipment checks internal physical structure of many materials. Spots on film indicate voids or other density variations. High voltage x-ray units produce hard radiation needed to penetrate thick specimens.

X-ray tube

Specimen

Figure 30-15. A schematic of an x-ray unit radiograph (photograph) of a weld.

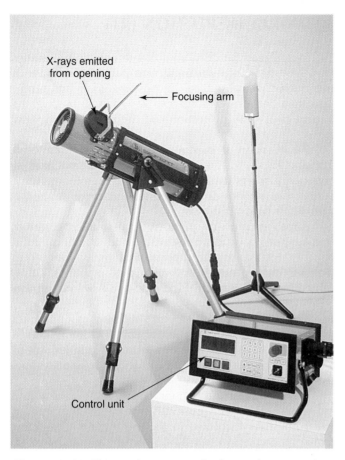

Figure 30-17. *The equipment required to make an on-site x-ray inspection. The radiating unit is shown on its stand and the control unit is on the floor at the right. (Seifert X-ray Corp.)*

Figure 30-18. *A portable x-ray unit used to inspect a jet engine nacelle. (Seifert X-ray Corp.)*

30.9 INSPECTING WELDS USING PNEUMATIC OR HYDROSTATIC PRESSURE

A common method of testing pressure-vessel welds (tanks and pipelines) for leaks is to use gas or air pressure. Carbon dioxide gas is well-suited for the purpose. It is nonexplosive when in contact with oils or greases. A small pressure of 25 psi to 100 psi (170kPa to 690kPa) is applied in the vessel or pipe, and a soap-and-water solution put on the outside of each weld. A leak will be indicated by the formation of bubbles. The ability of a vessel to hold pressure is also an indication of the soundness of its welds. A vessel to be inspected may be pressurized and the pressure noted on a gauge. The pressure is then turned off, and the gauge is again checked after 24 hours. Any drop in pressure indicates a leak. This test is easily applied. It is also *safe*, since the pressures used are usually under 100 psi (690kPa).

Another test for pressure vessels is to coat the surface with a lime solution. After the lime solution has dried, leaving a white coating, pressure is built up in the vessel. Where the lime flakes from the metal, a flaw is indicated as being present. **Hydrostatic pressure**, using water as the fluid, is the usual medium used in this test. This test may also be used to reveal the weakest portion of a welded vessel if enough pressure is created without destroying the vessel. Strain gauges that measure the slightest movement may be used to test for any flex points in a weld.

To inspect pressure vessels for leaks, water is still a popular method. However, water is not a reliable check for extremely small leaks. Pressure vessels may be filled with chlorine, fluorine, helium, or other nonoxygenated gases. These gases will flow through extremely small holes. A mass spectrometer pickup tube is placed on a suspected leakage area. A flow of one part of these gases in 1,000,000 parts of air may be detected to indicate a very small leak in a pressure vessel. **Follow all proper safety precautions when using chlorine gas.**

30.10 BEND TESTS

A popular method of destructively testing a weld which does not require elaborate equipment is the **bend test**. The method is fast and shows most weld faults quite accurately.

The most common bend tests are:
- Guided face and root bend test.
- Guided side bend test.
- Unguided bend test.
- Fillet weld bend test.

These tests will help determine:
- The physical condition of the weld, and thus confirm the weld procedure.
- The welder's qualifications.

The guided bend tests require that a test sample or test coupon be cut from parts that were welded together. Before bending, the edges of the test sample are filed or ground to form a slight radius. The radius prevents cracking along the edges. The test sample is bent with the weld and heat-affected zone being centered in the bend, as shown in Figures 30-19, 30-20, and 30-21.

Most bend tests are done with the weld going across the test sample. These tests are called *transverse bends,* as shown in Figure 30-19. If the face of the weld is on the outside of the bend, it is called a *transverse face bend*. If the root of the weld is on the outside, the bend is called a *transverse root bend*. When welding very thick metal, a face and root bend are not possible. In these cases, a *side bend* is performed. See Figure 30-20.

Bend tests can also be done with the weld going along the length of the test sample. These tests are called *longitudinal face and root bends*. Figure 30-21 shows longitudinal bend samples. The radius on the inside of the bend is often specified. Chapter 31 shows a bend jig specified by one code. Thicker weld samples require a larger radius than thinner weld samples. Figure 30-22 shows equipment used for a guided bend test.

After bending, each bend sample is examined for defects. An excellent bend sample will show no evidence

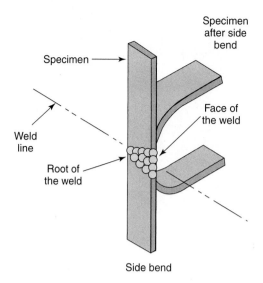

Figure 30-20. *Since a face bend test is not practical on thick welds, a side bend test is performed on a cross section of the weld. To perform this test, a sample is cut across a thick metal groove weld. The sample is machined to a uniform thickness.*

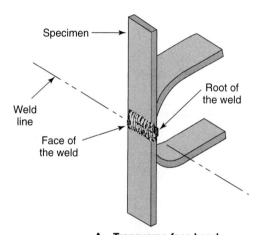

A – Transverse face bend
The weld face is on the outside of the bend

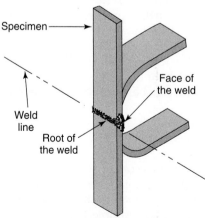

B – Transverse root bend
The root of the weld is on the outside of the bend

Figure 30-19. *Transverse face and root bend. The transverse sample is cut across the weld line.*

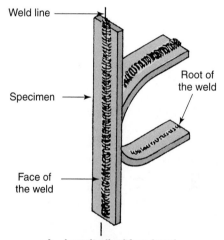

A – Longitudinal face bend
The weld face is on the outside of the bend

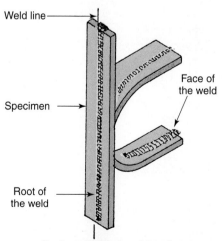

B – Longitudinal root bend
The root of the weld is on the outside of the curve

Figure 30-21. *Longitudinal face and root bend. The longitudinal sample is cut in the same direction as the weld line.*

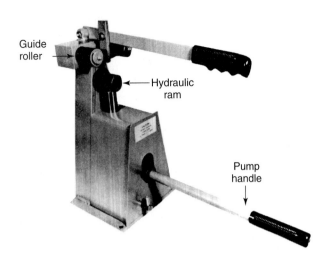

Figure 30-22. A hydraulically operated guided bend test machine. (Vega Enterprises)

of any cracking. However, some very small cracks may be visible even on a good bend sample. These small cracks are each looked at to make sure they are not caused by a welding defect.

A poorly made weld will have large cracks after bending. These are caused by defects like slag inclusions, lack of penetration, and lack of fusion. Some poorly made welds fracture into two pieces instead of bending into a U shape. The fractured surface can be examined to determine the reason for the failure.

Another bend test that is used is the *unguided bend test*. In this test, a welded piece is clamped in a vise or jig and one part of the metal is bent to an angle of about 90°. See Figure 30-23. This test will show most defects. It is not as controlled as the guided bend test, but it will give indications of the ductility and strength of the weld and base metal near the weld.

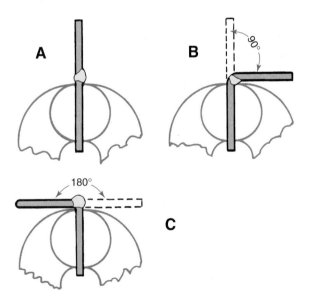

Figure 30-23. Testing a weld sample in a vise. A—Sample before bending. B—Sample after bending 90° (note the weld is closed in on itself, to test the penetration). C—Sample after bending it back 180°.

Caution: A screen or curtain should be used around the vise when doing an unguided bend test. This is to protect other workers from flying pieces if the welds should break.

A *fillet weld bend test* requires that a fillet weld be made on one side of a T-joint. The vertical piece is then bent over the weld until the weld fails, or until it is bent flat against the horizontal piece. Figure 30-24 illustrates a fillet weld bend test. There should be little or no evidence of defects along the weld joint. Chapter 31 refers to specific requirements of the ASME code concerning fillet weld tests.

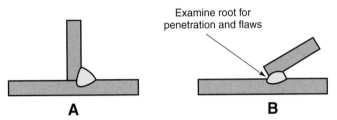

Figure 30-24. Bend testing a fillet weld. A—Sample before bending. B—Sample bent as far as possible in one direction. After bending, check the root area for penetration and flaws.

30.11 TENSILE TEST

A *tensile test* is a test in which a prepared sample is pulled in a tensile testing machine until it breaks. This test will determine different values of the sample. These values are useful to determine how a weld will perform in service. A welded structure must withstand all forces, or it will break. The tensile test helps determine how much pulling, or *tensile force*, a weld will withstand before it breaks.

To perform the test, a test sample is cut out of a weld. The test sample is further prepared to create a reduced area (weak area) where the weld is located. The test sample will always break in the reduced area, which is the weakest part of the test sample. This test, therefore, gives good information about the weld.

Three values can be obtained from a tensile test. These values are:
- Tensile strength.
- Yield point.
- Ductility.

Tensile strength is recorded as the number of pounds per square inch (psi) or kilopascals (kPa) required to break the sample. This value is obtained by measuring the width and the thickness of the sample before the test begins. Multiplying the width times the thickness yields the *area* of the sample. The area of a round sample is π times the diameter squared, divided by 4 ($\pi D^2 \div 4$). During the test, the maximum load applied to the sample is recorded. The tensile strength is determined by dividing the maximum tensile load by the cross-sectional area of the sample as shown in this formula:

$$\text{Tensile strength} = \frac{\text{tensile load}}{\text{area}}$$

As an example, let the sample be:
1/4" (6.4mm) thick
3/4" (19.1mm) wide.
Let the tensile load be: 11,250 lb (50,042 N), then

$$\text{Tensile strength} = \frac{11,250 \text{ lb}}{1/4" \times 3/4"} = 60,000 \text{ psi}$$

$$\text{Tensile strength} = \frac{50,042 \text{ N}}{6.4\text{mm} \times 19.1\text{mm}} = 409.5 \text{ N/mm}^2$$

The *elastic limit* of the metal is the *stress* (load) it can withstand and still return to its original length after the load is released. However, when more load is applied to the specimen after the elastic limit has been reached, the specimen exceeds its elasticity. The metal stretches and will not return to its original condition or size. The *yield point* or *yield strength* occurs when the specimen stretches or gives at a certain loading, but does not break. Figure 30-25 is a graph of the tensile load imposed on a metal sample.

The yield point is important. It is not desirable to load metal to the point where it will stretch and not return to its former shape. Metal structures are designed to be strong enough so the yield point is never reached.

Machines have been developed to test tensile strength and yield point. Figure 30-26 shows a tensile testing machine. Smaller machines and portable machines are used, but are limited in the size of the test sample that can be tested.

A tensile test will also determine the ductility of the test sample. Ductility is a measure of how much a sample will stretch or elongate before it breaks. Figure 30-27 shows tensile strengths and ductility of some common metals.

To measure the ductility of a sample, two points or two scribe lines are marked on the sample. The distance between the two points is recorded before the test begins. After the sample is tested and has broken, the two parts are fitted back together and the distance between the two points is measured again. The following formula is used to calculate the ductility:

$$\text{Ductility} = \frac{L_f - L_i}{L_i} \times 100\%$$

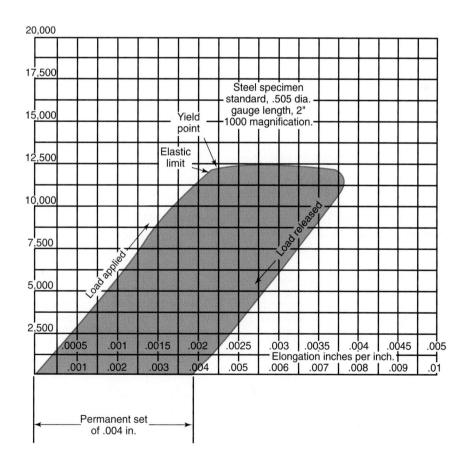

Figure 30-25. *Strength graph of a metal sample, showing the yield point and the permanent set (lengthening) left in the steel when returned to a no-load condition. (Tinius Olsen Testing Machine Co., Inc.)*

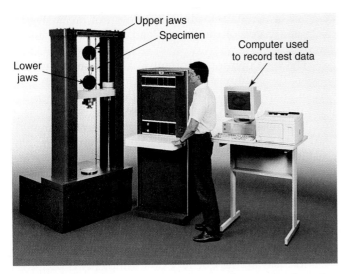

Figure 30-26. *A tensile test machine with its controls and a computer to record the test results. (Tinius Olsen Testing Machine Co., Inc.)*

As an example, let the initial length between the two points be 2" (50.8mm). After testing, the final length is 2 1/4" (57.15mm). Then,

$$\text{Ductility} = \frac{(2\ 1/4 - 2.00)}{2} \times 100\% = 12.5\%$$

$$\text{Ductility} = \frac{(57.15 - 50.8)}{50.8} \times 100\% = 12.5\%$$

30.12 LABORATORY METHODS OF TESTING WELDS

Most large companies that do welding have laboratories for determining the strength of welds. These laboratories are equipped with modern weld testing equipment. This equipment can be used to determine the physical and chemical properties of weld or base metal samples. Occasionally, the testing is performed by a metallurgical department. In some cases, a part of the shop is set aside for testing purposes.

Some of the tests performed are to determine or study the following:

- Impact strength.
- Hardness.
- Microstructure.
- Macrostructure.
- Chemical analysis.

These tests are usually in addition to the bend tests and tensile tests discussed in Headings 30.10 and 30.11.

The conditions under which the specimens are tested are kept identical. The specimens are all of a standard size specified by the welding code used. The length of the specimens need not be the same, but the cross-sectional area of the specimens must be the same. Samples must be taken from specified positions on the weld sample being tested.

The Society of Automotive Engineers (SAE), the American Society of Mechanical Engineers (ASME), and American Society for Testing Materials (ASTM), have adopted standards for laboratory tests of metals. These standards specify the physical qualities to be maintained for different types of welds and base metals. They also specify the size and location of various test specimens or samples. Some sample test pieces are shown in Figure 30-28. Specifying the location and size of the metal sample is necessary in order to compare test results. Test results of completed welds and base metal samples may be used to compare welds, welding operators, and welding procedures. Standardizing tests make it possible to build standard testing machines.

30.13 IMPACT TESTS

It is possible for a weld to show good results in a variety of tests, but to fail under a rapidly applied impact load. The *impact test* may be made by either the Izod method or the Charpy method. The methods are similar, but the shape and position of the notch on the sample varies. Test samples are taken from the weld, the heat-affected zone (HAZ), and the base metal.

Metal	Tensile Strength Annealed		Tensile Strength Heat-treated or Hardened		Percent Elongation
	psi	MPa	psi	MPa	
Low-carbon steel	55,000	379	—	—	25
Medium-carbon steel	76,000	524	—	—	18
Stainless steel	75,000	517	90,000	621	—
Chrome-moly steel	—	—	128,000	886	17
High tensile steel	—	—	262,000	1806	4.5
Duralumin-2017	26,000	179	62,000	427	22
Aluminum-6061	18,000	124	45,000	310	17
Commercial bronze	38,000	262	45,000	310	25-45

Figure 30-27. *Tensile strength and percent of elongation values for various metals.*

A test piece is notched in a specified manner and clamped in the jaws of an impact testing machine. A heavy pendulum is lifted to a given height and then dropped against the notched specimen. See Figure 30-29.

The indicator needle on the testing machine shows what force was exerted to break or *fracture* the test specimen. This test determines the impact strength of the weld specimen. The higher the force required, the greater the sample's toughness. A lower force indicates the sample does not withstand an impact load well and may be a brittle weld.

Figure 30-28. Samples of weld test specimens tested to destruction. These were machined to standardized measurements before testing. All pieces shown are tensile test specimens except the pair on the right. That pair is a Charpy impact test specimen shown after it has been fractured.

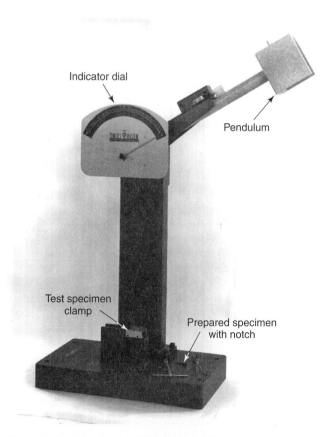

Figure 30-29. A Charpy impact test machine. The pendulum is lifted and dropped against the test weld, which is held in a clamp. The impact force is registered on the indicator dial. (Tinius Olsen Testing Machine Co., Inc.)

30.14 HARDNESS TESTING

Another important factor that should be determined when testing base metal or a weld is the hardness. *Hardness* may be defined as a resistance to permanent indentation.

Methods used to determine metal hardness have been standardized. The most popular method is the Rockwell hardness test. A Rockwell testing machine is shown in Figure 30-30. This machine works somewhat like a press. Testing is done by applying fixed weights operating through leverage to the sample being tested, as shown in Figure 30-31. An indenter is pressed into the sample. Two loads are applied. During the first load of 10kg (22 lb.), the machine is calibrated and the gauge on the machine is set to zero. Then, a final or major load is applied. The distance the indenter penetrates the metal between the first and second load indicates the hardness on a scale of 0 to 100. The depth of the indentation determines the hardness of the sample. Harder materials are indented less; softer materials are indented more.

There are different ranges or *scales* of Rockwell hardness testing. The most common scales are the Rockwell B and Rockwell C scales. The *Rockwell B test* uses a 1/16" (1.6mm) diameter ball indenter and a major load of 100kg (220 lb.). It is used on materials like copper alloys, soft

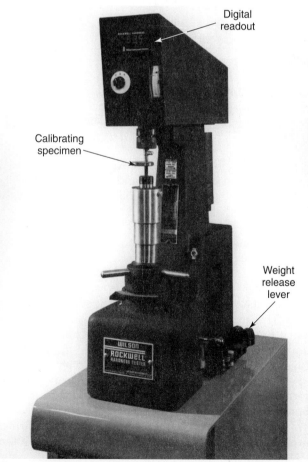

Figure 30-30. A laboratory-type digital Rockwell hardness tester. (Page-Wilson Corporation)

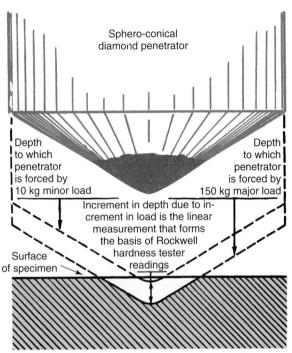

Figure 30-32. Drawing of the diamond penetrator used in the Rockwell hardness tester. Note the difference in depth of the depressions left in the metal by the diamond under the 10kg (22 lb) and the 150kg (331 lb) loads. (Shore Instrument Company)

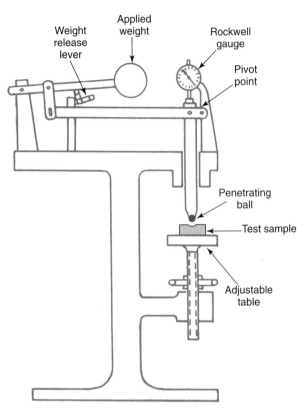

Figure 30-31. This diagrammatic drawing shows how leverage is used to apply weight to a penetrator in a Rockwell hardness tester.

steels, and aluminum alloys. The **Rockwell C test** uses a diamond ground to a 120° point and a major load of 150kg (331 lb). It is used on harder materials like steel, deep case-hardened steel, hard cast irons, and titanium. Figure 30-32 shows the preload and major load indentations.

Portable hardness testers are used to test large samples in the field or in production. Figures 30-33 and 30-34 show examples of portable hardness testers.

Another type of hardness test is called a **Scleroscope test**. A Scleroscope testing machine is shown in Figure 30-35. This machine is based on the impact or rebound from a test sample of a diamond-tipped hammer. A steel hammer with a diamond tip is dropped onto a sample. The distance that the hammer rebounds after it contacts the metal may be read on a scale. The higher the number, the harder the metal. A high-carbon steel will indicate approximately 95 on the Scleroscope scale.

A third method of testing hardness is to use a **Brinell hardness testing machine.** This type machine has a 10mm (0.394") diameter indenter built into the press, as shown in

Figure 30-36. The indenter is pressed (not dropped) onto the test sample for a given period of time. After the load is removed, the diameter of the indentation is measured using a microscope. The Brinell hardness number is calculated by dividing the applied load by the area of the indentation.

Figure 30-37 shows a table that compares the hardness numbers among the Rockwell, Brinell, and Scleroscope scales.

Microhardness testers have been developed which make it possible to test the hardness of a part without appreciably harming the part. An extremely small diamond penetrator is used to penetrate the surface to be tested. The loads on the diamond penetrator may be varied from 0.25N to 490N (0.06 lb to 110 lb). After the surface has been penetrated, the size of the indentation is measured, using a powerful microscope. This measurement indicates the hardness of the surface tested. Because the metal surface is marked microscopically, the tests can be performed more frequently over the surface of a metal object. Figure 30-38 shows a microhardness testing machine.

The length of an indentation made on hardened steel with the diamond penetrator under a load of 0.98N (0.22 lb) is about 0.0015" (0.038mm) and the depth of the penetration is about 0.00005" (0.001mm). Figure 30-39 illustrates the indentation made with the Knoop diamond penetrator.

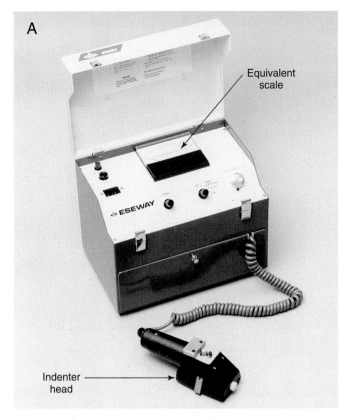

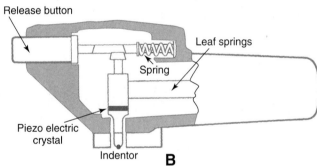

Figure 30-33. *An equivalent hardness testing device. A—The equivalent scale and indenter head. B—A schematic of the indenter head. (Engineering and Scientific Equipment, Ltd.)*

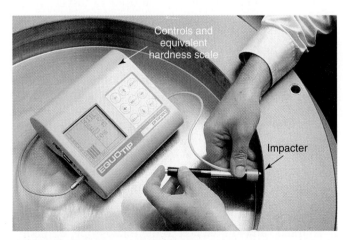

Figure 30-34. *A lightweight portable equivalent hardness tester. (Foerster Instruments, Inc.)*

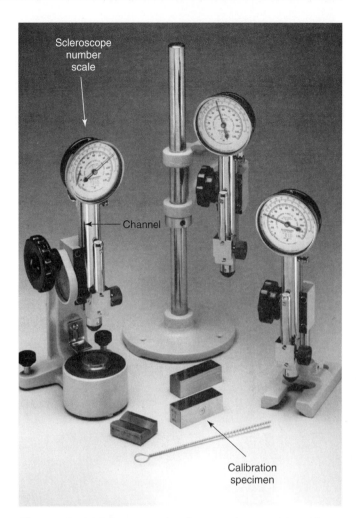

Figure 30-35. *Scleroscope hardness testing machine. The distance the small hammer rebounds up the channel after it is dropped on the test specimen is read on the dial gauge. Three different mounting arrangements are shown here. (Shore Instrument Company)*

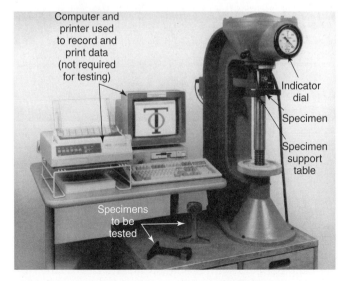

Figure 30-36. *A pneumatically operated Brinell hardness tester. (Tinius Olsen Testing Machine Co., Inc.)*

Rockwell C	Rockwell B	Brinell	Scleroscope
69	—	755	98
60	—	631	84
50	—	497	68
40	—	380	53
30	—	288	41
24	100	245	34
20	97	224	31
10	89	179	25
0	79	143	21

Figure 30-37. *This table compares hardness numbers in Rockwell, Brinell, and Scleroscope scales.*

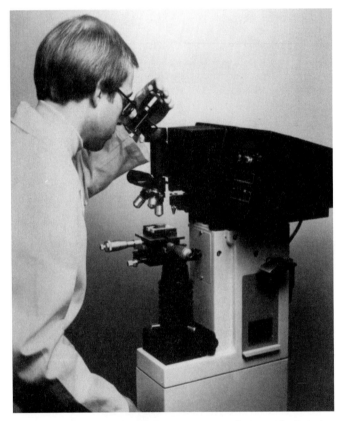

Figure 30-38. *Microhardness testing machine with digital reading. (Page-Wilson Corporation)*

30.15 MICROSCOPIC METHOD OF TESTING WELDS

A test commonly used in the metallurgical laboratory for testing a weld is to procure a sample of the weld and polish it to a very smooth, mirrorlike finish. After polishing, the sample should show absolutely no scratches on the surface. The sample is then placed under a microscope which magnifies the surface of the metal from 50 to 5000 times. The usual magnification is 100 to 500 times (100× to 500×).

The appearance of the metal under the microscope reveals such things as the amount of impurities, heat treat-

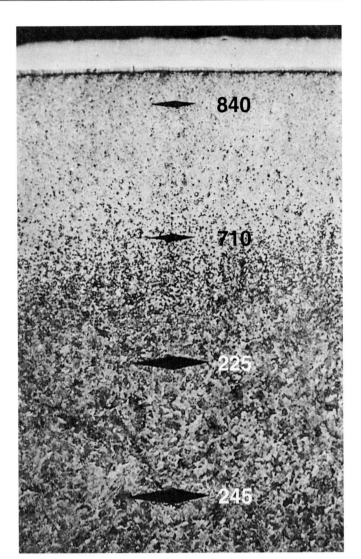

Figure 30-39. *The microhardness indentations made by a Knoop diamond penetrator. The white layer at the top of the photograph is chromium (75×).*

ment, and grain size. In most steels, a microscopic study of a sample can accurately determine the carbon content. Figure 30-40 shows a microscope used to inspect metal specimens.

Usually, the metal being worked on is treated with acid, or etched. To *etch* a sample after it has been polished, it is wiped with a weak acid solution. The acid solution used for steel is usually 4% nitric acid and 96% alcohol, and is called *nitrol*. The acid is allowed to remain on the metal for a time. It is then washed off. The metal is then studied under a microscope. Certain acids bring out features, such as the grain boundaries in the metal, the kind of impurities, and slag spots. Poor fusion and shrinkage cracks are easily seen under a microscope.

The most important feature of the micrographic study of the metal is that photographs may be taken. Photographs are taken of the metal by means of a specially adapted camera attached to the microscope. By taking photographs of each specimen, a very accurate comparison may be made between the samples studied. See Chapter 28 for microphotographs of common metals.

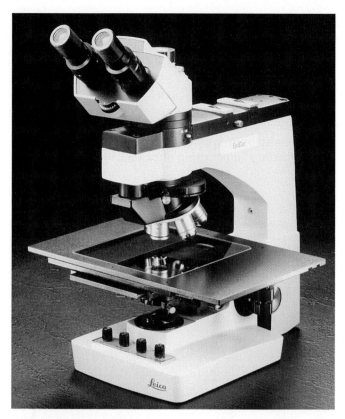

Figure 30-40. *A laboratory type microscope used to examine polished metal specimens. This microscope has lenses of four different powers mounted in a rotating turret. (Leica Optical Products Division)*

Adaptor

40× macroscope

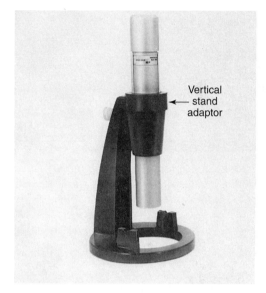

Vertical stand adaptor

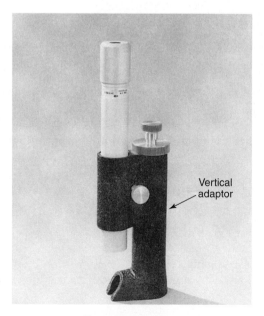

Vertical adaptor

30.16 MACROSCOPIC METHOD OF TESTING WELDS

A microscopic view of a weld does not cover enough area to obtain a picture of an entire weld for inspection purposes. *Macroscopic* pictures are only 10 to 40 magnifications (10× to 40×), and are better suited to this purpose. When the sample is deeply etched with hot nitric acid, the structure of the weld stands out more clearly. Figure 30-41 shows a macroscope with 40× magnifications. The crystalline structure of the metal is not so clearly revealed, but cracks, pits, and pin holes are clearly seen, as shown in Figure 30-42. Scale inclusions are easily detected by this method. This test also shows up the crystal grain size. A large grain size indicates improper heat treatment after or during welding.

30.17 CHEMICAL ANALYSIS METHOD OF TESTING WELDS

The complete investigation of welded material consists of a thorough *chemical analysis.* This type of test is made in a metallurgical laboratory. Most companies are

Figure 30-41. *A small pocket-size 40× macroscope is shown at A. At B, C, and D, various adaptors for the macroscope are shown. (Leica Optical Products Division)*

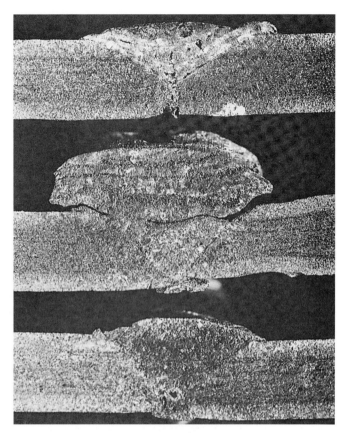

Figure 30-42. *A macroscope photograph of etched welds in cross section. Such magnified views allow the quality of the weld to be more easily judged than with the naked eye.*

strength of a resistance spot weld. All recommended machine settings are made on the spot welding machine before several spot welds are completed on test pieces.

The test pieces are then peeled apart. See Figure 30-43. If the spot weld nugget is of the correct diameter and is torn out of one piece, the spot weld is considered to be properly made. All machine settings are considered to be correct for the parts to be welded when the peel test is made satisfactorily. Spot welds may also be inspected in a *nondestructive* manner using ultrasonic test methods.

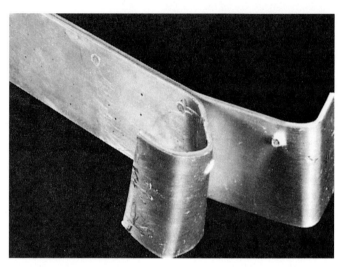

Figure 30-43. *Destructive peel test. Test welds are made to determine weld strength, size, and proper machine settings required for a given job.*

not equipped to perform this test. The chemical analysis may be both qualitative and quantitative. *Qualitative analysis* determines the different kinds of chemicals in the metals. The *quantitative analysis* determines the kind and amount of each chemical in the metal. This type of investigation is necessarily tedious and expensive. The tests are not of direct value to a welder or to a welding company. A chemical analysis is usually used when large quantities of metal are believed to be of the wrong composition. The weldability of a metal is dependent to a great extent on the impurities in the metal. Usually, the manufacturer of the metal can supply complete data on the physical and chemical properties of the metal.

30.18 THE PEEL TEST

Lap joints may be tested to destruction by means of the *peel test*. The peel test is most often used to check the

30.19 REVIEW OF SAFETY

Proper inspecting and testing of welds requires attention to certain safety factors. There is a danger of parts flying when the weld is broken. **A face shield should be worn to protect the operator. The operator should never stand in the likely trajectory of a flying test piece.**

You should avoid exposure to the types of radiation from x-ray machines used to inspect metals. Only trained experts should work with x-ray equipment. **Proper signs should be used to warn people that x-ray inspection is being performed in an area. Also, proper protection should be used by the operator to minimize any x-ray exposure.**

TEST YOUR KNOWLEDGE

Write your answers on a separate sheet of paper. Do not write in this book.

1. What is a discontinuity?
2. When does a discontinuity become a defect?
3. What does NDE mean? What does NDT mean?
4. List nine types of NDE.
5. What are the most popular methods of NDE?
6. What equipment may be used to visually inspect welds inside small pipes or cylinders?
7. When making a magnetic particle inspection, the test should be done twice. During the second inspection, the yoke or test prods should be turned _____° to the direction of the first inspection.
8. What are the steps used in a dye penetrant test?
9. Name two types of NDE tests that require little or no equipment and are most often used in the small shop.
10. In ultrasonic testing, what is a couplant and what is it used for?
11. Permanent records are often required for critical welds. Give three examples of industries or products where permanent records may be required.
12. What method is used to test for leaks as small as one part in a million?
13. The Izod and Charpy tests are examples of _____ tests.
14. Name three laboratory tests generally not done in a small shop or company.
15. Which test is made to check the elongation and the yield point of a material?
16. When a metal stretches at a certain loading or force, but does not break, this point is called the _____ point or _____ strength of the metal.
17. Ductility is the ability of a metal to _____ before it breaks.
18. Name three types of tests used to test for hardness.
19. Which type of hardness test may be used with minimum damage to the metal surface?
20. _____ pictures are 10 to 40 magnifications (10× to 40×).

A portable hardness tester with a digital readout. The "hardness" reading is converted to other measuring scales (such as Brinnell or Rockwell B) by using a conversion chart. (Foerster Instruments, Inc.)

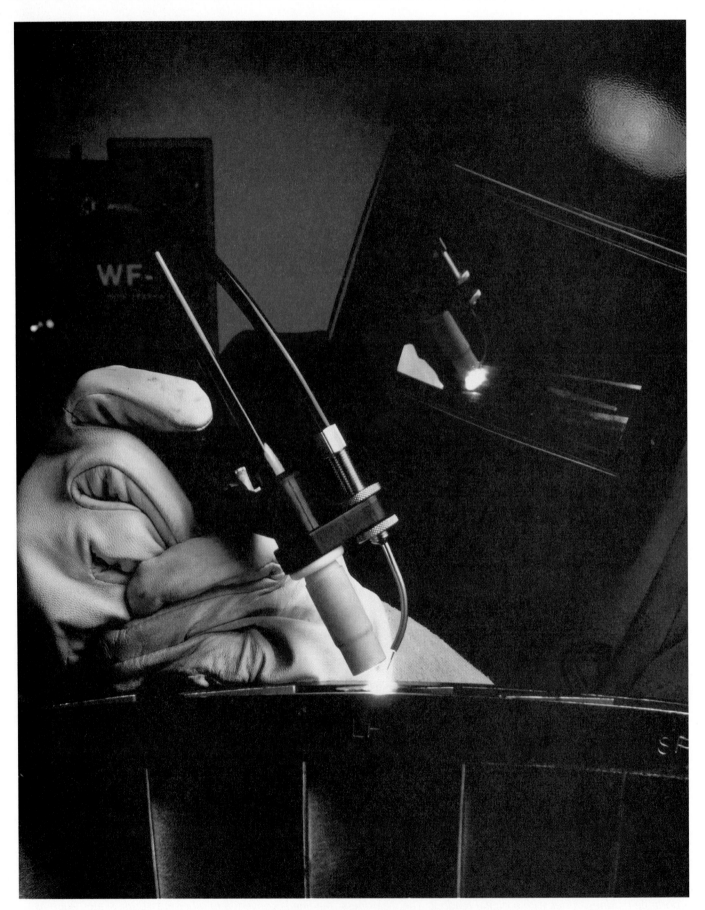

To meet the requirements of a Welding Procedure Specification, a welder must be tested to qualify for both the welding process that will be used and the welding positions that will be specified. (CK Worldwide)

Chapter 31

PROCEDURES AND WELDER QUALIFICATIONS

LEARNING OBJECTIVES

After studying this chapter, you will be able to:

* List several governmental agencies that use and produce welding codes of various types.
* Understand and relate the reasons for welding codes and specifications.
* List the essential and nonessential welding variables.
* Define a "Welding Procedure Specification" and a "Welding Procedure Qualification Record."
* Define a "Welding Performance Qualification."
* List the weld qualification test positions.
* Describe what test specimens are cut from a weld sample and describe the types of test made on each specimen.

Whenever an item such as a building, bridge, ship, or pressure vessel is to be manufactured or built, it is necessary that the manufacturer and buyer reach an agreement on the scope of the work involved. The agreement includes requirements for how the item is to be built, including requirements for the welding that will take place.

The manufacturer must prove that the welds made meet the requirements agreed upon. To prove this, the manufacturer writes out procedures that specify exactly how each weld will be done. The manufacturer makes and tests sample welds to show that the procedures being used meet the requirements for the job. Each welder doing work must be tested to ensure that he or she can follow the procedure and produce the required weld. Requirements for writing welding procedures and testing them are written in the form of a *code*.

31.1 WELDING CODES

To eliminate the necessity of writing a new code for each new job, a number of government agencies, societies,

and associations have developed codes that may be used. Federal, state, and local governmental agencies, for example, have written building and safety codes. Insurance companies that must insure welded structures have also prepared procedures and welder performance qualification codes. Codes are also written for soldered and brazed joints.

31.1.1 Government Agencies

Governmental codes and standards are established by the following partial list of agencies:

* Bureau of Mines, U.S. Department of the Interior.
* Federal Aviation Administration (FAA) — aircraft welding.
* Interstate Commerce Commission (ICC) — design, construction, content, and transportation of welding cylinders.
* Military Specifications (MIL).
* National Aeronautics and Space Administration (NASA).
* National Bureau of Standards.
* Occupational Safety and Health Administration (OSHA).
* U.S. Air Force.
* U.S. Army.
* U.S. Coast Guard.
* U.S. Navy.

31.1.2 Associations and Societies

A sampling of the associations and societies that have established codes, standards, and/or tentative standards for welding:

* American Society for Testing Materials (ASTM).
* American Institute of Steel Construction (AISC).
* American National Standards Institute (ANSI) — voluntary safety, engineering, and industrial standards.
* American Petroleum Institute (API) — standard for field welding for pipelines.
* American Society for Metals (ASM).

685

- American Society of Mechanical Engineers (ASME) — Boiler and Pressure Vessel Code.
- American Welding Society (AWS) — Structural Welding Code.
- Association of American Railroads (AAR), Operation and Maintenance Division.
- Institute of Electrical and Electronics Engineers (IEEE).
- Mechanical Contractors Association of America.
- Resistance Welder Manufacturers' Association (RWMA).
- Society of Automotive Engineers (SAE).
- Tubular Exchanger Manufacturers' Association, Inc. (TEMA).

31.1.3 Insurance Companies and Associations

Insurance companies or associations that have established standards for eligibility for insurance where welding is used include:

- Factory Assurance Corporation.
- The Hartford Steam Boiler Inspection and Insurance Co.
- Lloyd's of London.
- Lloyd's Register of Shipping Rules and Regulations.
- National Board of Fire Underwriters.
- National Fire Protection Association.
- Underwriters Laboratories.

31.2 IMPORTANCE OF CODES AND SPECIFICATIONS

The codes and specifications written by various organizations have been developed to aid manufacturers in producing a safe product. The codes and specifications that are available are only guidelines for a manufacturer. The codes and specifications do not have to be followed unless they are part of a contract agreement. The manufacturer or builder of a product must reach an agreement with the buyer as to how the product will be built. The agreement is called a *contract*.

The contract will refer to the codes and specifications that are to be used when constructing a product. The manufacturer or builder is required to follow those codes and specifications. If the manufacturer or builder does not follow the specifications in the contract, the firm has broken the contract.

Companies write procedures or practices that tell how they will perform the work. A company can write its own specifications, but usually uses established codes and specifications and refers to them in its own procedures and practices. Figure 31-1 shows the relationship between a contract, codes or specifications written by different companies and agencies, and company practices.

There are other examples of situations in which a code must be followed. The Interstate Commerce Commission (ICC) regulates the transportation of all gas

cylinders in the United States. Any manufacturer that produces cylinders for use in the United States must follow ICC specifications. In a similar way, some states have adopted certain codes as state laws. Any product produced or sold in that state must be manufactured according to the adopted code.

Some jobs may require the builder to follow a number of codes. If a bridge is constructed that also carries oil and pressure pipes over a span of water, several codes may be involved. The American Welding Society (AWS) *Structural Welding Code* may be used for the bridge structure. The oil pipelines would be welded using the American Petroleum Institute (API) *Standard for Field Welding of Pipe Lines*. Pressure pipes may be welded using the American Society of Mechanical Engineers (ASME) *Boiler and Pressure Vessel Code*.

All codes applying to a welding job must be available to the company engineers and quality department. The company practices must be available to the welders and inspectors. Copies of any welding code are available, for a fee, from the agency, society, or association that publishes them.

31.3 WELDING PROCEDURE SPECIFICATIONS

Procedure qualifications are limiting instructions, written by a manufacturer or contractor, which explain how welding will be done. The welding must be done in accordance with a code. These limiting instructions are listed in a document known as a *Welding Procedure Specification (WPS)*.

A welding procedure specification must list in detail:

- The various base metals to be joined by welding.
- The filler metal to be used.
- The range of preheat and postweld heat treatment.
- Thickness and other variables described for each welding process.

The variables are listed as essential or nonessential. These variables will be covered in Heading 31.3.1.

Each manufacturer or contractor must qualify the Welding Procedure Specification (WPS) by welding *test coupons* (samples) and by testing the coupons in accordance with the code. A test weld is made and test coupons are cut from it. The test coupons are used to make tensile tests, root bends, and face bends, as required by the code. The results of these tests are recorded on a document known as a *Welding Procedure Qualification Record (WPQR)*.

A different welding procedure specification is required for each change in an essential variable. Each Welding Procedure Specification (WPS) must have a Welding Procedure Qualification Record (WPQR) to document the quality of the weld produced. Figure 31-2 shows the relationship between the WPS and the WPQR. Each

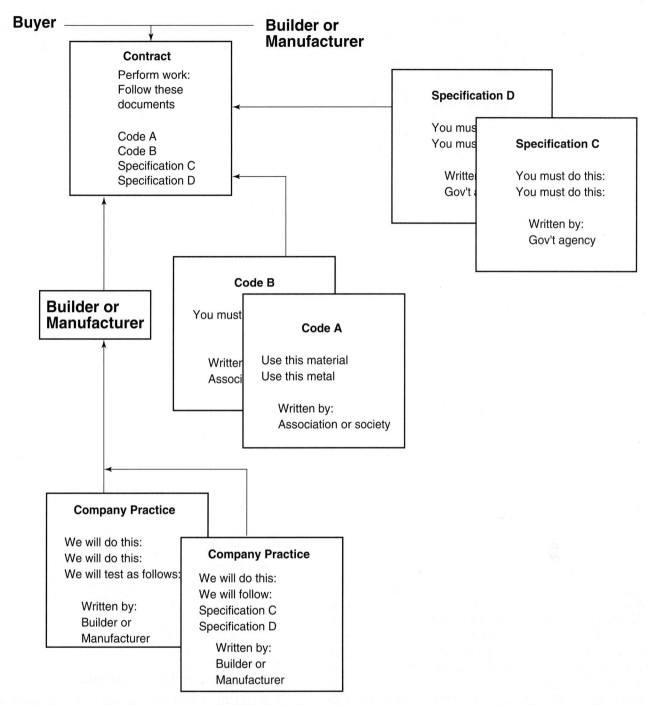

Figure 31-1. *A buyer and builder agree on what is to be built and sign a contract. Any codes and specifications called out by the contract are part of the contract. These codes and specifications help to assure the quality and performance of the item being produced. The builder writes company practices that will follow the contract and the codes and specifications called out by the contract.*

welding procedure must be written and the WPQR performed before the welding procedure can be used. Both the WPS and the WPQR must be kept on file by the manufacturer.

Figure 31-13, at the end of this chapter, shows a blank Welding Procedure Specification. Each of the fields on this form is filled in to complete the WPS. When complete, it clearly specifies what material can be welded, the welding process that will be used, and all the parameters that are required to perform the weld.

Figure 31-14, at the end of this chapter, shows a blank Welding Procedure Qualification Record. Information is recorded on this form about how the weld was done and on the test results. Notice that the first line of the form references the WPS used.

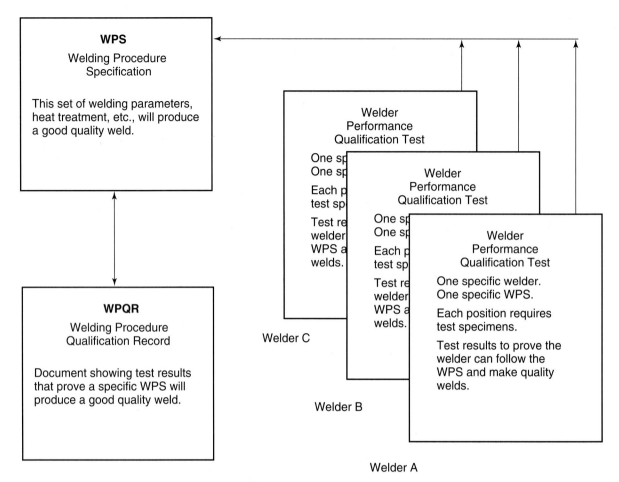

WPS

Welding Procedure
Specification

This set of welding parameters,
heat treatment, etc., will produce
a good quality weld.

Welder
Performance
Qualification Test

One sp
One sp

Each p
test sp

Test re
welder
WPS a
welds.

Welder
Performance
Qualification Test

One sp
One sp

Each p
test sp

Test re
welder
WPS a
welds.

Welder C

Welder
Performance
Qualification Test

One specific welder.
One specific WPS.

Each position requires
test specimens.

Test results to prove the
welder can follow the
WPS and make quality
welds.

Welder B

WPQR

Welding Procedure
Qualification Record

Document showing test results
that prove a specific WPS will
produce a good quality weld.

Welder A

Figure 31-2. *This diagram shows the relationship between the Welding Procedure Specification (WPS) and the Welding Procedure Qualification Record (WPQR).*

31.3.1 Welding Procedure Variables

Most welding codes list the welding procedure variables that are essential and nonessential. *Essential variables* are those which, when changed, will affect the mechanical properties of the weldment. Changes of essential variables require that the Welding Procedure Specifications (WPS) be requalified. *Nonessential variables* are those which, when changed from the approved welding procedure specifications, do not require requalification of the procedure. The changes made must be recorded on the WPS.

For example, some of the *essential variables* for welds made with the shielded metal-arc welding (SMAW) process are changes in:
- Base metal thickness (beyond accepted limits).
- Base metal strength or composition.
- Filler metal used (strength or composition).
- Preheat temperature.
- Postheat temperature.
- Thickness of postheat treatment sample.

For SMAW, some of the *nonessential variables* are changes in:
- Type of groove.
- Deletion of backing in single welded butt joints.

- Size of the electrode.
- Addition of other welding positions to those welding positions which are already qualified.
- Maintenance or reduction of preheat prior to postheat treatment.
- Current or polarity or the range of amperage or voltage.

31.3.2 Procedure Qualification Specimens

To qualify a Welding Procedure Specification (WPS), a contractor or manufacturer must use the WPS and weld a test plate. Several *test specimens* (samples) are taken from the plate for testing and approval.

Tension tests and bend tests are required. The number and thickness of the required tension test and transverse bend test specimens are shown in Figure 31-3. Also shown are the requirements for tension and longitudinal bend test specimens. The test results are recorded on a WPQR.

When a test plate is qualified, it will serve to qualify a range of thicknesses. If a test plate 1/8" (3.2mm) thick is qualified, it will serve to qualify all thicknesses from 1/16" to 1/4" (1.6mm to 6.4mm), as shown in Figure 31-3. The maximum thickness in the qualified range is "2t" or twice the test plate thickness.

Tension Test and Transverse Bend Tests

Thickness T of Test Coupon Welded, in. [Note (1)]	Range of Thickness T of Base Metal Qualified, in. [Note (2)]		Range of Thickness t of Deposited Weld Metal Qualified, in. [Note (2)]		Type and Number of Tests Required Tension and Guided Bend Tests — [Note (6)]			
	Min.	Max.	Min.	Max.	Tension QW-462.1[4]	Side Bend QW-462.2	Face Bend QW-462.3(a)	Root Bend QW-462.3(a)
Less than 1/16	T	2T	t	2t	2	—	2	2
1/16 to 3/8, incl.	1/16	2T	1/16	2t	2	—	2 (5)	2 (5)
Over 3/8, but less than 3/4	3/16 (8)	2T	3/16 (8)	2t	2	Note (3)	2 (5)	2 (5)
3/4 to less than 1 1/2	3/16 (8)	2T	3/16 (8)	2t when t < 3/4	2	4	—	—
3/4 to less than 1 1/2	3/16 (8)	2T	3/16 (8)	2t when t > 3/4	2	4	—	—
1 1/2 and over	3/16 (8)	8 (7)	3/16 (8)	2t when t < 3/4	2	4	—	—
1 1/2 and over	3/16 (8)	8 (7)	3/16 (8)	8 (7) when t > 3/4	2	4	—	—

Notes:
(1) When the groove weld is filled using two or more welding processes, the thickness t of the deposited weld metal for each welding process shall be determined and used in the "Range of thickness t" column. The test coupon thickness T is applicable for each welding process.
(2) See QW-403 (.2, .3, .6, .7, .9, .10) and QW-407.4 for further limits on range of thicknesses qualified.
(3) Four side bend tests may be substituted for the required face and root bend tests.
(4) The deposited weld metal of each welding process shall be included in the tension test of QW-462.1(a), (b), (c), or (e); and in the event turned specimens of QW-462.1(d) are used, the deposited weld metal of each welding process shall be included in the reduced section insofar as possible.
(5) Applicable for a combination of welding processes only when the deposited weld metal of each welding process is on the tension side of either the face or root bend.
(6) When toughness testing is a requirement of other Sections, it shall be applied with respect to each welding process.
(7) For the welding processes of QW-403.7 only; otherwise per Note (2) or 2T, whichever is applicable.
(8) When the weld metal thickness deposited by a process is 3/8" or less, this minimum shall be 1/16" for that process.

Tension Tests and Longitudinal Bend Tests

Thickness T of Test Coupon Welded, in. [Note (1)]	Range of Thickness T of Base Metal Qualified, in. [Note (2)]		Range of Thickness t of Deposited Weld Metal Qualified, in. [Note (2)]		Type and Number of Tests Required Tension and Guided Bend Tests — [Note (5)]		
	Min.	Max.	Min.	Max.	Tension QW-462.1[3]	Face Bend QW-462.3(b)	Root Bend QW-462.3(b)
Less than 1/16	T	2T	t	2t	2	2	2
1/16 to 3/8, incl.	1/16	2T	1/16	2t	2	2 (4)	2 (4)
Over 3/8	3/16 (6)	2T	3/16 (6)	2t	2	2 (4)	2 (4)

Notes:
(1) When the groove weld is filled using two or more welding processes, the thickness t of the deposited weld metal for each welding process shall be determined and used in the "Range of thickness t" column. The test coupon thickness T is applicable for each welding process.
(2) See QW-403 (.2, .3, .6, .7, .9, .10) and QW-407.4 for further limits on range of thicknesses qualified.
(3) The deposited weld metal of each welding process shall be included in the tension test of QW-462.1(a), (b), (c), or (e); and in the event turned specimens of QW-462.1(d) are used, the deposited weld metal of each welding process shall be included in the reduced section insofar as possible.
(4) Applicable for a combination of welding processes only when the deposited weld metal of each welding process is on the tension side of either the face or root bend.
(5) When toughness testing is a requirement of other Sections, it shall be applied with respect to each welding process.
(6) When the weld metal thickness deposited by a process is 3/8" or less, this minimum shall be 1/16" for that process.

Figure 31-3. ASME procedure qualification specimens. The QW numbers are article numbers in the ASME Code.

31.4 WELDING PERFORMANCE QUALIFICATIONS

Every welding procedure must be qualified. In addition, every welder and welding operator must be qualified to perform each welding procedure. The welder or welding operator is qualified by passing a welding performance qualification test.

The *Welding Performance Qualification Test* requires the welder or welding operator to complete a weld made in accordance with the Welding Procedure Specification (WPS). Some departures from the exact WPS may be allowed by the code used. Test specimens are cut from the test plate. Figure 31-15, at the end of this chapter, shows a form used to record information and document the Welding Performance Qualification Test.

The type and number of test specimens required for mechanical testing is specified by the code used. Also specified is the manner by which the specimens are removed from the weld sample. Specimens may also be tested by radiographic examination (x-ray).

The welder or welding operator is qualified by welding position. Welders may be qualified to perform a welding procedure in only one position, in several positions, or possibly in all positions. A welder qualified only in the flat position, for example, is not qualified to make welds in any other position.

Each welder or welding operator must be qualified for each process used. A welder qualified to weld in accordance with one qualified WPS may also be qualified to weld in accordance with another WPS. This is only true when the second WPS and the essential variables are within the limits of the code.

Welders may be required to *requalify* if they have not used a specific process for three months. They may also be required to requalify if there is a reason to question their ability to make welds that meet the welding procedure specification.

A welder must be requalified for a process whenever a change is made in one or more of the essential variables. The essential variables for Shielded Metal-Arc Welding (SMAW) listed in the ASME Boiler and Pressure Vessel Code are:

- The deletion of backing in single welded butt joints.
- A change in the ASME electrode specification number or AWS electrode classification number.
- The addition of welding positions other than those in which the welder has already qualified.
- A change from uphill to downhill, or from downhill to uphill, in the progress of the weld.

31.4.1 Welding Qualification Specimens

Welding positions and test specimen specifications illustrated in the following headings are taken from the American Society of Mechanical Engineers (ASME) Boiler and Pressure Vessel Code, Section IX.

This information is provided for those who may wish to practice making specimens similar to those that may be required for welder qualification tests. This information is only a small part of the complete code and should not be used as a code. The complete code should be obtained from the agency, society, or association that wrote the code.

31.4.2 Welding Test Positions

The weld positions shown in Figures 31-4 through 31-7 are typical of those used for many welding codes. A welder or welding operator must be tested to become qualified for each position in which welding is required. The welder or welding operator must follow the WPS for each position. A qualification in one position does not automatically qualify the welder to weld in any other position. When the welding position changes, the welder must be requalified for that position.

Tabulation of positions of welds

Position	Diagram reference	Inclination of axis	Rotation of face
Flat	A	0° to 15°	150° to 210°
Horizontal	B	0° to 15°	80° to 150° 210° to 280°
Overhead	C	0° to 80°	0° to 80° 280° to 360°
Vertical	D E	15° to 80° 80° to 90°	80° to 280° 0° to 360°

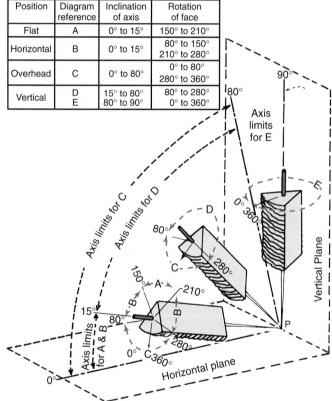

Figure 31-4. *Welding test positions. The horizontal reference plane is taken to lie always below the weld under consideration. Inclination of axis is measured from the horizontal reference plane toward the vertical. Angle of rotation of face is measured from a line perpendicular to the axis of the weld and lying in a vertical plane containing this axis. The reference position (0°) of rotation of the face invariably points in the direction opposite to that in which the axis angle increases. The angle of rotation of the face of the weld is measured in a clockwise direction from this reference position (0°) when looking at point P. (AWS A3.0)*

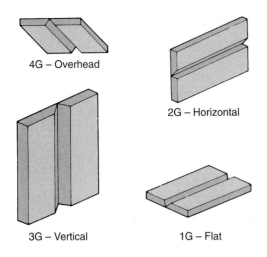

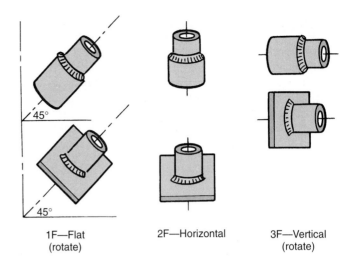

Groove welds in plate

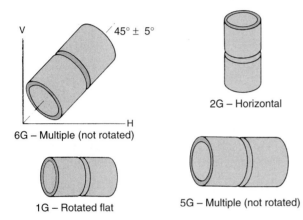

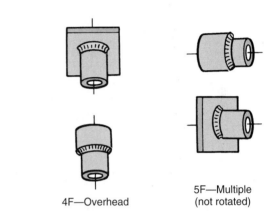

Figure 31-6. *Test positions for fillet welds in pipe joints. (ASME–Section IX)*

Groove welds in pipe

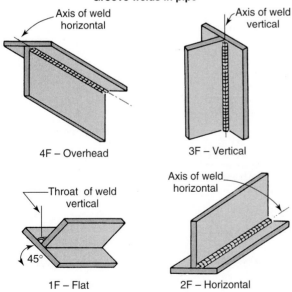

Fillet welds in plate

Figure 31-5. *Test positions for various joints. G stands for groove weld, F stands for fillet weld. Note that test positions 5G and 6G are not rotated, so the weld position changes as the weld is made. (AWS A3.0)*

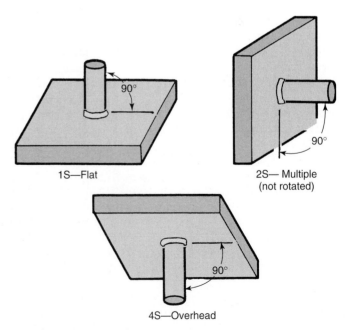

Figure 31-7. *Test positions for stud welds. (ASME–Section IX)*

One exception is that often (but not always) a code will state that a welder who passes a qualification test for a position other than flat is also qualified to weld in the flat position. For example, if a welder passes a Welder Performance Qualification Test in the horizontal position, he or she would be considered qualified to weld in both the horizontal and the flat positions.

31.5 METHODS OF TESTING SPECIMENS

There are five different weld specimen tests:
- Tension-reduced section (Heading 30.11).
- Side bend (Heading 30.10).
- Face and root bend – transverse (Heading 30.10).
- Face and root bend – longitudinal (Heading 30.10).
- Fillet weld – procedure (Heading 30.10).

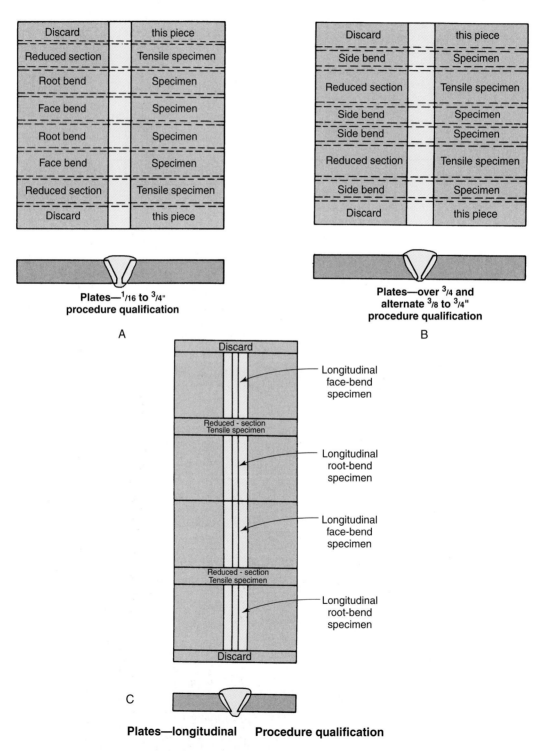

Figure 31-8. Order of removal of test specimens from the test weld. (ASME–Section IX)

Refer to the appropriate Heading for information on these tests. These tests are called out by different codes. The codes will specify or recommend what size test sample will be welded. Codes also specify where the samples to be tested will be cut from. Figure 31-8 shows test specimen locations for groove tests on plate. Figures 31-11 and 31-12 show locations for fillet weld tests.

31.5.1 Tension-Reduced Section Testing

The tension-reduced section test sample should be cut from the test weld from the area shown in Figure 31-8. Figure 31-9 shows how to prepare a sample for testing. Notice that the narrowest area is in the middle of the weld. This puts the most force into the weld.

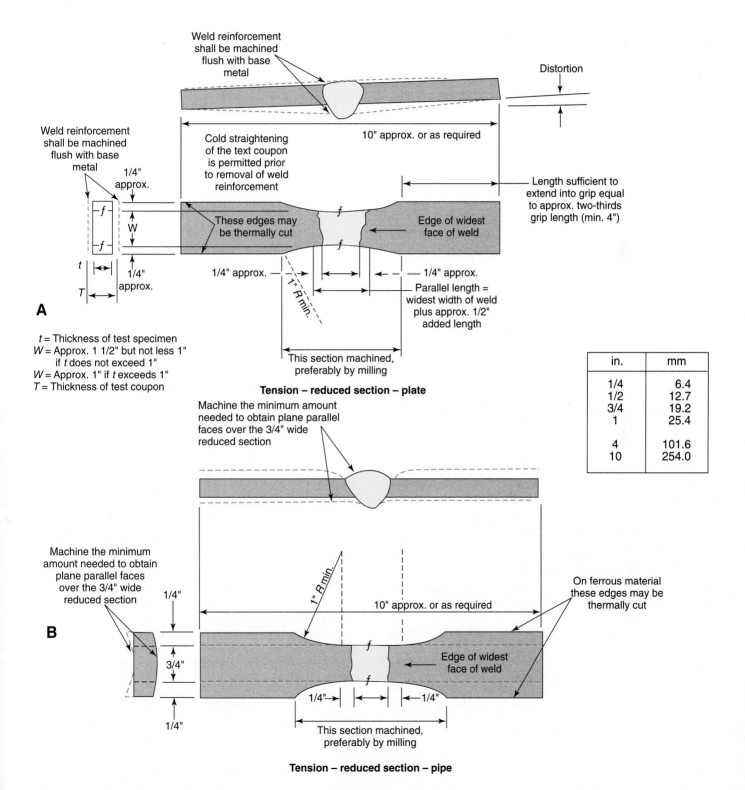

in.	mm
1/4	6.4
1/2	12.7
3/4	19.2
1	25.4
4	101.6
10	254.0

t = Thickness of test specimen
W = Approx. 1 1/2" but not less 1" if *t* does not exceed 1"
W = Approx. 1" if *t* exceeds 1"
T = Thickness of test coupon

Tension – reduced section – plate

Tension – reduced section – pipe

Figure 31-9. *Specifications for preparing tension-reduced section specimens. A—Specimens from plate. B—Specimens from pipe. (ASME–Section IX)*

Before putting the sample into the tensile testing machine, the thickness and width of the narrowest point are measured. This determines the cross-sectional area. Usually two points or lines are scribed onto the test sample. This allows the elongation or ductility to be measured.

The test sample is then mounted into the tensile testing machine and a tensile load (stretching force) is applied until the sample breaks. As discussed in Heading 30.10, the tensile strength and ductility are calculated.

31.5.2 Guided Face and Root Bends

Weld samples are generally welded from *one side* only. The face of the weld is the surface of the joint where the weld bead is applied. The root or bottom of the weld is the surface opposite the weld bead. After a test weld is completed, test specimens (samples) are cut from the test weld as shown in Figure 31-8.

After a bend sample is cut from the test weld, it is placed in a **bending jig**. The sample is placed so the weld and heat-affected zone will be in the center of the bend after bending. A guided bend jig is shown in Figure 31-10.

To be acceptable, the sample must bend a full 180° and must not break. After bending, the test sample may not show defects larger than those allowed by the code being used. The ASME code states the following concerning a bend test:

The weld and heat-affected zone of a transverse-weld bend specimen shall be completely within the bent portion of the specimen after testing.

The guided-bend specimens shall have no open defects exceeding 1/8" (3.2mm), measured in any direction on the convex surface of the specimen after bending, except that cracks occurring on the corners of the specimen during testing shall not be considered, unless there is definite evidence that they result from slag inclusions or other internal defects. For corrosion-resistant weld overlay cladding, no open defect exceeding 1/16" (1.6mm) measured in any direction shall be permitted in the cladding, and no open defects exceeding 1/8" (3.2mm) shall be permitted in the bend line.

31.5.3 Guided Side Bend Test

The **guided side bend test** is used on thick materials, where a face and root bend are not practical. The side bend test sample is removed from the test weld, as indicated in Figure 31-8B. Each side bend sample is bent using a jig, as shown in Figure 31-10.

To be acceptable, the side bend test sample must bend a full 180° and must not break. After bending, the test sample may not show defects larger than those allowed by the code being used. See the ASME code information in Heading 31.5.2.

31.5.4 Fillet Weld Procedure Test

A weld sample must be prepared according to the dimensions shown in Figure 31-11. This illustration also shows how specimens for the procedure test are cut from the test weld. In order to pass the ASME test, a specimen must satisfy the following:

Macro-Examination–Procedure Specimens

One face of each cross section shall be smoothed and etched with a suitable etchant to give a clear definition of the weld metal and heat-affected zone. In order to pass the test:

Visual examination of the cross section of the weld metal and heat-affected zone shall show complete fusion and freedom from cracks; and

There shall not be more than 1/8" (3.2mm) difference in the length of the legs of the fillet.

31.5.5 Fillet Weld Performance Test

The performance test sample for the fillet weld must be prepared and made as shown in Figure 31-12. Test specimens are removed from the weld sample as shown in the illustration.

The test specimens are placed under a load to bend the two parts (stems) flat on each other. To pass this performance test, the specimen must agree with the ASME code, which covers fillet weld performance test as follows:

Fracture Tests

The stem of the 4" (102mm) performance specimen center section (see Figure 31-12A) or the stem of the quarter section (see Figure 31-12B) shall be loaded laterally in such a way that the root of the weld is in tension. The load shall be steadily increased until the specimen fractures or bends flat upon itself.

If the specimen fractures, the fractured surface shall show no evidence of cracks or incomplete root fusion, and the sum of the lengths of inclusions and gas pockets visible on the fractured surface shall not exceed 3/4" (19.1mm) to pass the test.

A macro–examination is then made.

Macro-Examination - Performance Specimens

The cut end of one of the end sections from the plate or the cut end of the quarter section from the pipe, as applicable, shall be smoothed and etched with a suitable etchant to give a clear definition of the weld metal and heat-affected zone. In order to pass the test:

Visual examination of the cross section of the weld metal and heat-affected zone shall show complete fusion and freedom from cracks, except that linear indications at the root, not exceeding 1/32" (0.8mm) shall be acceptable; and

The weld shall not have a concavity or convexity greater than 1/16" (1.6mm); and

There shall be not more than 1/8" (3.2mm) difference in the lengths of the legs of the fillet.

Thickness of specimens		A		B		C		D		Material	Refer to
in.	mm	in.	mm	in.	mm	in.	mm	in.	mm		
3/8	9.5	1 1/2	38.1	3/4	19.1	2 3/8	60.3	1 3/16	30.2	All	
t	t	4t	4t	2t	2t	6t +1/8	6t + 3.2	3t + 1/16	3t +1.6	Others	
1/8	3.2	2 1/16	52.4	1 1/32	26.2	2 3/8	60.3	1 3/16	30.2	P-no. 23 and P-no.35 except as shown below	QW-422.23 QW-422.35
3/8	9.5	2 1/2	63.5	1 1/4	31.8	3 3/8	85.7	1 11/16	42.9	P-no.11, P-no.25, SB-148 alloys CDA-952 & 954 and SB-271 alloy CDA-952 & 954	QW-422.11 QW-422.25 QW-422.35
t	t	6 2/3t	169.3t	3 1/3t	84.7t	8 2/3t + 1/8	220t + 3.2	4 1/2t + 1/16	114.3t + 1.6		
1/16-3/8" incl	1.6-9.5	8t	8t	4t	4t	10t + 1/8	10t + 3.2	5t + 1/16	5t + 1.6	P-51	QW-422.51
1/16-3/8" incl	1.6-9.5	10t	10t	5t	5t	12t + 1/8	12t + 3.2	6t + 1/16	6t + 1.6	P-no. 52	QW-422.52
1/16-3/8" incl	1.6-9.5	10t	10t	5t	5t	12t + 1/8	12t + 3.2	6t + 1/16	6t + 1.6	P-no. 61	QW-422.61

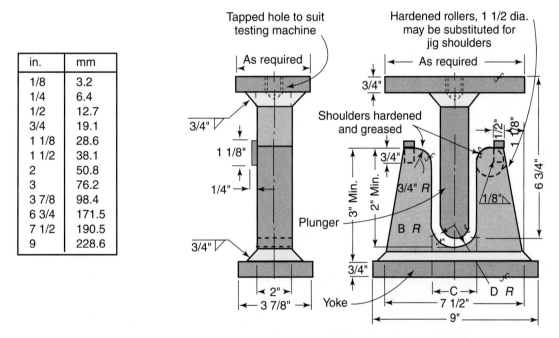

in.	mm
1/8	3.2
1/4	6.4
1/2	12.7
3/4	19.1
1 1/8	28.6
1 1/2	38.1
2	50.8
3	76.2
3 7/8	98.4
6 3/4	171.5
7 1/2	190.5
9	228.6

Figure 31-10. *Guided-bend jig specifications. Notice that a different jig is required for various thicknesses and metals. The QW-422 numbers referred to are ASME P number metal specifications. (ASME–Section IX). The millimeter equivalents are not shown as a part of the ASME–Section IX code. They are shown here for reference.*

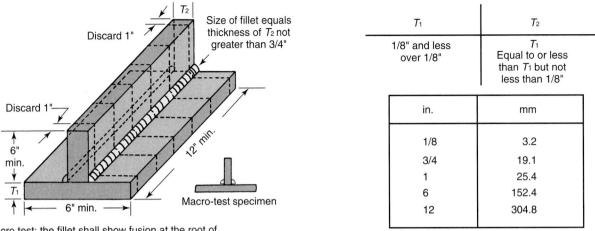

T₁	T₂
1/8" and less over 1/8"	T₁ Equal to or less than T₁ but not less than 1/8"

in.	mm
1/8	3.2
3/4	19.1
1	25.4
6	152.4
12	304.8

QW-462.4(a) Fillet welds – procedure

Macro test: the fillet shall show fusion at the root of the weld but not necessarily beyond the root. The weld metal and heat affected zone shall be free of cracks.

Figure 31-11. *Specifications for preparing a fillet weld procedure qualification weld. This figure also shows how the test specimen is removed from the weld sample. (ASME–Section IX) The millimeter equivalents are not shown as part of the ASME–Section IX code. They are shown here for reference.*

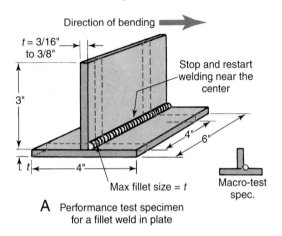

in.	mm
3/16	4.8
3/8	9.5
2	50.8
3	76.2
4	101.6
6	152.4

A   Performance test specimen for a fillet weld in plate

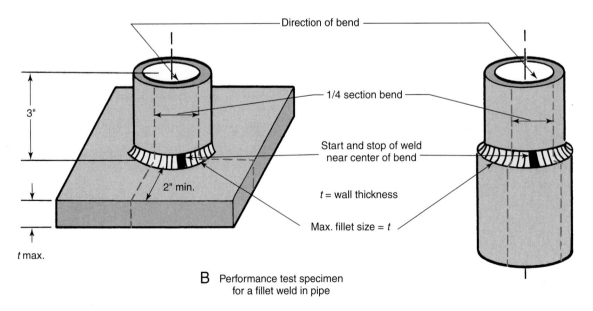

B   Performance test specimen for a fillet weld in pipe

Note: Either pipe to plate or pipe to pipe may be used as shown.

Figure 31-12. *Specifications for preparing a fillet weld performance qualification weld. This figure also shows how the test specimen is removed from the weld sample. (ASME–Section IX) The millimeter equivalents are not shown as a part of the ASME–Section IX code. They are shown here for reference.*

Suggested Welding
Procedure Specification (WPS)

Identification _____

Date _____ Revision _____

Company name _____

Supporting WPQR no.(s)_____ Type - Manual (   )   Semi-automatic (   )

Welding process(es) _____ Machine   (   )   Automatic (   )

Backing: Yes (   )   No (   )

Backing material (type) _____

Material number _____ Group _____ To material number _____ Group _____

Material spec. type and grade _____ To material spec. type and grade _____

Base metal thickness range: Groove _____ Fillet _____

Deposited weld metal thickness range _____

Filler metal F no. _____ A no. _____

Spec. no. (AWS) _____ Flux tradename _____

Electrode-flux (Class) _____ Type _____

Consumable insert: Yes (   )   No (   )   Classifications _____

Shape _____

Position(s) of joint _____ Size _____

Welding progression: Up (   )   Down (   )   Ferrite number (when reqd.) _____

Preheat: **Gas:**

  Preheat temp., min _____ Shielding gas(es) _____

  Interpass temp., max _____ Percent composition _____
  (continuous or special heating, where   Flow rate _____
  applicable, should be recorded)   Root shielding gas _____

Postweld heat treatment: Trailing gas composition _____

  Temperature range _____ Trailing gas flow rate _____

  Time range _____

Tungsten electrode, type and size _____

Mode of metal transfer for GMAW: Short-circuiting (   )   Globular (   )   Spray (   )

Electrode wire feed speed range: _____

Stringer bead (   )   Weave bead (   )   Peening: Yes (   )   No (   )

Oscillation _____

Standoff distance

Multiple (   )   or single electrode (   )

Other _____

Weld layer(s)	Filler metal Process	Class	Dia.	Type & polarity	Current Amp range	Volt range	Travel speed range	
								e.g., Remarks, comments, hot wire addition, technique, torch angle, etc.

Approved for production by _____

Employer

Note: Those items that are not applicable should be marked N.A.

Figure 31-13. A suggested Welding Procedure Specification (WPS) form. (American Welding Society)

Suggested Welding
Procedure Qualification Record (WPQR)

WPQR No. _____

WPS no. used for test _____

Company _____

Welding process(es) _____

Equipment type and model (sw) _____

Joint design used

Single ( )   Double weld ( )

Backing material _____

Root opening _____ Root face dimension _____

Groove angle _____ Radius (J-U) _____

Back gouging: Yes ( )   No ( ) Method _____

Base metals

Material spec. _____ To _____

Type or grade _____ To _____

Material no. _____ To material no. _____

Group no. _____ To group no. _____

Thickness _____

Diameter (pipe) _____

Surfacing: Material _____ Thickness _____

Chemical composition _____

Other _____

Filler Metals

Weld metal analysis A no. _____

Filler metal F no. _____

AWS specification _____

AWS classification _____

Flux class _____ Flux brand _____

Consumable insert: Spec. _____ Class. _____

Supplemental filler metal spec. _____ Class. _____

Non-classified filler metals _____

Consumable guide (ESW) Yes ( ) No ( )

Supplemental deoxidant (EBW) _____

Position

Position of groove _____ Fillet _____

Vertical progression: Up ( )   Down ( )

Preheat

Preheat temp., actual min _____

Interpass temp., actual max _____

Weld increment sequence

Postweld heat treatment

Temp. _____

Time _____

Other _____

Gas (2.6.7)

Gas type(s) _____

Gas mixture percentage

Flow rate _____

Root shielding gas _____ Flow rate _____

EBW vacuum ( ) Absolute pressure ( )

Electrical characteristics

Electrode extension _____

Standoff distance _____

Transfer mode (GMAW) _____

Electrode diameter tungsten _____

Type tungsten electrode _____

Current: AC ( ) DCEP ( ) DCEN ( ) Pulsed ( )

Heat input _____

EBW: Beam focus current _____ Pulse freq. _____

Filament type _____ Shape _____ Size _____

Other _____

Technique

Oscillation frequency _____ Weave width _____

Dwell time _____

String or weave bead _____ Weave width _____

Multi-pass or single pass (per side) _____

Number of electrodes _____

Peening _____

Electrode spacing _____

Arc timing (SW) _____ Lift ( )

PAW: Conventional ( ) Key hole ( )

Interpass cleaning:

Pass no.	Filler metal size	Amps	Volts	Travel speed (ipm)	Filler metal wire (ipm)	Slope induction	Special notes (process, etc.)

Note: Those items that are not applicable should be marked N.A.

Figure 31-14. A suggested *Welding Procedure Qualification Record (WPQR) form. (American Welding Society)*

Page 2 of 2

Tensile Test Specimens: Suggested Welding Procedure Qualification Record   WPQR No. ____

Type: _____ Tensile specimen size: _____

Groove (   )   Reinforcing bar (   )   Stud welds (   )

Tensile test results: (minimum required UTS _____ psi)

Specimen no.	Width, in.	Thickness, in.	Area, in.²	Max load lb.	UTS, psi	Type failure and location

Guided bend test specimens - specimen size: _____

Type	Result	Type	Result

Macro-examination results: Reinforcing bar (   )   Stud (   )

1. _____   4. _____
2. _____   5. _____
3. _____

Shear test results - fillets: 1. _____   3. _____
2. _____   4. _____

Impact test specimens:

Type: _____ Size: _____

Test temperature: _____

Specimen location: WM = weld metal;   BM = base metal;   HAZ = heat-affected zone

Test results:

Welding position	Specimen location	Energy absorbed (ft.-lbs.)	Ductile fracture area (percent)	Lateral expansion (mils)

If applicable: **Results**

   Hardness tests: (   )   Values _____   Acceptable (   )   Unacceptable (   )

   Visual (special weldments) (   )   Acceptable (   )   Unacceptable (   )

   Torque (   )   Acceptable (   )   Unacceptable (   )

   Proof test (   ) Method _____   Acceptable (   )   Unacceptable (   )

   Chemical analysis (   )   Acceptable (   )   Unacceptable (   )

Non-destructive exam (   ) Process _____   Acceptable (   )   Unacceptable (   )

Other _____   Acceptable (   )   Unacceptable (   )

Mechanical testing by (Company) _____   Lab no. _____

We certify that the statements in this Record are correct and that the test welds were prepared, welded, and tested in accordance with the requirements of the American Welding Society Standard for Welding Procedure and Performance Qualification (AWS B2.1).

Qualifier: _____

Date: _____

Suggested
Welding Performance Qualification Test Record

Name _____ Identification _____ Welder ( )   Operator ( )

Social security number: _____ Qualified to WPS no. _____

Process(es) _____ Manual ( )   Semi-automatic ( )   Automatic ( )   Machine ( )

Test base metal specification _____ To _____

Material number _____ To _____

Fuel gas (OFW) _____

AWS filler metal classification _____ F no. _____

Backing: Yes ( )   No ( )   Double ( ) or Single side ( )
Current: AC   ( )   DC ( )   Short-circuiting arc (GMAW) Yes ( ) No ( )
Consumable insert: Yes ( )   No ( )
Root shielding: Yes ( )   No ( )

Test Weldment   **Position Tested**   **Weldment Thickness (T)**

Groove:
  Pipe   1G ( )   2G ( )   5G ( )   6G ( ) 6GR ( )   Diameter(s) _____ (T) _____
  Plate   1G ( )   2G ( )   3G ( )   4G ( )   (T) _____
  Rebar   1G ( )   2G ( )   3G ( )   4G ( )   Bar size _____ Butt   ( )
                                                        Spliced butt ( )

Fillet:
  Pipe ( )   1F ( )   2F ( )   3F ( )   4F ( )   5F ( )   Diameter _____ (T) _____
  Plate ( )   1F ( )   2F ( )   3F ( )   4F ( )   (T) _____

  Other (describe) _____

Test results: Remarks

  Visual test results   N/A ( )   Pass ( )   Fail ( )
  Bend test results   N/A ( )   Pass ( )   Fail ( )
  Macro test results   N/A ( )   Pass ( )   Fail ( )
  Tension test   N/A ( )   Pass ( )   Fail ( )
  Radiographic test results N/A ( )   Pass ( )   Fail ( )
  Penetrant test   N/A ( )   Pass ( )   Fail ( )

Qualified for:
Processes
Groove: **Thickness**
  Pipe   1G ( )   2G ( )   5G ( )   6G ( )   6GR ( )   (T) Min _____ Max _____ Dia _____
  Plate   1G ( )   2G ( )   3G ( )   4G ( )   (T) Min _____ Max _____
  Rebar   1G ( )   2G ( )   3G ( )   4G ( )   Bar size   Min _____ Max _____

Fillet:
  Pipe   1F ( )   2F ( )   4F ( )   5F ( )   (T) Min _____ Max _____
  Plate   1F ( )   2F ( )   3F ( )   4F ( )   (T) Min _____ Max _____
  Rebar   1F ( )   2F ( )   3F ( )   4F ( )   Bar size   Min _____ Max _____

Weld cladding ( )   Position(s) _____   (T) Min _____ Max _____ Clad min _____

Consumable insert ( )   Backing type ( )

Vertical up ( )   Down ( )

Single side ( )   Double side ( )   No backing ( )

Short-circuiting arc ( )   Spray arc ( )   Pulsed arc ( )

Reinforcing bar - butt ( ) or Spliced butt ( )

The above named person is qualified for the welding process(es) used in this test within the limits of essential variables including materials and filler metal variables of the AWS Standard for Welding Procedure and Performance Qualification (AWS B2.1)

Date tested _____ Signed by _____
                                                        Qualifier

Figure 31-15. A suggested Performance Qualification Test Record form. (American Welding Society)

TEST YOUR KNOWLEDGE

Write your answers on a separate sheet of paper. Do not write in this book.

1. When a structure is to be built, the _____ and _____ must reach an agreement on how each weld will be made.
2. Name four organizations that have prepared welding codes.
3. What organization has prepared a code for structural welding, such as welding used for bridges and similar structures?
4. What organization has prepared a welding code for pressure vessels?
5. Why have different codes and specifications been written?
6. When must a code or specification be followed?
7. Which documents must be made available to welders: Company Practices, or the Codes and Specifications?
8. What do the letters WPS mean?
9. What do the letters WPQR stand for?
10. How is a WPS qualified?
11. What must be done if an essential variable on a WPS is changed?
12. What must be done if a nonessential variable on a WPS is changed?
13. List three essential variables and three nonessential variables for SMAW.
14. If a welder passes a qualification test on 1/8" (3.2mm) thick material, what is the range of material thicknesses the welder is qualified to weld according to the ASME code?
15. Under what conditions is it necessary to requalify a welder who is qualified in a given procedure?
16. Is it necessary that a welder be qualified for different welding positions?
17. What are three of the essential welding procedure variables for shielded metal arc welding?
18. What position of welding does a 3F signify? A 5G?
19. Name four specimen tests for qualifying welders.
20. An E7018 electrode is used to butt weld two 1/2" (12.7mm) plates together. A tension-reduced section specimen is prepared with a width of 3/4" (19.1mm). The specimen is tested. The maximum load is 27,000 lb. (120,100N). What is the tensile strength of this weld sample?

For safety and efficiency, the welding shop should be kept clean and neat. This welder is working in an environment free of clutter and scrap materials. (Lincoln Electric Co.)

Chapter 32

THE WELDING SHOP

Welding shops may be divided into three principal groups:
* The independent shop that specializes in repairing and fabricating metal structures by the various methods of welding.
* The welding shop, or department, attached to a factory or manufacturing establishment.
* The welding equipment shop that overhauls and repairs welding equipment and sells equipment and supplies.

The equipment found in each shop will vary, depending on the type and amount of repair or production that is being done.

Various types of supplies and equipment found in a welding shop will be explained and illustrated in this chapter. Welding machines and stations have been explained in other chapters, and will not be included in this chapter.

32.1 WELDING SHOP DESIGN

The architectural design of the welding shop is important. A typical shop would incorporate features such as these:

* Heavy-duty loadbearing floors, preferably of concrete.
* A fire-resistant building structure.
* Good general ventilation, with provision for localized exhaust ventilation.
* Large doors and some means of moving heavy equipment and material into and out of the shop.
* Heavy-duty electrical service.

The concrete floors should be at the ground level. The utilities should be arranged around the outside of the room and overhead. Equipment should be arranged to provide logical flow of work through the shop. Raw materials like plate, pipe, and sheet metal are often kept in one area. Wide aisles are necessary to allow fork lifts to move material around. A monorail or a double rail crane system may be installed to provide easy movement of equipment and material to any spot in the room.

If the business is of sufficient size to justify accessory rooms such as paint booths, storerooms, toilets, showers, locker rooms, and offices, these should be located in an annex to the shop for safety, cleanliness, and reduced noise levels.

32.2 WELDING SHOP EQUIPMENT

Equipment used in the welding shop depends to a considerable extent upon the kinds of work handled. Some of the common equipment used includes:
* Oxyfuel gas welding stations.
* Ac and dc SMAW arc welding stations.
* GMAW and GTAW arc welding stations.
* Resistance welding machines.
* Oxyfuel gas cutting equipment.
* Plasma cutting equipment.
* Preheating and postheating furnaces.
* Overhead crane and/or heavy-duty hoists.
* Fork lifts.
* Benches, vises, anvils.
* Jigs and fixtures.

- Heavy-duty power tools, such as saws, shears, brakes, mills, lathes, nibblers, drill presses, and grinders.
- Grit-blasting equipment.
- Weld inspecting and testing equipment.
- A paint booth.
- Weld positioners.
- Stationary and portable ventilation units.
- Portable screens to set up around shop floor welding jobs. See Figure 32-1.

Local and state safety and building codes must be followed when designing, building, and working in a welding shop. The shop must be adequately equipped with personal safety equipment, first aid supplies, and fire safety equipment (such as fire extinguishers and blankets).

Figure 32-1. A portable welding booth with filter-quality plastic screening material. This type screen can be set up anywhere to protect other workers from ultraviolet rays. (Singer Safety Company)

32.3 AIR VENTILATION AND CONDITIONING EQUIPMENT

Air contamination in the welding shop must be kept to a minimum for health and good housekeeping. The welders or equipment operators must have sufficient clean air and oxygen to eliminate respiratory problems and be comfortable. The temperatures and humidity conditions should be comfortable. Toxic gases and toxic dust particles must be held to an absolute minimum by the proper selection of the materials used. If dust and gases occur, they must be withdrawn and filtered. Welders who work in highly toxic areas should wear supplied air breathing equipment or at least filtered air gear.

Welding stations that produce dust particles and gases should have adequate exhaust ventilation. The amount of dust produced varies with the size of the equipment, kind of metal being heated, fluxes used, and the like. It is desirable to have an industrial hygiene technician take air samples under operating conditions, so that the ventilating system can be adjusted or modified until safe working conditions prevail.

Venting exhaust fumes from a welding station is shown in Figure 32-2. Figure 32-3 illustrates a portable fume extractor. This unit can be set up on the site of a large shop-floor welding job.

Air that is exhausted must be replaced. When the outdoor temperatures are comfortable, open doors and windows will provide the air replacement. However, during cold weather conditions, the replacement air must be both heated and humidified. Provisions must be included for seasonal shop requirements.

Figure 32-2. A fume extractor that is connected to a shop-wide extraction system. (Nederman, Inc.)

32.4 MECHANICAL METAL CUTTING EQUIPMENT

Metal may be cut to size and beveled to specifications by several different methods prior to welding. These methods include:
- Abrasive cutting wheel.
- Power hacksaw.
- Metal bandsaw.
- Shears.
- Nibblers.

Figure 32-3. *A portable fume extractor with an electronically controlled air filter, in use on a welding repair. (Nederman, Inc.)*

Power hacksaws, shears, and power cutoff saws are needed to cut standard-size stock to needed lengths. Figure 32-4 illustrates an abrasive wheel cutoff machine. The machine illustrated is enclosed to prevent accidents.

Figure 32-4. *A large abrasive wheel cutoff saw. The unit is enclosed to provide safety. Note the size of the pieces being cut. (W.J. Savage Co.)*

Where 90° or other standard angle cuts are needed, a reciprocating-type power hacksaw or a band-type power hacksaw is commonly used. Metal-cutting bandsaws are used for both straight and contour cuts.

Metal shears of all types, both manual and power-operated, find extensive use in welding shops. Floor-mounted, manually operated floor shears are sometimes used. The powered floor shear is more often found in shops. See Figure 32-5. Shears can be equipped with different blades that make it possible to cut flat stock, angle iron, and bar stock to length. Squaring shears are used for thin metal. Figure 32-6 shows a shear in use.

Figure 32-5. *This hydraulic shear is capable of cutting mild steel 1/4" (6.4mm) by 12' (3.6m). (Clearing-Niagara)*

Figure 32-6. *A metal shear in use.*

A nibbler machine, complete with attachments, provides a means of cutting irregular shapes for fabricating sheet metal and plate. This machine uses two small, sharp steel blades, one stationary and one powered. The cutting is done by shearing action, as shown in Figure 32-7. It is possible to cut straight or curved lines with this type of machine. Figure 32-8 shows a floor model and Figure 32-9, a portable model. Portable nibblers will cut metal up to 1/4" (6.4mm) thick. The nibbler is an excellent tool for the job of cutting curved shapes out of sheet metal. The internal construction of a nibbler, showing the mechanism that produces the shearing action, is illustrated in Figure 32-10.

Figure 32-7. *Enlarged view of a cut made by a nibbler machine.*

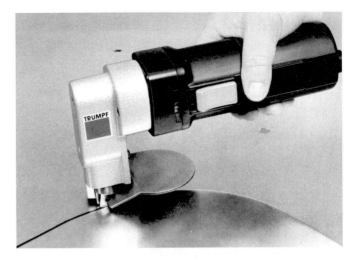

Figure 32-9. *A light-duty portable electric nibbler. The tool can cut on straight or curved lines. (Trumpf, Inc.)*

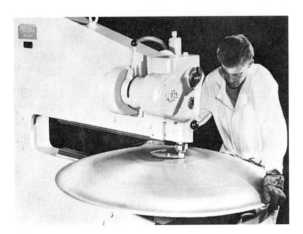

Figure 32-8. *Electrically powered floor model nibbler machine making a circular shearing cut. (Pullmax, Inc.)*

Tools have been developed for cutting tubing and pipe of all diameters and thicknesses to produce the special notched ends needed for welded or brazed joints. Figure 32-11 shows a manually operated notching machine. The different shapes that may be obtained for tubing ends are shown in Figure 32-12.

When metal to be gas or arc welded exceeds 1/8" (3.2mm) in thickness, it is usually beveled. The bevels may be made by grinding, by flame or arc cutting, or may be made by a beveling machine, as shown in Figure 32-13.

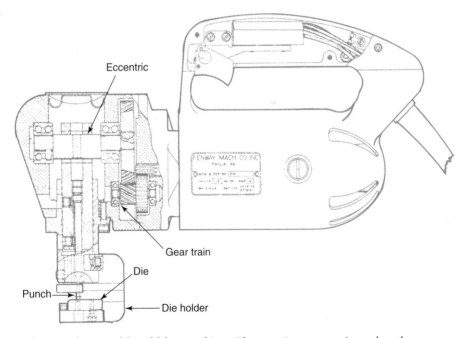

Figure 32-10. *A cross-sectional view of a portable nibbler machine. The rotating eccentric makes the upper punch or cutter go up and down. (Fenway Machine Co., Inc.)*

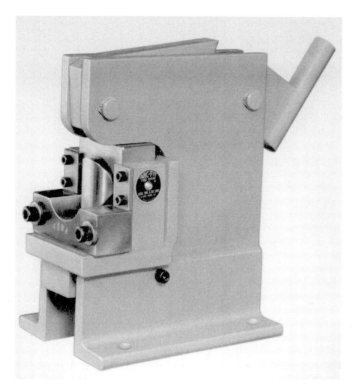

Figure 32-11. *A tube notching machine. The shape of the upper and lower dies can be changed to cut tubes of various sizes. (Vogel Tool & Die Corp.)*

32.5 METAL FORMING EQUIPMENT

After metal has been cut to size and shape, it must often be bent at various angles. This bending may be done in a shop vise if the parts are small and thin. However, most metal bending is done in a manual or powered piece of equipment called a *brake*. Figure 32-14 is a photo of a brake.

To form or bend a piece of metal in a brake, two tools are used. One is called a *punch* and is located above the metal to be bent. The second is called a *die* and is located below the metal to be bent. Figure 32-15 shows different punches and dies. A different punch and die set is required to obtain different radii and to obtain clearance while bending certain parts.

In operation, either the punch comes down or the die on the bottom comes up. As the punch and die come together, they will contact the metal to be bent. As they continue to close, the angle of the bend increases as shown

Figure 32-13. *A portable electric edge beveling machine. (Heck Industries, Inc.)*

in Figure 32-16. Figure 32-17 shows a brake being used to form cold-rolled steel parts.

32.6 FURNACES

A *preheating furnace* is a necessity in a number of welding shops. Some metals and practically all complicated metal structures are subject to excessive strain and perhaps breakage if heated or cooled unevenly. To minimize these conditions, the metal must be heated gradually to the correct preheat temperature before the necessary welding is performed. The structure is then allowed to cool slowly and evenly after the welding is completed. This practice makes it possible to minimize warpage. It also prevents excessive stresses and cracking of the metal as it cools to room temperature. Different styles of preheaters are:

- Portable torch type.
- Flat top open type.
- Enclosed furnace.
- Electrical resistance heaters. See Heading 29.1.1.
- Induction heaters. See Heading 29.1.1.

The torch type may be a large portable blowtorch such as shown in Figure 32-18. A preheat furnace may use the municipal gas supply, oil, gasoline, kerosene, or propane as the fuel. It may use air or oxygen as the combustion-supporting gas. These furnaces, or heaters, may be used for smaller metal structures or for localized preheating.

Figure 32-12. *Metal angles, square tubing, and round tubing all can be notched in various ways for making joints, as shown. (Vogel Tool & Die Corp.)*

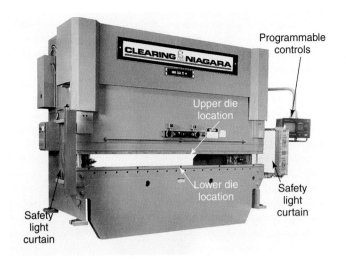

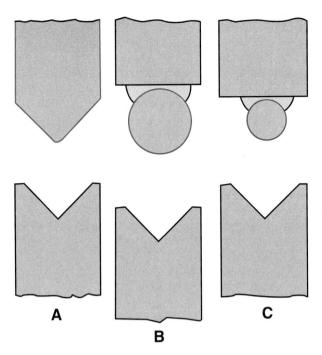

Figure 32-14. A hydraulic brake press. The upper and lower dies are not installed in this photo. Note the electronic controls at the right. This brake is equipped with a safety light curtain that will stop the brake if an unplanned object interrupts the light beam. (Clearing-Niagara)

Figure 32-15. Three upper and lower dies for a hydraulic brake press. A—90° angles with minimum bend radius. B and C—Round bars on the upper die vary the bend radius on the part.

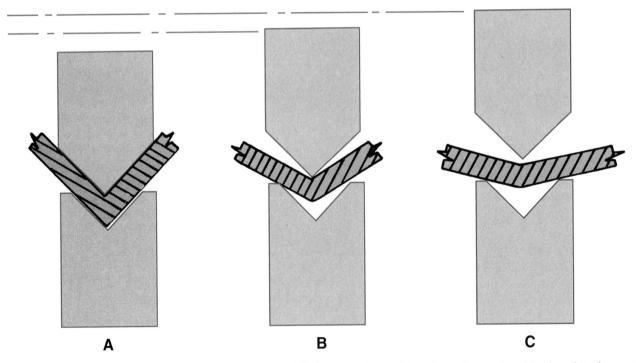

Figure 32-16. Hydraulic brake press dies at various full-down positions. Note how the angle of the bend in the metal is varied by changing the travel distance of the upper die. A—The upper die is adjusted for a 90° bend. B and C—The upper die is stopped at a higher setting for bends of less than 90°.

Figure 32-17. *A press brake being used to form cold-rolled steel pieces.*

Figure 32-18. *A large preheating torch in use as a welder heats a previously welded area. Such torches can be used for preheating, interpass heating, or postweld heat treating. (Belchfire Corp.)*

The open flat top preheater consists of a grate, or series of bars, with a gas burner underneath. The article to be preheated is placed on top of the bars or grates. Occasionally, a part too large for any existing shop furnace may be brought in for repair. In such cases, a preheat furnace may be built around the part using firebrick and heat-resistant materials. Asbestos materials should not be used. Heat is then produced under the part. In some cases, large gas-fired industrial space heaters are used to provide heat for improvised preheat furnaces like the one shown in Figure 32-19.

Clay and firebrick may be built up around the article to be preheated. This shield also will protect the welder from the furnace's heat, which is dissipated into the room. The furnace heat may make working conditions uncomfortable, especially in warm weather.

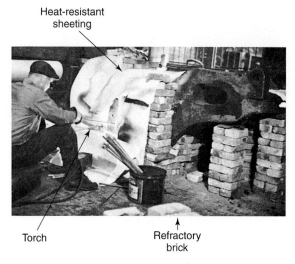

Heat-resistant sheeting

Torch

Refractory brick

Figure 32-19. *A shop-built furnace made of firebrick and heat-resistant sheeting.*

An advantage of this type of furnace is its flexibility. The article being preheated, or welded, is readily accessible if certain bricks or heat-resistant sheets are removed. The article then may be welded while it is still located in the furnace.

The enclosed type of preheating furnace is a typical furnace that may be used for heat treating, carburizing, or preheating. It consists of a firebrick-lined steel structure, and usually uses gas for fuel. A forced-air blower is generally used for supporting the combustion and for raising the temperature. An advantage of this type of furnace is its economy. It is able to cool an article slowly, for annealing purposes, and to vent the heat outdoors rather than into the shop. The enclosed furnace has the disadvantage of requiring that the preheated article be removed from the furnace before it is welded. If continual heating is required for welding, this furnace cannot be used.

32.7 OVERHEAD CRANE

The size of the work to be handled by the welding shop may require the use of a crane to move articles to be welded from place to place.

In the small shop, a chain fall, either hand- or motor-driven, may be suspended from rails or beams that run the length of the shop. A movable hoist can be used for lifting jobs which weigh only a few hundred pounds. Figure 32-20 illustrates a powered cable hoist. Figure 32-21 shows a similar hoist in use. Fork lifts are also often used to move material in a welding shop

32.8 JIGS AND FIXTURES

One of the greatest problems encountered in a welding shop is preventing the metal from warping or buckling during or after the welding operation. Many devices have been used to hold the metal during welding so that the

Figure 32-20. A 2200 lb. (4850kg) capacity overhead hoist. This hoist moves along an overhead I-beam. It is controlled electrically by the operator. (Jet Equipment and Tools)

Figure 32-21. A hoist in use in a large shop. (Harnischfeger Corp.)

been used to hold the metal during welding so that the warpage and bending will be reduced to a minimum.

Clamps, v-blocks, vises, and special holding fixtures are used extensively for this purpose. Figure 32-22 shows a fast-action clamp. This clamp has a beryllium copper alloy spindle that resists damage from arc welding spatter. Figure 32-23 shows a table-mounted type of fast-action clamp. Figure 32-24 shows how close you may arc weld to the clamp without damaging it. When a large quantity of parts of the same type are to be made, it is economical to design and make fixtures to hold the parts to be welded. *Fixtures* help each part have repeatable dimensions. Figure 32-25 shows such a fixture. When fixtures are designed, they often include the use of clamps similar to those shown in Figure 32-26. Refer also to Figures 25-24 and 25-25, which show fixtures being used.

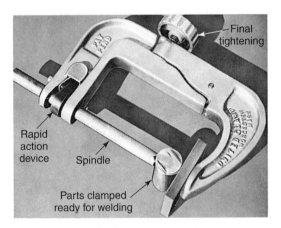

Figure 32-22. Fast-acting clamp. The shaft or spindle is made of beryllium copper alloy. This permits arc welding within 1/4" (6.4mm) of the spindle without damage from arc spatter. (United Clamp Mfg. Co.)

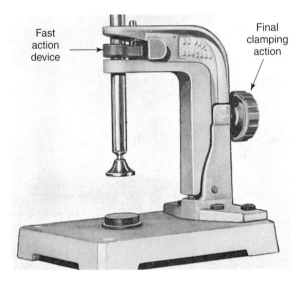

Figure 32-23. A table-mounted fast-action clamp used for production work. (United Clamp Mfg. Co.)

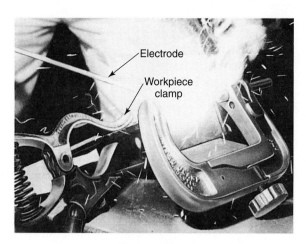

Figure 32-24. *An assembly being arc welded while the parts are held together with a fast-action clamp. Note the workpiece clamp. (United Clamp Mfg. Co.)*

Figure 32-25. *An assembly of toggle clamps mounted on a steel plate to form a clamping fixture. (De-Sta-Co)*

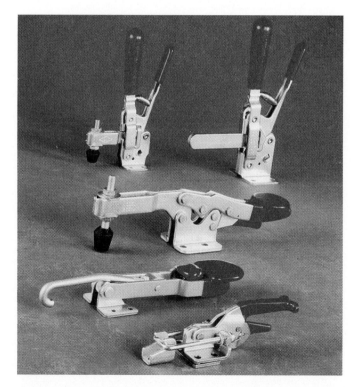

Figure 32-26. *Different types of clamps and toggle lock devices that may be used when designing and making welding fixtures. (De-Sta-Co)*

Large platens or table tops can be used for fixturing. Special clamping devices can be used to create fixturing for specific jobs. Figure 32-27 shows various accessories used to clamp and hold parts for welding.

Figure 32-27. *A faceplate (platen) with a variety of clamping devices shown installed on the faceplate. The material was being displayed at a welding industry trade show. (Weldsale Co.)*

Welding ***positioners*** are popular for production work in shops. They help to turn the work so that flat position welding is possible on all joints. Some positioners are moved manually, others are motorized. Some positioners, like the one shown in Figure 32-28, are controlled by a computer. This positioner moves the weldment into position as required to permit automatic robot welding. Figure 32-29 shows a very large positioner. For additional photos of positioners, refer to Figures: 25-19, 25-20, and 25-24.

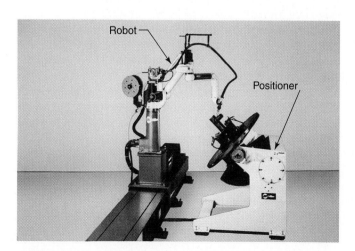

Figure 32-28. *A computerized welding positioner which moves the workpiece so that the robot can weld in the flat position. (Miller Electric Mfg. Co.)*

Figure 32-29. *Freight car welding positioner. The large assembly is rotated by positioner to achieve a flat or horizontal weld position as a means of producing quality welds economically under the production line conditions. Welders use semiautomatic gas shielded flux cored arc welding equipment to make finish welds. (Thrall Car Manufacturing Co.)*

32.9 POWER TOOLS

Power tools of various types are a necessity for high production welding. They help to cut overall production costs.

The drill press or portable drill is used to make holes in parts. Electrically or air powered wrenches can be used when disassembling large jobs. Power chisels save a great deal of labor when opening up a crack for a welding repair, or when removing scale from ferrous or nonferrous castings prior to welding. The power chisel may also be used when removing excess welding material from a surface to be ground.

The pedestal grinder may be used for rough grinding, for finishing parts, and for preparing small parts for welding. Figure 32-30 illustrates a pedestal grinder. The portable grinder finds wide use in preparing and finishing welded joints. It is especially useful for grinding welds on large weldments that cannot be brought to a pedestal grinder. Portable grinders may be used with rigid or flexible type abrasive discs. See Figure 32-31 and 32-32. An abrasive belt grinder is shown in Figure 32-33. A flexible shaft grinder is shown in Figure 32-34.

In all grinding operations, you must wear eye protection. Even if you are not doing the grinding, but are in an area where grinding is being done, you need to wear eye protection.

Abrasive wheels must be in good condition. They should be inspected for cracks, looseness, and general good condition before being used. **The operator should take precautions to stand in a safe and secure position.** Fasten small articles before grinding or sanding them.

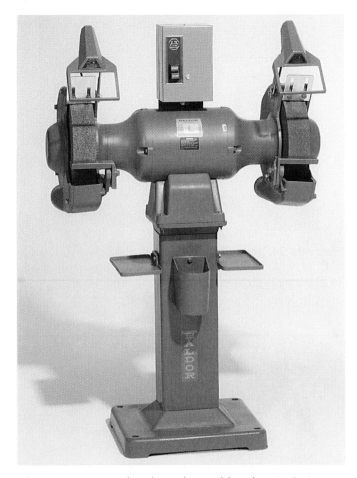

Figure 32-30. *A pedestal grinder. (Baldor Electric Co.)*

Figure 32-31. *A flexible shaft grinder being used to smooth the flange of a stainless steel weldment. (Suhner Industrial Products Corp.)*

Figure 32-32. *A portable grinder being used with a flexible grinding disc. (Norton Co.)*

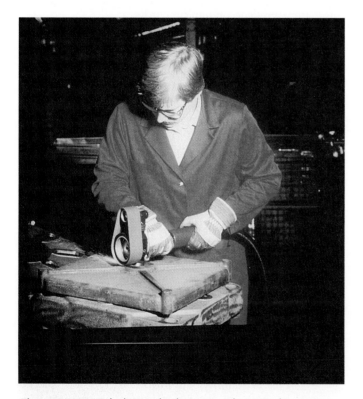

Figure 32-33. *A belt grinder being used to grind a casting flange area. Note the use of gloves and safety glasses. (Suhner Industrial Products Corp.)*

32.10 BLAST CLEANING EQUIPMENT

A blast-cleaning unit is handy when dirty or rusty metal stock must be cleaned before and/or after the welding has been done.

Air pressure is used to drive steel grit, shot, sand, or artificial abrasives against the surface to be cleaned. The material used depends on the metal and the condition of the surface being cleaned. The smoothness of the finished

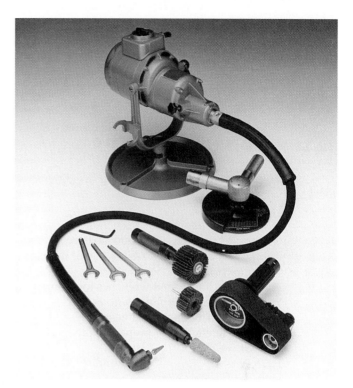

Figure 32-34. *A flexible shaft grinder with several accessory type grinders. (Suhner Industrial Products Corp.)*

surface will be affected by the material used for blasting. The blasting operation is performed in a special cabinet with a suction unit and a recovery chamber for the abrasive. The operator stands outside the cabinet and uses special built-in gloves to handle the blast nozzle and the parts, as shown in Figure 32-35.

Figure 32-35. *A blast cleaning cabinet. A nozzle which uses compressed air and grit is used to clean the metal before and after welding. (Ruemelin Mfg. Co.)*

The air pressure varies from 20 psig to 100 psig (140kPa to 690kPa). An exhaust system is used to remove the abrasive from the booth. Filters are mounted in the exhaust system to trap and remove the blasting material before the material can reach the exhaust fan.

32.11 WELDING SHOP TOOLS

Additional tools often needed in a welding shop include:
- Wrenches.
  - ◆ Welding equipment wrenches.
  - ◆ Socket wrench sets (up to 1" drive size) for large jobs. Metric sizes are useful in many applications.
  - ◆ Box and open-end wrenches for dismantling and assembling various articles.
  - ◆ Cylinder wrenches for acetylene cylinders.
  - ◆ Pipe wrenches for preparing pipe material for welding.
- Hammers of several sizes and types. These include chipping hammers and sledgehammers for straightening and bending heavy stock.
- Chisels (manual and power).
- Files of all types and sizes.
- Screwdrivers.
- Wire brushes (manual and power).
- Power grinder dresser.
- Grinder safety goggles.
- Squares.
- Levels.
- Clamps of all types.
- Mallets.
- Soldering irons.
- Hacksaws and blades (hand and power).
- Pedestal drill press.
- Portable hand drills (electric and air).
- Tape measures, scales, and calipers used for measuring.

32.12 WELDING SHOP SUPPLIES

It is recommended that you refer to Chapters 5, 7, 11, 13, 16, and 18 for information on welding supplies. A list of the more common supplies is as follows:

Gases
- Oxygen.
- Acetylene.
- Other fuel gases, such as natural gas, propane, or methylacetylene-propadiene (MAPP) for cutting or heating.
- Argon, helium, CO_2, or gas mixtures for GTAW, GMAW, and FCAW.

Filler metals and electrodes
Appropriate filler metals and electrodes should be kept for the types of welding jobs typically worked on. The common types of filler metals and electrodes include:
- Oxyacetylene filler metals for steel and cast iron and any special welding tasks.
- Brazing and soldering rods and fluxes for various metals.
- Covered electrodes for SMAW steel, stainless steel, and special applications. Different types and diameters are usually required.
- GMAW electrode wire for steel, stainless steel, and aluminum. Different types and diameter electrode wires are required.
- Various types and diameters of tungsten electrodes and appropriate filler metals for GTAW.

Replacement parts
Certain parts of a welding outfit wear during use. Other parts may sometimes fail. It is necessary to keep a supply of some items to allow welding operations to continue. Items to keep on hand include:
- Welding and cutting tips for oxyacetylene torch.
- Electrode holder for SMAW.
- Workpiece clamp.
- Welding cables.
- Collets, collet bodies, and nozzles for GTAW.
- Contact tips, adapters, and nozzles for GMAW.
- Foot pedals for GTAW and resistance welding.
- Different types of electrode tips and electrode holders for resistance welding.

Metal stock
- *Sheet and plate metal.* A welding shop should have on hand, at all times, a quantity of the various standard sizes of sheet metal and plate required for normal operation. See Figure 33-8 for a table of the Brown and Sharpe gauge sizes for metal thicknesses. Since a great variety of plain carbon, alloy steel, and other metals and alloys may be stored, it is essential that stored metal be properly identified. Often sheet and plate material will have a tag attached to the pallet, rack, or skid identifying the material type and thickness. The purchase order number or other identifying number may be used also.
- *Pipe.* A quantity of appropriate pipe stock should be kept on hand for normal operation. It is important to keep the material identified.
- *Angles and shapes.* Various sizes of angles, rectangular tubes, C-channels, rounds, and other shapes are necessary for some welding shops to keep on hand. As with all metals, it is important that stored angles and shapes be identified. Most suppliers ship bar and formed stock with one end painted. A specific color is used to identify the alloy and carbon content of the metal. If the supplier does not paint the metal, the shop should paint the ends. A shop-developed color

code should be kept on file. Employees should be instructed never to cut the end of the stock which is painted. If the type of metal becomes unknown, it is virtually worthless. Most jobs require a specific metal to be used.

Fluxes

Fluxes are used principally when oxyfuel gas welding, soldering, or brazing are done. The common fluxes that should be kept on hand are:

- Brazing flux.
- Cast iron welding flux.
- Cast iron brazing flux.
- Aluminum flux.
- Silver brazing (soldering) flux.
- Soldering fluxes.

It is important that these fluxes be kept in sealed containers and in a cool, dry place when not in use. When using the flux, only enough material to handle the job at hand should be transferred to a smaller container. This will keep the larger supply clean and fresh.

Vitrified firebrick

Firebrick is used for bench tops and for enclosing articles that are to be preheated. The brick is also used to slow the cooling rate of heated articles.

Miscellaneous

- Glycerin for lubricating oxyacetylene moving parts, and wiping cloths for cleaning purposes.
- Welding backing rings, and adhesive backing forms.
- First aid materials.
- Safety goggles.
- Gloves.
- Parts for normal repair and replacement.
- Miscellaneous fastening devices (bolts, nuts, washers) for assembly jobs.

32.13 WELD TEST EQUIPMENT

Welding shops may employ one or more of the following methods to check the quality of the welds and weldments produced:

- Visual inspection.
- Dye penetrants for surface flaws.
- Magnetic particle equipment for surface flaws.
- X-ray to detect internal flaws.
- Ultrasonic testers to locate internal flaws.
- Brinell, Rockwell, Scleroscope, and/or testers to test metal hardness.
- Tensile testers to check yield strengths in welds.
- Bend testers to determine qualifications of the welders.

These tests are described in Chapter 30. Special equipment and supplies are needed to perform most of these tests.

32.14 REPAIR OF WELDING EQUIPMENT

There are many specialty firms in business to repair oxyfuel gas regulators, gauges, and torches. Similar firms specialize in the repair of arc and resistance welding machines, flowmeters, gauges, electronic timers, and wire drives. The specialty tools, gauges, and meters required to repair the wide variety of makes and styles of equipment are very expensive.

It is recommended that repairs, other than very minor maintenance types of work, be done only by trained specialists. Large companies may employ such experts. Smaller shops may have to send small equipment items out for repair. For large equipment repair, the expert will have to be called in.

Oxyfuel gas tips can be cleaned out with the correct size drill. See Figure 6-7. The end of the tip may be reshaped with a tip dressing tool. The packing or seal around the torch valve stem can usually be replaced with manufacturer's replacement parts. Hose nipples, hose nuts, and hoses may be replaced easily and safely by shop personnel with some training.

Arc welding electrode and workpiece leads can be replaced, and end connections or lugs can also be easily installed. Figure 32-36 and Figure 32-37 show cable connections and one method used to install them. GMAW wire drives may be repaired in the shop, but major repairs may require a service call.

Calling in a repair specialist may be the most economical way of repairing an item when safety and downtime are considered.

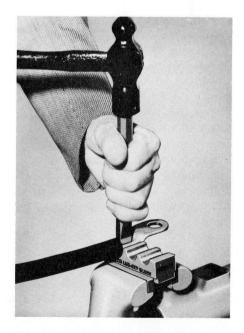

Figure 32-36. *Forming the barrel of a cable lug to obtain a connection between the cable and the lug. The completely assembled lug and cable is shown in Figure 32-37. (Tweco/Arcair, Division of Thermadyne Industries, Inc.)*

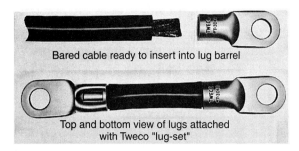

Bared cable ready to insert into lug barrel

Top and bottom view of lugs attached
with Tweco "lug-set"

Figure 32-37. A metal-forming method for making cable connections. (Tweco/Arcair, Division of Thermadyne Industries, Inc.)

32.15 WELDING SHOP POLICY

It is difficult to list all the policies that a welding shop should pursue in order to maintain correct business relationships. Work should all be done under a legal contract, with clearly and definitely defined provisions for all emergencies. The welding shop owner or manager should be extremely careful to make certain that all metal supplied meets the specifications required for the job. The shop should list the specifications of the metal when purchasing it, and should hold the supplier of the material to these specifications.

Welding shop personnel must estimate the final cost of fabricating an article prior to starting work. This takes considerable training and experience. To estimate it properly, these items must be considered:

- Type and quantity of material to be used.
- Preparation cost to get metal ready for welding. This includes sawing, cutting, shearing, bending, grinding, and other operations as needed.
- Length, depth, and shape of weld.
- Position of the weld: flat, vertical, horizontal, or overhead.
- Type of electrode or welding rod required.
- Shielding gas costs.
- Energy costs for oxygen, fuel gases, electricity.
- Labor costs.
- Costs after welding (grinding, heat treating, testing, plating, painting, inspection, and shipping).

Depreciation of equipment and other overhead costs also affect the final cost of the job. By calculation, the estimator can determine the operating costs. Accurate information may be obtained from welding supply houses. They may provide information on how much welding rod or electrode to use for a specific thickness and length of weld. They may also furnish information on how much gas or electricity may be involved in making welds of certain lengths and on various thicknesses of metal.

With this basic information, the estimator can determine the labor cost, the cost of material, and the power cost. Another item that should not be neglected in estimating costs is the matter of handling the material to be welded. When dealing with small articles, this may be

neglected. With large and cumbersome forms that must be moved from one part of the shop to another and turned in various positions, this item is important.

There should be a clear understanding about transportation costs. Either party may take this responsibility. Transportation costs involve moving the articles to the shop to be welded, and returning them to the person contracting for the work.

As with many other kinds of work, estimators can calculate the cost of a weld job after gaining considerable experience. This cost estimation of a job is vital to a shop's financial survival. An error in estimating the real cost of a job can cost the business a great deal of money. Inspection and quality testing costs should be included in any agreement. A welding shop should use carefully worded contract forms. A lawyer who specializes in drawing up industrial contracts should be employed to develop a standard form that may be applied to various types of welding jobs handled by the shop.

Many skilled welders have failed in operating their own welding businesses because they neglected the *business* aspects of the shop. Accurate records must be kept at all times. The total cost of operating a shop includes not only the actual operating cost but the overhead cost as well. Consultation with an experienced accountant is highly recommended.

It is important to keep an accurate record of each job. These records can then be used as a base for estimating future jobs similar in nature.

32.16 REVIEW OF SAFETY

All the safety procedures described throughout the text also apply to the various types of welding shops covered in this chapter.

Adequate ventilation, eye protection, protective clothing are all necessary. Safety shoes should be worn at all times when handling weldments and when repairing equipment.

Some of the precautions you must take when working in a welding shop are as follows:

- Cranes and hoists must be periodically inspected for condition of cables, hooks, clamps, rails, and the like.
- Ventilation ducts and air-handling equipment must be periodically inspected, cleaned as needed, and moving parts lubricated.
- Aisles in the shop should be clearly marked and kept clear of material and equipment.
- When heavy equipment is being moved, all personnel in the vicinity must be alerted and warned to move out of potentially dangerous locations.
- Periodic inspections should be made of all safety equipment and supplies in the shop.

TEST YOUR KNOWLEDGE

Write your answers on a separate sheet of paper. Do not write in this book.

1. List four important features that must be included in a well-designed welding shop.
2. For what reasons should the offices, paint booths, toilets, shower and locker rooms be kept separated from the welding shop?
3. List three reasons for having adequate ventilation in a shop.
4. During the winter, replacement air for the ventilation system must be _____ and humidified.
5. List four different mechanical processes used to cut metal in a welding shop.
6. Name three types of shears available.
7. What is a notching machine?
8. What type of tool or machine is used to make straight bends in metal? Name two.
9. What adjustment is made to change the angle of the bend in a hydraulic brake press?
10. Name two reasons why weldments may be preheated.
11. Name one disadvantage of an enclosed furnace.
12. Why is a jig or fixture used in welding?
13. Welding _____ are used to rotate weldments so most welds may be made in the flat position.
14. In all grinding operations, you should wear _____ _____.
15. List four materials used as cleaning abrasives when blast cleaning.
16. Why do smaller welding shops send most welding equipment out to specialty shops for repair?
17. When heavy equipment is being moved, all _____ in the vicinity should be alerted and moved out of potentially dangerous locations.
18. What is the cause of failure for most new welding shops?
19. Why is it a good practice to identify all metal stock which is kept in shop inventory?
20. Which type of mechanical metal cutting machine mentioned in this chapter is best for cutting curved lines on metal sheet?

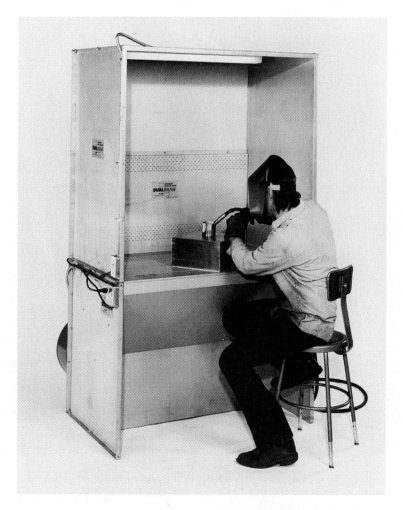

This welding workstation has more than 2000 tiny holes in the back and bottom of the work area to allow fumes and smoke to be drawn away from the welder. Good ventilation is an important consideration in designing any area used for welding. (Dualdraw Division of Industrial Manufacturing and Machining)

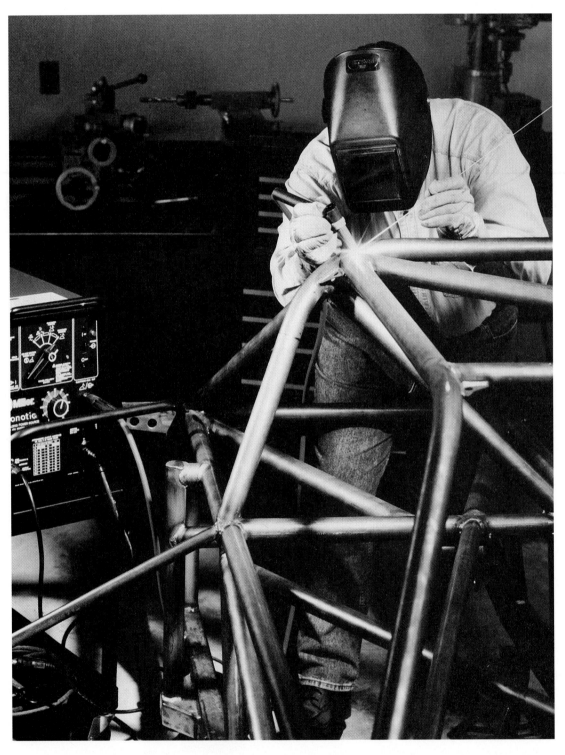

To get and hold a job in the welding industry today, you should always be ready to learn and refine new skills. This welder is using GTAW to fabricate a complex tubing assembly. (Miller Electric Mfg. Co.)

Chapter 33

GETTING AND HOLDING A JOB IN THE WELDING INDUSTRY

According to articles in welding trade magazines, the welding industry is growing, as indicated by the increasing sales of electrodes and electrode wires. As this growth occurs, more workers will be required in the welding industry.

There are many jobs and careers in the welding industry. Most jobs will require a high school education or an apprenticeship. Other jobs in the industry may require a junior college associate degree, a four-year degree, or an advanced degree.

Preparing for a career begins in high school and may require continued study throughout your lifetime. Getting a job is sometimes difficult, for various reasons that a job-seeker has control over. Losing a job also occurs for various reasons, many of which the employee has control over.

33.1 THE FUTURE OF THE WELDING INDUSTRY

The future of the welding industry looks bright. The industry has continued to use more and more electrodes and electrode wires year after year. Many welding processes have been developed and refined to weld newer and stronger metal alloys, complex parts, unusual joint designs, and unusual metal combinations. New welding processes will still need to be developed to weld new metals and to weld ceramic and plastic parts.

Robots are doing some of the repetitive-type production welding jobs in industry. However, welding products in low production, repair welding, and on-site welding do not lend themselves to robotic-type welding. Skilled welders will be required for most of the welding that will be done in the future.

33.2 WELDING OCCUPATIONS AND THE EDUCATION REQUIRED

Every career field has a variety of job levels and responsibilities that are required in order to produce a completed product or service. A number of welding and welding-related jobs, categorized by the educational level normally required, are listed below:

High School and Trade School
* Arc welder
* Assembler-brazer
* Assembler-welder
* Combination welder
* Fitter-welder
* Gas welder
* Gun welder
* Resistance welder
* Solderer
* Tack welder

Technical School or Community College
- Experimental welder
- Laser beam welding machine operator
- Robotic welding machine programmer
- Tool and die welder
- Ultrasonic welding machine operator
- Welding inspector
- Welding machine repair person
- Welding supervisor

Military Training School
- Combination welder
- Diver/welder
- Repair welder
- Specialty welder

Trade Apprenticeship
- Blacksmith for experimental parts
- Ornamental ironworker
- Pipe and steamfitter
- Sheet metal worker
- Structural ironworker

College or University
- Metallurgist
- Metallurgical engineer
- Welding engineer

Figure 33-1. *The ability to make and use sketches or mechanical drawings is a useful skill for a person entering the welding field or a welding-related occupation.*

33.3 SCHOOL SUBJECTS SUGGESTED FOR SUCCESS

Welding is a *technical trade*. In addition to a good knowledge of welding, a number of other school subjects are suggested as preparation for success in a welding career. The school subjects suggested, and the reason they are important, are discussed in this Heading.

33.3.1 Print Reading and Mechanical Drafting

Entry level welding jobs often require that a welder be able to read and understand the information given on a print. Also, the ability to draw or sketch a plan for the assembly and welding of an object is important to a welder who wishes to advance. See Figure 33-1. The ability to be able to read a complicated print to find sizes and special directions is essential to any welder who wishes to become a welder fitter, inspector, or supervisor. If you always must be told what to do and how to do it by someone else, you will not advance very far.

33.3.2 Electricity and Electronics

An arc welder should know something about electricity and electronics if he or she wishes to go into machine repair.

33.3.3 Metals Shop

At the middle school or high school level, a student should take metals shop so that he or she understands how to follow directions and read prints. Also, this class will help the student understand how metals are worked, connected, and finished. Students should learn the various metalworking processes, such as turning, milling, shaping, or boring. Students should learn shop safety and how to work with others in a shop situation.

33.3.4 Physics

A knowledge of *physics principles* is important to anyone going into a career in a trade or industry. They should understand friction, the function of threads, gears, levers, stresses, heat, energy, and energy conservation. They should be able to understand various mechanical, hydraulic, and pneumatic mechanisms. The fundamentals of electricity and electronics should be understood. Physics is important to anyone who intends to pursue education beyond high school, including apprenticeships.

33.3.5 Math, Algebra, Geometry, Trigonometry, and Calculus

Math is important to a welder if she or he is a fitter-welder, assembler-welder, inspector, supervisor, metallurgist, or welding engineer. Many problems, both simple and complex, must be solved before a product can be produced. Taking as much math as the school offers will benefit anyone who intends to go beyond high school or who hopes to advance in the welding industry.

33.3.6 Welding Courses

To enter the welding field as a welder, cost estimator, inspector, or welding engineer, a student should take welding courses. See Figure 33-2. The knowledge of what is actually occurring in a welding situation will be helpful to anyone working in the industry.

Figure 33-2. Classes taken in high school or vocational school will help prepare you for employment in jobs requiring welding skills. (American Welding Society)

33.4 PERSONAL TRAITS SOUGHT BY EMPLOYERS

A study of employers was made recently to determine the personal traits that they look for in an employee. The top ten personal traits named were:

- Dependability.
- Ability to follow instructions.
- Getting along with peers.
- Doing work well.
- Self-confidence.
- The ability to take on responsibility.
- Initiative.
- Getting along with supervisors.
- The ability to communicate written ideas.
- The ability to communicate ideas orally.

As viewed by future *employers*, school is the young person's "first job." While still in school, the student can demonstrate to prospective employers that he or she possesses the traits that an employer wants. These desired traits can be demonstrated in the following ways:

Dependability
- Have a good attendance record.
- Be in class on time.
- Complete all assigned tasks on time and without supervision.
- Come prepared to work.

Ability to follow instructions
- Observe all school and classroom rules and guidelines.
- Follow oral and written instructions.
- Verify and proof (check) your work.

Getting along with peers
- Work well with other students.
- Support the group in which you are working.
- Respect the property of others.
- Participate willingly in group activities.

Doing work well
- Do all work neatly and accurately.
- Plan and organize your time.
- Use equipment properly and carefully.
- Always work up to your highest potential.

Self-confidence
- Try to appear calm, poised, and comfortable when performing any task or assignment.
- Show pride in your work.
- Be able to verbally present your ideas to a group.

Taking responsibility
- Be responsible for your actions.
- Seek positive leadership positions in school, clubs, church, and other organizations.
- Follow school, class, and organization rules.
- Be honest.
- Show concern for school or others' property.

Initiative
- Begin new assignments without being asked.
- Volunteer new ideas.
- Demonstrate positive leadership.
- Volunteer to do extra tasks or assignments.

Getting along with supervisors
- Follow written and oral instructions and directions.
- Follow all school and classroom rules.
- Demonstrate loyalty to school and superiors.
- Complete assigned tasks without complaint.
- Graciously accept constructive criticism.

Ability to communicate written ideas
- Use correct grammar.
- Spell words correctly.
- Write and print neatly and legibly.
- Write reports, instructions, and descriptions that are well thought out.

Ability to communicate ideas orally
- Speak using correct grammar.
- Use a tone of voice that indicates confidence, enthusiasm, eagerness, and harmony.
- Speak clearly, concisely, and convey complete thoughts.

You may wish to request a letter of recommendation from your school to indicate to a potential employer that you have demonstrated these traits while a student.

33.5 ACADEMIC SKILLS SOUGHT BY EMPLOYERS

Employers were also surveyed to identify those *academic skills* a person should possess to successfully get and keep a good job. These skills include the ability to:

- Read and understand written materials.
- Write and understand the technical terms and language of the trade or business in which the person is working.
- Read, understand, and interpret graphs and charts.
- Understand basic mathematics.
- Use mathematical skills to solve problems.
- Use research and library skills.
- Use the tools and equipment involved in the business.
- Use the technical language in which business is conducted for both spoken and written communication.
- Use the scientific method for solving problems.

33.6 PERSONAL MANAGEMENT SKILLS SOUGHT BY EMPLOYERS

The following *personal management skills* were identified by employers as necessary for an employee to obtain and keep a good job:

- Daily, on-time attendance at school or work.
- Demonstrated self-control.
- A developed career plan.
- The ability to follow verbal instructions.
- The ability to follow written instructions.
- Knowing personal strengths and weaknesses.
- The ability to identify and suggest new ways to get a job done.
- Capability of learning new skills.
- The ability to meet deadlines in school and work.
- Paying attention to details.
- Working without supervision.

33.7 APPLYING FOR A JOB

When filling out a *job application*, write or print clearly, as directed. Be sure to fill out the application completely. If you are asked for information for which there is no answer, enter "NA" or "Not applicable." For example, if you were asked "In what branch of service did you serve?" and you had not served in the armed forces, answer "not applicable" or "NA." If the answer space is left blank, a potential employer might think that you had overlooked the question and are not a careful worker.

If applying for your first job, list all your classes and projects that relate to the job that you are applying for. List all your educational experiences, even short training courses that are related to the job.

When you go in for an *interview*, it is important to be a little early. Prepare for the interview by thinking about the job for which you are applying. Mentally list all the skills and abilities you have that meet the job requirements. Review Headings 33.5 and 33.6 and be ready to answer questions to show you have what the employer is looking for.

It is important to look your best for an interview. If you are male, shave, get a fresh haircut, and dress well, but appropriately. A suit is not necessary if interviewing for a welder's job. A suit *would* be appropriate, however, if you are applying for a job as an inspector, foreman, or supervisor. If you are female, use only necessary makeup, and wear a neat and appropriate hair style. Dress appropriately for the job. *Never wear jeans or work clothes to an interview, since you will not normally be expected to go to work immediately.*

33.8 FACTORS THAT LEAD TO REJECTION FOR EMPLOYMENT

Personal factors or traits that would cause a person to be rejected for a job were identified by executives of 186 companies in a recent survey. These factors are:

- A poor scholastic record.
- Inadequate personality.
- Lack of goals.
- Lack of enthusiasm.
- Lack of interest in the business where employment is being sought.
- Inability to express oneself.
- Unrealistic salary demands.
- Poor personal appearance.
- Lack of maturity.
- Failure to get information about the company where employment is sought.
- Excessive interest in security and benefits.

33.9 FACTORS THAT CAN LEAD TO TERMINATION FROM A JOB

The factors or causes that can lead to failing to get a promotion, or even to being fired (terminated) from a job, include the following:

- Poor attendance without good reason.
- Habitually coming to work late.
- Alcohol abuse.
- Illegal drug use.
- Inability to perform the tasks required.
- Inability to work as a team member.
- Inability to work with peers.
- Fighting with or making threats to peers or supervisors.

- Insubordination to directions from a supervisor.
- Talking with others too much and too often.
- Lack of respect for others.
- Lack of respect for others' property.
- Always making excuses.
- Constant complaining.
- Inability to make independent decisions within the scope of the job.

TEST YOUR KNOWLEDGE

Write your answers on a separate sheet of paper. Do not write in this book.

1. Name three welding jobs or careers that require only a high school diploma.
2. List three welding jobs or careers that require a four-year college degree.
3. Explain why a student should take drafting or print reading in high school.
4. Explain why a student should take physics in high school.
5. List the top ten personal traits that employers would like to have in an employee.
6. List four ways you can demonstrate to an employer the fact that you are dependable.
7. What are the personal management skills listed by employers as desirable?
8. List at least six factors which, according to employers, may lead to a job applicant's rejection.
9. Describe how you could show an employer that you have an interest in his or her business.
10. List at least six reasons for which an employee might be fired.

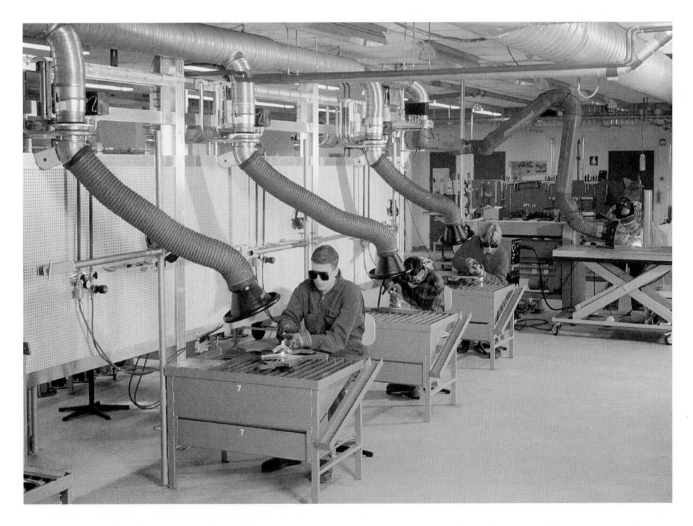

This welding classroom is well-equipped with a gas manifold system and well-arranged workspaces. Note that it has efficient fume extractors to protect the health of the student welders. (Nederman, Inc.)

Welders must be ready to work under a variety of weather conditions. This worker is warmly dressed in coveralls while working on a construction site on a winter day following a light snowfall. (Miller Electric Mfg. Co.)

Chapter 34

TECHNICAL DATA

LEARNING OBJECTIVES

After studying this chapter, you will be able to:
* Calculate, or determine from charts and tables provided, the needed pressures, sizes, and specifications to complete a job in the welding shop.
* Calculate the correct tap drill to use to tap a given size thread and perform the drilling and tapping operations.
* Use and convert various US Conventional to SI Metric units and vice versa.
* Create a list of twenty hazards that may be present in a welding shop.

Material covered in this chapter will deal with technical understanding and useful reference tables. Information in the chapter will include:
* The effect of hose diameter and length on gas flow and pressure.
* Various oxyfuel gas flame temperatures.
* The chemistry of welding.
* The chemical composition of pickling and metal cleaning solutions.
* Properties of metals.
* Sheet metal gages.
* Judging the temperature of mild steel by its color.
* Drill, metal gage, and tap sizes.
* Metric practices and conversions for use in welding.
* Understanding the meaning of the terms temperature and heat.

34.1 THE EFFECT OF TEMPERATURE ON CYLINDER PRESSURE

The pressure within a cylinder or tank will vary as the outside temperature changes. Gas in a confined space, such as a cylinder, cannot expand. Therefore, if it is heated, the pressure of the gas inside the cylinder will rise. If the temperature of the cylinder cools, the pressure of the gas in the cylinder will go down.

Figure 34-1 uses a full 244 ft^3 (6.91m^3) cylinder at 2200 psig (15 168kPa) and 70°F (21.1°C) to illustrate how the pressure changes as the temperature changes.

Temperature of Oxygen		Gauge Reading	
°F	°C	psig	kPa
100	37.8	2325	16 030
90	32.2	2283	15 741
80	26.7	2242	15 458
70	21.1	2200	15 168
60	15.6	2158	14 879
50	10.0	2117	14 596
40	4.4	2075	14 307
30	-1.1	2034	14 024
20	-6.7	1992	13 734
10	-12.2	1951	13 452
0	-17.8	1909	13 162

Figure 34-1. *The effect of temperature on the oxygen pressure within a fully charged 244 ft^3 (6.91m^3) oxygen cylinder at 2200 psig and 70°F (15 168kPa at 21.1°C).*

34.2 THE EFFECT OF HOSE DIAMETER AND LENGTH ON GAS FLOW AND PRESSURE AT THE TORCH

The volume of gas that reaches the torch valves is determined by the inside diameter of the hose, the length of the hose, and the gas pressure. In oxyfuel gas welding and cutting, an insufficient volume of oxygen or fuel gas will affect the energy output of the flame. It may also cause a flashback accident to occur. An insufficient supply of oxygen for cutting will cause a poor cut.

Insufficient volumes of shielding gases when GTAW, GMAW, GTAC, GMAC, or other gas-using processes will cause poor welds and cuts. Therefore, the volume of gas delivered to the torch valves is very important.

Pressure propels the gas. A low pressure will deliver lower volumes of gas; a high pressure, higher volumes of gas. Four factors affect the gas pressure that reaches the torch valves. One factor is the *pressure set on the regulator*. The second is the *diameter of the nozzle installed in the regulator*. The third is the *length of the hose* used in the system. The fourth is the *inside diameter of the hose*. The inside surface of the hose offers resistance to the flow of the gases.

As the length of hose increases, a higher regulator setting is required to obtain the desired pressure to the torch valve. Two pressures may be measured and shown on the working pressure gauge. The working pressure or low-

Hose diameter	Hose length	Cutting tip size	Pressures in lbs./in² (psig)				Flow cfh
			Reg. psi static	Reg. psi flowing	Inlet psi torch	psi drop in hose	
3/16	50	3	50	47	37$\frac{1}{2}$	9$\frac{1}{2}$	169
3/16	100*	3	51	47	26	21	129
3/16	50	5	84$\frac{1}{2}$	78	44	34	370
3/16	100*	5	83$\frac{1}{2}$	78	22	56	215
3/16	50	7	108	100	24	76	510
3/16	100*	7	106$\frac{1}{2}$	100	9	91	270
3/16	50	9	138$\frac{1}{2}$	130	19$\frac{1}{2}$	110$\frac{1}{2}$	735
3/16	100*	9	136$\frac{1}{2}$	130	7	123	405
1/4	50	3	50$\frac{1}{2}$	47	44$\frac{1}{2}$	2$\frac{1}{2}$	194
1/4	100*	3	50	47	42$\frac{1}{2}$	4$\frac{1}{2}$	188
1/4	50	5	86	78	68$\frac{1}{2}$	9$\frac{1}{2}$	540
1/4	100*	5	85	78	58$\frac{1}{2}$	19$\frac{1}{2}$	470
1/4	50	7	114	100	68	32	1140
1/4	100*	7	110	100	49	51	870
1/4	50	9	149$\frac{1}{2}$	130	65	65	2010
1/4	100*	9	144	130	36$\frac{1}{2}$	93$\frac{1}{2}$	1290
1/4	100**	3	50	47	36	11	164
1/4	100**	5	84$\frac{1}{2}$	78	42	36	360
1/4	100**	7	108	100	25	75	560
1/4	100**	9	140	130	18	112	795
3/8	50	3	51	47	46	1	190
3/8	50	5	86	78	74$\frac{1}{2}$	3$\frac{1}{2}$	580
3/8	50	7	117	100	86	14	1400
3/8	50	9	163$\frac{1}{2}$	130	89$\frac{1}{2}$	40$\frac{1}{2}$	2700
3/8	100*	3	51	47	46	1	198
3/8	100*	5	86	78	72	6	570
3/8	100*	7	115	100	77	23	1280
3/8	100*	9	155	130	75	55	2280

*—Two 50 ft. lengths of hose connected together with standard hose unions
**—Four 25 ft. lengths of hose connected together with standard hose unions

Caution: Do not exceed the maximum working pressures (regulator setting) listed for the hoses during welding and cutting operations.
Oxygen..150 psig maximum
Acetylene...15 psig maximum
Propane, propylene, MAPP, and all other fuel gasses......................40 psig maximum

Caution: HOSES. Do not use excessively long hoses or hoses with many hose unions. Either will restrict gas flow and pressure causing lower cutting efficiency and possibly leading to dangerous operating conditions.

MANIFOLDING CYLINDERS. When required flows in cubic feet per hour exceed the recommended withdrawal rate from one cylinder then additional cylinders must be manifolded to provide safe and efficient operation. Acetylene must not be withdrawn at more than 1/7 of the cylinder capacity (47 cubic feet per hour for a 330 cu. ft. cylinder). Consult your gas supplier for manifolding instructions for the gasses and cylinders supplied to you.

Figure 34-2. *The effect of hose diameter and length on the flow and pressure at the torch valves. (Smith Equipment, Div. of Tescom Corp.)*

pressure gauge will read a higher pressure when the gas is not flowing (*static pressure*) than when the gas is flowing (*dynamic pressure*). This occurs because some of the energy in the gas is used up in moving through the hose and overcoming friction.

Careful thought must be given to the hose diameter, hose length, and regulator pressure settings when an outfit is prepared for welding. Figure 34-2 illustrates the effects of hose diameter and length on the flow and pressure at the torch valves.

34.3 FLAME CHARACTERISTICS

Not all gases burn at the same temperature. Some gases are well-suited for preheating; others, because of their high flame temperatures, are best for welding. Figure 34-3 lists the flame temperatures for a number of fuel gases.

Fuel Gas	Chemical Formula	Btu/ft³	Mj/m³	Neutral Flame Temperature with Oxygen	
				°F	°C
Acetylene	C_2H_2	1470	55	5589	3087
Hydrogen	H_2	325	12	4820	2660
Methyl-acetylene-propadiene	C_3H_4	2460	91	5301	2927
Natural gas (methane)	CH_4	1000	37	4600	2538
Propane	C_2H_3	2498	104	4579	2526
Propylene	C_3H_6	2400	89	5250	2899

Figure 34-3. *The heat units and flame temperatures of various fuel gases.*

34.4 THE CHEMISTRY OF OXYACETYLENE GAS WELDING

The chemical symbols for the products involved in the combustion of the oxyacetylene flame are:
- Acetylene = C_2H_2
- Oxygen = O_2
- Carbon monoxide = CO
- Carbon dioxide = CO_2
- Water (vapor) = H_2O

The chemical formula for the combustion of oxygen and acetylene is:

$$2C_2H_2 + 3O_2 \rightarrow 4CO + 2H_2O + heat$$

In this equation, a chemical change takes place as the welding gases are ignited at the torch tip. This is what happens: two molecules of acetylene (C_2H_2) combine with three molecules of oxygen (O_2) and are ignited. The result of the burning of this combination of molecules is four molecules of carbon monoxide (CO), plus two molecules of water vapor (H_2O), plus heat.

Carbon monoxide (CO) is a very unstable gas. It will unite readily with oxygen to form carbon dioxide (CO_2). The carbon dioxide accounts for the fact that, surrounding the cone of the welding flame, there is an area where the flame is less intense.

It is in this area that the carbon monoxide is mixing with atmospheric oxygen to form carbon dioxide. The layer of carbon monoxide tends to keep the molten weld metal from oxidizing. The carbon monoxide absorbs any free oxygen present.

The chemical action in the outer flame becomes:

$$2CO + CO_2 \rightarrow 2CO_2 + heat$$

The oxygen, in this case, comes from the atmosphere surrounding the welding flame.

This principle must be remembered when welding in a confined space where a free movement of air does not exist around the torch tip. Under these conditions, more oxygen must be fed to the torch tip to maintain the neutral flame.

34.5 CHEMICAL CLEANING AND PICKLING SOLUTIONS

Chemical cleaning, degreasing, and pickling solutions often must be used before brazing some nonferrous metals. Magnesium, magnesium alloys, and copper and copper alloys require chemical cleaning prior to brazing. The following solutions are used on magnesium. Solutions for preparing copper are also given.

Degreasing is generally done with an alkaline solution of sodium bicarbonate, sodium hydroxide, and water. The solution is mixed at a temperature of 190°F to 212°F (88°C to 100°C) follows:

> 3 oz. (85g) sodium bicarbonate
> 2 oz. (56.7g) sodium hydroxide
> 1 gal. (3.785L) water

Bright chrome *pickling* or modified chrome pickling is done to remove oxides and scale from the metal surfaces. Bright pickle solution may be made by mixing these chemicals at 60°F to 100°F (15.6°C to 37.8°C):

> 24 oz. (684.4g) chromic acid
> 5.3 oz. (150.3g) ferric nitrate
> 0.47 oz. (0.013 kg) potassium fluoride
> 1 gal. (3.785L) water

The chrome pickling solution is made by mixing the following chemicals at 70°F to 90°F (21.1°C to 32.2°C):

> 24 oz. (684.4g) sodium dichromate
> 24 fl. oz. (71cc) concentrated nitric acid
> 1 gal. (3.785L) water

A modified chrome pickling mixture is made by combining the following at 70°F to 100°F (21.1°C to 37.8°C):

> 2 oz. (56.7g) sodium acid fluoride
> 24 oz. (684.4g) sodium dichromate
> 1.3 oz. (36.9g) aluminum sulfate
> 16 fl. oz. (47.3cc) nitric acid, 70%
> 1 gal. (3.785L) water

To remove oil, grease, and dirt from copper and its alloys, the following solution can be prepared. The solution should be used at 125°F to 180°F (51.7°C to 82.2°C).

      6.8 oz. (192.8g) sodium orthosilicate
      0.8 oz. (22.7g) sodium carbonate
      0.4 oz. (11.3g) sodium resinate
      1 gal. (3.785 L) water.

A pickling solution for copper is a 10% solution of sulfuric acid used at 125°F to 150°F (51.6°C to 65.6°C). Scale can be removed by using the following solution at room temperature:

      40% concentrated nitric acid
      30% concentrated sulfuric acid
      0.5% concentrated hydrochloric acid
      29.5% water
      (Percent by volume)

34.6 THERMIT REACTION CHEMISTRY

The welding of parts or making of castings using the exothermic Thermit® process has been done for many years. The chemical reaction of this steel welding or casting process is:

    $8Al + 3Fe_3O_4 \rightarrow 9Fe + 4Al_2O_3 + heat$

The aluminum and iron oxide mixture must be heated to approximately 2200°F (1204°C) to start the reaction described above.

Copper, nickel, and manganese have also been welded or cast using this process. The word "Thermit" is a trade name commonly used to identify this process.

34.7 BLAST FURNACE OPERATIONS

The blast furnace is used to convert iron ores to pig iron. See Heading 27.1.2. The blast furnace, in production of pig iron, must:
1. Deoxidize the iron ore.
2. Melt the iron.
3. Melt the slag.
4. Carburize the iron
5. Separate the iron from slag.
    The following raw materials are fed to the furnace:
ORES
Hematite (red iron) Fe_2O_3 — 70% iron
Magnetite (black) Fe_3O_4 — 72.4% iron
Limonite (brown) $Fe_2O_3H_2O$ — 63% iron
Siderite (iron carbonate) $FeCO_2$ — 48.3% iron
FLUX
A good flux must melt and unite with impurities, called **gangue,** and carry them away in the form of slag. Sand (silica) and alumina are chief impurities. The basic flux used in the blast furnace is limestone.
FUEL
    The fuel used in the blast furnace must melt the charge and furnish heat for the reactions in the furnace. It must be low in phosphorus and sulfur. Coke is ideal for this purpose.

AIR
Preheated compressed air is forced into the lower part of the furnace. The main chemical reactions in the blast furnace are:
1. $C + O_2 \rightarrow CO_2 + heat$
2. $CO_2 + C \rightarrow 2CO + heat$
3. $Fe_2O_3 + 3CO \rightarrow 3CO_2 + 2Fe + heat$
(1-3) In the lower zone of the furnace heat, the CO acts as a reducing agent.
4. $MnO + CO \rightarrow CO_2 + Mn + heat$
5. $SO_2 + 2CO \rightarrow 2CO_2 + S + heat$
(4-5) In the upper zone of the furnace.
6. $Fe_2O_3 + 3C \rightarrow 3CO_2 + 2Fe + heat$
7. $CaCO_3 + heat \rightarrow CaO + CO_2$
(6-7) In the lower zone, as much heat is required as in the upper part and the process goes practically to completion at about 1500°F (816°C).
8. $MgCO_3 + heat \rightarrow MgO + CO_2$
9. $CaSO_4 + 2C \rightarrow CaS + 2CO_2$
10. $CaO + Al_2O_3 \rightarrow CaO + Al_2O_3$
11. $CaO + SiO_2 \rightarrow CaO + SiO_2$
(8-11) In the lower zone and only partly complete.

The Al_2O_3, CaO, and MnO go through both zones unchanged, since not enough heat is furnished to cause the reactions in (4) and (5) to go to completion.

Pure iron melts at 2700°F (1482°C). With impurities, it melts at lower temperatures. Iron and slag separate in the bottom of the blast furnace because the slag is lighter and floats on top of the molten iron. Impurities such as silicon, manganese, and carbon are soluble in iron and remain in the iron. Iron sulfide and iron phosphide are also soluble in iron, so their quantities must be kept small.

34.8 PROPERTIES OF METALS

Figure 34-4 lists most metals used in the welding industry. The chemical symbol of the metal is given, along with its melting temperatures and some other properties.

When designing a structure, the type of material to be used and its weight must be considered. To prevent distortions in the completed weldment, the expansion rate of metal must be considered. This is especially true when jigs and fixtures are used. Figure 34-5 is a table which lists the weights and expansion rates of various metals.

34.9 STRESSES CAUSED BY WELDING

Stress is a force which causes or attempts to cause a movement or change in the shape of parts being welded (strain).

During welding, the heat created by the welding process causes the metal to expand. When the welded metal cools, it contracts (shrinks). Usually, it does not return to the original shape or position. If the metal does not return to its original shape, *distortion* has occurred. Distortion can be minimized by clamping the parts into a fixture while welding.

Metal	Symbol	Melting Temperature		Specific Gravity	Weight per cu./ft.	Grams per cu./cm	Specific Heat	
		°F	°C				Btu/lb/°F	Cal/g/°C
Aluminum	A1	1218	659	2.7	166.7	2.67	0.212	0.226
Antimony	Sb	1166	630	6.69	418.3	6.6	0.049	0.049
Armco iron	. .	2795	1535	7.9	490.0	7.85	0.115	0.108
Barium	Ba	1600	870	3.6	219.0	. . . .	. . .	0.068
Beryllium	Be	2348	1285	1.84	. . . .	1.845	. . .	0.46
Bismuth	Bi	520	271	9.75	612.0	. . . .	. . .	0.029
Boron	B	3990	2200	2.29	143.0	. . . .	. . .	0.309
Brass (70Cu 30Zn)	. .	1652-1724	900-940	8.44	527.0	. . . .	0.092	. . .
Brass (90Cu 10Zn)	. .	1868-1886	1020-1030	8.60	540.0	. . . .	0.092	. . .
Bronze (90Cu 10Sn)	. .	1562-1832	850-1000	8.78	548.0	. . . .	0.092	. . .
Cadmium	Cd	610	321	8.64	550.0	. . . .	. . .	0.055
Carbon	C	6,510	3600	2.34	219.1	3.51	0.113	0.165
Cast pig iron	. .	2012-2282	1100-1250	7.1	443.2	. . . .	0.13	. . .
Cerium	Ce	1184	640	6.8	432.0	. . . .	. . .	0.05
Chromium	Cr	2770	1520	6.92	431.9	6.92	0.104	0.12
Cobalt	Co	2700	1480	8.71	555.0	. . . .	. . .	0.099
Columbium	Cb	3124	1700	7.06	452.54	7.25	. . .	. . .
Copper	Cu	1981	1100	8.89	555.6	8.9	0.092	. . .
Gold	Au	1900	1060	19.33	1205.0	19.2	0.032	0.031
Hydrogen	H	-434.2	-259	0.070	0.00533	. . . .	. . .	3.415
Iridium	Ir	4260	2350	22.42	1400.0	22.4	0.032	0.032
Iron	Fe	2790	1530	7.865	490.0	7.85	0.115	0.108
Lead	Pb	621	327	11.37	708.5	11.32	0.030	0.030
Lithium	Li	367	186	.534	. . . .	32.8	. . .	0.79
Magnesium	Mg	1204	651	1.74	108.5	. . . .	. . .	0.249
Manganese	Mn	2300	1260	7.4	463.2	7.40	0.111	0.107
Mercury	Hg	-38	-39	13.55	848.84	13.6	0.033	0.033
Molybdenum	Mo	4530	2500	10.3	638.0	. . . .	. . .	0.065
Nickel	Ni	2650	1450	8.80	555.6	8.9	0.109	0.112
Plain carbon steel	. .	2462-2786	1350-1530	7.8	486.9	. . . .	0.115	. . .
Osmium	Os	4890	2700	22.48	1405.0	. . . .	. . .	0.031
Palladium	Pd	2820	1550	12.16	750.0	. . . .	. . .	0.059
Platinum	Pt	3190	1750	21.45	1336.0	21.4	0.032	0.032
Rhodium	Rh	3540	1950	12.4	776.0	. . . .	. . .	0.060
Ruthenium	Ru	4440	2450	12.2	762.0	. . . .	. . .	0.061
Silenium	Se	424	218	4.8	300.0	. . . .	. . .	0.084
Silicon	Si	2590	1420	2.49	131.1	2.10	0.175	0.176
Silver	Ag	1800	960	10.5	655.5	10.5	0.055	0.056
Tantalum	Ta	5160	2800	16.6	1037.0	. . . .	. . .	0.036
Tellurium	Te	846	452	6.23	389.0	. . . .	. . .	0.047
Thallium	Ti	576	302	11.85	740.0	. . . .	. . .	0.031
Thorium	Th	3090	1700	11.5	717.0	. . . .	. . .	0.028
Tin	Sn	450	232	7.30	455.7	7.30	0.054	0.054
Titanium	Ti	3270	1800	5.3	218.5	3.50	0.110	0.142
Tungsten	W	5430	3,000	17.5	1186.0	19.0	0.034	0.034
Uranium	U	. . . .	. . . .	18.7	1167.0	18.7	0.028	0.028
Vanadium	V	3130	1720	6.0	343.3	. . . .	0.115	. . .
Wrought iron bars	. .	2786	1530	7.8	486.9	. . . .	0.11	. . .
Zinc	Zn	787	419	7.19	443.2	. . . .	0.093	. . .
Zirconium	Zr	3090	1700	6.38	398.0	. . . .	. . .	0.066

Figure 34-4. *Metal properties. This table lists the metals used in welding, the chemical symbol for each, and other useful information.*

Metal	Weight per ft³ (lbs.)	Weight per m³ (kg)	Expansion per °F rise in temperature (0.0001")	Expansion per °C rise in temperature (0.0001mm)
Aluminum	165	2643	1.360	62.18
Brass	520	8330	1.052	48.10
Bronze	555	8890	0.986	45.08
Copper	555	8890	0.887	40.55
Gold	1200	19 222	0.786	35.94
Iron (Cast)	460	7369	0.556	25.42
Lead	710	11 373	1.571	71.83
Nickel	550	8810	0.695	31.78
Platinum	1350	21 625	0.479	21.90
Silver	655	10 492	1.079	49.33
Steel	490	7849	0.689	31.50

Figure 34-5. *Weight and expansion properties of various metals.*

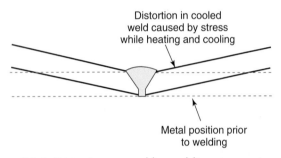

Figure 34-6. *Distortion caused by welding stresses in an unclamped part. Residual stress after distortion may be zero.*

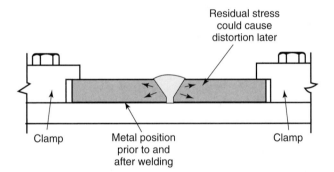

Figure 34-7. *Residual stress remaining in a clamped part after welding. After the weld has cooled, the distortion may be near zero, but the internal residual stress may be considerable. Residual stress may be removed by stress relieving.*

When unclamped parts cool down, they are often distorted (changed in shape). This is due to the stresses remaining in them. Stresses that remain in parts after welding is completed are known as *residual stresses*. See Figure 34-6.

When a part is clamped into a jig or fixture for welding, expansion, contraction, and resulting distortion are held to a minimum. Even though the distortion has been reduced, residual stresses remain in the metal after cooling. See Figure 34-7. These residual stresses may cause the metal to distort at any time if not removed. To reduce the residual stresses caused by welding, parts should be given a stress-relieving heat treatment.

34.10 WIRE AND SHEET METAL GAGES

Two systems are in use in the United States to measure the thickness of metal. The thickness of metal is measured in gage sizes. The two systems are the *Brown and Sharpe gage (B&S)* and the *Steel Sheet Manufacturers Standard (SSMS)*. The B&S gage system is usually used to measure nonferrous sheet and wire. Sheet steels are measured using the Steel Sheet Manufacturers Standard gage system. Figure 34-8 lists the B&S and SSMS gage sizes.

34.11 JUDGING TEMPERATURE BY COLOR

Chapter 29 explains the need for and procedure for the proper heat treatment of a metal after welding or brazing. Proper heat treatment and/or mechanical working will maximize the desired physical properties of a metal.

For proper heat treatment, the metal must be heated to specific temperatures and then cooled. As it is heated to higher and higher temperatures, steel turns different colors. These colors give a fairly accurate indication of the temperature of the steel and may be used as a guide when tempering. Refer to Figure 4-38. The approximate temperature at which each color occurs is shown in Figure 34-9.

Gage	Brown and Sharpe Gage (inch)	Steel Sheet Mfrs. Std. (inch)	Gage	Brown and Sharpe Gage (inch)	Steel Sheet Mfrs. Std. (inch)
6-0	0.5800	—	17	0.0453	0.0538
5-0	0.5165	—	18	0.0403	0.0478
4-0	0.4600	—	19	0.0359	0.0418
3-0	0.4096	—	20	0.0320	0.0359
2-0	0.3648	—	21	0.0285	0.0329
0	0.3249	—	22	0.0253	0.0299
1	0.2893	—	23	0.0226	0.0269
2	0.2576	—	24	0.0201	0.0239
3	0.2294	0.2391	25	0.0179	0.0209
4	0.2043	0.2242	26	0.0159	0.0179
5	0.1819	0.2092	27	0.0142	0.0164
6	0.1620	0.1943	28	0.0126	0.0149
7	0.1443	0.1793	29	0.0113	0.0135
8	0.1285	0.1644	30	0.0100	0.0120
9	0.1144	0.1495	31	0.0089	0.0105
10	0.1019	0.1345	32	0.0080	0.0097
11	0.0907	0.1196	33	0.0071	0.0090
12	0.0808	0.1046	34	0.0063	0.0082
13	0.0720	0.0897	35	0.0056	0.0075
14	0.0641	0.0747	36	0.0050	0.0067
15	0.0571	0.0673	37	0.0045	0.0064
16	0.0508	0.0598	38	0.0040	0.0060

Figure 34-8. A table of the Brown and Sharpe (B&S) and Steel Sheet Manufacturers Standard (SSMS) gage sizes for metal thickness. The decimal equivalents of the US Conventional system are also shown. B&S gages are usually used with nonferrous metals. SSMS gages are generally used for steel thicknesses.

Color	°F	°C
Faint straw	400	205
Straw	440	225
Deep straw	475	245
Bronze	520	270
Purple	540	280
Light blue	590	310
Blue	640	340
Black	700	370
Dark red	1000	600
Dull cherry red	1200	650
Cherry red	1300	700
Bright cherry red	1400	750
Orange red	1500	800
Orange yellow	2200	1200
Yellow white	2370	1300
White heat	2550	1400

Figure 34-9. The approximate temperature at which various color changes occur when clean carbon steel is heated to the melting temperature.

34.12 DRILL SETS AND SIZES

Drill bits are produced in a variety of sizes, and are offered in four different size marking systems. These four sizing systems are:

- Fractional drills.
- Number drills.
- Letter drills.
- Metric drills.

The shank of each drill (when large enough) carries a stamped identification of the drill size. The size is given as a number, fraction, or letter. Metric drills are marked with a number. It is important to know if the drill set being used is a number drill set or a metric drill set. The diameters are different, even though the *number* used may be the same.

Drill bits are available in either high-carbon steel (least expensive) for general use, or alloy steel, marked HSS (high-speed steel), for high speed use.

Fractional drill sets may begin with size 1/64" and go to 1/2" in steps of 1/64". Larger fractional sizes are available. See Figure 34-10.

Number drill sets begin with No. 1 (0.228") and go to No. 80 (0.0135"). Most commonly used number drill sets include No. 1 through No. 60. In number drills, the higher

Fractional Size	Decimal Size	Fractional Size	Decimal Size
1/64	0.015625	17/64	0.265625
1/32	0.03125	9/32	0.28125
3/64	0.046875	19/64	0.296875
1/16	0.0625	5/16	0.3125
5/64	0.078125	21/64	0.328125
3/32	0.09375	11/32	0.34375
7/64	0.109375	23/64	0.359375
1/8	0.125	3/8	0.375
9/64	0.140625	25/64	0.390625
5/32	0.15625	13/32	0.40625
11/64	0.171875	27/64	0.421875
3/16	0.1875	7/16	0.4375
13/64	0.203125	29/64	0.453125
7/32	0.21875	15/32	0.46875
15/64	0.234375	31/64	0.484375
1/4	0.250	1/2	0.500

Figure 34-10. *Fractional drill sizes from 1/64" to 1/2" with their decimal equivalents.*

the number, the smaller the drill. Figure 34-11 lists the number drill sizes.

Letter drill sets come in sizes "A" (0.234") to "Z" (0.413"); see Figure 34-12. *Metric drill sets* include sizes from 0.100mm (0.0039") to 25.50mm (1.004"). See Figure 34-13.

34.13 TAPPING A THREAD

When a thread is cut in a hole, a tool called a *tap* is used. Before the tap can be used, a hole of the correct size must be drilled. Regular letter, fraction, number, or metric drills are used. When these drills are used prior to tapping a hole, they are called *tap drills*.

Figure 34-14 lists a number of different National Fine (NF) and National Coarse (NC) thread sizes and their proper tap drill sizes. If the tap drill used is too small in diameter, the worker may break the tap. If the tap drill is too large, the threads will be too shallow. They will have little holding power.

34.14 METRIC PRACTICES FOR WELDING

The entire world, with the exception of the United States, uses the metric system of measurement. At one time, countries using the metric system were not using the same unit as their standard for measurement. In weight, for example, the unit might have been grams or kilograms; and lengths could have been in millimeters or meters.

Drill Number	Decimal Size	Drill Number	Decimal Size	Drill Number	Decimal Size
1	0.2280	28	0.1405	55	0.0520
2	0.2210	29	0.1360	56	0.0465
3	0.2130	30	0.1285	57	0.0430
4	0.2090	31	0.1200	58	0.0420
5	0.2055	32	0.1160	59	0.0410
6	0.2040	33	0.1130	60	0.0400
7	0.2010	34	0.1110	61	0.0390
8	0.1990	35	0.1100	62	0.0380
9	0.1960	36	0.1065	63	0.0370
10	0.1935	37	0.1040	64	0.0360
11	0.1910	38	0.1015	65	0.0350
12	0.1890	39	0.0995	66	0.0330
13	0.1850	40	0.0980	67	0.0320
14	0.1820	41	0.0960	68	0.0310
15	0.1800	42	0.0935	69	0.02925
16	0.1770	43	0.0890	70	0.0280
17	0.1730	44	0.0860	71	0.0260
18	0.1695	45	0.0820	72	0.0250
19	0.1660	46	0.0810	73	0.0240
20	0.1610	47	0.0785	74	0.0225
21	0.1590	48	0.0760	75	0.0210
22	0.1570	49	0.0730	76	0.0200
23	0.1540	50	0.0700	77	0.0180
24	0.1520	51	0.0670	78	0.0160
25	0.1495	52	0.0635	79	0.0145
26	0.1470	53	0.0595	80	0.0135
27	0.1440	54	0.0550		

Figure 34-11. *A table of number drills with their decimal sizes. If the size cannot be seen on the drill, a micrometer can be used to measure the size.*

Drill Letter	Decimal Size	Drill Letter	Decimal Size
A	0.234	N	0.302
B	0.238	O	0.316
C	0.242	P	0.323
D	0.246	Q	0.332
E	0.250	R	0.339
F	0.257	S	0.348
G	0.261	T	0.358
H	0.266	U	0.368
I	0.272	V	0.377
J	0.277	W	0.386
K	0.281	X	0.397
L	0.290	Y	0.404
M	0.295	Z	0.413

Figure 34-12. *A table of letter drills with their decimal equivalents.*

Metric Drills		Metric Drills		Metric Drills		Metric Drills	
mm	Equiv. Inch	mm	Equiv. Inch	mm	Equiv. Inch	mm	Equiv. Inch
0.40	0.0157	2.0	0.0787	4.1	0.1614	7.3	0.2874
0.45	0.0177	2.05	0.0807	4.2	0.1654	7.4	0.2913
0.50	0.0197	2.1	0.0827	4.3	0.1693	7.5	0.2953
0.58	0.0228	2.15	0.0846	4.4	0.1732	7.6	0.2992
0.60	0.0236	2.2	0.0866	4.5	0.1772	7.7	0.3031
0.65	0.0256	2.25	0.0886	4.6	0.1811	7.8	0.3071
0.70	0.0276	2.3	0.0906	4.7	0.1850	7.9	0.3110
0.75	0.0295	2.35	0.0925	4.8	0.1890	8.0	0.3150
0.80	0.0315	2.4	0.0945	4.9	0.1929	8.1	0.3189
0.85	0.0335	2.45	0.0965	5.0	0.1968	8.2	0.3228
0.90	0.0354	2.5	0.0984	5.1	0.2008	8.3	0.3268
0.95	0.0374	2.55	0.1004	5.2	0.2047	8.4	0.3307
1.00	0.0394	2.6	0.1024	5.3	0.2087	8.5	0.3346
1.05	0.0413	2.65	0.1043	5.4	0.2126	8.6	0.3386
1.10	0.0433	2.7	0.1063	5.5	0.2165	8.7	0.3425
1.15	0.0453	2.75	0.1083	5.6	0.2205	8.8	0.3465
1.20	0.0472	2.8	0.1102	5.7	0.2244	8.9	0.3504
1.25	0.0492	2.85	0.1122	5.8	0.2283	9.0	0.3543
1.30	0.0512	2.9	0.1142	5.9	0.2323	9.1	0.3583
1.35	0.0531	2.95	0.1161	6.0	0.2362	9.2	0.3622
1.40	0.0551	3.0	0.1181	6.1	0.2402	9.3	0.3661
1.45	0.0571	3.1	0.1220	6.2	0.2441	9.4	0.3701
1.50	0.0591	3.2	0.1260	6.3	0.2480	9.5	0.3740
1.55	0.0610	3.25	0.1280	6.4	0.2520	9.6	0.3780
1.60	0.0630	3.3	0.1299	6.5	0.2559	9.7	0.3819
1.60	0.0630	3.4	0.1339	6.6	0.2598	9.8	0.3868
1.65	0.0650	3.5	0.1378	6.7	0.2638	9.9	0.3898
1.70	0.0669	3.6	0.1417	6.8	0.2677	10.0	0.3937
1.75	0.0689	3.7	0.1457	6.9	0.2717	10.1	0.3976
1.80	0.0709	3.75	0.1476	7.0	0.2756	10.2	0.4016
1.85	0.0728	3.8	0.1496	7.1	0.2795	10.3	0.4055
1.90	0.0748	3.9	0.1535	7.2	0.2835	10.4	0.4094
1.95	0.0768	4.0	0.1575	7.25	0.2854	10.5	0.4134

Figure 34-13. A table of metric drill sizes with their decimal equivalents in US Conventional units.

An international group developed a metric system, called the International System of Units or SI Metric, that was agreed to worldwide. Industries in the United States must use the SI Metric system if they are to compete successfully in producing products for sale overseas.

The American Welding Society has published a booklet entitled *Metric Practice Guide for the Welding Industry.* This publication describes the manner in which SI measurements are used in the welding industry. SI Metric is an easily used system, since it never deals with fractions. Parts of a unit are expressed as decimals. The SI system uses only one unit for each physical quantity.

The base units used in the International System of Units (SI Metric system) are as follows:

Measure	Unit	Symbol
Length	meter	m
Mass (weight)	kilogram	kg
Time	second	s
Electric current	ampere	A
Thermodynamic temperature	degree Kelvin	°K
Luminous intensity	candela	cd
Amount of a substance	mole	mol
Temperature	degree Celsius	°C

In the SI system, the following terms are used to express parts or multiples of a base unit:

- milli = 1/1000 or 0.001 of the base unit.
- micro = 1/1,000,000 or 0.000001 of the base unit.

Tap		
Dia. of screw	Threads per inch	Tap Drill
6	32	No. 36
6	40	No. 33
8	32	No. 29
8	36	No. 29
10	24	No. 25
10	32	No. 21
12	24	No. 16
12	28	No. 14
1/4	20	No. 7
1/4	28	No. 3
5/16	18	F
5/16	24	I
3/8	16	5/16
3/8	24	Q
7/16	14	U
7/16	20	25/64
1/2	13	27/64
1/2	20	29/64

Figure 34-14. *A table of NF and NC tap drill sizes.*

Examples:

millisecond = 1/1000 of a second or 0.001 seconds.

microampere = 1/1,000,000 of an ampere or 0.000001 amperes.

These terms are used to express large numbers of the base unit:

- kilo = 1000 × 1 unit.
- mega = 1,000,000 × 1 unit.

Examples:

kilogram = 1000 grams.

megavolt = 1,000,000 volts.

Figure 34-15 is a list of conversion factors for use with the common measurement units used in welding. Figure 34-16 is a conversion table for changing electrode diameters and fillet weld sizes to SI Metric units. Figure 34-17 shows how to convert frequently used technical measurements from the US Conventional system to SI Metric units.

34.15 TEMPERATURE SCALES

Many means have been used to measure the temperature level. In the SI Metric system, the temperature scale is in degrees Celsius. On this scale, ice melts at 0°C and water boils at 100°C. The Fahrenheit scale is based on a scale of 180 equal points with ice melting at 32°F and water boiling at 212°F.

One other temperature scale is used in scientific work. This scale measures temperatures in degrees Kelvin. On the Kelvin scale, the temperature 0°K is equivalent to –273°C. This is the point, known as **absolute zero**, at which all molecular motion stops. The point at which ice melts is 273°K. Water boils at 373°K.

To convert °F to °C, use the formula below:

(°F – 32) × 0.555 = °C

Example:

80°F = ? °C

(80°F – 32) × 0.555 = ? °C

48 × 0.555 = 26.6°C or 27°C

To convert °C to °F, use this formula:

(°C × 1.8) + 32 = °F

Example:

100°C = ? °F

(100°C × 1.8) + 32 = ? °F

180°C + 32 = 212°F

or 40°C = ? °F

(40 × 1.8) + 32 = ? °F

72 + 32 = 104°F

34.16 PHYSICS OF ENERGY, TEMPERATURE, AND HEAT

The welding industry is interested in the energy in heating gases. It is also interested in turning solids to liquids. Therefore, welders should know some of the theory of molecular energy.

The **molecular theory of heat** is generally accepted by chemists and engineers as the best explanation of heat energy. The three most important forms of energy are heat energy, mechanical energy, and electrical energy. One form of energy can be converted into another. For example: an electric motor may be used to turn electrical energy into mechanical energy. The bearings of a motor will warm up a little while it is running. This shows that some of the mechanical energy is being turned into heat.

One horsepower (*mechanical energy*) = 2545.6 Btu (2684kJ) per hour (*heat energy*).

One horsepower (*electrical energy*) = 746 watts.

746W = 2545.6 Btu (2684kJ) per hour.

1W = 3.412 Btu (3.497kJ) per hour.

1 kW = 3412 Btu (3497kJ) per hour.

The molecular theory of heat, briefly explained, is as follows: All matter consists of molecules and atoms. The atoms contain protons, electrons, neutrons, and short-lived particles. It is believed that molecules and atoms are always in motion.

For examples, molecules in a sheet of paper or a piece of steel are continually in motion. The rapidity with which these molecules move determines the heat level and is known as **temperature**. Whether one atom is moving at a certain speed, or whether a thousand atoms are moving at that speed, the temperature will be the same.

The number of molecules or atoms in a substance at a given temperature is what determines the **heat** or heat energy of a substance. This, in brief, is the difference between temperature and the amount of heat. The *speed of the motion of the molecules* determines the temperature. The *amount of the substance* determines the energy.

Conversions for Common Welding Terms*			
Property	**To convert from**	**To**	**Multiply by**
Area dimensions (mm^2)	in^2	mm^2	$6.451\ 600 \times 10^2$
	mm^2	in^2	$1.550\ 003 \times 10^{-3}$
Current density (A/mm^2)	A/in^2	A/mm^2	$1.550\ 003 \times 10^{-3}$
	A/mm^2	A/in^2	$6.451\ 600 \times 10^2$
Deposition rate** (kg/h)	lb/h	kg/h	0.045**
	kg/h	lb/h	2.2**
Electrical resistivity ($\Omega \cdot m$)	$\Omega \cdot cm$	$\Omega \cdot m$	$1.000\ 000 \times 10^{-2}$
	$\Omega \cdot m$	$\Omega \cdot cm$	$1.000\ 000 \times 10^2$
Electrode force (N)	pound-force	N	4.448 222
	kilogram-force	N	9.806 650
	N	lbf	$2.248\ 089 \times 10^{-1}$
Flow rate (L/min)	ft^3/h	L/min	$4.719\ 475 \times 10^{-1}$
	gallon per hour	L/min	$6.309\ 020 \times 10^{-2}$
	gallon per minute	L/min	3.785 412
	cm^3/min	L/min	$1.000\ 000 \times 10^{-3}$
	L/min	ft^3/h	2.118 880
	cm^3/min	ft^3/h	$2.118\ 880 \times 10^{-3}$
Fracture toughness ($MN \cdot m^{-3/2}$)	$ksi \cdot in.^{1/2}$	$MN \cdot m^{-3/2}$	1.098 855
	$MN \cdot m^{-3/2}$	$ksi \cdot in.^{1/2}$	0.910 038
Heat input (J/m)	J/in.	J/m	$3.937\ 008 \times 10$
	J/m	J/in.	$2.540\ 000 \times 10^{-2}$
Impact energy	foot pound force	J	1.355 818
Linear measurements (mm)	in.	mm	$2.540\ 000 \times 10$
	ft	mm	$3.048\ 000 \times 10^2$
	mm	in.	$3.937\ 008 \times 10^{-2}$
	mm	ft	$3.280\ 840 \times 10^{-3}$
Power density (W/m^2)	$W/in.^2$	W/m^2	$1.550\ 003 \times 10^3$
	W/m^2	$W/in.^2$	$6.451\ 600 \times 10^{-4}$
Pressure (gas and liquid) (kPa)	psi	Pa	$6.894\ 757 \times 10^3$
	lb/ft^2	Pa	$4.788\ 026 \times 10$
	N/mm^2	Pa	$1.000\ 000 \times 10^6$
	kPa	psi	$1.450\ 377 \times 10^{-1}$
	kPa	lb/ft^2	$2.088\ 543 \times 10$
	kPa	N/mm^2	$1.000\ 000 \times 10^{-3}$
	torr (mm Hg at 0°C)	kPa	$1.333\ 22\ \times 10^{-1}$
	micron (μm Hg at 0°C)	kPa	$1.333\ 22\ \times 10^{-4}$
	kPa	torr	$7.500\ 64\ \times 10$
	kPa	micron	$7.500\ 64\ \times 10^3$
Tensile strength (MPa)	psi	kPa	6.894 757
	lb/ft^2	kPa	$4.788\ 026 \times 10^{-2}$
	N/mm^2	MPa	1.000 000
	MPa	psi	$1.450\ 377 \times 10^2$
	MPa	lb/ft^2	$2.088\ 543 \times 10^4$
	MPa	N/mm^2	1.000 000
Thermal conductivity ($W/[m \cdot K]$)	cal/(cm·s·°C)	W/(m·K)	$4.184\ 000 \times 10^2$
Travel speed, wire feed speed (mm/s)	in./min	mm/s	$4.233\ 333 \times 10^{-1}$
	mm/s	in./min	2.362 205

*Preferred units are given in parentheses.
**Approximate conversion.

Figure 34-15. *Conversions for common welding terms. To convert from SI Metric to US Conventional, divide the SI Metric unit by the factor shown in Column 4.*

Electrode Sizes		Fillet Sizes	
in.	mm	in.	mm
0.030	0.76	1/8	3
0.035	0.89	5/32	4
0.040	1.02	3/16	5
0.045	1.14	1/4	6
1/16	1.59	5/16	8
5/64	1.98	3/8	10
3/32	2.38	7/16	11
1/8	3.18	1/2	13
5/32	3.97	5/8	16
3/16	4.76	3/4	19
1/4	6.35	7/8	22
		1	25

Figure 34-16. *Conversion table for changing the US Conventional system to SI Metric. The table gives equivalents for electrodes and fillet sizes in SI units.*

All substances may exist in four forms: solid, liquid, gas, or plasma. It is usually conceded that there is no change in the chemical composition of a substance in any of these four forms. Therefore, it is explained that when a substance is in the *solid* form, the molecules have a vibrating motion. They stay in the same position relative to each other, but are vibrating. When energy is applied to the substance, the molecules vibrate faster. This is indicated by an increase in temperature. However, only a certain amount of heat energy may be put into a solid substance, and only a certain temperature rise can be obtained in a solid.

After this amount of heat has been absorbed, any additional energy that is added will cause the molecules to travel at such a rate that they cannot stay within their vibrating bonds. An internal change of structure occurs within the molecule. This is accompanied by the absorption of a large amount of energy, and the substance changes slowly from the solid to the *liquid* state. The substance will absorb heat during the change, but there is *no temperature rise*. All the heat being applied produces internal structural change, rather than increasing the motion of the molecule.

The theory of the energy in a liquid substance is that the molecule, instead of vibrating, now travels in a straight line until it comes into contact with another molecule. This necessarily means that the substance will now have no definite shape and will have no rigidity. It must, therefore, be kept in a container. However, the individual molecules still have considerable attraction for one another. One molecule will attract another strongly enough to divert its path and also to prevent it from traveling too far away.

As energy is applied to the liquid molecule, the rate of molecular travel increases and is indicated by a temperature rise. After a certain amount of energy has been absorbed by the liquid, an internal structure change of the molecule will take place. The result of this is that the liquid now turns into a *gas*. While a liquid is being turned into a gas, the temperature cannot rise, since the heat being

applied results only in a structural change. Upon becoming a gas, molecules lose their attraction one for the other and travel in straight paths until they contact another molecule or some other substance. This means that a gas must be confined in a sealed container. Further heating will change the gas into the *plasma* state.

Ice, water, and steam are the best examples of the above theory. All three conditions are easily obtained. Changing from one to the other does not reveal any change in chemical composition.

The heat that turns solids to liquids, and liquids to gases is called *latent heat* (hidden heat). This is because the thermometer gives no indication of the amount of this heat. For example, it requires 970 Btu to change one pound of water at 212°F, to steam at 212°F (or 2255 joules to change one gram of water at 100°C to steam at 100°C).

The oxyacetylene flame temperature is the same for all tip sizes. The amount of energy delivered to the metal, however, increases as the tip size and gas volume increases.

34.17 HEALTH HAZARDS

As a part of instruction in welding, it is important to be familiar with, and to be able to recognize, conditions that may be hazardous to health. It is also important to remember that the best way to attack a hazard is to eliminate or control the conditions causing it.

The oxyfuel gas flame is generally safe, but carbon monoxide (CO) can build up to dangerous levels in a poorly ventilated place. The use of carbon dioxide as a shielding gas when arc welding creates a carbon monoxide (CO) problem as the CO_2 breaks down in the welding arc.

Reducing furnaces may have carbon monoxide (CO) emissions. Such furnaces should be well-vented at both the charging end and the discharging end. When vacuum furnaces or welding chambers are used, the pump exhaust must be vented away from people.

Inert gases are dangerous to use in confined spaces. Inert gases displace air and oxygen. **Workers entering spaces or tanks that are filled with an inert gas must wear an air-supplied purifier or self-contained breathing equipment.**

Coatings on metals are a problem when the metal is heated. Nitrogen dioxide, coated electrode fumes, and iron oxide, cadmium, and zinc coating fumes are examples.

For many years, red lead paint was commonly used outdoors for metal finishing and protection. Lead oxide fumes caused by burning lead paint coatings can produce acute lead poisoning. Terne plate is a metallic lead coating and is dangerous when heated.

Cadmium plating is frequently used on small parts. At low levels, cadmium oxide fumes produce a chronic condition. At high levels, the fumes are very harmful to lungs and liver.

Beryllium is toxic in even very small amounts. Therefore, any operation involving beryllium must be contained. Cobalt should be handled about the same as

General Conversions			
Property	**To convert from**	**To**	**Multiply by**
Acceleration (angular)	revolution per minute squared	rad/s^2	1.745 329 x 10^{-3}
Acceleration (linear)	in./min^2	m/s^2	7.055 556 x 10^{-6}
	ft/min^2	m/s^2	8.466 667 x 10^{-5}
	in./min^2	mm/s^2	7.055 556 x 10^{-3}
	ft/min^2	mm/s^2	8.466 667 x 10^{-2}
	ft/s^2	m/s^2	3.048 000 x 10^{-1}
Angle, plane	deg	rad	1.745 329 x 10^{-2}
	minute	rad	2.908 882 x 10^{-4}
	second	rad	4.848 137 x 10^{-6}
Area	in.^{2}	m^2	6.451 600 x 10^{-4}
	ft^2	m^2	9.290 304 x 10^{-2}
	yd^2	m^2	8.361 274 x 10^{-1}
	in.^{2}	mm^2	6.451 600 x 10^2
	ft^2	mm^2	9.290 304 x 10^4
	acre (U.S. Survey)	m^2	4.046 873 x 10^3
Density	pound mass per cubic inch	kg/m^3	2.767 990 x 10^4
	pound mass per cubic foot	kg/m^3	1.601 846 x 10
Energy, work, heat, and impact energy	foot pound force	J	1.355 818
	foot poundal	J	4.214 011 x 10^{-2}
	Btu*	J	1.054 350 x 10^3
	calorie*	J	4.184 000
	watt hour	J	3.600 000 x 10^3
Force	kilogram-force	N	9.806 650
	pound-force	N	4.448 222
Impact strength	(see energy)		
Length	in.	m	2.540 000 x 10^{-2}
	ft	m	3.048 000 x 10^{-1}
	yd	m	9.144 000 x 10^{-1}
	rod (U.S. Survey)	m	5.029 210
	mile (U.S. Survey)	km	1.609 347
Mass	pound mass (avdp)	kg	4.535 924 x 10^{-1}
	metric ton	kg	1.000 000 x 10^3
	ton (short, 2000 lbm)	kg	9.071 847 x 10^2
	slug	kg	1.459 390 x 10
Power	horsepower (550 ft lbf/s)	W	7.456 999 x 10^2
	horsepower (electric)	W	7.460 000 x 10^2
	Btu/min*	W	1.757 250 x 10
	calorie per minute*	W	6.973 333 x 10^{-2}
	foot pound-force per minute	W	2.259 697 x 10^{-2}
Pressure	pound force per square inch	kPa	6.894 757
	bar	kPa	1.000 000 x 10^2
	atmosphere	kPa	1.013 250 x 10^2
	kip/in.^{2}	kPa	6.894 757 x 10^3
Temperature	degree Celsius, t°C	K	$t_K = t°C + 273.15$
	degree Fahrenheit, t°F	K	$t_K = (t°F + 459.67)/1.8$
	degree Rankine, t°R		$t_K = t°R/1.8$
	degree Fahrenheit, tF		$t°C = (t_F - 32)/1.8$
	kelvin, t$_K$		$t°C = t_K - 273.15$
Tensile strength (stress)	ksi	MPa	6.894 757
Torque	inch pound force	N·m	1.129 848 x 10^{-1}
	foot pound force	N·m	1.355 818

Figure 34-17. Conversions for generally used technical measurements. To convert from SI Metric to US Conventional, divide the SI Metric unit by the factor shown in Column 4.

Property	To convert from	To	Multiply by
Velocity (angular)	revolution per minute degree per minute revolution per minute	rad/s rad/s deg/min	1.047 198 x 10^{-1} 2.908 882 x 10^{-4} 3.600 000 x 10^2
Velocity (linear)	in./min ft/min in./min ft/min mile/hour	m/s m/s mm/s mm/s km/h	4.233 333 x 10^{-4} 5.080 000 x 10^{-3} 4.233 333 x 10^{-1} 5.080 000 1.609 344
Volume	in.^3 ft^3 yd^3 in.^3 ft^3 in.^3 ft^3 gallon	m^3 m^3 m^3 mm^3 mm^3 L L L	1.638 706 x 10^{-5} 2.831 685 x 10^{-2} 7.645 549 x 10^{-1} 1.638 706 x 10^4 2.831 685 x 10^7 1.638 706 x 10^{-2} 2.831 685 x 10 3.785 412

Figure 34-17. *Continued*

beryllium. Thorium is also toxic. When thoriated electrodes are used, an alpha emission is produced and causes an ionization effect. Ventilate the area well.

Silver brazing alloy fumes can be quite dangerous. Fluorides are common in fluxes. Fluoride fumes are harmful. Manganese dioxide is not too toxic, but it may cause trouble if ventilation is poor.

When gas tungsten arc welding aluminum, there is a possibility of real trouble as the ultraviolet frequency is right to form ozone. This gas is most toxic and may cause severe lung and body damage. It is irritating and causes coughing. **It is advisable to keep the welding area well ventilated. When welding under toxic conditions, it is also advisable to wear an appropriate air supplied purifier, a powered air purifier, or a respirator. Equipment choice depends on the fumes involved.** Refer to Chapter 1.

Oil smoke is a problem. The aromatics produced can be dangerous. Figure 34-18 shows a table of safe limits for some welding fumes.

Ionizing radiations markedly increase the possibility of eye cataracts. Even small amounts of ultraviolet may generate ionization and accelerate cataracts. Workers who are operating welding equipment, as well as helpers and other people in the area, must take precautions at all times to prevent eye injury by wearing safety goggles. Ultraviolet rays are harmful to the welder's vision. Safety goggles with proper lenses will provide protection.

Never fail to protect the eyes, skin, and respiratory system. Provide adequate ventilation during all welding

Limits of Air Contaminants				
Substances	PEL		STEL	
	ppm	g/m^3	ppm	mg/m^3
Acetylene tetrabromide	1	14	—	—
Beryllium and beryllium compounds	—	2μg	—	5μg
Cadmium oxide fumes	—	0.10	—	0.3
Cadmium dust	—	0.2	—	0.6
Carbon dioxide	5000	9000	30,000	54,000
Copper fumes	—	0.1	—	—
Iron oxide	—	10	—	—
Manganese fumes	—	5	—	3
Nitrogen dioxide	5	9	1	1.8
Oil mist	—	5	—	—
Ozone	0.1	0.2	0.3	0.6
Titanium dioxide	—	15	—	—
Zinc oxide	—	5	—	10

PEL	= permissible exposure limit
STEL	= short-term exposure limit
ppm	= parts per million
mg/m^3	= milligrams per cubic meter
μg	= micrograms or millionths of a gram

Figure 34-18. *A table showing the permissible exposure limits (in an 8-hour period) and short-term exposure limits (15 minutes) for various gases and air contaminants, as recommended by the Michigan Occupational Health Standards Commission.*

Never fail to protect the eyes, skin, and respiratory system. Provide adequate ventilation during all welding cutting, brazing, and soldering operations.

TEST YOUR KNOWLEDGE

Write your answers on a separate sheet of paper. Do not write in this book.

1. What is the pressure in a 244 ft³ oxygen cylinder at 30°F?
2. If an oxyfuel gas torch with a 1/4" inside diameter hose and a number 7 tip is used 100' from the cylinders, what pressure drop occurs in the hose?
3. Which fuel gas has the highest flame temperature? What is that temperature?
4. When mixing an alkaline solution for cleaning, _____ ounces of sodium bicarbonate are mixed with one gallon of water.
5. What chemicals are mixed to make a bright pickle solution?
6. A pressure of 250 psi is equal to what SI Metric pressure?
7. What material is used in the blast furnace as a flux?
8. At what temperature (in °F and °C) does pure iron melt?
9. Pure nickel melts at what temperature?
10. How much heavier than aluminum is steel?
11. Which two metal thickness gages are used in the United States?
12. At what temperature is clean carbon steel if heated to a color of cherry red?
13. A number 44 drill is how large in diameter (in decimals of an inch)?
14. What is the decimal size of a 7/32" drill?
15. Which letter drill is larger in diameter, a G or an H?
16. Before tapping a 10-32NF thread, what size tap drill should be used? What is the drill's decimal size?
17. When tapping a hole, what may happen if the tap drill is too small?
18. What do the letters SI mean?
19. In SI Metric, how large is a microampere?
20. What precautions must be taken before entering a space or tank that is filled with an inert gas like argon, helium, or CO_2?

A two-wire submerged arc welding outfit with automated flux recovery. The equipment is being used to make a fillet weld on a T-joint in the fixture on the floor. (Koike-Aronson, Inc.)

A student practicing plasma arc cutting on a light truck cab. (Miller Electric Mfg. Co.)

GLOSSARY OF TERMS

This Glossary of Terms explains the meaning of terms most used by welders. Technical engineering terms have been simplified. For additional definitions, refer to the publication *Standard Welding Terms and Definitions* (ANSI/AWS A3.0). It is available from the American Welding Society.

A

Abrasion: Wear produced by rubbing.

Ac or alternating current: Kind of electricity that reverses its direction of electron flow in regular intervals.

Acetylene (C_2H_2): Gas composed of two parts of carbon and two parts of hydrogen. When acetylene is burned in an atmosphere of oxygen, it produces one of the highest flame temperatures obtainable.

Acetylene cylinder: Specially built container manufactured according to ICC standards. Used to store and ship acetylene (sometimes called "tank" or "bottle").

Acetylene hose: See *Hose.*

Acetylene regulator: A device used to control the flow of acetylene. The regulator reduces acetylene cylinder pressures to torch pressures and keeps the pressures constant.

Acoustic emission: Sound produced by a material as it undergoes some change.

Actual throat: Shortest distance from the root of the weld to the face of the weld.

Actuator: A device (usually an electric motor or a hydraulic cylinder) that is used to move a robot about its base or to move the robot arm.

Adhesion: Act of sticking or clinging.

Air carbon arc cutting (AAC): Arc cutting process that uses a carbon arc to heat the metal and an air blast to remove the molten metal and form the cut or gouge.

Air purifying respirator: A device worn over the face to filter harmful or toxic particles or chemicals from the air breathed. This device is not powered.

Alloy: An alloy is a pure metal that has additional metal or nonmetal elements added while molten. The alloy has mechanical properties that are different (usually improved) from the pure metal.

Ampere: Unit of electrical current. One ampere is required to flow through a conductor having a resistance of one ohm at a potential (pressure) of one volt.

Ampere-turns: A term used with electromagnets. It is equal to the number of turns of wire in a coil times the number of amperes flowing in the coil.

Annealing: Softening metals by heat treatment.

Anode: Positive terminal of an electrical circuit.

Arc: Flow of electricity through a gaseous space or air gap.

Arc blow: Wandering of an electric arc from its normal path because of magnetic forces.

Arc cutting: Making a cut in metal using the energy of an electric arc.

Arc gouging: Arc cutting process used to cut a groove in a surface.

Arc length: The distance from the welding electrode to the weld pool surface.

Arc spraying (ASP): Process using the heat of the arc and a compressed gas to atomize and propel a coating material onto the base material.

Arc voltage: Electrical potential (pressure or voltage) across the arc.

Arc welding: A group of welding processes using the heat of an electric arc to melt and weld metal. Arc welding may be done with or without filler metal.

Articulated motion robot: A robot that rotates around the X, Y, and Z axes and swings in arcs in those axes.

Assembly drawing: A drawing that shows a product completely assembled, with all subassemblies and parts.

Atom: Smallest whole part of an element. Its nucleus is made up of protons and neutrons.

Austenite: High-temperature form of steel. It has a face-centered cubic structure.

Automatic welding: Welding with equipment that needs only occasional or no monitoring and no manual adjustments during the process.

Axis of a weld: An imaginary line through the center of the weld metal from the beginning to the end of the weld.

B

Backfire: Short "pop" of the torch flame followed by extinguishing of the flame or continued burning of the gases.

Backhand welding: Moving the weld in opposite direction to which gas flame is pointing.

Backing: Material that is placed on the root side of a weld to aid in the control of penetration.

Backing ring: A metal ring placed inside a pipe before butt welding. The backing ring ensures complete weld penetration and a smooth inside surface.

Back-step welding: Welding small sections of a joint in a direction opposite the progression of the weld as a whole.

Backward welding: See *Backhand welding.*

Base metal: Metal to be welded, cut, or brazed.

Bauxite: Ore from which aluminum is obtained. It consists mostly of hydrated alumina ($Al_2O_3 3H_2O$).

Bead: The appearance of the finished weld. Also, the metal added in welding.

Bevel: An angle cut on the edge of the base metal in a weld joint to create a groove form.

Bird's nest: A tangle of electrode wire that may occur in a wire feeder.

Black light: Light waves that are below the visible range of light. The wavelength of the light reacts with certain dye materials. This causes the dye to fluoresce in a color range visible to the eye.

Blasting: A method of cleaning or surface roughening by projecting a stream of grit (sharp angular abrasives) against the material.

Blowhole: See *Porosity.*

Blowpipe: A nonstandard term for an oxyfuel gas torch.

Blueprint: See preferred term *Print.*

Bond line: In thermal spraying, the junction between the thermal spray deposit and the base metal.

Braze welding: Making an adhesion groove, fillet, or plug weld above 840°F (450°C). The metal is not distributed by capillary action.

Brazement: An assembly joined by brazing.

Brazing: Making an adhesion connection using a minimum amount of an alloy that melts above 840°F (450°C). The alloy flows by capillary action between close-fitting parts.

Brinell hardness: An accurate measure of hardness of metal, made with an instrument. Measurement is made as a hard steel ball is pressed into the smooth surface at standard conditions.

Brittleness: Quality of a material that causes it to develop cracks with only a small degree of bending (deformation) of the material.

Bronze welding: A nonstandard term for *braze welding.*

Buildup: The amount that the weld face extends above the surface of the base metal.

Burning. A nonstandard term. See *Flame cutting.*

Butt joint: An assembly in which the two pieces joined are in the same plane, with the edge of one piece touching the edge of the other.

Buttering: A surfacing deposit that is applied to provide a metallurgically compatible surface for a weld metal to be applied later.

Button: Part of a resistance weld torn out in destructive testing of a spot, seam, or projection weld.

C

Cable: A nonstandard term. See *Lead.*

Capillary action: Property of a liquid to move into small spaces if it has the ability to "wet" those surfaces.

Carbon: The element that, when combined with iron, forms various kinds of steel. In solid form, it is used as an electrode for arc welding. As a mold, it will hold weld metal. Motor brushes are made from carbon.

Carburizing flame: A reducing oxyfuel gas flame in which there is an excess of fuel gas, resulting in a carbon-rich flame zone beyond the inner cone.

Casehardening: Adding carbon to the surface of a mild steel object and heat treating it to produce a hard surface.

Castings: Metallic forms that are produced by pouring molten metal into a shaped container or cavity called a mold.

Cast-weld assembly: An assembly in which cast parts fixed in place by welding.

Cathode: Electrical term for a negative terminal.

Caustic solution: A solution that is capable of eating away a material by chemical action.

Celsius: Temperature scale in SI Metric.

Cementite: The compound also known as iron carbide (Fe_3C). It contains 6.67% carbon.

Chamfer: See *Bevel.*

Charpy: A type of impact testing machine which strikes the specimen with a swinging hammer. The specimen is placed against an anvil with supports 40mm apart.

Check valve: A valve that allows a liquid or gas to flow in only one direction. It closes when flow in the opposite direction is attempted.

Chemical flux cutting (FOC): Oxyfuel gas cutting process that uses a chemical flux to help in cutting some materials.

Chill: To cool rapidly.

Chlorination: A process in which dry chlorine gas is passed through a molten aluminum alloy to remove trapped oxides and dissolved gases.

Circuit: The path of electron flow from a source through components and connections back to the source.

Cladding: A somewhat thick layer of material applied to a surface to improve resistance to corrosion or other agents that tend to wear away the metal.

Clearance: The gap or space between adjoining or mating surfaces.

Coalescence: The intermixing or growing together of materials into one body while being welded.

Coated electrode: A nonstandard term. See *Covered electrode.*

Coating: A nonstandard term. See *Thermal spray deposit.*

Cohesion: Sticking together through the attraction of molecules.

Cold welding (CW): A process that makes use of high pressure and no outside heat to force metal parts to fuse.

Cold work: A metal part on which a permanent strain has been placed by an outside force while the metal is below its recrystallization temperature.

Cold working: Bending (deforming) metal at a temperature lower than its recrystallization temperature.

Combined stresses: A stress type that is more complex than simple tension, compression, or shear.

Combustible: Flammable, easily ignited.

Complete fusion: Fusion that has occurred over the entire fusion surface of a base metal being welded.

Complete joint penetration: A situation in which weld metal completely fills the groove and fuses with the base metal through its entire thickness.

Compressive strength: The greatest stress developed in a material under compression.

Concave weld face: A weld with the center of its face below the weld edges.

Conductivity: The ability of a conductor to carry current.

Conductor: A substance capable of readily transmitting electricity or heat.

Cone: The inner visible flame shape of a neutral or nearly neutral flame.

Constricting: Reducing in size or diameter, as in a constricted arc or constricting orifice.

Continuous casting: A method of casting metal in an open-ended mold so that metal is fed into and cools in the mold in a continuous form.

Continuous weld: Making the complete weld in one nonstop operation.

Controller: The heart of a machining or robotic system. It contains the program that controls the welding or machining variables.

Convex weld: A weld with the face above the weld edges.

Cooling stresses: Stresses resulting from uneven distribution of heat during cooling.

Corner joint: The junction formed by edges of two pieces of metal touching each other at angle of about 90° (a right angle).

Corrosion: The chemical and electrochemical interaction of a metal with its surroundings that causes it to deteriorate.

Corrosion embrittlement: The loss of ductility or workability of a metal due to corrosion.

Corrosion fatigue: The effect of repeated stress in a corrosive atmosphere, characterized by shortened life of the part.

Couplant: A material placed between the transducer and metal surface when doing ultrasonic testing.

Coupon: A piece of metal used as a test specimen. Often an extra piece, as in a casting or forging.

Cover lens or plate: A removable pane of clear glass or plastic used to protect the expensive filtering welding lens.

Covered electrode: A metal rod used in arc welding. It has a covering of materials to aid in the arc welding process.

Crack: A break or separation in rigid material running more or less in one direction.

Cracking: The action of opening a valve slightly, then closing it immediately. Cracking is used to blow out any dust in the valve orifice.

Crater: A depression in the face of a weld, usually at the termination of a weld. The crater is visible after a weld has cooled.

Creep: Permanent deformation caused by stress or heat or both.

Crevice corrosion: Deterioration of a metal caused by a concentration of dissolved salts, metal ions, oxygen, or other gases in pockets that are not disturbed by the fluid stream. This buildup eventually causes deep pitting.

Crown: The curved or convex surface of finished weld.

Cryogenics: Term for the study of physical phenomena at temperatures below –50°F (–46°C).

Cup: A nonstandard term. See *Nozzle.*

Cupola: A blast furnace, in the shape of a vertical cylinder, that is used in making gray iron.

Cutting head: The part of a cutting machine or cutting equipment to which a cutting torch or tip is attached.

Cutting process: An action which causes separating or removal of metal.

Cutting tip: The part of an oxygen cutting torch from which the gases are released.

Cutting torch: Nozzle or device that controls and directs the gases and oxygen needed for cutting and removing the metal in oxyfuel gas cutting.

Cycle: A set of repeating events which occur in order. Also, one complete reversal of alternating current.

Cylinder: A container holding the supply of high-pressure gas used in welding. See *Oxygen, Acetylene.*

Cylinder manifold: See *Manifold.*

D

Dc or direct current: Electric current that flows only in one direction.

DCEN: Direct current that flows from the electrode to the work.

DCEP: Direct current that flows from the work to the electrode.

DCRP: A nonstandard term. See *Direct current electrode positive.*

DCSP: A nonstandard term. See *Direct current electrode negative.*

Decalescence: Transformation which takes place during superheating of iron or steel. The metal surface darkens due to the sudden decrease of temperature during the rapid absorption of the latent heat of transformation.

Decarburization: Carbon loss from the surface of a ferrous alloy when the alloy is heated in a medium that reacts with the surface carbon.

Deep etching: The eating away (as with the corrosive action of acid) of a metal surface . This allows examining of the surface with a magnifier to detect features such as grain flow, cracks, or porosity.

Defect: An imperfection which, by its size, shape, location, or makeup, reduces the useful service of a part.

Degasifier: A substance that can be added to molten metal to draw off soluble gases before they are trapped in the metal.

Degassing: Process of removing gases from liquids or solids.

Degree: One unit of a temperature scale.

Demagnetization: Removal of existing magnetism from a part.

Demurrage: Charge made by a gas supplier to the gas user as rent on the gas cylinder. The user is allowed free use for a number of days. Then, a daily charge is made for additional use.

Dendrite: A crystal development often found in cast metals as they are slowly cooled through the solidification range. The crystal shows a tree-like branching pattern.

Deoxidizer: A substance which, when added to molten metal, removes either free or combined oxygen.

Deoxidizing: Process of removing oxygen from molten metals with a deoxidizer. Also, the removal of other undesirable elements using elements or compounds that readily react with them; in metal finishing, removing oxide films with chemicals or electrochemical processes.

Deposition rate: Weight of material applied in a unit of time. Usually expressed in lbs./hr. or kg/hr.

Depth of fusion: Depth to which base metal is melted during welding.

Destructive testing (DT): The process of testing a sample until it fails. Used to determine how large a discontinuity can be before it is considered a flaw.

Detail drawing: A drawing that shows the shape and size of each small part of an assembly.

Detonation flame spraying: Thermal spraying process that uses controlled explosions of a mixture of fuel gas and oxygen to propel a powdered coating material onto the surface of a workpiece.

Dewar flask: A cylinder designed to hold gases that have been reduced in temperature until they become liquid in form.

Die casting: Production of parts by forcing molten metal into a metal die or mold. Also, the part made by this process.

Die forging: A part shaped by the use of dies in a forging operation.

Diffusion: Spreading of an element throughout a gas, liquid, or solid, so that all of it has the same composition.

Diode: A device used in an electrical circuit that permits electricity to flow in only one direction.

Direct current electrode negative (DCEN): Direct current flowing from the electrode (cathode) to the work (anode).

Direct current electrode positive (DCEP): Direct current flow from the work (cathode) to the electrode (anode).

Direct current reverse polarity: A nonstandard term. See *Direct current electrode positive.*

Direct current straight polarity: A nonstandard term. See *Direct current electrode negative.*

Discontinuity: Any abrupt change or break (cracks, seams, laps, bumps, or changes in density) in the shape or structure of a part. The usefulness of the part may or may not be affected.

Distortion: Warping of a part of a structure.

Downhill: Welding with a downward progression.

Drag: The offset distance between the actual and theoretical exit points of the cutting oxygen stream, measured on the exit side of the material.

Drawing: A shop term sometimes used mistakenly for the term "tempering" in the heat treatment of metal.

Dross: Oxidized metal or impurities that may form on molten metals. It often forms on the other side of metal that is flame or arc cut.

Ductile crack propagation: Slow development of cracks and noticeable warping or deformation caused by an outside pressure.

Ductility: The ability of a material to be changed in shape without cracking or breaking.

Duty cycle: The percentage of time in a 10-minute period that an arc welding machine can be used at its rated output without overloading. A resistance welding machine duty cycle is usually calculated over a 1-minute period.

E

Edge joint: A joint between the edges of two or more parallel parts.

Edge weld: Weld produced on an edge joint.

Effective penetration: (A term used in ultrasonic testing.) The greatest depth at which ultrasonic transmissions can effectively detect discontinuities.

Effective throat: On a fillet weld, the least distance from the root of a weld to the weld face, not counting any reinforcements. See also *Joint penetration.*

Elastic deformation: A temporary change of dimensions caused by stress. The part returns to its original dimensions when the stress is removed.

Elastic limit: The greatest stress to which a structure may be subjected without causing permanent deformation.

Elasticity: Ability of a material to regain its original size and shape after deformation.

Electric arc spraying: A nonstandard term. See *Arc spraying.*

Electrochemical corrosion: Corrosion caused by current flow between the contact areas of two dissimilar metals.

Electrode: Terminal point to which electricity is brought in the welding operation and from which the arc is produced to do the welding. In most electric arc welding, the electrode is usually melted and becomes a part of the weld.

Electrode extension: The length of unmelted electrode extending beyond the end of the collet.

Electrode lead: Electrical conductor between the welding machine and the electrode holder.

Electrode skid: Sliding of an electrode along the work surface during spot, seam, or projection welding.

Electromagnet: A device in which a magnetic field is produced by a coil carrying an electric current. The coil surrounds a mass of ferrous material which also becomes magnetized.

Electromotive force: Energy, measured in volts, which causes the flow of electric current.

Electron: One of the fundamental parts of an atom. It has a small negative electrical charge.

Electron beam welding (EBW): Focused stream of electrons that heats and fuses metals.

Electronic controller: A device which controls the variables in a welding process, such as in automatic welding or when using robotics.

Electroslag welding (ESW): A process using molten slag to melt the base metal. As the weld progresses vertically, the molten metal, slag, and flux are held in place by water-cooled moving shoes.

Element: A chemical substance that cannot be divided into simpler substances by chemical action.

Elongation: The percentage increase in the length of a specimen when it is stressed to its yield strength.

Embrittlement: Reducing the normal ductility of a metal by a physical or chemical change.

Erosion: A reduction in size of an object as a result of a liquid or gas impacting on it.

Eutectic alloy: A mixture of metals that has a melting point lower than that of any of the metals in the mixture, or of any other mixture of these metals.

Exothermic: The ability to release heat. The exothermic cutting process uses rods that release a great deal of heat as they burn.

Explosion welding (EXW): A process that joins metal as powerful shock waves create pressure to cause metal flow and resultant fusion.

Expulsion weld: Resistance spot weld that squirts (expels) molten metal.

F

Face of a weld: The exposed surface of the weld.

Fahrenheit: The temperature scale used in the United States. Its symbol is °F.

Fatigue: A condition of metal leading to cracks under repeated stresses below the tensile strength of the material.

Fatigue limit: Stress limit below which a material can be expected to withstand any number of stress cycles.

Fatigue strength: The greatest amount of stress that a metal will withstand without cracking for a certain number of stress cycles.

Feed rate: Speed at which a material passes through the welding gun in a unit of time.

Feedback: Information about welding variables that is fed back to the welding controller during welding.

Feedback control: Adjustments made to welding variables or mechanical devices (such as robots or positioners) as a result of feedback received by the controller.

Ferrite: Iron which contains very small amounts of carbon. It has a body-centered cubic structure.

Ferrite banding: Bands of free ferrite which line up in the direction the metal was worked. Bands are parallel.

Ferrous metal: A metal containing iron.

Filler metal: Metal that is added when making welded, brazed, or soldered joints.

Filler rod: A nonstandard term. See *Welding rod.*

Filler wire: A nonstandard term. See *Welding wire.*

Fillet weld: Metal fused into a corner formed by two pieces of metal whose welded surfaces are approximately 90° to each other.

Filter plate: An optical material that protects the eyes from ultraviolet, infrared, and visible radiation.

Fixture: A device designed to hold a specific part and accurately locate it during production. Parts are often located in the fixture by pins and held by clamps.

Flame cutting: Cutting performed by an oxyfuel gas torch flame which has a second oxygen jet.

Flame spraying (FLSP): A thermal spraying process in which an oxyfuel gas flame is the source of heat for melting the coating material. Compressed gas may be used to propel the coating onto the base material.

Flash: Impact of electric arc rays against the human eye. Also, the surplus metal formed at the seam of a resistance weld.

Flash welding (FW): Process using electric arc in combination with resistance and pressure welding.

Flashback: A burning back of the gases into the oxyfuel gas torch, hoses, and possibly into the regulator and cylinder. This is a very dangerous situation.

Flashback arrestor: A type of check valve, usually installed between torch and welding hose to prevent flow of burning fuel gas and oxygen mixture back into hoses and regulators.

Flat position weld: Horizontal weld on the upper side of a horizontal surface.

Flaw: Discontinuity in a weld that is larger than the accepted limit.

Fluorescent dye: Dye that gives off or reflects light when exposed to short-wave radiation.

Fluorescent penetrant: Penetrating fluid with a fluorescent dye added to improve the visibility of discontinuities.

Flux: Material used to prevent, dissolve, or help remove oxides and other undesirable surface substances.

Flux cored arc welding (FCAW): Welding method in which heat is supplied by an arc between a hollow, flux-filled electrode and the base metal.

Flux cored arc welding electrogas (FCAW-EG): A type of flux cored arc welding process. Similar to electroslag welding with shielding gas and flux cored electrode added.

Flux oxygen cutting: A nonstandard term. See *Chemical flux cutting.*

Focal spot: In EBW and LBW, the spot where a beam's energy level is most concentrated and where it has the smallest cross-sectional area.

Forehand welding: Welding in the same direction that the flame is pointing.

Forging: The process of making metallic shapes by either hammering or squeezing the original piece of metal. The resulting shape is also called a forging.

Forming: Changing the shape of a metal part without changing its thickness.

Free bend test: Bending a weld specimen without using a fixture or guide.

Free carbon: In steel or cast iron, that part of the total carbon content which is present in the form of graphite or temper carbon.

Friction welding (FRW): Welding process in which the welding heat is generated by revolving one part against another part, under very heavy pressure to create friction.

Front view: The view on a working drawing that shows the greatest amount of detail about an object's shape and size.

Full-wave rectified single-phase ac: Alternating current in which the reverse half of the cycle is made to travel the same direction as the other half of the wave. It produces pulsating direct current with no interval between pulses.

Full-wave rectified three-phase ac: Conversion of alternating current by a rectifier in which there is little pulsation in the resulting direct current.

Fusion: Intimate mixing or combining of molten metals.

Fusion face: The surface of the base metal which is melted during welding.

Fusion welding: Any type of welding that uses fusion as part of the process.

G

Gamma rays: Electromagnetic radiation given off by the nucleus of an atom. Gamma rays always accompany fission.

Gas holes: Holes created by gas escaping from molten metal. These holes are round or elongated, smooth edged dark spots. The holes appear in clusters or individually.

Gas lens: A device used in a GTAW nozzle to straighten the flow of the shielding gas.

Gas metal arc welding (GMAW): Arc welding using a continuously fed consumable electrode and a shielding gas. Sometimes incorrectly called "MIG welding."

Gas pockets: Cavities in weld metal caused by entrapped gas.

Gas tungsten arc welding (GTAW): Arc welding using a tungsten electrode and a shielding gas. The filler metal is added using a welding rod. Sometimes incorrectly called "TIG welding."

Gauss: Unit used to measure magnetic flux density or *induction.* One gauss is one line of flux per square centimeter.

Generator: Mechanism that generates electricity or produces some substance; for example an electric generator or an acetylene generator.

Gouging: Cutting a groove in the surface of a metal using an oxyfuel gas or arc cutting process.

Graphitization: Forming of graphite in iron or steel either during solidification or later during heat treatment.

Groove joint: A joint that has one or both edges cut or machined to form a bevel, V, U, or J groove. This preparation allows the weld to be made into the bottom of the joint.

Groove weld: Welding rod fused into a joint which has the base metal removed to form a V, U, or J trough at the edge of the metals to be joined.

Gross porosity: Large pores, gas holes, or globular voids in a weld or casting. Called *gross* because they are bigger and more numerous than would be found in good practice.

Guided bend test: Bending a specimen in a definite way by using a fixture.

H

Half-wave rectified ac: Simplest manner of rectification in which the reverse half of the cycle is blocked out completely. Result is a pulsating direct current with intervals when no current is flowing.

Hammer forging: Process involving deforming of the workpiece by repeated blows.

Hand shield: An approved arc welding helmet designed to be hand-held. Often used by instructors or observers.

Hardening: Making metal harder by a process of heating and cooling.

Hardfacing: Surfacing material deposited on a surface so that it will resist abrasion, erosion, wear, corrosion, galling, or impact wear.

Hardness: Ability of metal to resist plastic deformation. The same term may refer to stiffness or temper, resistance to scratching or abrading.

HAZ: See *Heat-affected zone.*

Heat: Molecular energy of motion.

Heat-affected zone (HAZ): That part of the base metal altered by heat from welding, brazing, or cutting operations.

Heat-checking: Crazing of a die surface, especially one subjected to alternate heating and cooling.

Heat conductivity: Speed and efficiency of heat energy movement through a substance.

Heat-treatable alloys: Aluminum alloys that reach maximum strength by solution heating and quenching.

Helium (He): Inert, colorless, gaseous element used as a shielding gas in welding.

Helmet: Protective hood that fits over the arc welder's head, provided with an approved filtering lens through which the operator may safely observe the electric arc.

Hertz: Unit of frequency equal to one cycle per second.

Hold time: Time that force continues to be applied to electrodes in resistance welding after the current is turned off.

Home point: See *Set-up point.*

Horizontal position: Weld performed on a horizontal seam that is at least partially on a vertical surface.

Horn: The arm of a resistance welding machine that carries the current and applies the electrode pressure.

Hose: A flexible, usually reinforced, rubber tube used to carry pressurized gases or water to a torch.

Hot forming: Term for operations performed on metal while it is above its recrystallization temperature. Hot forming operations may include bending, drawing, forging, heading, piercing, and pressing.

Hot shortness: A weakness of metal that occurs in the hot forming temperature range.

Hot working: Shaping of metal at a temperature and rate that does not cause strain hardening.

Hydrogen: A gaseous element. When combined with oxygen, it forms a very clean flame. The oxyhydrogen flame does not produce a very high temperature, however.

Hydrogen embrittlement: A low-ductility condition that occurs in metals due to absorption of hydrogen.

I

Impact energy: The amount of energy that must be exerted to fracture a part. The measurement is usually made in an Izod or Charpy test.

Impact strength: Material's ability to resist shock.

Impact test: Careful measurement of how materials behave under heavy loading, such as bending, tension, or torsion. Charpy or Izod tests, for example, measure energy absorbed in breaking a specimen.

Impedance: Total resistance to flow of alternating current as a result of resistance and reactance.

Impurities: Undesirable elements or compounds in a material.

Inclusion: Foreign matter introduced into and remaining in welds or castings.

Incomplete fusion: Less-than-complete fusion of weld material with the base metal or with the preceding bead.

Incomplete joint penetration: Lack of fusion between metals, appearing as elongated darkened lines. May occur in any part of a weld groove.

Incomplete penetration: Incomplete root penetration or failure of two weld beads to fuse.

Indentation: A depression left on the surface of the base metal after spot, seam, or projection welding.

Indentation hardness: The degree of resistance of a material to indentation. This is the standard test used to measure a material's hardness.

Indication, magnetic: Magnetic particle pattern held magnetically on the surface of a material being tested.

Indication, penetrant: Visual evidence of a discontinuity (the penetrant can be seen in the crack).

Indication, ultrasonic: Signal on ultrasonic equipment indicating a crack or discontinuity in a material being tested.

Induced current: Secondary current that is set in motion when a second conductor in the shape of a closed loop is placed in the magnetic field around another current-carrying conductor.

Inductance: In the presence of a varied current in a circuit, the magnetic field surrounding the conductor generates an electromotive force in the circuit itself. If another circuit is next to the first, the changing magnetic field of the first circuit will cause voltage in the second.

Induction: Magnetism that is induced in a ferromagnetic material by some outside magnetic force.

Induction hardening: Process of quench hardening using electrical induction to produce the heat.

Inductive reactance: Force that opposes flow of alternating current through a coil. This force is independent of the resistance of a conductor to a flow of current.

Inert gas: A gas which does not normally combine chemically with the base metal or filler metal.

Infrared rays: Heat rays coming from either an arc or a welding flame.

Inside corner weld: Two metals placed at an angle to each other (usually 90°), and fused at the vertex of the angle.

Interface: Surface that forms a common boundary between two bodies.

Intergranular corrosion: Corrosion occurring, for the most part, between grains or on the edges of the grain in a ferrous material.

Intermittent weld: Method of joining two pieces, with sections of the weld joint left unwelded.

Ion: An atom or a group of atoms that becomes positively or negatively charged as a result of having gained or lost one or more electrons. Also, a free electron.

Ionization: Adding or removing electrons from atoms or molecules to create ions.

Iron-carbon diagram: A diagram that shows the critical temperatures for the varying amounts of carbon in iron which form steels or cast irons.

Izod test: Type of test for impact strength that is made by striking the test piece with a measured downstroke of a pendulum. The specimen, usually notched, is held by one end in a vise. Energy absorbed, which is measured by the upward swing of the pendulum, indicates the impact strength of the specimen.

J

Jig: A device that holds the work and guides a drill or tool during a cutting operation.

Joint: Line or area where two pieces are joined in an assembly.

Joint efficiency: Strength of a welded joint, given as a percentage of the strength of the base metal.

Joint penetration: The depth of weld metal and base metal fusion in a welded joint.

K

Kerf: Width of cut produced by a cutting operation.

Keyhole: A welding technique in which concentrated heat penetrates the workpiece, leaving a hole at the leading edge of the weld. The keyhole, while visible, ensures that the joint has proper penetration.

Killed steel: Steel that has been deoxidized (treated with a strong agent to reduce the oxygen content). Deoxidizing is carried to the point where no reaction takes place between carbon and oxygen as the metal solidifies.

Kilopascal (kPa): The unit of pressure in SI Metric (one thousand pascals). See *Pascal*.

L

Lack of fusion: A type of weld defect caused by lack of union between the weld metal and the base metal.

Laminate: Sheets or bars made up of two or more metal layers built up to form a structural member. Also, the process of forming a metallic product with two or more bonded layers.

Lap joint: A joint in which the edges of the two metals to be joined overlap.

Laser beam welding (LBW): Process in which a single-frequency light beam concentrates a small spot of heat to fuse small, light metal materials.

Layer: A certain weld metal thickness made of one or more passes.

Lead: A wire carrying electricity from the power source to the electrode holder or to the ground clamps.

Left-side view: The view on a working drawing that shows the shape and size of an object's left side.

Leg of fillet weld: Distance from the point where the base metals touch at the root of the joint to the toe of the fillet.

Lens: Specially treated glass or plastic through which a welder may look at an intense flame without being injured by the harmful rays or glare.

Light metal: A low-density metal such as aluminum, magnesium, titanium, beryllium, or their alloys.

Liquidus: Lowest temperature at which a metal or alloy is completely liquid.

M

Macro-etch: Eating away of the metal surface to make gross structural details stand out so that they can be observed with the naked eye or with magnification of up to 10 times.

Macrograph: Photographic or graphic reproduction of the surface of a prepared specimen that has been magnified up to 10 times normal size.

Macrostructure: The physical makeup or structure of metals revealed under magnification of not more than 10 diameters.

Malleability: The ability of a metal to be deformed without breaking.

Malleable cast iron: Cast iron made by annealing white cast iron while the metal undergoes decarburization, graphitization, or both. The process eliminates all or most of the cementite.

Malleable castings: Cast forms of metal that have been heat treated to reduce their brittleness.

Manifold: A pipe or cylinder with several inlet and outlet fittings. It is designed so that several gas or oxygen cylinders can be connected together and the material piped to a number of locations or stations.

MAPP: Stabilized methylacetylene-propadiene fuel gas. Also referred to as MPS.

Martensite: A very hard and brittle form of steel. It is formed by rapid cooling from the austenite phase.

Mechanical properties: Descriptions of a material's behavior when force is applied for purpose of determining the material's suitability for mechanical use. Properties described, for example, are modulus of elasticity, elongation, fatigue limit, hardness, and tensile strength.

Mechanized welding: Welding that requires manual adjustment of controls and variables and visual observation while the torch or gun is held in a mechanical device.

Megapascal (Mpa): One million pascals; a unit of measurement for pressure in SI Metric. See *Pascal.*

Metal powder cutting (POC): An oxygen cutting process that uses a powder, such as iron or aluminum, to improve the cutting of metal.

Metallic: Term indicating that a material contains metal.

Metallography: The scientific study of the constitution and structure of metals and alloys, as observed by the naked eye or with the aid of magnification and x-ray.

Metallurgy: The science and technology of metal.

MIG: A nonstandard term. See *Gas metal arc welding.*

Mixing chamber: Part of the welding or cutting torch in which the welding gases are mixed prior to combustion.

Modulus of rupture: In a bend or torsion test, the stress at which fracture occurs, expressed as a constant.

N

NEMA: Acronym for the National Electrical Manufacturers' Association.

Neutral flame: The flame resulting from combustion of perfect proportions of oxygen and the welding gas.

Newton: The SI Metric unit of force. A force of 9.8 newtons is required to lift a mass of 1 kilogram.

Nitriding: A casehardening process that involves adding nitrogen to a solid ferrous alloy by keeping the alloy at a suitable temperature while in touch with a material rich in nitrogen.

Nodular cast iron: Cast iron containing primary graphite that is in a ball-like or globular form, rather than in flakes as in gray cast iron. Also known as spheroidal graphite iron, it is more ductile and has greater strength than ordinary iron.

Nondestructive evaluation: A nonstandard term. See *Nondestructive examination.*

Nondestructive examination (NDE): The act of determining the suitability of materials or parts, using techniques that do not affect the serviceability of the part or material.

Nonferrous metal: A metal that contains no iron.

Nonmetallic: Term that indicates a material does not contain metal.

Normalizing: The process of heating steel above the temperature used for annealing and then cooling it in still air at room temperature; used as a preparation for further heat treatment. Normalized steel has a uniform unstressed condition with a grain size and refinement that makes the metal more suitable for heat treating.

Notch brittleness: Tendency of a material to break at points where stress is concentrated.

Notch sensitivity: A measure of the extent of reduction of strength in a metal after introducing stress concentration (by notching).

Nozzle: A device that directs a shielding medium or gas.

O

Offtime: In resistance welding, the time that the electrodes are off the work. It also can be defined as the time between repeating cycles.

Optical pyrometer: Temperature-measuring device that compares the incandescence (white, glowing heat) of a heated object with an electrically heated filament whose incandescence can be regulated.

Orifice: An opening through which gases flow.

Orthographic projection: A method of making a working drawing and projecting the sizes from one view to another.

Outside corner weld: A fused joint of two pieces that are at an angle (often 90°) to each other. The weld is performed on the side opposite the vertex of the corner.

Overhead position: Weld made on the underside of joint with the face of the weld in a horizontal position.

Overheating: Damaging the properties of a metal by applying too much heat. When original properties cannot be restored, the overheating is known as "burning."

Overlap: Extension of the weld face metal beyond the toe of the weld.

Oxidation: Combining of a substance with oxygen. Rapid oxidation is called "burning."

Oxidizing: The process in which oxygen combines with any other substance.

Oxidizing flame: The flame produced by an excess of oxygen in the torch mixture, which leaves some free oxygen that tends to burn the molten metal.

Oxyacetylene cutting: An oxyfuel gas cutting process that uses an oxygen and acetylene flame for heat and a jet of oxygen to oxidize the molten metal and form a cut.

Oxyacetylene welding (OAW): Method of oxyfuel gas welding in which oxygen and acetylene are combined and burned to provide the heat.

Oxyfuel gas cutting: Cutting metal using an oxygen jet and a preheating flame that combines oxygen and a fuel gas.

Oxyfuel gas welding: Method of welding that combines and burns oxygen and a fuel gas to create the required heat.

Oxygen (O_2): Gas that makes up 21% of the composition of air. A gas used to support combustion in oxyfuel gas welding and cutting.

Oxygen cylinder: A specially built container manufactured according to ICC standards and used to store and ship certain quantities of oxygen.

Oxygen hose: A reinforced, multilayered flexible tube, usually of rubber, that is used to carry high-pressure gases.

Oxygen lance cutting (LOC): An oxyfuel gas cutting process that heats base metal and then blows away the molten metal with jet of oxygen from an iron pipe.

Oxygen regulator: Automatic valve used to reduce cylinder pressures to torch pressures and to keep the pressures constant.

Oxyhydrogen flame: The chemical combining of oxygen with the fuel gas hydrogen.

Oxypropane gas flame: The chemical combining of oxygen with the fuel gas propane (also referred to as LP, or liquefied petroleum, gas).

P

Pascal: The SI Metric unit for measuring pressure.

Pass: See *Weld pass*.

Pearlite: A form of steel that has alternating layers of ferrite and cementite.

Peel test: A destructive test that mechanically separates a resistance welded lap joint by peeling one piece away from the other.

Penetrant: Either a liquid or a gas which, when applied to the surface of a metal, enters cracks (discontinuities) to make them visible.

Penetration: The extent to which the weld metal combines with the base metal, as measured from the surface of the base metal.

Percussion welding (PEW): A type of resistance welding in which the heat comes from an arc produced by an electrical discharge and instantaneous pressure applied during or immediately following the heating.

Physical properties: Properties or qualities (other than mechanical properties) that have to do with the physics of a material. Examples are density, ability to conduct electricity, ability to conduct heat, and thermal expansion.

Physical testing: Examination of a material to find out its physical properties.

Pickling: A method of removing surface oxides from metals by chemical or electrochemical reaction.

Plasma: Temporary physical condition of a gas after it has been exposed to and has reacted to an extremely high temperature.

Plasma arc cutting (PAC): A metal cutting process that uses an electric arc and fast flowing ionized gases.

Plasma spraying (PSP): Thermal spraying process that uses a nontransferred plasma arc to melt and propel the surfacing material onto the base.

Plastic welding: Process in which heated air softens and fuses synthetic plastic materials.

Plasticity: Ability of a metal to bend without breaking (rupturing).

Platen: A smooth, extremely flat surface that is built so that it will not warp. Usually made of iron or granite, a platen is used to check measurements and to arrange parts for large assemblies. In *resistance welding*, a large, generally flat surface through which current flows and to which projection welding dies are attached.

Plug weld: A weld made through and in a round hole that has been cut into one piece of metal that is lapped over another piece.

Polarity: The direction of flow of electrons in a closed direct current welding circuit.

Porosity: Gas pockets or voids in a metal.

Positioner: A device used to hold a weldment and move it or rotate it into the best position for welding. It may be moved manually or automatically.

Postflow: The timed flow of shielding gas after the arc is extinguished. This is done to protect the hot electrode and weld area from oxidation.

Postheating: Application of heat to a metal after a welding or cutting operation has been performed.

Powder metallurgy: The art and technology of producing powdered metal and of utilizing it in the production of parts.

Powered air purifier: A device, worn by a welder, that has a motorized blower and filters to remove harmful or toxic particles and chemicals from the recirculated air.

Preflow: The flow of shielding gas that begins before the arc is struck.

Preheating: Application of heat to a metal before a welding or cutting operation has been performed.

Print: A copy of a drawing or plan. Previously referred to as a "blueprint."

Procedure qualification record (PQR): A nonstandard term. See *Welding procedure qualification record (WPQR).*

Prods: Two hand-held electrodes that are pressed against the surface of a part to pass a magnetizing electric current through it; used to find defects with magnetic particles.

Program: A series of movements, actions, times, pressures, welding sequences, and other variables that must occur during an automatic welding procedure. Usually placed into a computer memory.

Projection welding (PW): A type of resistance welding in which current flow is concentrated at predetermined points by projections, embossments, or intersections.

Psia: Pounds per square inch atmospheric.

Psig: Pounds per square inch gauge.

Puddle: A nonstandard term. See *Weld pool.*

Pulsed arc welding: A welding arc in which the current is interrupted or pulsed as the welding arc progresses. GMAW-P and GTAW-P are two pulsed arc processes.

Purging: Passing the proper gas through a system to ensure that there is no air or foreign substances in the system. Purging is recommended before welding.

Purified air breathing apparatus: A device worn by a welder that will provide pure, clean air that contains no undesirable particles or chemicals.

Pyrometer: Device for determining temperatures over a wide range.

Q

Quench aging: A change in metal produced by rapid cooling after heat treating.

Quench annealing: The process used to soften austenitic ferrous alloys by solution heat treatment.

Quench hardening: Hardening an iron alloy by austenitizing followed by rapid cooling, so that some or all of the austenite becomes martensite.

Quenching: Rapid cooling of metal in a heat treating process.

R

Radiograph: Photograph made by passing x-rays or gamma rays through the object to be photographed and recording the variations in density on a photographic film.

Radiographer: Person who performs radiographic operations.

Radiographer's exposure device: Instrument containing the x-ray source for making radiographic records on sensitized film.

Radiographic interpretation: "Reading" of the films to determine cause and significance of discontinuities below the surface of the material being examined. One of the determinations of the reading is the suitability of the material for use.

Radiographic screens: Sheets, either metallic or fluorescent, that are used to intensify the radiation effect on film.

Radiography: Use of radiant energy found in x-rays or gamma rays to examine opaque objects and make a record of the examination.

Reactance: Opposition to the flow of alternating current as a result of inductance or capacitance.

Recarburize: To add carbon to molten cast iron or steel. Also, the process of adding more carbon to a surface that has lost some carbon in processing.

Rectified alternating current: Alternating current made to flow in one direction only by use of a device like a diode to stop normal reversing.

Rectifier: A device, such as a diode or a circuit, that acts like a one-way valve. It converts one half of a waveform of alternating current to useful current flowing in the same direction as the other half of the waveform.

Rectilinear motion robot: A robot that moves in the X, Y, or Z axes. Rotation around these axes is also possible.

Reducing flame: Oxyfuel gas flame with a slight excess of fuel gas.

Reduction of area: Difference in cross-sectional area of a specimen after fracture as compared to original cross-sectional area.

Refractory: A material that is resistant to heat or difficult to melt.

Regulators: See *Acetylene regulator, Oxygen regulator.*

Reinforcement of weld: Excess metal on the face of a weld.

Residual stress: Stress that is still present in a body freed of external forces or thermal gradients.

Resistance seam welding (RSEW): A resistance welding process that typically uses round, rotating electrodes to make a continuous seam or overlapping spot welded seam.

Resistance spot welding (RSW): A resistance welding process that uses resistance to the flow of electricity to create the heat for fusion. A small spot is welded on two overlapping pieces between two electrodes.

Resistance welding (RW): Process that uses the resistance of the metals to the flow of electricity as the source of heat.

Right side view: The view on a working drawing that shows the shape and size of an object's right side.

Robot: A mechanical device that can perform various movements at the direction of an automatic controller.

Robot work cell: The volume of space in which the robot, positioner, controller, and all other related components are located.

Robotics: The use of robots and automatic controllers, along with tools like welding torches or guns, to perform welds, cuts, or machining operations.

Rockwell hardness tester: Tester that measures the hardness of materials, based on depth of penetration of a standardized force.

Root crack: Crack in either the weld or the heat-affected zone at the root of a weld.

Root of joint: Point at which metals to be joined by a weld are closest together.

Root of weld: That part of a weld farthest from source of weld heat and/or from the side where filler metal is added.

Root opening: The space at the bottom of the joint between the pieces being welded.

Root penetration: Depth to which weld metal extends into the root of a welded joint.

Rosette weld: See *Plug weld.*

S

Scleroscope test: Hardness test that uses the height of rebound of a falling piece of metal to determine how much energy is absorbed by the material being tested.

SCR: See *Silicon-controlled rectifier.*

Secondary hardening: The process of tempering of some alloy steels at a higher temperature than normally used for hardening. The result is a hardness greater than is achieved by tempering at the lower temperature for the same period of time.

Semiautomatic welding: Manual welding with one or more of the welding variables controlled by automatic devices.

Semikilled steel: Incompletely deoxidized steel that contains sufficient dissolved oxygen to react with the carbon, so that carbon monoxide is formed. This offsets solidification shrinkage.

Sequence: The order in which operations take place.

Servomotor: A motor that can move in small increments according to electrical signals. Used for steering mechanical devices, robots, and welding guns.

Set-up point: A specific reference point in space from which all movements of a robot arm or electrode tip are measured.

Shear: Force that causes two parts of the same body touching each other to slide parallel to their contacting surfaces.

Shear fracture: Break in which crystalline material separates by sliding under action of shear stress.

Shear strength: Stress required to fracture a part in a cross-sectional plane when two forces being applied are parallel and opposite, but are offset somewhat.

Shield: An eye and face protector. It enables a person to look directly at the electric arc through a special lens without being harmed.

Shielded metal arc cutting (SMAC): An arc cutting process in which metal is cut by melting it with the heat of an arc between a covered electrode and the base metal.

Shielded metal arc welding (SMAW): Arc welding process that melts and fuses the metals using the heat of an arc between a covered electrode and the base metal. The electrode wire also acts as the filler metal.

Short arc: Gas metal arc process that uses a low arc voltage. The arc is continuously interrupted as the molten electrode metal bridges the arc gap.

Shot peening: Working the surface of a metal by bombarding it with metal shot.

Silicon-controlled rectifier (SCR): A semiconductor device with three terminals that can be switched from conducting to nonconducting through the use of signals controlled by logic gates.

Sinusoidal wave: In alternating current, a plot of time against amperage flow. An ac sine wave pattern.

Slag: Nonmetallic byproduct of smelting and refining made up of flux and nonmetallic impurities. Also, material that forms on the underside of an oxyfuel gas or arc cut.

Slag inclusions: Nonfused, nonmetallic substances in the weld metal.

Soldering: A means of fastening metals together by adhering to another metal at a temperature below 840°F (450°C). Only the filler metal is melted.

Solenoid: An electromagnetic device used to open or close switches or valves.

Solid pellet oxyfuel gas welding: A portable welding system using fuel gas with pellets. Pellets produce oxygen and eliminate need for bulky oxygen cylinders.

Solid-state controller: An electronic controller that uses transistors, diodes, and other semiconductor devices.

Solidus: The highest temperature at which a metal or alloy is completely solid.

Spheroidizing: The process of heating and cooling steel to produce a spheroidal or globular form of carbide in the metal.

Spray arc: A gas metal arc process with an arc voltage high enough to continuously transfer the electrode metal across the arc in small globules.

Stepdown transformer: A device used to reduce a higher voltage to a lower voltage. As the voltage decreases, amperage increases, and vice-versa.

Stepping motor: An electric motor that has the ability to rotate in increments of a few degrees at a time.

Stickout: See *Electrode extension.*

Strain: Reaction of an object to a stress.

Stress: A load imposed on an object.

Stress relieving: A process involving even heating of a structure to a temperature below the critical temperature followed by a slow, even cooling.

Submerged arc welding (SAW): Process in which electric arc is submerged in granular flux.

Superficial Rockwell hardness test: Test for determining surface hardness of thin sections or small parts, or where a large hardness impression might be harmful.

Surfacing: Depositing material on a surface to obtain desired properties or dimensions.

T

Tack weld: Small weld used to temporarily hold components together.

Tank: Thin-walled container of fluids or gases. Usually weaker than a cylinder for pressurized gases.

Temper: To heat-treat hardened steel or hardened cast iron by heating to a temperature below its melting point for purpose of decreasing the hardness and increasing the toughness.

Tempering: Reheating hardened or normalized ferrous alloys and then cooling at any desired rate.

Tensile strength: Maximum pull stress in pounds per square inch or megapascals (newtons per square millimeter) that a specimen will withstand.

Thermal spraying processes: Processes used to apply molten surfacing materials. These processes include arc spraying, flame spraying, plasma arc spraying, and detonation spraying.

Throat of a fillet weld: See *Actual throat.*

TIG: A nonstandard term. See *Gas tungsten arc welding* .

Time-temperature-transformation diagram: Diagram that shows how time and temperature affect steel transformation when cooling from the austenite phase.

Tinning: In soldering, a thin coating of solder placed on the metals to be joined.

Tip: End of the torch where the gas burns, producing the high-temperature flame. In resistance welding, the electrode ends.

T-joint: Joint formed by placing one metal against another at an angle near 90° to form a "T" shape.

Toe crack: A crack in the base metal at the toe of a weld.

Toe of weld: Junction between the face of a weld and the base metal.

Top view: The view on a working drawing that shows the size and shape of an object's top side.

Torch: Mechanical device that a welder holds during gas welding and cutting and from which issue the gases that are burned to produce heat. The device held during some arc welding processes is also known as a torch.

Torsion: Twisting motion resulting in shear stresses and strains.

Toughness: Metal's ability to absorb energy and deform before breaking.

Toxic: Term used to describe material that is poisonous and possibly deadly.

Transducer: Device used to send and receive sonic (sound) waves when doing ultrasonic testing.

Transistor: A semiconductor device similar to an SCR. It can be turned on or off by a control signal.

T-T-T diagram: See *Time-temperature-transformation diagram.*

Tuyere: Opening in a blast furnace through which air is forced to support combustion.

U

Ultimate compressive strength: The maximum compression stress that a material can stand under a gradual and evenly applied load.

Ultimate strength: The greatest conventional stress, tensile, compressive, or shear that a material can stand.

Ultrasonic: Mechanical vibrations in a frequency above the range of humanly audible sound.

Ultrasonic testing (UT): Nondestructive examination method that transmits and receives high-frequency sound waves through the material being examined.

Ultrasonic welding (USW): Process using high sound frequencies to produce metal fusion.

Ultraviolet rays: Energy waves that come from electric arcs and welding flames at such a frequency that these rays are in the short light wavelength known as the ultraviolet spectrum.

Underbead crack: Crack in the base metal near the weld and beneath the surface.

Undercut: Depression at the toe of the weld that is below the surface of the base metal.

Uphill: Welding with an upward progression.

V

Vertical position: Type of weld in which the welding is done in a vertical seam and on a vertical surface.

Vertically down: A nonstandard term. See *Downhill*

Vertically up: A nonstandard term. See *Uphill.*

W

Weld bead: Deposited line of filler metal from a single welding pass.

Weld crack: Crack in weld metal.

Weld metal: The fused portion of base metal or fused portion of the base metal and the filler metal.

Weld nugget: Weld metal in a spot, seam, or projection weld.

Weld pass: Single progression of a weld or surfacing operation. The result of a pass is a bead, layer, or spray deposit.

Weld pool: Portion of the weld that is molten due to the heat of welding.

Weld size: The depth of penetration in a groove weld. Also, the nominal length of a leg in a fillet weld.

Welder: A person who performs a weld. (The term is sometimes incorrectly used to describe a welding machine.)

Welder performance qualification test (WPQT): A test of a welder's ability to perform a weld according to a particular welding procedure specification.

Welding: A joining method that produces coalescence of materials by heating them to the welding temperature. The process may or may not use pressure or filler material.

Welding arc: See *Arc.*

Welding procedure qualification record (WPQR): A record of the welding variables and tests conducted to qualify a specific welding procedure specification.

Welding procedure specification (WPS): A document that lists all the variables and procedures required to perform a specific weld. This is done to ensure the acceptable repeatability of the weld when performed by trained and qualified professional welders.

Welding rod: Metal rod that is melted into the weld metal.

Welding sequence: Order in which the component parts of a structure are welded.

Welding wire: A filler metal that is normally in a coil form. It may or may not conduct electricity.

Weldment: Assembly of component parts joined together by welding.

Working drawings: Drawings that are used to produce a part or product. They may be detail or assembly drawings.

Working pressure: The gas pressure set on the low-pressure regulator. Also called the torch pressure.

Working volume: The volume of space that can be covered by a robot at the full extent of its reach in all directions.

Workpiece: The part that is to be welded, brazed, cut, or surfaced.

Workpiece lead: Electrical conductor that carries the current between the welding machine and the workpiece.

WPQR: See *Welding procedure qualification record.*

WPQT: See *Welder performance qualification test.*

WPS: See *Welding procedure specification.*

Wrought iron: Commercial iron made up of slag (also known as iron silicate) fibers entrained in a ferrite matrix.

Y

Yield point: The lowest stress to which a material or body can be subjected. The point at which strain increases without an appreciable or proportionate increase in stress.

Yield strength: Stress value in psi or kPa at which a specimen assumes a specified limiting permanent set.

Z

Zero point: See *Set-up point.*

A technician using a teach pendant to program a welding robot. The robot is being set up to make GMA welds on computer furniture. (ABB Flexible Automation)

This huge weldment is being held on a positioner as submerged arc welding is done by a machine mounted on the end of the long crane arm. (Harnischfeger Corp.)

ACKNOWLEDGMENTS

The publishing of a book of this nature would not be possible without the assistance of the many segments of the Welding Industry.

The authors gratefully acknowledge the cooperation of the following:

ABB Flexible Automation, Inc.
ACRO Automation Systems, Inc.
Aeroquip Corporation
AGA Gas, Inc.
Air Products and Chemicals, Inc.
Airco Gases, Div. of the BOC Group
Airmatic-Allied, Inc.
Airmatic/Beckett-Harcum
Ajax Magnethermic Corporation
The Aluminum Association
American Iron and Steel Institute
American Society of Mechanical Engineers
American Tool Companies, Inc.
American Torch Tip Company
American Welding Society
Anchor Research Corporation
Applied Power, Inc.
Arc Machines, Inc.
Arcair Company, Division Thermadyne Industries
Argopen
Atlas Welding Accessories, Inc.
Baldor Electric Company
Belchfire Corporation
Bernard Welding Equipment Co., A Dover Industries Company
Bethlehem Steel Corporation
British Federal
Broco, Inc.
Bug-O Systems
Carol Cable Company, Inc.
Central Foundry, Division of General Motors Corporation
Century Manufacturing Company
CK Worldwide, Inc.
Clearing Niagara
CMS Gilbreth Packaging Systems
CMW, Inc.
CONCOA (Controls Corp. of America)
Convergent Energy
CPS Products, Inc.
Craftsweld Equipment Corporation
Dearman, a Division of Cogsdill Tool Products, Inc.
De-Sta-Co., A Dover Resources Company
Dow Chemical Company
Dualdraw Division of Industrial Manufacturing and Machining.
Duffers Scientific
Duraline Div. of J.B. Nottingham & Company, Inc.
Econospect Corporation
Electrovert
Engineering & Scientific Equipment Ltd.
Englehard Corporation
Entron Controls, Inc.
ESAB AB
ESAB Welding & Cutting Products
Eureka Welding Alloys, Inc.
Falstrom Company
FANUC Robotics North America, Inc.
Fenway Machine Co., Inc.
The Fibre-Metal Products Co.
Foerster Instruments, Inc.
Fusion, Inc.
The Gasflux Company
Goss, Inc.
Gullco International Limited
Handy & Harman/Lucas Milhaupt, Inc.
Harnischfeger Corporation
Harris Calorific Div. of the Lincoln Electric Co.
J.W. Harris Company, Inc.
Heck Industries, Inc.
Hercules Welding Products
Hobart Brothers Company
Hornell Speedglas, Inc.
Hughes Aircraft Company
Hypertherm, Inc.
Inco Alloys International, Inc.
Ind. Schweiztechnic E. Jankus
Ingersoll-Rand Waterjet Cutting Systems

Invincible AirFlow Systems
Jackson Products, Div. Thermadyne Industries, Inc.
Jet Equipment & Tools
Jordon Controls, Inc.
Kamweld Products Company, Inc.
Kedman Company
Kelsey-Hays Company
Kemper Purification Systems
Koike Aronson, Inc.
Kolene Company
Laramy Products Company, Inc.
Leica Optical Products Division
Lenco-NLC, Inc.
The Lincoln Electric Company
LORS Machinery, Inc.
Magnaflux Corporation
Magnatech Limited Partnership
Maitlen & Benson, Inc.
Manufacturing Technology, Inc.
Marc-L-Tec, Inc.
Mathey/Leland International, Ltd.
McDonnell & Miller, Inc.
Medar, Inc.
Michigan Seamless Tube Company
Miller Electric Mfg. Co.
Miller Thermal, Inc.
M.K. Products, Inc.
Modern Engineering Co., Inc.
Motoman, Inc.
Nederman, Inc.
Niagara Machine and Tool Works
Norton Company
Charles Albright, The Ohio State University
The Ohio Nut and Bolt Company
Olympus America, Inc., Industrial Fiber Optics Division
Osram Sylvania, Inc.
P & H Material Handling
Page-Wilson Corporation
Panametrics, Inc.
Cecil P. Peck Company
Pertron Controls Corporation
Phoenix Products Company Inc.
Physical Acoustics, Inc.
Plasma-Giken
Pow-Con, Inc.
Pressed Steel Tank Co.
PTR-Precision Technologies, Inc.
Pullmax, Inc.
Racal Health & Safety, Inc.
Raytheon Company
Resistance Welder Corporation

J.W. Rex Company
Rexarc, Inc.
Robotron Corporation
Rocketdyne Division, Rockwell International Corporation
Royco Instruments, Inc.
Ruemelin Mfg. Company, Inc.
W.J. Savage Company
Sciaky, Inc. subsidiary of Phillips Service Industries
Seifert X-Ray Corporation
Sensor Developments, Inc.
Shore Instrument Company
Singer Safety Company
Smith Equipment, Division of Tescom Corporation
Sonobond Ultrasonics, Inc.
Dipl. Ing. Ernest Spirig and Solder Absorbing Technology, Inc.
Stoody Company
Strippit, Inc., Unit of IDEX Corporation
Suhner Industrial Products Corporation
Sulzer Metco
Swanstrom Centerline
TAFA, Inc.
Taylor-Wharton Cylinders
The Taylor Winfield Corporation
Tempil Division, Air Liquide America Corp.
Thermacote-Welco Company
Thermadyne Industries, Inc.
Thermco Instrument Corporation
Thrall Car Manufacturing Company
Tinius Olsen Testing Machine Company, Inc.
Tri Tool, Inc.
Trumpf, Inc.
TRW Nelson Stud Welding Division
Tuffaloy Products, Inc.
Tweco/Arcair, Division of Thermadyne Industries, Inc.
United Clamp Mfg. Company
U.S. Steel Corp.
Vacuum Atmospheres, Inc.
Vega Enterprises
Veriflo Corporation
Victor Equipment Co., Division Thermadyne Industries, Inc.
Vogel Tool & Die Corporation
E.H. Wachs Company
Wall Colmonoy Corporation
Weldcraft Products, Inc.
Welding & Fabrication Data Book
Welding Design & Fabrication
Weldma Company
Weldsale Company
Wilton Corporation

INDEX

FEB 1999

52 MOD

dern welding

DATE DUE

The Joint Free Public Library
of
Morristown and Morris Township
1 Miller Road
Morristown, New Jersey 07960